Lecture Notes in Computer Science 1625
Edited by G. Goos, J. Hartmanis and J. van Leeuwen

Springer

Berlin
Heidelberg
New York
Barcelona
Hong Kong
London
Milan
Paris
Singapore
Tokyo

Bernd Reusch (Ed.)

Computational Intelligence

Theory and Applications

International Conference, 6th Fuzzy Days
Dortmund, Germany, May 25-28, 1999
Proceedings

Volume Editor

Bernd Reusch
University of Dortmund, Computer Science I
D-44221 Dortmund, Germany
E-mail: reusch@cs.uni-dortmund.de

Cataloging-in-Publication data applied for

Die Deutsche Bibliothek - CIP-Einheitsaufnahme

Computational intelligence : theory and applications ; international conference ; proceedings / 6th Fuzzy Days, Dortmund, Germany, May 25 - 27, 1999. Bernd Reusch (ed.). - Berlin ; Heidelberg ; New York ; Barcelona ; Hong Kong ; London ; Milan ; Paris ; Singapore ; Tokyo : Springer, 1999
(Lecture notes in computer science ; Vol. 1625)
ISBN 3-540-66050-X

CR Subject Classification (1998): I.2.3, F.4.1, F.1.1, I.2, F.2.2, I.4, J.2

ISSN 0302-9743
ISBN 3-540-66050-X Springer-Verlag Berlin Heidelberg New York

Typesetting: Camera-ready by author
SPIN: 10703278 06/3142 – 5 4 3 2 1 0 Printed on acid-free paper

Preface

Fuzzy Days in Dortmund were held for the first time in 1991. Initially, the conference was intended for scientists and practitioners as a platform for discussions on theory and application of fuzzy logic. Early on, synergetic links with neural networks were included and the conference evolved gradually to embrace the full spectrum of what is now called Computational Intelligence (CI). Therefore, it seemed logical to launch the 4th Fuzzy Days in 1994 as a conference for CI—one of the world's first conferences featuring fuzzy logic, neural networks and evolutionary algorithms together in one event. Following this successful tradition, the 6th Fuzzy Days' aim is to provide an international forum for reporting significant results on the theory and application of CI-methods.

Once again, we have received a remarkable number of papers. I would like to express my gratitude to all who have been interested in presenting their work within the framework of this conference and to the members of the programme committee for their valuable work (in this edition each paper was reviewed by five referees). In particular, I wish to thank all keynote and tutorial speakers for their commitment. Last but not least, I am obliged to the Deutsche Forschungsgemeinschaft and Kommunalverband Ruhrgebiet for their financial support.

March 1999

Bernd Reusch

Organisation

The 6th Fuzzy Days in 1999 are organised by the following institutions:

University of Dortmund, Computer Science I,
VDE/VDI-Society of Microelectronics, Micro- and Precision Engineering (GMM),
Information Technology Society within VDE (ITG),
VDI/VDE-Society on Measurement and Control (GMA),

in co-operation with German Informatics Society (GI).

Honorary Chairmen

D. B. Fogel, USA
T. Kohonen, Finland
Z. Pawlak, Poland
L. A. Zadeh, USA

General Chairman

B. Reusch, Dortmund

Programme Committee

Chairmen

K. Goser, Dortmund
H. Kiendl, Dortmund
B. Reusch, Dortmund
H.-P. Schwefel, Dortmund

Members

Aguilar-Martin, J.; France
Aizenberg, I.; Ukrain
Albrecht, R.; Austria
Bäck, Th.; Germany
Baldwin, J.; UK
Banzhaf, W.; Germany
Berenji, H. R.; USA
Borisov, A. N.; Latvia*
Bosc, P.; France
Bouchon-Meunier, B.; France
Brauer, W.; Germany
Carlsson, C.; Finland
Chan, L.; Hong Kong*
Damiani, E.; Italy*

* regional co-ordinator

De Jong, K.; USA
Di Nola, A.; Italy
Drechsler, R.; Germany
Dubois, D.; France
Eklund, P.; Sweden*
Esteva, F.; Spain*
Fathi, M.; Germany
Fedrizzi, M.; Italy
Frank, P. M.; Germany
Freksa, Ch.; Germany
Fukuda, T.; Japan*
Glesner, M.; Germany
Gottwald, S.; Germany
Grauel, A.; Germany
Gupta, M. M.; USA
Grzymala-Busse, J.; USA
Hajek, P.; Czech Republic
Halgamuge, S.; Australia*
Hata, Y.; Japan
Hirota, K.; Japan
Homaifar, A.; USA
Isermann, R.; Germany
Jamshidi, M.; USA*
Kacprzyk, J.; Poland*
Kasabov, N.; New Zealand*
Kaynak, O.; Turkey*
Kerre, E.; Belgium
Klement, E. P.; Austria
Koczy, L.; Hungary
Kruse, R.; Germany
Larsen, H. L.; Denmark
Liu, Y. M.; China*
Mamdani, E. H.; UK
Mansour, M.; Switzerland
Mesiar, R.; Slovak Republic
Mohammadian, M.; Australia
Moraga, C.; Germany
Nauck, D.; Germany
Negoita, M. Gh.; New Zealand
Novak, V.; Czech Republic
Oh, K.-W.; South Korea
Ollero, A.; Spain
Orlowska, E.; Poland
Pedrycz, W.; Canada*
Reuter, M.; Germany
Ruspini, E. H.; USA
Sipper, M.; Switzerland
Tanaka, H.; Japan
Trillas, E.; Spain
Valverde, L.; Spain
Varachiu, N.; Romania*
Verdegay, J. L.; Spain
Yager, R. R.; USA
Yamakawa, T.; Japan

Industry Track Committee

Bastian, A.; VW AG
Georg, E.; Endress + Hauser GmbH & Co.
Grollmann, J.; Siemens AG
Schäfer, R.; Roche Diagnostics GmbH
Schröder, H.-J.; VDO Adolf Schindling AG
Schwärtzel, H. J.; FAST
Schubert, A.; Dow Deutschland Inc.
Splettstößer, W.; Siemens AG
Tresp, V.; Siemens AG
Wagner, R.; Thomson Multimedia GmbH

* regional co-ordinator

Sponsoring Organisations

Arbeitsgemeinschaft Fuzzy Logik und Softcomputing Norddeutschland (AFN)
Asian Pacific Neural Network Assembly (APNNA)
Berkeley Initiative in Softcomputing (BISC)
City of Dortmund
Dachverband Medizinische Technik (DVMT)
Deutscher Verband für Schweißtechnik e. V. (DVS)
Deutsche Gesellschaft für Logistik e. V. (DGfL)
Dortmund Chamber of Industry and Commerce (IHK zu Dortmund)
European Society for Fuzzy Logic and Technologies (EUSFLAT)
Federation of International Robot-Soccer Association (FIRA)
Forschungsgemeinschaft Bekleidungsindustrie e. V.
Forschungskuratorium Gesamttextil
International Fuzzy Systems Association (IFSA)
Neural Network Council Member Societies (IEEE-NNC)

Organising Committee

Dr. N. Jesse, University of Dortmund, Computer Science I
Dr.-Ing. V. Schanz, ITG – Information Technology Society within VDE
Dipl.-Ing. R. Theobald, GMM – VDE/VDI-Society of Microelectronics, Micro- and Precision Engineering

Table of Contents

Poster Abstracts

Topological Theory of Fuzziness

Rudolf F. Albrecht

Faculty of Natural Science, University of Innsbruck
Technikerstr. 25, A-6020 Innsbruck, Austria
Rudolf.Albrecht@uibk.ac.at

Abstract. The intention of this article is to show how fuzzy set theory fits into classical topology. The basic concepts are filter and ideal bases and morphisms on these. Filter bases are used to define abstract distances. Then compositions and comparisons of filter and ideal bases are considered and uniform neighborhood measures between such bases are introduced. On filter bases homomorphisms and antimorphisms can be defined as set functions, in particular, they can be given by the set extension of point functions. With these concepts, fuzzy set theory can be expressed in terms of topology. For some applications we consider contractive mappings, roundings, hierarchical filter bases ("pyramids"), adaptable networks.

1 Naive fuzzy set theory

Classical general set theory started 1874 with the fundamental work of G. Cantor [1], was later on in many ways axiomized (e.g. the axiom systems of Russell, Zermelo-Fraenkel, von Neumann-Bernays-Gödel), and became the basis of several mathematical disciplines like algebra, topology, measure theory. In his approach to measure theory, H. Lebesgue [2] introduced 1904 the "characteristic function" $\chi : X \to \{0, 1\} \subset \mathbf{R}_+$ of a subset X of a set S by $\chi(s) = 1$ if $s \in X$, and $\chi(s) = 0$ if $s \in S \setminus X$, and defined the measure of X by the integral $\int_S \chi(s) ds$ if existing.

In 1965 L. A. Zadeh [3] generalized the characteristic function χ to any function $f: S \to C \subseteq [0, 1] \subset \mathbf{R}$ with $1 \in C, 1 \in f(S)$ (i.e. f is normalized) and considered this to be an extension of the set concept, named "fuzzy set", with the intuitive meaning that $f(s)$, the "membership" function, expresses a weight or a measure of the membership of s in the fuzzy set $F = (s, f(s))_{s \in S}$. Then numerous definitions for "union" $\cup$, "intersection" $\cap$, "complement" $\mathbf{C}$ of fuzzy sets were invented as generalizations of classical set operations, preserving at least part of their classical properties. Let $G = (s, g(s))_{s \in S}$ be another fuzzy set. Then for example $F \cup G =_{def} (s, \max(f(s), g(s)))$, $F \cap G =_{def} (s, \min(f(s), g(s)))$, $\mathbf{C}F = (s, (1-f(s)))_{s \in S}$, where in general $F \cup \mathbf{C}F \neq (s, 1)_{s \in S}$, $F \cap \mathbf{C}F \neq (s, 0)_{s \in S}$.

For any normalized f and $\alpha \in C$, $C_{[\alpha]} =_{def} \{c \mid c \in C \wedge c \geq \alpha\}$, $S_{[\alpha]} =_{def} f^{-1}(C_{[\alpha]}) \subseteq S$ denotes an "α-cut" of the fuzzy set. We notice the following structural properties: $S_{[\alpha]}$ and $C_{[\alpha]}$ are in correspondence, for $\alpha \to 1$ $S_{[\alpha]}$ and $C_{[\alpha]}$ are monotonously decreasing with non-empty limits. In case $C = [0, 1]$, $S = [a, b] \subset \mathbf{R}$, $a < b$, and for f integrable over S, to $S_{[\alpha]}$ and to $C_{[\alpha]}$ are assigned $\int_{S_{[\alpha]}} (f(x) - \alpha)dx$ and $\int_{C_{[\alpha]}} S_{[y]}dy$ respectively, both values are equal.

Fuzzy set theory and generalizations of it have been introduced to express uncertainty, vagueness, approximating properties of objects in engineering science. On the other hand, topology is the earlier and well developed mathematical discipline dealing with neighborhoods and approximations. We are going to show the relationship between both.

2 Some set theoretical and topological concepts

2.1 Filters and ideals

If ind: $I \to S$ is an indexing of elements of S, we use the notation $ind(i) = s(i) = s_{[i]}$, $s_i = (i, s_{[i]})$. $\bigwedge$, $\bigvee$ denote the universal and the existential quantifier.

Let there be given a lattice $(\mathscr{L}, \leq, \sqcap, \sqcup)$, $\sqcap$, $\sqcup$ lattice meet and join, **o** the zero and **e** the unit element if in $\mathscr{L}$, and a non-empty subset $\mathscr{B} = \{B_{[k]} \mid k \in K\} \subset \mathscr{L}$, the indexing bijective, with the following properties:

$\bigwedge k \in K\, (B_{[k]} \neq o) \wedge \bigwedge k', k'' \in K\, (\bigvee k''' \in K\, ((B_{[k''']} \leq B_{[k']}) \wedge (B_{[k''']} \leq B_{[k'']})))$. Then $\mathscr{B}$ is a "filter base" on $\mathscr{L}$. If in addition $\bigwedge k \in K \bigwedge L \leq e\, (B_{[k]} \leq L \Rightarrow L \in \mathscr{B})$ then $\mathscr{B}$ is a "filter". We define $\lim \mathscr{B} = \sqcap \mathscr{B}$. The dual notions to filter base and filter are "ideal base" and "ideal". Filter bases were introduced by L. Vietoris in 1921 [4]. A filter base can also be an ideal base (see Fig. 1).

If S is a non-empty set, then the above applies to the complete, atomic, boolean lattice $(\text{pow}\, S, \subseteq, \cap, \cup, \varnothing, S)$. For a filter base $\mathscr{B} = \{B_{[k]} \mid k \in K\}$ on pow S, $B^* =_{def} \lim \mathscr{B} = \bigcap_{k \in K} B_{[k]}$. For $B^* \neq \varnothing$ (then we name $\mathscr{B}$ a "proper" filter base) the neighborhood of any $s \in B$, $B =_{def} \bigcup_{k \in K} B_{[k]}$, to the elements of B^* can be expressed by membership or non-membership of s in certain $B_{[k]}$:

Let $\bigwedge s \in B\, ((K(s) =_{def} \{k \mid k \in K \wedge s \in B_{[k]}\}) \wedge \overline{K}(s) =_{def} K \setminus K(s))$, $\mathscr{B}_\cap(s) =_{def} \{B_{[k]} \mid k \in K(s)\}$, $\mathscr{B}_\cup(s) =_{def} \{B_{[k]} \mid k \in \overline{K}(s)\}$. We have $s \in \bigcap_{k \in K(s)} B_{[k]} \cap \bigcap_{k \in \overline{K}(s)} \mathbf{C}B_{[k]}$, **C** the complement with respect to B. Let $K_{min}(s) =_{def} \{k \mid k \in K(s) \wedge \neg \bigvee k' \in K(s)\, (B_{[k']} \subset B_{[k]})\}$, $\overline{K}_{max}(s) =_{def} \{k \mid k \in \overline{K}(s) \wedge \neg \bigvee k' \in \overline{K}(s)\, (B_{[k']} \supset B_{[k]})\}$, then $\mathscr{B}_{\cap min}(s) =_{def} \{B_{[k]} \mid k \in K_{min}(s)\}$, $\mathscr{B}_{\cup max}(s) =_{def} \{B_{[k]} \mid k \in \overline{K}_{max}(s)\}$. General "distance" / "similarity" *relations* of s from / to $s^* \in B^*$ are then given by $\bigwedge s \in B\, (D_\cap(s^*, s) =_{def} \mathscr{B}_{\cap min}(s) \wedge D_\cup(s^*, s) =_{def} \mathscr{B}_{\cup max}(s))$. $D_\cap(s^*, s) = D_\cap(s^*, s')$ and $D_\cup(s^*, s) = D_\cup(s^*, s'')$ define equivalence relations $s \sim_\cap s'$ and $s \sim_\cup s''$. In particular, if $\mathscr{B}$ is itself a complete lattice then $d_\cap(s^*, s) =_{def}$

$\bigcap D_{\cap}(s^*, s) \in \mathscr{B}$ and $d_{\cup}(s^*, s) =_{def} \bigcup D_{\cup}(s^*, s) \in \mathscr{B}$ are *functional* in s and $d_{\cup}(s^*, s) \subset d_{\cap}(s^*, s)$. Dual results hold for $\mathscr{B}$ being an ideal base. For illustration see Fig. 2.

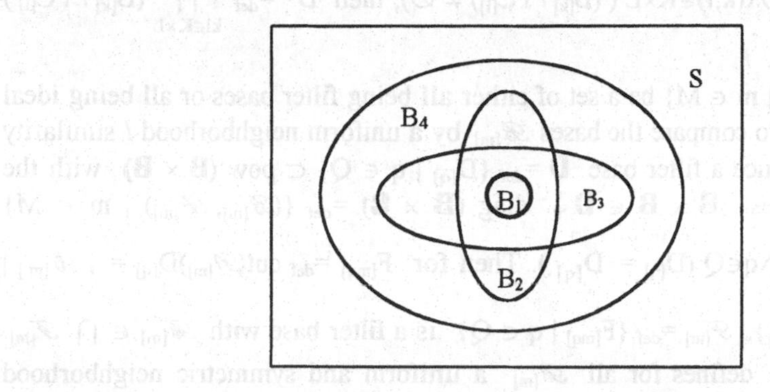

Fig. 1. $\{B_1,B_2,B_3,B_4\}$ is a filter base and an ideal base. $\{B_1,B_2,B_3\}$ is a filter base but not an ideal base.

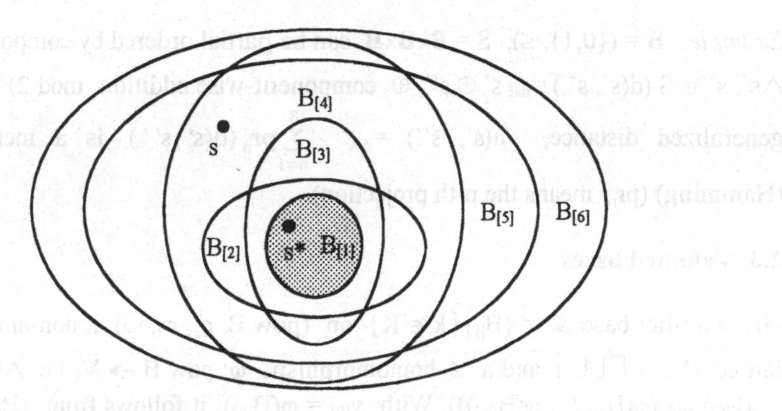

Fig. 2. $\mathscr{B} = \{B_{[1]}, B_{[2]}, B_{[3]}, B_{[4]}, B_{[5]}, B_{[6]}\}$, $\lim\mathscr{B} = B_{[1]}$, $D_{\cap}(s^*,s) = \{B_{[4]}, B_{[5]}\}$, $D_{\cup}(s^*,s) = \{B_{[2]}, B_{[3]}\}$

2.2 Comparison and composition of bases

Given a non-empty set S and two filter bases $\mathscr{B} = \{B_{[k]} \mid k \in K\} \subset \text{pow } S$ with $B =_{def} \bigcup_{k\in K} B_{[k]}$, $B^* =_{def} \bigcap_{k\in K} B_{[k]}$, $\mathscr{C} = \{C_{[l]} \mid l \in L\} \subset \text{pow } S$ with $C =_{def} \bigcup_{l\in L} C_{[l]}$, $C^* =_{def} \bigcap_{l\in L} C_{[l]}$, all indexings bijective. We say $\mathscr{B}$ is "finer" than $\mathscr{C}$, $\mathscr{B} \prec \mathscr{C} \Leftrightarrow_{def} \bigwedge l\in L \bigvee k\in K\ (B_{[k]} \subseteq C_{[l]})$, $\mathscr{B}$ is "equivalent" $\mathscr{C}$, $\mathscr{B} \sim \mathscr{C} \Leftrightarrow_{def} \mathscr{B} \prec \mathscr{C} \wedge \mathscr{C} \prec \mathscr{B}$, and for finite cardinalities, $\mathscr{B}$ "finer granulated" than $\mathscr{C}$ if card $\mathscr{B}$ > card $\mathscr{C}$.

$\mathscr{S} =_{def} \{B_{[k]} \cup C_{[l]} \mid (k,l) \in K \times L\}$ is a filter base on pow S with $S^* =_{def} \bigcap_{kl \in K \times L} (B_{[k]} \cup C_{[l]}) = B^* \cup C^*$, $\mathscr{D} =_{def} \{B_{[k]} \cap C_{[l]} \mid (k,l) \in K \times L\}$ is a filter base on pow S only if $\wedge(k,l)\in K\times L\ ((B_{[k]} \cap C_{[l]}) \neq \varnothing)$, then $D^* =_{def} \bigcap_{kl \in K \times L} (B_{[k]} \cap C_{[l]}) = B^* \cap C^*$.

Let $\mathbf{B} = \{\mathscr{B}_{[m]} \mid m \in M\}$ be a set of either all being filter bases or all being ideal bases on pow S. To compare the bases $\mathscr{B}_{[m]}$ by a uniform neighborhood / similarity measure we introduce a filter base $\mathbf{D} =_{def} \{D_{[q]} \mid q \in Q\} \subset \text{pow}\ (\mathbf{B} \times \mathbf{B})$ with the following properties: $\mathbf{B} \times \mathbf{B} \in \mathbf{D}$, $\text{diag}\ (\mathbf{B} \times \mathbf{B}) =_{def} \{(\mathscr{B}_{[m]}, \mathscr{B}_{[m]}) \mid m \in M\} \subseteq \bigcap_{q \in Q} D_{[q]}$, and $\wedge q\in Q\ (D_{[q]} = \overset{-1}{D_{[q]}})$. Then for $F_{[mq]} =_{def} \text{cut}(\mathscr{B}_{[m]})D_{[q]} = \{\mathscr{B}_{[m']} \mid (\mathscr{B}_{[m]}, \mathscr{B}_{[m']}) \in D_{[q]}\}$, $\mathscr{F}_{[m]} =_{def} \{F_{[mq]} \mid q \in Q\}$ is a filter base with $\mathscr{B}_{[m]} \in \bigcap \mathscr{F}_{[m]}$. Consequently, $\mathbf{D}$ defines for all $\mathscr{B}_{[m]}$ a uniform and symmetric neighborhood system. Thus according **2.1**, for any pair $(\mathscr{B}_{[m]}, \mathscr{B}_{[m']})$ the neighborhood / similarity measures $D_{\cap}(\mathscr{B}_{[m]}, \mathscr{B}_{[m']})$ and $D_{\cup}(\mathscr{B}_{[m]}, \mathscr{B}_{[m']})$ can be applied.

Example: $\mathbf{B} = (\{0,1\}, \leq)$, $S = \mathbf{B}\times\mathbf{B}\times\mathbf{B}$ can be partial ordered by component-wise $\leq$, $\wedge s', s'' \in S\ (d(s', s'') =_{def} s' \oplus s''$, $\oplus$ component-wise addition mod 2) is a uniform generalized distance, $h(s', s'') =_{def} \sum_{n=1}^{3} pr_n(d(s', s''))$ is a metric distance (Hamming) (pr_n means the n-th projection).

2.3 Valuated bases

Given a filter base $\mathscr{B} = \{B_{[k]} \mid k \in K\}$ on $(\text{pow}\ B, \subseteq, \cap, \cup)$, a non-empty complete lattice $(V, \leq, \sqcap, \sqcup)$, and a $\leq$–homomorphism φ: pow B $\to$ V, i.e. $\wedge k,k' \in K\ ((B_{[k]} \subseteq B_{[k']}) \Rightarrow (\varphi(B_{[k]}) \leq \varphi(B_{[k']})))$. With $v_{[k]} = \varphi(B_{[k]})$ it follows from $(B_{[k]} \subseteq B_{[k']}) \wedge (B_{[k]} \subseteq B_{[k'']})$ that $(v_{[k]} \leq v_{[k']}) \wedge (v_{[k]} \leq v_{[k'']})$, hence $\varphi(\mathscr{B})$ is a filter base on V if all $v_{[k]} \neq \mathbf{o}$, and we have $\varphi(\lim \mathscr{B}) \leq \lim \varphi(\mathscr{B})$. In the function $(B_{[k]}, v_{[k]})_{k\in K}$ the elements $B_{[k]}$ of base $\mathscr{B}$ with "support" $B = \bigcup \mathscr{B}$ are valuated by $v_{[k]}$.

We consider $\overset{-1}{\varphi}: V \to \text{pow}\ B$ defined by $\wedge v\in V\ (\overset{-1}{\varphi}(v) =_{def} \bigcup_{\varphi(U)\leq v} U)$. Then $\overset{-1}{\varphi}$ is a homomorphism. If $\mathscr{V} = \{v_{[l]} \mid l \in L\}$ is a filter base on V and $\wedge v\in\mathscr{V}$ $(\overset{-1}{\varphi}(v) \neq \varnothing)$, then $\overset{-1}{\varphi}(\mathscr{V})$ is a filter base on pow B.

For $U \subseteq S$ and for $v \in V$ we have $U \subseteq \overset{-1}{\varphi}\varphi(U)$ and $v \geq \varphi\overset{-1}{\varphi}(v)$, thus $\mathscr{B} \prec \overset{-1}{\varphi}\varphi(\mathscr{B}))$ and $\varphi\overset{-1}{\varphi}(\mathscr{V}) \prec \mathscr{V}$.

An example is: φ the set extension of a function $f: B \to V$, $\varphi(U) = \bigsqcup(f(u))_{u \in U}$, in particular, $V = \text{pow } C$, C a non-empty set. This case can be found in textbooks, e.g. [5] in 1951.

φ being a homomorphism corresponds to the "neighborhood to B^*" interpretation. Choosing φ as antimorphism, $\varphi(\mathscr{B})$ is an ideal base, which corresponds to the "similarity to B^*" interpretation.

The presentation in this section followed the one given in [6].

3 Fuzziness in terms of topology

The basic concept is a filter bases $\mathscr{B} = \{B_{[k]} \mid k \in K\} \subset (\text{pow } S, \subseteq)$, $B_{[k]} \neq \varnothing$, with $\bigcap \mathscr{B} = B^*$, or dual, an ideal base. $\overline{\mathscr{B}} =_{\text{def}} \mathscr{B} \cup \{B^*\}$ is partially ordered by $\subseteq$ which can semantically express a mutual "more or less" relationship between the elements of the $B_{[k]}$, B^*, like neighborhood, similarity, certainty etc. To formulate this quantitatively, a "valuation" $\varphi: \overline{\mathscr{B}} \to (V, \leq, \sqcap, \sqcup)$, a lattice, is applied, with the set function φ isotone with $\subseteq$. φ maps $\mathscr{B}$ onto a filter base $\mathscr{V}$ on V if $\bigwedge k \in K\ (\varphi(B_{[k]}) \neq \mathbf{o})$.

We assume φ is extended to a $\leq$-homomorphism $\text{pow } S \to V$. $(\text{pow } S, \leq, \cap, \cup)$ is a complete, atomic, boolean lattice. If $\mathscr{V}$ is a filter base on V and $\bigwedge v \in \mathscr{V}(\overset{-1}{\varphi}(v) \neq \varnothing)$ then $\overset{-1}{\varphi}$ maps $\mathscr{V}$ onto a filter base on pow S.

Let there be given two filter bases $\mathscr{B} = \{B_{[k]} \mid k \in K\}$, $\mathscr{C} = \{C_{[l]} \mid l \in L\}$ on pow S with valuations $\varphi_{\mathscr{B}}: \mathscr{B} \to V$, $\varphi_{\mathscr{C}}: \mathscr{C} \to V$. $\bigcup \mathscr{B} \neq \bigcup \mathscr{C}$ is admitted. Then $\mathscr{S} = \{B_{[k]} \cup C_{[l]} \mid (k,l) \in K \times L\}$ is a filter base with $\lim \mathscr{S} = \lim \mathscr{B} \cup \lim \mathscr{C}$, $\mathscr{D} = \{B_{[k]} \cap C_{[l]} \mid (k,l) \in K \times L\}$ is a filter base with $\lim \mathscr{D} = \lim \mathscr{B} \cap \lim \mathscr{C}$ if $\bigwedge(k,l) \in K \times L\ (B_{[k]} \cap C_{[l]} \neq \varnothing)$ ("union" and "intersection" of filter bases).

Valuations of the elements of $\mathscr{S}$ and of $\mathscr{D}$ are given by
$\varphi_{\mathscr{S}}(B_{[k]} \cup C_{[l]}) =_{\text{def}} \varphi_{\mathscr{B}}(B_{[k]}) \sqcup \varphi_{\mathscr{C}}(C_{[l]})$ and by
$\varphi_{\mathscr{D}}(B_{[k]} \cap C_{[l]}) =_{\text{def}} \varphi_{\mathscr{B}}(B_{[k]} \cap C_{[l]}) \sqcup \varphi_{\mathscr{C}}(B_{[k]} \cap C_{[l]})$.

A particular case is: $\Sigma \subseteq \text{pow } S$ a σ-algebra, $V = (\overline{\mathbf{R}}_+ = \mathbf{R}_+ \cup \{\infty\}, \leq)$ which is a lattice with $\mathbf{o} = 0$ and $\mathbf{e} = \infty$, $\mu: \Sigma \to \overline{\mathbf{R}}_+$ a σ-additive measure ($\mu(\varnothing) = 0$). If for the filter base $\mathscr{B} = \{B_{[k]} \mid k \in K\}$ holds all $B_{[k]} \in \Sigma$, then μ is a $\leq$-homomorphism $\mathscr{B} \to \overline{\mathbf{R}}_+$.

According **2.3**, the set extension of a function $f: B \to V$, $B = \bigcup \mathscr{B}$, $\mathscr{B} = \{B_{[k]} \mid k \in K\}$ being a filter base, yields a homomorphism $\varphi: \mathscr{B} \to V$. However, a homomorphism of $\mathscr{B}$ need not be given by the set extension of a point function. Similarly, for $\mathscr{B}$ a filter base on Σ, an integrable point function f can exist such that the integrals of f over the $B_{[k]}$ yield a homomorphism μ of $\mathscr{B}$ into $\overline{\mathbf{R}}_+$.

We assume, an algebraic composition law $\bullet$ is defined on the lattice V, $\bullet: V \cup (V \times V) \to V$ with $\bullet$ being the identity on V and $\bullet$ compatible with $\leq$. Let be $\varnothing \neq$

$\mathscr{B} \subseteq \text{pow } S$, $\emptyset \neq \mathscr{C} \subseteq \text{pow } S$, and $\varphi: \mathscr{B} \to V$, $\psi: \mathscr{C} \to V$ *any* two functions. Then the families $(B, \varphi(B))_{B\in\mathscr{B}}$ and $(C, \psi(C))_{C\in\mathscr{C}}$ can be concatenated to a family (function) $(K, \phi(K))_{K\in\mathscr{K}}$ on $\mathscr{K} = \mathscr{B} \cup \mathscr{C}$, $\mathbf{K}(\bullet)((B, \varphi(B))_{B\in\mathscr{B}}, (C, \psi(C))_{C\in\mathscr{C}}) = (K, \phi(K))_{K\in\mathscr{K}}$, with $\phi(K) = \varphi(K)$ for $K\in\mathscr{B}\backslash\mathscr{C}$, $\phi(K) = \psi(K)$ for $K\in\mathscr{C}\backslash\mathscr{B}$, $\phi(K) = \varphi(K) \bullet \psi(K)$ for $K\in\mathscr{B}\cap\mathscr{C}$. Examples are $\bullet = \sqcup$, $\bullet = \sqcap$, and for $V = \overline{\mathbf{R}}_+$, $\bullet = +$. All are isotone with $\leq$.

We apply this to the case: V a boolean lattice, $\mathscr{B} = \mathscr{C} = \{B_{[k]} \mid k\in K\}$ a filter base, φ a homomorphism, $\psi = \overline{\varphi}$ the antimorphism dual to φ. Complementary to the filter base $\mathscr{V} = \{v_{[k]} = \varphi(B_{[k]}) \mid k\in K\}$ is the ideal base $\overline{\mathscr{V}} = \{\overline{v}_{[k]} = \overline{\varphi(B_{[k]})} \mid k\in K\}$. Concatenation of these families by $\sqcup$ and by $\sqcap$ results in $\mathbf{K}(\sqcup) = (B_{[k]}, e)_{k\in K}$ and $\mathbf{K}(\sqcap) = (B_{[k]}, o)_{k\in K}$, respectively.

We distinguish operations on bases from operations on functions defined on bases.

Revisiting the naive fuzzy set theory of section 1, we see that it is embedded in the theory of morphisms of filter and ideal bases: V is $\text{pow } C$, $\emptyset \neq C \subseteq (\mathbf{R}, \leq)$ a discrete or continuous subset, for example the interval $[0,1]$, $f: S \to C$, $\mathscr{V} = \{V_{[k]} \mid k\in K\}$ a filter base on V, for example the monotone filter base $\mathscr{V} = \{[\alpha, 1] \mid \alpha\in C\}$ for which $\lim \mathscr{V} = \{1\}$. Then for φ the set extension of f, the "cuts" $\overset{-1}{\varphi}(V_{[k]})$ are considered, for example the cuts $\overset{-1}{\varphi}([\alpha, 1])$, which, if all of them are $\neq \emptyset$, generate a filter base on $\text{pow } S$. The latter condition holds if f maps S *onto* V.

An illustration for union and intersection of filter bases is shown in Fig. 3.

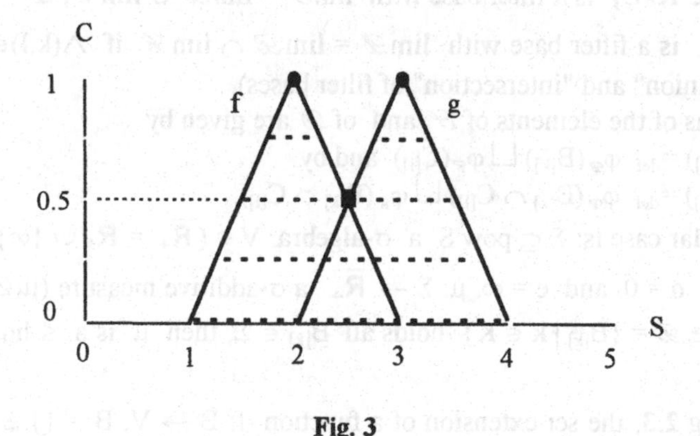

Fig. 3

$S = [0, 5]$, $C = [0, 1]$, $f: [1, 3] \to C$, $g: [2, 4] \to C$, $f(s) = s - 1$ for $1 \leq s \leq 2$, $f(s) = -s + 3$ for $2 \leq s \leq 3$, $g(s) = f(s - 1)$. $\mathscr{V} = \{[0, 1], [.25, 1], [.5, 1], [.75, 1], \{1\}\}$. $\overset{-1}{f}(\mathscr{V}) = \{[1, 3], [1.25, 2.75], [1.5, 2.5], [1.75, 2.25], \{2\}\}$, $\lim \mathscr{V} = \{1\}$, $\lim \overset{-1}{f}(\mathscr{V}) = \{2\}$, similar for $\overset{-1}{g}(\mathscr{V})$. The union filter base $\mathscr{S}(\overset{-1}{f}(\mathscr{V}), \overset{-1}{g}(\mathscr{V}))$ is equivalent to the base $\mathcal{S}' = \{[1, 3] \cup [2, 4], [1.25, 2.75] \cup [2.25, 3.75], ..., \{2\} \cup \{3\}\}$ where only *equally* valuated sets are joined, $\mathcal{S} \sim \mathcal{S}'$. $\lim \mathcal{S} = \lim \mathcal{S}' = \{2, 3\}$. There is no intersection filter base as image of $\mathcal{V}$, but as image of the filter base $\tilde{\mathcal{V}} = \{[0, .5], [.25, .5], \{.5\}\}$. We have $\lim \overset{-1}{f}(\tilde{\mathcal{V}}) = \{1.5, 2.5\}$, $\lim \overset{-1}{g}(\tilde{\mathcal{V}}) = \{2.5, 3.5\}$, and for the intersection filter base $\mathcal{D}(\overset{-1}{f}(\tilde{\mathcal{V}}), \overset{-1}{g}(\tilde{\mathcal{V}}))$, $\lim \mathcal{D} = \{2.5\}$. Notice, the functions f, g do not have the same support.

Now we consider two filter bases $\mathcal{V}_f = \{[.75, 1], \{1\}\}$, $\mathcal{V}_g = \{[0, .25], \{0\}\}$ on V with filter bases $\overset{-1}{f}(\mathcal{V}_f) = \{[1.75, 2.25], \{2\}\}$, $\overset{-1}{g}(\mathcal{V}_g) = \{[2, 2.25] \cup [3.75, 4], \{2, 4\}\}$. The union $\mathcal{S}(\overset{-1}{f}(\mathcal{V}_f), \overset{-1}{g}(\mathcal{V}_g)) = \{[1.75, 2.25] \cup [3.75, 4], [1.75, 2.25] \cup \{4\}, [2, 2.25] \cup [3.75, 4], \{2, 4\}\}$. $\lim \mathcal{S} = \{2, 4\}$, $\lim \overset{-1}{f}(\mathcal{V}_f) = \{2\}$, $\lim \overset{-1}{g}(\mathcal{V}_g) = \{2, 4\}$. Valuation of the limits yields $\varphi_f(\lim \overset{-1}{f}(\mathcal{V}_f)) = \{1\}$, $\varphi_g(\lim \overset{-1}{g}(\mathcal{V}_g)) = \{0\}$, $\varphi(\lim \mathcal{S}) = \{1\} \cup \{0\} = \{0, 1\}$. The intersection $\mathcal{D}(\overset{-1}{f}(\mathcal{V}_f), \overset{-1}{g}(\mathcal{V}_g)) = \{[2, 2.25], \{2\}\}$. $\lim \mathcal{D} = \{2\} = \{2\} \cap \{2, 4\}$, valuated by $\varphi(\lim \mathcal{D}) = \{0, 1\}$.

4 Some applications

4.1 Contractive mappings

Let $X = X_{[0]}$ be a non-empty set and let $\mathcal{F} = (f_{[n]})_{n \in \mathbf{N}}$ be a family of functions $f_{[n]}$: $X \to X$, $X_{[n]} =_{def} f_{[n]}(X)$. If $\mathcal{X} = (X_{[n]})_{n \in \mathbf{N}}$ forms a filter base on pow X, we say, $\mathcal{F}$ is a "filter generating" or a "contractive" function sequence on X. We assume $\emptyset \neq X^* = \bigcap \mathcal{X}$ and have $X^* \subseteq X$, $f_{[n]}(X^*) \subseteq X_{[n]}$, $\bigcap_{\mathbf{N}} f_{[n]}(X^*) \subseteq X^*$.

If in particular f: $X \to X$, i.e. $f(X) \subseteq X$, and $\wedge n \in \mathbf{N}$ $(f_{[n]} =_{def} f^n)$, we have iterations, monotone $\mathcal{X}$ and $f(X^*) \subseteq X^*$. For $X^* = \{x^*\}$, x^* is a fixpoint.

We assume, a function h: $\mathbf{N} \to X$ exists such that for the filter base $\{B_{[n]} = \{z \mid n \leq z \wedge z \in \mathbf{N}\} \mid n \in \mathbf{N}\}$ $X_{[n]} = \{h(z) \mid z \in B_{[n]}\}$. If h is known on $\{1,2,..m\}$, extrapolation $\tilde{h}$ of h to $\{m+1, ...\}$ can be used to find an approximation $\tilde{X}_{[m+1]}$ to $X_{[m+1]}$. This is a well known practice in numerical analysis.

4.2 Roundings

For $n = 1,2,....$ we consider idempotent mappings (roundings) $\rho_{[n]}$: $X_{[n-1]} \to X_{[n]}$, $\emptyset \neq X_{[n]} = \rho_{[n]}(X_{[n-1]}) \subset X_{[n-1]}$. The set extensions of roundings are idempotent $\subseteq$-homomorphisms: for $\emptyset \neq X' \subseteq X'' \subseteq X_{[n-1]}$ we have $\rho_{[n]}(X') \subseteq \rho_{[n]}(X'')$, $\rho_{[n]} \circ \rho_{[n]}$

$(X') = \rho_{[n]}(X')$, and $\rho_{[n]}(X') \subseteq \rho^{-1}_{[n]}\rho_{[n]}(X')$. The set $\{\rho^{-1}_{[n]}\rho_{[n]}(X'), \rho_{[n]}(X')\}$ is a filter base. Given roundings $\rho_{[k]}: X_{[k-1]} \to X_{[k]}$ for stages $k = 1,2,\ldots n$, we define $f_{[k]} = \rho_{[k]} \circ \rho_{[k-1]} \ldots \circ \rho_{[1]}$ to obtain a contractive sequence $(f_{[k]})_{k=1,\ldots n}$.

Example: Rounding of the reals $\mathbf{R}$ onto integers $\mathbf{Z}$ by $\rho(r) \to \{z\}$ for $r \in (z - 0.5, z + 0.5]$ and $z \in \mathbf{Z}$. On the next stage, $\mathbf{Z}$ can be rounded by another rounding rule, for example onto $10 \times \mathbf{Z}$, and so on.

4.3 Valuated filter hierarchies ("pyramids")

We generalize multi-stage rounding: Let be $X = X_{[0]}$, $\Sigma \subseteq \text{pow}\, X$ a σ-algebra, $\mu: \Sigma \to \mathbf{R}_+$ a σ-additive measure, $X_{[0]}$ partitioned into a finite number of non-empty subsets $X_{[0,k]}$, $k = 1,2,\ldots K_{[0]}$, with $X_{[0,k]} \in \Sigma$. For all k let there be given roundings $\rho_{[0k]}: X_{[0k]} \to \rho_{[0k]}(X_{[0,k]}) \subset X_{[0k]}$. We define $\mu_{[0k]} =_{def}$ measure of $\rho_{[0k]}(X_{[0,k]}) =_{def} \mu(X_{[0,k]})$, $\mu_{[0]} =_{def} +(\mu_{[0k]})_{k \in K_{[0]}}$.

For $n = 0,1,2,\ldots$ let $K_{[n]}$ be partitioned into $I_{[ni]}$, $i = 1,\ldots,K_{[n+1]}$. For $k = 1,\ldots,K_{[n+1]}$ we define

$X_{[n+1,k]} =_{def} \bigcup\{\rho_{[ni]}(X_{[ni]}) \mid i \in I_{[nk]}\}$,

$\mu_{[n+1,k]} =_{def}$ measure of $X_{[n+1,k]} =_{def} +(\mu_{[ni]})_{i \in I_{[nk]}}$,

$X_{[n+1]} =_{def} \bigcup\{X_{[n+1,k]} \mid k \in K_{[n+1]}\}$,

$\mu_{[n+1]} =_{def}$ measure of $X_{[n+1]} =_{def} +(\mu_{[n+1,k]})_{k \in K_{[n+1]}}$,

roundings $\rho_{[n+1,k]}: X_{[n+1,k]} \to \rho_{[n+1,k]}(X_{[n+1,k]}) \subset X_{[n+1,k]}$.

We have $\mu_{[0]} = \mu_{[1]} = \ldots$ (conservation of total measure). An illustration is given in Fig.4.

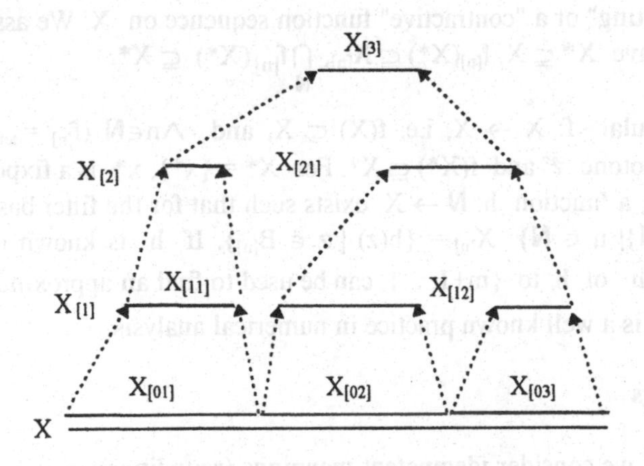

Fig. 4

An application to valuated picture pyramids is shown in Fig. 5. The grid is $\{-2, -1, 0, +1, +2\} \times \{-2, -1, 0, +1, +2\}$, additive valuations are $\{0, .25, .5, ...15\}$, not all grid points are valuated. The example shows for level 0 how overlap of $X_{[n-1,k]}$'s can be handled.

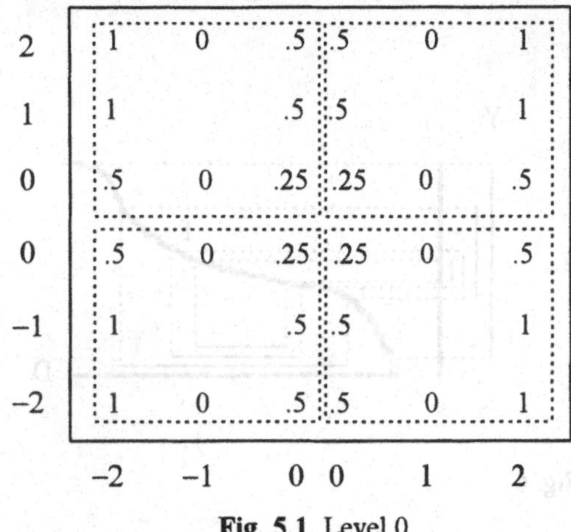

Fig. 5.1. Level 0

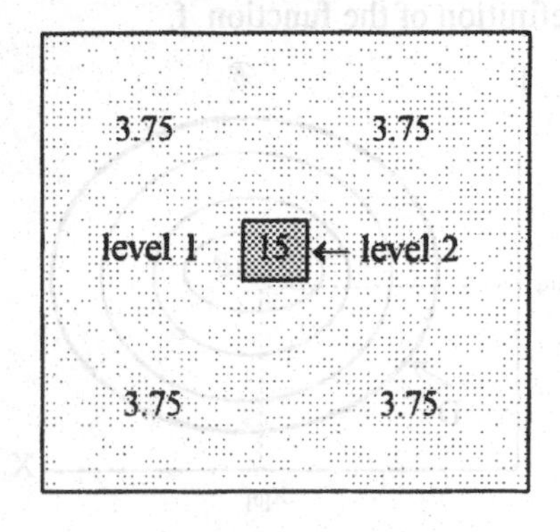

Fig. 5.2. Level 1 and 2

4.4 Adaptable networks

We first consider a function $f: X \to Y$. X and Y can be multi-dimensional, $X \subseteq X_{[1]} \times ... \times X_{[n]}$, $Y = Y_{[1]} \times ... \times Y_{[m]}$. We write $y = f(x)$, $x = (x_{[i]})_{i=1,2,...,n}$, $y = (y_{[j]})_{j=1,2,...,m}$. Let there be given $x_{[0]} \in X$ and let $y_{[0]} = f(x_{[0]})$. To deal with values $\tilde{x}$, $\tilde{y} = f(\tilde{x})$ which are approximations to $x_{[0]}$, $y_{[0]}$ we introduce a neighborhood system to $x_{[0]}$ by a filter base $\mathscr{B}$ on pow X with $x_{[0]} \in \bigcap \mathscr{B}$. The set extension of f maps $\mathscr{B}$ on a filter base $f(\mathscr{B})$ with $y_{[0]} \in \bigcap f(\mathscr{B})$. If each base is isomorphically and injectively valuated by values v of a filter base $\mathscr{V}$ on a complete lattice V and if we assume $v(\bigcap \mathscr{B}) = v(\bigcap f(\mathscr{B})) = \bigcap \mathscr{V}$, we can measure generalized distances of $\tilde{x}$ to $x_{[0]}$ and $\tilde{y}$ to $y_{[0]}$ according **2.2** by their values $v(\tilde{x})$ and $v(\tilde{y})$. A simple case is $\mathscr{B}$ monotone.

Example (interval arithmetic on $\mathbf{R}$): $f \in \{+,-,*,/\}$, $X = \mathbf{R} \times \mathbf{R}$ for $\{+,-,*\}$, $X = \mathbf{R} \times (\mathbf{R} \setminus \{0\})$ for /, $Y = \mathbf{R}$. $x_{[0]} = (x_{[01]}, x_{[02]})$, $\mathscr{B} = \{\text{interval } I = (x_{[01]}-a, x_{[01]}+b] \times (x_{[02]}-a', x_{[02]}+b'], X\}$, assuming $0 \notin (x_{[02]}-a', x_{[02]}+b']$ for / . a, b, a′, b′ positive reals. $\lim \mathscr{B} = I$, $f(\mathscr{B}) = \{f(I), \mathbf{R}\}$. All $\tilde{x} \in I$ and all $\tilde{y} \in f(I)$ are equivalent. $v(I) = v(f(I)) = \{1\}$, $v(X) = v(Y) = \{0, 1\}$.

Now we consider a parameterized function $f: D \to Y$ with $D \subseteq X \times P$, P a non-empty set of parameters $p = (p_{[l]})_{l \in L}$. Changing p results in changing the function and thus in general in changing the function value for fixed x. We say, f is "controllable" or "programmable" and p is a "control parameter" or a "program" to f. To $(x_{[0]}, p_{[0]}) \in D$ a filter base (neighborhood system) $\mathscr{B} = \{B_{[k]} \mid k \in K \wedge B_{[k]} \subseteq D\}$, $D = \bigcup \mathscr{B}$, can be introduced and then mapped by the set extension of f onto a filter base $f(\mathscr{B})$. A visualization is given in Fig. 6.

Notice, in general $D \neq X \times P$, not all values (x,p) belong to the domain of definition of the function f.

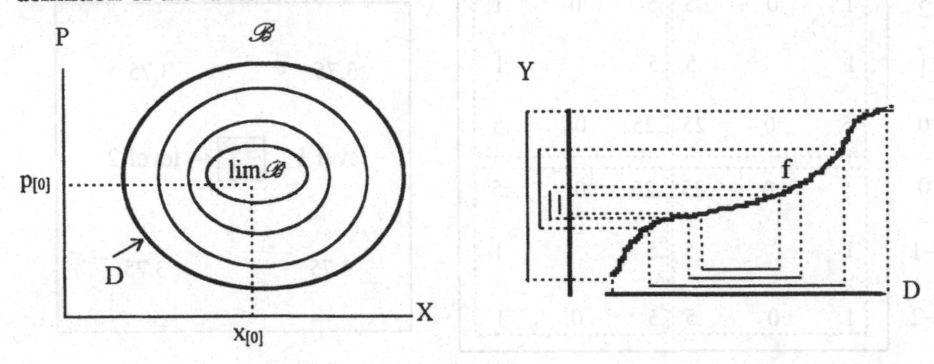

Fig. 6

The parameters are used to adapt the function f to a certain behavior: For fixed $x_{[0]}$, the cut of D along $x_{[0]}$ is a filter base $\mathscr{B}(x_{[0]})$. In practice a sequence of approximate parameters $\tilde{p}$ is constructed converging to $\lim \mathscr{B}(x_{[0]})$ by trying to minimize the deviation of $f(x_{[0]}, \tilde{p})$ from $f(x_{[0]}, p_{[0]})$ ("learning").

Finally, let there be given a family $\mathscr{F}$ of parameterized functions, $\mathscr{F} = (y_s = f_s(x_{[s]}, p_{[s]}))_{s \in S}$, S finite. Notice, for $s \neq s'$ $f_{[s]} = f_{[s']}$ is possible. Composing functions $f_s, f_q \in \mathscr{F}$ by identifying ("connecting") some ("output" with "input") components $y_{[sj]} = x_{[qi]}$, can (but need not) result in a composite function F. A theory on this is for example given in articles of Albrecht [6] and of Pearson and Dray [7]. To adapt an F to a wanted behavior is in general a complex multi-parameter problem which has two variants:

selection of $\mathscr{F}$ and of the interconnections for given parameters p (numerical analysis),

adaptation of the parameters p for given $\mathscr{F}$ and given interconnections (artificial neural networks).

Example (simple classifier): We assume $\bigwedge s \in S\ (X_{[s]} = X \wedge Y_{[s]} = Y = [0, 1] \subset \mathbf{R} \wedge f_s(X) = Y)$. Some patterns $x = (x_{[1]}, ..., x_{[n]}) \in X$ are considered to belong more or less to the same class s. We take a sample of such x to group them in a filter base $\mathscr{B}(s)$ on pow X. We try to adapt the (multi-dimensional) parameter in $f_s(x,p_{[s]})$ to a parameter $p_{[s0]}$ for which $f_s(x, p_{[s0]})$ maps only $\mathscr{B}(s)$ isomorphically on a filter base $\mathscr{V}(s)$ on pow Y with $\bigcap \mathscr{B}(s)$ mapped on $\{1\}$, but not a $\mathscr{B}(s')$ with $s' \neq s$. If this has been done for all s, then to any given $\hat{x} \in X$, we obtain a family h =

$(\hat{y}_s)_{s\in S}$ of values $\hat{y}_{[s]}$, representing a function on S. The reciprocal $\overset{-1}{h}$ can be used for cuts along any $U \subset Y$, e.g. to find an s with maximal $\hat{y}_{[s]}$. A visualization is given in Fig. 7.

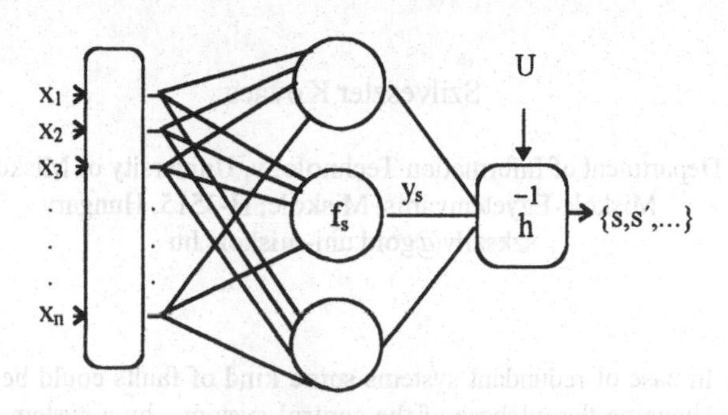

Fig. 7

Frequently used is $X = \mathbf{R}^n$, $f_s = \exp(-\frac{1}{2}(x-x_{[s]}^{(0)})^T\ K_{[s]}^{-1}\ (x-x_{[s]}^{(0)}))$, x the input vector, parameters are the center vector $x_{[s]}^{(0)}$ and the elements of the inverse $K_{[s]}^{-1}$ (if not singular) of the covariance matrix $K_{[s]}$.

References

1. Cantor G.: Über eine Eigenschaft des Inbegriffs aller reellen algebraischen Zahlen. Jour. Reine u. Angew. Mathematik (1874)
2. Lebesgue H.: Leçons sur l'intégration et la recherche des fonctions primitives. Paris (1904), 2[ed] ed. Paris (1928)
3. Zadeh L. A.: Fuzzy Sets. Information and Control **8** (1965)
4. Vietoris L.: Stetige Mengen. Monatshefte f. Mathematik u. Physik **31** (1921)
5. Bourbaki N.: Topologie Général. Act. Sc. Ind., Vol. 1142. Hermann & Cie, Paris (1951) 40-41
6. Albrecht R. F.: On mathematical systems theory. In: R. Albrecht (ed.): Systems: Theory and Practice. Springer-Verlag, Vienna New York (1998) 33-86
7. Pearson D. W., Dray G.: Applications of artificial neural networks. In: R. Albrecht (ed.): Systems: Theory and Practice. Springer-Verlag, Vienna New York (1998) 235-252

Similarity Based System Reconfiguration by Fuzzy Classification and Hierarchical Interpolate Fuzzy Reasoning

Szilveszter Kovács

Department of Information Technology, University of Miskolc
Miskolc-Egyetemváros, Miskolc, H-3515, Hungary
szkszilv@gold.uni-miskolc.hu

Abstract. In case of redundant systems some kind of faults could be tolerated simply by changing the rulebase of the control system - by a system (strategy) reconfiguration. One solution for this kind of system reconfiguration could be the combination of the fuzzy clustering based symptom evaluation and the hierarchical interpolate fuzzy reasoning - *the similarity based system reconfiguration*. Its main idea is the following: More similar the actual symptom to one of the system behaviour classes, more similar must be the final conclusion to the decision done by the strategy handling that system behaviour class. A similarity based system reconfiguration method and as an example of its practical application, fault diagnosis and reconfiguration of a simplified three-tank benchmark system is introduced in this paper.

1 Introduction

The main idea of the similarity based system reconfiguration -*more similar the actual symptom to one of the system behaviour classes, more similar must be the final conclusion to the decision done by the strategy handling that system behaviour class* - is based on the premise, that all the possible behaviours of a given system can be handled by a combination of control strategies belonging to some special known system behaviours. (Handling the relevant system behaviours by special control strategies and generating all the other strategies as a combination of the relevant ones.)

The two main tasks of the similarity based system reconfiguration are the following:

- decision about the level of similarities the actual system behaviour to the studied relevant system behaviours (the level of necessity and the type of the rulebase needed to handle the actual system behaviour),
- generating the correct control action, as a combination of the control actions belonging to the known relevant strategies.

One solution for these tasks could be the combination of the fuzzy clustering based symptom evaluation and the hierarchical interpolate fuzzy reasoning. The conclusion of the fuzzy clustering based symptom evaluation is a set of similarity values. These are the levels of similarities of the actual system behaviour to all the known behaviour classes (the normal and all the handled known faulty system behaviours). Having all the conclusions of the different strategies (handling the different kind of system behaviours), as an upper level *interpolate reasoning*, we could simply combine these conclusions in a function of the corresponding similarities to get the actual final conclusion.

2 Fuzzy clustering in symptom evaluation

The symptom (or residual) evaluation is a very important and in many cases a very heuristic step of fault diagnosis. The task of symptom classification, is basically a series of similarity checking between an actual symptom (or residual) and a series of known symptoms (characterising the normal and the faulty behaviour of the studied system).

The task of fuzzy clustering in symptom evaluation could be viewed as a kind of information compressing step. It can reduce the number of the observed symptoms (characterising the main classes of faulty or normal system behaviour of the monitored process) dramatically, to a few relevant symptoms patterns (these relevant symptoms are forming fuzzy partitions). Based on these patterns, evaluating the actual symptom is nothing else than calculating the similarity values of the actual symptom (the actual system behaviour) to all the relevant symptoms patterns (the known - normal and faulty - system behaviours).

This case the task of "generating the relevant symptom patterns" is a two stage clustering hierarchy. The first level is based on a priori human knowledge about the possible fault classes. The main goal of this step is to classify the system behaviours to normal and some faulty behaviour classes. The second one is an automatic information processing, or data compression stage. Its main task is to characterise the observed symptoms of a system behaviour class (faulty or normal) only by a few relevant symptoms (characterising all the observed symptoms belonging to the same behaviour class).

2.1 Evaluating the actual symptom by fuzzy classification

The actual symptom is basically a point in the symptom space. Its evaluation, is nothing else than calculating the similarity (membership) values between this point and the member fuzzy sets of the fuzzy partition characterising the relevant system behaviours. These membership values can be calculated the same way as the relevant symptom patterns generation has calculated the fuzzy partition matrix (e.g. applying functions of the Fuzzy c-Means fuzzy clustering algorithm [1]). Having more cluster centres for the same system behaviour class, the membership value u_i of a given point **x** to the ith system behaviour class, characterised by a set of cluster centres C_i, could be calculated by the following formula [1]:

$$u_i = \sum_{\forall k \in C_i} \frac{1}{\sum_{j=1}^{c} \left(\frac{d_k}{d_j} \right)^{\frac{2}{m-1}}}, \qquad (1)$$

where $d_k = \|\mathbf{x} - \mathbf{v}_k\|$, the distance (measure of dissimilarity) of the point **x** and the kth cluster centre $\mathbf{v}_k$, m is a weighting exponent (usually m=2), c is the number of the clusters (cluster centres) and C_i is the ith subset of cluster centres belonging to the same i[th] system behaviour class.

The result of the symptom evaluation is a set of membership value, a set of similarity measures between the studied symptom, and the system behaviour classes.

3 Similarity based system reconfiguration

The main idea of the similarity based system reconfiguration is the following: more similar the actual symptom to one of the system behaviour classes, more similar must be the final conclusion to the decision done by the strategy handling that system behaviour class.

The conclusion of the fuzzy clustering based symptom evaluation is a set of similarity values, the level of similarities of the actual system behaviour to all the studied ones. Having all the conclusions of the different strategies (handling the different kind of relevant system behaviours FLC_i), the actual conclusion could be simply combined from them in the function of the corresponding similarities (u_N, u_{Fi}), as an upper level *interpolate fuzzy reasoning* [2] (see Fig.1a.).

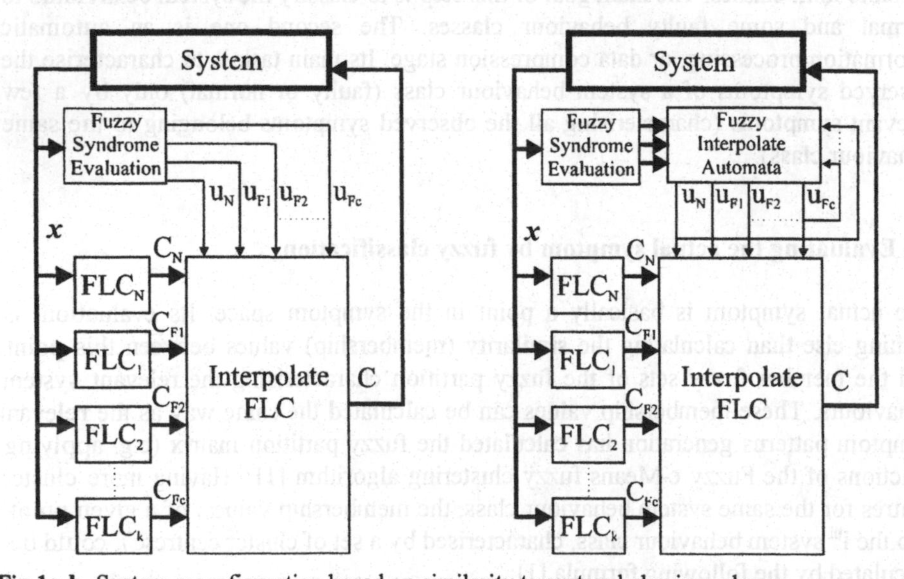

Fig.1a,b. System reconfiguration based on similarity to system behaviour classes

3.1 Evaluating the actual symptom

Having similarity based system reconfiguration the syndrome evaluation is more complicated. In most cases the syndromes are strongly dependent on the actual behaviour (normal or faulty) of the system. Each system behaviour class with the corresponding control strategy has its own syndrome structure (these syndrome structures are characterising the system behaviour classes). It means, that the relevant symptom patterns (characterising the normal behaviour and all the studied fault classes of the system) must be determined separately in all the studied system behaviour classes (see example on Fig. 3 and Fig. 4).

Having all these relevant symptoms the fuzzy syndrome evaluation works as a *fuzzy automata.* The actual system behaviour is characterised by a set of similarities, the similarities of the actual system behaviour to the studied relevant system behaviours. Depending on the actual system behaviour (a set of similarities) the fuzzy syndrome evaluation evaluates the actual symptom to get the next system behaviour as a conclusion (see Fig. 1b.). More precisely, considering the actual system behaviour to be the state variables (a set of similarities - fuzzy membership values), it determines the new state as a function of the actual state and the conclusions of the syndrome evaluation. In other words the conclusions of the syndrome evaluation are the state-transition functions of the fuzzy automata. The automata is a fuzzy automata, because of the state variables are fuzzy membership values (similarities – infinite states) and the state-transitions are driven by fuzzy rules.

The rulebase applied for the state-transitions of the fuzzy automata (rules for interpolate fuzzy reasoning) for the ith state S_i (R_{Ai}):

If S_i = One **And** S_i-S_i = One **Then** S_i = One
If S_i = One **And** S_i-S_k = One **Then** S_i = Zero
If S_k = One **And** S_k-S_i = One **Then** S_i = One
If S_k = One **And** S_k-S_i = Zero **Then** S_i = Zero

where S_i-S_j is the conclusion of the syndrome evaluation about the state-transition from state i to j.

The structure of the state-transition rules is similar for all the states (relevant system behaviours).

The reason of the interpolate way of fuzzy reasoning is the incompleteness of state-transition rulebase [3].

3.2 Combining the conclusions of the different strategies

Having all the conclusions of the different strategies (handling the different kind of relevant system behaviours FLC_I - C_i), the actual conclusion can be generated as their combination (in a function of the corresponding similarities (u_N, u_{Fi})) (see Fig. 1b.). The simplest way for such a combination is the application of the *interpolate fuzzy reasoning* [2] (as an upper level reasoning). The main idea of the similarity based system reconfiguration - *more similar the actual symptom to one of the system behaviour classes, more similar must be the conclusion to the decision done by the strategy belonging to that system behaviour* - can be directly translated to an interpolate fuzzy rulebase. (Applying interpolate fuzzy reasoning the completeness off the fuzzy rulebase is not necessary.)

The rulebase applied for the upper level interpolate fuzzy reasoning to combine the conclusions in a function of the corresponding similarities (see Fig.1b. for notation) is the following:

If u_N=One **And** u_{F1}=Zero **And** ... **And** u_{Fc}=Zero **Then** C=C_N
If u_N=Zero **And** u_{F1}=One **And** ... **And** u_{Fc}=Zero **Then** C=C_{F1}
...
If u_N=Zero **And** u_{F1}=Zero **And** ... **And** u_{Fc}=One **Then** C=C_{Fc}

4 Fault diagnosis and reconfiguration of the three-tank example

As a simple demonstration of the proposed similarity based system reconfiguration, fault diagnosis and reconfiguration of a simplified configuration (two tanks only) of the three tank benchmark (see Fig.3.) is introduced in the followings.

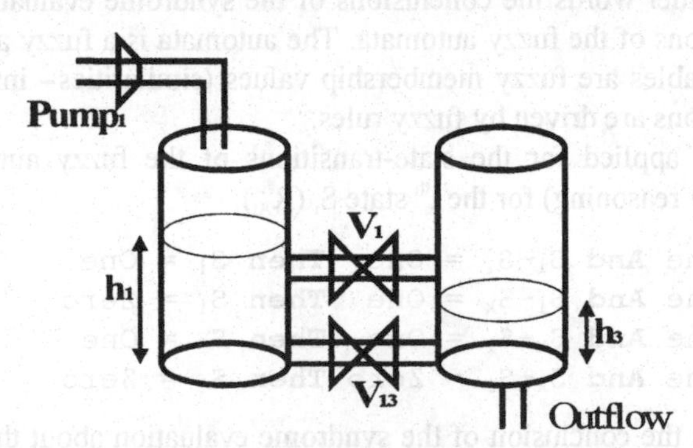

Fig.2. Simplified configuration of the three tank benchmark

The goal of the control system is to keep the water levels in $tank_1$ and $tank_3$ h_1=0.5 and h_3=0.1 by controlling the $valve_{13}$ and the $pump_1$ at a constant value of outflow from $tank_3$ (normal behaviour of the system).

The example is concentrating of the faults of the $valve_{13}$. Were this valve opened and blocked, the water level in $tank_3$ h_3=0.1 could be controlled by $pump_1$ (this case h_1 is changed) - this is the fault condition no.1. Were this valve closed and blocked, the water levels in $tank_1$ and $tank_3$ h_1=0.5 and h_3=0.1 could be controlled by the $valve_1$ and the $pump_1$ - this is the fault condition no.2.

The first step of the proposed method is to handle the separate system behaviours by separate controllers (one controller for the normal, one for fault 1 and one for fault 2). The controllers used for handling the different system behaviours are interpolate fuzzy logic controllers [2].

The simulated results of the different system behaviours direct handled by their own specific controllers are introduced on Fig.3.

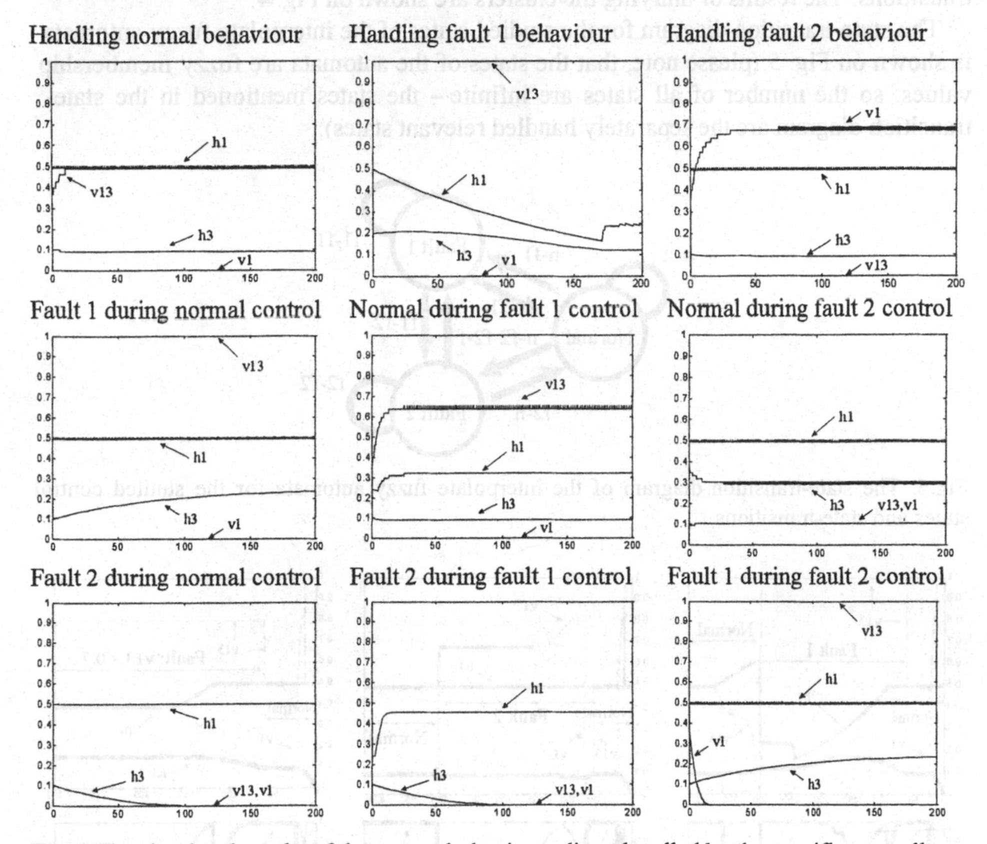

Fig.3. The simulated results of the system behaviours direct handled by the specific controllers

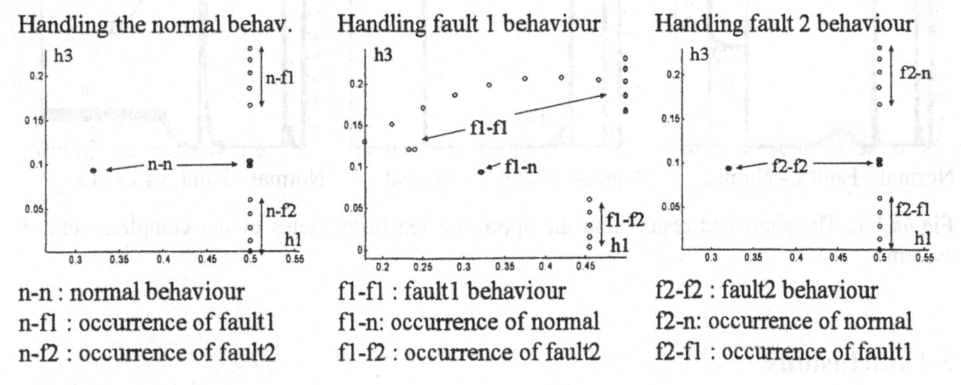

Fig.4. The unified clusters characterising all the studied control states and state-transitions.

The next step is the clustering. We have to characterise all the system behaviour classes by different relevant symptoms - by cluster centres. Having the relevant symptom patterns, the next step is unifying the cluster centres (fuzzy partitions), to characterise the system behaviour classes in all the studied control states and state-transitions. The results of unifying the clusters are shown on Fig. 4.

The state-transition diagram for the studied states of the interpolate fuzzy automata is shown on Fig. 5 (please note, that the states of the automata are fuzzy membership values, so the number of all states are infinite – the states mentioned in the state-transition diagram are the separately handled relevant states).

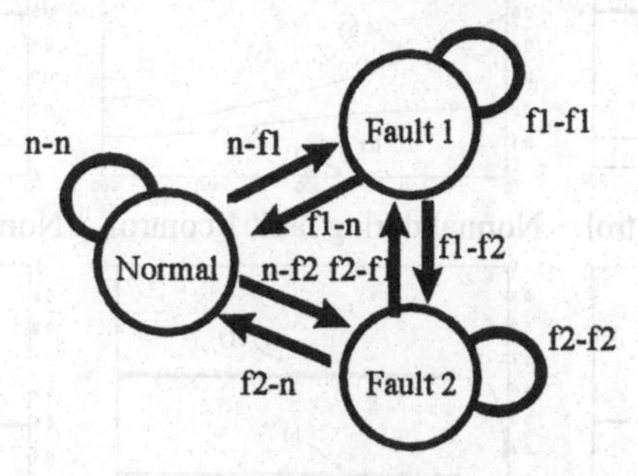

Fig.5. The state-transition diagram of the interpolate fuzzy automata for the studied control states and state-transitions.

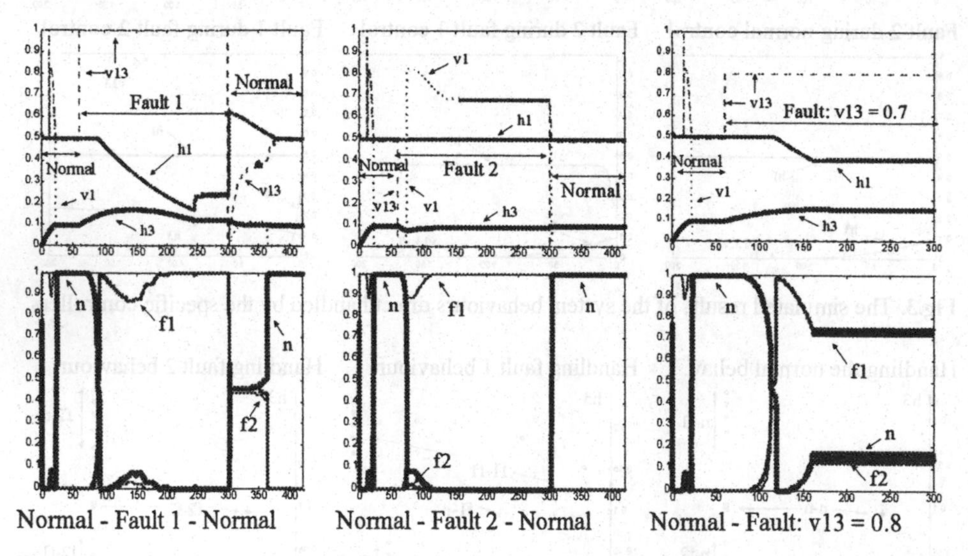

Normal - Fault 1 - Normal   Normal - Fault 2 - Normal   Normal - Fault: v13 = 0.8

Fig.6a,b,c. The simulated results and the approximated fuzzy states of the complete control system

5 Conclusions

The simulated example application demonstrated, that the fuzzy automata is able to follow the studied relevant states and state-transitions (normal, fault1 and fault2 behaviour). Moreover, because of the interpolate properties of the fuzzy automata in some cases it is able to approximate the unstudied states too (its states are the similarities of the actual system state to the studied ones). Based on these similarities the hierarchical interpolate fuzzy logic controller can combine the existing (studied) control strategies to handle unknown (unstudied) fault situations (see e.g. Fig. 6c, where v13=0.8 open is an unstudied fault situation).

This approximated control strategy could be correct only in the case if the unknown fault situation can be handled as a slight modification or combination of the studied ones.

One simple practical application of the proposed controller structure could be the following: study and handle only the relevant characteristic fault situations of a system (faults needs significantly different handling) and let the proposed similarity based system reconfiguration to handle all the others by approximation. (Handling the relevant system behaviours by special control strategies and generating all the other strategies as a combination of these relevant ones.)

6 Acknowledgement

This research was partly supported by the Hungarian National Scientific Research Fund grant no: F 029904 and T019671, the IQQFD COPERNICUS project No. PL-96-4383, and the Hungarian Ministry of Culture and Education grant no 0422/1997.

References

1. Bezdek, J.C.: Pattern Recognition with Fuzzy Objective Function, Plenum Press, New York, (1981)
2. Kovács, Sz., Kóczy, L.T.: Approximate Fuzzy Reasoning Based on Interpolation in the Vague Environment of the Fuzzy Rule base as a Practical Alternative of the Classical CRI, Proceedings of the 7th International Fuzzy Systems Association World Congress, Prague, Czech Republic, (1997) 144-149
3. Kóczy, L. T., Hirota, K.: Interpolative reasoning with insufficient evidence in sparse fuzzy rule bases, Information Sciences, Vol. 71, (1992) 169-201

A Fuzzy System for Fetal Heart Rate Assessment

J.F. Skinner, J.M. Garibaldi, E.C. Ifeachor

School of Electronic, Communication & Electrical Engineering,
University of Plymouth,
Plymouth, PL4 8AA, UK.
Tel +44-1752-232573
Email: jonski@cis.plym.ac.uk
Web: http://www.cis.plym.ac.uk/cis/home.html

Abstract. The clinical interpretation of fetal heart rate traces is a difficult task that has led to the development computerised assessment systems. These systems are limited by their inability to represent uncertainty. This paper describes the first stage in the development of a fuzzy expert system for fetal heart rate assessment. A preliminary evaluation study comparing the initial fuzzy system with three clinicians and an existing crisp expert system is presented. The fuzzy system improved on the crisp system and achieved the highest overall performance. The use of fuzzy systems for analysis of fetal heart rate traces and similar time varying signals is shown to have potential benefit.

1 Introduction

A continuous recording of fetal heart rate and maternal contractions, the cardiotocogram (CTG), is routinely used to monitor the health of the fetus during labour. The outcome of labour is usually good for the fetus, however problems may occur that can result in permanent fetal brain damage or even death. Changes in the fetal heart rate, and their timing relative to maternal contractions provide an indication of 'fetal distress', and are used to identify those cases which require clinical intervention. CTG interpretation is a difficult task, requiring clinical experience and significant expertise. Studies have shown that this is often lacking in the clinical setting, with CTG misinterpretation implicated in a large number of preventable fetal deaths [1] and unnecessary interventions [2].

As a result, a number of computerised systems have been developed to provide decision support for CTG interpretation. These range from simple feature extraction and classification programs [3,4,5,6], to intelligent expert systems that assess the CTG along with clinical information to provide management advice [7,8,9,10]. Preliminary evaluation of these systems has shown some promising results, but the inherent uncertainty in CTG interpretation has limited their performance.

Much of the knowledge relating to CTG interpretation is based on empirical observations made in the 1960s that certain heart rate features are associated with poor fetal outcome. An example of 15-minutes of CTG trace, with the important features labelled, is shown in Fig. 1.

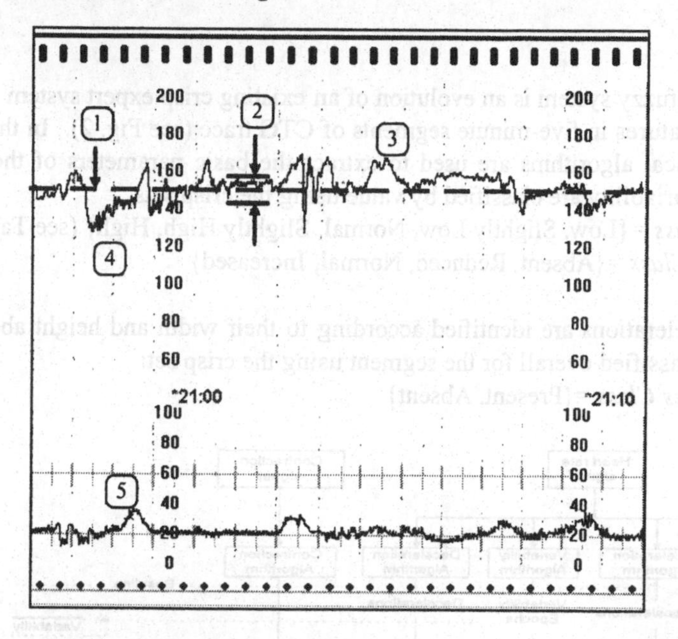

Fig. 1. An example 15 minute segment of cardiotocogram with important features. The top trace is fetal heart rate and the bottom trace maternal contractions. The features of interest are: *Baseline-* the basic heart rate value about which the heart rate pattern fluctuates *(1)*, *Variability-* peak to peak amplitude of high frequency perturbations about the baseline *(2)*, *Accelerations-* relatively long term transient increases in heart rate from the baseline *(3)*, *Decelerations-* relatively long term transient decreases in heart rate from the baseline *(4)*, *Contractions-* maternal contractions *(5)*

Feature classification guidelines have been suggested by many researchers, and those published by FIGO are now generally accepted [11]. Definitions extracted from such guidelines are often used to form the knowledge base in computerised systems [7,8,10]. However, while the classification and meaning of features is defined in precise terms, the identification of features is linguistic and inexact. For example, a normal baseline is defined as 110-160 beats per minute (bpm), however there are no precise definitions of how to identify the baseline. In consequence, any classification based on these guidelines will have inherent uncertainty. This uncertainty cannot easily be represented or manipulated in traditional rule based systems. The use of fuzzy techniques in such systems may improve performance by allowing uncertainty to be explicitly represented [12].

In this paper we present the development of a fuzzy system for assessment of the cardiotocogram. In section 2 the design of the fuzzy system is explained. Section 3 describes the preliminary evaluation of the fuzzy system compared with a previous crisp system and three clinicians.

2 Methods

The developed fuzzy system is an evolution of an existing crisp expert system [7] that assesses the features in five-minute segments of CTG trace (see Fig. 2). In this crisp system, numerical algorithms are used to extract the basic parameters of the CTG. Baseline and variability are classified by value using the crisp sets:

Baseline Class ={Low, Slightly Low, Normal, Slightly High, High} (see Table 1.)
Variability Class ={Absent, Reduced, Normal, Increased}

Individual accelerations are identified according to their width and height above the baseline and classified overall for the segment using the crisp set:

Accelerations Class ={Present, Absent}

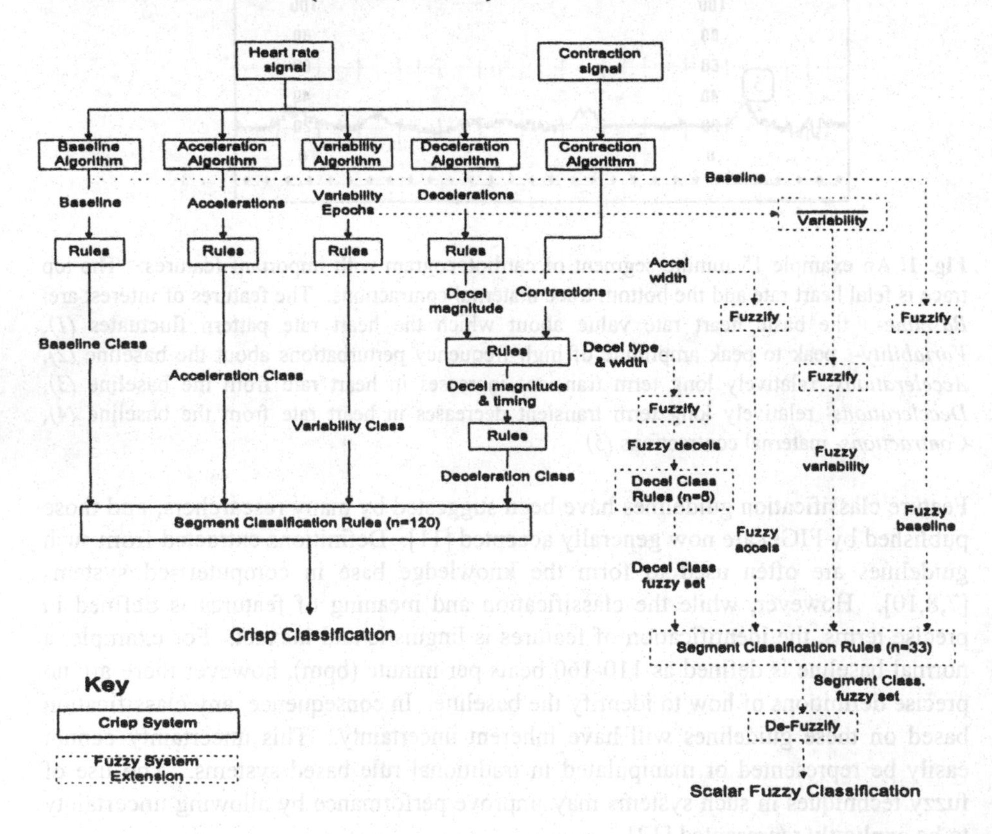

Fig. 2. The architecture of the fuzzy CTG assessment system and its relationship to the existing crisp CTG system.

Table 1. Crisp classification ranges of baseline heart rate

Baseline value (beats per minute)	Crisp Classification
<90	Low
90-109	Slightly low
110-159	Normal
160-179	Slightly high
>180	High

Individual decelerations are identified according to their width, depth and area below the baseline. Their magnitude is classified using the crisp set:

Deceleration Size ={Absent, Present, Severe}

Identified decelerations are further classified by their timing relative to a contraction to give the set:

Deceleration Type ={Absent, Early, Late, Severe Early, Severe Late}

Classified decelerations in the segment are assessed using a small rule set (n=5) to provide the overall crisp deceleration classification set for the segment:

Decelerations Class ={Absent, Present, Severe}

The classifications of the four basic features are assessed using a second rule set (n=120) to provide a crisp overall classification for every possible combination of crisp input features. The crisp classification output set is:

Segment Class ={Normal, Intermediate, Abnormal, Severely abnormal}

Such a crisp classification provides little information regarding *how* abnormal a segment actually is, and is also unstable for small changes in the input parameters near classification boundaries.

2.2 The fuzzy system

The fuzzy system builds on and extends the feature extraction algorithms in the crisp system. The relationship between the two system architectures can be seen in Fig. 2. Features identified by the crisp system are fuzzified and assessed using new rule sets. The values of baseline and variability calculated by the existing algorithms are fuzzified onto their respective termsets:

Baseline ={Very low, Low, Normal, High, Very high}

Variability ={Absent, Reduced, Normal, Increased}

These fuzzy sets are based on the existing crisp classification sets. (See Table 1 and Fig. 3 for the relationship between crisp and fuzzy baseline) The fuzzy classification of accelerations is found by considering the total duration of identified accelerations as a proportion of the segment length. The acceleration classification set is:

Accelerations ={Absent, Present} (Fig. 4)

Each of the four types of deceleration are similarly fuzzified by their proportional duration in the segment, using the sets:

Deceleration Type ={Absent, Present}.

A small rule set (n=5) based on the equivalent crisp rule set is used to provide an overall fuzzy classification of the decelerations in the segment. The overall deceleration classification set is:

Decelerations Class ={Absent, Present, Severe}.

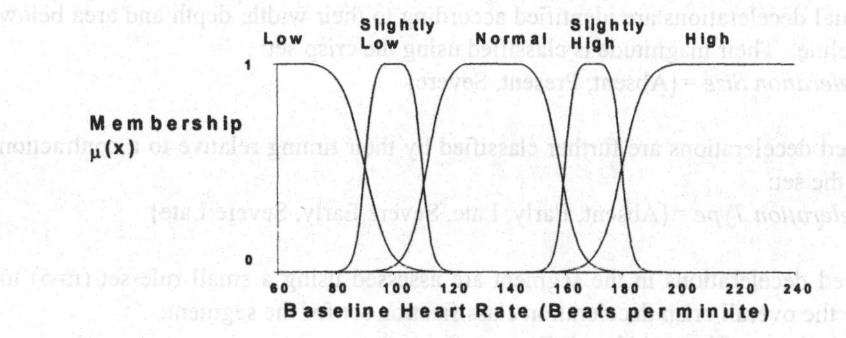

Fig. 3. Fuzzy membership sets for baseline heart rate.

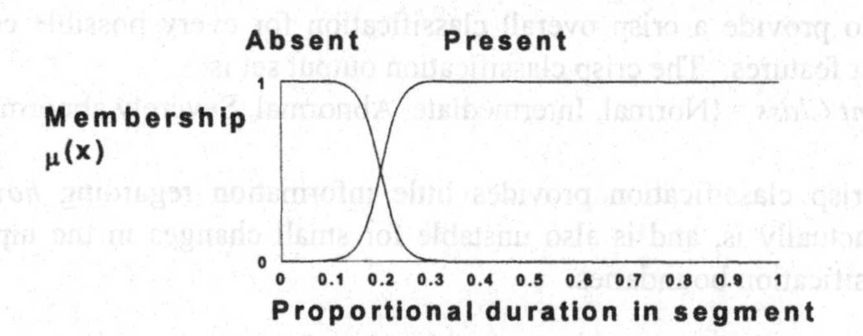

Fig. 4. Fuzzy membership sets for accelerations.

A second rule set provides overall segment classification based on the fuzzy variables *Baseline*, *Variability*, *Accelerations* and the fuzzy set *Deceleration Class*. In this manner the deceleration classification output set is forward chained directly into the overall segment classification. The final output set is:

Segment Class ={Normal, Intermediate, Abnormal, Severely abnormal}.

This rule set was derived from the existing crisp system, but was reduced from n=120 to n=33 using the ID3 rule induction algorithm [13]. This algorithm automatically searches for the most efficient combination of rules that will achieve identical overall behaviour in a rule set, thereby removing redundancy and reducing the total number of rules required to represent knowledge.

All membership functions in the system are normal Gaussian and MaxMin inference is used throughout. The segment classification output set is defuzzified using the centroid method to give a scalar value (0-100), to provide an overall index of the segment abnormality. The performance of the system on a number of example cases was examined and minor changes to the position and width of fuzzy membership functions were made.

2.3 Testing

Ninety five 15-minute segments of CTG trace were chosen from a database of approximately 6,500 hours of digitised CTG data. The segments selected had not been used in system development and were chosen to represent as many of the 120 potential combinations of crisp features as possible.

The selected segments were independently evaluated by three clinicians. Two were senior grade obstetricians, the third a Professor of clinical physiology. All three are currently involved in fetal monitoring research and have significant experience of CTG interpretation. The reviewers were required to assess each 15-minute segment and allocate a score from 0-39, indicating how normal or abnormal they considered the heart rate pattern to be. The relationship between scores and overall linguistic categorisation of the trace is shown in Table 2. These linguistic categories mirror the output of the crisp system and were used to guide the reviewers' scores.

Table 2. Scoring protocol for reviewers showing the relationship between score and linguistic description of the trace

Score	Linguistic Classification
0-9	Normal
10-19	Intermediate
20-29	Abnormal
30-39	Severely abnormal

So that the two systems could be directly compared with the reviewers, the system assessments of three consecutive 5-minute segments were combined as follows:- For the fuzzy system the score for a 15-minute segment was calculated as the mean of the scores for three 5-minute segments. For the crisp system, each possible combination of 5-minute classifications were ranked from most nomal to most abnormal. This ranking is arbitrary, but is considered to be the most appropriate order (Table 3).

Table 3. Ranking of three 5-minute segment combinations from crisp system output classification. n = normal, i = intermediate, a = abnormal, s = severely abnormal

Classifications			Rank	Classifications			Rank
n	n	n	1	n	a	a	11
n	n	i	2	i	a	a	12
n	n	a	3	n	a	s	13
n	n	s	4	i	a	s	14
n	i	i	5	a	a	a	15
i	i	i	6	a	a	s	16
i	i	a	7	n	s	s	17
i	i	s	8	i	s	s	18
n	i	a	9	s	s	a	19
n	i	s	10	s	s	s	20

3. Results

The mean scores for each reviewer are shown in Table 4. ANOVA has shown that A's mean is higher than B's ($p < 10^{-13}$) and C's ($p < 10^{-7}$) at the 5% significance level. B and C's means are not significantly different ($p > 0.05$).

Table 4. Mean and SD of scores for each reviewer.

Reviewer	A	B	C
Score Mean(SD)	24.23 (10.17)	12.44 (9.39)	15.42 (11.04)

The traces were then ranked, by score for each reviewer and by the output scores derived in 2.3 for the two systems. The agreement in case ranking between each reviewer and the systems was calculated by the Spearman rank correlation statistic, corrected for ties. The calculated agreement and significance are shown in Table 5.

4 Discussion

The use of scoring systems to assess CTG traces has been known to be problematic for many years, often with wide inter-observer variation. This is highlighted in the difference in scores between reviewer A and reviewers B and C. Overall, reviewer A has assigned significantly higher scores to the traces than reviewers B and C. If such a scoring system were used as the basis for clinical decision making, the scores assigned by A would result in approximately twice as many interventions as those of B and C, for a given threshold.

Table 5. Agreement in case rankings for each reviewer and the two systems. Significance (p value) is given at the 95% confidence level

Reviewer	A	B	C	Crisp System	Fuzzy System	Mean Agreement
A		0.80 $p < 10^{-21}$	0.37 $p < 0.0003$	0.51 $p < 10^{-7}$	0.73 $p < 10^{-16}$	0.60
B	0.80 $p < 10^{-21}$		0.59 $p < 10^{-9}$	0.50 $p < 10^{-6}$	0.65 $p < 10^{-12}$	0.64
C	0.37 $p < 0.0003$	0.59 $p < 10^{-9}$		0.46 $p < 10^{-5}$	0.47 $p < 10^{-5}$	0.47
Crisp System	0.51 $p < 10^{-7}$	0.50 $p < 10^{-6}$	0.46 $p < 10^{-5}$		0.82 $p < 10^{-24}$	0.57
Fuzzy System	0.73 $p < 10^{-16}$	0.65 $p < 10^{-12}$	0.47 $p < 10^{-5}$	0.82 $p < 10^{-24}$		0.67

Remark 1. Values of agreements in Table 5 may range from -1.00 to 1.00. An agreement of 1.00 indicates complete agreement, -1.00 complete disagreement, and 0.00 random agreement. Mean agreement is the mean value of each reviewer and system's agreement.

However, when trace ranking is used to examine relative assessment, reviewer A achieves good correlation with the others, particularly reviewer B. This implies that common criteria are being used for relative assessment of traces, but the assignment of an overall linguistic description to a trace is inconsistent. The difference in meaning that domain experts attach to linguistic classifications is a significant problem for knowledge engineers and system designers.

Reviewer C has lower agreement in ranking than the other reviewers. It is interesting to note that this reviewer, while experienced in CTG assessment, is not involved in CTG interpretation and decision making in the clinical setting. The two practising clinicians, A and B, have very high agreement in ranking. It is likely that different criteria for assessment have been used, e.g. reviewer C assessed the traces according to the normality of physiological response shown, while the clinicians assessed the heart rate patterns according to their relative clinical frequency.

Each reviewer has expertise and experience in trace assessment, but an individual perspective. Further discussions with the reviewers will allow their specific criteria of assessment to be determined. It is impossible to ascertain which of the reviewers is the 'best' from this study, as the limited amount of data in 15-minute segments prevents meaningful comparison with clinical outcome, the available gold standard.

4.2 System performance

The agreement of the fuzzy system with each of the reviewers, particularly the two practising clinicians, is higher than that of the crisp system. The fuzzy system also achieves the highest mean overall agreement. These results are extremely encouraging and demonstrate the potential of fuzzy techniques to improve on the performance of crisp rule-based systems. The fuzzy system assessment has a very high correlation with that of the crisp system. This is likely to be because the current fuzzy system is an evolution of the crisp system, using the same rules and with membership functions that are largely based on existing crisp classifications.

It is difficult to produce management recommendations for a case from a single 15-minute segment of trace, but the abnormality index generated by the system allows temporal changes in the CTG trace to be examined. This is not possible using the crude classification provided by the existing crisp system. The next stage of development will be to consider multiple segments of trace and extend the system to assess temporal changes. The knowledge acquired from the study can also be used in further development and system tuning prior to testing on a larger number of traces.

5 Conclusions and further work

A complex model to assess a biological system has been successfully produced using a hybrid of advanced data processing algorithms and fuzzy rule based techniques. Initial evaluation of the system has shown the potential benefit of fuzzy computerised CTG assessment over crisp CTG classification. The continuous measure of abnormality produced by the fuzzy system will allow temporal changes in the trace to be evaluated. The high performance of the feature extraction and classification functions of the preliminary fuzzy system provide a very solid platform for the next stages of development, towards a fuzzy system that can interpret entire CTG records.

6 Acknowledgements

The authors would like to thank Prof Karl Rosen, Dr Mark Davies, and Dr Roberto Luzietti for their enthusiastic participation in this study and ideas for future work.

References

1. CESDI (1995) Confidential Enquiry into Stillbirths and Deaths in Infancy. Annual report for 1 January - 31 December 1993. Part 1: Summary of methods and main results. Department of Health, UK

2. Neilsen JP, and Grant AM. (1993) The randomised trials of intrapartum electronic fetal heart rate monitoring. In *Intrapartum fetal surveillance.* JAD Spencer & RHT Ward, (eds) RCOG Press, London, 77-93
3. Chung TKH, Mohajer MP, Yang ZJ, et al. (1995) The prediction of fetal acidosis at birth by computerised analysis of intrapartum cardiotocography. *Br J Obstet Gynaecol* **102**:454-460
4. Dawes G S, Moulden M, Redman CWG. (1991) System 8000: Computerized antenatal FHR analysis. *J Perinat Med* **19**:47-51
5. Ulbricht C, Doffner G, Lee A. (1998) Neural networks for recognising patterns in cardiotocograms. *Artificial Intelligence in Medicine* **12**:271-84.
6. Krause W. (1990) A computer aided monitoring system for supervision of labour. In: *Computers in perinatal medicine.* Maeda K (ed) Elsevier Science Publishers BV Amsterdam 103-111
7. Keith RDF, Beckley S, Garibaldi JM, et al. (1995) A multicentre comparitive study of 17 experts and an intelligent computer system for managing labour using the cardiotocogram. *Br J Obstet Gynaecol* **102**:688-700.
8. Bernardes J, Ayres-de-Campos D, Costa-Pereira A, et al. (1998) Objective computerized fetal heart rate analysis. *Int J Gynecol Obstet* **62**:141-147
9. Alonso-Betanzos A, Devoe LD, Castillo RA, et al. (1989) FOETOS in clinical practice: A retrospective analysis of its performance. *Artificial Intelligence in Medicine* **1**:93-99
10. Hamilton E, Kiminani EK. (1994) Intrapartum prediction of fetal status and assessment of labour progress. In: Analysis of Complex Data and Artificial Intelligence in Obstetrics. Chang and Rogers (eds) *Bailliere's Clinical Obstetrics and Gynaecology* (International Practice and Research), Bailliere Tindall, London 567-581
11. Rooth GA, Huch A, Huch R. (1987) Guidelines for the use of fetal monitoring. FIGO News. *Int J Gynecol Obstet* **25**:159-167
12. Zadeh LA (1983) The role of fuzzy logic in the management of uncertainty in expert systems. *Fuzzy Sets and Systems* **11**:199-227
13. Quinlan JR (1986) Induction of Decision Trees, *Machine Learning* **1**, 81-106.

Efficient Graph Coloring by Evolutionary Algorithms

Nicole Drechsler, Wolfgang Günther, Rolf Drechsler

Institute of Computer Science
Albert-Ludwigs-University
79110 Freiburg, Germany
email: {ndrechsler,guenther,drechsler}@informatik.uni-freiburg.de

Abstract. In this paper we present an Evolutionary Algorithm (EA) for the *Graph Coloring Problem* (GCP). GCP has many important applications in VLSI CAD. The problem in general is known to be NP-hard. Our EA approach incorporates problem specific knowledge, i.e. we make use of different heuristics to obtain efficient solutions. We perform a large set of experiments and compare our EA to state-of-the-art heuristics. It turns out that the combination of EA methods with "classical" heuristics clearly outperforms previous approaches.

1 Introduction

One area of large interest in general are graph problems. Many "real world" problems can be modeled as a subclass of these. This is especially true in the area of VLSI CAD. In this paper we have a closer look at the *Graph Coloring Problem* (GCP). To each node in a graph a color has to be assigned such that two adjacent nodes have different colors. The GCP is the underlying graph problem in many approaches in VLSI CAD, like logic synthesis, register allocation, state minimization, routing (see e.g. [13, 11, 12, 3, 4]). For this there is a need for efficient algorithms solving this problem with high quality. Several methods for the GCP have been published during the last years: simple greedy heuristics [10] and heuristics with polynomial time bounds [2] have been presented. Furthermore, alternative approaches based on *Evolutionary Algorithms*[1] (EAs) for the GCP have also been studied.

Recently, many successful applications of EAs have been reported in the literature [7]. In [6] an order-based EA is presented, where the number of colors used for a feasible coloring, is minimized. The EA determines a sequence in which the vertices are colored by a greedy heuristic. In [9, 8] further EA approaches for graph coloring are presented. There EAs are used to find a coloring of a graph using a fixed number k of colors. Notice, that this method does not guarantee a feasible solution.

We present a hybrid EA for the GCP, i.e. heuristics are directly integrated in the EA and are also used as genetic operators [6]. In our application we minimize

[1] In the following we do not consider *genetic algorithms*, *evolution strategies*, *genetic programming*, etc. separately; instead we use EA as a global term.

the number of colors which are used for a feasible coloring of a given graph. As mentioned above the EA makes use of problem specific knowledge. Heuristics are applied to subgraphs to perform a local search on the graphs. We also investigate the influence of several mutation methods which are proposed in this paper. A large set of experimental results is reported that demonstrates the efficiency of our approach. We compare our EA to heuristics from [2] and [4], and to the EA approach reported in [6].

The paper is structured as follows: In Section 2 we describe the GCP in more detail. The EA approach including problem specific heuristics and genetic operators is presented in Section 3. In Section 4 experimental results are discussed. Finally, the paper is summarized.

2 Graph Coloring Problem

We start with some basic notation and definitions to make the paper self-contained.

Definition 1. *A* graph $G = (V, E)$ *consists of a finite set of nodes* $V \neq \emptyset$ *and a set of edges* E.

Notice, that throughout the paper we use undirected graphs.

Definition 2. *Two nodes* $v, w \in V$ *in a graph* $G = (V, E)$ *are called* adjacent *iff there exists an edge* $e \in E$ *such that* $e = \{v, w\}$.

For each node $v \in V$ the number of edges $e \in E$ with $v \in e$ is denoted as $deg(v)$.

Definition 3. *A* coloring *of graph* G *is an assignment of colors from a finite set* C *to nodes* $v \in V$ *such that each two adjacent nodes have different colors.*

The minimal number of colors needed to color a graph G is denoted as $\chi(G)$. $\chi(G)$ is called the *chromatic number* of graph G. In the following we assume that each color $c \in C$ is assigned to an integer value from $\{1, 2, \cdots, |C|\}$. $|C|$ is given by an upper bound for the number of colors. This upper bound is shown in Theorem 1 below.

We now consider the following problem:

How can we determine a coloring of graph G that needs a minimum number of colors?

It has been shown that *graph k-colorability*, i.e the question if a graph is colorable with k colors ($k \geq 3$), is NP-complete [10]. Since we are only interested in feasible solutions, i.e. solutions that color the complete graph, we optimize the coloring of a graph with respect to the number of colors that have to be used.

An upper and a lower bound for the number of colors are shown in the following theorem. They are used in basic heuristics and in the genetic operators of the approach.

Theorem 1. *For each graph G it holds*

$$\omega(G) \leq \chi(G) \leq \Delta(G) + 1,$$

with

$$\omega(G) := \textit{maximum size of a clique in } G$$

and

$$\Delta(G) := \textit{maximum } deg(v) \textit{ for all } v \in V.$$

Proof. Obviously, each node in a clique has its own color, since each node is connected with each other. Thus, a lower bound is given by the maximum size of a clique in G and the first inequation holds.

The second inequation is proven by construction: an algorithm is described that makes use of at most $\Delta(G)$ colors.

(1) $c(v_1) := 1;$
(2) **for** $i = 2$ **to** n
(3) $\quad c(v_i) := min\{j \ : \ c(v_h) \neq j$ for all $h < i$, h and i adjacent $\}$

Since each node v_i that is colored in (3) has at most $\Delta(G)$ adjacent predecessors, at most $\Delta(G)$ colors cannot be chosen. For this the algorithm finds a coloring with $\leq \Delta(G) + 1$ colors.

Since the proof of the theorem is constructive we directly obtain that each graph G can be completely colored with $\leq \Delta(G) + 1$ colors.

Using this result we directly derive the heuristic from Brelaz [2]: For each uncolored node $v \in V$ in graph $G = (V, E)$ we determine two integer values $A(v)$ and $B(v)$:

$A(v)$: The number of colors that has already been assigned to the adjacent nodes.
$B(v)$: The number of neighbors that have not been colored so far.

Remark 1. Let G be a completely uncolored graph. Then for all nodes $v \in V$ it holds:

1. $A(v) = 0$
2. $B(v) = deg(v)$

The algorithm of [2] starts with an (arbitrarily chosen) node. The next considered node is always the adjacent node with the largest value $A(v)$. If several nodes have the same value the one with the largest value $B(v)$ is chosen. For this node the smallest color (=integer number) possible is chosen (analogously to Theorem 1). In the following this heuristic is denoted by BRE.

3 Evolutionary Algorithm

To the problem defined above we apply a standard EA [7]. We make use of ***roulette-wheel selection*** and ***steady-state reproduction***.

In the following sections we describe some internals of our approach.

3.1 Representation and Objective Function

To store individuals in population $\mathcal{P}$ we use integer valued strings of length $|V|$. The entry of position i represents the color of node v_i ($i \in \{1,\ldots,|V|\}$). Genetic operators are carried out directly on the graph. Thus the operators do not work on a string encoding. This extremely simplifies the use of problem specific knowledge in the operators.

For a given coloring of graph G the *objective function* (*fitness*) is determined by counting the number of colors that have been used. For an individual $p \in \mathcal{P}$ it is denoted by $fitness(p)$.

3.2 Initialization

As mentioned above our EA does not allow infeasible solutions. Each individual in a population represents a feasible coloring of the graph. To create an initial population we make use of heuristic BRE to generate a first valid element. Then the algorithm from Theorem 1 is applied to generate further individuals. The node sequence for coloring is determined at random. This strategy guarantees variation in the population, and additionally all elements in the initial population represent feasible solutions.

3.3 Genetic Operators

In this section the genetic operators are introduced. We restrict to mutation operators, since in the studies it turned out that they are working well in this application and crossover operators have no significant effect. All operators make use of domain specific knowledge of the GCP. They all create feasible solutions, i.e. each coloring is a correct solution.

Most often in EAs a mutation consists of changing the value of a single element. In our case this would result in changing the color of only one single node. Obviously, this does not result in large gains.

Mutation operators generate one new solution from one parent element. The main idea is to recolor a whole subgraph $G_S = (V_S, E_S)$ from graph $G = (V, E)$ in one step. The operators consist of two phases:

1. selection of a subgraph to be recolored (by MUT1, MUT2, and MUT3 described below)
2. coloring the subgraph using heuristic BRE

The mutation strategies have in common that the number of nodes S in G_S is determined by uniform distribution in the intervall $[1, |V|]$. G_S is then recolored using heuristic BRE. The mutations differ in the strategy how G_S is determined starting from a randomly chosen node. These strategies are described in the following:

First, one parent element $p \in \mathcal{P}$ is selected using *roulette-wheel selection*. Let $A = \{v | \{v, v_i\} \in E, v \neq v_i, v \notin V_S, v_i \in V_S\}$ be the set of nodes that are not in V_S and that are adjacent to nodes in V_S. Initially, V_S represents the empty set ($V_S = \emptyset$); Then V_S is enlarged by different strategies:

MUT1 (Random Mutation): A node $v_i \in V$ is randomly chosen and inserted in V_S. The adjacent nodes v_j, $\{\{v_i, v_j\} \in E\}$ $(1 \leq j \neq i \leq |V|)$ are inserted in A. Then randomly chosen nodes $v \in A$ are deleted from A and inserted in V_S iteratively. In a next step A is enlarged by adjacent nodes from v. The algorithm terminates if $|V_S| = S$.

MUT2 (Color Priority Mutation): A node $v_i \in V$ is randomly chosen and inserted in V_S. Again, the adjacent nodes v_j of v_i, $\{\{v_i, v_j\} \in E\}$ $(1 \leq j \neq i \leq |V|)$ are inserted in A. A color $c < \mathit{fitness}(p)$ is randomly chosen. A node $v \in A$ is deleted from A and inserted in V_S, if v has color c. (If several nodes have the same color, one of them is chosen at random.) Again, the algorithm terminates if $|V_S| = S$.

MUT3 (Degree Neighbor Mutation): A node $v_i \in V$ is randomly chosen and inserted in V_S. Again, the adjacent nodes v_j, $\{\{v_i, v_j\} \in E\}$ $(1 \leq j \neq i \leq |V|)$ are inserted in A. A node $v \in A$ is deleted from A and inserted in V_S, if v has the most adjacent nodes in V_S. (If several nodes have the same number of neighbors if V_S, one of them is chosen at random.) Again, the algorithm terminates if $|V_S| = S$. (The idea is to find subgraphs that are strongly connected. In the best case this subgraph is a clique.)

The selection of the subgraph largely influences the performance of the mutation operators. Using for example *breadth first search* for determination of G_S does not result in any gain.

After V_S has been determined these nodes are recolored using BRE. (Notice, that all individuals then represent feasible solutions.) Examples of mutations MUT1 and MUT2 are given in Figure 1. In the first (second) row MUT1 (MUT2) is illustrated.

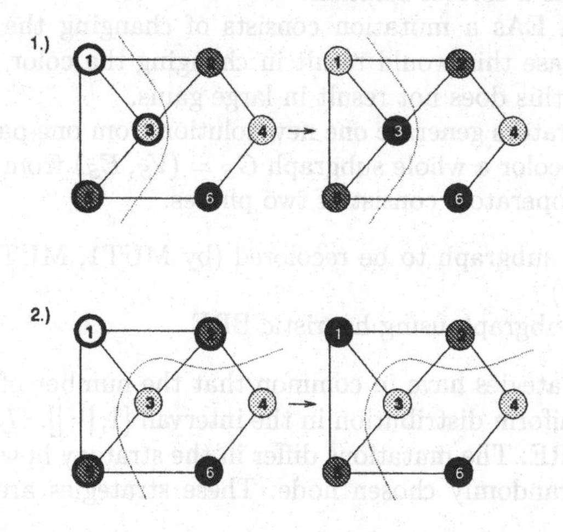

Fig. 1. Illustration of mutations

3.4 Heuristics

Often it is useful to combine EAs with problem specific heuristics. Concerning the GCP we make use of a heuristic to improve the quality during the EA run. The heuristic is used like a genetic operator and is described briefly in the following:

Iterated Greedy Heuristic (IG) [5]: A greedy heuristic is used to recolor the graph. The nodes are considered in a specific order. This order is determined such that the fitness, i.e. the number of colors used in the present solution, cannot increase. Dependent on the starting point and the selection of the nodes the algorithm can also "escape" from local minima. Thus the method profits from several calls of the core algorithm.

Heuristic IG is applied to an already colored graph G. When using IG as genetic operator, the graph is recolored only once, i.e. we do not apply it iteratively.

3.5 The Algorithm

We briefly describe the underlying basic EA:

- First an initial population of size $|\mathcal{P}|$ is generated. The individuals are created as described above.
- A fitness has to be assigned to each individual by counting the number of colors that are used by the corresponding coloring of the graph.
- Then $|\mathcal{CH}|$ children are created by performing the genetic operators. They are applied with corresponding probabilities.
- The best $|\mathcal{P}|$ individuals are selected from the pool of $|\mathcal{P}| + |\mathcal{CH}|$ individuals.
- This is continued for a fixed number of generations.

The algorithm is illustrated in Figure 2.

```
evolutionary_algorithm (graph) {
 generate_initial_population ();
 fitness_calculation () ;
 for( k iterations ){
     apply_operators () ;
     offspring_fitness_calculation () ;
     update_population () ;
 }
 return best_element ;
}
```

Fig. 2. Sketch of the algorithm

3.6 Parameter Setting

To determine parameter settings for the EA a large set of experiments were performed. Both, the quality of the results and the runtime of the EA were taken into account. Variations of the population size, the termination criterion, operator probabilities, and the number of offsprings were considered. Concluding these studies we have chosen the following parameters:

The size of the populations is set to $|\mathcal{P}| = 50$ and the number of children is set to $|\mathcal{CH}| = 10$. Each mutation operator is carried out with the same probability. The algorithm stops after 10000 iterations.

4 Experimental Results

In this section we present experimental results that have been carried out on a *SUN Ultra 1-140* workstation. A part of the benchmark set is taken from [1]. Benchmarks denoted by *gen1* and *gen2* are generated by a quasi-random graph generator[2]. These graphs have in common that they have a high edges-per-node count. The best heuristic solutions in our experiments are printed in bold.

4.1 Parameter Study

In a first series of experiments we investigated the influence of the genetic operators defined in Section 3.3. First, each type of mutation operators is considered separately as a genetic operator. The results are given in Table 1. In column *graph* the considered graph benchmarks are given. We consider *large* graphs with up to 1000 nodes and nearly 450000 edges (see columns *nodes* and *edges*). The chromatic number $\chi(G)$ of graph G (if it is known) is given in column χ. The remaining columns denote heuristic BRE and the different mutation operators. Notice, that each mutation operator improves on the results obtained by heuristic BRE. As can be seen mutation MUT2 performs best on the given benchmark set. Experiments have shown that using two or more mutations in the EA (we show MUT1+MUT2 in Table 1) does not result in large gains.

In a next series of experiments we give an impression on the influence of the iterated greedy heuristic IG. The EA also makes use of a mutation operator, since otherwise it gets stuck in a local minimum very early. It has been shown that MUT2 works well in this application, thus we apply MUT2 and IG as operators in the EA. The influence of the EA using IG is reported in Table 2.

In columns BRE and IG the number of colors after applying the corresponding heuristic is given. Heuristic IG is iterated 10000 times. Column EA_{IG} shows the results obtained by the EA using MUT2 and IG as operators.

A comparison from EA_{IG} to BRE and IG shows that the EA improves the heuristics for most examples. EA_{IG} is more powerful than using MUT2 only as genetic operator. For this, the number of colors could be improved for 50% of the considered graph benchmarks.

[2] The *C-code* is available via ftp from:
ftp://ftp.cs.ualberta.ca/pub/joe/GraphGenerator/generate.tar.gz

graph	*nodes*	*edges*	χ	BRE	MUT1	MUT2	MUT3	MUT1+MUT2
le450_5a	450	5714	5	10	**5**	6	8	**5**
le450_5d	450	9757	5	11	**5**	**5**	**5**	**5**
le450_15c	450	16680	15	24	**22**	23	23	23
le450_15d	450	16750	15	24	**23**	**23**	**23**	**23**
le450_25c	450	17343	25	28	**27**	**27**	28	28
le450_25d	450	17425	25	29	**27**	**27**	**27**	**27**
gen1	300	21611	≤25	42	38	38	**37**	**37**
gen2	200	15464	≤30	55	**30**	**30**	**30**	**30**
flat300_20_0	300	21375	20	42	**36**	37	38	38
flat300_26_0	300	21633	26	42	39	**38**	39	**38**
flat300_28_0	300	21695	28	41	**38**	**38**	39	**38**
flat1000_50_0	1000	245000	?	114	110	**107**	110	109
flat1000_60_0	1000	245830	?	112	110	**108**	110	109
flat1000_76_0	1000	246708	?	115	111	**108**	111	109
DSJC1000.5	1000	249826	?	114	111	**110**	112	111
DSJC1000.9	1000	449449	?	301	**267**	**267**	280	271

Table 1. Comparison of different mutation operators

graph	*nodes*	*edges*	χ	BRE	IG	MUT2	EA_{IG}
le450_5a	450	5714	5	10	7	6	**5**
le450_5d	450	9757	5	11	6	**5**	**5**
le450_15c	450	16680	15	24	24	**23**	**23**
le450_15d	450	16750	15	24	25	**23**	**23**
le450_25c	450	17343	25	28	29	**27**	**27**
le450_25d	450	17425	25	29	29	**27**	**27**
gen1	300	21611	≤25	42	**33**	38	**33**
gen2	200	15464	≤30	55	**30**	**30**	**30**
flat300_20_0	300	21375	20	42	**20**	37	**20**
flat300_26_0	300	21633	26	42	39	38	**36**
flat300_28_0	300	21695	28	41	37	38	**36**
flat1000_50_0	1000	245000	50	114	108	107	**103**
flat1000_60_0	1000	245830	60	115	108	108	**105**
flat1000_76_0	1000	246708	76	114	108	108	**105**
queen11_11	121	1980	11	15	14	**13**	**13**
queen13_13	169	3328	13	17	**16**	**16**	**16**
DSJC1000.5	1000	249826	?	114	109	110	**105**
DSJC1000.9	1000	449449	?	301	249	267	**248**

Table 2. Influence of heuristic IG

graph	*nodes*	*edges*	χ	BRE	EA_{seq}	EA_{IG}	time per edge
le450_5a	450	5714	5	10	12	**5**	0.02
le450_5d	450	9757	5	11	10	**5**	0.01
le450_15c	450	16680	15	24	28	**23**	0.12
le450_15d	450	16750	15	24	28	**23**	0.14
le450_25c	450	17343	25	**28**	33	**27**	0.13
le450_25d	450	17425	25	29	33	**27**	0.13
gen1	300	21611	$\leq$25	42	42	**33**	0.11
gen2	200	15464	$\leq$30	55	51	**30**	0.09
flat300_20_0	300	21375	20	42	43	**20**	0.09
flat300_26_0	300	21633	26	42	42	**36**	0.09
flat300_28_0	300	21695	28	41	43	**36**	0.10
DSJC1000.5	1000	249826	?	114	120	**105**	0.10
DSJC1000.9	1000	449449	?	301	305	**248**	0.10

Table 3. Comparison to other EA

4.2 Comparison

First, we compare our best results EA_{IG} to the approach from [6] denoted by EA_{seq} as summarized in Table 3. This method optimizes the sequence how the nodes should be colored. It has been reimplemented using the operators given there. Thus the methods are compared in the same environment.

As can be seen the results from EA_{seq} are in the range of the results obtained by BRE. The EA method from [6] does not make use of problem specific knowledge, i.e. it uses the genetic operators typically used for sequencing problems. This demonstrates the necessity of incorporating problem specific methods into evolutionary methods. As can be seen the combination of heuristics BRE, IG and "pure" EA is a promising approach for solving GCP. The average execution time of our proposed method denoted by EA_{IG} is in the range of 0.1 CPU seconds per edge in a graph. This property shows the linear scalability of EA_{IG} with respect to the number of edges.

Finally, a comparison of the quality of heuristics recently summarized in [4] is given as can be seen in Table 4. These heuristics are denoted by H1, H2, and H3, respectively, and are applied to large instances. Notice, that for these benchmarks the exact results could not be computed [4]. Again, EA_{IG} is given in the last column. It turns out that the evolutionary based approach outperforms several "standard" heuristics, like BRE and the algorithms given in [4]. As can be seen EA_{IG} profits from evolutionary techniques in combination with problem specific heuristics.

5 Conclusions

In this paper we focused on the *Graph Coloring Problem*. It has been demonstrated by experimental results that our EA approach outperforms the state-of-the-art heuristics and can often determine the optimal results. Due to the

graph	*nodes*	*edges*	χ	BRE	H1	H2	H3	EA_{IG}
le450_15d	450	16750	15	24	31	25	25	**23**
le450_25c	450	17343	25	28	38	30	28	**27**
flat1000_50	1000	245000	50	114	104	110	113	**103**
flat300_20	300	21375	20	42	40	41	42	**20**
queen11_11	121	3960	11	15	16	16	14	**13**
queen13_13	169	6656	13	17	18	18	17	**16**

Table 4. Comparison to heuristics from VLSI CAD

problem specific operators that are used in our evolutionary approach it turns out that our method is very robust, i.e. it computes best heuristical solutions on the considered benchmark set. If the optimal solution can not be found the problem instances turned out to be "hard" in the sense that also all other heuristics failed.

References

1. J.E. Beasley. OR-Library: Distributing test problems by electronic mail. *Journal of the Operational Research Society*, 41(11):1069–1072, 1990.
2. D. Brélaz. New methods to color vertices of a graph. *Comm. of the ACM*, 22:251–256, 1979.
3. S. Chiusano, F. Corno, P. Prinetto, and M. Sonza Reorda. Hybrid symbolic-explicit techniques for the graph coloring problem. In *European Design & Test Conf.*, pages 422–426, 1997.
4. O. Coudert. Exact coloring of real-life graphs is easy. In *Design Automation Conf.*, pages 121–126, 1997.
5. J.C. Culberson. Iterated greedy graph coloring and the difficult landscape. Technical Report TR92-07, University of Alberta, Department of Computing Science, Edmonton, Alberta, Canada, 1992. ftp://ftp.ualberta.ca/pub/TechReport.
6. L. Davis. *Handbook of Genetic Algorithms.* van Nostrand Reinhold, New York, 1991.
7. R. Drechsler. *Evolutionary Algorithms for VLSI CAD.* Kluwer Academic Publisher, 1998.
8. A.E. Eiben and J.K. van der Hauw. Graph coloring with adaptive genetic algorithms. Technical Report 11/96, Leiden University, Leiden, Netherlands, 1996.
9. C. Fleurent and J.A. Ferland. *Annals of Operations Research.* Genetic and Hybrid Algorithms for Graph Coloring. edited by G. Laporte, I. H. Osman, and P. L. Hammer, 1994.
10. M.R. Garey and D.S. Johnson. *Computers and Intractability - A Guide to NP-Completeness.* Freemann, San Francisco, 1979.
11. T. Sasao. *Logic Synthesis and Optimization.* Kluwer Academic Publisher, 1993.
12. C. Scholl, S. Melchior, G. Hotz, and P. Molitor. Minimizing ROBDD sizes of incompletely specified functions by exploiting strong symmetries. In *European Design & Test Conf.*, pages 229–234, 1997.
13. W. Wan and M.A. Perkowski. A new approach to the decomposition of imcompletely specified multi-output functions based on graph coloring and local transformations and its application to FPGA mapping. In *European Design Automation Conf.*, pages 230–235, 1992.

Determination of Decision Rules on the Basis of Genetic Algorithms

Arita Takahashi[1] and Arkady Borisov[2]

[1] Institute of Information Technology, Technical University of Riga,
1 Kalkyu St., LV-1658 Riga, Latvia
a.takahasi@konts.lv

[2] Institute of Information Technology, Technical University of Riga,
1 Kalkyu St., LV-1658 Riga, Latvia
aborisov@egle.cs.rtu.lv

Abstract. The paper considers generation of decision rules with the help of genetic algorithms. The performed experiments are described, and the results are analyzed. The instances are compared in which different object training sets are used for the pair comparison, i.e. for the evaluation of solutions generated by the genetic algorithm. There are compared the selection operator based on dominance features (Pareto set) and that of scalar function based.

1 Introduction

Genetic algorithms have widely been used in optimization tasks, and there are only a few works that use genetic algorithms to find rules. One of these studies is described [2] as follows: a set of solutions consists of various encoded rules:

$$attribute1 < valueA; \quad attribute2 >= valueB; \quad attribute3 = valueC. \tag{1}$$

One symbol string may encode several rules. One should foresee whether they will be combined with a conjunction or with a disjunction (conjunctive or disjunctive method [3]). The task of a genetic algorithm is to determine the preferences according to which a decision maker divided patterns into acceptable and unacceptable ones.

2 Filled Squares Problem

The present work deals with the following issues:

1. how to find the preferences by using only a small part of solution space;
2. how to find the preferences for fuzzy evaluations;
3. how to generate hypotheses for *the best ideal* and *the worst ideal* solutions.

Elements of the solution space are shown in Fig. 1 and they may be evaluated with the 10-mark scale. The decision maker will evaluate phenotypes over a small area of the solution space. The darkness of filled squares can change within the limits from 1 to 9 (see Fig. 1). Unfilled square is encoded as 0. The code of a string will be formed by passing along the squares in the direction from left to right and passing through the rows in the direction from top to bottom.

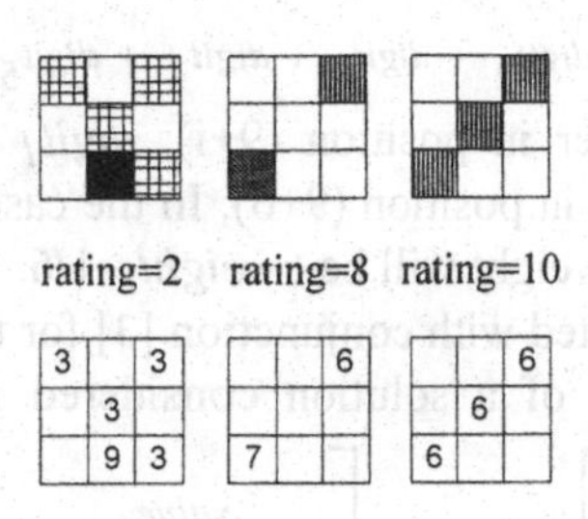

Fig. 1. Phenotypes and their encoding

Strings with evaluation " 10" are connected by the following rules (see Table 1):

1. The configuration of filled squares forms a diagonal
2. The number of filled squares is 3.
3. The degree of filling darkness is 6 or 7.
4. The difference between lightest filled and darkest filled areas should not exceed two darkness levels.

Table 1. Rating of patterns made by the expert

String (of pattern)	Rating
303030093	2
006060600	10
700060006	9

By using the accumulated data base, the genetic algorithm should generate additional objects and rule weights for each object. The rules themselves will look as the hypotheses either affirmed or denied by the genetic algorithm:

Rule 1: Best objects have only certain squares filled.
Rule 2: Best objects have a certain number of filled squares, no matter which ones.
Rule 3: Best objects should have a certain maximum darkness degree.
Rule 4:The difference between the lightest and the darkest squares of a certain size.
Rule 5: Best objects have the first position unfilled or vice versa.
Rule 6: Best objects have the last position unfilled or vice versa.

If any rule is not to be taken into account, the genetic algorithm should indicate a low weight coefficient for this rule. If the rule should be taken into account to a certain extent, then for this rule the genetic algorithm should generate a large coefficient.

3 Weighs and Encoding

Thus, a solutions will be formed from strings of length 15 (9 positions for structure and 6 - for rule weights) -see Table 2. Then the weight coefficient for the i-th rule (i-th criterion) could be computed as follows:

$$weight_i = \frac{digit_i}{digit_1 + digit_2 + digit_3 + digit_4 + digit_5 + digit_6} \qquad (2)$$

,where $digit_i$ is the number in position (9+i), $digit_1$ is the number in position (9+1),..., $digit_6$ is the number in position (9+6). In the case when the last six symbols are zeros, i.e. "000000", the weight will be : $weight_i$=1/6.

All the six rules are connected with conjunction [3] for the selection mechanism: where $level_1$ is the value of a solution considered in the previous cycle, i.e.

$$if \left[\left(\frac{value_1}{1 + weight_1}\right) >= level_1 \right] and \ldots and \left[\left(\frac{value_6}{1 + weight_6}\right) >= level_6 \right] then\ select \qquad (3)$$

$level_1 = (value_1/(1+weight_1))_{iteration-1}$, $value_1$ is the value that expresses the quality of a solution with regard to *Rule1*.

Although in the task a complete Pareto set is not determined, those solutions which dominate are being chosen for the next generation.

Table 2. The weight coefficients for the rules are computed by equation (2)

String (of solution)	Weight1	Weight2	Weight3	Weight4	Weight5	Weight6
000014300**980260**	0,36	0,32	0	0,08	0,24	0
062210310**043382**	0	0,2	0,15	0,15	0,4	0,1
002068080**787315**	0,226	0,258	0,226	0,097	0,032	0,161
001780100**961107**	0,375	0,25	0,042	0,042	0,0	0,292

From the expressions it follows that the greater the meaning of $weight_i$ is, the higher requirements are imposed on the objects by the i-th rule which is to participate in forming the next generation, whereas if there is generated a small weight, the meaning of $value_i$ might also be small. The experiments have proved that coefficients converge to certain values.

4 Genetic Operations: Mutation and Inversion

In a binary string, mutation is treated as transformation of zero into 1 and vice versa [1], whereas in this decimal encoding any code should be allowed to transform into 0. If the example under consideration is encoded to a binary code, then each darkness degree would be characterized by a 4-position code. Say, mutation from "7" can be implemented in three different directions:

$$0111 \rightarrow 0011 \qquad 7 \rightarrow 3$$
$$0111 \rightarrow 0101 \qquad 7 \rightarrow 5$$
$$0111 \rightarrow 0110 \qquad 7 \rightarrow 6$$

That is why, also in the decimal system there are allowed four mutations from "7" with equal probabilities, namely, the three above and $7 \rightarrow 0$. For any string, an inverse string is obtained in the same way by mutating all elements of the former string.

5 Phenotypes Characterizing Criteria

Each phenotype will be characterized by criterion1 *Similarities_of_structure_ij* showing whether the sequence of filled and empty squares fits or not:

```
Similarities_of_structure_j=0
Do_In_cycle_122_times:(i=1 to 122)
  Similarities_of_structure_ij=0
  Do_In_cycle_9_times: If (Digit_in_pattern_i = 0 and Digit_in_solution_j = 0) or
  (Digit_in_pattern_i ≠ 0 and Digit_in_solution_j ≠ 0)
     then Similarities_of_structure_ij= Similarities_of structure_ij+1
  Endif;Enddo; Store Similarities_of_structure_ij
  Similarities_of_structure_j=Similarities_of_structure_j+
  +rating*Similarities_of_structure_ij
Enddo; Store Similarities_of_structure_j
```

where j is the index of solution, i is the index of pattern and 9 is maximal number of digit position in string (positions corresponding to certain squares of design).

Other five criteria are shown in Table 3. They are computed as follows:

Criterion2 shows the number of filled squares in design:

$$FilledSquaresNumber=NumberOfSquares-Occurs("0", substring_{pos1-9}) \quad (4)$$

Criterion3 shows the degree of shading for the darkest colour used in design:

$$Darkest=max(colour_1 \ldots colour_n) \quad (5)$$

Criterion4 shows the difference between the darkest colour and the lightest one:

$$\text{Dispersion\_of\_colours=Darkest-Lightest} \quad (6)$$

Criterion5 shows whether the first square is filled or not:

$$Presence\_at\_first=0, \text{ if } digit_1=0; \quad else \; Presence\_at\_first=1 \quad (7)$$

Criterion6 shows whether the last square is filled or not:

$$Presence\_at\_last=0, \text{ if } digit_9=0; \quad else \; Presence\_at\_last=1 \quad (8)$$

Thus the differences and then coincidences of each generated solution with each pattern can be found. For criterion2 *FilledSquaresNumber* (see equation (4)) the following holds:

$$DifferenceCriterion2_{ij}=abs(FilledSquares_{InPattern\ i}-FilledSquares_{InSolution\ j}) \quad \textbf{(9)}$$

Table 3. Criteria computed by equations (4) - (8)

	Phenotype				
Number of criterion:	2	3	4	5	6
String (of solution)	Filled squares (number)	Darkest colour	Dispersion of colour	Presence at first	Presence at last
200014000980260	3	4	3	1	0
062210310043382	6	6	5	0	0
200068008787315	4	8	6	1	1
001780003961107	4	8	7	0	1

After six criteria and six differences are evaluated, one should determine coincidences. For the second criterion the following holds:

$$CoincidenceOfFilledSquares_{ij}=9-DifferenceOfFilledSquares_{ij} \quad \textbf{(10)}$$

6 Performance of the Generated Solution

For each generated solution the coincidences are being summed up until comparisons with all the examples evaluated by the decision maker are finished up. The ratings of patterns are considered by multiplying them with coincidence values. The total sum of coincidences is then stored and the next generated solution is included into comparison. Thus the i-th solution receives an evaluation with regard to the k-th criterion:

$$value_{i,k} = \sum_j (CoincidenceOfCriterion\_k_i * rating_j) \quad \textbf{(11)}$$

where i is the number of solution: $1 \le i \le$PopulationSize, j is the number of pattern: $1 \le j \le$Quan.ofPatterns. The fitness for i-th solution with regard to the k-th criterion is:

$$FitnessOfCriterion\_k_i = \frac{value_{k,i}}{1 + weight_{k,i}} \quad \textbf{(12)}$$

where $value_{k,i}$ is calculated from equation (11) and $weight_{k,i}$ is calculated from equation (2).

From this expression it can be seen that the selection does not occur according to one *fitness* function value only, but to all six values simultaneously.

7 Results

The following experiments have been performed:

1. Comparisons with patterns, whose evaluation is rating>7 (see Fig. 2, Appendix);
2. Comparisons with patterns whose evaluation is rating=10 ;
3. Comparisons with all examples of the training set ;
4. Comparisons with "bad" patterns whose evaluation is rating<4 ;
5. Comparisons with patterns whose evaluation is rating=1 (see Fig. 3, Appendix);

The rate of good solutions in the population with regard to each of criterion:

$$RateOfGoodSolutionsOnCriterion\_i = \frac{NumberOfGoodSolutionsOnCriterion\_i}{NumberOfSolutionsOnPopulation} \quad \mathbf{(13)}$$

8 Conclusions

As compared with patterns that have ratings 8, 9, and 10 there are strings in the training set that are closer to the objective, that is why this searching path is the shortest (see Fig. 2).

As compared with patterns that have rating 1, one can see that convergence takes no place at all. This may be explained by the fact that contrary to the best ideal, the worst ideal is fuzzy.

To compare, an experiment has also been performed with function dependent selection (see Fig. 4). In this experiment, one criterion (criterion 1) was sufficiently advanced in comparison with other criteria since for it there was chosen a large coefficient in the function. In this case the criteria compensate each other due to the evaluation with regard to the same criterion. Hence, the final solution will nearly always be more favorable for this criteria than for the others. However, searching for results can be performed in such a way that more difficult developing features can be more favored, whereas it was not possible in the domination based (Pareto) selection in these experiments.

References

1. Goldberg, D.E. (1989). *Genetic Algorithms in Search, Optimization, and Machine Learning*. Reading/MA, USA: Addison-Wesley.

2. Oliver, J. Finding Decision Rules with Genetic Algorithms. (1994). *AI Expert*, March, 33-39.

3. Hwang, C-L. and Yoon, K. (1981). *Multiple Attribute Decision Making. Methods and Applications*. Lecture Notes in Economics and Mathematical Systems, 186.

Appendix: Results

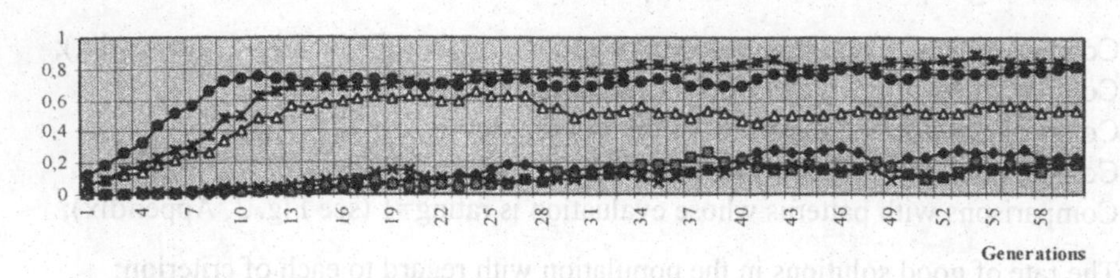

Fig. 2. *Solutions, compared with patterns which have Rating=8; 9; 10. The best string="006000400 900300":* If a training sample contains not only best patterns, searching can proceed faster

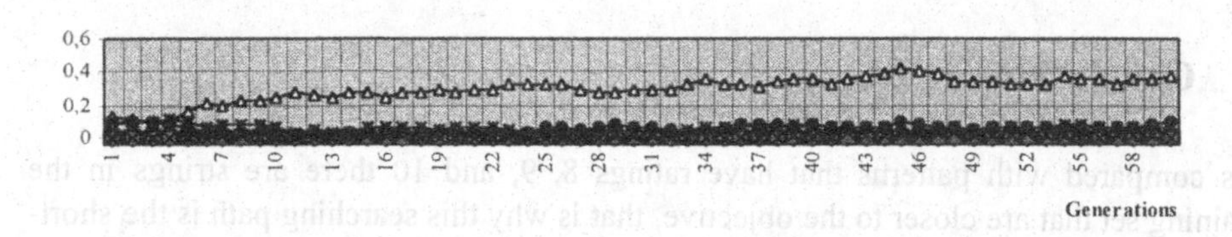

Fig. 3. *Solutions, compared with patterns which have Rating=1. The best string="254507447 720900":* It is impossible to find *the worst ideal* solution at all since there is no any particular attribute to characterize it

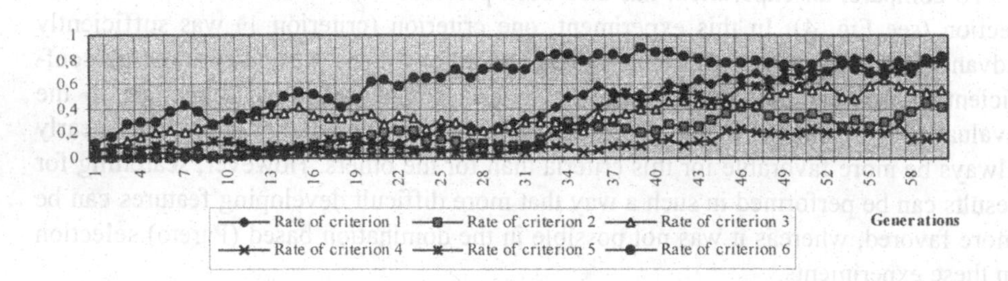

Fig. 4. *Function-based selection. The best string="006070500 900000":* Improvement of *easier* criteria owing to *more difficult* ones is not favoured since the *most difficult* criterion (criterion 1) contains a coefficient in the function that helps to improve its value

Modeling a Refrigeration System Using Recurrent Neural Networks

Ressom Habtom

Institute of Process Automation, University of Kaiserslautern, Postfach 3049, D-67653 Kaiserslautern, Germany
habtom@e-technik.uni-kl.de

Abstract. Refrigeration systems are characterized by the fact that they posses a large number of state variables that are coupled non-linearly with each other. Hence, the use of neural networks for modeling such processes is more appealing than employing traditional modeling approaches. This paper presents the results of the experiments conducted on a medium-sized laboratory setup of a refrigeration system. Using data collected from the setup, recurrent multilayer perceptron networks are trained to mimic the behavior of the system. The networks are validated not only with test data collected under similar external conditions but also with those that are gathered when the measurements of the external temperature are beyond the range inspected during the collection of the training data. Despite a significant change in external conditions, the validation results showed a fairly good performance in a multi-step prediction of the temperature and relative humidity inside the refrigerator.

1. Introduction

Neural networks have the ability to learn sophisticated nonlinear relationships. Hence, they provide an ideal means of modeling complicated nonlinear dynamic systems. The fact that neural networks demand no detailed a priori knowledge about the internal events of the process for identification is one of the appealing factors. Feedforward neural networks (FNNs) with tapped delay lines are commonly used for the identification of nonlinear dynamic systems. This could happen because of the availability of standard learning algorithms like back-propagation for multilayer perceptrons (MLPs) and fast as well as efficient training methods like least squares for radial basis function (RBF) networks. On the other hand, due to the computational intensity required for training, recurrent neural networks (RNNs) are rarely utilized in spite of their important capabilities, which are not found in FNNs. To mention few, RNNs have attractor dynamics, have inherent dynamic nature (i.e., they have the ability to store information for later use), have the ability to deal with time-varying input or output through their own natural temporal operation [1].

One of the typical architectures of RNNs is the so-called Williams-Zipser [1] architecture, by which all units are fully interconnected. A cascade of such fully connected layers arranged in a multilayer feedforward fashion form an architecture described as recurrent multilayer perceptron (RMLP) in [2]. One may ask here the advantage of having several layers if all units within a layer are connected. The point is that an RMLP has fewer connections compared to a fully connected recurrent network of the same units yielding a faster and efficient training. This is not only due to the few number of connections but also due to the inherent feedforward structure, which allows the network to enjoy the efficient learning algorithms developed for MLPs with minor modifications. Besides, RMLP networks have the effect of modeling plants, in which the plant's nonlinear model is assumed to be described by a set of difference equations, where the plant's states are coupled non-linearly with the imposed controls [3].

Fig. 1 shows an RMLP network with a single hidden layer and an external recurrence as applied for modeling a nonlinear dynamic system. At time t, $\boldsymbol{u}(t)$, $\boldsymbol{y}(t)$, and $\hat{\boldsymbol{y}}(t)$ denote the input vector, the observed output vector, and the network output vector, respectively. Based on the error between the observed and the network output vectors, the parameters of the network can be adjusted. The dynamic back-propagation (DBP) is a learning algorithm that takes the internal and external recurrences into account.

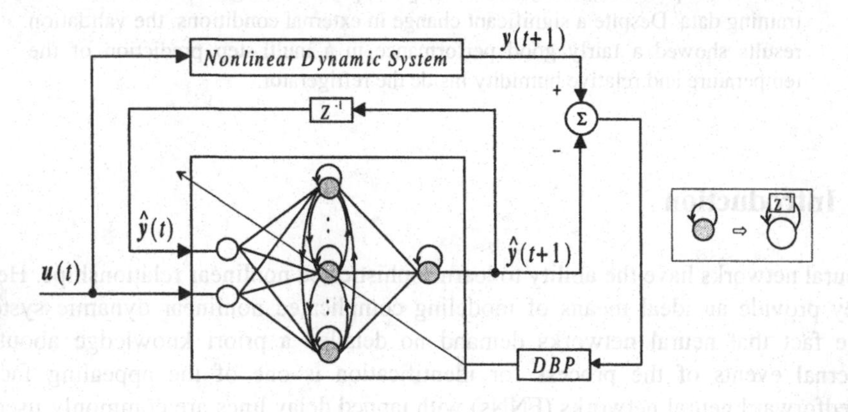

Fig. 1. RMLP for modeling a dynamic system

2. The Refrigeration System

Refrigeration is defined as a process of removing heat from an enclosed space or material, and maintaining that space or material at a temperature lower than its surroundings, [4]. A mechanical refrigeration system consists of an arrangement of components, which contribute to transferring heat. Refrigerant is one of the key components necessary to make the mechanical refrigeration work. It is a chemical compound that is alternately compressed and condensed into a liquid, and then permitted to expand into a vapor or gas as it is pumped through the mechanical refrigeration cycle. Fig. 2 illustrates the basic refrigeration cycle. It is based on the physical principle that a liquid extracts heat from the surrounding area as it expands (boils) into a gas. The refrigerant is circulated through the system by a compressor, which increases the pressure and temperature of the vaporous refrigerant and pumps it into the condenser. In the condenser, refrigerant vapor is cooled by air or water until it condenses into a liquid. The liquid refrigerant then flows to the flow control device, or expansion valve, where flow is metered and the pressure is reduced, resulting in a reduction in temperature. After the expansion valve, refrigerant flows into the lower pressure evaporator, where it boils by absorbing heat from the space being cooled, and changes into a vapor. The cycle is completed when the compressor draws the refrigerant vapor from the evaporator and, once again, compresses the gas so that the cycle can continue.

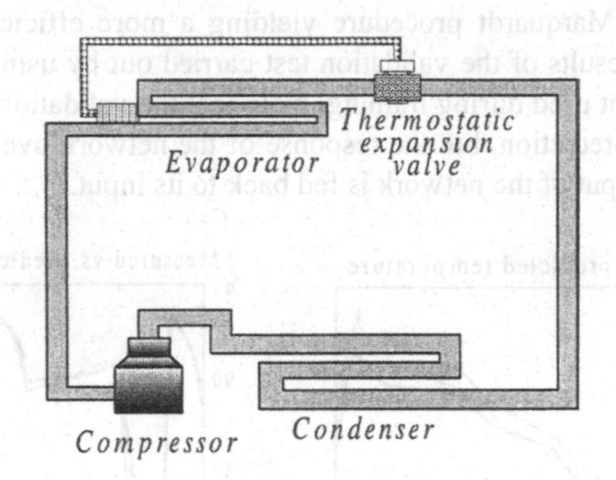

Fig. 2. The basic refrigeration cycle

In this paper, a laboratory setup that is fully equipped with the necessary components of a medium-sized commercial refrigeration is employed to conduct the modeling experiment. The setup is mainly made up of a compressor, a condenser with a blower, an evaporator with a fan, and an expansion valve. The compressor's speed is adjusted by a dc-potential (u_1) that is supplied to a static frequency changer. Similarly, the speed of the evaporator's fan is regulated using a dc-potential (u_2) fed to a thyristor-driven motor to effect a phase control. While the condenser's blower is driven at a constant speed, the temperature and the relative humidity inside the refrigerator are regulated by varying the two dc inputs, whose voltages range from 0 to 10 volts.

The temperature outside the room is an external disturbance. This temperature has a remarkable effect on the system's behavior since the air-cooled condenser is mounted on the side of a wall that permits a direct contact to the outside environment.

3. Experimental Results

The aim of the experiment is to establish a model that predicts the temperature (y_1) and relative humidity (y_2) inside the refrigeration system from the two input variables u_1 and u_2. To simplify the modeling task, only specific operating conditions like a given temperature range and a specified start-up procedure are considered. Thus, on 22 July 1998, 4500 samples were collected from the laboratory setup for a temperature range between -5 and 5°C at a sampling rate of 5 sec. After reducing the measurement noise by filtering the data, the first 4000 samples are used to train two RMLP networks NN_1 and NN_2 for the prediction of y_1 and y_2, respectively. Each network has three inputs (i.e., u_1, u_2, and an external recurrence), one hidden layer with 11 hidden units, and one output unit. The networks are trained using the dynamic back-propagation algorithm based on the framework of the real-time recurrent learning method introduced in [1]. The algorithm is extended with the help of the Levenberg-Marquardt procedure yielding a more efficient performance [5]. Fig. 3 shows the results of the validation test carried out by using the remaining 500 samples that are not used during training. Note that the validation results do not show a one-step-ahead prediction, but the response of the network over the whole range of time where the output of the network is fed back to its input.

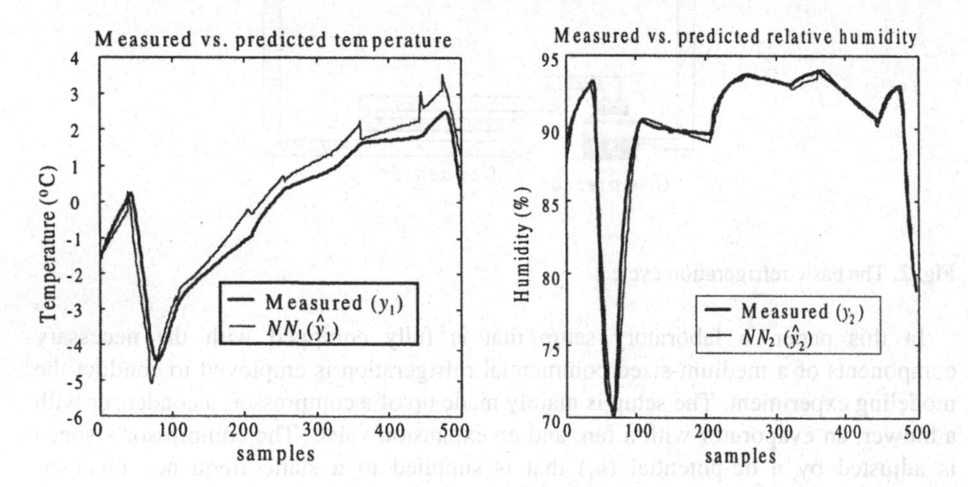

Fig. 3. Measured vs. predicted values of temperature and relative humidity

Needless to say, validating the networks using samples gathered on some other day when the outside temperature is far different from the day on which the training data are collected, is an interesting issue. The measurements of the external temperature recorded for two days are depicted in Fig. 4. Fig. 5 and Fig. 6 show the validation tests carried out by using test data that were collected on August 21, 1998. These figures demonstrate the robustness of the networks which performed very well despite a significant difference between the measurements of the external temperature inspected during the collection of the training data and during validation.

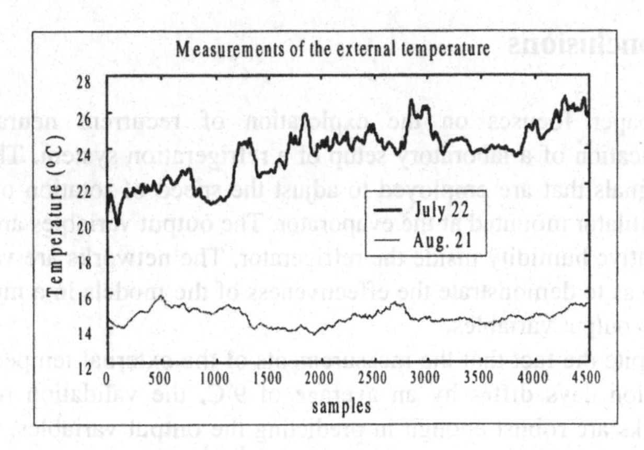

Fig. 4. Measurements of the external temperature on July 22 and August 21, 1998

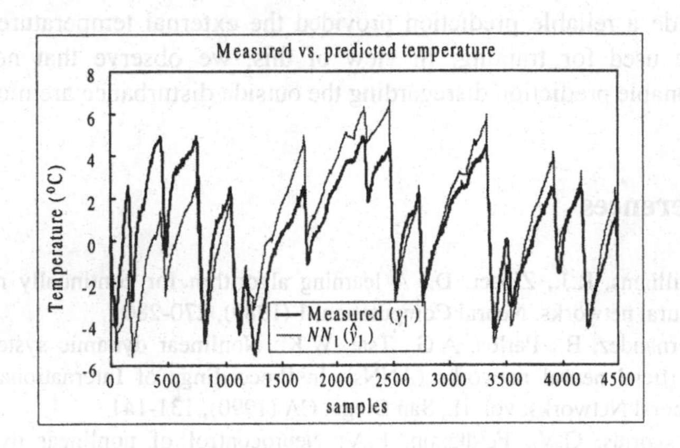

Fig. 5. Validation result for NN_2 with samples collected on August 21, 1998

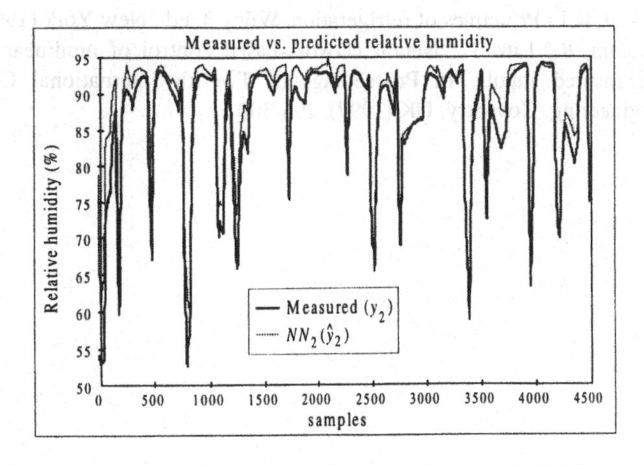

Fig. 6. Validation result for NN_2 with samples collected on August 21, 1998

4. Conclusions

This paper focuses on the exploration of recurrent neural networks for the identification of a laboratory setup of a refrigeration system. The input variables are two signals that are employed to adjust the speed of rotation of the compressor and the ventilator mounted at the evaporator. The output variables are the temperature and the relative humidity inside the refrigerator. The networks are validated using unseen data so as to demonstrate the effectiveness of the models in a multi-step prediction of the two output variables.

Despite the fact that the measurements of the external temperature on training and validation days differ by an average of 9°C, the validation results reveal that the networks are robust enough in predicting the output variables. Evidently, the results could be improved by incorporating the measurements of the external temperature as an additional input to the networks. In this case, however, the resulting networks can provide a reliable prediction provided the external temperature remains within the range used for training. In view of this, we observe that networks providing a reasonable prediction disregarding the outside disturbance are more appealing.

References

1. Williams, R.J., Zipser, D.: A learning algorithm for continually running fully recurrent neural networks. Neural Computation, 1 (1989), 270-280
2. Fernandez, B., Parlos, A.G., Tsai, W.K.: Nonlinear dynamic system identification using artificial neural networks (ANNs). In Proceedings of International Joint Conference on Neural Networks, vol. II., San Diego, CA (1990), 131-141
3. Puskorius, G.V., Feldkamp, L.A.: Neurocontrol of nonlinear dynamical systems with Kalman filter trained recurrent networks. IEEE Trans. on Neural Networks. 5 (1994), 279-297
4. Dossat, R.J.: Principles of refrigeration. Wiley 3. ed., New York (1991).
5. Habtom, R., Litz, L.: Neural network-based control of nonlinear dynamic systems with unmeasured inputs. In Proceedings of Twelfth International Conference on Systems Engineering. Coventry, UK (1997), 296-300

Evaluating Nugget Sizes of Spot Welds by Using Artificial Neural Network

Tan Yiming, Fang Ping, Zhang Yong, and Yang Siqian

Materials Science and Engineering College of
Northwestern Polytechnical University,
710072, Xi'an, P.R. China
duyin@nwpu03.nwpu.edu.cn

Abstract. Spot welding is widely used for numerous industrial applications, especially in automobile, aerospace and electronic manufacturing applications. One of the most important issues is how to evaluate the nugget diameter of spot weld, upon which the quality of joint was found to be dependent, with non-destructive inspection method. This paper introduces the investigation on evaluating quality of spot weld using back-propagation network. The network is configured by learning the pattern sets which consist of electrical parameters detected during the welding process and the relevant nugget sizes (diameter and height) measured after welding. The electrical parameters composed of welding current and voltage between the electrodes are selected as the input of the network. The output of the network is the nugget sizes of the weld which are compared with the experimental data. Results showed that the neural network system is valid for evaluating application in spot welding.

Introduction

network was mostly utilised in engineering applications. The essence of back-propagation learning is to encode an input-output relation, represented by a set of examples {x,d}, with a multilayer perceptron (MLP) well trained in the sense that it learns enough about the past to generalise to the future.
This study aims to establish such a system with artificial Spot welding has been used successfully for many years to join different kinds of components of sheet metals. Because of its high level of productivity, this technique has become one of the main welding methods. However, there is no guarantee that each weld will have the required quality determined normally by using destructive testing, because the welding quality is affected by a lot of factors, such as electrical dips or surges, mechanical compliance, electrode deformation, workpiece fit-up, material surface roughness and cleanliness etc.
Therefore, developing a non-destructive inspection system seems to be always one very important task for quality control of spot welding. Usually the weld quality can be supposed to depend upon the nugget sizes formed at the faying surface between work-pieces. So it is very useful to estimate the nugget diameter and height of spot weld with non-destructive method.
Artificial neural networks provide a useful tool for analysing many industrial formulation, pattern recognition and process control problems. Among of them the

back-propagation neural network for the quality evaluation and control in spot welding.

Experimental conditions and data detection

The mild steel was selected as the experimental material and the thickness was 1.0mm. A single-phase 200 KVA AC welding machine was used. The welding parameter setting was so designed according to cross test method that the experiments could examine weld quality under a wide range of welding conditions. These factors of the cross test were: (A)Welding heat; (B)Electrode tip diameter; (C)Electrode force: (D)Ferromagnetic materials inserting and (E)Secondary voltage. Their relationship were shown in table 1.

Table 1. Factors and levers of cross test (welding time 12 cycle)

Factor	Level		
	1	2	3
A: Welding heat (%)	70	60	50
B: Tip diameter (mm)	8	6.4	5
C: Electrode force (N)	2610	3800	3920
D: Ferromagnetic inserting * (Block)	0	1	2
E: Secondary power supply (V)	6.35	5.50	4.93

* The size of block: 170×310×3 (mm).

Under every parameter setting 4 specimens were welded, one of which was used as for evaluation purpose, and others as for training purpose. The electrical data(e.g., welding current and voltage between the electrodes) were sampled at each cycle throughout the welding process. After the process the corresponding nugget diameter and height were measured.

A data acquisition system was developed to detect the electrical data from the spot welding process. The hardware structure of the data acquisition system is shown in Figure 1. A single-chip microcomputer MCS-51 was selected as the main control unit of the system. The chip 2817 was for storing the application program of the system. The chip 6264 was used as data dynamic memory of the system. The sample circuit chip in the system was chip 0809. The sample rate of the system was determined as 120 μs per time. During the welding process the current and voltage were in turn sampled at every electric cycle. These detected data were treated as effective value through the software of the system and later used as the input patterns for the training of the neural network.

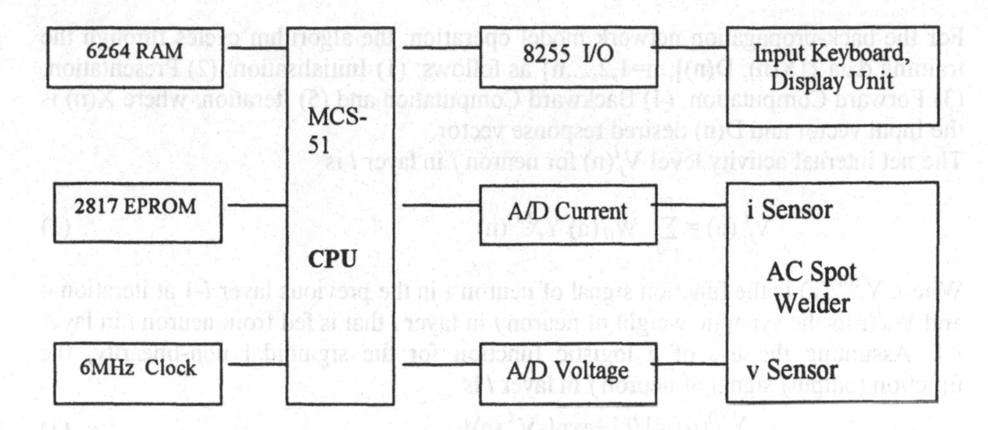

Fig. 1. Block diagram of spot welding data acquisition system.

Structure model and training process

The back-propagation network with a single hidden layer can usually map inputs to outputs for arbitrarily complex non-linear systems, static or dynamic. In the study this kind of neural network was chosen as the evaluating model. The network is composed of three separate layers, including an input layer, a hidden layer and an output layer, as shown in Fig.2. The input layer has 24 processing elements (units) which deal with the electrical data from the welding process. The output layer has 2 processing elements (giving data on the nugget diameter and nugget height). The units in the hidden layer will be so determined that during the evaluations its sum-squared error is the most minimum under the whole learning conditions.

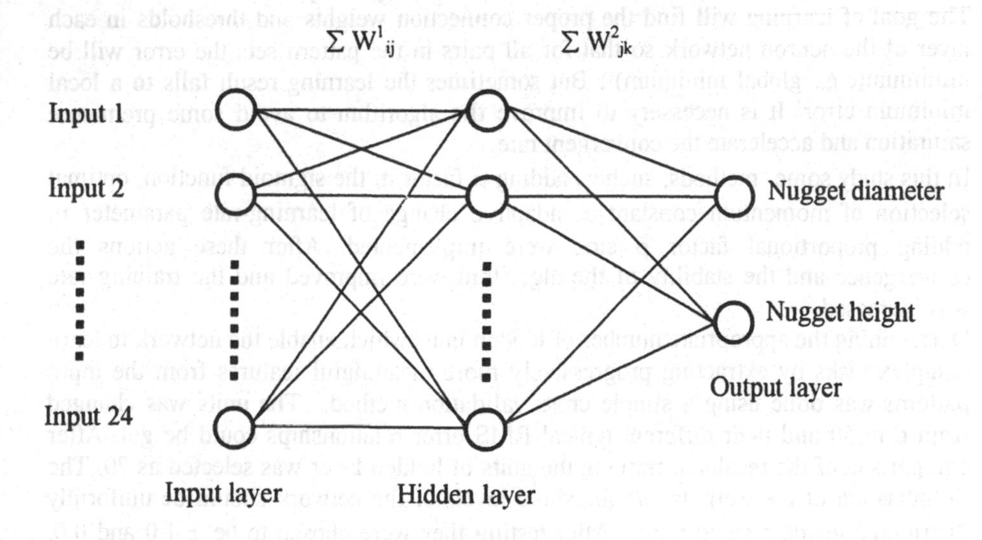

Fig. 2. Structure diagram of neural network.

For the back-propagation network model operation, the algorithm cycles through the training data {[X(n), D(n)]; n=1,2,...,n} as follows: (1) Initialisation; (2) Presentation; (3) Forward Computation; (4) Backward Computation and (5) Iteration, where X(n) is the input vector and D(n) desired response vector.
The net internal activity level $V_j^l(n)$ for neuron j in layer l is

$$V_j^l(n) = \sum_{i=0}^{p} W_{ji}^l(n)\, Y_i^{(l-1)}(n) \qquad (1)$$

Where $Y_i^{(l-1)}(n)$ is the function signal of neuron i in the previous layer l-1 at iteration n and $W_{ji}^l(n)$is the synaptic weight of neuron j in layer l that is fed from neuron i in layer l-1. Assuming the use of a logistic function for the sigmoidal non-linearity, the function (output) signal of neuron j in layer l is

$$Y_j^{(l)}(n) = 1/\{1+\exp[-V_j^l(n)] \qquad (2)$$

If neuron j is in the output layer(i.e., l=L), set

$$Y_j^{(L)}(n) = O_j(n)$$

So, computer the error signal

$$E_j(n) = D_j(n) - O_j(n)$$

where $D_j(n)$ is the jth element of the desired response vector.
The local gradients of the network is computed as follows:
For neuron j in the output layer L

$$\delta_j^L(n) = E_j^L(n)\, O_j(n)[1-O_j(n)] \qquad (3)$$

For neuron j in hidden layer l

$$\delta_j^l(n) = Y_i^l(n)\, [1-Y_j^l(n)] \sum_k \delta_k^{(l+1)}(n) W_{kj}^{(l+1)}(n) \qquad (4)$$

Hence, adjust the synaptic weights of the network in layer l according to the generalised delta rule:

$$W_{ji}^l(n+1) = W_{ji}^l(n) + \alpha[\, W_{ji}^l(n) - W_{ji}^l(n-1)] + \eta\, \delta_j^l(n) Y_i^{(l-1)}(n) \qquad (5)$$

where α is the momentum constant and η is the learning-rate parameter.
The goal of learning will find the proper connection weights and thresholds in each layer of the neuron network so that for all pairs in the pattern sets the error will be minimum(e.g., global minimum) . But sometimes the learning result falls to a local minimum error. It is necessary to improve the algorithm to avoid some premature saturation and accelerate the convergent rate.

In this study some methods, such as adding μ factor in the sigmoid function, optimal selection of momentum constant α, adaptive change of learning-rate parameter η, adding proportional factor ρ etc., were implemented. After these actions the convergence and the stability of the algorithm were improved and the training rate was increased.

Determining the appropriate number of hidden units which enable the network to learn complex tasks by extracting progressively more meaningful features from the input patterns was done using a simple cross validation method. The units was changed from 6 to 50 and their different typical RMS error relationships could be got. After comparison of the results of training the units of hidden layer was selected as 20. The initialisation of the weights and threshold levels of the network should be uniformly distributed inside a small range. After testing they were chosen to be ± 1.0 and 0.0. The order in which the training pattern sets were presented to the network was randomised from one epoch to the next and the pattern data were all generalised in the range from 0.1 to 0.9999.

Under the conditions of the optimal fixed learning parameters the network was trained by the pattern sets. The number of iterations in training was 60000 and the final weights matrix was got. The training results indicated that the sum-squared error of the evaluation pattern sets was 0.0169434, the average relative error of nugget diameter was 4.66468% and the average relative error of nugget height was 6.48885%.

Validity of the neural network

The network which had been properly trained was checked with non-learning pattern sets in order to find whether the network had the ability to evaluate the nugget sizes of spot weld. The network's performance confirmed that during the 22 pattern sets the average relative error of nugget diameter and height was respectively less than 5% and 8%, the highest error value being respectively 10.22% and 16.15%, as shown in table 2.

Table 2. Experimental results of neural network evaluation.

Test-No.	Measured value d (mm)	Measured value 2h (mm)	Estimated value d (mm)	Estimated value 2h (mm)	Absolute error Δd (mm)	Absolute error Δ2h (mm)	Relative error d (%)	Relative error 2h (%)
1	2.95	0.96	2.812	0.834	0.138	0.126	4.67	13.12
2	3.10	0.95	2.778	0.805	0.305	0.145	9.83	15.26
3	5.88	1.24	5.841	1.214	0.039	0.026	0.67	2.07
4	4.16	1.06	4.101	1.036	0.059	0.024	1.41	2.29
5	4.98	1.16	5.053	1.121	0.073	0.039	1.47	3.39
6	3.32	1.04	3.472	0.944	0.152	0.096	4.57	9.22
7	4.24	1.14	4.585	1.185	0.345	0.045	8.14	3.95
8	4.70	1.37	4.515	1.259	0.185	0.111	3.94	8.12
9	4.93	1.24	4.431	1.179	0.499	0.061	10.12	4.96
10	4.83	1.40	5.043	1.386	0.213	0.014	4.42	1.01
11	3.04	0.91	2.895	0.880	0.145	0.030	4.77	3.28
12	5.47	1.33	5.205	1.358	0.365	0.028	4.85	2.11
13	4.05	1.22	4.026	1.161	0.024	0.059	0.59	4.87
14	4.60	1.55	4.533	1.299	0.067	0.251	1.45	16.15
15	3.81	0.98	3.771	0.979	0.039	0.001	1.01	0.05
16	3.73	1.02	3.718	1.011	0.012	0.009	0.33	0.88
17	5.03	1.08	5.544	1.205	0.514	0.125	10.22	11.57
18	3.92	1.09	3.844	1.059	0.076	0.031	1.94	2.79
19	5.22	1.18	5.064	1.220	0.156	0.040	2.99	3.41
20	5.73	1.47	5.609	1.432	0.121	0.038	2.09	2.59
21	2.51	0.67	2.509	0.730	0.001	0.060	0.05	9.01
22	5.65	1.52	5.682	1.366	0.032	0.154	0.57	10.11

Conclusions

From the investigation it is concluded that the principal of evaluating nugget sizes of spot weld based on neural networks is valid, and this technique can be utilised in monitoring quality of spot welding by some further researches and improvements.

References

1. Haykin, S.: Neural network. First edn. Macmillan College Publishing Company, New York Toronto Oxford Singapore Sydney (1993)
2. Zhang, L.: Models and applications of artificial neural networks. First edn. Fu Dan University Press, Shanghai (1993)
3. Javed, M.A.:and Sanders, S.A.C.: Neural networks based learning and adaptive control for manufacturing systems. IEEE/RSJ int. Workshop on intelligent Robots and Systems IROS'91. IEEE Cat. 91TH0375-6,Osaka, Japan (1991) 242-246
4. Takuma, M., Shinke, N. and Motono, H.: Evaluation of function of spot-welded joint using ultrasonic inspection. JSME International Journal, Series A. Vol.39, No.4. (1996) 626-632.
5. Rojas, R.: Theorie der Neuronalen Netze. Springer-Verlag, Berlin Heidelberg New York (1993)

Spotting Relevant Information in Extremely Large Document Collections

Teuvo Kohonen

Helsinki University of Technology
Neural Networks Research Centre
P.O. Box 2200, FIN-02015 HUT, Finland
e-mail teuvo.kohonen@hut.fi

Abstract. The self-organizing map (SOM) converts statistical relationships between high-dimensional data into geometric relationships on a low-dimensional grid. It can thus be regarded as a projection and a similarity graph of the primary data. The most important topological and metric relationships of the primary data elements are shown on the display. The SOM may also be thought to produce some kinds of abstraction. These two aspects, visualization and abstraction, can be utilized in many complex tasks: in this presentation the automatic organization of very large document collections and searching relevant information from them are discussed.

Among the neural-network architectures and algorithms, the Self-Organizing Map (SOM) [1] is in a special position, because it is able to form abstract, but ordered images of large and often high-dimensional data sets. It converts complex, nonlinear statistical relationships between high-dimensional data elements into simple geometric relationships between points on a low-dimensional display.

The central idea in this algorithm is to use a large number of relatively simple and structurally similar, but interacting statistical submodels. Each submodel then describes only a limited domain of observations, but as the submodels are able to communicate, they can mutually decide what and how large a domain belongs to which submodel. By virtue of such collective interactions it becomes possible to span the whole data space nonlinearly, thereby minimizing the average overall modeling error. As the SOM implements a characteristic nonlinear projection from the input space to a visual display, it can be used, e.g., to reveal process states that otherwise would escape notice.

The SOM usually consists of a two-dimensional regular grid of nodes. The SOM algorithms described below compute the models so that they optimally describe the domain of input data. The models will be organized in a meaningful two-dimensional order such that similar models become closer to each other in the grid than the more dissimilar ones. In this sense the resulting SOM is also a similarity graph, and a clustering diagram, too. Its computation is a nonparametric, recursive regression process.

It is not possible to survey the whole range of applications of the SOM method in more detail here. Let it suffice to refer to the list of 3343 research papers on the SOM [2] that is also available at the Internet address http://www.icsi.berkeley.edu/~jagota/NCS/vol1.html.

The basic SOM carries out a clustering in the Euclidean vector space. Surprisingly, the same vector-space clustering methods sometimes apply even to entities that are basically symbolic by their nature. We shall point out in the following that it is possible to perform the clustering of free-text, natural-language documents, if their contents are described statistically by the usage of different words in them. Document word statistics can be shown to be very powerful for the discrimination between different documents and their topic areas.

The histograms, although having a vast dimensionality, are usually very sparsely occupied: in one document one may use only, say, a few dozen to a couple hundreds of different words, depending on its length. Therefore a simple but still effective method to reduce the dimensionality of the representation vectors, without essentially decreasing their discriminatory power, is to project them randomly onto a much lower-dimensional Euclidean space.

Without going into all computational details, the document-clustering SOM called the WEBSOM produces the visual display of the document collection in the following steps: 1. Some preprocessing of the texts is first carried out, removing nontextual symbols and very rare words. Eventually, a stemmer is used to transform all word forms into their most probable stem words. 2. The word histogram of each document is projected randomly onto a space of dimensionality 300 to 500, thereby obtaining a representation vector $\mathbf{x}$ for each document. 3. A SOM is formed using the $\mathbf{x}$ as input data. 4. The models $\mathbf{m}_i$ formed at the nodes of the SOM are labeled by all those documents that are mapped onto the said node. In practice, the nodes are linked to the proper document data base by address pointers (indexing). 5. Standard browsing software tools are used to read the documents mapped to the SOM nodes.

The starting points of browsing can be determined by a set of keywords or a complete new document, which are first processed as described at points 1 and 2 above. The models that match best with this projected „document„ are then marked on the SOM display.

We have already implemented WEBSOM systems for the following applications:

Internet Usenet newsgroups; the largest experiment consisted of 85 newsgroups, with a total of over 1 million documents. The size of the SOM was thereby 104 448 nodes.

News bulletins (in Finnish) of the Finnish News Agency (Finnish Reuter).

Patent abstracts (English) that are available in electronic form. The largest demonstration, being finished during the writing of this report, consists of seven million patent abstracts from the U.S., Japan and European patent offices and the SOM array consists of 1 002 240 nodes.

Demonstrations of various WEBSOM displays are available on the Internet at the address http://websom.hut.fi/websom/, where they can be browsed with standard www browsers.

References

[1]T. Kohonen, 1995. *Self-organizing maps*. Series in Information Sciences, Vol. 30. Springer, Heidelberg. 2nd ed. 1997.

[2] S. Kaski, J. Kangas, and T. Kohonen, 1998. Bibliography of self-organizing map (SOM) papers: 1981-1997. *Neural Computing Surveys*, 1(3&4), 1-176. http://www.icsi.berkeley.edu/~jagota/NCS/vol1.html.

Fuzzy Controller Generation with a Fuzzy Classification Method

István Borgulya
Janus Pannonius University, Faculty of Business and Economics
H-7621 Pécs, Rákóczi út 80.
e-mail: Borgulya@ktk.jpte.hu

Abstract. Fuzzy controller generating procedures when using crisp input-output data produce the necessary system in two steps: first they produce a starting rule set and then they tune the parameters that influence the approximation with a learning algorithm. Other solutions work under special conditions as hybrid neuro-fuzzy systems improving the approximation with a gradient based learning algorithm (e.g. in the case of monotonous membership functions), or use the methods of the genetic algorithms to generate the fuzzy controller. This article demonstrates a new method which reduces the problem to a classification task and carries out the generation of the rules and the tuning of the system in a single step.

1. Introduction

It is a common task to reveal the relations among data. Recognition of rules is needed when analyzing data or modeling industrial controllers or relations in economy. The recognized rules, when, can be viewed as a mapping, which in special cases can be treated as a function approximation problem.

In addition to the different approximation methods of mathematics and statistics, new methods have recently appeared, applying the techniques of artificial intelligence: connections between crisp input and output data are determined with neural networks, or fuzzy systems. Both neural networks and fuzzy systems are suitable to approximate functions: some neural network models and fuzzy systems are universal approximators [10], [18]. They are the fuzzy controllers that can be applied to produce mappings. Mostly the controller of Mamdani type is used and several manual and automatic methods are known for generating the rules and fuzzy sets constituting the controller [4], [9], [16], [21].

Mostly the automatic rule generation is used. The basic idea of these approach is to estimate fuzzy rules though learning from input and output data. Usually the controller is formed in two steps. First, the initial rule set is defined, then the system is optimized (self-tuning). For example, in the first step the data are divided into clusters by fuzzy cluster analysis and rules are formed projecting each class [11], [16], [21]. The fuzzy sets are also either obtained as projections of the classes or are chosen from the standard fuzzy set types (like Gaussian functions). The second tuning step means changing the fuzzy set parameters or the rules in order to make the system more accurate. The tuning is usually performed with a neural network or a genetic algorithm [13], [15].

Though methods based on fuzzy cluster analysis are most frequent, other approaches exist as well. For example Zhang et al. propose a fuzzy controller for modeling or function approximation which apply B-spline basis functions as a fuzzy set [19]. They help the optimization of the controller by supplementing the input data with additional points (peak support points), and the learning algorithm changes the location of these points so as to improve the approximation. We can see a few approaches, which under special circumstances, improve the approximation with gradient descent learning algorithms as a hybrid neural-fuzzy system (e.g. choosing monotonous membership functions [6]). Additionally, there are a lot of methods using genetic algorithms to generate fuzzy controllers ([3], [5], [7]).
In the following, I demonstrate a new method that is based on an extended fuzzy classification algorithm. The method generates a controller that applies Gaussian functions as membership functions of the fuzzy sets and the learning algorithm performs the generation of the rules and the tuning of the system in a single step.

2. Fuzzy controller generation based on fuzzy classification

Before describing the new method, let us take the Yager-Filev method (based on cluster analysis) as a point of comparison [21]. The fuzzy controller generation method of Yager and Filev involves two steps. In the first step it divides the crisp data into m classes with the mountain-clustering procedure, and then it generates m rules:

IF k_{i1} is B_{r1} AND k_{i2} is B_{r2} AND ... AND k_{in} is B_{rn} THEN y is D_r (1)
r=1,2,...,m and

$$B_{rl}(k_{il}) = \exp(-1/\sigma_{rl}^2 (k_{il}-c_{rl})^2)$$
$$D_r(y_i) = \exp(-1/\sigma_{rl}^2 (y_i-c_r)^2)$$

where $k_{i1}, k_{i2}, \ldots, k_{in}, y_i$ is the input data row number i with the corresponding output value, $c_{r1}, c_{r2}, \ldots, y_r^*$ is the centre of class number r, and $\sigma_{r1}\ \sigma_{r2} \ldots \sigma_{rn}\ \sigma_r$ are the spreads of the Gaussian functions. For an arbitrary $x_1, x_2, \ldots, x_n$ input data row the output of the controller is

$$y = \frac{\sum_{r=1}^{m} y_r \cdot \exp\left(-\sum_{j=1}^{n} \frac{1}{2\sigma_{rj}^2}(x_j - p_{rj})^2\right)}{\exp\left(-\sum_{j=1}^{n} \frac{1}{2\sigma_{rj}^2}(x_j - p_{rj})^2\right)} \quad (2)$$

In the second step a learning algorithm tunes the parameters (the class centres, spreads) of the membership functions using the gradient method.
The new method uses this principle in part only. It looks for the controller in the form (1) and it determines the output of the system with the formula (2). The generation of the rules is carried out by an extended version of a fuzzy classification algorithm instead of by cluster analysis. The classification algorithm, let us term it FCR (Fuzzy Classification based on Ranking), uses a learning algorithm and generates a fuzzy

system that characterizes each class with one or more elements (these are the prototypes) and incorporates the input elements into classes according to their similarity to the prototypes [2]. To solve the controller generation problem, the FCR algorithm should be extended so that any output value could be used as a class identifier, and in addition to providing the classification it would improve the accuracy of the approximate value of formula (2).
The differences compared to the classification problem and their treatments are the following:

- The classes are not identified by k different numbers, but by a "rho-alpha" environment. Only those elements belong to the same class that are both sufficiently similar to the class prototype (similarity value > rho) and possess an output value given close to the output value of the prototype (difference between the output values < alpha).
- The forming of the classes are incidental.
- The prototype defines the centres of the Gaussian functions of the controller described by formula (1) in addition to the classes.
- The result given by formula (2) is the approximation of the output value. The learning algorithm has to track the error of this approximation and decrease it.
- Stopping the algorithm depends on the sufficiently accurate approximation instead of learning the correct classes. If the sum of the error squares drops under a threshold vale, the learning process can be stopped.

3. The extended FCR algorithm (EFCR)

The FCR algorithm can process crisp or fuzzy input data and it produces crisp output data as class values. Although our task in view has crisp input values, we reserve the possibility of fuzzy input. As the FCR processes the elements to be classified with a fuzzy ranking method as well, we have to interpret the crisp input data also as fuzzy data. (Symmetric triangles or Gaussian functions are suitable membership functions.)
Let us review the original FCR algorithm concept before introducing the extended one.

3.1 The FCR algorithm concept

Let us treat the elements to be classified alternatives of a ranking problem and choose a fuzzy ranking method. If we accord a value to each alternative (element) by the fuzzy ranking-method as by a function, the sequence of alternatives can be checked by means of these values (The function values depend on both the criterion values and the criterion weights). The elements of the classes are ordered into one or more groups by their similarity (analogy) and we can then accord a centre and a prototype to each group. The centre is defined as the mean of the function values of elements in the given class (group) and the prototype is an element of the class whose function value most closely approaches the value of the centre. The algorithm learns the prototypes and centres of each classes and groups, and the weight numbers of criteria. In order to recognize the classes with relatively high intrinsic scatter, it is possible to define more than one group within a class automatically, depending on the fault sequence.

In the case of a classification the class is known for each element. The fuzzy system has to recognize this correct class by the criteria values and it has to assign this to the element. Because the classes are unambiguously characterized by their centres and prototypes, an algorithm has to be formulated which assigns that class to each element whose prototype is the closest to the given element and which is identical to the correct class.

The system learns the classification step by step. It takes the elements randomly one by one and it assigns each element into a class (group) on the basis of the similarity measure. In order that the centres may converge to a stable value, the system modifies the weight numbers of the criteria in such a way that it lets the new function value of the element approach the proper centre. The weight numbers are modified in every case when either the classification is incorrect or the sequence of the function values of the last two elements differs from the sequence of their class-centres. The modification of weight numbers is carried out by means of a formula similar to the delta-rule of the neural networks. The process will be continued until the classes are recognized with the prescribed accuracy and the centres became stable, i.e. the scale of their alteration decreases to something below a threshold value.

Let us see the heuristic algorithm in more detail. Denote $a_1, ..., a_n$ the elements to be classified and the serial numbers $ho_1, ho_2... ., ho_n$ their known classes. Let us consider each fuzzy criterion K_j $(1 \leq j \leq m)$ as a language variable and let $K_j = \{L_{j1}, . . ., L_{jpj}\}$, where $L_{j1}, . . ., L_{jpj}$ are the values of the language variable. An a_i element is characterized by the $k_{i1}, k_{i2}, ..., k_{im}$ and ho_i input data. The chosen fuzzy ranking method (FRM) orders a real number, y_i (i=1, . ., n) to each elements a_i .

Three important points have to be decided to formulate the algorithm: the handling of analogies, the rule of the weight-modifications and the choosing of the appropriate FRM.

Similarity measure There are different kinds of similarity measures and distance measures [12]. A similarity measure is defined in terms of distance measure as

$$H(a_i,a_j)=1/(1+d(a_i,a_j)) \qquad (3)$$

where $H(a_i,a_j)$ and $d(a_i,a_j)$ represent the similarity and the distance measures of the elements, respectively. In the formula (3) we use the "semantic distance" proposed by Munda [14]. There is a generalization of the Minkowski p-metric with given R crisp criteria and S stochastic and/or fuzzy criteria (m=R+S). In this case, the distance of two arbitrary elements a_i and a_j is given by

$$d(a_i,a_j) = |a_i - a_j| = \left[\sum_{r=1}^{R} w_r{}^p \left|x_r - y_r\right|^p + \sum_{s=1}^{S} w_s{}^p \left(\iint \left|x-y\right| \; f_s(x) g_s(y) dy dx\right)^p\right]^{1/p}$$

where continouos and convex membership functions are supposed and

$$f_s(x)=d1_s*\mu_{Lis}(x) \text{ and } g_s(y)= d2_s*\mu_{Ljs}(y),$$

$d1_s$ and $d2_s$ are real numbers and f_s and g_s fulfils:

$$\int f_s(x)dx = \int g_s(y)dy = 1$$

(While the weights of the criteria are very important, I modified the semantic distance: all criteria are multiplied by their weight (w_s^p).)
The FCR algorithm establishes the belonging of an element to a class by the similarity measure (p=2). The element will be ordered into the class whose prototype is the most analogous with it. Considering, that in the case of classes with relatively high intrinsic scatter not all the elements can be properly characterized by one prototype only, more than one prototypes need to be used in such a class.

Modification of weights. The algorithm checks the ranking of classes based on centres by means of arbitrarily chosen elements. When the sequence of the two latest investigated elements is not correct, it alters the y numbers ordered to the element into the direction of the correct sequence by "small steps". To do this, all those criterion-weights will be changed - increased into the correct and decreased into the wrong direction - whose criterion-values are different for the two elements investigated.
A fault sequence may occur when classifying by analogy in two ways:

- The sequence of class-centres of two arbitrary, to different classes belonging elements is opposite to the sequence of the function-values ordered to the element by FRM
- The class obtained by the algorithm is not the correct class for the element.

When the classification of the elements a_i and a_j is correct, and only the sequence of the function-values obtained by FRM is incorrect, then the weights have to be modified considering the crisp or fuzzy criteria. The range of the modification of weight numbers depends on

- the difference of the individual values of the identical criteria at the last two elements
- the similarity value of the last element with its correct prototype
- the "learning rate" which is the unit of alteration of weight-numbers.

When the sequence of the a_i, a_j pair investigated is not correct the correction comprises even more than one criterion-weight. When the class obtained by the algorithm differs from the correct one, then the analogy between the element and the nearest prototype of the correct class should be increased. The rule of the modification of weights is the same even in the case, but the prototype of the correct class has to be chosen as previous element.
The rule of modification of weights differs for crisp and fuzzy criteria, respectively. In the case of crisp criteria the formula of modification:

$$\Delta w_s = -\eta * \text{sign}(y_i - y_j)*(k_{si} - k_{sj})* H(a_j, p_m) \qquad (4)$$

where η is the learning rate, k_{si} and k_{sj} are the latest and previous criterion-values of the s-th criterion, and p_m is the prototype of the correct class.
In the case of fuzzy criteria, one can not infer the necessary direction and range of the modification of weights from the real values k_{si}, k_{sj}. However, the less-greater relations between the criteria - as fuzzy sets - can be determined for each criterion, as suggested by Munda [14]. When the sign of the integral

$$int_s = \iint (x-y)\, f_s(x)\, g_s(y)\, dy\, dx$$

is positive then the greater, else the less relation is true between the actual values of criteria. Using this to modify the weights in eq. (4) we obtain

$$\Delta w_s = -\eta * \text{sign}(y_i - y_j)*(int_s) * H(a_j, p_m) \qquad (5)$$

Note: A simplification has been carried out in the code of the algorithm: the continuous semantic distance values were approached by discrete ones, i.e. instead of int_s the formula

$$\mu_{Lis}(k_{si})^* \mu_{Ljs}(k_{sj})^*)^*(k_{si} - k_{sj}) \text{ was taken.}$$

The ranking method. The max-min method [20] and the "mark-based method" [1] (see the appendix) were chosen as fuzzy ranking methods when testing the algorithm. Both methods produced similar results.

3.2 The extended FCR algorithm (EFCR)

For fuzzy controller generation we define class more generally in the FCR algorithm, extend it with the approximation given by formula (2) and we use the error of the approximation in the modification of the criteria weight numbers.
The EFCR algorithm allows elements not only with the same class value (output value) to form a class, but also the elements within a certain environment of the prototype. The "*rho-alpha environment*" of the prototype includes all the elements which differ from the class value of the prototype only with *alpha* and sufficiently similar to the prototype (similarity value > *rho*).
The spread value has to be estimated to get the approximation value by formula (2). The spreads are estimated for each prototype (that is the centre of the (1) Gaussian function in the same time) with the distances of the nearest prototypes by criteria, and the criteria weight numbers.
The σ_{is} spread belonging to the p_{is} criterion of the prototype number i is estimated as

$$\sigma_{is} = abs(p_{is} - p_{js})/(2\, w_s) \qquad (6)$$

where the prototype number j is the nearest to the prototype number i.
The *ykoz* approximation of the output value, which is calculated for each element, is also considered by the algorithm in the creation of the classes. The *(ykoz - class value)* error is considered also; it is used as a multiplication factor in the weight number modification formula:

$$\Delta w_s = - \eta^* sign\,(y_i - y_j)^*(k_{si} - k_{sj})^*\,(ykoz - ho_i)^* H(a_i, p_z) \qquad (7)$$

$$\Delta w_s = - \eta^* sign(y_i - y_j)^*(int_s)\,^*(ykoz - ho_i)^* Hd(a_i, p_z) \qquad (8)$$

In the above η is the learning rate, y_i and y_j are the function values corresponding to the a_i and a_j elements (they are determined by the fuzzy ranking method), k_{is}, k_{js} are the preceding and last criteria values at the sth criterion, ho_j is the correct output value, and p_z is the prototype of the correct class.

The steps of the heuristic EFCR algorithm. Let the number of classes be t. Let ρ and α be given values of the rho-alpha environment, ε a threshold number when checking the stability of centres and γ the allowed maximal value of the sum of the squared errors of the approximation. Let $ho_1, \ldots, ho_n$ denote the correct classes of individual elements (the output values).

1. Let us define the initial empty clusters (sets) of classes as $C = \{C_i\}$, $i = 1,2,\ldots,t$, further $c_1, c_2, \ldots, c_t$ and $p_1, p_2, \ldots p_t$ denote the centres and prototypes of classes, respectively. The centre, prototype and class value of the cluster C_i are then c_i, p_i, o_i. Let us choose $c_i := p_i := o_i := 0$ for $i = 1, 2, \ldots, t$ at the beginning. The initial values of the σ_{ij} $(i=1,2, \ldots, t,\ j=1, 2, \ldots, m)$ spreads of the prototypes are randomly generated from the interval [0, 1].
2. Let us take a randomly generated series of weight numbers $w_1, \ldots, w_m$. and two integer indices i and j randomly generated from the interval [1, m]. Let the beginning pair be (a_i, a_j). Let $oc:=0$, $y_i:=0$ and $a_i:=0$.
3. Let FRM: $a_j \rightarrow y_j$.
4. Checking of analogies
   If $H(a_j, p_z) = \max_q H(a_j, p_q)$; $q, z \in \{1,2, \ldots, t\}$ using eq. (3), then $a_j \in C_z$, FRM: $p_z \rightarrow y$.
5. The approximation
   Calculate the *ykoz* approximating value using the formula (2).
6. Modification of prototypes.
   a.) If $(|ho_j - o_z| < \alpha) \wedge (c_z - y_j) < (c_z - y) \wedge (H(a_j, p_z) \geq \rho)$, then $p_z := a_j$, $((\exists x)\ (a_j = p_x)\ (x \in \{1,2, \ldots, z-1, z+1, \ldots, t\} \rightarrow c_x := o_x := 0)$.
   b.) If $((|ho_j - o_z| > 2\alpha) \vee (H(a_j, p_z) < \rho)) \wedge (\exists\, ii)\ (c_{ii} = 0)$, $ii \in \{1,2, \ldots, t\}$, then $a_j \in C_{ii}$, $p_{ii} := a_j$, $c_{ii} := y$, $o_{ii} := ho_j$, $z:=ii$, $((\exists x)\ (a_j = p_x)\ (x \in \{1,2, \ldots, ii-1, ii+1, \ldots, t\} \rightarrow c_x := o_x := 0$
7. Modification of weights
   If $(|ho_j - o_z|) < \alpha) \wedge ((oc < c_z) \wedge (y_i > y_j)) \vee ((oc > c_z) \wedge (y_i < y_j))$, then in the case of crisp criteria eq. (7), in the case of fuzzy criteria eq. (8) is used for modification of weights.
8. The normalization of weights

$$\|W\| = \sum_{l=1}^{m} w_l \quad and \quad w_l = \frac{w_l}{\|W\|},\ i = 1,2,\ldots, m$$

9. Let $i := j$, $oc:=c_z$ and let us choose a new a_j element randomly. The new pair is now (a_i, a_j).
10. Let us check the result in every n-th step of the iteration. Recomputed the values of centres: $uc_1, uc_2, \ldots uc_t$,

$$uc_z = \left(\sum_{q=1}^{db_z} y_q\right) / db_z$$

   where db_z is the number of the elements of the z-th class and y_q is the value ordered to the elements of the class by FRM: $a_q \rightarrow y_q$. If the error function fulfils the condition

$$\left(\sum_{l=1}^{s+t} |c_l - uc_l| < \varepsilon\right) \wedge (err < \gamma)$$

   where err is sum of the squared errors of the approximation, then the iteration is finished, else $c_l := uc_l$ for $l = 1,2, \ldots, t$, recalculates the spreads of formula (2) with formula (6) and the iteration is continued at point 3.

The algorithm converges if suitable parameters are chosen. Each of the centres of all the classes converge to stable values, and at the same time the approximation of the system (1) converges to a minimal error. At this point the system works similarly to the learning of a neural network. For example the convergence may be slower or faster depending on the learning rate or the learning process and it might even be non-convergent. However, the "*rho-alpha environment*" is important also; both accuracy and convergence change, depending on *rho* and *alpha*. If a proper *rho-alpha environment* is not defined, the learning will not be convergent or it will approximate to the required output value only with a significant error. This means that the appropriate parameter values differ from task to task.

Further usage of the algorithm. In case of successful learning, the EFCR algorithm results in a given number of prototypes and spreads. By substituting the results into system (1) the required fuzzy controller can be obtained, and by using the formula (2) the output of the controller can be determined on any input data.

4. Examples

Fuzzy controller generations are usually achieved based on input-output data that are not describable by a precise mathematical function. Noisy, accidental data have to be taken as a starting point, where the relationship between the input and output data can be characterized with a nonlinear system. Let us demonstrate the estimating capability of the EFCR algorithm in such examples and let us examine the generation of some controllers expressed in two dimensions, and then a few in three. In both cases the nonlinear systems were derived from precise functions and the EFCR algorithm correctly approximated these functions, although the accuracy cannot be increased to any serious extent. (Uniformly for these examples a randomly generated data set from the interval [0,1] of 200 elements were used. In the nonlinear systems the volatile value of the noise fluctuated with the maximal amplitude of 0.2.)
The examples demonstrate approximations of 14 nonlinear systems or functions. The first five examples discuss nonlinear system or function approximation in detail. (Fig. 4. summarizes the results of the examples.)

Two dimensional space. Examples 1-2. Let us generate the input-output data with the $Y=0.9/(X+0.2)+\text{noise}$
nonlinear system, where X is from the interval [0, 1]. Yager and Filev used 3 clusters for the approximation [21]. The controller produces by the EFCR gave similar results as the method of Yager and Filev (Fig. 1. part A).
By the parameter values of

alpha = 0.12, *rho* = 0.8, max number of prototypes = 4, learning rate = 0.1

the algorithm gave a stable result after 293 learning steps (the sum of approximation error squares is 0.97). The results that have to be substituted into the (1) system (the x, y coordinated and spreads of the 4 prototypes) are the following:

X_i	0.9	0.06	0.01	0.24
Y_i	0.7	0.24	0.01	0.5
σ_i	0.37	0.03	0.03	0.08

(Hereinafter I pass over the data of the prototypes and spreads.)
If we leave the noise variable from the nonlinear system, the result will be good function regression. Setting *alpha* = 0.01, *rho* = 0.8, max number of prototypes = 20 the algorithm gave a stable result after 292 learning steps with an error of 0.01 (Fig. 1 part B).

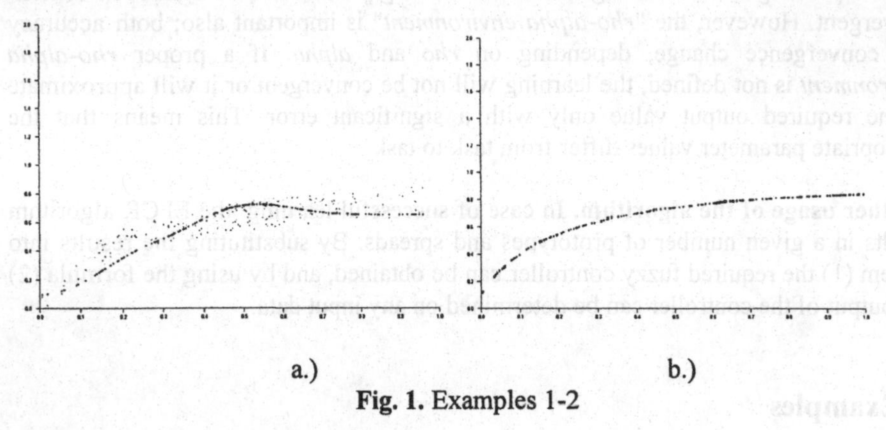

a.) b.)

Fig. 1. Examples 1-2

Examples 3-4. Let us now generate the input-output data with the

$$Y=2.5 (X-0.45)^2 + \text{noise}$$

where $X=0.5 \sin(\pi k) +0.5$
nonlinear system (k=0, 1, ...). Yager and Filev used 7 clusters to approximate this example [21]. The EFCR algorithm by choosing

alpha = 0.12, *rho* = 0.7, max number of prototypes = 7, learning rate = 0.1

gave a stable result after 391 learning steps (the sum of approximation error squares is 1.45). The function regression gives good results again, when leaving the noise variable. Setting *alpha* = 0.01, *rho* = 0.8, max number of prototypes = 20 the algorithm gave a stable result after 245 learning steps with an error of 0.04. Fig. 2 part A demonstrates the approximation of the nonlinear system, part B demonstrates the approximation of the function without the noise variable.

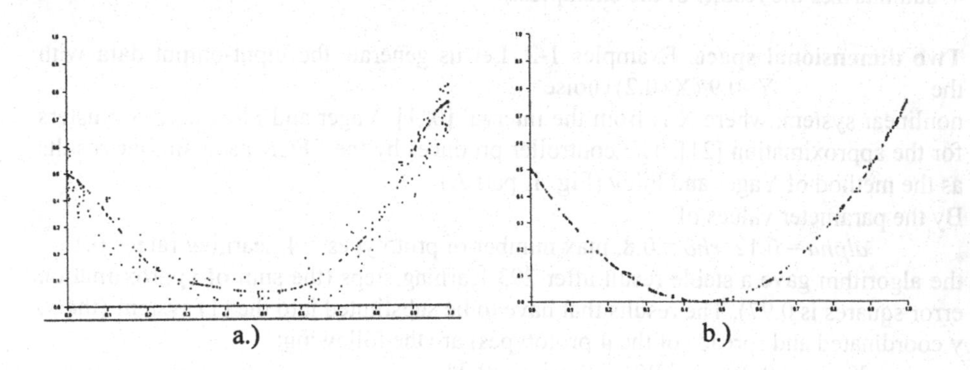

a.) b.)

Fig. 2. Examples 3-4

Example 5. Let us end the row of the two dimensional examples with a function regression. Approximate the

$$Y = -0.5 \sin(7\pi x^3) + x - 0.5$$

function over the interval [0, 1]. By the parameter values of

alpha = 0.01, *rho* = 0.99, max number of prototypes = 70, learning rate = 0.1

the EFCR algorithm gave a stable result after 293 learning steps (the sum of squared approximation error is 0.06). As the function regression demonstrates (Fig. 3.), the accuracy is satisfactory and the original and estimated function values coincide at the most of the points. Naturally, if we decreased the allowed number of prototypes, the accuracy would decrease as well.

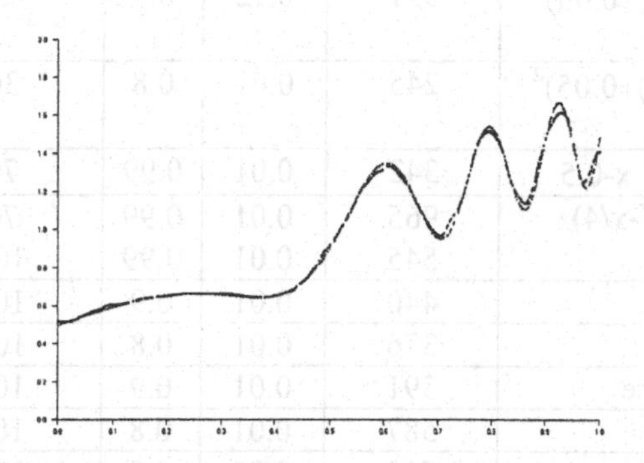

Fig. 3. Function regression of example 5

Three dimensional space. The further examples for fuzzy controller generation are three dimensional nonlinear systems or functions (examples 7-14.). The results are tabulated in Fig. 4.

In a three-dimensional space, the mappings needed the application of significantly more prototypes. Uniformly letting 100 prototypes, the approximations are of acceptable accuracy. (If we increased the intervals of the variables or decreased the number of the prototypes, the accuracy could naturally weaken.) However, this accuracy does not mean that the function curves could be approximated to an arbitrary precision, because accuracy can not be improved to any extent (if we consider a given number of prototypes). Nevertheless, for nonlinear systems, where the output values are points of a band the precision is satisfactory.

The variables of the demonstrated examples take their values from the interval [0, 1], and the target values are in the interval [0, 2]. Additional studies are needed to examine under which conditions the width of the intervals could be broadened in a way that the precision would remain acceptable.

As the parameters chosen for the different examples show, the "*rho-alpha environment*" and the number of prototypes are crucial from the point of view of the solution of the task. The appropriate combination of parameter values is easy to find and with a few systematic attempts we can obtain the ideal parameter values. However, these depend very much on the task itself, and mostly on the relative number of the data in the input space and the mapping or the shape of the surface given by the output values. However the appropriate parameter combination can be

found with the application of a simple search strategy (The graph of the input-output data can help in the search of the parameter).
This method is primarily capable of the approximation of nonlinear systems, but it can also provide function approximations of adequate accuracy.

	Example	Learning steps	alpha	rho	Number of prototypes used	error
1	$y=0.9x/(x+0.2)$+noise	293	0.12	0.8	4	0.97
2	$y=0.9x/(x+0.2)$	292	0.01	0.8	20	0.01
3	$y=2.5*(0.5 \sin(\pi k)+0.05)^2$ +noise $k=1,2,\ldots$	391	0.12	0.7	7	1.45
4	$y=2.5*(0.5 \sin(\pi k)+0.05)^2$ $k=1,2,\ldots$	245	0.01	0.8	30	0.04
5	$y=-0.5 \sin(7\pi x^3)+ x-0.5$	342	0.01	0.99	70	0.06
6	$y=\sin((x/4)^2) \exp(-x/4)$	965 545	0.01 0.01	0.99 0.99	70 100	1.85 0.002
7	$z= x^2+y^2$+noise	440	0.01	0.9	100	1.78
8	$z= x^2+y^2$	376	0.01	0.8	100	0.8
9	$z=\sin x \cos y$ +noise	391	0.01	0.9	100	0.98
10	$z=\sin x \cos y$	587	0.01	0.8	100	0.01
11	$z= \sin (2 \pi x)\cos (\pi y)$ + noise	734	0.05	0.9	100	0.7
12	$z= \sin (2 \pi x)\cos (\pi y)$	530	0.05	0.9	100	1.5
13	$z=1/(1+\|(x+iy)^6-1\|)$	545	0.01	0.9	100	0.7
14	$z=(200-(x^2+y-11)^2-(x+y^2-7)^2)/100$	441	0.01	0.99	100	0.28

Fig. 4 Tabulation of the results of the examples (where the error is the sum of the squared errors)

5. Conclusion

The demonstrated fuzzy controller generating method produces the unknown controller in a single step, using learning. The EFCR fuzzy classification algorithm produces the classes by gradual improvements of the parameters of the classes, keeping track of the accuracy of the approximation. This method is mainly suitable for approximating nonlinear systems, but it could also functions approximations of sufficient precision. Although the EFCR algorithm was tested on crisp input-output data, in principle it would similarly produce the unknown fuzzy controller in case of fuzzy input data.

Acknowledgment
The study was supported by the research OTKA T 18562.

Appendix : The mark-based method

This classification method ranks such multi-criterion alternatives where the criteria are weighted and the criterion values are fuzzy sets. It approaches the solution of the problem in two alternative ways:

- in a special case, the criterion values are supposed to be given by marks like the ones used in education, or they are transformable into such marks.
- in a general case, the criterion values themselves are considered to be marks, without which they possess the properties of marks used in education.

The method orders an individual number, an "extra mark" to each alternatives in both of the cases, on whose basis they can be ranked [1].

The general case. Let $X = \{a_1, ..., a_n\}$ be a finite set of alternatives and $K = \{k_1, . . ., k_m\}$ finite set of fuzzy criteria and let $0 \leq g_1, ..., g_m \leq 1$ weight numbers of the criteria. Let us consider each fuzzy criterion kj ($1 \leq j \leq m$) as language variable over X and let $k_j = \{S_{j1}, . . . S_{jpj}\}$, where $S_{j1}, . . ., S_{jpj}$ are the values of the language variable. The alternative a_i ($1 \leq i \leq m$) is evaluated by the fuzzy sets $S_{j1}, . . ., S_{jpj}$ of the criterion k_j. Let us further define the result-set E as the union of the criterion-sets:

$$E = \bigcup_{j=1}^{m} k_j$$

Let us define the FAM rules below:

$$(g_j^2) \quad \textbf{IF} \ k_j = S_{js} \quad \textbf{THEN} \ E = S_{js} \ (s=1,2, ...,p_j)$$

where (g_j^2) $j = 1, 2, . . ., m$ are the weight numbers of the rules.
Aggregating the constitutions of the rules (inference operator: algebraic product, aggregation operator: sum) the fuzzy system constructed this way orders to each individual a_i alternative a y_i value by a centre-defuzzification process and the alternatives can be ranked on the basis of the y_i values.

The special case. As a special case of the general classification method, let the criterion values of alternatives be ranked uniformly as fuzzy sets interpreted marks not necessarily of identical degree. Let the marks $(S_1, . . ., S_p)$ fuzzy sets given by symmetric triangles or by Gaussian - functions over the [mark-1, mark+1] interval where the mark $\in \{1, 2, . . . ,p\}$ (The set E consists now of the union of the possible $S_1, . ., S_p$ values). In the special case, the y_i values ordered to the alternatives can be interpreted as marks and the alternatives can be ranked by mean of them.

References

1. Borgulya, I.: A Ranking Method for Multiple Criteria Decision Making. International Journal of Systems Science 28, (1997) 905-912
2. Borgulya, I.: Két fuzzy osztályozó módszer. Szigma XXIX. (1998) No. 1-2. 7-28

3. Chin T.C., Qi X.M.: Genetic algorithms for learning the rule base of fuzzy logic controller. Fuzzy Sets and Systems 97. (1998) 1-7
4. Cho K.B., Wang B.H.: Radial basis function based adaptive fuzzy systems and their applications to system identification and predection. Fuzzy Sets and Systems 83, (1996) 325-339
5. Cordón O., Herrera F.: A Hybrid Genetic Algorithm-Evolution Strategy Process for Learning Fuzzy Logic Controller Knowledge Bases In: F. Herrera, L.J. Verdegay (eds): Genetic Algorithms and Soft Computing. Physica-Verlag, Heidelberg (1996) 250-278
6. Fullér R.: Egy fuzzy-neurális megközelítés a keresztárfolyamoktól függő portfóliók kiértékelésére. XXIII. Magyar Operációkutatási Konferencia Pécs. (1997)
7. Geyer-Schulz A.: Fuzzy genetic programming and dynamic decision making Proc. ICSE'96 (1996) 155-172
8. Halgamuge S.K., Glesner M.: Neural networks in designed fuzzy systems for real world applications. Fuzzy Sets and Systems 65, (1994) 1-12
9. Höppner F., Klawonn F., Kruse R.: Fuzzy-Clusteranalyse. Vieweg, Braunschweig / Wiesbaden (1997)
10. Hornik K., Stinchgombe M.: Multilayer Feedforward Networks Are Universal Approximators. In: White H.: Artificial Neural Network. Approximation and Learning Theory. Blackwell Pub. Cambridge (1992) 12-28
11. Kosko B.: Neural Networks and Fuzzy Systems. Englewood Cliffs, NJ. Prentice Hall (1992)
12. E.S. Lee, Q. Zhu: Fuzzy and evidence Reasoning. Physica-Verlag Heidelberg (1995)
13. Mitra S., Pal S.K.: Fuzzy multi-layer perceptron inferencing and rule generation. IEEE Trans. Neural Networks 6, (1995) 51-63
14. Munda G.: Multicriteria Evaluation in a Fuzzy Environment. Physica-Verlag Heidelberg (1995)
15. Shimojima K., Kubota N., Fukuda T.: Virus-Evolutionary Algorithm for Fuzzy Controller Optimization. In: F. Herrera, L.J. Verdegay (eds): Genetic Algorithms and Soft Computing. Physica Verlag Heidelberg (1996) 367-388
16. Sugeno M., Yakusawa T.: A fuzzy logic-based approach to qualitative modeling. IEEE Trans. On Fuzzy Systems, Vol. 1. (1993)
17. Wang L.X., Mendel J.M.: "Back-propagation fuzzy system as nonlinear dynamic system Identifiers". Proc. First IEEE International Conference on Fuzzy Systems, (1992) 1409-1416
18. Wang L.X.: Universal approximation by hierarchical fuzzy systems. Fuzzy Sets and Systems. 93, (1998) 223-230
19. Zhang J., Knoll A., Le K.V.: A New Type of Fuzzy Logic System for Adaptive Modelling and Control. In: Reusch (ed): Computation Intelligence. Springer Berlin (1997) 363-380
20. Yager R.R.: Fuzzy Decision Making Including Unequal Objectives. Fuzzy Sets and Systems. 1, (1978) 87-95
21. Yager R.R., Filev D.: Generation of Fuzzy Rules by Mountain Clustering. Journal of Intelligent & Fuzzy systems, Vol. 2, (1994) No. 3, 209-219

Transformation and Optimization of Fuzzy Controllers Using Signal Processing Techniques

Felipe Fernández[1] and Julio Gutiérrez[2]

Dep. Tecnogía Fotónica, Facultad de Informática U.P.M.
Campus de Montegancedo
28660 Madrid, Spain
[1]felipe.fernandez @es.bosch.com
[2]jgr@dtf.fi.upm.es

Abstract. This paper proposes an eclectic approach for the efficient computation of fuzzy rules based on fuzzy logic and signal processing techniques. The rules $\{R_r\}$ of the MISO zero-order Takagi-Sugeno fuzzy system considered, are given in the form of R_r: If X_1 is A_{r1} and ... and X_N is A_{rN} then z is c_r , where Xj are fuzzified input variables, A_{rj} are standard fuzzy sets which belong to the corresponding partition of unity $\{A_{rj}\}$ and c_r is a nonfuzzy singleton term of output variable z. A relevant feature of this approach is a quantitative, signal processing based, transformation of uncertainty (imprecision) of each input Xj into an additional uncertainty (vagueness) on the corresponding fuzzy partition $\{A_{rj}\}$. This transformation greatly simplifies the involved matching computation. Moreover, this fuzzification transformation gives a new set of linguistic terms $\{A_{rj}'\}$ which is also a partition the unity.

1 Introduction

Fuzzy engineering provides a systematic procedure of transforming a rule-based system into a nonlinear mapping and it is, at the present time, strongly related to the function approximation theory. Fuzzy control discipline is also a subfield of general system theory; but the fact that fuzzy controllers excel in nonlinear systems may produce a misleading impression that the relationship between classical and fuzzy control is very limited. Those may also lead to the erroneous concussion that most of the tools of linear time invariant system theory completely lose their validity in the nonlinear domain.

The approach of this paper is based on the application of some additional mathematical tools from related disciplines, in particular: functional analysis, multidimensional signal processing theory and spline approximation theory [3][5][9] [10]. This paper presents a model for compiling, optimising and implementing fuzzy controllers using these related techniques. The main external differential characteristic from other analogous neuro-fuzzy controllers [1][11][14][15] is the additional inclusion of a general input-fuzzification function (shift invariant operator) which transforms observed inputs (nonfuzzy singletons x) into symmetrical fuzzy variables (fuzzy sets X). The use of a standard cross-correlation/convolution operation to characterize the corresponding matching process (X is A) gives a powerful framework to derive interesting properties for the simplification of the corresponding fuzzy rules. Also, a deeper consideration of the spline approximation theory allows us a simpler decomposition of the elements of fuzzy rules.

From a theoretical point of view our system-theory/signal-processing orientation supplies a method for the undivided consideration of uncertainty of input variables and uncertainty of antecedent linguistic terms of fuzzy rule antecedents. This way, a fuzzy proposition (X is A) with two fuzzy components (X , A) can be orthogonally decomposed into a three-element fuzzy system (x , A^1, Φ): a variable singleton input (x), a crisp linguistic term (A^1) and an additional fuzzy set (Φ) which globally captures the imprecision of X and the vagueness of A.

The rest of the final paper is organized as follows: Section 2 reviews some related fuzzy basic concepts and introduces new ones; Section 3 summarizes the main characteristics of the fuzzy controller model used; Section 4 introduces the cross correlation/convolution matching operation on fuzzy sets and describes the preservation theorem on fuzzy partitions; Section 5 presents *F*-splines, a new general spline model specifically adapted to fuzzy control; some conclusions are drawn in Section 6.

2 Basic Definitions and Terminology

The main concept of fuzzy set theory is the notion of fuzzy sets. Many authors found different ways of defining fuzzy sets; in this paper we consider the following particular definition adapted from Geering [7]. A fuzzy set A is an ordered pair (U,A) where U is the domain or universe of discourse (usually the real vector space R^n) and A is a set membership function mapping U onto the interval [0,1] of the real line R, i. e., $A: U \rightarrow [0,1]$. Every fuzzy set can be represented by its membership function: if the reference set is discrete, membership functions are discrete domain functions; if the referential set is a continuous set, we can represent analytically the corresponding continuous domain membership functions.

In this section, we review some main concepts of discrete and continuous fuzzy sets [6][8], for clarifying the notation used and for convenience of the ulterior development. For a general fuzzy set A defined on the universe U, the following standard definitions are considered:

- Support of fuzzy set A is an interval defined by

$$\mathrm{Supp}(A) = \{u \mid A(u)>0 \ , \ u \in U\} \quad \textbf{(1)}$$

- Core of fuzzy set A is an interval defined by

$$\mathrm{Core}(A) = \{u \mid A(u)=1 \ , \ u \in U\} \quad \textbf{(2)}$$

- Cardinality of fuzzy set A, on a discrete domain U, is defined by

$$|A| = \sum_{u \in U} A(u) \tag{3}$$

- Area of fuzzy set *A*, on a continuous domain *U*, is defined by

$$|A| = \int_U A(u)\,\mathrm{d}u \tag{4}$$

In this paper some general concepts from functional analysis [5][10][12] are also considered. For functions defined on the same domain as the corresponding fuzzy sets, the following standard definitions are used:

- Shift by *p* of a general function *F*, on a domain *U*, is a new function defined by

$$F_p(u) = F(u-p) \tag{5}$$

- An even function *F* on a domain *U*, is a function which verifies

$$F(u) = F(-u) \quad \forall\, u \in U \tag{6}$$

- Restriction of general function *F*, on a domain *R*, onto an interval $[\alpha,\beta] \subset R$ is defined by (using conditional notation of Mc Carthy [2])

$$F_{[a,\,b]} = \{(u \in [a,\, b]) \rightarrow F(u);\, 0\} \tag{7}$$

- An inner product of two functions of two continuous functions *F* and *G*, on a continuous domain *U*, is customarily defined by

$$\langle F,G \rangle = \int_U F(u) \cdot G(u)\, du \tag{8}$$

  Analogously for functions *F* and *G* defined on a discrete domain *U*, a common inner product is defined by

$$\langle F,G \rangle = \sum_{u \in U} F(u) \cdot G(u) \tag{9}$$

A central point in this paper is a general functional consideration of the family of linguistic terms used in the antecedent description of a fuzzy controller. These linguistic terms can be seen as qualitative values used to describe fuzzy rules. Usually, a set of *M* linguistic terms: $\{A\}=\{Aj\}=\{A_1, A_2 \ldots A_M\}$ is defined in the domain *U* of a given scalar variable *u*. It is generally required that the linguistic terms satisfy the property of coverage, i.e. for each domain element is assigned at least one fuzzy set *Aj* with nonzero membership degree.

In this paper, for practical reasons, the linguistic terms are normally constrained to be fuzzy numbers [8]. Also the model here considered, and presently in many other fuzzy control models [1][8][11], imposes stronger conditions on the linguistic terms used:

- Partition of unity $\{Aj\}$ of a domain *U* is defined by the following constraints

$$Aj(u) \in [0,1] \quad \forall u \in U \quad 1 \leq j \leq M \tag{10}$$

$$\sum_{j=1}^{M} A_j(u) = 1 \quad \forall u \in U; \tag{11}$$

$$\left(\sum_{u \in U} A_j(u) \right) > 0 \quad 1 \leq j \leq M \tag{12}$$

Equation (10) is the standard membership constraint; equation (11) means that for each u, the sum of all membership degrees $A_j(u)$ equals one, and Equation (12) implies that none of the fuzzy sets $A_j(u)$ is empty. This fuzzy segmentation of domain U is also called *fuzzy partition* [1][2]. A hard or *crisp partition* $\{Aj\}=\{Aj^1\}$ can be considered as a particular case: $Aj^1(u) \in \{0,1\}$. For discrete domains $U=\{u \in \mathrm{Z} \mid (1 \leq u \leq Q)\}$ a fuzzy partition can be conveniently represented in matrix notation: $[A]= [A_{j,u}]$ where $[A]$ is a $M \times Q$-matrix whose rows $[A_j]$ define point-wise the membership functions of the corresponding fuzzy terms.

Example 1. The sequence $T= \{(t_0, t_1, \ldots; t_M) \mid (a=t_0 < t_1 < \ldots < t_M=b)\}$ defines a set of M intervals $\{A_0, A_1, \ldots, A_M\}$ which is a crisp partition of the interval $[a,b] \in U$ (Figure 1.a). Analogously, the same sequence T defines a piecewise linear fuzzy partition $\{A_0, A_1, \ldots, A_M\}$ composed of piecewise linear triangular functions $\{Aj(u)\}$, with $A_j(t_j)=1$ and $A_j(t_k)=0$ if $k \neq j$. (Figure 1.b). □

In this paper, a particular extension of a fuzzy partition is usually considered, in order to avoid bound problems in window-convolution operations applied on fuzzy partitions, which are used to model fuzzification transformations:

- Constant extension of a fuzzy partition $\{Aj\}$ $(1 \leq j \leq M)$ defined on a domain $U=[\alpha,\beta]$ onto a bigger domain $U^+= [\alpha-w, \beta+w] \mid (w>0)$ is characterized by

$$A_j(u)^+ = \{ u \in [\alpha, \beta] \rightarrow Aj(u);\ u \in (\alpha-w, \alpha) \rightarrow Aj(\alpha);\ u \in (\beta, \beta+w) \rightarrow Aj(\beta)\} \tag{13}$$

In Figure 2, it is shown an example of constant extension of a fuzzy partition.

- An important parameter of a fuzzy partition is the *overlapping factor o*, which

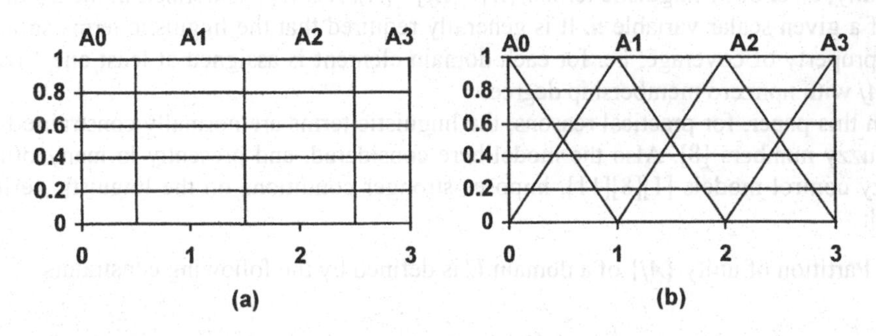

Fig. 1. Fuzzy partitions: (a) Crisp partition. (b) Triangular fuzzy partition.

is the maximum number of membership functions with non-null membership degrees for each element u of domain U.

$$o(\{A_j\}) = \max_u \left(\sum_j \lceil A_j(u) \rceil \right) \quad u \in U \quad 1 \leq j \leq M \tag{14}$$

where $\lceil \alpha \rceil$ denotes the ceiling function (smallest integer $\geq \alpha$). For example, the overlapping factor of fuzzy partition that appears in Figure 2 is equal tree.

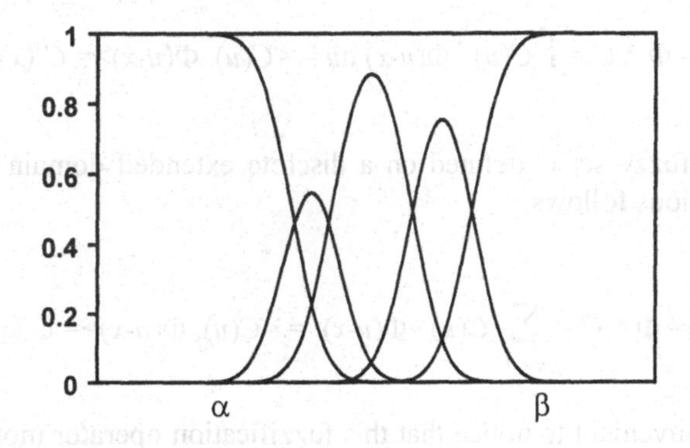

Fig. 2. Constant extension of a fuzzy partition defined on [α,β]

We also introduce in this section, a non-standard fuzzy set that we use as a conceptual bridge between fuzzy and signal processing systems, to accomplish some standard signal operations. The $[0,1]$ membership boundedness condition is relaxed to allow general correlation operations on fuzzy sets.

- A c-normal S-fuzzy set $X(u)$ (signal-fuzzy set with cardinality unity) is defined as

$$X(u) = \frac{1}{|Y|} \cdot Y(u) \tag{15}$$

where Y(u) is the membership function of a standard nonempty fuzzy set Y defined in a discrete or continuous domain U, and $|Y|$ is the scalar cardinality [2] or area of Y. Therefore, $X(u)$ is a unit-area fuzzy signal set. It is convenient to notice, that unit impulse functions (for discrete domain)

$$\delta_x = \delta(u\text{-}x) = \{(u{=}x) \rightarrow 1;\ 0\} \tag{16}$$

or informally Dirac pulses (for continuous domain) can be included into this type of s-fuzzy sets.

Finally, in this section, we introduce the concept of *fuzzification operator*, which is a generalization of the concept of fuzzification function [8][13]. This operator allows a unified consideration of the uncertainty associated to the input variables or to the corresponding antecedent terms of a fuzzy controller.

- The fuzzification operator applies an even fuzzification function Φ (c-normal S-fuzzy set) onto a fuzzy set C, defined on a continuous extended domain U, and gives a new fuzzy set C' defined by means of the following cross-correlation/convolution operation

$$\Phi(C) = \Phi * C = \int_U C(u) \cdot \Phi(u-x)\, du = \langle C(u), \Phi(u-x) \rangle = C'(x) \tag{17}$$

For a fuzzy set C defined on a discrete extended domain U, an analogous definitions follows

$$\Phi(C) = \Phi * C = \sum_{u \in U} C(u) \cdot \Phi(u-x) = \langle C(u), \Phi(u-x) \rangle = C'(x) \tag{18}$$

It is convenient to notice that this fuzzification operator modifies, in general, the core and support of the corresponding fuzzy set. A Particular interesting case is the singleton fuzzification operator δ:

$$\delta * C = C \tag{19}$$

3 Fuzzy Controller Model

In this section we summarize the main characteristics of the fuzzy implementation model considered. It is assumed that the fuzzy controller has N fuzzified inputs: $(X_1, X_2, \dots X_N)$, defined within a universal set U^N, and one nonfuzzy-singleton output z, defined within a universal set V.

This MISO fuzzy controller is a zero-order Takagi-Sugeno fuzzy system and it is defined by a set of rules $\{R_r\}$, which are given in a canonical conjunctive normal form:

$$R_r\text{: If } (X_1 \text{ is } A_{r1} \text{ and } \dots \text{ and } X_N \text{ is } A_{rN}) \text{ then } (z \text{ is } c_r) \tag{20}$$

where X_i is a fuzzified input variable (c-normal S-fuzzy set), $A_{r,i}$ is a standard fuzzy set which belongs to a corresponding partition of unity $\{Arj\}$ and c_r is a nonfuzzy singleton term associated to output variable z of the r-th rule.

Before describing the implementation model proposed, it is convenient to notice that the smoothness of the control surface of a fuzzy system $\{R_r\}$ directly depends on the smoothness of antecedent membership functions (X and A). This fact restricts the choice of the type of membership functions that we can use. For instance, the frequently used trapezoidal membership functions for linguistic terms $\{A_j\}$, and singletons for inputs X, result in a nonsmooth output.

The evaluation of output z of the T-S fuzzy controller referred, with N fuzzy set input variables Xi on a discrete domain U, and M rules R_r, is determined via a sequence of a tree-stage procedure:

- **Matching.** The standard procedure computes, for each fuzzy set variable X_i and each fuzzy set term A_{ri}, the degree of cross correlation u_{ri} given by

$$\mathrm{T}(X_i \text{ is } A_{ri}) = u_{ri} = \underset{u \in U}{\mathrm{S}} \left(\mathrm{T}\left(X_i(u), A_{ri}(u)\right) \right) = \sum_{u \in U} \left(X_i(u) \cdot A_{ri}(u) \right) \tag{21}$$

where S and T mean t-norm and s-norm respectively. In this paper T is implemented by the algebraic product (which is a differentiable function) and S by algebraic addition operation (which is also a differentiable function on the partitions of unity considered).

For a continuous domain U the following analogous inner product is considered

$$\mathrm{T}(X_i \text{ is } A_{ri}) = \langle X_i, A_{ri} \rangle = \int_U \mathrm{T}\left(X_i(u), A_{ri}(u) \right) du = \int_U X_i(u) \cdot A_{ri}(u)\, du \tag{22}$$

In our approach, this matching operation is previously transformed, during the compilation time of fuzzy rules, into a standard singleton computation. This transformation is accomplished by applying input fuzzification operator Φ, not to the original singleton input, but to fuzzy terms of the corresponding fuzzy partition (Φ: $\{A_{ij}\} \to \{A_{ij}'\}$) (this point is more deeply described in Section 4). The new obtained partition $\{A_{ij}'\}$ is also a fuzzy partition (this property is formally expressed by the theorem of fuzzy partition preservation in Section 4). Furthermore, the new singleton computation only is accomplished for active rules. If o is the common overlapping factor of fuzzy partitions $\{A_{ij}\}$ then the number of active rules is equal to o^N.

- **Antecedent conjunction.** For each active rule r compute its activation degree by

$$u_r = \underset{i \in N}{\mathrm{T}} \left(u_{ri} \right) = \prod_{i \in N} \left(u_{ri} \right) \tag{23}$$

where T means a t-norm. In this paper, T is also implemented by algebraic product (also differentiable).

- **Rule aggregation.** The output z is computed as follows

$$u_{ri} = \frac{\sum_{r \in R\,\text{active}} \left(u_r \cdot c_r \right)}{\sum_{r \in R\,\text{active}} \left(u_r \right)} = \sum_{r \in R\,\text{active}} \left(u_r \cdot c_r \right) \tag{24}$$

where R_{active} is the set of active rules. Last expression does not use any division, since in a fuzzy partition, the denominator of previous fraction equals one for every point of domain U^N.

4 Fuzzy Cross Correlation

Fuzzy propositions are sentences with truth-value from the interval [0,1]. They can generally be expressed, in a canonical form, by

$$\text{p}(x)\text{: } X \text{ is } A \tag{25}$$

or in equivalent form

$$\text{p(x): } \Phi_x \text{ is } A \tag{26}$$

where Φ_x is a fuzzified input variable: $\Phi_x=\Phi(u\text{-}x)$, and A is a fuzzy predicate: $A=A(u)$ which characterises a term of the considered variable. $\Phi(.)$ is also an additional even fuzzification function (shift invariant) applied to nonfuzzy singleton $\delta_x=\delta(u\text{-}x)=\{(u=x)\rightarrow 1;0\}$ in order to consider the imprecision related to x:

$$\Phi_x = \Phi * \delta_x \tag{27}$$

where * stands for the corresponding cross-correlation/convolution operation. Input $\Phi_x(.)$ is also defined as a c-normal S-fuzzy set. It is the result of a c-normal fuzzification function (of unit-area) applied to a singleton input variable δ_x of a fuzzy controller, to express the associated uncertainty.

In this paper, the truth degree of a fuzzy proposition, in a continuous domain R, is defined by means of the cross-correlation operation of fuzzy set $A(.)$ and a c-normal S-fuzzy set $\Phi(.)$

$$\text{T}(X \text{ is } A) = \text{T}(\text{p}(x)) = \int_R A(u) \cdot \Phi(u\text{-}x)\,\text{du} = A * \Phi\ (x) = \langle A\,,\Phi_x\rangle = A'(x) \tag{28}$$

The cross-correlation integral is similar to convolution integral except that the fuzzy set $X(u)$ is simply displaced to $X(u\text{-}x)$ without reversal. For the case considered here, where $X(.)$ is a even S-fuzzy set, convolution and correlation are equivalent; this follows because an even fuzzy set and its image are identical. With this restriction, convolution and correlation are loosely interpreted as identical.

For discrete domains the integral is replaced with the sum of products:

$$T(X \text{ is } A) = T(p(x)) = \sum_{u \in U} \big(\Phi(u\text{-}x) \cdot A(u)\big) = A * \Phi\ (x) = \langle A, \Phi_x \rangle = A'(x) \quad \textbf{(29)}$$

The result of a cross-correlation/convolution of an even c-normal S-fuzzy set $\Phi(.)$ on a set of linguistic terms of a fuzzy partition $\{A_j\}$ can also be considered a new set of linguistic terms $\{A'_j\}$; this new set of linguistic terms includes the uncertainty of fuzzy variable $X=\Phi_x$ and of linguistic terms $\{A_j.\}$.

An example of this partition transformation is shown in Figure 3. The big advantage of this transformational method is that it only carries out a standard singleton fuzzification in execution time, but has the same expressive power of a general input fuzzification [8].

Note that we have substituted the standard max-min matching operation with a sum-product matching operation. A comparative advantage of this product-sum matching operation, as we will see below, is that it is a partition of unity preserving transformation. Unfortunately, this property is not accomplished by max-min matching operation usually considered [4][8][13].

Theorem 1. The partition of unity condition of a fuzzy partition $\{A_j\}$ is preserved under the cross-correlation/convolution:

$$\{A_j\} * \Phi = \{A_j * \Phi\} = \{A_j'\} \quad \textbf{(30)}$$

where Φ is an even c-normal S-fuzzy set (fuzzification operator).

Proof. Is a consequence of distributive property of convolution operation:

$$\sum_{j=1}^{M} A'_j(u) = \sum_{j=1}^{M} (A_j(u) * \Phi) = (\sum_{j=1}^{M} (A_j(u)) * \Phi = 1 * \Phi = 1 \quad \forall u \in U \quad \textbf{(31)}$$

5 F-Spline Partition

The first step of a fuzzy controller design is the division of input space into fuzzy regions. The simplest solution is the rectangular crisp partition, i. e. to divide each input domain $[a,b] \in U$ into a ordered set of M intervals: $a=t_0 < t_1 < \ldots < t_M = b$, and assign each region to a crisp linguistic term A_j^1 defined by (using conditional notation of Mc Carthy):

$$A_j^1(u) = \{(t_j < u < t_{j+1}) \rightarrow 1; 0\} = 1_{[tj,\, tj+1]} \qquad j = 0,\ldots,M-1 \quad \textbf{(32)}$$

A possible second step is the introduction of a global fuzzification function Π to express the associated uncertainty of the corresponding input (imprecision), the uncertainty of predicates considered (vagueness), and the preferred uncertainty of the control surface involved (smoothness).

In one dimension space, the initial basis considered for fuzzification functions is the even rectangular c-normal S-fuzzy set (of unit-area and base σ) $\Pi_0^1[\sigma]$:

$$\Pi^1[\sigma](u) = \{(|u|<\sigma/2) \rightarrow 1/\sigma \,;\, 0\} = (1/\sigma)_{[-\sigma/2,\, \sigma/2]} \quad \textbf{(33)}$$

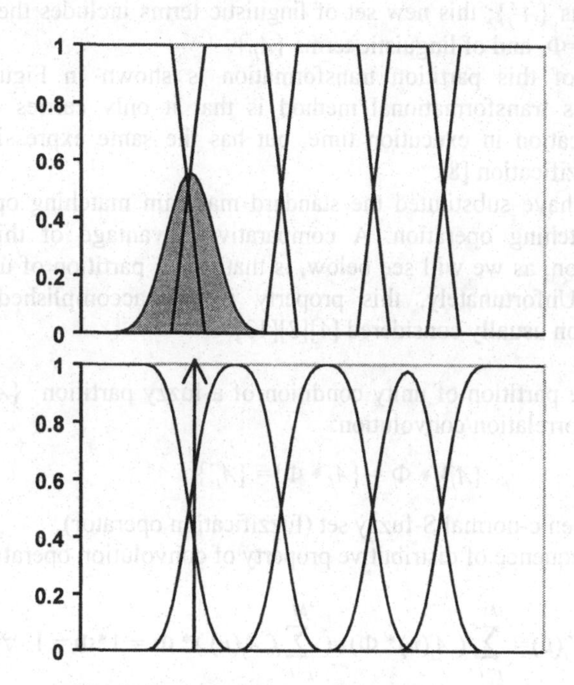

Fig. 3. Equivalent fuzzy matching computations:
a) T(X is {A}) b) T(x is {Φ (A)})

The matching between the function $\Pi^1[\sigma]$ and a crisp domain partition $\{Aj\}=\{Aj^1\}$ is a new fuzzy partition $\{A_j^2[\sigma]\}$ defined by the multiple correlation operation:

$$\{A_j^2[\sigma]\} = \Pi^1[\sigma] * \{A_j^1\} \quad \textbf{(34)}$$

The resulting trapezoidal fuzzy terms $\{A_j^2[\sigma]\}$ also form a partition of unity (fuzzy partition).

An n-order B-spline fuzzification function can be defined by the following recurrence equation

$$\Pi^n[\sigma] = \{(n=1) \rightarrow \Pi^1[\sigma] \,;\, \Pi[\sigma] * \Pi^{n-1}[\sigma]\} \quad \textbf{(35)}$$

In general, if an n-order B-spline fuzzification function $\Pi^n[\sigma]$ is correlated with a crisp partition $\{A_j^1\}$:

$$\Pi^n[\sigma] * \{A_j^1\} = \{\Pi^n[\sigma] * A_j^1\} = \{A_j^{n+1}[\sigma]\} \quad \textbf{(36)}$$

it gives a new fuzzy partition $\{A_j^{n+1}[\sigma]\}$ which is called, in this paper, *F-spline partition* of order n+1. The parameters n (order) and σ (uniform-averaging window size) control the smoothness of the corresponding control surface. For example, in Figure 4 it is shown a nonuniform F-spline fuzzy partition of order n=8 and uniform-averaging window size σ=11: $\{A_j^8[11]\}$, derived from the corresponding nonuniform crisp partition $\{A_j^1\}$:

$$\{A_j^8[11]\}=\Pi^7[11] * \{A_j^1\} \quad \textbf{(37)}$$

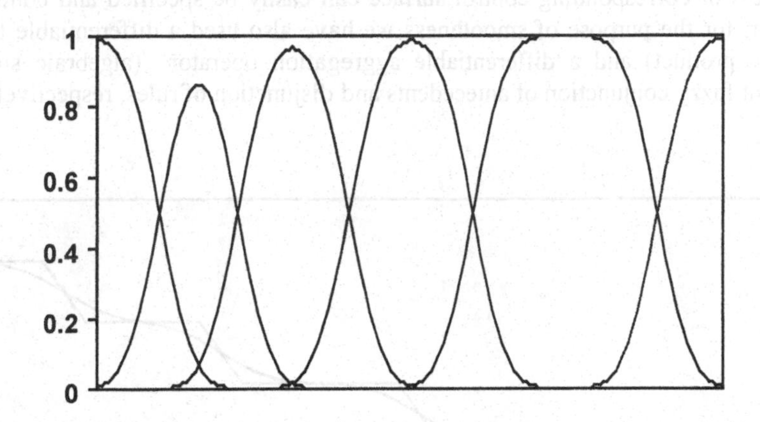

Fig. 4. F-spline partition of eight-order

This way, the generation of complex linguistic fuzzy partitions for controlling the smoothness of a fuzzy controller, it is not always necessary, since the user can factorize the regular vagueness of a linguistic fuzzy partition, and specify it as an additional uncertainty of the corresponding input. Therefore, in regular vagueness domains, it is also possible to specify, in an equivalent form, crisp linguistic partitions for antecedents and fuzzified inputs

Example 2. As a simple application example, Figure 5 shows the corresponding output function of the SISO zero-order Takagi-Sugeno fuzzy controller:

$$\{R_r\text{: If } X \text{ is } A_r \text{, then } z \text{ is } c_r\} \quad r = 1,2,...,6. \quad \textbf{(38)}$$

where $\{c_r\}$ ={1.5, 2, 5, 6, 7, 8} and fuzzy partition $\{A_r\}$ is initially a typical trapezoidal partition , also shown in the same figure. If a singleton fuzzification function is used, the corresponding output function is not smooth (it does not belong to C^1). If a suitable fuzzification function is applied to the corresponding input X (a B-spline of order 10 in this case: $\Pi^{10}[13]$), a new equivalent F-spline partition is obtained, which gives the smoother output function indicated. Notice that it is also equivalent in this case, to specify a nonuniform crisp partition $\{A_r^1\}$, and a similar B-spline fuzzification function, but of an additional order: $\Pi^{11}[13]$.

6 Conclusions

In this paper, we have presented a new efficient approach to implement real time fuzzy controllers. The computation model used is mainly based on the additional application of standard cross correlation/convolution techniques for the matching stage of antecedents.

By means of the additional fuzzification function on involved inputs, the smoothness of corresponding control surface can easily be specified and controlled. Moreover, for the purpose of smoothness we have also used a differentiable t-norm (algebraic product) and a differentiable aggregation operator (algebraic sum) to implement fuzzy conjunction of antecedents and disjunction of rules, respectively.

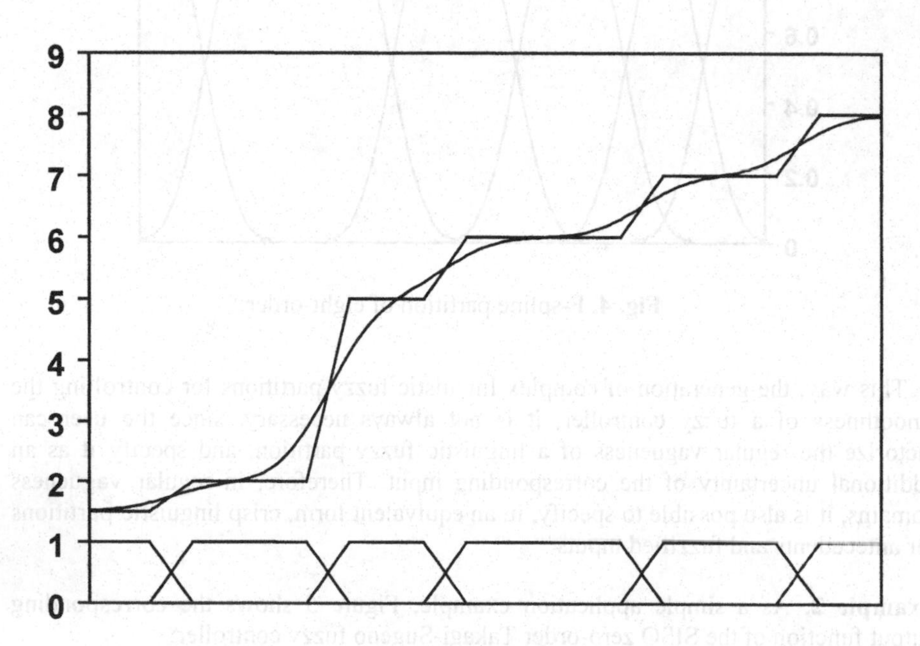

Fig. 5. Output function of a SISO zero-order T-S fuzzy controller using a fuzzified input (B-spline)

The utilization of fuzzified input variables in the description of fuzzy controllers, gives more expressive power to the fuzzy source language, allows direct specification of control-surface smoothness and does not involve an additional computation, in execution time, using the referred transformation.

Furthermore, the introduction of F-spline fuzzy partitions for defining linguistic terms provides a direct method to specify and implement high performance fuzzy systems.

References

1. R. Babuska, *Fuzzy Modeling for Control*, Kluwer Academic Publisher, 1998.
2. D. W. Barron, Recursive Techniques in Programming, Macdonald, London, 1976
3. K. R. Castleman, Digital Image Processing, Prentice Hall, 1996.
4. T. Chiueh, "Optimisation of Fuzzy Logic Inference Architecture", Computer, IEEE, vol. 25, No. 5, May 1992, pp. 67-71.
5. C. K. Chui, Wavelets: A mathematical Tool for Signal Analysis, SIAM, 1997.
6. D. Driankov, H. Hellendoorn, M. Reinfrank, An Introduction to Fuzzy Control, Springer-Verlag, 1993.
7. H. Geering, Introduction to Fuzzy Control, Institut fuer Mess- und Regeltechnik. IMRT-Bericht Nr. 24, October 1993.
8. G. Klir, B. Yuan, Fuzzy Sets and Fuzzy Logic, Prentice Hall, 1995.
9. S. Mallat, A wavelet Tour of Signal Processing, Academic Press, 1998.
10. A. N. Michel & C. J. Herget, Applied Algebra and Functional Analysis, Dover, 1981.
11. T. Miyoshi H. Ichihashi and F. Tanaka, Fuzzy Projection Pursuit ID3, IEEE International Conference on Fuzzy Systems, 1997, pp. 1301-1306.
12. A. Pinkus & S. Zafrany, Fourier Series Integral Transforms, Cambridge University Press, 1997.
13. H. Watanabe, W. Dettloff, K. Yount, "A VLSI Fuzzy Logic Controller with Reconfigurable, Cascadable architecture" IEEE Journal of Solid-State Circuits, Vol. 25, No. 2, April 1990, pp. 376-381.
14. J. Zhang and A. Knoll, Constructing Fuzzy Controllers with B-spline Models, IEEE International Conference on Fuzzy Systems, 1996.
15. J. Zhang and A. Knoll, Unsupervised Learning of Control Surfaces Based on B-spline Models, IEEE International Conference on Fuzzy Systems, 1997, pp. 1725-1730.

Fuzzy-Control Design Tool for Low-Cost Microcontrollers (FHFC-Tool)

Prof. Dr.-Ing. Klaus-Dietrich Kramer
Dipl.-Ing. Jochen Kirschner
Sven Wöhlbier
Hochschule Harz, University of Applied Sciences
Dep. of Automation and Informatics
Friedrichstraße 57-59
D-38855 Wernigerode
Phone +49 (0) 3943/659-317
Fax +49 (0) 3943/659-107
eMail: kkramer@fh-harz.de

1. Introduction

The CASE process of Fuzzy systems typically consists of the partial processes:

- graphical editing of the project
- generation of code (in general via C-intermediate code)
- and, if necessary, verification or simulation of the system

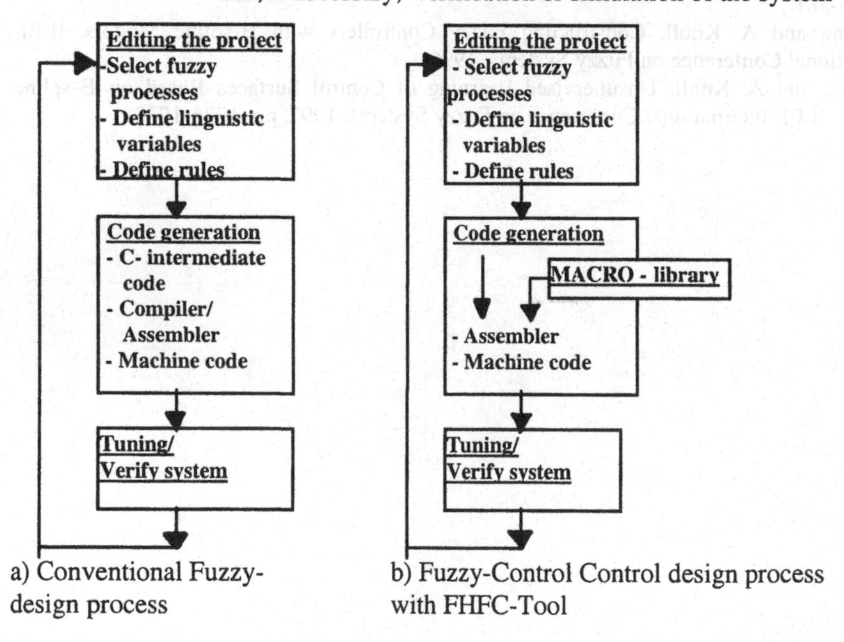

a) Conventional Fuzzy-design process

b) Fuzzy-Control Control design process with FHFC-Tool

Fig. 1: Fuzzy-Control design processes

Using the C-intermediate code for the realization of a target-independent development is favourable. The disadvantage, however, of this concept is that such a solution is not optimal regarding the resources of the system (time, memory).

For a fuzzy design tool for low-cost microcontrollers it is necessary to realize an optimized (with reference to the reaction time and the memory requirements) and flexible (with reference to different targets) development system.

2. Structure of Fuzzy Control Processing

A Fuzzy Processing System consists of the sensory mechanism (INPUT periphery), the processing system (fuzzy controller) and the actuatory mechanism (OUTPUT periphery).

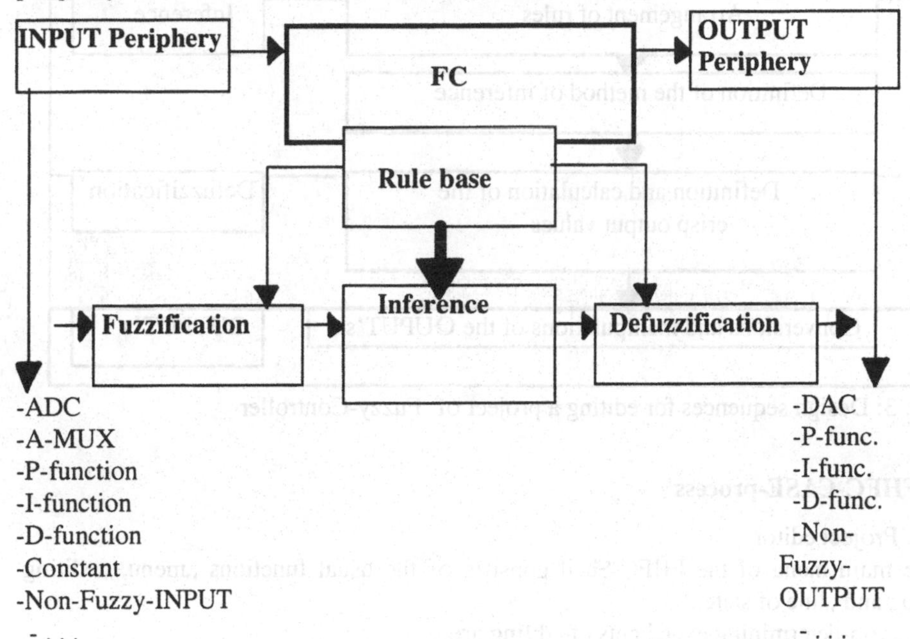

Fig. 2: Fuzzy process

From the point of view of the design process, all these functions are suboperators, of the complete process. Both NON-Fuzzy and Fuzzy operations are parts there of. These operators are deposited as optimized Assembler Macros in a library.

3. Optimized Fuzzy-Design-Process

3.1. Goals of development

The FHFC-Tool has been developed to realize the following approaches:

- target-optimized generation of fuzzy- and NON-fuzzy algorithms for different microcontrollers
- system design by using a uniform graphic platform
- verification of the result system (fuzzy debugging, fuzzy monitoring, fuzzy benchmarks)

3.2. Design sequences for editing a fuzzy system

The structure of the design process of a fuzzy system is realized analogously to the fuzzy control process shown in Fig. 2. All the definitions of the structure and the algorithms of the fuzzy process must be determined, including all data conversions and data manipulations.

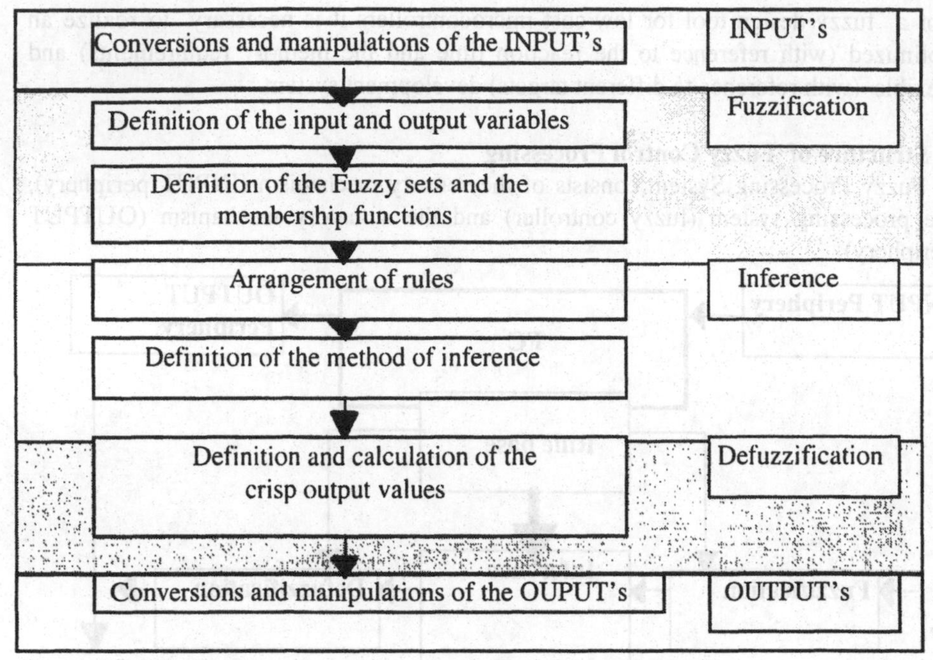

Fig. 3: Design sequences for editing a project of Fuzzy-Controller

4. FHFC-CASE-process

4.1. Project editor

The main menu of the FHFC-Shell consists of the usual functions „menu„„„editing area„ and „line of state„.

Function-determining expedients of editing are:

- Treatment of an element
- Word element
- Input elements
- Output elements
- Manipulation elements
- Fuzzy elements
- In-Line Assembler (module element)

4.2. Generation of code

After structuring, the Fuzzy Control process starts the phase of generating the code. For this purpose the MACROs which correspond to the edited functions are linked by the structure of the Fuzzy system (generation of an ASM-File). In addition to this, variables between the functions have to be defined to realize communication in the system. In order to make it possible to realize different applications, these variables are dynamically defined, also under the aspect of an optimized system (with reference to optimizing the memory and process time).

Different MACROs must be available accordingly for different controllers or derivates of controllers. The dynamic generation of code must be oriented to the characteristics of the target controllers.

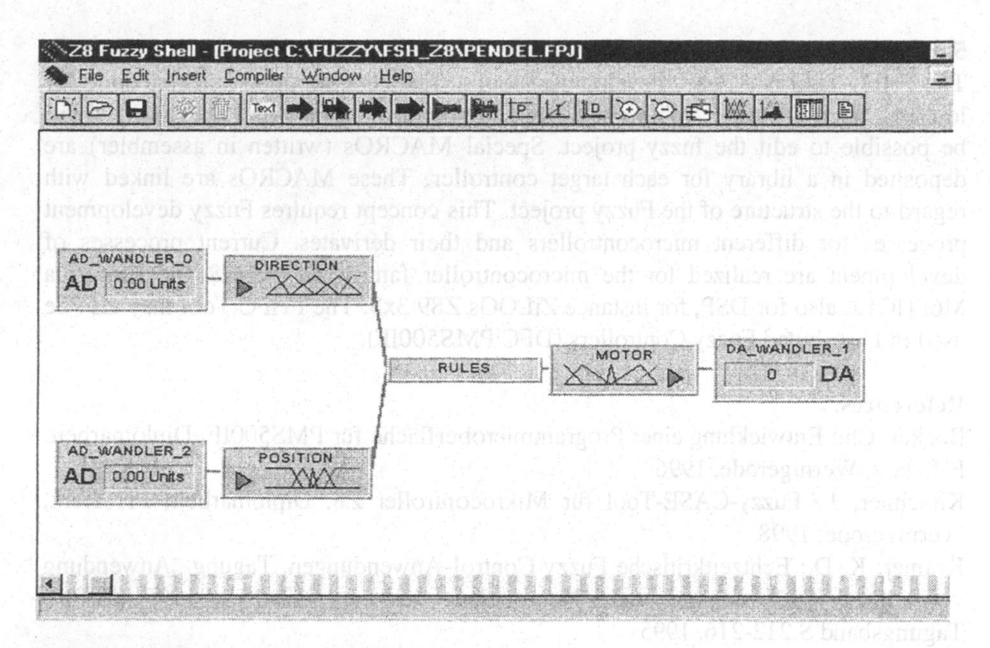

Structure of the source program:

MACRO **„ADU_0„** (driver for ADU-channel 0 - direction)
MACRO **„DIRECTION„** (fuzzification - INPUT 0 - direction)
MACRO **„ADU_1„** (driver for ADU-channel 1 - position)
MACRO **„POSITION„** (fuzzification - INPUT 1 - position)
MACRO **„RULES„** (rule base and inference)
MACRO **„MOTOR„** (defuzzification - voltage of motor)
MACRO **„DAU_1„** (driver for DAU 1)

Fig. 4 : Editing a project with the structure of the source program (example)

4.3. Verification of the system

In addition to the design process a verification and tuning phase is necessary. This verification and tuning phase can be realized directly (On-line or Off-line) or indirectly (Off-line or with a simulator). The modification of parameters or rules has to be carried out in this phase. It must be possible to change functions of the system. Different goals exist:

- improvement of the dynamic of the system
- improvement of the exactness of the system
- general optimization of the system

All these processes, which are described as tuning and optimized processes require efficient system support:

- On-line debugging (realized by the system debugger)
- On-line debugging with management of the INPUT /OUTPUT data (generally realized by a real-time system)
- complex analysis of performance (measuring methods, monitoring, Fuzzy benchmarks)

In the case of using low-cost microcontrollers it will be of advantage to use On-line debugger, to indicate relevant data and manipulate them.

5. Conclusions

The FHFC-Tool is a Fuzzy development tool to realize an optimal machine code for low-cost microcontroller. With a standardised graphical surface (project editor) it will be possible to edit the fuzzy project. Special MACROs (written in assembler) are deposited in a library for each target controller. These MACROs are linked with regard to the structure of the Fuzzy project. This concept requires Fuzzy development processes for different microcontrollers and their derivates. Current processes of development are realized for the microcontroller family ZiLOG Z8 and Motorola M68HC12, also for DSP, for instance ZiLOGs Z89 3x3. The FHFC Tool may also be used in Distributed Fuzzy Controllers (DFC:PMS500IF).

References:

Becker, Ch.: Entwicklung einer Programmieroberfläche für PMS500IF, Diplomarbeit, FH Harz, Wernigerode, 1996

Kirschner, J.: Fuzzy-CASE-Tool für Mikrocontroller Z8, Diplomarbeit, FH Harz, Wernigerode, 1998

Kramer, K.-D.: Echtzeitkritische Fuzzy-Control-Anwendungen, Tagung „Anwendung von Fuzzy-Technologien und Neuronalen Netzen„ 13.-15.12.1995 Erfurt, Tagungsband S.212-216, 1995

Wöhlbier, S.: Fuzzy-CASE-Prozess für ZILOG-DSP (Z89xx3), Projektarbeit, Hochschule Harz, Wernigerode, FB A/I, 1999

Search of Optimal Error Correcting Codes with Genetic Algorithms

Jérôme Lacan and Pascal Chatonnay

IUT Belfort-Montbéliard, LIFC,
Rue Engel-Gros, BP 527, 90016 Belfort Cedex, France*

Jerome.Lacan@utbm.fr, Pascal.Chatonnay@iut-bm.univ-fcomte.fr

Abstract. In this paper, it is described one of the most important open problem in the area of the Error Correcting Codes: the search of optimal codes. It is explained why and how the Genetic Algorithm can be used for this problem. Some promising results of computations are presented.

1 Optimal Error Correcting Codes

Error Correcting Codes (ECC) are a technique to improve the quality of some communications (deep space communications, networks communications, compact disc, ...). Their principle consists in transmitting a redundant information with the initial information. By this way, if some errors occur during the transmission, it is possible to recover the original information by using the redundancy.

There exists principally two classes of ECC: convolutional codes and block codes. In this paper, we only consider block codes.

Let us explain more precisely the principle of these codes.

Let us suppose now that the information to transmit is a sequence of elements which take q possible values, where q is a power of a prime number p. We then can consider these elements as elements of the finite field of cardinal q, which is denoted by $\mathbb{F}_q$. In general, transmitted symbols are binary, then $q = 2$.

Principle of block codes is the following: the initial message is cut out into blocks of length k. The length of the redundancy is $n - k$ and thus, the length of the transmitted blocks is n. Main block codes are linear codes. In this case, redundancy is computed in such a way as concatenation of information and of redundancy is an element of vector space (a code) of dimension k of $(\mathbb{F}_q)^n$.

The operation of coding can be computed by multiplying the information (considered as a vector of length k) by a $k \times n$ systematic generator matrix of this vector space. Note that a generator matrix is called systematic if its k first column form the identity matrix. For a given code, there exists at most one systematic generator matrix. Some codes does not admit a systematic generator matrix. On the other hand, all these codes are equivalent (modulo a permutation of the positions) to a code which admit one.

* The first author is now with the Université de Technologie de Belfort-Montbéliard, Rue du château - Sévenans F-90010 BELFORT Cedex

For linear codes, k and n are respectively called the dimension and the length of the code.

Example 1. Let C be a code of length 7 and of dimension 4 over $\mathbb{F}_2$ characterized by its systematic generator matrix

$$G = \begin{pmatrix} 1\,0\,0\,0 & 1\,0\,1 \\ 0\,1\,0\,0 & 0\,1\,1 \\ 0\,0\,1\,0 & 1\,1\,0 \\ 0\,0\,0\,1 & 1\,1\,1 \end{pmatrix}.$$

The transmitted codeword corresponding to the information (1101) is :

$$(1101) \times G = (1101\,001)$$

The last parameter of a code is its minimum distance d. It is the smallest Hamming distance (the number of distinct components) between two codewords. As the considered codes are linear vector spaces, their minimum distance is also the Hamming weight (the number of nonzero components) of the codeword of smaller Hamming weight. The correction capability (the maximum number of error that can be corrected per word of length n) of the code is then equal to $t = (d-1)/2$.

A code is said optimal if, for a given length and a given dimension, its minimum distance is maximum. The search of optimum codes is one of the main open problems in coding theory. A table of the best known codes is regularly updated on [3]. For each pair of parameters n and k, this table contains the distance (d) of the best known code and its theoretical upper bound.

2 Utilization of Genetic Algorithms to Find Optimal ECC

In this part, we explain why and how we used the genetic algorithms to solve the problem of optimal ECC search.

2.1 Why Using Genetic Algorithms ?

Usually, used techniques to find some good codes are mathematical techniques. For example, the well-known theorem of BCH [1] permits to compute the minimum distance of some cyclic codes and gives a lot of optimal codes. It is based on the consecutive roots of the cyclic code. The family of geometric codes [2], which contains some best codes, is derived form algebraic geometry.

But there is no mathematical technique which can be applied to all lengths and all dimension of codes.

But, as it is explained in the first section, the linear codes are characterized by their generator matrix, which can be explained as a sequence of numbers. So rationally, we tried to use the genetic algorithms (GA) to solve this optimization problem (see [4]).

2.2 Implementation

For the implementation of GA, we only consider the binary case $q = 2$, but the following ideas can be applied for $q > 2$.

We first fixed the length and the dimension of the codes. We wanted to represent a population of codes and applied onto this population the principles of GA.

In the first section, it is explained that every linear code which does not admit a systematic generator matrix is equivalent (modulo a positions permutation) to a linear code which admits one. Thus for each length and each dimension, there is optimal codes in the set of linear codes admitting a systematic generator matrix. Moreover such linear codes are entirely characterized by their systematic generator matrix.

So, the population contains only codes admitting a systematic generator matrix. Any systematic matrix can be represented under the form $G = I \mid R$ where I is the $k \times k$ identity matrix and R is a $k \times (n-k)$ matrix representing the redundancy. In the population, each code is represented by its corresponding matrix R.

The operations on the different individuals are the following : an evaluation of the population member, a classical crossover, a classical probabilistic mutation and a directed mutation.

The evaluation of a code consists in computing its minimum distance and in determining the number of codewords of which Hamming weight is equal to the minimum distance. The final fitness is a combination of these two elements. Clearly, the most important is the minimum distance. Many formulas were used for calculate this fitness, but the best results were given by the following formula:

$$f(C) = E^{2*d(C)} * (1 + (E-1) * \frac{tabNbMax[d(C)] - x}{tabNbMax[d(C)]})$$

where:

- E is a constant (most of times, we used $E = 2$)
- $d(C)$ is the minimum distance of C
- $tabNbMax$ is an array containing in rank d the number of codewords of the first found code of minimum distance d

For the operation of crossover, we chose to concatenated the rows of the matrices R into a sequence and to applied an usual crossover (see the figure 1).

We used a classical probabilistic mutation operator which leads to digit modification in a code. this operateur do not change the matrix structure.

We also introduced an operator which can be considered as a directed mutation. It is an optimization of each new individual in function of their codewords of least Hamming weight.

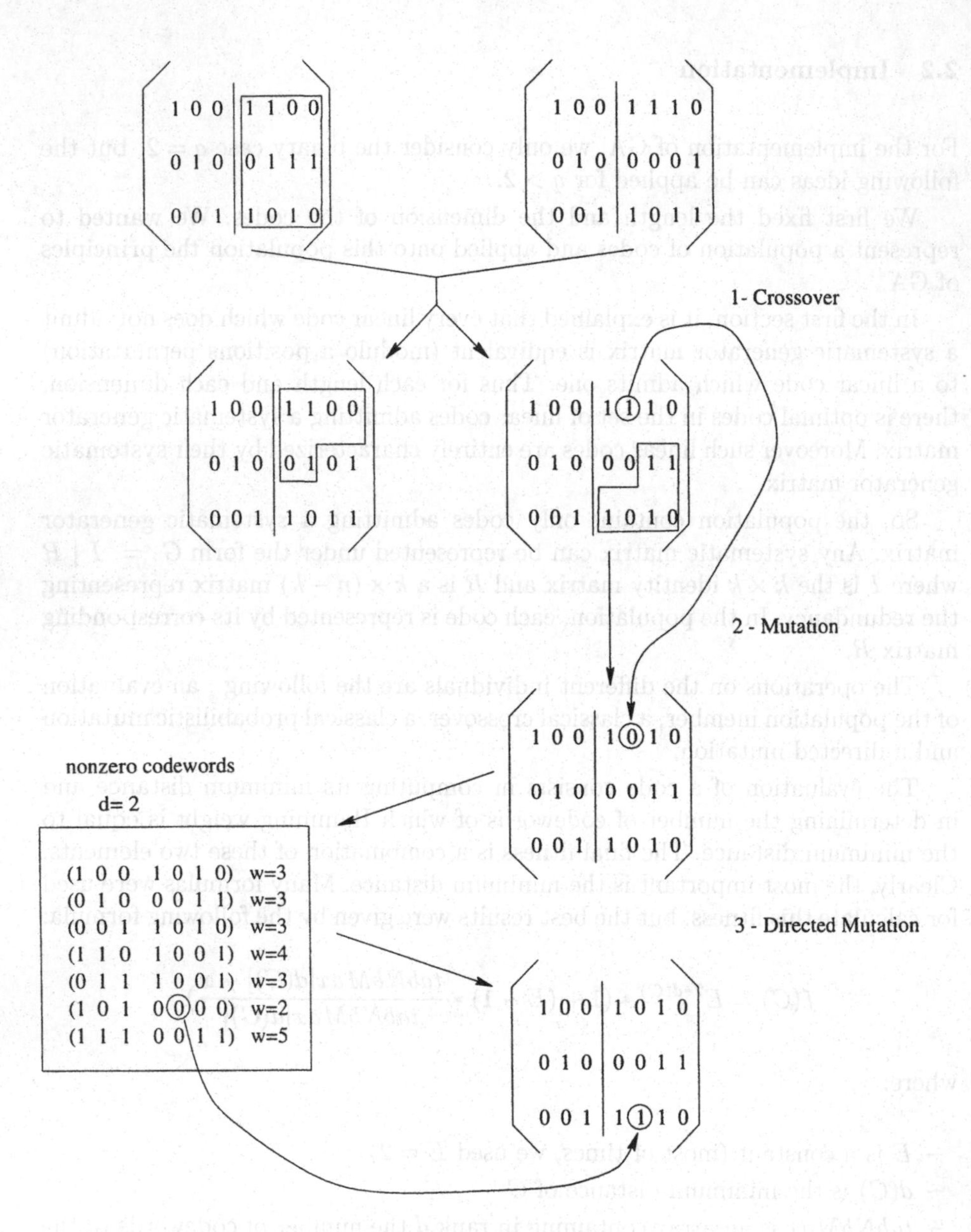

Fig. 1. Operators

Its principle consists first in computing the codewords of minimal weight. As all codewords, they are combinations of rows of the generator matrix. So, we compute the number of times that each row of the matrix is used to generate the set of codewords of minimal weight d. We consider the row which is used the most of times. Then we consider the set of codewords of minimal weight which use this row. If these codewords contain a same position in which they are all equal to zero, then this position is changed in the considered row. The aim of this operator is to decrease the number of codewords of minimal weight. Sometimes, as in the example described in the figure 1, the minimum distance can be increased. Note that this kind of modification can decrease of 1 the weight of codewords of weight upper than d, and thus, some new codewords of weight d can result from words of weight $d+1$. But the computations proved that in most cases, this directed mutation provides a better code.

2.3 Difficulties, Results and Perspectives

During this work, we met some difficulties.

The first one is the time needed for computing the fitness of each code. To find the minimum distance and the codewords of minimum distance,the simplest way is to compute all the codewords. For example, for binary codes of dimension 25 and length n, $n \times 2^{25} = n \times 33554432$ operations are needed. It is easy to reduce this number but it is known that there no exists algorithms in polynomial time to compute the minimum distance of a linear code.

Another problem is the determination of some schemes which have to be preserved by the crossover. We did not determined if it is better to keep the columns or the rows. The computations did not indicate a solution better than others.

We implemented the computations at the computer science department of the IUT of Belfort-Montbliard on a PC Pentium 200. Clearly, we were limited by the capabilities of the computer.

Nevertheless, the results are very interesting. Actually, for all the cases in which the computations are possible i.e. when $n < 25$, the maximum known was reached and some new codes with these parameters were produced. This results are promising because the program can easily be parallelized. Actually for larger values, the difference between the minimum distances of the best known codes and the theoretical upper bounds are more important [3] and it seems to be possible to obtain some new codes with minimum distance better than the known codes.

The work we present here is on going. In the future we want to introduce three main improvements:

First of all, the directed mutation technique we present could actually leads to a worse code than its generator. To solve this problem we are introducing the codewords of weight $d+1$ in the directed mutation. By this way we will never generate a worse code, but always a better one, if it is possible.

The second improvement is to parallelize, at least, the minimum distance computation. We hope that the execution of our method on a cluster of 16 Alpha Station will give significant benefit in performance and will permit to compute larger codes.

Finally, an important part of our job can be apply in the domain of non-binary codes. This kind of codes were not studied as many as binary codes. Generally, for two given parameters, the difference between the best known code and the theoretical upper bound is larger.

References

1. McWilliams, F.J., Sloane N.J.A, The Theory of Error-Correcting Codes, North Holland, 1983.
2. Goppa V.D., Algebraico-geometrical Codes, Math. USSR Izvesiya, **21**, pp. 75-91, 1983.
3. EIDMA, Linear Code Bounds, `www.win.tue.nl/math/dw/voorlincod.html`.
4. Goldberg D. E., Genetic Algorithms in Search, Optimization, and Machine Learning, Addison-Wesley, 1989.

An Unsupervised Clustering with Evolutionary Strategy to Estimate the Cluster Number

[1,2]Katsuki IMAI, [2]Naotake KAMIURA and [2]Yutaka HATA

[1]SHARP Corporation,
22-22, Nagaike, Abeno, Osaka, 545-8522, Japan

[2]Department of Computer Engineering, Himeji Institute of Technology,
2167, Shosha, Himeji, 671-2201, Japan
imai@comp.eng.himeji-tech.ac.jp

Abstract. Clustering is primarily used to uncover the true underlying structure of a given data set. Most algorithms for fuzzy clustering often depend on initial guesses of the cluster centers and assumptions made as to the number of subgroups presents in the data. In this paper, we propose a method for fuzzy clustering without initial guesses on cluster number in the data set. Our method assumes that clusters will have the normal distribution. Our method can automatically estimates the cluster number and achieve the clustering according to the number, and it uses structured Genetic Algorithm (sGA) with graph structured chromosome.

1 Introduction

Database users must treat large amounts of stored data that generally forms complex distribution. Clustering is primarily used to discover the true underlying structure of a given data set [1][2]. The well-known clustering methods require some information about databases in advance. Especially in Fuzzy C-Means (FCM for short) [3][4] the number of clusters is one of the most important factor. An unsupervised clustering, which can automatically estimate the important factor, are also proposed [5][6].

In this paper, in order to automatically estimate the cluster number, we propose an unsupervised clustering method, which forms the clusters according to a given standard deviation. In it, we assume that every cluster has the normal distribution. Genetic Algorithm (GA for short) based search is used to avoid falling into local minimum. For every datum, the probabilistic like degree is calculated by the normal density curve formed by the given standard deviation. An objective function in GA evaluates the degree. Then our method with GA can achieve a clustering according to the standard deviation.

2 Preliminaries

Let $\boldsymbol{X} = \{\boldsymbol{x}_1, \boldsymbol{x}_2, \cdots, \boldsymbol{x}_n\}$ be a set of all data. Each element of $\boldsymbol{X}$ has an N-dimensional vector specified by values of variables x_1, x_2, ... and x_N. Let c be the number of clusters. The objective function Q of FCM is defined as a sum of squared errors. Namely,

$$Q = \sum_{i=1}^{c} \sum_{k=1}^{n} u_{ik}{}^{m} \|\boldsymbol{x}_k - \boldsymbol{v}_i\|^2 \tag{1}$$

where $\boldsymbol{v}_i$ is the center of the i-th cluster and $\|\boldsymbol{x}_k - \boldsymbol{v}_i\|$ is a distance function between $\boldsymbol{x}_k$ and $\boldsymbol{v}_i$. u_{ik} is the grade of the i-th membership function on $\boldsymbol{x}_k$, namely

$$\begin{cases} u_{ik} \in [0,1] \\ \sum_{i=1}^{c} u_{ik} = 1 \quad \text{for any } k \end{cases} \tag{2}$$

When m in Equation (1) equals 1, u_{ik} takes 0 or 1 for optimum solution. In this case, FCM method converges in the theory to the traditional k-means solution. Already known fuzzy clustering method including FCM usually set m to the value over 1 [6][7] to constrain u_{ik} to be fuzzy value. In FCM, local minimal solution of Equation (1) is calculated as follows.

$$u_{ik} = \left[\sum_{j=1}^{c} \left(\frac{\|\boldsymbol{x}_k - \boldsymbol{v}_i\|^2}{\|\boldsymbol{x}_k - \boldsymbol{v}_j\|^2} \right)^{\frac{1}{m}} \right]^{-1}$$

$$\boldsymbol{v}_i = \frac{\sum_{k=1}^{n} u_{ik}{}^{m} \boldsymbol{x}_k}{\sum_{k=1}^{n} u_{ik}{}^{m}}$$

In conventional clustering method, as the number of clusters increase, u_{ik} approaches 0 or 1, and the minimal value of Equation (1) approaches 0. When the number of clusters equals the number of data, Equation (1) becomes 0 and u_{ik} takes 0 or 1 for optimum solution. Therefore, we can't estimate the number of clusters from

Equation (1). Moreover, another problem exists. Because FCM method is consider only local minimal solutions, we don't obtain global minimal solution.

3 Fuzzy Clustering Based on the Normal Distribution

3.1 Overview

To keep u_{ik} from taking 0 or 1, we assume that each cluster has the normal distribution. To avoid falling into local minimum, we use Genetic Algorithm (GA for short) as search algorithm without initial guesses of the centers.

Figure 1 shows our flow diagram. In advance, we set the standard deviation of every cluster: σ_0, that determines the distribution of cluster. Then GA searches the number of clusters and the center of each cluster. According to them, we calculate the fitness function of GA. As a result, we obtain the clustering result.

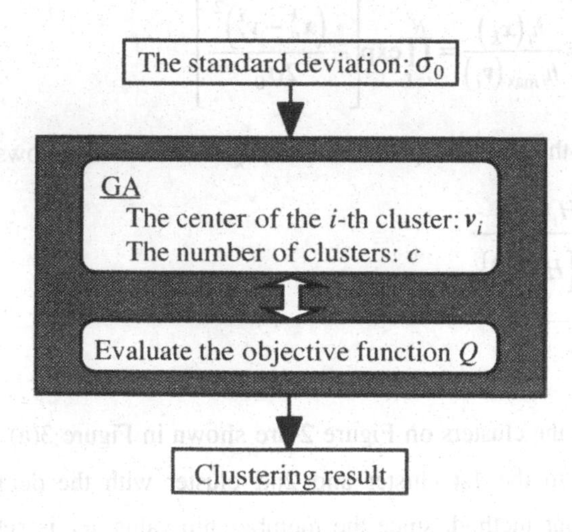

Fig. 1. Flow diagram.

3.2 Our Clustering Method

Let $H_i \in [0,1]$ be the membership function for i-th cluster. The belongingnesses of the all points are calculated based on the normal density curve. The distribution of each clusters at least overlaps with the adjacent ones. So, the degree u_{ik} never takes 0 and 1.

In the following discussion, we denote $\boldsymbol{x}_k\ (1 \le k \le n)$ by $\left(x_1^k, x_2^k, \cdots, x_N^x\right)$ and denote $\boldsymbol{v}_i\ (1 \le i \le c)$ by $\left(v_1^i, v_2^i, \cdots, v_N^i\right)$. We derive a normal distribution h_i for given standard deviation σ_0 and calculate the membership function of the i-th cluster H_i for h_i. We show the following formula $h_i(\boldsymbol{x}_k)$ that expresses the degree of normal distribution of $\boldsymbol{x}_k$ to the i-th cluster.

$$h_i(\boldsymbol{x}_k) = \prod_{t=1}^{N} \frac{1}{\sqrt{2\pi}\sigma_0} \exp\left[-\frac{\left(x_t^k - v_t^i\right)^2}{2\sigma_0^{\,2}}\right] \tag{3}$$

The maximal value of h_i is

$$h_{i\max}(\boldsymbol{v}_i) = \left(\frac{1}{\sqrt{2\pi}\sigma_0}\right)^N$$

We calculate the fuzzy membership function $H_i(\boldsymbol{x}_k)$ from $h_i(\boldsymbol{x}_k)$ as follows.

$$H_i(\boldsymbol{x}_k) = \frac{h_i(\boldsymbol{x}_k)}{h_{i\max}(\boldsymbol{v}_i)} = \prod_{t=1}^{N} \exp\left[-\frac{\left(x_t^k - v_t^i\right)^2}{2\sigma_0^{\,2}}\right] \tag{4}$$

The value of the i-th membership function on $\boldsymbol{x}_k$ is defined as follows.

$$u_{ik} = \frac{H_i(\boldsymbol{x}_k)}{\sum_{i=1}^{c}\left\{H_i(\boldsymbol{x}_k)\right\}} \tag{5}$$

[Example 1]

The $H_i(\boldsymbol{x}_1)$ of the clusters on Figure 2 are shown in Figure 3(a). The data marked with $\boldsymbol{x}_k$ belongs to the 1st cluster and 2nd cluster with the degrees 0.9 and 0.2, respectively. In our method, since the membership value u_{ik} is set to Equation (2), u_{1k} and u_{2k} are calculated as follows.

$$u_{1k} = \frac{0.9}{0.9 + 0.2} = 0.82$$

$$u_{2k} = \frac{0.2}{0.9 + 0.2} = 0.18$$

Figure 3(b) shows the membership value u_{ik}.

[End of Example]

As the number of clusters increases, the overlap with the adjacent ones increases and hence the variance of u_{ik}s becomes smaller. In Figure 3(b), the overlap is shown by gray area. If the overlap increases, we can not clearly recognize each cluster. As the number of clusters decreases, $u_{ik}\|x_k - v_i\|^2$ increases. So, the objective function must decrease depending on the overlap and $u_{ik}\|x_k - v_i\|^2$. We define the objective function Q as follows.

$$Q = \sum_{i=1}^{c}\sum_{k=1}^{n} u_{ik}^{\frac{1}{2}}\|x_k - v_i\|^2 \tag{6}$$

In Equation (6), u_{ik} never takes 0 and 1 according to Equations (4) and (5).

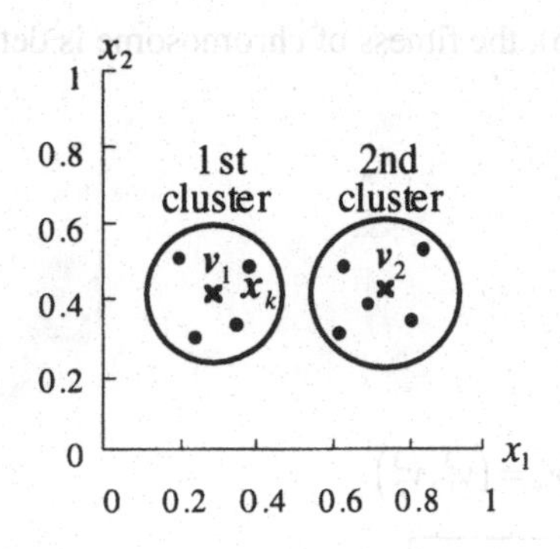

Fig. 2. An example of clustered data.

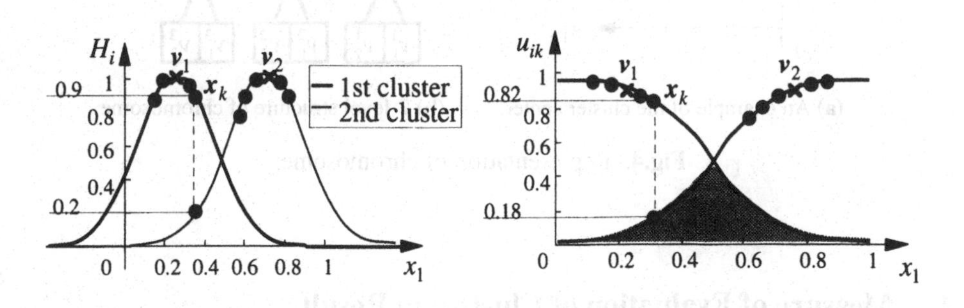

(a) Membership function of the clusters: H_i. **(b)** Membership value on each data: u_{ik}.

Fig. 3. Membership functions.

3.3 Evolutionary Strategy on Our Method

We use structured Genetic Algorithm (sGA for short) to globally search the optimum solution. An sGA utilizes chromosomes with a largely hierarchical directed graph genetic structure [8][9].

In our method, the chromosome has a two-level hierarchy directed graph structure of genes. Figure 4 shows the chromosome that represents the center of cluster in Figure 4(a).

The high level of chromosome expressed by a set of binary strings searches the number of clusters. When the *i*-th cluster can not be found, we assign 0 to the *i*-th gene, otherwise we assign 1. Namely, a sum of all genes equals the number of cluster *c*. The lower level that expressed by a set of real number searches the centers of the clusters. The length of the upper level is the number of data *n*. The length of the lower level is $N \times n$.

According to Equation (6), the fitness of chromosome is defined as follows.

$$fitness = \frac{1}{Q}$$

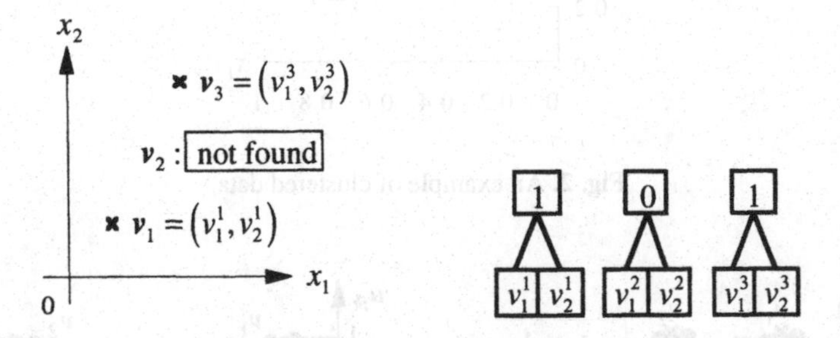

(a) An example of the cluster center. (b) 2-level structure of chromosome.

Fig.4. Representation of chromosome.

4 Measure of Evaluation of Clustering Result

The result obtained by this method depends on the standard deviation σ_0 which is given in advance. We consider a measure based on the probability for comparison with different result on various σ_0.

Let i_{max} be i where u_{ik} takes the maximum value for i. Then $\boldsymbol{x}_k$ should belong to the i_{max}-th cluster. The value of the normal density curve $H_{i_{max}}(\boldsymbol{x}_k)$ gives the probability of selecting the i_{max}-th cluster for the given k-th data when the probability for $\boldsymbol{v}_{i_{max}}$ equals 1. A good standard deviation needs to give high probability to a cluster that the data should belong to.

We define the measure M_{sd} as the average of probability as an indicator to express the clustering result.

$$M_{sd} = \frac{\sum_{k=0}^{n} H_{i_{max}}(\boldsymbol{x}_k)}{n} \qquad (c > 1)$$

The higher M_{sd} includes better clustering results.

5 Experimental Result

Figure 5 shows artificially synthesized three clusters with 68 data specified by two-variable vectors using Neyman-Scott's method [10]. We apply our method to these data.

Table 1 shows the number of clusters for different σ_0 where $1 \le c \le 4$. Figure 6(a) shows M_{sd}. When σ_0 equals 0.68, the number of clusters c equals 3 and M_{sd} becomes the maximum value 0.82. Figure 6(b) shows the clustering result under the assumption of $\sigma_0 = 0.68$. The result specified the artificial cluster very well. Our method can thus estimate correctly the number of clusters.

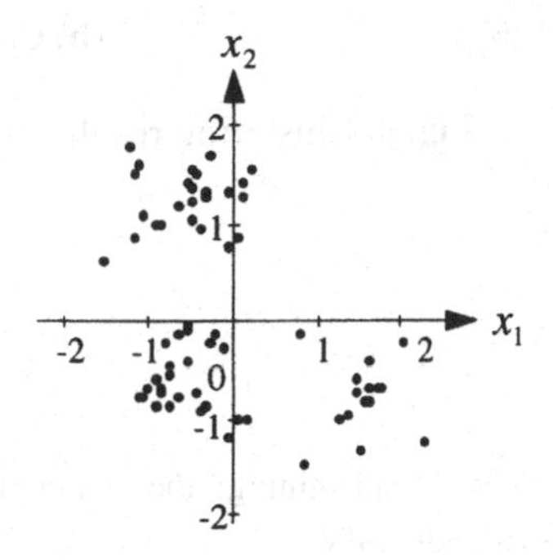

Fig. 5. Artificial data.

6. Conclusion

In this paper, we have proposed the unsupervised fuzzy clustering method. We use GA to search optimum results without initial cluster centers under the assumption that the clusters are distributed according to the normal distribution. Moreover, we define the measure of the standard deviation of each cluster. The experimental results show that our clustering method is useful in searching the optimum number of clusters.

Table 1. The number of clusters.

standard deviation	the number of clusters
$0 \le \sigma_0 < 0.12$	4
$0.12 < \sigma_0 \le 0.68$	3
$0.68 < \sigma_0 \le 0.81$	2
$0.81 < \sigma_0$	1

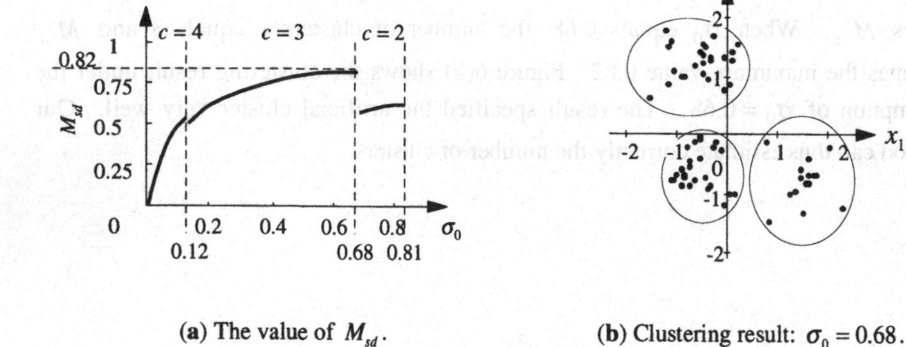

(a) The value of M_{sd}. (b) Clustering result: $\sigma_0 = 0.68$.

Fig. 6. Clustering result.

References

1. M. Holsheimer and A. Siebes, "Data mining: the search for knowledge in databases," Report CS-R9406, The Netherlands, 1994.

2. U. Fayyad, G. P. Shapiro, P. Smyth and R. Uthurusamy, *Advances in knowledge discovery and data mining*, AAAI/MIT Press, Boston, MA, 1996.
3. W. Pedrycz, "Data mining and fuzzy modeling," *Proc. of IEEE Biennial Conference of the North American Fuzzy Information Processing Society - NAFIPS*, pp.263-267, 1996.
4. J. C. Bezdek and S. K. Pal, *Fuzzy models for pattern recognition*, IEEE Press, 1992.
5. P. Cheeseman and J. Stutz, "Bayesian classification (AutoClass): Theory and results," *Advances in knowledge discovery and data mining*, ed. U. M. Fayyad, pp.153-180, 1996.
6. R. P. Li and M. Mukaidono, "A maximum entropy approach to fuzzy clustering," *Proc. of the 4th IEEE Intern. Conf. on Fuzzy Systems (FUZZ IEEE/ IFES'95)*, Yokohama, Japan, March 20-24, pp.2227-2232, 1995.
7. T. Ohta, M. Nemoto, H. Ichihashi, T. Miyoshi, "Hard clustering by fuzzy c-means," *Journal of Japan Society for Fuzzy Theory and Systems*, Vol. 10, No. 3, pp. 532-540, 1998, (in Japanese).
8. R. J. Bauer, Jr., *Genetic algorithms and investment strategies*, John Wiley & Sons Inc., N. York, 1994.
9. D. Dasgupta and D. R. McGregor, "Designing application-specific neural networks using the structured genetic algorithm," in *Proc. of COGANN-92 International Workshop on Combinations of Genetic Algorithms and Neural Networks*, Baltimore, Maryland, June 6, pp.87-96, 1992.
10. T. Kato and K. Ozawa, "Non-hierarchical clustering by a genetic algorithm," *Information processing society of Japan*, Vol.37, No.11, pp.1950-1959, 1996, (in Japanese).

Multi-objective Optimization in Evolutionary Algorithms Using Satisfiability Classes

Nicole Drechsler, Rolf Drechsler, Bernd Becker

Institute of Computer Science
Albert-Ludwigs-University
79110 Freiburg im Breisgau, Germany
email: {ndrechsler,drechsler,becker}@informatik.uni-freiburg.de

Abstract. *Many optimization problems consist of several mutually dependent subproblems, where the resulting solutions must satisfy all requirements.*

We propose a new model for Multi-Objective Optimization (MOO) in Evolutionary Algorithms (EAs). The search space is partitioned into so-called Satisfiability Classes (SC), where each region represents the quality of the optimization criteria. Applying the SCs to individuals in a population a fitness can be assigned during the EA run. The model also allows the handling of infeasible regions and restrictions in the search space. Additionally, different priorities for optimization objectives can be modeled. Advantages of the model over previous approaches are discussed and an application is given that shows the superiority of the method for modeling MOO problems.

1 Introduction

Evolutionary Algorithms (EAs) become more and more important as a tool for search and optimization. Especially for hard combinatorial problems they often have successfully been applied (see e.g. [10,3]). Multiple, competing criteria have to be optimized subject to a large number of non-trivial constraints. One strategy is to artificially divide a problem into a number of subproblems, which are then solved in sequence. Obviously, this is not a promising strategy if the objectives are conflicting. EAs are well suited for solving this kind of problems, when mutually dependent subproblems are considered in parallel. One problem arises when using EAs is to rank a pool of solutions.

Traditionally, the evaluation is done by an objective function which maps a solution of multiple objectives to a single value. A classical aggregative method is the *linear combination by weighted sum* where the value of each objective is weighted by a constant coefficient. The values of the weights determine how strong the specific objective influences the value of the single fitness value. Disadvantages are that the weights have to be known in advance to find good solutions or have to be determined by experiments. Obviously, this is time consuming and not desirable, since the parameters resulting from different runs may vary ending in "in-stable" algorithms [6].

1.1 Previous Work

Advanced methods for ranking solutions with multiple objectives have been developed over the years. If priorities exist between the objectives, a simple *lexicographic* order can be used[1].

In [8] a method is described where solutions with multiple objectives without preferences can be compared. This is realized by a relation, called *dominate.* A solution x *dominates* y, if x is equal or better for each objective than y, and x is for at least one component strongly better than y. Thus, the solutions in the search space can be ranked by the relation *dominate.* Implementations of this approach are presented in [9, 11].

In [7] another approach is proposed, where the search space is divided in a *satisfiable*, *acceptable*, and *invalid* range. This model has successfully been applied to a specific problem in the area of *chip design*, but it requires a user interaction. The designer has to specify the limits between *satisfiable*, *acceptable*, and *invalid* solutions.

1.2 Results

In this paper we propose a new model for ranking solutions of *Multi-Objective Optimization* (MOO) problems. A relation *favor* is defined analogously to *dominate* in [8], but this relation has several advantages, e.g. *favor* is able to compare solutions which are not comparable using relation *dominate.* The search space is divided into several *Satisfiability Classes (SCs).* Thus our approach can be seen as a generalization of the approach in [7] using a finer granularity. By this technique no user interaction is required any more. Furthermore, handling of priorities is also supported. Infeasible solutions are assigned to their own *SCs* depending on the objectives which are in an infeasible range. The *SCs* can be efficiently manipulated using operations on graphs. The model has been implemented and application to an example from the domain of *heuristic learning* [5] demonstrate the efficiency of our approach.

2 Multi-objective Optimization Problems

In general, many optimization problems consist of several mutually dependent subproblems. MOO problems can be defined as follows: Let $\Pi \subset \mathbf{R}^n$ be the feasible range of solutions in a given search space, where $\mathbf{R}$ denotes the real valued numbers. The objective function

$$f_c : \Pi \longrightarrow \mathbf{R}_+^n$$

assigns a cost to each objective of a solution, where

$$f_c(x) = (f_{c_1}(x), f_{c_2}(x), \cdots, f_{c_n}(x)), x \in \Pi.$$

[1] Informations on *lexicographic* sorting of vectors can be found in standard mathematical literature.

To compare several multi-objective solutions ***superior points*** in the search space can be determined. In the following we restrict to minimization problems. The relation *dominate* is defined as proposed in [8]. Let $x, y \in \mathbf{R}_+^n$ be the costs of two different solutions. x ***dominates*** y ($x <_d y$), if no component of y is smaller than any component of x. More formally:

Definition 1.

$$x <_d y \Longleftrightarrow (\quad \exists i : f_{c_i}(x) < f_{c_i}(y)) \\ \wedge(\forall j \neq i : f_{c_j}(x) \leq f_{c_j}(y))$$

$$x \leq_d y \Longleftrightarrow \quad (\forall i : f_{c_i}(x) \leq f_{c_i}(y))$$

A non-dominated solution is called a *Pareto-optimal* solution. $\leq_d$ defines the set of all Pareto-optimal solutions, called the *Pareto-set*, and additionally $\leq_d$ is a partial order. All elements $x \in \Pi$ in the Pareto-set are equal or *not comparable*. Usually, all points in this set are of interest for the decision maker or designer.

3 The Model

Let $\Pi \subset \mathbf{R}^n$ be the set of all feasible solutions. The objective function $f_c : \Pi \longrightarrow \mathbf{R}_+^n$ assigns a cost to each $x \in \Pi$ as defined above. To classify solutions in *satisfiability classes* we define the relation *favor* ($\prec_f$):

Definition 2. Let $x, y \in \mathbf{R}_+^n$ and let $\prec_f$ be a relation on $\mathbf{R}^n$ which is defined as follows:

$$x \prec_f y \Longleftrightarrow \exists i, j : j < i : | \{x_k < y_k | 1 \leq k \leq n\}| = i \\ \wedge | \{x_l > y_l | 1 \leq l \leq n\}| = j$$

Using Definition 2 we are able to compare elements $x, y \in \Pi$ pairwise more precisely. x is *favored* to y ($x \prec_f y$) iff i ($i \leq n$) components of x are smaller than the corresponding components of y and only j ($j < i$) components of y are smaller than the corresponding components of x.

We use a graph representation for the relation, where each element is a node and "preferences" are given by edges. Relation $\prec_f$ is not a partial order, because it is not transitive, as can be seen as follows:

Example 1. Consider some solution vectors from $\mathbf{R}^3$:

(8,7,1) (1,9,6) (7,0,9) (1,1,2) (2,1,1)

The relation graph of $\prec_f$ is given in Figure 1a). Vectors (1,1,2) and (2,1,1) are preferred to all other vectors, but they are not comparable. The remaining three vectors (8,7,1), (1,9,6), and (7,0,9) are pairwise comparable. But as can be seen in the relation graph they describe a "cycle". Thus relation $\prec_f$ is not transitive.

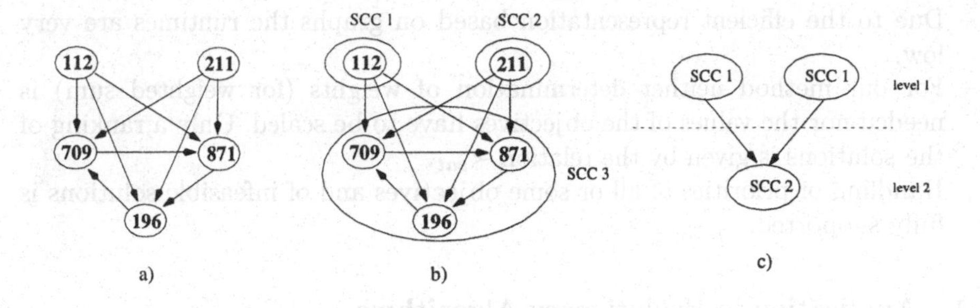

Fig. 1. Relation graph

To get some more insight in the structure of our model we briefly focus on the meaning of the cycles in the relation graph: We rank elements equally which are included in a cycle, because no element is superior to all the others. Elements that describe a cycle are denoted as not comparable. The determination of the *Strongly Connected Components* (SCC) of the relationgraph groups all elements which are not comparable in one SCC. The SCCs are computed by a DFS-based linear time graph algorithm [2]. The transitive closure is computed. A directed graph G_{SC} is constructed by replacing each SCC in G by one node representing this SCC. Thus, all cycles in G are eliminated.

In the following, we denote the relation that is represented by G_{SC} by $\prec_{sat}$. Relation $\prec_{sat}$ is asymmetric and transitive. It is not a partial order, since $\prec_{sat}$ is not reflexive. Nevertheless, the relation properties antisymmetry and transitivity are sufficient for our purposes. Elements are successfully ranked by $\prec_{sat}$. For each SCC in G_{SC} we define a *Satisfiability Class* (*SC*). Level sorting of the nodes in G_{SC} determines the ranking of the SCCs; each level contains at least one SCC. Then each level corresponds to a *SC*.

Example 2. In Figure 1b) SCC 1, SCC 2, and SCC 3 are illustrated. As can be seen the elements of SCC 1 and SCC 2 are superior to the elements of SCC 3. Figure 1c) shows the relation graph G_{SC} after level sorting. Level 1 corresponds to *SC* 1 and level 2 corresponds to *SC* 2.

Our model has several advantages over "classical" approaches.

- Relation $\prec_{sat}$ can also handle infeasible solutions. An infeasible component of the solution is considered as the worst possible value.
- Dependent on the population the model dynamically adapts the relation graph that performs the comparison of the fitness function. In each generation the relation graph is recomputed and the ranking is updated. (This is done totally automatic and no user interaction is required.) Thus the granularity of the *SCs* is dynamically adapted to present conditions.
- If the structure of the search space changes during the EA run these changes are directly included in the relation that is updated online.
- No problem specific knowledge is needed for the choice of the weights of the fitness function, like in the case of weighted sum.

- Due to the efficient representation based on graphs the runtimes are very low.
- For our method neither determination of weights (for weighted sum) is needed nor the values of the objectives have to be scaled. Only a ranking of the solutions is given by the relation $\prec_{sat}$.
- Handling of priorities of all or some objectives and of infeasible solutions is fully supported.

3.1 Application in Evolutionary Algorithms

The relation $\prec_{sat}$ describes a ranking of solution elements in the search space Π. Using $\prec_{sat}$ the individuals of a population are ordered. Then, for selection processes, a single fitness value *sf* is assigned to each individual. Elements of one *SC* have the same fitness value, i.e. the elements of SC 1 have the single fitness *sf*= 1. Doing so, the ***tournament selection*** as selection method is suggested, since *sf* does not take the "distances" of the solutions' fitness into account.

4 A Case Study: Heuristic Learning

In [5] a new approach to minimization of ***Binary Decision Diagrams*** (BDDs), the state-of-the-art data structure in VLSI CAD, has been proposed. They are very sensitive to the chosen variable ordering, i.e. the size may vary from linear to exponential. An EA has been used to learn heuristics for BDD minimization. The heuristics are given by sequences of basic operations, i.e. local search operations. Multiple optimization criteria have been considered in parallel in [5]. The evaluation was done by a weighted sum, where the parameters have been finetuned "by hand", i.e. many experiments had to be performed.

The model proposed in this paper has been applied to the optimization problem introduced above. Experiments are performed and a comparison to [5] is given. Many experiments were needed to obtain the high quality results in [5], since the weights of the weighted sum had to be determined. In the new approach no finetuning is needed and in only one program run the same optimization results are obtained.

Furthermore, the efficiency of this approach is demonstrated by a comparison to the method of ***nondominated sorting*** (NDOM) [11]. By our experiments it turns out that the new model for MOO is superior to NDOM.

4.1 The Learning Model

To make the paper self-contained the basic learning model is briefly reviewed [4]. It is assumed that the problem to be solved has the following property: A non empty set of optimization procedures can be applied to a given (non-optimal) solution in order to further improve its quality. (These procedures are called ***Basic Optimization Modules*** (BOMs).) Each heuristic is a sequence of BOMs. The length of the sequence depends on the problem instances that are considered.

The goal of the approach is to determine a good (or even optimal) sequence of BOMs such that the overall results obtained by the heuristic are improved.

The set of BOMs defines the set H of all possible heuristics that are applicable to the problem to be solved in the given environment. H may include problem specific heuristics, local search operators but can also include some random operators.

The individuals of our EA make use of multi-valued strings. The sequences of different length are modeled by a variable size representation of the individuals.

To each BOM $h \in H$ we associate a *cost function cost*: $H \to \mathbf{R}$. *cost* estimates the resources that are needed for a heuristic, e.g. execution time of the heuristics. The quality fitness measures the quality of the heuristic that is applied to a given example.

For more details about the learning model see also [4, 5], where the complete model, using a weighted sum for measuring the fitness, has been presented.

4.2 Binary Decision Diagrams

As well-known each Boolean function $f : \mathbf{B}^n \to \mathbf{B}$ can be represented by a *Binary Decision Diagram* (BDD), i.e. a directed acyclic graph where a Shannon decomposition is carried out in each node.

We make use of reduced and ordered BDDs [1]. For briefness they are called OBDDs in the following.

Example 3 shows the importance of good variable orderings for OBDDs:

Example 3. In Figure 2 the OBDDs of the function $f = x_1x_2 + x_3x_4 + x_5x_6$ with variable orderings $x_1x_2x_3x_4x_5x_6$ and $x_1x_3x_5x_2x_4x_6$ are illustrated. As can be seen the choice of the variable ordering largely influences the size of the OBDDs. The left (right) outgoing edge of each node x_i denotes $f_{x_i=1}$ ($f_{x_i=0}$).

We now consider the following problem that will be solved using EAs:

> ***How can we develop a good heuristic to determine variable orderings for an OBDD representing a given Boolean function f such that the number of nodes in the OBDD is minimized?***

Notice we do *not* optimize OBDDs by EAs. Instead we optimize the heuristic that is applied to OBDD minimization.

Dynamic Variable Ordering The algorithms that are used as BOMs in the EA are based on dynamic variable ordering. *Sifting* (S) is a local search operation for variable ordering of OBDDs which allows hill climbing. *Siftlight* (L) is a restricted form of sifting where hill climbing is not allowed. The third BOM is called *inversion* (I) which inverts the variable ordering of an OBDD. For more details see [5].

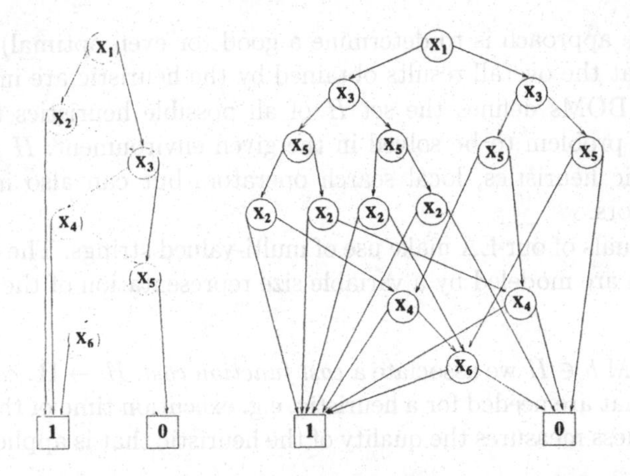

Fig. 2. OBDDs of $f = x_1x_2 + x_3x_4 + x_5x_6$

4.3 Evolutionary Algorithm

Representation In our application we use a multi-valued encoding, such that the problem can easily be formulated. Each position in a string represents an application of a BOM. Thus, a string represents a sequence of BOMs which corresponds to a heuristic. The size of the string has an upper limit of size l_{max} which is given by the designer and limits the maximum running time of the heuristic.

In the following we consider three-valued vectors: S (L, I) represents sifting (siftlight, inversion).

Objective Function As an *objective function* that measures the *fitness* of each element we apply the heuristics to k benchmark functions in a *training set* $T = \{b_1, \cdots, b_k\}$. The quality of an individual is calculated by constructing the OBDD and counting the number of nodes for each $b_i, 1 \leq i \leq k$. Additionally, the execution time (measured in CPU seconds) that is used for the application of the newly generated heuristic is minimized. Then, the objective function is a $(k+1)$-tuple and it is given by:

$$f_c(T) = (\#nodes(b_1), \cdots, \#nodes(b_k), \sum_{i=1}^{k} time(b_i)),$$

where $\#nodes(b_i), 1 \leq i \leq k$, denotes the number of OBDD nodes that are needed to represent function b_i. $time(b_i), 1 \leq i \leq k$ is the execution time of the newly created heuristic applied to function b_i.

The choice of the fitness function largely influences the optimization procedure. It also possible to chose a fitness function as a tuple of length $2 \cdot k$ by considering each $time(b_i)$ separately, instead of using the sum. By our choice the EA focuses more on quality of the result than on the runtime needed.

solutions		OBDD size					
		add6	addm4	cm85a	m181	risc	$\sum$time
dom	1.	28	181	27	60	65	0.17
	2.	28	163	27	59	65	0.33
	3.	68	163	35	60	66	0.17
	4.	68	163	35	54	70	0.18
	5.	28	236	37	60	65	0.16
	6.	28	175	27	60	65	0.28
	7.	68	163	35	60	66	0.17
	8.	54	163	35	56	75	0.17
	9.	68	177	38	63	76	0.12
	10.	28	165	36	61	66	0.28
	11.	68	163	35	54	65	0.21
favor	1.	28	163	27	54	65	0.29
	2.	28	163	27	60	65	0.26
exact		28	163	27	54	65	*

Table 1. Training set 1

Algorithm

1. The initial population $\mathcal{P}$ is randomly generated.
2. Then $\frac{\mathcal{P}}{2}$ elements are generated by the genetic operators ***reproduction*** and *crossover*. The newly created elements are then mutated.
3. The offsprings are evaluated. Then the new population is sorted using relation $\prec_f$.
4. If no improvement is obtained for 50 generations the algorithm stops.

For more details of the algorithm see also [5].

4.4 Experimental Results

In this section we present results of experiments that were carried out on a *SUN Ultra 1* workstation. All running times are given in CPU seconds. We performed our experiments using examples of Boolean functions that are taken from LGSynth91 [12].

In a first series of experiments we compare the model proposed in this paper to the method of nondominated sorting [8]. The results are summarized in Table 1. The EA is applied to a training set of size five that is used to develop heuristics for OBDD minimization. The size of the population is set to 20. In the first row of Table 1 the names of the five benchmark functions and the execution time are given. The next block of rows show the best solutions resulting from nondominated sorting (*dom*). All these solutions are not comparable with respect to relation *dominate*, i.e., this set of solutions is the Pareto-set. As can be seen no solution hits the minimum OBDD size for each function (see

	training set 2	
	dom	*favor*
# solutions	18	2
# SCs	3	2
time EA	4h 17m s	3h 50m

Table 2. Comparison of *dom* and *favor*

row exact). Analogously, the best solutions of the EA using the new model of ranking MOO solutions can be seen in the next two rows (*favor*). In this case the EA comes up with only two concurring solutions. Both are superior to the solutions from *dom*. Notice that solution 1 even hits the minimum OBDD size in each case. Each component of solution *favor* 2 is even better or equal than the corresponding components from some solutions in *dom*, e.g. solution *dom* 10. The granularity of the search space partition of relation $\prec_f$ is clearly better in comparison to *dominate*. Only two instead of 11 solutions are ranked in the same level. Additionally, the overall quality of the EA is improved by relation $\prec_f$ as can be seen by the quality of solutions *favor* 1 and *favor* 2.

In a next series of experiments a larger training set consisting of 8 functions of varying sizes has been considered to give some more insight in the effect of using our approach. The results are summarized in Table 2.

For training set 2 the number of solutions in the final solution set is shown. Again it can be seen in row ***# solutions*** that the method of nondominated sorting gives 18 best solutions in the Pareto-set. Using the technique proposed in this paper again the differentiation of the elements is better, i.e. only 2 best elements, and thus the convergence of the EA is much stronger. To show the effect of our method we apply the concept of SCs to the 18 elements resulting from nondominated sorting. As can be seen in row *# SCs* the number of elements can be drastically reduced. (Obviously, for *favor* the number does not change.) But still both solution elements from experiment *favor* are better than the best solution from *dom*, if the comparison is based on relation $\prec_f$. Clearly, this results in significant reductions in runtime of the EA. In [3] it has been observed that most of the runtime of EAs applied in VLSI CAD is needed for evaluation of the fitness function. Using the model presented in this paper, the number of evaluations can be significantly reduced by applying operations on graphs. These operations have polynomial runtime and can be carried out efficiently. For our example the runtime is given in the last row for both approaches. As can be seen *favor* is even faster and also gives better results.

5 Conclusion

A new model for *Multi-Objective Optimization* has been proposed. We could overcome the limitations of classical EA approaches that often require several runs to determine good starting parameters. Furthermore, our model gives very robust results since the number of parameters is reduced without reducing the quality of the result. Only "non-promising" candidates (for which can be guaranteed that they are not optimal and better individuals already exist) are not considered. An additional advantage is the fact that components with different measures have not to be scaled. This is done automatically comparing elements with relation $\prec_f$. This pruning of the search space even may result in significant speed-ups in EAs and simplifies the handling of the algorithms.

A comparison to the approach of *nondominated sorting* has shown that relation $\prec_f$ performs a ranking of finer granularity. In the case study it turned out that the overall quality of the EA increased using the new model for MOO based on satisfiability classes and furthermore speeds up the computation since less evaluations of the fitness function are needed. This is especially useful in cases where the evaluation of the fitness function dominates the overall execution time, as it is often the case in applications in VLSI CAD.

References

1. R.E. Bryant. Graph - based algorithms for Boolean function manipulation. *IEEE Trans. on Comp.*, 35(8):677–691, 1986.
2. T.H. Cormen, C.E. Leierson, and R.C. Rivest. *Introduction to Algorithms.* MIT Press, McGraw-Hill Book Company, 1990.
3. R. Drechsler. *Evolutionary Algorithms for VLSI CAD.* Kluwer Academic Publisher, 1998.
4. R. Drechsler and B. Becker. Learning heuristics by genetic algorithms. In *ASP Design Automation Conf.*, pages 349–352, 1995.
5. R. Drechsler, N. Göckel, and B. Becker. Learning heuristics for OBDD minimization by evolutionary algorithms. In *Parallel Problem Solving from Nature, LNCS 1141*, pages 730–739, 1996.
6. H. Esbensen. Defining solution set quality. Technical report, UCB/ERL M96/1, University of Berkeley, 1996.
7. H. Esbensen and E.S. Kuh. EXPLORER: an interactive floorplaner for design space exploration. In *European Design Automation Conf.*, pages 356–361, 1996.
8. D.E. Goldberg. *Genetic Algorithms in Search, Optimization & Machine Learning.* Addision-Wesley Publisher Company, Inc., 1989.
9. J. Horn, N. Nafpliotis, and D. Goldberg. A niched pareto genetic algorithm for multiobjective optimization. In *Int'l Conference on Evolutionary Computation*, 1994.
10. Z. Michalewicz. *Genetic Algorithms + Data Structures = Evolution Programs.* Springer-Verlag, 1994.
11. N. Srinivas and K. Deb. Multiobjective optimization using nondominated sorting in genetic algorithms. *Evolutionary Computation*, 2(3):221–248, 1995.
12. S. Yang. Logic synthesis and optimization benchmarks user guide. Technical Report 1/95, Microelectronic Center of North Carolina, Jan. 1991.

Neural Network Approach to Design of Distributed Hard Real-Time Systems

Jerzy Martyna

Jagiellonian University, Dept. of Computer Science
ul. Nawojki 11, 30-072 Kraków, Poland

Abstract. This paper examines the utility of neural networks for optimization problems occuring in the design of distributed hard real-time systems. In other words, it describes how neural networks may also be used to solve some combinatorial optimization problems, such as: computer locations in distributed system, minimalization of overall costs, maximization of system reliability and availability, etc. All requested parameters and constraints in this optimization process fullfil the conditions for design of distributed hard real-time systems. We show that the neural network approach is useful to obtain the good results in the optimization process. Numerical experimentation confirms the appropriateness of this approach.

1 Introduction

Fault tolerant flight control system, autopilot systems, radar for tracking missiles, systems for monitoring patients in critical conditions, nuclear power plants and process control are inherently distributed and have several real-time constraints, such as reliability and availability requirements. These constraints add considerable complication to the design of these systems whose correctness depends not on their logical and functional behaviour, but also on the temporal properties of this behaviour. They can be classified as *distributed hard real-time systems* [6], [26], in which the consequences of missing deadline may be catastrophic. These systems are usually working in the typical or specialized local area networks. Therefore, the design process must take into consideration different conflicting objectives such as low investment and operating cost, high reliability and availability of the system, minimum response time, etc. This designing process can be treated as nonlinear goal programming model which may be solved by heuristic methods [21]. The application of genetic algorithms to computer assignment problem in distributed hard real-time systems was presented in the paper [22]. However, there is no guarantee that these methods would reach a global optimal solution in finite time.

The solving of this problem is also possible by the use of new, alternative way of addressing optimization problems. These methods include artificial neural networks studied in many handbooks [9], [10], [16]. These networks are also used to solve specific optimalization problems, such as: job-shop scheduling [30], linear programming [28], traveling salesman problem [5], [12], [29], simulated annealing [15], etc. The global optimization issue was presented among others in [13]. The idea of the use of neural networks in global optimization was presented in [7], [14].

The remainder of this paper is organized as follows. In Section 2, the problem is formulated as a nonlinear combinatorial optimization problem. In Section 3, we describe the Hopfield neural networks. In Section 4, we present simulated annealing to solve the submitted problem and use the stopping rule in global optimization. In Section 5, we report computational experience with our neural network. Section 6 provides summary and conclusions.

2 Computer assignment problem in distributed hard real-time systems

The assignment problem of computers in distributed hard real-time systems has been studied in the paper [21]. There, having shortly described this problem, we formulated the following model of distributed hard real-time systems.

We assumed that the typical distributed system for the processing of time-constrained tasks had a given location of nodes in the local area network. Let S be the set of all nodes and s represents an individual node. A set of all classes of processors available in this model was defined by M. An individual class of processor was here described by m. To simplify, we assumed that only the processor of single class would be located at a node. In our model we had only two classes of tasks ($J = 2$): time constrained tasks ($j = 1$) and tasks without time constraints ($j = 2$).

In problem statement, the following parameters are also used: $SPEED_s$ is the speed of processor at node s (in millions of instructions per second); $MAINMEM_s$ is the size of main memory required at node s (in Megabytes); $SECMEM_s$ is the secondary memory required at node s (in Gigabytes).

The computer assignment problem in distributed hard real-time systems is stated as follows: given the locations of all nodes in computer network, assign computers with the defined processor class m to nodes so as to optimize all costs. The costs belong to designed objectives and are as follows:

1) minimum investment cost of located computers;
2) minimum operating cost;
3) maximum reliability and availability of distributed system;
4) minimum response time of distributed system.

The investment cost consists of summarized costs of processor, main and secondary computer memory allocated in each node. The operating cost of distributed system is the sum of mean costs of processing and task transmission [21]. The reliability and availability of distributed hard real-time can be computed as in the papers [19], [20]. The response time of distributed hard real-time was defined in [21]. All objectives have constraints. Accordingly, the first constraint is as follows:

$$\sum_{s \in S} FCOST_s \leq INVCOST \tag{1}$$

where $FCOST_s$ is the investment cost of computer of at node s and $INVCOST$ includes the admissible value of total investment cost specified by the designer.

The second constraint is stated as follows:

$$\sum_{s \in S} FOPC_s \leq OPCOST \tag{2}$$

where $FOPC_s$ is the operating cost of computer at a node s and $OPCOST$ is the maximal operating cost of the system specified by the designer.

The third constraint is given by:

$$\prod_{s \in S} R_s \leq RELCON \text{ and } \forall R_s > 0 \tag{3}$$

where R_s is the reliability of computer at node s and $RELCON$ is the minimal reliability of the system specified by the designer.

The fourth constraint is as follows:

$$\prod_{s \in S} A_s \leq AVALCON \text{ and } \forall A_s > 0 \tag{4}$$

where A_s is the availability of computer at node s and $AVALCON$ is the allowed minimal value of system availability specified by the system designer.

The fifth constraint is provided by the relation:

$$\sum_{j \in J} \sum_{s \in S} RESP_{sj} \leq RESPCON \tag{5}$$

where $RESP_{sj}$ is the response time for a task of class j on node s and $RESPCON$ is the admissible maximal response time of the system specified by the designer.

The crucial parameter for the design of hard real-time systems is given by p_{dyn} [17]. It defines the probability of dynamic failures of system in time-constrained environment caused by the hardware and software failures, system reconfiguration, interference with the communication network as well as other causes such as missing the hard deadline. Besides, we take into consideration such parameters as maximum mission lifetime, maximum processing power, etc. The threshold of any constraints and parameters mentioned above is associated with changing or rejecting of the project [21]. The outline of the program for the design distributed hard real-time system is given in Fig. 1.

The above program operates as follows: first the topology of designed system is settled. For a given topology and some parameters, p_{dyn} - the probability of dynamic failures of the system is computed. If the calculated value is less than

```
program design_of_distributed_hard_real-time_systems;
const p_dyn = ?;  {maximum value of dynamic failure probability}
      INVCOST = ? ; {max. value of investment cost}
      RELCON = ?; {min. value of system reliability}
      AVALCON = ?;  {min. value of system availability}
      mission_lifetime = ?; {max. length of mission lifetime}
      processing_power = ?; {max. number of task processed per unit time}

procedure computation_of_system_parameters(p_dyn^(com): real);
var best_result: boolean;
begin
  best_result := false;
  compute_actual_parameters(parameters);
  while p_dyn^(com) <= p_dyn do
    begin
      simulated_annealing_to_solve_the_problem;
      compute(p_dyn^(com));
    end;
  if best_result then write('parameters');
end;

begin {main program}
  repeat
    settle_the_initial_topology;
    compute(p_dyn^(com));
    if p_dyn^(com) <= p_dyn then computation_of_system_parameters(p_dyn^(com));
    else change_preliminary_foundations
  until all_possible_conditions
end.
```

Fig. 1. Outline of the program *design_of_distributed_hard_real_time_systems*

the assumed value of p_{dyn}, then the procedure to computation the rest parameters is called. In this procedure we use the continous Hopfield neural network to solve the problem. This procedure finds the global optimum of the actual parameters to improve the solution. If our distributed system does not meet the dynamic failure requirements, we print out all obtained values of the parameters.

3 The Hopfields neural networks

The original model described by J. J. Hopfield [11] is a neural network with binary input variables. Each of the N elements (neurons) may acquire one of the two states: $x_i = +1$ or $x_i = -1$. For such network it is possible to introduce for the special symmetric case $w_{ij} = w_{ji}$ a so-called global energy function given as follows:

$$E = -\frac{1}{2}\sum_{i=1}^{N}\sum_{j=1}^{N} w_{ij}x_i x_j - \sum_{i=1}^{N}\theta_i x_i \tag{6}$$

where w_{ij} is the weight of the connection from the jth element (neuron) to the ith neuron and θ_i is a fixed threshold applied externally to neuron i.

There is an essential analogy between this neural network and the spin system in physics, where E has then the meaning of energy and an expression analogous to Eq. (6) is valid for the deterministic system at absolute zero temperature.

J.J. Hopfield has suggested to use the local minima of the global energy function E as attractors for the storage of input patterns presented to the neural network. With the use of the so-called Hebbian learning rule he has deduced an expression for the pertinent connection weights:

$$w_{ij} = \frac{1}{N}\sum_{p=1}^{P} u_i^p u_j^p, \quad i \neq j \tag{7}$$

where $\{u_i^p\}$ for $i = 1, \ldots, N$ is the pth pattern from the ensemble of P memorized input patterns.

Besides the symmetry condition $w_{ij} = w_{ji}$, there are further conditions for the validity of Eq. (7):

a) the presented input patterns are random and mutually uncorrelated;
b) the connection weights are the result of learning only (the so-called premiss of tabula rasa at the beginning);
c) each element (neuron) is connected with all the others;
d) the number N is very large (for a thermodynamic analogy $lim\ N \to \infty$ would even be required).

4 A Hopfield neural network approach to distributed hard real-time system design

We recall that a Hopfield neural network is a fully interconnected network of artificial neurons whose connections (excitatory and inhibitory) embody a *"computational energy"* function in the following sense: a relaxation algorithm is employed to alter the activation levels of individual neurons repeatedly until a stable state is reached. Every stable state represents a minimum of the energy function. Therefore, to solve a problem using a Hopfield neural network, an appropriate energy function from which to derive the connection weights was suggested.

We assume that a Hopfield neural network is represented as a matrix of units. Each row in the matrix corresponds to a node of distributed system. Each column of this matrix represents a class of computer. An assignment of computers to a node can be here defined by the activation of exactly one unit in each row.

Our energy function obtains a minimum, when only one computer is assigned to a node. In other words, if the sum of activation levels of the neurons in each row is equal to 1, for M class of computers and S nodes there is only one computer of one class on each node. Thus, we define the energy function of our neuronal network as follows:

$$E(\mathbf{d}) = A \sum_{s \in S} \left(\sum_{m \in M} V_s^{(m)} - 1 \right)^2 + \sum_{i=1}^{5} \Phi[B_i * h_i(\mathbf{d})] \tag{8}$$

where V_{ms} represents the activation level of the neuron at the intersection of row s and column m. $\mathbf{d} = [d_s^{(m)}]_{|M| \times |S|}$ is a set of all possible computer locations in the designed network. A and B_i are constant weights expressing the relative importance of both terms of energy functions. The function Φ is continuous and so that

$$\Phi[h_i(\mathbf{d})] \begin{cases} = 0, \text{ if } h_i(\mathbf{d}) \geq 0 \\ > 0, \text{ if } h_i(\mathbf{d}) < 0 \end{cases} \tag{9}$$

Bounded conditions, $h_i(\mathbf{d})$, $(1 \leq i \leq 5)$, are in the forms:

$$h_1(\mathbf{d}) = \sum_{s \in S} \sum_{m \in M} FCOST_s^{(m)} - INVCOST \geq 0 \tag{10}$$

$$h_2(\mathbf{d}) = \sum_{s \in S} \sum_{m \in M} FOPC_s^{(m)} - OPCOST \geq 0 \tag{11}$$

$$h_3(\mathbf{d}) = \left(\prod_{s \in S} \sum_{m \in M} R_s^{(m)} - RELCON \right)^{-1} \geq 0 \tag{12}$$

$$h_4(\mathbf{d}) = \left(\prod_{s \in S} \sum_{m \in M} A_s^{(m)} - AVALCON \right)^{-1} \geq 0 \tag{13}$$

$$h_5(\mathbf{d}) = \sum_{j \in J} \sum_{s \in S} RESP_{sj} - RESPCON \geq 0 \tag{14}$$

where $FCOST_s^{(m)}$, $FOPC_s^{(m)}$, $R_s^{(m)}$, $A_s^{(m)}$ are respectively: investment cost, operating cost, reliability, availability of computer of class m at node s. $RESP_{sj}$ is the response time of task of class j at node s. We assume that $\forall R_s^{(m)} > 0$ and $\forall A_s^{(m)} > 0$.

The first term of energy function (Eq. 8) is the objective function to be minimized. The remaining terms, $h_1(\mathbf{d})$, $h_2(\mathbf{d})$, ... represent here the parameters of penalty functions whose minimization satisfies the constraints. The B_1, B_2, ... are constant weights assigned to the respective penalty functions.

```
procedure simulated_annealing_to_solve_the_problem;
begin
initialize_the_parameters;
randomly_generating_S0;
C(S0); {computation of the initial solution for S0};
E(S0); {computation of the mean investigation and operational costs
        for the point S0}
while not_satisfied_the_stopping_conditions_in_global_optimization do
  begin
    while not_available_the_optimum do
      begin
        generate_next_point(S0';
        Δ E := E(S0') - E(S0);
        if (Δ E < 0) or (RANDOM < exp (-Δ E / C(S0')))
        then   S0 = S0'
      end;
    C(S0);
  end;
  write( 'results');
end;
```

Fig. 2. The proposed neural network approach as the procedure *simulated_annealing_to_solve_the_problem*

4.1 Simulated annealing to solve the problem

The algorithm of simulated annealing investigates the behaviour of an object which is a subject-matter of optimization handled as a set of physical molecules reach their balance in temperature T. For the assumed temperature T which a sequence of steps called Metropolis technique [23] is executed.

In each step, for the random chosen molecule its dislocation and the associated ΔE change of energy is defined. It is accepted with the probability 1, if $\Delta E \leq 0$, or with the probability equal to $e^{-\Delta E/k_B T}$, if $\Delta E > 0$, where k_B is the Boltzmann constant. If the number of steps in Metropolis algorithm is large, then for given T system reaches the Boltzmann distribution which is a global state of termical balance of molecules. Next, the temperature is decreased and it's made by new set of steps of Metropolis algorithm.

Theoretically, the simulated annealing is guaranteed to converge to global optimum with probability 1, in practice, this method for finding solution around a predefined error rate and within a reasonable computation time is difficult to design [1]. None the less, as it was shown in [24], if the object is defined with the energy function, then due to the use of simulated annealing, this function converges to global minimum.

In typical implementation of simulated annealing (see Fig. 2) a pair of nested loops is used. In each step of the internal loop, a unit of a system (here it's a computer of a noted class) is subjected to a small random displacement, and the resulting change ΔE in the energy of the system is computed. If we find that the change $\Delta E \leq 0$, the displacement is accepted, and the new system configuration with the displaced unit is used as the starting point for the next step of algorithm. In other words, if we find that the change $\Delta E > 0$, the algorithm proceeds in a probabilistic manner as described next. The probability of accepting the configuration with the displaced unit is given by

$$P(\Delta E) = \exp(-\frac{\Delta E}{T}) \tag{15}$$

where T is the temperature. In our implementation we chose the sum of investment and operating costs as the temperature. To implement the probabilistic part of the algorithm, we may use a generator of random numbers distributed uniformly in the interval (0, 1). One such number is selected and compared with the probability $P(\Delta E)$ of Eq. (15). If the random number is smaller than the probability $P(\Delta E)$, the new configuration with the displacement unit is accepted. Otherwise, the original system configuration is reused in the next step of the algorithm.

4.2 Stopping rule in global optimization

In this section the problem of stopping for global optimization is considered. The crucial point of the rule proposed is the decision connected with sequential sampling and cost associated with this sampling.

We adopted here the approach given in the papers [3], [4]. In this solution k-step *look-ahead* (sla) rule for sampling with recall from an unknown distribution of function underlying the observations is used. We recall, that k-sla rule calls for stopping the sampling process as soon as the current cost defined as

$$L(t_1, t_2, \ldots, t_n; c) = -t_n^* + nc \tag{16}$$

is not greater than the cost expected if at least k more observations are taken, where t_i is the function value at the local optimum found during the ith search, $t_n^* = \max_{i \in [1,\ldots,n]} t_i$ and $c > 0$ is a fixed constant. The t_i's are realizations of a discrete random variable. For F parametric used in a model the 1-sla is found to be the rule which calls for stopping as soon as

$$\int_{t_n^*}^{\infty} (t - t_n^*) d\widehat{F}_n(t) \leq c \tag{17}$$

where $\widehat{F}_n(t) = E(F(t) \mid t_1, \ldots, t_n)$ is the Baeyes estimate of $F(t)$ or, equivalently integrated by parts [8]

$$\int_{t_n^*}^{\infty} (1 - F(t)) dt \leq c \tag{18}$$

$\widehat{F}_n(t)$ can be interpreted as the predictive distribution of the next observation given the sample so far collected, so that (17) it is said that stopping is prescribed by the 1-sla rule as the improvement in the best sampled value given by a further observation is not greater than c was expected.

It is obvious that the cost c can be interpreted as an acceptability threshold for the expected improvement in continuing the sampling process.

Behind Betro and Schoen [2] was chosen

$$\widehat{F}_n(t) = 1 - \frac{m(t)+\lambda}{n+\lambda}\exp\{-\sum_{j=1}^{n(t)} \frac{\gamma(t_{(j)}) - \gamma(t_{(j-1)})}{m(t_{(j-1)})+\lambda}\} * \exp\{-\frac{\gamma(t) - \gamma(t_{(n(t))})}{m(t_{(n(t))})+\lambda}\} \tag{19}$$

where $n(t)$ is the number of distinct observations not greater than t, $m(t)$ is the total number of observations strictly greater than t, $\gamma(t) = -\lambda\log(1-F_0(t))$ and $t_{(j)}$, $j = 1,\ldots,n(t_n^*)$ are the increasingly distinct observations. It was assumed that

$$\int t^2 d\widehat{F}_0(t) < \infty \tag{20}$$

and according to [3], the optimal stopping rule does exist.

It should also be noticed that the use of stopping rule in global optimization problem implies the choice of some parameters. We have chosen as in [3] the following parameters:

$$\widehat{F}_0(t) = \begin{cases} (1-(1-\frac{1}{\sqrt{2}})exp(\frac{t-a}{b})^2, & t \leq a \\ 0.5exp(-2(\sqrt{2}-1)(\frac{t-a}{b})), & t > a \end{cases} \tag{21}$$

and $\lambda = 1$. In Eq. (21) a is the median of the distribution and b is a shape parameter, regulating the concentration of $\widehat{F}_0(t)$ arround a. In our experience the parameter a was set to 0, $b = 10$, $c = 0.05$.

5 Numerical experiments and results

An example of application of Hopfield neural network to computer assignment problem in distributed hard real-time has been presented. The Hopfield network consists here of 76 neurons. The output of each neuron in the network is fed back to all other neurons. There is no self-feedback in the neural network (i.e., $w_{ij} =$ 0). The neural network was trained using samples that span the allowable ranges of the input variables. A total of 30 samples were used for training. The input variables consist of the control variables, such as the class of processor allocated at the defined node.

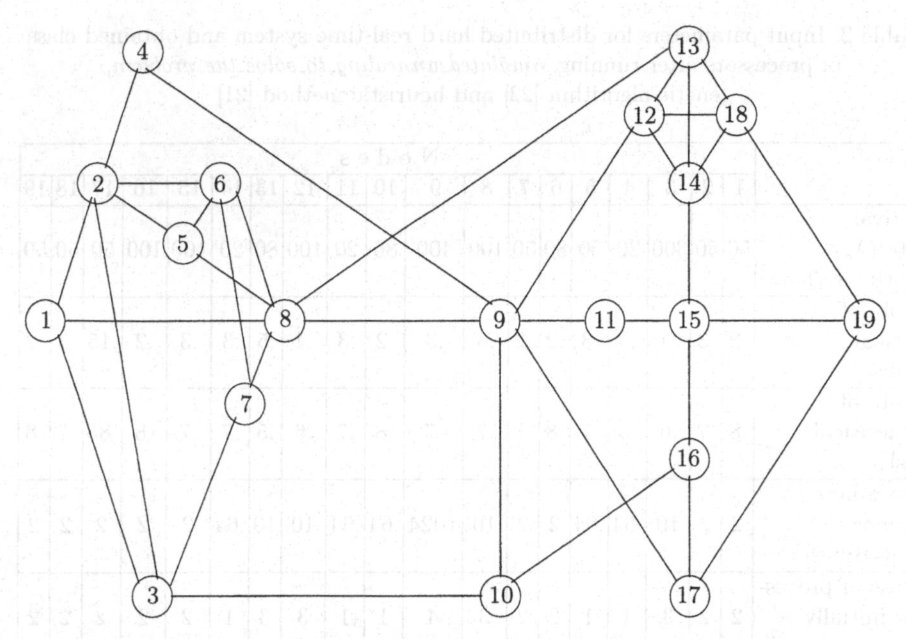

Fig. 3. Exemplary physical topology of distributed computer systems

Physical topology of studied, distributed computer system is shown in Fig. 3. This topology was also used in the paper of [18] for solving resource allocation problem in distributed computer system. The structure consists of 19 computers interconnected through a communication network. The computer network is assumed to be logically fully connected. It means that every computer can communicate with every other computer.

In our model it was assumed that there are 4 classes of processors: microcomputers, workstations, network servers, and mainframe. Hewlett Packard computer family was here selected for the computation, because these computers (except for the microcomputers) use the same operating system HP-UX and according to the producer information the $MTBF$ for the electronical units belongs to the highest and exceeds 6 years. The HP computer family is represented here by: microcomputer HP VE2 5/75, workstation HP J200, network server HP E55, mainframe HP T500. More parameters for these computers are given in Table 1.

Table 1. Exemplary parameters of processor classes

		Processor class			
		1 *Micro-computer*	2 *Work-station*	3 *Network server*	4 *Main-frame*
Availability	A_m	0.99	0.98	0.98	0.92
Reliability	R_m	0.99	0.98	0.97	0.96
Price*	in thousand USD	1.368	20.0	22.0	200.0
Main memory	in MBytes	4.0	64.0	256.0	512.0
Secondary memory	in GBytes	0.64	2.0	10.0	1024.0

*Source: HP price catalog, 1998 year..

Table 2. Input parameters for distributed hard real-time system and obtained class of processors after running *simulated_annealing_to_solve_the_problem*, genetic algorithm [22] and heuristic method [21]

	Nodes																		
	1	2	3	4	5	6	7	8	9	10	11	12	13	14	15	16	17	18	19
Arrival rate (λ_s) in [thous./hour]	50	50	300	20	50	80	50	100	400	80	20	100	80	20	100	100	50	50	50
Prob. of critical tasks	.2	.3	.4	.5	.3	.2	.2	.3	.3	.2	.3	.4	.5	.3	.3	.2	.15	.3	.2
Prob. of noncritical tasks	.8	.7	.6	.5	.7	.8	.8	.7	.7	.8	.7	.6	.5	.7	.7	.8	.85	.7	.8
Secondary memory in [GBytes]	2	2	10	.64	.64	2	2	10	1024	.64	.64	10	10	.64	2	2	2	2	2
Class of processor initially located at node	2	2	3	1	1	2	2	3	4	1	1	3	3	1	2	2	2	2	2
Class of processor after running sim. ann.	1	2	3	1	1	2	1	2	3	2	1	1	2	2	3	1	1	1	1
Class of processor after running gen. alg.	1	2	3	1	2	1	1	1	3	3	1	1	3	2	3	3	1	1	1
Class of processor after run. heur. meth.	1	2	3	1	2	3	1	3	3	3	1	3	3	1	3	3	1	2	1

Exemplary initial parameters for each: assumed arrival rate per hour, probabilities of both classes of tasks arrivals and the found maximal neighbouring of first order for each node (e.g. the number of nodes which are connected with given node by immediate channel) are shown in Table 2. These parameters allow to determine the initial value of processor class located at each node. The input parameters for neural network are presented in Table 3. These values characterize the time constrained environment in which the designed distributed system should function. Change of the values of constant weights A, B_i, ($1 \leq i \leq 5$), allows to investigate in which way individual preferences of the designer influence each solution.

Table 3. Assumed data for neural networks

No.	Parameter	Value
1	Constant weight A	1.24
2	Constant weight B_1	7.065
3	Constant weight B_2	5.152
4	Constant weight B_3	8.319
5	Constant weight B_4	10.175
6	Constant weight B_5	14.036

Table 4. Initial data and final solutions obtained after running the procedure *simulated_annealing_to_solve_the_problem*, genetic algorithm [22], heuristic method [21]

	Initial data	Final solution		
		heuristic method	genetic algorithm	neural network
Probability of dynamic failure of distr. system	0.1	0.05	0.02	0.02
Reliability of distributed system	0.653	0.666	0.709	0.724
Availability of distributed system	0.687	0.731	0.739	0.754
Mean response time in [sec]	$1.05 * 10^{-1}$	$9.4 * 10^{-2}$	$9.0 * 10^{-2}$	$9.0 * 10^{-2}$
Mean investment cost in thous. USD	473.5	266.73	206.00	201.68
Mean operating cost	$8.83 * 10^{6}$	$6.0 * 10^{6}$	$6.0 * 10^{6}$	$6.0 * 10^{6}$

The values obtained as a result of the program for designing hard real-time distributed system with the specially neural network prepared for project optimization and computer location is given in Table 2. To compare the solutions obtained before and after using the neural network, the final results of computation are presented in Table 4. These results indicate an advantage of using the neural network to computer assignment problem in the reasonable amount of time. Although, the results obtained are only approximated solutions, they seem sufficient for the assumed computation.

6 Conclusions

The computational effort required to solve the combinatorial optimization problems, such as computer locations in distributed system with minimalization of overall costs, maximization of system reliability and availability, etc., is substantial. We have demonstrated that neural network can be used to solve this global optimization problem occuring in the design of distributed hard real-time systems. This approach would permit a straightforward amendment of heuristic method [21] or genetic algorithm solution [22] to this problem.

References

1. Aarts, E., van Laarhoven, P.: Simulated Annealing: Theory and Practice, John Wiley and Sons, New York, 1989
2. Betro, B. Schoen, F.: Sequential Stopping Rules for the Multistart Algorithm in Global Optimization, Mathematical Programming, Vol. 38 (1987) 271 - 286
3. Betro, B., Schoen, F.: Optimal and Sub-Optimal Stopping Rules for the Multistart Algorithm in Global Optimization, Mathematical Programming, Vol. 57 (1992) 445 - 458
4. Boender, C.G.E., Rinnooy Kan, A.H.G.: On When to Stop Sampling for the Maximum, Journal of Global Optimization, 1 (1991) 331 - 340
5. Burke, L.I.:Neural Methods for the Traveling Salesman Problem: Insights From Operations Research, Neural Networks, Vol. 7, No. 4 (1994) 681 - 690
6. Burns, A.: Distributed Hard Real-Time Systems: What Restrictions Are Necessary, in: H.S.M. Zedan (Ed.): *Real-Time Systems, Theory and Applications*, North-Holland, Amsterdam, 1990, 297 - 303
7. Cichocki, A., Unbehauen, R.: Neural Networks for Optimization and Signal Processing, John Wiley, New York, 1993
8. Feller, W.: An Introduction in Probability Theory and Its Applications, Vol. II, John Wiley, New York, 1966
9. Haykin, S.: Neural Networks: A Comprehensive Foundation, IEEE Computer Society Press and Macmillan, New York, 1994
10. Hertz, J., Kroch, A., Palmer, R.G.: Introduction to the Theory of Neural Computation, Addison-Wesley, Reading, 1991
11. Hopfield, J.J.: Neural Networks and Physical Systems with Emergent Collective Computational Abilities, Proc. Natl. Acad. Sci. USA, Vol. 79 (1982) 2554 - 2558
12. Hopfield, J.J., Tank, D.W.: Neural Computation of Decisions in Optimization Problems, Biological Cybernetics, Vol. 52 (1985) 141 - 152
13. Horst, R., Pardalos, P.M.: Handbook of Global Optimization, Kluwer Academic Publ., Dordrecht, 1995
14. Karhunen, J., Joutsensalo, J.: Generalization of Principal Component Analysis, Optimization Problems, and Neural Networks, Neural Networks, Vol. 8, No. 4, (1995) 549 - 567
15. Kirkpatrick, S., Gelatt Jr., C.D., Vecchi, M.P.: Optimization by Simulated Annealing, Science, Vol. 220 (1983) 670 - 680
16. Korbicz, J., Obuchowicz, A., Uciński, D.: Artificial Neural Networks, Akademicka Oficyna Wydawnicza PLJ, Warszawa, 1994 (*in Polish*)
17. Krishna, C.M., Shin, K.G.: Performance Measures for Multiprocessor Controllers, in: A. K. Agrawala, S. K. Tripathi (Eds.), *Performance'83*, North-Holland Publ. Comp., Amsterdam, 1983, 229 - 250
18. Kurose, J.F., Simha, K.: A Microeconomic Approach to Optimal Resource Allocation in Distributed Computer Systems, IEEE Trans. on Computers, Vol. 38, No. 5 (1989) 705 - 717
19. Martyna, J.: Reliability Analysis of Distributed Systems for the Hard Real-Time Environment, Archiwum Informatyki Teoretycznej i Stosowanej, Vol. 6, No. 1 - 4 (1994) 89 - 100
20. Martyna, J.: Functional Availability Analysis of Hard Real-Time Distributed Systems, Archiwum Informatyki Teoretycznej i Stosowanej, Vol 6, No. 1 - 4 (1994) 101 - 114

21. Martyna, J.: A Methodology for the Design of Hard Real-Time Distributed Systems, Archiwum Informatyki Teoretycznej i Stosowanej, Vol 7, No. 1 - 4 (1995) 105 - 124
22. Martyna, J.: Application of Genetic Algorithms to Computer Assignment Problem in Distributed Hard Real-Time Systems, Prace Naukowe UJ, Zeszyty Informatyczne, Vol. 8 (1998) 45 - 62; and also in: B. Reusch (Ed.): *Computational Intelligence. Theory and Applications*, Lecture Notes in Computer Science, Vol. 1226, Springer-Verlag, Berlin, 1997, 564
23. Metropolis, N., Rosenbluth, A.W., Rosenbluth, M.N., Teller, A.H.: Teller, E.: Equations of State Calculations by Fast Computing Machines, Journal of Chemical Physics, Vol. 21 (1953)
24. Mitra, D., Romeo, F., Sangiovani-Vincentelli, A.: Convergence and Finite-time Behavior of Simulated Annealing, Advances in Applied Probability, Vol. 18, No. 3 (1986)
25. Shang, Y., Wah, B.W.: Global Optimization for Neural Network Training, IEEE Computer, No. 3 (1996) 45 - 54
26. Stankovic, J.A.: Misconceptions About Real-Time Computing. A Serious Problem for Next-Generation Systems, IEEE Computer, Vol. 21, No. 10 (1988) 10 - 19
27. Tang, Z., Koehler, G.J.: Deterministic Global Optimal FNN Training Algorithms, Neural Networks, Vol. 7, No. 2 (1994) 301 - 311
28. Tank, D.W., Hopfield, J.J.: Simple "Neural" Optimization Networks: An A/D Converter, Signal Decision Circuit, and Linear Programming Circuit, IEEE Trans. on Circuits and Systems, Vol. CAS-33, No. 5 (1986) 533 - 541
29. Xu, X., Tsai, W.T.: Effective Neural Algorithms for the Traveling Salesman Problem, Neural Networks, Vol. 4 (1991) 193 - 205
30. Zhao, D.N., Cherkassky, V., Baldwin, T.R., Olson, D.E.: A Neural Network Approach to Job-Shop Scheduling, IEEE Trans. on Neural Networks, Vol. 2, No. 1 (1991) 175 - 179

Supporting Traditional Controllers of Combustion Engines by Means of Neural Networks

Claudia Ungerer[1], Dirk Stübener[2], Clemens Kirchmair[1], and Michael Sturm[1]

[1] Institut für Informatik
Technische Universität München
80290 München
Germany

[2] Kratzer Automation GmbH
Carl-von-Linde-Straße 38
85716 Unterschleißheim
Germany

Abstract In this paper it is investigated whether neural networks are able to improve the performance of a PI controller when controlling a combustion engine. The idea is not to replace but to assist a PI controller by a neural co-controller. Three different neural approaches are investigated for this use: Dynamic RBF (DRBF), Adaptive Time-Delay Neural Network (ATNN), and Local Ellipsoidal Model Network (LEMON).

1 Introduction

Developing combustion engines or aggregates of cars like gearboxes, catalytic converters and control boxes means testing them on a testbed. The ability to follow a prescribed speed table with a combustion engine on a testbed with high precision, is one of the main characteristic features of a controller for a combustion engine.

Presently throttle angle based control of combustion engines is done using traditional controllers and inverse engine maps. The inverse engine map is used to predict the throttle angle dependent on the demanded torque. One of the major drawbacks of using inverse engine maps is the high measurement effort, resulting in high stress for the combustion engine.

A problem of traditional controllers is the lack of constant control conditions in steady state. There is always an error between the reference and the measured value due to unsteady torque present in a combustion engine. Another problem using static inverse engine maps are changes within the control box. Gear shifts for instance may cause disturbances of speed and torque.

In this paper we investigate whether the use of neural networks instead of inverse engine maps can help to overcome these problems. As development environment a Matlab/Simulink simulation is used.

In the following, the paper is divided into three parts. First the simulation model is introduced. Then we will describe the applied control strategy as well as the used neural architectures. Finally, simulation results are presented in the last section.

2 The Simulation

The Simulink simulation model consists of a three phase AC induction machine, a spark ignition engine described by a mean value method and a car model. The spark ignition engine is directly coupled with the induction machine. That means, that there is no real gear-box at the test bed. (See figure 1)

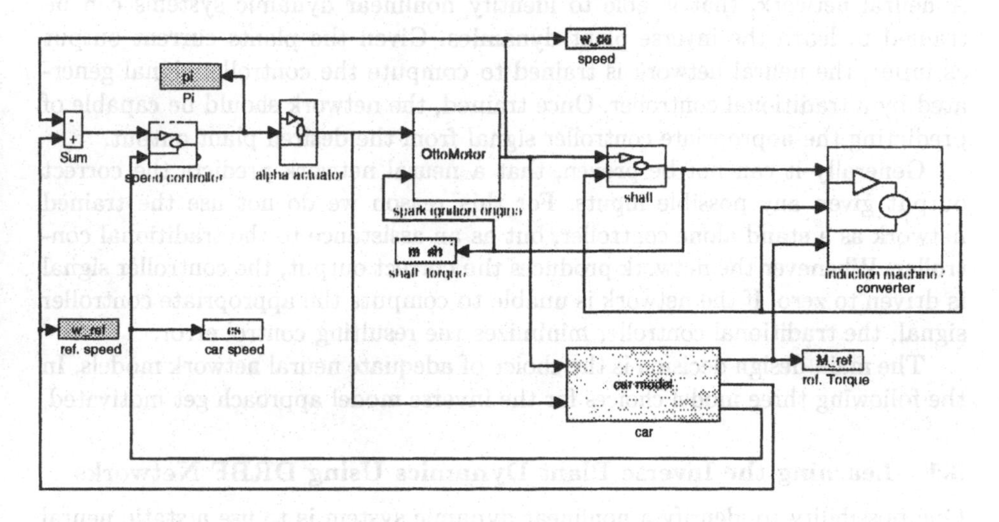

Figure1. The Simulink simulation of the testbed and the combustion engine

Inputs to the spark ignition engine include: throttle angle, an external load torque and ambient conditions (i.e., atmospheric temperature and pressure). The engine model is physically based and captures the major dynamics (lags and delays) inherent in the spark ignition torque production process. The model takes into account the air and fuel dynamics but it does not attempt to generate torque pulsations due to individual cylinder filling events. The reference value of the spark ignition engine is a speed table, that is included in the car model block.

The AC induction machine is supplied by a three phase current converter with field orientated control (direct flux orientation). The function of the induction machine is to simulate the load of a car, inclusive clutch and gear shifts.

The load of a car, which is the reference value for the induction machine, is generated by the car model. The car model simulates the quasi-stationary torque (rolling resistance, air resistance and break) and the spring- and damping torque of the motor-gearbox system with high torque transients during gear shifts and coupling processes. Inputs to the car model are throttle angle and engine speed. The simulated car velocity is used as feedback to the speed controller. Parameters of the car model are damping constant and spring constant of the motor-gearbox system, car inertia, gear transmissions and efficiency, time for engage and disengage the clutch, speed for up shifting and speed for down shifting. If the engine speed passes over the speed for up shifting the clutch model shifts up automatically, else if the engine speed falls the speed for down shifting the model shifts down.

3 Neural Control Strategies

A neural network, that is able to identify nonlinear dynamic systems can be trained to learn the inverse plant dynamics. Given the plants current output as input, the neural network is trained to compute the controller signal generated by a traditional controller. Once trained, the network should be capable of predicting the appropriate controller signal from the desired plant output.

Generally it can not be proven, that a neural network predicts the correct output given any possible inputs. For this reason we do not use the trained network as a stand alone controller, but as an assistance to the traditional controller. Whenever the network produces the correct output, the controller signal is driven to zero. If the network is unable to compute the appropriate controller signal, the traditional controller minimizes the resulting control error.

The main design decision is the choice of adequate neural network models. In the following three model choices for the inverse model approach get motivated.

3.1 Learning the Inverse Plant Dynamics Using DRBF Networks

One possibility to identify a nonlinear dynamic system is to use a static neural network model and to provide current and delayed process inputs and outputs as inputs to the model ([4], [6]). Obviously this approach suffers from the course of dimensionality as the dimension of the input space of the model is proportional to the number of delayed process inputs and outputs and the dimension of the input and output spaces of the process. Also the order of the system has to be known.

To avoid these problems, Ayoubi and Isermann propose neural networks with distributed dynamics ([1], [2]). They suggest to integrate a second order ARMA filter in each neuron, such that the output of a neuron is not only dependent on its current input, but also on its current state. Using neural networks with distributed dynamics to identify unknown systems there is no need to use past measurements of the process input and output as network inputs, because the neurons store a representation of the process history in their local memories. Ayoubi and Isermann report good results identifying a turbocharger using RBF networks with distributed dynamics (DRBF).

Dynamic Radial Units: A DRBF network is a RBF network with dynamic radial units as basic components. A dynamic radial unit consists of a multidimensional Gaussian basis function and a second order ARMA filter. Given P

neuron inputs $u_p(k)$ at time instant k the filter input $x(k)$ is computed using the Gaussian basis function:

$$x(k) = \exp\left(-\frac{1}{2}\sum_{p=1}^{P}\left(\frac{u_p(k) - c_p}{\sigma}\right)^2\right) \tag{1}$$

The parameters of the Gaussian basis function are given by $c_1, \ldots, c_P$ and σ respectively.

The output $y(k)$ at time instant k of a dynamic radial unit can be calculated by

$$y(k) = \sum_{i=0}^{2} b_i\, x(k-i) - \sum_{i=1}^{2} a_i\, y(k-i) \tag{2}$$

where b_0, b_1, b_2, a_1, a_2 are the coefficients of the filter function.

The filter transfer function can be defined using the linear time-shifting operator $q^i[x(k)] = x(k-i)$ as

$$[y(k)] = \frac{b_0 + b_1\, q^{-1} + b_2\, q^{-2}}{1 + a_1\, q^{-1} + a_2\, q^{-2}}\,[x(k)] \tag{3}$$

The output of the whole DRBF network at time instant k is the weighted sum of the outputs of all dynamic radial units at time instant k.

The Adaptation Algorithm: In order to identify a nonlinear dynamic system using a DRBF network, the adaptable parameters of the network have to be adjusted. The adaptable parameters are the parameters of the Gaussian basis functions and the filter coefficients of the dynamic radial units as well as the weights which are used to compute the output of the entire network.

The parameters are adjusted to a set of measurements using a gradient descend procedure. This means that the partial derivatives of an error function with respect to all adjustable parameters of the network must be computed, which can be done applying the chain rule. We skip the details and focus on calculating the partial derivatives of the dynamic radial units with respect to their parameters. The details can be found in [1] and [2].

The partial derivatives of the output $y(k)$ of a dynamic radial unit with respect to the filter coefficient b_i and a_i are given by

$$\frac{\partial[y(k)]}{\partial b_i} = \frac{q^{-i}}{1 + a_1\, q^{-1} + a_2\, q^{-2}}\,[x(k)] \tag{4}$$

$$\frac{\partial[y(k)]}{\partial a_i} = \frac{-q^{-i}}{1 + a_1\, q^{-1} + a_2\, q^{-2}}\,[y(k)] \tag{5}$$

For the parameters c_p and σ of a Gaussian basis function the derivatives can be written as

$$\frac{\partial[y(k)]}{\partial c_p} = \frac{b_0 + b_1\, q^{-1} + b_2\, q^{-2}}{1 + a_1\, q^{-1} + a_2\, q^{-2}}\left[\frac{u_p(k) - c_p}{\sigma_p^2}\, x(k)\right] \tag{6}$$

$$\frac{\partial[y(k)]}{\partial\sigma} = \frac{b_0 + b_1\, q^{-1} + b_2\, q^{-2}}{1 + a_1\, q^{-1} + a_2\, q^{-2}} \left[\frac{(u_p(k) - c_p)^2}{\sigma_p^3}\, x(k) \right] \tag{7}$$

3.2 Learning the Inverse Plant Dynamics Using ATNN Networks

Looking at the structure of the control system under concern yields to a second model choice. The system contains some time delay components, i.e., components, whose input affects the output not immediately but after a certain time delay (e.g. the delay between intake and power production). Therefore, a neural network is chosen that is capable of identifying such systems: the Adaptive Time-Delay Neural Network (ATNN).

Traditional feedforward neural networks immediately generate a specific output to a particular input. There is no memory incorporated for information from past time steps. Within the ATNN however, internal delay lines explicitly store recent information from the input signal.

ATNN Architecture: The Adaptive Time-Delay Neural Network, investigated in [10], is a multilayer feedforward neural network. Beside the weight there is a time delay defined on each connection. Each layer is fully connected and there may be more than one connection between two neurons. Consequently, different time delays can be defined between two neurons and therefore information from different time steps can be processed together. Figure 2 shows two neurons i and j of an ATNN having n connections between each other. The total input to neuron i at time t is defined as

$$x_i(t) = \sum_j \sum_k w_{ijk}(t)\, y_j(t - \tau_{ijk}(t)) \tag{8}$$

where $w_{ijk}(t)$ is the weight of the k-th connection from neuron j to neuron i at time t, $\tau_{ijk}(t)$ is the time delay on that connection at time t and y_j is either the output of neuron j, $y_j(t) = f_j(x_j(t))$, or a network input. The activation function f is sigmoidal. The connections of the network can be seen as the memory of the network. They operate as buffers holding the neurons outputs. The time delay values define which past output gets processed to the next neuron.

ATNN Learning Algorithm: The learning algorithm is an extended form of the backpropagation algorithm for traditional multilayer feedforward neural networks. Therefore, to minimize the error function, the weights and time delays must adapt with changes proportional to the negative gradients

$$\Delta w_{ijk}(t) = -\mu \frac{\partial E(t)}{\partial w_{ijk}(t)} \tag{9}$$

$$\Delta \tau_{ijk}(t) = -\eta \frac{\partial E(t)}{\partial \tau_{ijk}(t)} \tag{10}$$

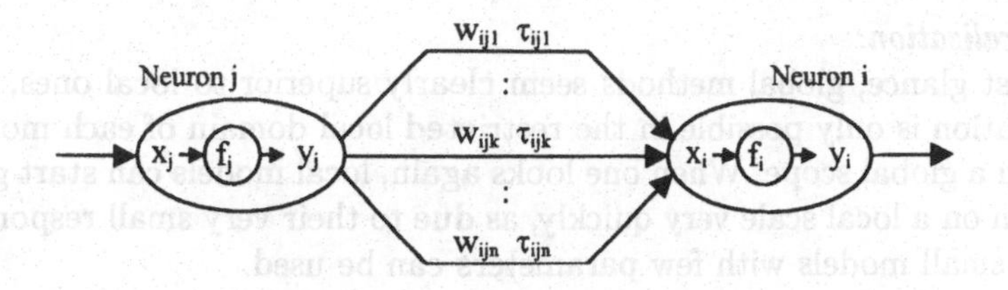

Figure2. Section of an Adaptive Time-Delay Neural Network showing two neurons i and j having n connections between each other. Each connection holds a weight w_{ijk} and a time delay τ_{ijk}

where μ and η are learning rates. The learning algorithm is defined by means of the Jacobian matrix. The derivation of this algorithm is described in detail in [11]. We summarize the update rules for the weights and time delays as

$$\Delta\tau_{ijk}(t) = \eta\ opt_a\ jac\_\delta_i^a(t-a_i)\ w_{ijk}(t-a_i)\ y'(t-\tau_{ijk}(t-a_i)-a_i) \quad (11)$$

$$\Delta w_{ijk}(t) = -\mu\ opt_a\ jac\_\delta_i^a(t-a_i)\ y(t-\tau_{ijk}(t-a_i)-a_i) \quad (12)$$

where opt_a is the Jacobian matrix of the error function E calculated by

$$opt_a = \frac{\partial E(t)}{\partial y_a(t)} \quad (13)$$

where $y_a(t)$ is the output of the a-th output neuron. $jac\_\delta_i^a(t-a_i)$ is the error signal associated with each neuron i and is defined as

$$jac\_\delta_i^a(t-a_i) = \frac{\partial y_a(t)}{\partial x_i(t-a_i)} \quad (14)$$

The parameters a_i are the Aging parameters of the network associated to each layer due to causality constraints. The Aging is described in detail in [10].

3.3 Learning the Inverse Plant Dynamics Using the Local Ellipsoidal Model-Network LEMON

A third approach is used due to considerations about gear switching and different dynamics in different driving stages. With LEMON[1] the input space is splitted into several model regions, each containing a model responsible for that region. By this strict partitioning different models for different dynamics may be used. Using local models has several advantages over using global models as stated by [5] and [3]:

- *Changing of global information:*
  Often learning of new data sets by global methods means modifying already gained knowledge in an unpredictable way. Using local learning methods will guarantee that only local modifications in the knowledge base are made.

[1] Local Ellipsoidal Model-Network

- *Generalization:*
  At first glance, global methods seem clearly superior to local ones, as generalization is only possible in the restricted local domain of each model and not on a global scope. When one looks again, local models can start generalization on a local scale very quickly, as due to their very small responsibility areas small models with few parameters can be used.
- *Computational costs:*
  Two main computational tasks have to be accomplished using local models. First, the responsible local model must be determined. Second, its model prediction must be calculated. Global methods skip the first step and directly calculate the prediction. As several investigations have shown, the additional costs of step one are much less than the costs introduced by using one global model in a complex scenario instead of several local ones.
- *Model selection:*
  Depending on the actual operating point the model structure can be changed. Different complex models may be used. The selection is done examining model properties guided by a set of fuzzy rules.

In the special case of aiding a controller we may in addition simply switch the prediction off, if we enter a state-space region where no model has been placed yet. In other words: we make a fallback to traditional control.

The LEMON Architecture: As shown in figure 3, the modeling process of LEMON is a hierarchical one. The main modeling system consists of an ellipsoidal input clustering mechanism similar to [9] and [8]. One main feature of LEMON is its capability to handle multiple different models. The user may provide the system with several prototype models in conjunction with a selection rule base. Every time a new model is needed, the rule base is queried to decide on which model to choose. The rulebase may contain references to several static and dynamic model properties, like model complexity or model quality on current data.

During the online training process, model quality is augmented. If it keeps small or even decreases, the learning process starts to put submodels into the main models responsibility region according to the obviously badly mapped input/output pairs. Again the model selection is guided by the rule base mentioned above. If at the end the model ceases to work, thus model quality goes to zero, the model but not the submodels, gets deleted.

LEMON Learning Algorithm: As LEMON uses different kinds of models, no unique learning algorithm may be provided here. In fact, if using a dynamic learning paradigm like RProp [7], even each model derived from the same prototype model may possess different learning parameters.

But as model quality is a result of learning, the used learning mechanism is weaved into the online clustering mechanism. On the one hand this means model regions get expanded only if model prediction is good regarding the new

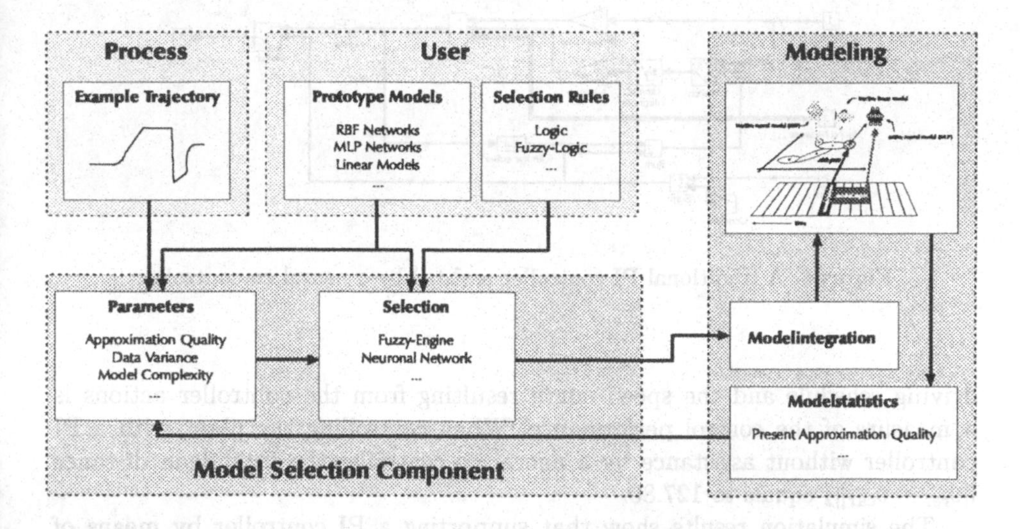

Figure3. The LEMON Architecture: User provided prototype models get combined by also given selection rules. The rule base may use parameters specific to the prototype models or their present modeling performance due to collected model statistics. The selected model gets integrated into the main modeling system after an initialization with already known input/output pairs has been done.

aquired region. On the other hand bad performing models get substituted by new models.

4 Simulation Results

As described above, ATNN, DRBF and LEMON neural networks are used to identify the inverse plant dynamics of the combustion engine. The training data set was produced by running the simulation with a predefined driving schedule and consists of the torque and the speed of the combustion engine as network inputs and the measured controller signal as target. The measurements were taken over a time interval of 50 seconds, one sample every 0.01 second. Both inputs as well as the target were scaled to the interval [0; 1].

To incorporate the trained neural networks into a controller of PI type we choose the adding of the network output to the proportional and integral part as shown in figure 4. As one can see, the anti-windup and limiter mechanism is located *behind* this point to take the network input into account as well. If the network is unable to produce a suitable output, the controller nevertheless minimizes the resulting control error by using the PI part. By this architecture we automatically provide a fallback mechanism if the network fails.

The control task is to follow a prescribed speed schedule as good as possible. The aim is to yield higher precision with the assistance of a neural co-controller. The Euklidean distance between the Matlab vectors v_{ref} and v_{car} describing the

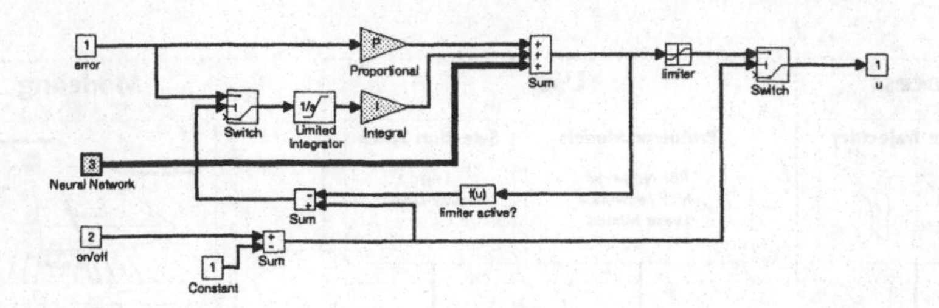

Figure4. **A traditional PI controller assisted by a neural co-controller**

driving schedule and the speed curve resulting from the controller actions is a measure of the control performance. When controlling the plant with a PI controller without assistance by a neural co-controller the Euklidean distance $\|v_{ref} - v_{car}\|_2$ equals to 127.80.

The simulation results show that supporting a PI controller by means of neural networks leads to a higher precision when following a prescribed speed table with a combustion engine on a testbed. In all cases we obtained a smaller Euklidean distance, that is $\|v_{ref} - v_{car}\|_2 = 118,48$ for the DRBF network, $\|v_{ref} - v_{car}\|_2 = 124.05$ when using an ATNN network and $\|v_{ref} - v_{car}\|_2 = 123.92$ in case of a LEMON network. Figures 5 and 6 show the target speed and the actual speed curve without neural co-controller and with neural co-controller of DRBF type respectively. The results are similar for the use of ATNN and LEMON networks.

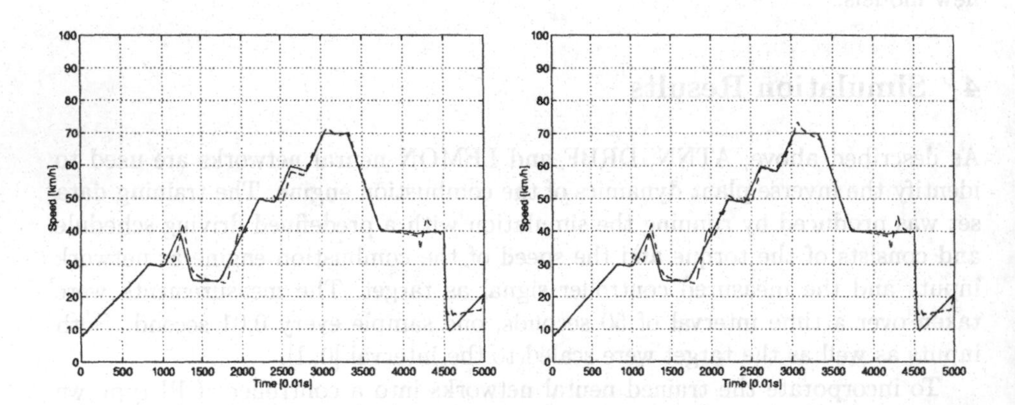

Figure5. **v_{ref} (solid line) and v_{car} (dashed line) when controlled with a PI controller only**

Figure6. **v_{ref} (solid line) and v_{car} (dashed line) when controlled with a PI controller assisted by a DRBF network**

The fact that the simulation results improve using neural networks of all three types investigated, proves that the architecture of a neural co-controller presented in this paper leads to better performance.

References

1. M. Ayoubi. Das dynamische Perzeptronmodell zur experimentellen Modellbildung nichtlinearer Prozesse. *Informatik Forschung und Entwicklung*, 12:14–22, 1997.
2. M. Ayoubi and R. Isermann. Radial Basis Function Networks with Distributed Dynamics for Nonlinear Dynamic System Identification. In *Third European Congress on Intelligent Techniques and Soft Computing (EUFIT)*, 1995.
3. Walter L. Baker and Jay A. Farrell. An introduction to connectionist learning control systems. In David A. White and Donald A. Sofge, editors, *Handbook of Intelligent Control*, chapter I, pages 35–63. VAN NOSTRAND REINHOLD, New York, 1992.
4. S. Chen, S. Billings, and P. Grant. Nonlinear System Identification using Neural Networks. *Int. J. Control*, 51(6), 1990.
5. Roderick Murray-Smith, editor. *Multiple model approaches to modelling and control*. Taylor & Francis, London, 1997.
6. K.S. Narendra, S. Kumpati, and K. Parthasarathy. Identification and Control of Dynamical Systems Using Neural Networks. *IEEE Trans. Neural Networks*, 1(1):4–27, 1990.
7. Martin Riedmiller and Heinrich Braun. A direct adaptive method for faster backpropagation learning: The rprop algorithm. In *Proceedings of the IEEE International Conference on Neural Networks 1993 (ICNN 93)*, 1993.
8. M. Sturm and K. Eder. Self-organizing process state detection for on-line adaptation tasks. In A. B. Bulsari, S. Kallio, and D. Tsaptsinos, editors, *Solving Engineering Problems with Neural Networks, Proceedings of the International Conference EANN ´96*, pages 33–36, London, 6 1996. Systeemitekniikan seura ry.
9. Michael Sturm and Till Brychcy. On-Line Prozeßraumkartierung mit ellipsoider Vektorquantisierung zur lokalen Modellbildung. In A. Grauel/W.Becker/F.Belli, editor, *Fuzzy-Neuro-Systeme ´97 - Computational Intelligence*, Proceedings in Artificial Intelligence, pages 463–470. Infix Verlag, St. Augustin, 1997.
10. C. Ungerer. *Neuronale Modellierung von Totzeiten in nichtlinearen dynamischen Systemen*. Diplomarbeit, Institut für Informatik, Technische Universiät München, 1998.
11. C. Ungerer. Identifying Time-Delays in Nonlinear Control Systems: An Application of the Adaptive Time-Delay Neural Network. In Masoud Mohammadian, editor, *Computational Intelligence for Modelling, Control and Automation (CIMCA)*. IOS Press, 1999.

Controlling Biological Wastewater Treatment Plants Using Fuzzy Control and Neural Networks

Michael Bongards

Cologne University of Applied Sciences, Campus Gummersbach,
Faculty of Electrical Engineering
Am Sandberg 1, D-51643 Gummersbach, Germany
bongards@gm.fh-koeln.de
http://www.bongards.de

Abstract. Improving the performance of wastewater treatment plants by optimising the control systems is a very cost efficient method but it involves risks caused by the time variant and non linear nature of the complex biochemical processes. Key problem is the removal of nitrogen combined with an optimal processing of the sludge water.
Different control strategies are discussed. A combination of neural network for predicting outflow values one hour in advance and a fuzzy controller for dosing the sludge water are presented. This design allows the construction of a highly non-linear predictive controller adapted to the behaviour of the controlled system with a relatively simple and easy to optimise fuzzy controller. The system has been successfully tested on a municipal wastewater treatment plant of 60.000 inhabitant equivalents.

1 Introduction

Reducing the cost of operation of industrial and municipal wastewater treatment plants plays a more and more important role in the design of automation systems. Furthermore the quality of effluent water must not decrease. Modifications of German wastewater quality regulations now under discussion will result in new boundary conditions:

- Until today it is sufficient to keep the quality of effluent wastewater below fixed limit values which are defined by governmental authorities.
- In the next future this procedure shall be replaced by new regulations in which the plant has to pay "pollution units" depending from the amount of biological and chemical load in the effluent water.

This new situation will force technical improvements of many wastewater treatment works which can be done by investments (1) in concrete, (2) in machinery, or (3) in intelligent control systems. The results presented in this paper show that option (3) is generating substantial improvements that can be achieved by moderate investments.

The complexity of a modern wastewater treatment plant can be compared with a biotechnological production plant. But there is an fundamental difference [1]: The industrial plant is producing a product and the operator tries to optimise quality and the requirement of time, energy and space by varying the process parameters.

Wastewater works are not producing but removing substances with purified water and sludge as "final products", so that the major design target is the security of operation. Furthermore the variation of process parameters like temperature or pH-value is not possible for economical and ecological reasons. The quality of the inflow is not controllable and its quantity can only be varied in a relatively small range by buffer basins. For an optimal operation wastewater works have to react on very dynamic variations of inflow parameters which shows the relevance of a well adapted closed loop control.

The biological wastewater treatment plant has to degrade the dissolved organic carbon and it has to remove the nitrogen and the phosphorus. Removing the nitrogen is the most difficult part of the process, if it works well then in most cases the effluent concentration of organic carbon is sufficiently low[2]. Phosphorus is in most cases eliminated by chemical precipitation [3, 4].

In this paper a method is presented to control the removal of nitrogen by combining fuzzy control with the capability of prediction by neural networks. At first the biochemical principles are described followed by the discussion of the controller system and by a presentation of first experimental results.

2 Removal of Nitrogen

The organic nitrogen is part of human and animal's excrements. In the sewer system it is transferred into ammonium (NH_4). Micro-organisms are modifying the substance firstly into nitrate (NO_3) and then into a biochemical neutral gaseous state (N_2) by a two step reaction [5]:

- First step – Nitrification:

$$2\,NH_4^+ + 3\,O_2 \rightarrow 2\,NO_2^- + 4\,H^+ + 2\,H_2O$$

$$2\,NO_2^- + O_2 \rightarrow 2\,NO_3^-$$

The nitrification process requires oxide conditions. The concentration of dissolved oxygen has to be in the range of 0.5 to 2 mg/l.

- Second step - Denitrification:

$$2\,NO_3^- + 2\,H^+ + 10\,H \rightarrow N_2\uparrow + 6\,H_2O$$

The denitrification needs anoxic conditions, so that the concentration of dissolved oxygen has to be as low as possible. Sewage works use different methods for the technical implementation like cascaded, alternating or simultaneous denitrification [6]. In the plant optimised during this project reactions of denitrification and nitrification are proceeded sequentially in consecutive basins.

Unfortunately only 2/3 of the inflowing nitrogen is diffusing in a gaseous state into the atmosphere. 1/3 is remaining in the sludge which has to be treated separately [7]. The sludge must be dewatered for further treatment in landfill or combusting. Water is removed by sedimentation, filter-pressing or centrifugation. This sludge water has to be recirculated into the inflow of the plant because it has an extremely high concentration of nitrogen. If the plant is working at the limits of its capacity this is a very critical step which has to be controlled by an appropriate strategy.

3 Control Strategies

If the recirculation of the sludge water into the inflow of the plant is done in the daytime during normal hours of operation the ammonium load at the inflow can be increased by 30% or more. This results in an increase of the nitrogen outflow value so that the limit values can be exceeded.
To handle this problem a buffer tank with a capacity of several production days of sludge water is necessary. Furthermore an appropriate strategy of dosing the sludge water into the inflow has to be developed. Fig. 1 shows a sketch of the main elements of the controlled system.

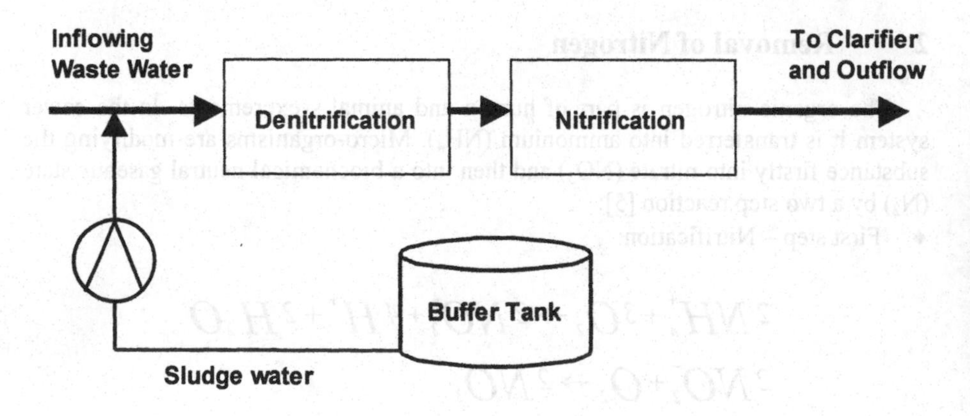

Fig. 1. Water way and sludge water way in the controlled system – part of the wastewater works.

The behaviour of the controlled system, the biological part of the wastewater treatment plant, is only very roughly predictable: The activity of the micro-organisms depends on the concentration and biochemical composition of the biologically active sludge and on the amount of biologically degradable biomass which is of central importance for the denitrification. In a more general view the "history" of all these parameters has a decisive influence on the reaction because it determines the composition of the biomass which removes the nitrogen. A typical growth rate of the micro-organisms varies between 1/day to 6/day [8]. The higher value indicates that a doubling of the population of micro-organisms takes place in

less than 4 hours so that during a day the micro-organisms will continuously adapt to changes in their environment.
The simplest control approach is an open loop control using time controlled switches for dosing of the sludge water in times of low biomass load into the inflow. But this solution has two major disadvantages:

- Caused by the time dependent fitness of the complete system the switching program has to be adapted manually from day to day to the capacity of the plant – a time consuming and often neglected work.
- At night in times with low inflowing load the concentration of the natural carbon sources can be too low so that the process of denitrification only works well if additional carbon sources like ethanol are added, causing additional costs of operation.

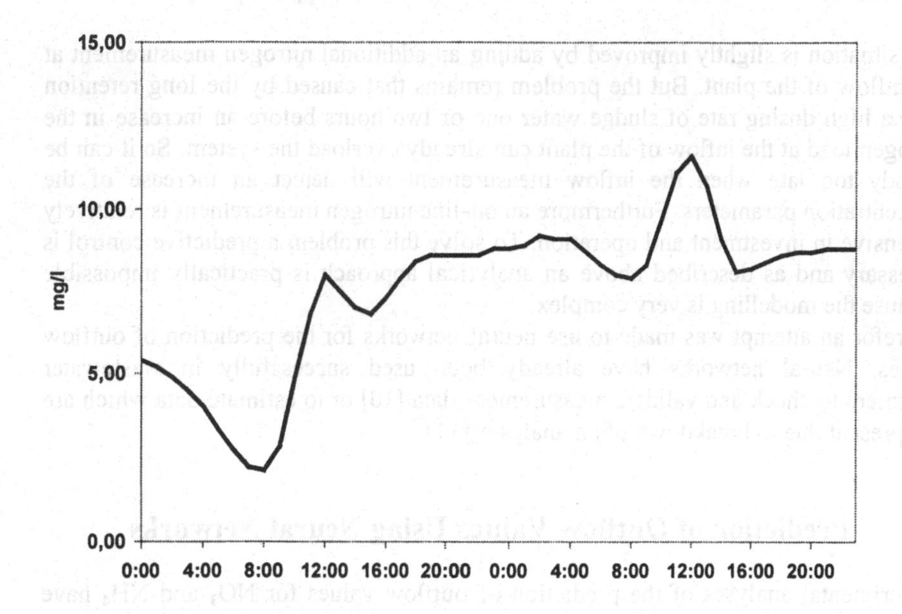

Fig. 2. Concentration of NH4 measured over 2 days at the outflow (18.03. – 19.03.98)

Fig. 2 shows a typical graph of the concentration of ammonium in the outflow over two days. As it can be seen the behaviour is highly non-linear and there is no periodicity from day to day. To handle this system the simple open-loop-control must be tuned to a very careful and conservative mode of operation, resulting in problems in the operation of the plant: Insufficient amounts of sludge water are pumped out of the buffer tank so that the sludge pressing has to be reduced.

A more sophisticated approach uses a closed loop control by on-line measurement of the concentration of nitrogen at the outflow of the aeration basin. The control-system normally is a standard PID-controller or a fuzzy controller.
This solution cannot convince in practical applications:

- Due to the continuous changes in the behaviour of the biomass the controller has to be adapted continuously to the process, a very complex task in every day operation.
- Optimising a controller by simulation using an analytical model has made a lot of progress in the last years [9], but it remains a nearly endless job of high complexity caused by the various model parameters to be adapted and by the continuous modifications in system behaviour.
- The retention time of the system is in the range of one to ten hours mainly depending from the amount of inflow into the plant. So especially in the night times a high retention time makes an effective control impossible. The dosing of sludge water has to be reduced one or several hours before an increase of the nitrogen load in the inflow and without integrating a prediction into the control system an increase of the outflow concentration will happen.

The situation is slightly improved by adding an additional nitrogen measurement at the inflow of the plant. But the problem remains that caused by the long retention time a high dosing rate of sludge water one or two hours before an increase in the nitrogen load at the inflow of the plant can already overload the system. So it can be already too late when the inflow measurement will detect an increase of the concentration parameters. Furthermore an on-line nitrogen measurement is relatively expensive in investment and operation. To solve this problem a predictive control is necessary and as described above an analytical approach is practically impossible because the modelling is very complex.
Therefor an attempt was made to use neural networks for the prediction of outflow values. Neural networks have already been used successfully in wastewater treatment to check and validate measurement data [10] or to estimate data which are not present due to breakdown of an analyser [11].

4 Prediction of Outflow Values Using Neural Networks

Experimental analyses of the prediction of outflow values for NO_3 and NH_4 have been made in 1998 at the wastewater treatment plant Krummenohl in Gummersbach (Germany) [12]. The plant is purifying mainly municipal and some industrial wastewater of 60.000 inhabitant equivalents. In the year 2001 the wastewater works has to guarantee new and sharper effluent values which was the initial motivation for the research presented in this paper.
On the process control system of the plant 8 on-line measurements of the water way were continuously monitored and stored at a 5 minutes time interval: Temperature, conductivity, flow-rate and pH-values at the inflows, the oxygen-concentration in the aerated nitrification area and the ammonium and nitrate concentration at the outflow.
In a first step all these data were used to train a neural network for predicting the ammonium and nitrate concentration at the outflow one hour in advance.
A commercially available neural network software was integrated in Microsoft-Excel-worksheets which made its application easy to use. Multilayer perceptrons

(MLPs) were trained with static backpropagation. Also Jordan / Elman networks were used but they showed no significant improvement of the prediction quality.
A sensitivity analysis of the input data over a complete month of operation showed that most of the input data are of only small relevance to the predicted outflow values. So the network could be simplified by only using the flow-rate at the inflows, the oxygen-concentration in the aerated nitrification area and the ammonium and nitrate concentration at the outflow as input data. Best results were achieved by configuring the neural network with two hidden layers of 15 respectively 10 neurones.
Fig. 3 shows the prediction of nitrate concentration using a trained neural network in comparison with the measured values. The error of prediction is below 0.3 mg/l which is a well acceptable value because the measurement error of the on-line-monitoring system in everyday operation is a value in the magnitude of 0.5 mg/l.
The predicted values of ammonium are of similar precision so that the neural network could be installed as part of a predictive control system which is described in the following chapter.

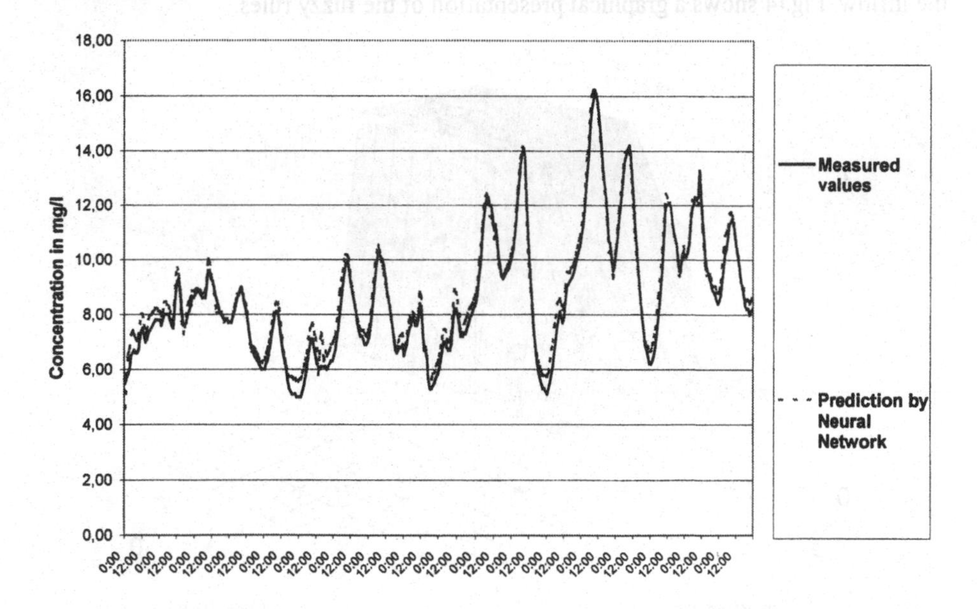

Fig. 3. Measurement and neural-network-prediction of nitrate (NO_3). Outflow values over 17 days of operation in March 1998

5 Fuzzy Control for Dosing the Sludge Water

Fuzzy controllers have become a well accepted element in controlling the highly non-linear systems in a wastewater treatment plant [13]. A general problem in the application of this technology is the experimental "try and error" approach in design and optimisation of the fuzzy rules. The adaptation of a controller to the process is a time-consuming affair which becomes even more complex in systems with many

rules. Controlling the removal of nitrogen and phosphorus can require a complex fuzzy controller of 47 rules [14].
Concerning the dosing of sludge water the complex non-linear behaviour of the elements of the wastewater works is integrated already in the neural network which predicts the outflow values. Therefor a more simple fuzzy controller could be generated out of 14 rules that is based on the following principal ideas:

- If the buffer tank is nearly full, as much water as possible shall be dosed, otherwise the dosing can be done in a more conservative low-risk-strategy.
- If the predictions for ammonium (NH_4) and total nitrogen, the sum of ammonium and nitrate (NO_3), are low, the dosing should be high, otherwise it should be low.

The controller needs as input only three parameters: the on-line measured level of the buffer tank and the one hour prediction of ammonium and total nitrogen. Output of the controller is a setting value for the amount of sludge water to be pumped into the inflow. Fig. 4 shows a graphical presentation of the fuzzy rules.

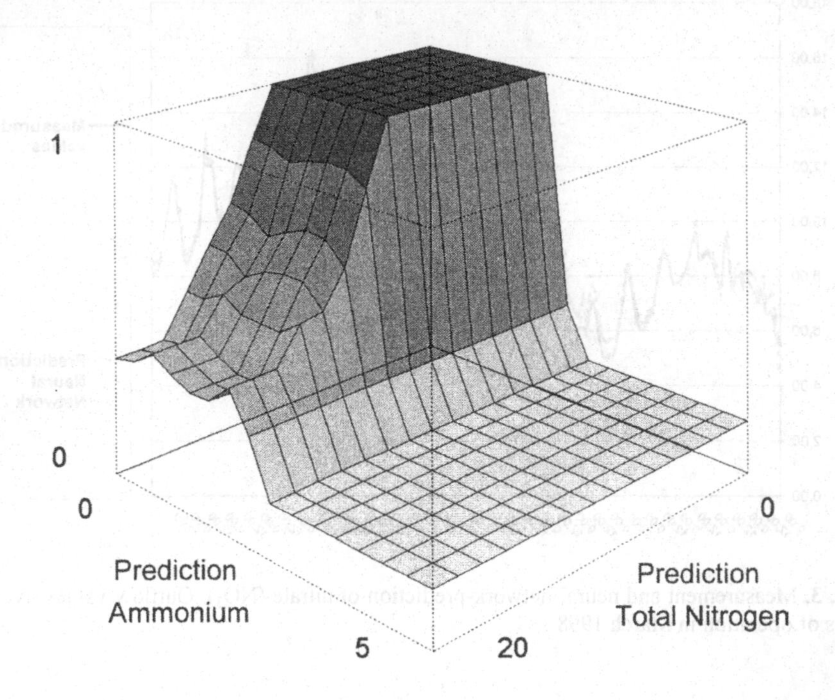

Fig. 4. Fuzzy rules in graphic form. Vertical axis is the amount of pumped sludge water out of the buffer tank (Graphic calculated for a level of 85% in the buffer tank).

6 Experimental Results

The combination of neural network and fuzzy controller was installed as a prototype on the wastewater treatment plant Krummenohl and was operated for several weeks. The main results can be summarised as following:

- Before installation the new controller every two days the time switches for controlling the pump of the buffer tank had to be adjusted. This work has become obsolete.
- The amount of pumped sludge water remained the same because this is only determined by the amount of sludge pressed which did not change. But the buffer tank was mostly empty or half filled. Before installation of the new controller the buffer tank often was full at the end of the week so that sludge only could be pressed in the evening or at the weekend.
- The oxygen concentration in the nitrification tank remained low during the night. Formerly it increased up to 4 times its nominal value because the organic load in the night was very low. Now high amounts of sludge water are pumped into the inflow during night hours which equilibrates the load.
- The outflow concentration of nitrogen was always kept clearly within the governmental requirements.

The system has proved as a very reliable tool for controlling the complex behaviour of the wastewater works. The laboratory equipment installed on the plant will now be replaced by an industrial version.

7 Conclusions

The combination of neural network and fuzzy controller has proved as a robust and highly adaptive tool to control the complex and extremely non-linear processes at a wastewater treatment plant. Integrating the neural network avoids the necessity of designing complex fuzzy controllers with many input parameters and rules. Furthermore the neural network allows a reliable prediction of process values so that predictive control is possible. This predictive control is especially valuable on wastewater works where strongly varying retention times of several hours occur.

In the near future the neural network will be applied to other wastewater works with partially different control tasks to test its applicability.

8 Acknowledgement

We like to thank the "Ministerium fuer Schule und Weiterbildung, Wissenschaft und Forschung des Landes Nordrhein-Westfalen" for its grant which made this work possible. We are grateful to the Aggerverband (Gummersbach) for the permission to operate our test equipment on their wastewater works.

References

1. Svardal, K.: Meß-, Regel- und Steuerungsstrategien (Measurement, Instrumentation and Control Strategies), Korrespondenz Abwasser **9** (1994) 1586-1596
2. Poepel, H. J.: Einfluss der Betriebsfuehrung von Abwasserbehandlungsanlagen auf die Ablaufguete (Influence of the Operation of Wastewater Treatment Plants on the Quality of the Effluent), Schriftenreihe WAR - Darmstadt **75** (1994) 13-39
3 Automatisierung der chemischen Phosphatelimination (Automation of Chemical Phosphate Elimination), ATV-Merkblatt M 206 – Hennef (1994)
4 Poepel, H. J.: Grundlagen der chemisch-physikalischen Phosphorelimination (Principles of the Chemical and Physical Elimination of Phosphorus), Schriftenreihe WAR - Darmstadt **51** (1991) 15-49
5 Mudrack, K., Kunst, S.: Biologie der Abwasserreinigung (Biology of Wastewater Treatment) G. Fischer Verlag Stuttgart (1991)
6 Köhne, M.: Analyse und Regelung biologischer Abwasserreinigungsprozesse in Kläranlagen (Analysis and Control of Biological Wastewater Treatment Processes in Sewage Plants) at-Automatisierungstechnik **46** (1998) 215-234
7 Cornel, P., Ante, A.: Reduzierung der Stickstoffrueckbelastung durch separate Schlammwasserbehandlung (Reducing the Load of Recirculated Nitrogen by a Separate Treatment of the Sludge Water), Schriftenreihe WAR - Darmstadt **108** (1998) 173-197
8 Henze, M. et al.: Activated Sludge Model No. 2. IAWQ - International Association on Water Quality London (1995)
9 Jumar, U., Alex, J., Tschepetzki, R.: Zur Nutzung von On-line-Modellen für die Regelung von Kläranlagen (Using On-line Models to Control Wastewater Treatment Plants), at-Automatisierungstechnik **46** (1998) 235-244
10 Haeck, M., Koehne, M.: Lokale und globale Validierung der Meßsignale kontinuierlich arbeitender Prozeßanalysatoren (Local and Global Validation of Measurement Signals of Continuously Working Process Analyzers), at-Automatisierungstechnik **46** (1998) 315-325
11 Haeck, M., Rahrbach, B., Koehne, M.: Schaetzung wichtiger Prozessgroessen der Abwasserreinigung mit Hilfe neuronaler Netze (Estimating Important Process Parameters in Wastewater Treatment with the help of Neural Networks), Korrespondenz Abwasser **3** (1996) 368-378
12 Graner, M., Weise, M.: Entwicklung und Optimierung einer praediktiven Regelung mit kuenstlichen Neuronalen Netzen und Fuzzy-Control zur Verfahrensoptimierung auf Klaeranlagen (Development and Optimisation of a Predictive Control using Artificial Neural Networks and Fuzzy Control to Optimise the Processes of a Wastewater Treatment Plant); Diplomarbeit, Cologne University of Applied Sciences, unpublished (1998)
13 Schmitt, T.G., Hansen, J.: Fuzzy Control und Neuronale Netze zur Optimierung der Stickstoff- und Phosphorelimination (Fuzzy Control and Neural Networks used for Optimising the Nitrogen and Phosphorus Elimination), Schriftenreihe Universität Kaiserslautern, **11**, (1998)
14 Hansen, J.: Fuzzy-Regelungen zur Optimierung der Nährstoffelimination in kommunalen Kläranlagen (Fuzzy Controls for the Optimization of Nutrients Removal in Municipal Sewage Treatment Plants) Korrespondenz Abwasser **12** (1998) 2259-2268

A New Fuzzy Character Segmentation Algorithm for Persian / Arabic Typed Texts

Mohammad Bagher Menhaj , Farbod Razzazi

Electrical Engineering Department of Amirkabir University of Technology ,
Hafez Ave. No.424 Tehran 15914 Iran
{menhaj,r7623915}@cic.aku.ac.ir

Abstract. This paper introduces a new off-line fuzzy segmentation algorithm to separate cursive Persian / Arabic typed texts. Precise boundary detection of typed characters is the main feature of the proposed method. The method is applied into some sample texts. The simulation results are very promising indicating 95% error free segmentation performance.

1 Introduction

Automatic recognition of typed characters with low probability of error is one of the main concerns in OCR systems. Almost all official texts and books are print formatted, hence automatic machine storing of the above information becomes very important.

As writing has been invented to convey information among mankind, computerized writing recognition algorithms try to reproduce intelligent human techniques. One of the human algorithm characteristics is that they are tolerant to pattern variations, and noise or distortion. To model these characteristics, fuzzy recognition methods seem to be effective.[1]-[3].

A general text recognition procedure block diagram is shown in Fig. 1. The input image is preprocessed first. Preprocessing includes classical image enhancement and noise reduction methods. The simplest model for this block is a 2 dimensional low pass filter. In the next step, recognizing elements (e.g. characters) are separated in the segmentation block. The required features for the decision making block are then extracted. Finally, a decision making algorithm is applied into each segmented element based on the extracted features. The segmentation block of the above procedure is the main concern of this paper.

Text recognition methods are categorized in some aspects. One aspect is typed text recognition versus handwritten text recognition. Although the recognition principles are the same in both cases, but due to the variety and complexity of handwritten texts, the algorithms used for handwritten text recognition are more complicated and the segmentation process principles are different.

Another classification is on-line recognition methods versus off-line recognition methods. On-line methods need faster algorithms and have more features to recognize characters (e.g. tracking pen motion and emphasis's). On the other side off-line methods can be more time consuming while using less features to distinguish characters.

The most important classification in segmentation problem is cursive character recognition versus recognition of separated characters. In non-cursive texts, characters are written separately and the segmentation process is limited to track separated contours of an image. On the other hand, in a cursive text the boundaries of each character is not definitely determined and more complex segmentation algorithms are needed.

Two segmentation approaches are common in the literature. The first is based on language standard characters.[8] This approach is common when isolated characters are attached to make a cursive text and connection points are practically distinguishable (e.g. typed texts). Another approach is based on separating smaller writing units than characters which are the distance between two pen emphasis's or breaking points.[4-7] This approach is used for separating cursive handwritten characters.

Persian/Arabic texts are different from Latin texts in many aspects that cause some difficulties in the segmentation process. Some of these characteristics are different character shapes due to character location on the word, variety of Persian / Arabic fonts, cursive writing in both typed and handwritten texts.

This paper presents a new off-line fuzzy segmentation algorithm for separating cursive Persian/Arabic typed characters. Most of the literature in this area are concentrated on cursive handwritten English texts [1]-[3], and almost all of them employ structural features of handwriting. Therefore their segmentation and recognition lexicons are not natural language characters. Persian texts are naturally written cursively; hence the first step in Persian recognition is to separate the cursive characters[4]-[8]. Because the typed characters have specific fixed shapes, the lexicons selected here are the characters themselves. This, in turn, simplifies the recognition procedure.

In [8] a segmentation algorithm has been presented for Persian typed texts. This algorithm, because of its crisp nature, cannot locate the character boundaries precisely . This problem makes the recognition phase more complicated. Since this phenomena is inherently qualitative requiring soft decision making process. The main feature of our method is that it increases detection precision of character boundaries using fuzzy reasoning.

The remainder of this paper is organized as follows. Segmentation algorithm, it's ideas and concepts are described in section 2. Section 3, presents fuzzy reasoning implementation, and experimental results and practical considerations are presented in section 4. Finally section 5 concludes the paper.

2 Segmentation Algorithm

Segmentation is employed as a preprocessing stage to prepare recognizable objects for recognition stage. Figure 2 presents a block diagram representation of the segmentation algorithm used in this paper.

As observed in the figure, text lines are first separated using vertical histogram and the other steps such as subword segmentation, dot elimination, ... are then applied to separated text lines. Vertical histogram is calculated by counting black pixels in each image horizontal line. Here, the picture is considered black and white, however if it is in a gray-scale format, it should be thresholded in order to achieve a black and white image. It is expected to have no black pixels between two subsequent lines. However because of noise effects, an experimental threshold level is determined to detect line boundaries. The baseline of each text line is also detected in this step. Baseline in each text line is determined by finding the maximum density image line. In the next step, horizontal histogram is calculated and in a similar manner, non-overlapped words (or subwords) are separated in each text line.

Most of the Persian / Arabic characters have some dots. These dots in most cases are the only difference between characters. But because there exist no role in segmentation procedure for dots, upper or lower dots of subwords are omitted using their surrounding contour lengths considered as recognition features. The contour length threshold is extracted experimentally. Text dots are restored in the recognition phase to distinguish similar characters.

After the above preprocessing, cursive characters in subwords are disconnected in the first cursive segmentation step, and then in the second step, overlapped non-connected characters are separated using a standard contour tracing algorithm. In the first step, the following features are used to recognize character boundaries:

1- Average height
2- Baseline distance

Average height is calculated window-wise. It demonstrates the average width of the character in that window. The average window height is used instead of point height to make the algorithm more robust against image noise and distortion. The window is moved on the subword to determine the average height in all breaking point candidates. To avoid the error caused by overlapped connected characters, the average height is calculated in a window whose height is much smaller than that of the character itself. The height of this window is determined experimentally. If no candidate point is detected for a giver maximum length of a character, the window size is reduced while the window height is increased and the procedure is performed again. After detecting a boundary point, the new boundary is searched after finding an over threshold point in this line of the image. In this manner, it is expected to measure the pen writing widths at the character boundaries.

Baseline distance is indeed the distance from pen writing line to baseline. It is expected to have small values for both features at the character boundaries. In the ideal case, the average height should be minimum and the pen tracking contour has to cross the baseline (i.e. baseline distance = 0). In the next section, the reasoning procedure is described.

3 Fuzzy Reasoning

Fuzzy set theory is used to characterize and model ambiguous phenomena. Hence noisy and non deterministic pattern recognition problems are well fitted to this theory. Some papers have employed this theory in character recognition. In [3] Zimmermann presented a practical example of fuzzy isolated typed character recognition with about 8% probability of error. Fuzzy Farsi / Arabic handwritten character recognition are described in [4] - [7]. In [8] a Fuzzy Latin handwritten segmentation method is proposed. Hence it's features are based on handwritten structure and nonstandard lexicons.

In classical segmentation algorithms, reasoning is based on finding a threshold for both baseline distance and average height conditions[8]. If baseline distance and average height is smaller than a given threshold, then a character boundary will be detected. Whereas the algorithm presented in this paper uses a fuzzy reasoning to find the characters boundaries. The smaller the average height and baseline distance is, the greater will be the membership of the testing point to the character boundaries set. The membership functions used in this paper are shown in figure 4. Figures 4.a and 4.b are crisp intervals used in classical algorithm. figure 4.c and 4.d are the proposed membership functions. The ideal case is that the pen writing line is on the baseline and the average height is zero. Deviation from these ideal cases cause the membership values to be reduced. Parameters of the membership functions are determined by experiment. The importance of this algorithm is the use of crisp features which has dramatically reduced the required computations.

A membership value is calculated for each point of the text line in each subword as the minimal of the membership values of the two features. Candidate boundary points are selected if the membership values of those points are greater than a given threshold. The character boundaries are selected from local maxima in the set of candidate boundary points membership values. To avoid spurious boundary detections, some conditions on minimum and maximum size of characters have been developed.

4 Experimental Results

The proposed method has been implemented on a Pentium 166 PC and has been tested on a 300 dpi scanned normal typed texts. The normalized threshold value for line and word segmentation is determined equal to 0.02. The contour length for recognizing texture dots is assumed 20 dots. It is evident that this parameter is strongly font size dependent. Using this parameter, no error is observed in the text. Furthermore, no error in line and subword segmentation stage has been occurred

In the connected subword character segmentation stage, the necessary parameters to run the algorithm are estimated for the scanned text and experimentally corrected based on some trials and errors. The typical parameters are given in table 1. These parameters are indeed font dependent.

In the last stage, overlapped non-connceted characters should be separated. A recursive painting algorithm is implemented for this part. After segmentation of each character, a predetermined length is added up to both sides of the character. This leads to reduction of the probability of missing a part of the character.

In the test described above, the algorithm has shown 95% error free segmentation performance. The errors are in some special character pairs, like(áÂ¡àÂ) and in some cases the character (Ï) is segmented into two parts . Because these errors are not widely distributed, they can be separated using some other special features, or even they can be considered as a new lexicon in the recognition phase.

A sample of the segmented text has been presented in figure 5. Segmentation errors were underlined. All other characters have been correctly segmented.

5 Conclusion

In this paper, a fuzzy character segmentation algorithm for typed Farsi / Arabic texts was presented. The algorithm has been tested on some Persian sample texts The simulation results illustrate high performance of the proposed algorithm. The method is easily applicable to OCR systems used in formal writtings. Automatic evaluation of font dependent parameters and suitable fuzzy membership function selection are currently under development by the authors and will be reported in near future.

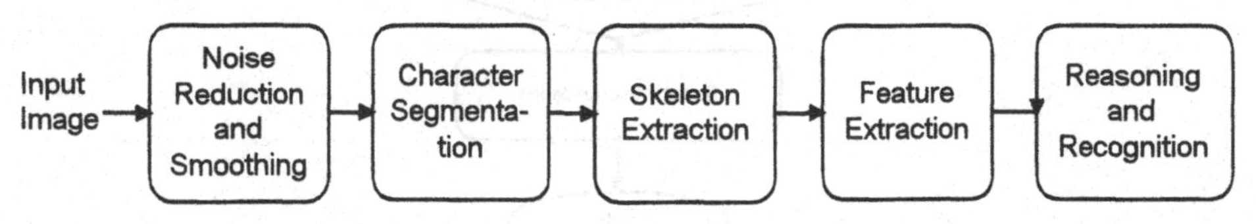

Figure 1 - Block Diagram of a Typical Character Recognition System

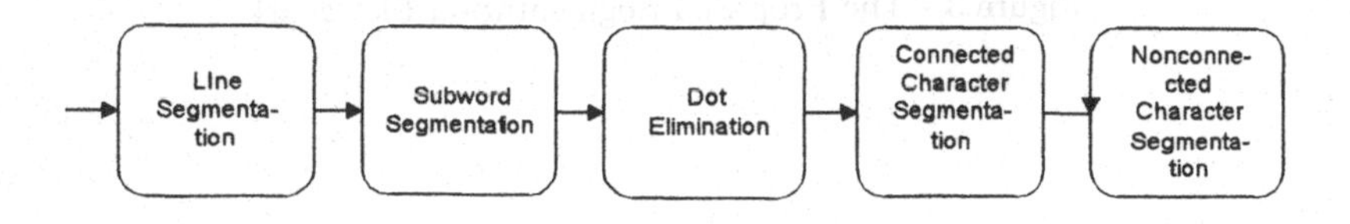

Figure 2 - Proposed Segmentation Process Block Diagram

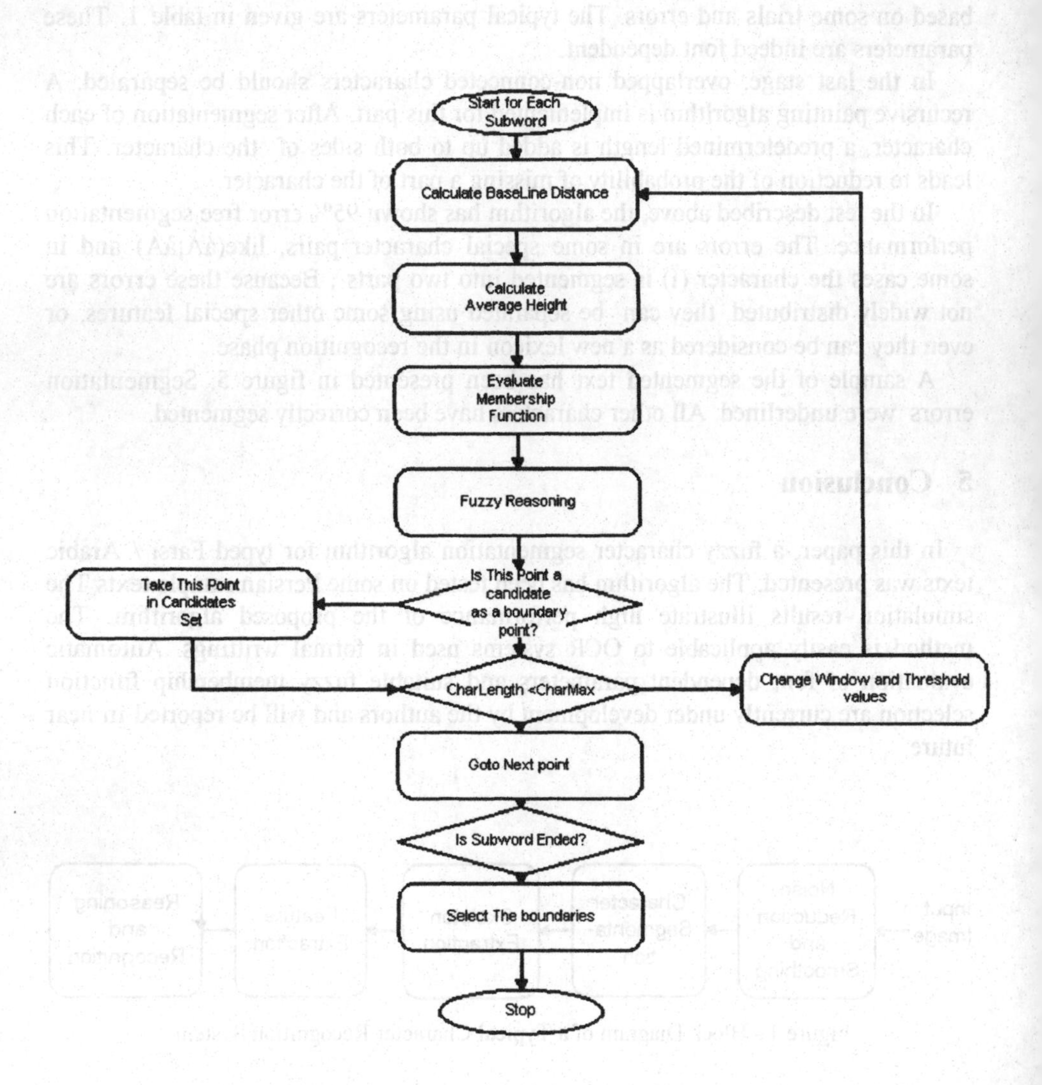

Figure 3 - The Proposed Segmentation Flowchart

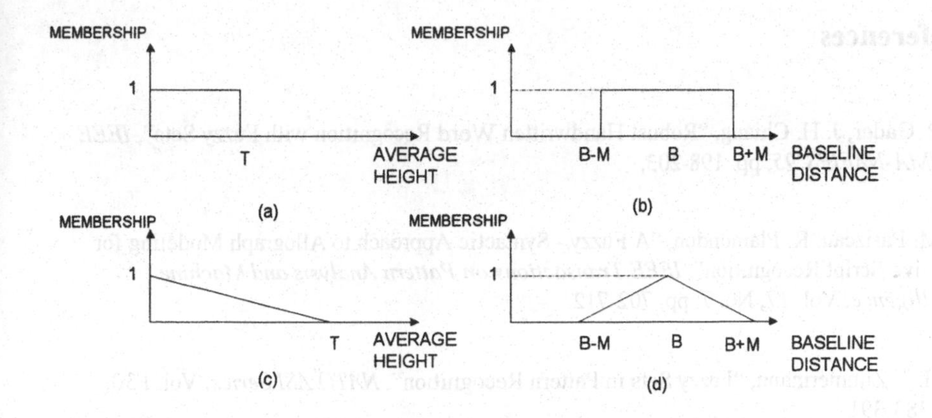

Figure 4 - Membership function(a) Crisp interval for average height (b) Crisp interval for baseline distance (c) Fuzzy membership for average height (d)Fuzzymembership for baseline distance

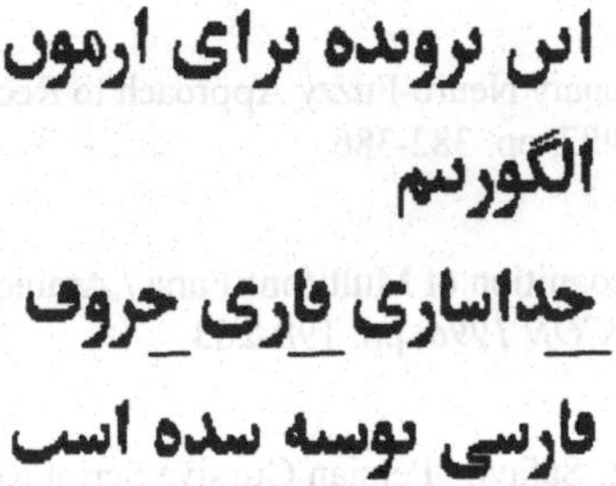

اں برونده برای ارمون

الگورتم

حداساری فاری حروف

فارسی بوسه سده اسب

Figure 5: Result of segmentation algorithm on a sample text (Errors are underlined)

Table 1-Typical Method Parameters

Parameter	Value	Description
W	2	Window Width
T	5.4	Membership Parameter
M	2	Membership Parameter
S	11	Window Height
CharMax	15	Maximum Length of a Character
CharMin	10	Minimum Length of a Character
L	3	Character length increasement

References

1- P. Gader, J. H. Chiang, "Robust Handwritten Word Recognition with Fuzzy Sets", *IEEE ISUMA-NAFIPS'95*, pp. 198-203.

2- M. Parizeau, R. Plamondon, "A Fuzzy - Syntactic Approach to Allograph Modeling for Cursive Script Recognition", *IEEE Transactions on Pattern Analysis and Machine Intelligence*, Vol. 17, No. 7, pp. 702-712.

3- H. J. Zimmermann, "Fuzzy Sets in Pattern Recognition", *NATO ASI Series*, Vol. F30, pp. 383-391.

4- H. Gorain, M. Usher, S. Al. Emami, "Off-line Arabic Character Recognition", *IEEE Computer Magazine*, June 1992, pp.71-74.

5- A. M. Alimi, "A Neuro - Fuzzy Approach to Recognize Arabic Handwritten Characters", *IEEE ICNN 1997*, pp. 1397-1400.

6- A. M. Alimi, "An Evolutionary Neuro-Fuzzy Approach to Recognize On-Line Arabic Handwriting", *IEEE IAPR 1997*, pp. 382-386.

7- M. Namazi, K. Faez, "Recognition of Multifont Farsi / Arabic Characters Using a Fuzzy Neural Network", *IEEE TENCON 1996*, pp. 198-203.

8- M. Hashemi, O. Fatemi, R. Safavi, "Persian Cursive Script Recognition", *IEEE Document Analysis and Recognition Conference 1995*, pp. 869-873.

RePART: A Modified Fuzzy ARTMAP for Pattern Recognition

Anne Canuto, Gareth Howells and Michael Fairhurst

University of Kent at Canterbury, Canterbury CT2 7NT, Kent - UK
{amdc1,W.G.J.Howells and M.C.Fairhurst}ukc.ac.uk

Abstract. Fuzzy ARTMAP has been proposed as a neural network architecture for supervised learning of recognition categories and multidimensional maps in response to arbitrary sequences of analog or binary input vectors [12]. In this paper, RePART, a proposal for a variant of Fuzzy ARTMAP is analysed. As in ARTMAP-IC, this variant uses distributed code processing and instance counting in order to calculate the set of neurons used to predict untrained data. However, it additionally uses a reward/punishment process and takes into account every neuron in the calculation process.. . .

1 Introduction

Fuzzy Logic [1][2] and Neural Networks [3][4] are complementary technologies in the design of intelligent systems. The combination of these two technologies into an integrated system appears to be a promising path toward the development of Intelligent Systems capable of capturing qualities characterizing the human brain. Neural fuzzy systems [5][6] have attracted the growing interest of the researches in various scientific and engineering areas. Specially in the area of Pattern Recognition hybrid neuro-fuzzy systems seems to be gaining a lot of interests [7][8][9].

The theory of adaptive resonance began with an analysis of human cognitive information processing and stable coding in a complex input environment [15], starting with ART 1 model [10]. Subsequently, several other ART-based models have emerged, such as: ARTMAP [11], Fuzzy ARTMAP [12], ART-EMAP [13], ARTMAP-IC [14], PROBART[16].

In this paper, RePART, a variant of Fuzzy ARTMAP is investigated. Essentially, RePART adds distributed code processing, instance counting (as in ARTMAP-IC) and a reward/punishment process to the standard fuzzy ARTMAP. This paper has been sub-divided as follows. The first section introduces the main aim of this paper. The second section describes ARTMAP and fuzzy ARTMAP models. The third section presents some extensions of the ARTMAP model as well as RePART. The fourth section analyses some experiments that have been performed to investigate the behaviour of the RePART model in comparison with the standard Fuzzy ARTMAP and ARTMAP-IC. Finally, the last section presents the final remarks of this paper.

2 ARTMAP and Fuzzy ARTMAP models

An ARTMAP network is used for supervised learning, self-organising a mapping from input vectors to output vectors. The original binary ARTMAP [11] incorporates two ART1 modules, ARTa (the input vector is received as module input) and ARTb (the desired output vector is received as module input), that are linked by a map field module. This module forms associations between categories from both ART Modules via a technique known as *outstar learning.*

The ARTMAP model works as follows. Input and desired output are clamped into ARTa and ARTb respectively. Once both ART modules produce their output, the map field model forms the association between both winner categories (from ARTa and ARTb). In case of failure in making the correct prediction, the search is triggered and a match tracking process occurs. The match tracking process increases the ARTa vigilance parameter in order to avoid a subsequent activation of the same category in ARTa. That is, a reset in ARTa category is performed and a search for another category is activated [11].

Fuzzy ARTMAP is a variant of the ARTMAP model which has the same abstract structure (two ART modules and a map field link) as ARTMAP, but makes use of the operations of fuzzy set theory [1][2], instead of those of classical set theory to govern the dynamics of ARTMAP [12]. The fuzzy ARTMAP system is especially suited to classification problems and is capable of learning autonomously in a non-stationary environment. Fuzzy ARTMAP has been widely applied in several tasks, mainly in the Pattern Recognition area [17],[18],[19],[20],[21],[22].

3 Some Extensions of the ARTMAP model

There are several extensions of ARTMAP model, such as: fuzzy ARTMAP [12], ART-EMAP [13], ARTMAP-IC [14], dART [23], PROBART [16], and so on. Almost every extension employs the same learning phase as the ARTMAP model, with improvements made in the recall phase. For instance, ART-EMAP and ARTMAP-IC (that are also extensions of Fuzzy ARTMAP), among other changes, employ a distributed code in order to define the winner node in the recalling phase. In distributed code, instead of only one winner (winner-take-all method), several neurons are described as winners and are used, along with the correspondent map field weights, to produce the final result. The main motivation for the use of distributed code is that in a winner-take-all method, a situation where other nodes, apart from the winner, can also satisfy the vigilance criterion for a given input I. Therefore, a better performance could be reached if all the nodes that satisfy the vigilance criterion are used to predict outputs, see [14].

Along with distributed code, ARTMAP-IC also uses a category instance counting which distributes prediction according to the number of training set inputs classified by each F_2 node. In other words, It is the frequency of activation of a F_2 node in ARTa during learning phase. During recall, instance counting is used along with y_i (output of ith neuron of the F_2 layer in ARTa) to the winner node choice process. In other words, instance counting is one more parameter to take into account during the winner node choice process.

One of the main problems of the distributed code method is to define the number of neurons that compose the winner's group. That is, the number of neurons that must co-operate in order to make a prediction of the best category in ARTa. Several methods to define the number of nodes can be used, such as: a Q-max rule or a rule of thumb [14]. However, in some situations, mainly in a pattern recognition task, a good prediction of an untrained pattern is not done just with the chosen set of neurons (an input pattern can be equally similar to more than one category). Thus, one question arises, why does it not take into account every F_2 node in the prediction process when every node was important in the training phase?

3.1 RePART

The RePART model employs distributed code processing and instance counting in order to calculate the set of neurons used to predict untrained data. However, it takes into account every neuron in the calculation process, using a reward/punishment process. If a neuron belongs to the set of winner neurons, its correspondent map field node will be rewarded, otherwise it will be punished. The main idea of this work is the following:

- To decreasingly rank the neurons, according to their output, from the most similar to the least similar;
- To divide the neurons into two groups: winners and losers;
- To calculate the map field output node rewarding correspondent nodes which belong to winners group and punishing correspondent nodes which belong to the losers group.

The proposed modification can be explained as follow.

1. According to the ranking of neurons, choose the N first neurons to belong to the set of winners. After that, calculate map field node output, taking into account the set of winners.

$$U_k = \sum_{j \in Wi} \left(\frac{w_{jk} c_j T_j}{c_j T_j} + w_{jk} R_j\right), \tag{1}$$

where:
w_{jk} is the weight from the jth F_2 layer neuron to the kth map field layer;
c_j is the instance counting of the jth F_2 layer neuron;
T_j is the output of the F_2 layer neuron;
Wi represents the set of winners;
R is the reward parameter.
The F_2 layer neuron output (T) is taken from its own output after a contrast enhancement function. The R parameter is used as a reward for the map field neuron because its corresponding neuron is in the set of winners. The

intensity of the reward depends on the position of the neuron in the set of winners, the first winner (with the biggest output) will have the biggest reward. The reward equation is defined as follows:

$$R_j = (T_j * \frac{NumWin}{NumWin + k_j^w})^2, \quad (2)$$

where:
T_j is the enhanced node output;
NumWin: is the number of neurons in the set of winners;
k_j^w: is the ranking of the jth neuron in the set of winners (its position in the set of winners).

2. An analogous process to that of equation 1 is performed, but using the remaining nodes of the rank (set of losers) and the punishment parameter.

$$U_k = \sum_{j \in L} (\frac{w_{jk} c_j T_j}{c_j T_j} + w_{jk} P_j), \quad (3)$$

where: L represents the set of losers.
Instead of a reward parameter, equation 3 uses a punishment parameter. Again, the intensity of punishment of a neuron depends on its position in an inverse sense, the first losers will have the smallest punishment. The equation of punishment is defined as follows:

$$P_j = ((W - l_j) * \frac{NumLos + k_j^l}{NumLos})^2, \quad (4)$$

where:
W is the enhanced node output of the last winner (with smallest output);
l_j is the enhanced output of the jth loser;
NumLos: is the number of neurons in the set of losers;
k_j^w: is the ranking of the jth neuron in the set of losers (its position in the set of losers).

3. After performing the above steps, the biggest map field output node is selected as the winner. Its output is passed to the ARTb output neuron which produces the overall network output.

Using the above proposed modification, it is possible to use the output as a measure of the similarity of the presented input for every neuron. The best class is chosen through the biggest map field output node and the other classes can have similarities to the input pattern, which can be defined by their output values.

4 Simulation

In this section, an analysis of the RePART model, compared with Fuzzy ARTMAP and ARTMAP-IC is presented. A handwritten numeral recognition task will be applied in the comparative analysis. The main purpose of using these three neural network is that Fuzzy ARTMAP is one of the first extension of ARTMAP, ARTMAP-IC uses instance counting and the RePART model was based upon this model.

In order to build a fuzzy ARTMAP neural network to be used in the simulation, it is necessary to use a preprocessing method in order to transform the binary set into a real one. The preprocessing method employed in this analysis is Histogram. The main idea here is to divide the pattern into rectangles and extract histogram information in the four possible directions (horizontal, vertical, right and left diagonals). In other words, the projections of a defined size window onto four representative orientations. Along with the histogram, the number of pixels which contains the value 1 (The number of positives pixels is calculated). Therefore, the number of pixels of the feature vector is the number of rectangles (windows) multiplied by 5 (the four projections and the number of pixels). The size of the window plays a very important role and the network performance will be analysed according to the size of the window.

Variations of the following parameters are performed in the comparative analysis: size of the training set, size of the window for the pre-processing method, the vigilance parameter, the learning rate.

4.1 The size of the window for the pre-processing method

As was mentioned above, the size of the window plays a very significant role in a histogram pre-processing method and the network performance was analysed according to the size of the window. In order to define the size of the window, fuzzy ARTMAP, ARTMAP-IC and RePART were used to analyse their behaviour according to variations in the size of the window.

In order to analyse the behaviour of the neural networks, the following configuration has been chose: 125 training patterns, 100 recalling patterns, the vigilance parameter = 0.9, the learning rate = 0.2 and the gaussian parameter = 0.02. Table 1 shows the performance of the neural networks with variations in the window size.

Table 1. Comparative analysis of Fuzzy ARTMAP, ARTMAP-IC and RePART while varying the size of window for histogram pre-processing

Window	Fuzzy ARTMAP	ARTMAP-IC	RePART
2x2	89,8%	92,2%	94,5%
3x3	89%	91,7%	93,9%
4x4	89,1%	91,6%	93,5%
5x5	87,6%	90%	92,2%
6x6	86,6%	89,1%	91,2%
7x7	81,6%	84%	85,8%
8x8	82,5%	86,9%	88,9%
9x9	79%	81,9%	83,9%

It can be observed from table 1 that for every neural network, higher recognition rates were reached using a small window. This is explained by the fact that a small window leads to a big input pattern and consequently a complex neural network (using a 2x2 window, the third layer of ARTa has 1176 neurons). On the other hand, it contains more information about the pattern and in turn leads to a higher performance. When using a big window the problem of complexity of the neural network is overcome. However, it leads to a decrease in the performance of the neural network.

In relation to the difference in performance between RePART and ARTMAP-IC and Fuzzy ARTMAP, it was almost constant, in an average of 2%, reaching to 2.8% when using a 7x7 window. It means that the variation in the size of window does have the same importance for both ARTMAP-IC and RePART.

The choice of the ideal size of window must take into account two factors: complexity and performance. For instance, the number of neurons in the third layer of ARTa are 1176, 1125, 1073, 1037, 1033, 652, 818 and 740 for the following size of window: 2x2, 3x3, 4x4, 5x5, 6x6, 7x7, 8x8 and 9x9 respectively. In analysing these factors, the best choice is a 3x3 since it has one of the best performance and a not very complex network.

4.2 Size of the Training Set

The size of the training set defines the number of patterns which will be used during training of the neural network. That is, the number of patterns which will be presented to the network during the training phase. This factor is very important for the performance of the neural network as it characterises the main resource for the neural network task. That is, a bad training set leads to poor performance of the neural network. While, on the other hand, a good training tends to lead to a good performance of the neural network. The table 2 shows the performance of fuzzy ARTMAP, ARTMAP-IC with variations of the size of training size.

Table 2. Comparative analysis of Fuzzy ARTMAP, ARTMAP-IC and RePART while varying the size of the training set

Size of training set	Fuzzy ARTMAP	ARTMAP-IC	RePART
10	77%	81%	80.2%
25	79.6%	83.1%	83.3%
50	82.1%	84%	85.1%
75	84.2%	86.9%	88.2%
100	87%	89.5%	90.9%
125	88.6%	91%	92.5%
150	89%	91.7%	93.9%

In analysing table 2, it can be observed that when few training patterns are used, ARTMAP-IC had a higher recognition rate than RePART. However,

RePART still had a higher rate recognition than fuzzy ARTMAP. As a result of increasing the size of the training set, RePART had a higher performance than ARTMAP-IC, reaching a 2.2% higher recognition rate. According to table 2, it can be deduced that RePART is more dependent on training pattern size than ARTMAP-IC. Since RePART employs a set of winners and a set of losers, it is necessary to acquire more information about the task to be performed and in turn to use a reasonable number of training patterns in order to achieve a good performance. This fact can be confirmed when using more training patterns where RePART achieved a higher recognition rate than ARTMAP-IC.

The overall performance of the neural network constantly increased, with the size of the training set. The best performance was reached with 150 training patterns for each class. However, in using 150 training patterns, the neural networks become very complex, having 1338 neurons in the third layer of ARTa. Therefore, the best choice for the size of training set must be a tradeoff between recognition rate and complexity of the neural network. 125 training patterns proved a good compromise.

4.3 Vigilance parameter

Vigilance is a parameter which is used in the match step of the ART model. It defines the minimum similarity allowed between the input pattern and the template (weights) of the chosen neuron in layer F_1 and it is used in both ART (ARTa and ARTb). This parameter defines whether a learning step or a reset will take place.

The vigilance parameter is one of the most important parameters in governing the performance of a ARTMAP-based network. It can range from 0-1 and the higher vigilance is the more sensitive to changes in the input the system becomes and in turn the more neurons in layer F_2 are required. On the other hand, the lower vigilance, the less sensitive the network becomes. However, the network becomes simpler and faster.

In this section, a comparative behaviour of Fuzzy ARTMAP, ARTMAP-IC and RePART is performed. The main purpose of this section is to analyse whether one of the networks is more affected by variations of the vigilance parameter than others. The same configuration as the above subsections is used. Table 3 shows the performance of ARTMAP, ARTMAP-IC and RePART while varying the vigilance parameter.

In analysing table 3, it can be observed that the performance of the neural networks increases, as the vigilance parameter increases. This is due to the fact that the networks become more sensitive to changes on the input and they do not group similar patterns (the higher the vigilance is, the more strict in grouping training patterns the neural network becomes). On the other hand, the neural networks becomes more complex.

For all vigilance parameter settings, RePART had a better performance than ARTMAP-IC and fuzzy ARTMAP. However, when using a low vigilance, the difference was almost non-existent. This is because there is a grouping of patterns which are not very similar when using low vigilance and this affects the proposed

Table 3. Comparative analysis of Fuzzy ARTMAP, ARTMAP-IC and RePART while varying the vigilance parameter

Vigilance parameter	Fuzzy ARTMAP	ARTMAP-IC	RePART
0.5	85.4%	85.8%	86%
0.6	86.2%	87%	89%
0.7	87.5%	88.1%	90.5%
0.8	87.8%	89.8%	92.3%
0.85	88.6%	90.8%	93%
0.9	89%	91.7%	93.9%

model to a greater extent since it is more dependent on a good learning process. According to increasing of the vigilance parameter, RePART started having a better performance than the other two neural networks, reaching a difference of 2.5% in the performance in comparison with ARTMAP-IC, when using vigilance 0.8. After that, a slight decrease occurs and the difference stands in 2.2% when using vigilance 0.9.

Despite the fact that the best performance difference was reached using vigilance 0.8, the best choice is vigilance 0.9 since the neural networks had the best performance. This is due to the fact that the networks become more sensitive to changes in input and it is a good factor to be taken into account.

4.4 Learning rate

Learning rate is a parameter which controls the amount of learning of a input pattern. In this experiment, a comparative analysis is performed, in which the behaviour of Fuzzy ARTMAP, ARTMAP-IC and RePART is analysed while the learning rate is varied. The same configuration used above with a 0.9 vigilance parameter is used. The results are displayed in Table 4.

Table 4. Comparative analysis of Fuzzy ARTMAP, ARTMAP-IC and RePART while varying learning rate

Learning rate	Fuzzy ARTMAP	ARTMAP-IC	RePART
0.2	89.4%	91.7%	93.9%
0.3	89.7%	91.6%	93.8%
0.4	89.7%	91.6%	93.7%
0.6	89.2%	90.9%	93%
0.8	88.9%	90.6%	92.6%

In examining table 4, it can be deduced that the best performance, for every neural networks, was reached using a learning rate of 0.2 or 0.3. This is due to the fact that there is a forgetting of previous learned patterns when using a high learning rate. As a consequence, there is a degradation in the learned information of the neural networks which leads to a decrease in their performance.

In relation to the difference of performance, RePART always had a better performance than ARTMAP-IC and fuzzy ARTMAP. The difference in comparison with ARTMAP-IC was always constant, at an average of 2%. Therefore, it can be concluded that both networks, ARTMAP-IC and RePART, were equally affected by a variation in the learning rate.

5 Final Remarks

In this work, RePART, a variant of fuzzy ARTMAP has been proposed. It is based upon ARTMAP-IC [14] and uses a set of winners and a set of losers in order to define the winner neuron in ARTa network. In these sets, an instance counting is used along with a reward or punishment parameter (reward for the set of winners and punishment for the set of losers) which is defined according to the ranking of the neurons in these sets.

In order to investigate the ability of RePART, a handwritten character recognition task was performed. In this simulation, RePART was compared with Fuzzy ARTMAP and ARTMAP-IC and always had a better performance. However, it is more dependent on training patterns (when few training patterns were used, RePART had a worse performance than ARTMAP-IC) in comparison with ARTMAP-IC. Although more processing time is required for RePART, the expectation that, in pattern recognition task, it is more suitable to use all the necessary information to determine the output of the neural network was confirmed.

It is important to emphasise that the results obtained in this work are from one pattern recognition task (handwritten character recognition). For future work, these networks will be analysed in other pattern recognition tasks.

References

1. Zadeh, L.A.: Fuzzy Sets. Inf Cont.8:338-353 (1965)
2. Ruspini, E., Bonissone, P., Pedrycz, W.: Handbook of Fuzzy Computation. Ed. Iop Pub/Inst of Physics (1998).
3. Haykin, S.: Neural Networks: A Comprehensive Foundation. Ed. Prentice Hall. 2nd Ed. (1998).
4. Mehrotra, K., Mohan, C.K., Ranka, S.: Elements of Artificial Neural Net-works. Ed. The MIT Press, (1997).
5. Lin, C-T., Lee, C.S.G.: Neural Fuzzy Systems: A neuro-fuzzy synergism to intelligent systems. Prentice hall P T R (1996).
6. Jang, J-S R, Sun, C.-T., Mizutani: Neuro-fuzzy and Soft Computing: A Computational Approach to Learning and Machine Intelligence. Ed. Prentice Hall, 1997.
7. Alimi, A.M.: A Neuro-Fuzzy Approach to Recognize Arabic Handwritten Characters. International Conference on Neural Networks, (1997), pp.1397-1400.
8. Baraldi, A., Blonda, P.: Fuzzy Neural Networks for Pattern Recognition. Tech Report, IMGA-CNR, Italy, 1998.
9. Meneganti, M., Saviello, F. Tagliaferri, R.: Fuzzy Neural Networks for Clas-sification and Detection of Anomalies. IEEE Transactions on Neural Networks, Vol.9(2), 1998, pp.848-861.

10. Carpenter, G., Grossberg, S.: A massive parallel architecture for a self-organizing neural pattern recognition machine. Computer Vision, Graphics and Image Processing, 37 (1987), 54-115.
11. Carpenter, G., Grossberg, S., Reynolds, J.H.: ARTMAP: Supervised real-time learning and classification of nonstationary data by a self-organizing neural network. Neural Networks, 4 (1991), 565-588.
12. Carpenter, G., Grossberg, S., Markunzo, N., Reynolds, J.H. , Rosen, D.B.: Fuzzy ARTMAP: A neural network architecture for incremental supervised learning of analog multidimensional maps. IEEE Transactions on Neural Net-works, 3 (1992), 698-713. 1
13. Carpenter, G., Ross, W.: ART-EMAP: A Neural Network Architecture for Object Recognition by Evidence Accumulation. IEEE Transactions on Neural Networks, 6:4 (1995), 805-818.
14. Carpenter, G., Markuzon, S.: ARTMAP-IC and Medical Diagnosis: instance counting and inconsistent cases. Neural Networks, 11 (1998), 323-336.
15. Grossberg, S.: Adaptive pattern classification and universal recording II: Feed-back, expectation, olfaction and illusions. Biological Cybernetics, 23 (1976), 187-202
16. Srinivasa, N.: Learning and Generalization of Noisy Mappings Using a Modified PROBART Neural Network. IEEE Transaction on Signal Processing, vol.45, No.10 (1997), pp. 2533-2550.
17. Ham, F.M., Han, S.: Classification of Cardiac Arrhythmias Using Fuzzy ARTMAP. IEEE Transactions on Biomedical Engineering. 43(4) (1996), 425-429.
18. Carpenter, G., Grossberg, S., Iizuka, K.: Comparative Performance Measures of Fuzzy ARTMAP, Learned Vector Quantization, and Back Propoagation for Handwritten Character Recognition. Int. Joint Conf. On Neural Networks, (1992) (I) 794-799.
19. Asfour, Y.R., Carpenter, Grossberg, S.: Landsat Satellite Image Segmentation Using the Fuzzy ARTMAP Neural Network. Tech. Report CAS/CNS-95-004, (1995).
20. Grossberg, S., Rubin, M.A., Streilein, W.W.: Buffered Reset Leads to Improved Compression in Fuzzy ARTMAP Classification of Radar Range Profiles. Tech Report CAS/CNS-96-014. (1996).
21. Murshed, N.A., Bortolozzi, F., Sabourin, R.: A Fuzzy ARTMAP-Based Classification System for Detecting Cancerous Cells, Based on the One-Class Problem Approach. Proceedings of Int. Conf. On Pattern Recognition (1996). 478-482.
22. Murshed, N., Amin, A., Singh, S.: Off-line Handwritten Chinese Character Recognition based on Structural Features and Fuzzy ARTMAP Proc. Interna-tional Conference on Advances in Pattern Recognition (ICAPR'98), Plymouth, UK, Springer, pp. 334-343 (23-25 November 1998).
23. Carpenter, G.: Distributed Learning, Recognition, and Prediction by ART and ARTMAP Neural Networks. Neural Networks, 10(8) (1997) 1473-1494.

An Adaptive C-Average Fuzzy Control Filter for Image Enhancement

Farzam Farbiz , Mohammad Bagher Menhaj , Seyed Ahmad Motamedi

Electrical Engineering Department of Amirkabir University of Technology ,
Hafez Ave. No.424 Tehran 15914 Iran
{f7223924, menhaj, motamedi}@cic.aku.ac.ir

Abstract. In this paper a new fuzzy control based filter called by Adaptive C-average Fuzzy Control Filter 'ACFCF' is introduced. The aim of this filter is removing impulsive noise, smoothing Gaussian noise out while at the same time preserving edges and image details efficiently. To increase the run-time speed, the floating point operations are avoided. This filter employs the idea that each pixel is not fired uniformly by each of the fuzzy rules which are adaptively tuned. To show how efficient our filtering approach is, we perform several different test cases on image enhancement problem. From these results we may list the concluding remarks of the proposed filtering as: i) no need of floating point calculations, ii) very fast performance of the filter in compared with those of the other recently proposed filters, iii) high quality of filtering, iv) high edge preserving ability.

1 Introduction

Image enhancement is an important step in most of image processing applications. The problem of image enhancement can be stated as that of removing impulsive noise, smoothing non-impulsive noise, and enhancing edges or certain salient structures in the input image. Noise filtering can be viewed as replacing the gray-level value of every pixel in the image with a new value depending on the local context. Ideally, the filtering algorithm should vary from pixel to pixel based on the local context. For example, if the local region is relatively smooth, then the new value of the pixel is worth being determined by averaging neighboring pixels values. On the other hand, if the local region contains edge or impulse noise pixels, a different type of filtering should be used. However, it is extremely hard, if not impossible, to set the conditions under which a certain filter should be selected, since the local conditions can be evaluated only vaguely in some portions of an image. Therefore, a filtering system needs to be capable of performing reasoning with vague and uncertain information; this is a clear justification of fuzzy logic common usage.

Noise smoothing and edge enhancement are inherently conflicting processes, since smoothing a region might destroy an edge while sharpening edges might lead to unnecessary noise. A plethora of techniques for this problem have been proposed in the literature [1-16].

In this paper, we propose a new filter, based on fuzzy logic control [6,7], that removes impulsive noise and it, furthermore, smoothes Gaussian noise while edges and image details are efficiently preserved.

2 The Proposed Method

This section presents the architecture of our proposed rule-based image processing system. In this system, we adopt the general structure of fuzzy if-then-else rule mechanism. In contrast to the original technique of this mechanism, our approach is mainly based on the idea of not letting each point in the area of concern being uniformly fired by each of the basic fuzzy rules. This idea is widely used in fuzzy control applications [4]. To furnish this goal, the following fuzzy rules and membership functions are proposed for image filtering:

R1: IF (**more** of x_i are **NB**) THEN y is **NB**
R2: IF (**more** of x_i are **NM**) THEN y is **NM** (1)
R3: IF (**more** of x_i are **NS**) THEN y is **NS**
R4: IF (**more** of x_i are **PS**) THEN y is **PS**
R5: IF (**more** of x_i are **PM**) THEN y is **PM**
R6: IF (**more** of x_i are **PB**) THEN y is **PB**
R0: ELSE :

IF(**fairly more** of x_i are **Z**) THEN y is $ave(x_i)$

where

$$ave(x_i) = \text{average}(x_i : x_i \in \text{support}(Z)) \quad (2)$$

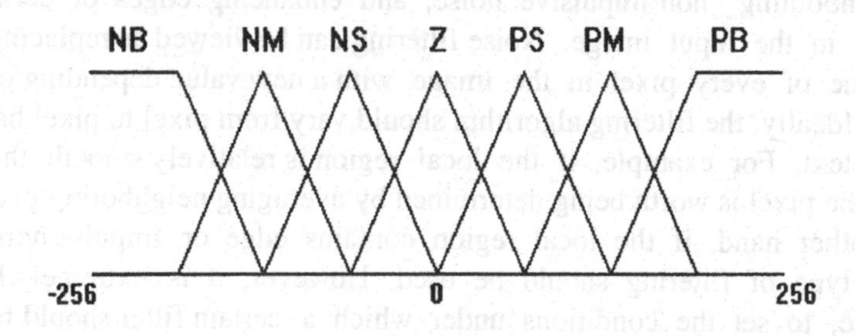

Fig. 1. General shape of the membership functions NB,NM,NS,Z,PS,PM,PB

In the above, x_i's are the luminance differences between neighboring pixels, P_i (located in a window of size N N), and the central pixel, P, : $x_i = P_i - P$. The output variable y is the quantity which is added to P to yield the resulting pixel luminance, P'.

The linguistic terms, **more** and **fairly more**, which is in fact a smoothed version of the **more** function, represent S-type fuzzy functions. To avoid floating point calculations and to increase the speed of algorithm, we adopted two Look-Up-Tables (LUT) with 5 bit integer values for these fuzzy functions.

The functions NB,NM,NS,Z,PS,PM,PB are membership functions with general shapes of figure 1. These functions are adaptively tuned based on the local information in the neighboring window which is fully described in [23].

2.1 Calculation of the Activity degree of the Rules

The activity degree of R1 in Eq.(1) is computed by the following relationship (the other if-then rules degree of activities are computed similarly)

$$\lambda_1 = Function\{\mu_{\text{NB}}(x_i) : x_i \in \text{support(NB)}\} \times \mu_{\text{more}}\left[\frac{\text{number of } x_i \text{ in support(NB)}}{\text{total number of } x_i}\right] \tag{3}$$

The output y is calculated by :

$$y = k.ave(x_i) + (1-k).\sum_{i=1}^{6} C_i \lambda_i \tag{4}$$

where C_i is the membership function center of the ith rule and

$$k = \mu_{\text{fairly more}}\left(\frac{\text{number of } x_i \text{ in support(Z)}}{\text{total number of } x_i}\right) \tag{5}$$

The *Function* term in Eq.(3) is responsible for selecting the membership degree of one of the most proper x_i's belonging to support(NB). To do so, we may take one of the following different choices. Each choice has its own advantages and disadvantages explained below:

1.

$$\text{Function}(.) = \frac{\sum \mu_{\text{NB}}(x_i) : x_i \in \text{support(NB)}}{\text{Total no. of } x_i\text{'s which are in support (NB)}} \tag{6}$$

Indeed, it computes the average of membership degrees of x_i's belonging to support(NB)

2.

$$\text{Function}(.) = \min[\mu_{\text{NB}}(x_i) : x_i \in \text{support(NB)}] \tag{7}$$

3.

$$\text{Function(.)} = \text{median}[\mu_{NB}(x_i) : x_i \in \text{support(NB)}] \tag{8}$$

Note that the above three formulae for computing the first term in Eq.(3) possess the following advantages and disadvantages:

- Although method 1, in compared with the other two methods , perform fast, it makes the output image get some blurred.
- Method 2, like method 1, performs fast but it shows poor filtering performance. To remedy this problem, we should apply it on the input image in more iterations which in turn weakens its fast computing property.
- The third method is slower than the others, but it performs well.

To remedy the disadvantages of the above methods while keeping the fast computation property, we came up with the following formula named by C-average:

$$Function(.) = \mu_{NB}(x_c) \tag{9}$$

where

$$x_c = \arg\left[\min_{x_i}\left(\left|\gamma - \mu_{NB}(x_i)\right|\right)\right] \tag{10}$$

and

$$\gamma = \frac{\sum \mu_{NB}(x_i) : x_i \in \text{support(NB)}}{\text{Total no. of } x_i\text{'s which are in support (NB)}} \tag{11}$$

Table 1 summarizes the performance of the four methods explained above. As seen in the column 3, the C-average takes 1.5 micro second to evaluate for a string with 9 integer elements on a PC Pentium-166 and at the same time shows good performance on the image restoration problem (column 2).

Table 1.

Method	Filtering Ability[1]	run-time (micro second)[2]
average	blur	0.425
minimum	poor	0.575
median	good	5.25
C-average	good	1.5

[1]-Here the Lena image is considered as a test case for a comparison study.

[2]-The run times are computed for a string with 9 integer elements. Programs are written on PC AT Pentium-166 with the software of BorlandC 4.5 for windows.

3 Experimental Results

In order to reemphasize the effectiveness of our method in a noisy environment, several test images were considered. In the Table 1 the results of some of them are depicted in the sense of MSE for different filters: the proposed filter (ACFCF), the median, ENHANCE (proposed in [17]) with two window sizes of 3x3 and 5x5, FWM (proposed in [18]) , EPS (proposed in [19]) with two different window sizes of 5x5 and 7x7, IFCF (proposed in [20]) with 1 and 6 iterations, SFCF (proposed in [21]), FFCF (proposed in [22])and AFCF (proposed in [23]).

The last row of Table 1 summarizes the behavior of the proposed filter. By comparing the columns 2-4 of Table1, it can be obviously seen that the proposed filter has high capability of image enhancement for complex and non complex images.

To show the ability of the proposed filter in image edge preservation, the MSEE (defined in [22]) for the image given in figure 2 is computed for all the above filters. The results are summarized in Table 1. By looking at the column 5 of Table 1, it can be easily observed that our proposed filter has the very good image edge preservation among the other ones.

Fig. 2. A complex test image used for computing the MSEE of filters

The noise removing performance of ACFCF for different variances of Gaussian noise compared with some other methods are depicted in figures 3 & 4. These figures justify the noise reducing performance of the proposed filter.

The last column of Table 1 shows the run-time of each filter on a PC-AT Pentium-166 machine. It can be seen that our proposed filter is much faster than the others. This ability together with the capability of being free of floating point calculations, make the proposed filter quite efficient for hardware implementation in real-time tasks.

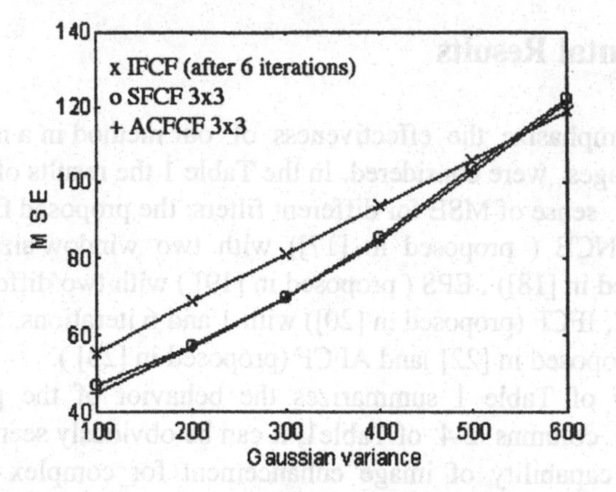

Fig. 3. MSE as a function of variance of Gaussian noise computed for the Lena image.

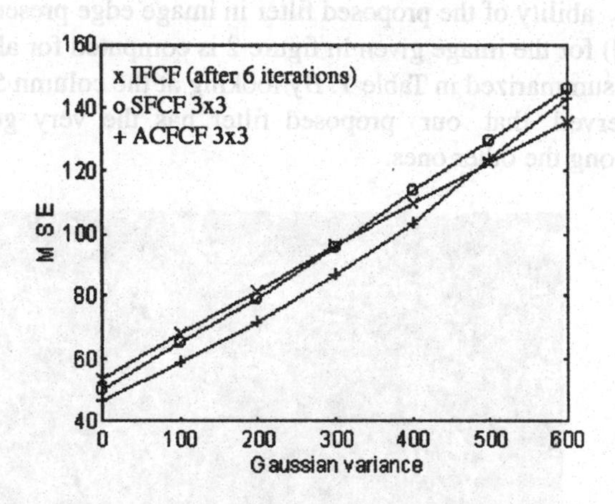

Fig. 4. MSE of restored images of different filters for the Lena image corrupted by mixed noise (Gaussian and [%2.5, %2.5] impulsive noise)

4 Conclusion

In this paper we presented a new filtering method based on fuzzy logic control to image enhancement. The goal of filtering process was to satisfy simultaneously the three tasks of image edge preserving, impulsive noise removal, and Gaussian noise smoothing with the highest speed as possible. We performed some experiments in order to demonstrate the effectiveness of the proposed filtering approach. The results of the proposed filter on different types of images (Lena image and figure 2) are compared with those of filters listed in table 2. From this table and figures 3&4, it can be concluded that i) the proposed filter performs very fast in compared with the other proposed filters such as FWM, EPS, ..., ii) in compared with the methods listed in Table 2, the proposed method has high performance on edge preserving , iii) the proposed filter shows high capability of image restoration in a noisy environment.

Table 2. The MSE MSEE and run-time of Filters

Filter Type	Image 1[1]	Image 2[2]	Image 3[3]	MSEE	Run time (second)[4]
median 3×3	127.9	529.1	111.1	610	1
median 5×5	134	907.6	127.1	1682	4
FWM	130.3	427.8	113.4	506	440
EPS 5×5	136.9	460.3	87.61	847	78
EPS 7×7	129.7	488.5	98.67	1053	151
ENHANCE 3×3	107.6	430.9	96.12	577	30
ENHANCE 5×5	110.8	588	102.4	1102	110
IFCF 3×3 (after 1 iteration)	49162	684.1	187.8	76	10
IFCF 3×3 (after 6 iterations)	109.8	441.1	93.61	359	60
SFCF 3×3	113.8	359.6	85.11	421	15
FFCF 3×3	110	342.8	86.45	389	2.1
AFCF 3×3	105.9	365	87.02	411	2
ACFCF 3×3	103.9	373	85.33	433	1.5

[1] Image 1: the Lena image contaminated by [%2.5, %2.5] impulsive noise and Gaussian noise with $\mu=0$, $\sigma^2=100$.

[2] Image 2: Figure 2 contaminated by [%2.5, %2.5] impulsive noise and Gaussian noise with $\mu=0$, $\sigma^2=100$.

[3] Image 3: the Lena image contaminated by Gaussian noise with $\mu=0$, $\sigma^2=100$.

[4] All filters are implemented by a PC AT Pentium-166 with the software of BorlandC 4.5 for windows. In the simulations, all variables take on real floating point values.

References

1. Pitas, I., Venetsanopoulos, A.N. (eds): Nonlinear Digital Filters: Principles and Applications. Kluwer Academic Publishers, (1990)
2. Haralick, R.M., Shapiro, L.G. (eds): Computer and robot vision. Addison Weseley, vol. 1, (1992)
3. Mastin, G.A.: Adaptive filters for digital image noise smoothing: an evaluation. Computer Vision, Graphics, Image Processing, vol. 31, (1985) 103-121
4. Pal, S.K.: Fuzzy sets in image processing & recognition. In Proc. First IEEE Int. Conf. Fuzzy System (1992) .119-126.
5. Krishnapuram, R., Keller, M.: Fuzzy sets theoretic approach to computer vision :an overview. In Proc. First IEEE Int. Conf. Fuzzy System (1992) 135-142.
6. Kosko, B. (ed): Neural networks and fuzzy systems. Prentice-Hall, (1992)

7. Bezdeck, J.C., Pal, S.K. (eds): Fuzzy Models for Pattern Recognition. New York: IEEE Press, (1992)
8. Russo, F., Ramponi, G.: Edge detection by FIRE operators. In Proc. Third IEEE Int. Conf. Fuzzy System (1994) . 249-253.
9. Russo, F., Ramponi, G.: Combined FIRE filters for image enhancement. In Proc. Third IEEE Int. Conf. Fuzzy System (1994) 261-264.
10. Russo, F., Ramponi, G.: Fuzzy operator for sharpening of noisy images. IEE Electron Lett., vol. 28, Aug. (1992) , 1715-1717
11. Russo, F.: A user-friendly research tool for image processing with fuzzy rules. In Proc. First IEEE Int. Conf. Fuzzy System (1992) 561-568.
12. Russo, F., Ramponi, G.: Nonlinear fuzzy operators for image processing. Signal Processing, vol. 38, Aug (1994) , 129-140
13. Russo, F., Ramponi, G.: A noise smoother using cascaded FIRE filters. In Proc. Fourth IEEE Int. Conf. Fuzzy System, vol.1 . (1995) , 351-358
14. Russo, F., Ramponi, G.: An image enhancement technique based on the FIRE operator" In Proc. Second IEEE Int. Conf. Image Processing, vol.1, (1995) , 155-158
15. Russo, F., Ramponi, G.: Removal of impulsive noise using a FIRE filter. In Proc. Third IEEE Int. Conf. Image Processing, vol.2, (1996) , 975-978
16. Russo, F., Ramponi, G.: A fuzzy filter for images corrupted by impulse noise. IEEE Signal Processing Letters, vol.3, no. 6, (1996) , 168-170
17. Choi, Y., Krishnapuram, R.: A robust approach to image enhancement based on fuzzy logic. IEEE Trans. Image Processing, vol. 6, no. 6, June (1997) , 808-825
18. Taguchi, A.: A design method of fuzzy weighted median filters. In Proc. Third IEEE Int. Conf. Image Processing, vol.1,. (1996) .123-126
19. Muneyasu, M., Wada, Y., Hinamoto, T.: Edge-preserving smoothing by adaptive nonlinear filters based on fuzzy control laws. In Proc. Third IEEE Int. Conf. Image Processing, vol.1,.(1996). 785-788
20. Farbiz, F., Menhaj, M.B., Motamedi, S.A.: An Iterative Method For Image Enhancement Based On Fuzzy Logic. In Proc. IEEE Int. Conf. ICASSP-98 (1998)
21. Farbiz, F., Menhaj, M.B., Motamedi, S.A.: Edge Preserving Image Filtering Based On Fuzzy Logic. In Proc. *EUFIT-98* vol.2 (1998) 1417-21
22. Farbiz, F., Menhaj, M.B., Motamedi, S.A.: Fixed Point Filter Design For Image Enhancement Using Fuzzy Logic. In Proc. IEEE Int. Conf. Image Processing ICIP-98 (1998)
23. Farbiz, F., Menhaj, M.B., Motamedi, S.A.: An Adaptive Fixed Point Filter Design For Image Enhancement Using Fuzzy Logic. In Proc. *EUFIT-98* vol.2 (1998).1432-6

Pareto-optimality in Scheduling Problems

Antonio F. Gómez-Skarmeta, Fernando Jiménez, and Jesús Ibáñez*

Dept. Informática, Inteligencia Artificial y Electrónica.
Facultad de Informática, Universidad de Murcia.
Apartado 4021. 30001 Murcia, Spain
{skarmeta, fernan, jesus}@dif.um.es

Abstract. Recent techniques in multi-mode project scheduling and multiobjective optimization with Evolutionary Algorithms are combined in this paper. We propose a single step Evolutionary Algorithm to find multiple Pareto-optimal solutions to the multi-mode resource-constrained project scheduling problem which make it possible that the decision maker can be able to choose the most appropiate solution according to the current decision environment.

1 Introduction

Within the classical resource-constrained project scheduling problem (RCPSP) and the more realistic multi-mode resource-constrained project scheduling problem (MRCPSP), the activities of a project have to be scheduled such that the makespan of the project is minimized. Precedence constraints have to be observed as well as limitations of the renewable resources required to accomplish the activities.

Minimization of the makespan is the aim of most genetic algorithms for the RCPSP and MRCPSP problems. This seems enough. However, when we try to implant our algorithms in real enterprises, the decision makers use to be critical with this limitation. They usually propose optimize another criterias (cost, human resources required, etc), or a combination of criterias.

In the last years a lot of research in the field of scheduling with Evolutionary Algorithms (EA) has been done [3]. Curiously, the genotypes used [1][2]in the EA usually developed, contain the information required to optimize the criterias proposed by the decision makers, although it's not exploited.

One of the major difficulties which appear in this context is finding the aggregation mechanism of the different criteria which should be taken into account to measure the fitness of a solution of the problem. In the classical multi-objective methods, multiple objectives are combined to form one objective by using some knowledge of the problem which is normally obtained by experimentation. A drawback of these methods is that a decision maker must have a prior knowledge of the priority of each objective before aggregating in a single objective

* The author thanks the *Instituto de Fomento de la Region de Murcia* for its financing through the *Seneca* Program.

the set of objectives. Thus, for different situations, different priorities need to be assigned and the same problem requires to be solved a number of times. A more realistic method would be one that can find multiple Pareto-optimal (non-dominated) solutions simultaneously so that a decision maker may be able to choose the most appropiate solution for the current situation. A pareto-optimum is a solution which offers the least objective conflict, i.e., a point in the search space that is optimally placed from the individual optimum of each objective and cannot be improved for one objective without making it worse for another. The knowledge of many Pareto-optimal solutions is also useful for later use, when the current situation has changed and a new solution is required. Since EA deal with a population of points, multiple Pareto-optimal solutions can be captured in the population in a single run. The idea of assigning equal probability of reproduction to all nondominated individuals in the population by means of a nondominated sorting procedure was first proposed by Goldberg [6]. Next, several authors (Fonseca and Fleming [4], Horn and Nafpliotis [8], Srinivas and Deb [11]) developed variants based on this concept, and successfully applied it to some problems. An overview of EA in multiobjective optimization can be found in [5].

In this paper we present an EA based approach to find multiple pareto-optimal solutions to the MRCPSP problem.

2 Evolutionary Algorithm

In the MRCPSP problem each activity can be performed in one out of several modes. Each mode of an activity represents an alternative way of combining different levels of resource requirements with a related duration. Renewable and nonrenewable resources are distinguished. Renewable resources have a limited per-period availability (manpower, machines, etc) and nonrenewables resources are limited for the entire project (budget).

The classical objective is to find a mode and a start time for each activity, such that the schedule is makespan minimal and feasible with respect to the precedence and resources constraints. Our algorithm, more realistic, is aimed to find multiple pareto-optimal solutions.

We use a EA to solve the MRCPSP. The search space, that is the set of possible genotypes, consists of the job sequences feasible with respect to the precedence and all the combinations of modes.

We use the notation used by Hartmann in [7] (as well as the idea for the representation and some variation operators). We consider a project which consists of J activities labeled $j = 1, ..., J$. The activities are partially ordered (there are precedence relations between some of the jobs).

These precedence relations are given by sets of inmediate predecessors P_j, indicating that an activity j may not be started before all of its predecessors are completed. We consider additional activities $j = 0$ representing the source and $j = J + 1$ representing the sink activity.

Each activity (with the exception of the source and the sink) requires certain amounts of resources to be performed. The set of renewable resources is referred to as R. For each renewable resource $r \in R$ the per-period-availability is constant and given by K_r^p. The set of nonrenewable resources is denoted as N. For each nonrenewable resource $r \in N$ the overall availability for the entire project is given by K_r^v.

Each activity can be performed in one of several different modes of accomplishment. A mode represents a combination of different resources with a related duration. Once an activity is started in one of its modes, it is not allowed to be interrupted, and its mode may not be changed. Activity j may be executed in M_j modes labeled $m = 1, ..., M_j$. The duration of job j being performed in mode m is given by d_{jm}. Activity j executed in mode m uses k_{jmr}^p units of renewable resource r each period it is in process, where we assume $k_{jmr}^p \leq K_r^p$ for each renewable resource $r \in R$. It consumes k_{jmr}^v units of nonrenewable $r \in N$. We assume that the source and the sink activity have only one mode each with a duration of zero periods and no request for any source.

2.1 Representation

An individual I is represented by a pair of an activity sequence and a mode assignment and is denoted as

$$I = ((j_1^I, ..., j_J^I), m^I)$$

The job sequence $j_1^I, ..., j_J^I$ is assumed to be a precedence feasible permutation of the set of activities. The mode assignment m^I is a mapping which assigns to each activity $j \in \{1, ..., J\}$ a mode $m^I(j) \in \{1, ..., M_j\}$.

Each genotype is related to an uniquely determined schedule which is computed using the serial schedule generation scheme (see below).

The schedule related to an individual is feasible with respect to the precedence relations and the renewable resource constraints, but not necessarily w.r.t. the nonrenewable resource constraints.

2.2 Initial Population

The procedure of generation of the initial population obtains a complete population of *popsize* individuals which satisfies the precedence constraints. Each individual $I = ((j_1^I, ..., j_J^I), m^I)$ is generated by means of J iterations. In each iteration i an activity j_i^I and a mode $m^I(j_i^I)$ are selected. The activity j_i^I is randomly selected, from the set of activities that fulfill two conditions:

1. $j_i^I \notin (j_1^I, ..., j_{i-1}^I)$, that is, the activity has not been selected in previous iterations.
2. There's no activity x such that $x \in P_{j_i^I}$ and $x \notin (j_1^I, ..., j_{i-1}^I)$.

The mode $m^I(j_i^I)$ is a random integer with $1 \leq m^I(j_i^I) \leq M_{j_i^I}$.

2.3 Evaluation

Each genotype is related to an uniquely determined schedule which is computed using a schedule generation scheme. Each individual is evaluated w.r.t each considered criteria (makespan, cost, human resources required,etc). Some evaluation functions are applied to the genotype directly, but most are applied to the related schedule. The complete process to evaluate the population is as follows:

1. Generate the schedule related to every individual, using a schedule generation scheme.
2. Apply every evaluation function to all the individuals (genotype or schedule).

Schedule generation schemes (SGS) are the core of most heuristic solution precedures for the RCPSP and the MRCPSP. SGS start from scratch and build a feasible schedule by stepwise extension of a partial schedule. A partial schedule is a schedule where only a subset of the $J+2$ activities have been scheduled. There are two different SGS avaiable. They can be distinguished with respect to the incrementation into activity- and time- incrementation. The so-called serial SGS performs activity-incrementation and the so-called parallel SGS performs time-incrementation. We use a serial SGS as described in [10].

Every decision maker will be interested in a particular set of evaluation criterias. We describe two evaluation functions we have used: the makespan and the total consumption of a renewable resource.

The *minimization of the makespan* is the clasical objective in the MRCPSP. The evaluation should be applied to the schedule, because the genotype lacks required information. To know the order of the activities and their durations is not sufficient to calculate the makespan. After apply the serial SGS we can easily calculate the makespan of the project, just selecting the latter finish time of the activities.

The evaluation function can be more complex. The schedule related to an individual is feasible with respect to the precedence relations and the renewable resource constrints, but not necessarily w.r.t. the nonrenewable resource constraint. The evaluation can take it into account.

Let T be the upper bound on the project's makespan given by the sum of the maximal durations of the activities. Moreover, let $L_r^v(I)$ denote the leftover capacity of nonrenewable resource $r \in N$ w.r.t the modes selected by the genotype of individual I, that is,

$$L_r^v(I) = K_r^v - \sum_{j=1}^{J} k_{jm^I(j)r}^v$$

If there is a nonrenewable resource $r \in N$ with $L_r^v < 0$, then the mode assignment of individual I is infeasible w.r.t. the nonrenewable resource constraints. In this case the fitness of I is given by

$$f(I) = T + \sum_{\substack{r \in N \\ L_r^v(I) < 0}} |L_r^v(I)|$$

Otherwise, if individual I is feasible w.r.t. the nonrenewable resources, the fitness of individual I is given by the makespan of the related schedule. Observe that a feasible individual always has a lower fitness than an infeasible one.

The *minimization of the total consumption of a particular renewable resource* is another possible objective. The evaluation in this case can be applied to the genotype. Let r be the renewable resource whose total consumption we want to minimize. In this case the fitness of I is given by

$$f(I) = \sum_{j=1}^{J} (d_{jm^I(j)} k^p_{jm^I(j)r})$$

Other possible objectives to consider are: minimization of the total consumption of a particular nonrenewable resource, minimization of the average consumption of a particular renewable resource, minimization of the average consumption of a particular nonrenewable resource, minimization of the highest consumption of a particular renewable resource, minimization of the highest consumption of a particular nonrenewable resource, etc.

2.4 Identification of Nondominated Individuals

Let f_k $(k = 1, \dots, M)$ be the objective functions which represent the criteria considered in the scheduling. We assume, without loss of generality, that all objectives have to be minimized. An individual H_i $(i = 1, \dots, \mathit{popsize})$ is said to be a *nondominated individual* in the population if there is no other individual H_j in the population such that:

$$f_k(H_j) \leq f_k(H_i), \text{ for all } k = 1, \dots, M$$

and

$$f_k(H_j) < f_k(H_i), \text{ for at least one } k.$$

Note that nondominated individuals are not necessary Pareto-optimal solutions.

2.5 Selection Mechanism and Generational Replacement

We propose to use the *Pareto domination tournament* (Horn and Nafpliotis [8]). In this method, first a comparison set comprising a specific number (t_{dom}) of individuals is picked at random from the population at the beginning of each selection process. Next, once we have the comparison set, two random individuals are picked from the population and compared with the members of the comparison set for domination with respect to the objective function. If one is nondominated and the other is dominated, then the nondominated is selected. If both are either nondominated or dominated, a niche count is found for each individual by counting in the entire population the number of points within a certain distance (σ_{share}) from it. The individual with the least niche count within the two, is selected.

Variation operators are applied to the selected individuals and the offspring are copied to the next population. This process is repeated until the whole new population is generated (*complete generational replacement*). Moreover we use *elitism*, selecting in each generation a random nondominated individual from the entire population.

2.6 Variation Operators

We have used a crossover operator and two mutation operators.

Crossover. We assume that two indivuals, M and P are selected as parents for the crossover operation. Then we produce two random integers p_1 and p_2 with $1 \leq p_1, p_2 \leq J$. Two new children, A and O, are produced from the parents. A is defined as follows: In the sequence of jobs of A, the positions $i = 1, ..., p_1$ are taken from the first parent, that is,

$$j_i^A = j_i^M$$

The sequence $i = p_1 + 1, ..., J$ in A is taken from the second parent. But the jobs already taken from the first parent may not be considered again. Using the general technique for permutation based genotypes we obtain:

$$j_i^A = j_k^P \text{ where } k \text{ is the lowest index such that } j_k^P \notin \{j_1^A, ..., j_{i-1}^A\}$$

This definition ensures that the relative positions in the parent's job sequences are preserved. Observe that the resulting job sequence is precedence feasible.

The modes of the activities on the positions $i = 1, ..., p_2$ in A are defined by the first parent's mode assignment m^M, that is,

$$m^A(j_i^A) = m^M(j_i^A)$$

The modes of the remaining jobs on the positions $i = p_2 + 1, ..., J$ in A are derived from the second parent's mode assignment m^P:

$$m^A(j_i^A) = m^P(j_i^A)$$

The child O of the individuals M and P is computed similarly. However, the positions $1, ..., p_1$ of the child's job sequence are taken from the second parent and the remaining positions are determined by the first parent. Analogously, the first part up to the position p_2 of the mode assignment of O is taken from P while the second part is derived from M.

Mutation. We have used two mutation operators. The first is defined as follows: Given an individual I of the current population, we draw two random integers q_1 and q_2 with $1 \leq q_1 < J$ and $1 \leq q_2 \leq J$. q_1 is used to modify the job sequence by exchanging activities $j^I_{q_1}$ and $j^I_{q_1+1}$ if the result is a job sequence which fulfills the precedence constraints. Each of the changed activities keeps its assigned mode, that is, this modification does not change the mode assignment. Then we randomly choose a new mode for the activity on position q_2, that is, we recalculate $m^I(j^I_{q_2})$.

The second mutation operator is defined as follows: Given an individual I of the current population, we draw two random integers q_1 and q_2 with $1 \leq q_1, q_2 \leq J$. Let Sec_1 be a section of the job sequence taken from I and defined as

$$Sec_1 = (j^I_{q_1}, ..., j^I_{q_2}), \text{ if } q_1 \leq q_2$$
$$Sec_1 = (j^I_{q_2}, ..., j^I_{q_1}), \text{ if } q_2 < q_1$$

We apply to the set of activies in Sec_1 the procedure described above to generate a new individual in the initial population procedure, and obtain a new section Sec_2. Sec_2 replace to Sec_1 in I. Observe that the new section and the new individual are precedence feasible.

2.7 Parameters

The parameters to be considered by the EA are maximal generation number (*maxgen*), population size (*popsize*), probability of crossover (p_c), probability of mutation (p_m), number of members in comparison sets (t_{dom}) and maximum phenotypic distance between individuals (σ_{share}). Note that p_m is the sum of the probabilities of the two mutation operators described above.

3 Results

We have used a subset of the standard test problems constructed by the project generator ProGen (developed by Kolisch et al. [9]) and used by Hartmann [7] and other authors. They are available in the project scheduling problem library PSLIB from the University of Kiel. In particular we have used the multi-mode problem sets containig instances with 10 and 12 non-dummy activities.

Firstly, we have evaluated the algorithm in the optimization of one objective: Minimization of the makespan. After this, we have tested the algorithm in the optimization of two objectives: Minimization of the makespan and minimization of the total consumption of a particular renewable resource. The parameters used in these tests are given in Table 1.

We have compared the values of the makespan in the bicriteria optimization with the ones in the monocriteria optimization. We have obtained good results. The values are comparable in general. In some problems the values of the makespan is the same in the two cases, in the remainder is almost the same. Logically, the results in this sense depends on the concrete problem considered.

Table 1. Values of the parameters used in the tests

Parameter	Value
maxgen	3000
popsize	20
p_c	0.6
p_m	0.8
t_{dom}	10
σ_{share}	0.5

Finally, we can not to take out definitive conclusions from the tests carried out up to now, however our objective is not to compare our performance with other results in the literature, but to show how the pareto-optimality can be used to solve real scheduling problems.

4 Conclusions and Future Trends

In this paper we have presented an evolutionary algorithm to find multiple pareto-optimal solutions in the context of scheduling. The results obtained up to now with the application of this algorithm although not definitive, show that its use can be of great interest for the decision makers as it offers a collection of solutions that can be used in a supervised environment.

References

1. Bierwirth, C.: A Generalized Permutation Approach to Job Shop Scheduling with Genetic Algorithms. Department of Economics, University of Bremen, Germany
2. Bierwirth, C., Mattfeld, D.C., Kopfer H.: On Permutaction Representations for Scheduling Problems. Department of Economics, University of Bremen, Germany
3. Fang, H.: Genetic Algorithms in Timetabling and Scheduling. PhD. Department of Artificial Intelligence, University of Edinburgh (1991)
4. Fonseca, C.M, Fleming, P.J.: Genetic algorithms for multi-objective optimization: formulation, discussion and generalization, in S. Forrest (ed.), Genetic Algorithms: Proc. of the Fifth Intern. Conf, Morgan Kaufmann, San Mateo (1993) 416-423
5. Fonseca, C.M., Fleming, P.J.: An overview of evolutionary algorithms in multiobjective optimization, Evolutionary Computation, vol. 3, no. 1, MIT Press (1995)
6. Goldberg, D.E.: Genetic Algorithms in Search, Optimization, and Machine Learning. Addison-Wesley (1989)
7. Hartmann, S.: Project Scheduling with Multiple Modes: A Genetic Algorithm. Technical Report 435. Instituten fr Betriebswirtschaftslehre. Universitt Kiel, Germany (1997)
8. Horn, J., Nafpliotis, N.: Multiobjective optimization using the niched pareto genetic algorithm, IlliEAL Report No. 93005 (1993)
9. Kolish, R., Sprecher, A., Drexl, A.: Characterization and generation of a general class of resources-constrained project scheduling problems. Management Science, Vol. 41, 1693-1703

10. Kolish, R., Hartmann, S.: Heuristic Algorithms for Solving the Resource-Constrained Project Scheduling Problem: Classification and Computational Analysis. Handbook on Recent Advances in Project Scheduling, edited by J.Weglarz and published by Kluwer (1998)
11. Srinivas, N., Deb, K.: Multiobjective optimization using nondominated sorting in genetic algorithms, Evolutionary Computation, vol. 2, no. 3 (1995) 221-248

Controlled Markov Chain Optimization of Genetic Algorithms

Yijia Cao[1] and Lilian Cao[2]

[1]Intelligent Computer Systems Centre
University of the West of England
Bristol, BS16 1QY, UK
[2]Dept. of Industrial & Manufacturing Engineering
New Jersey Institute of Technology
Newark, NJ 07102, USA

Abstract. Identifying the optimal settings for crossover probability p_c and mutation probability p_m is of an important problem to improve the convergence performance of GAs. In this paper, we modelled genetic algorithms as controlled Markov chain processes, whose transition depend on control parameters (probabilities of crossover and mutation). A stochastic optimization problem is formed based on the performance index of populations during the genetic search, in order to find the optimal values of control parameters so that the performance index is maximized. We have shown theoretically the existence of the optimal control parameters in genetic search and proved that, for the stochastic optimization problem, there exists a pure deterministic strategy which is at least as good as any other pure or mixed (randomized) strategy.

1 Introduction

Genetic Algorithms (GAs) are efficient stochastic search techniques for global optimization problems and have been found many potential engineering applications [1, 2, 3]. They start with a randomly generated population and evolve towards better solutions by applying genetic operators (crossover, mutation, inversion, etc). By imitating the principles of survival and evolution from nature, GAs process a set of solution proposals (called population), gradually enhancing the average performance and attempting to achieve the global optimum. In the course of genetic evolution, more fit specimens are given greater opportunities to reproduce, this selection pressure is counterbalanced by mutations and random crossover operations that add stochasticity to the evolution of the population. However, the population might undergo an undesired process of diversity loss, constantly decreasing the variety of its specimens, which results in the population likely degenerating to the local optimum. This is referred as a premature convergence problem in the genetic search.

In order to overcome the premature convergence problem and improve the convergence property of GAs, the idea of adapting crossover and mutation operators has been employed earlier. The significance of selections of crossover probability p_c and mutation probability p_m in controlling the GA performance has been acknowledged in the GA research. Identifying the optimal settings for p_c and p_m is of an important problem to improve the convergence performance of GAs. Grefenstette has performed a large amount of experiments to search for the optimal GA's parameter settings using a set of numerical optimization problems. It is shown that optimising GA control parameters based on the experimental data is possible, and a very good performance can be obtained within a range of control parameter settings [4]. Schaffer and Morishma have discussed a crossover mechanism in which distribution of crossover points is adapted based on the performance of generated offspring. The distribution information is encoded into each string using additional bits. Selection and recombination of the distribution bits occurs in the normal fashion along with the other bits of solutions [5]. L. Davis has presented an effective method of adapting operator probabilities based on the performance of the operators. The adaptation mechanism provides for the alteration of probabilities in proportion to the fitness of strings created by the operators [6]. T.E. Davis has proposed the SA-like strategy in which the mutation probability is reduced from $\frac{1}{2}$ to 0 with the generation changes [7]. Fogarty has studied the effects of varying the mutation rate over generations with integer encodings. Specially, a mutation rate that decreases exponentially with generations has demonstrated superior performance for a single application [8]. Srinivas and Patnaik recommend the use of adaptive probabilities of crossover and mutation to realise the twin goals of maintaining the diversity of population and sustaining the convergence capacity of the GA. The probabilities of crossover and mutation are varied depending upon the fitness value of solutions. High fitness solutions are protected, while the solutions with subaverage fitness are totally disrupted [9]. Bäck presents some experimental results which indicate that environmentally dependent self-adaptation of approximate settings for the mutation rate are useful for GAs [10]. Hinterding implements self-adaptation Gaussian mutation, which allows the GA to vary the mutation strength in the search process, and gives further improvement on some of the functions [11].

In this paper, we further the above research and proved theoretically the existence of the optimal control parameters in the genetic search. In this study, a GA is modelled as a controlled Markov chain, whose transition depends on control parameters (probabilities of crossover and mutation). The basic dynamic equation for GA search behavior is derived. A stochastic optimization problem is formed based on the performance index of populations during the genetic search. We have proved that, for the stochastic optimization problem, there exists a pure deterministic strategy which is at least as good as any other pure or mixed (randomized) strategy.

2 Vose's Complete Markov Model

Vose developed a complete Markov chain model with no special assumptions and derived an exact formula for the transition matrix [12]. Here we introduce his model to show that a GA could be considered as a stochastic system, whose behaviour depends on the control parameters, such as probabilities of crossover and mutation, etc.

Let S be the collection of length l binary strings, and let $r = |S| = 2^l$ be the number of possible strings. Let P be a population of elements from S, let $n = |P|$ be the population size, and then there are $N = \binom{n+r-1}{r-1}$ possible populations. A GA can be modelled with a Markov chain having N possible populations as states. Introducing Z as an $r \times N$ matrix whose columns represent the possible populations of size n. The ith column $\phi_i =< z_{0,i}, \cdots, z_{r-1,i} >^T$ of Z is the incidence vector for the ith population P_i. where $z_{y,i}$ is the number of occurrences of string y in P_i and integers y are identified with their binary representations of all individuals in a population, indexing from 0 to $r-1$. Therefore, the columns of Z, $\phi_1, \phi_2, \cdots, \phi_N$, describe the states of GA model.

The GA can be modelled as a discrete-time Markov chain $\{X^{(k)}\}$ ($k = 0, 1, 2, \cdots$), with *finite* state space

$$\Omega = \{\phi_1, \phi_2, \cdots, \phi_N\}$$

its kth step transition matrix is denoted as $Q(k)$, whose (i,j) element is

$$q_{ij}(k) = q_{ij}^{(k-1,k)} = \Pr\{X^{(k)} = \phi_j | X^{(k-1)} = \phi_i\}, \qquad i,j = 1, \cdots, N \tag{1}$$

and the initial probability distribution is given as

$$\alpha_i^{(0)} = \Pr\{X^{(0)} = \phi_i\},\ i = 1, 2, \cdots, N \tag{2}$$

$$\alpha_i^{(0)} \geq 0, \sum_{i=1}^{N} \alpha_i^{(0)} = 1. \tag{3}$$

As in the standard GA (SGA), the probabilities of crossover and mutation, p_c and p_m, remain fixed at each generation, the Markov chain could be homogeneous, i.e., $\forall\ s, t \in \{1, 2, \cdots, \}$, $q_{ij}(t) = q_{ij}(s)$, for all $i, j \in \Omega$. In this case, the $N \times N$ transition matrix $Q(k)$ is time independent, i.e. $Q(k) = Q = (q_{ij})$, $k = 1, 2, \cdots, i, j = 1, 2, \cdots, N$, where q_{ij} is the probability that the kth generation will be P_j given that the $(k-1)$th generation is P_i. The transition probabilities may be calculated by considering how the population incidence vector ϕ_j describes the composition of the next generation [12]. Let $p_i(y)$ be the probability of producing string y in the next generation given that the current population is P_i. If the next generation is P_j, then there must be $z_{y,j}$ occurrences of string y produced. The probability of this is almost given by $\{p_i(y)\}^{z_{y,j}}$, the probability of producing population P_j from population P_i can be written as a multinomial distribution with parameters n, $p_i(0), \ldots, p_i(r-1)$:

$$
\begin{aligned}
q_{ij} &= \frac{n!}{z_{0,j}! z_{1,j}! \cdots z_{r-1,j}!} \prod_{y=0}^{r-1} \{p_i(y)\}^{z_{y,j}} \\
&= n! \prod_{y=0}^{r-1} \frac{\{p_i(y)\}^{z_{y,j}}}{z_{y,j}!}. \qquad (4)
\end{aligned}
$$

Given a vector x, let $|x|$ denote the sum of its coordinates, $\oplus$ and $\otimes$ be *exclusive-or* and *logical-and* on integers, respectively, and $m_{ij}(k)$ be the probability that k results from the recombination process based on the parents i and j. If the recombination is 1-point crossover with crossover rate p_c and the mutation with mutation rate p_m, then

$$
\begin{aligned}
m_{ij}(0) &= \frac{(1-p_m)^l}{2} \{\eta^{|i|}(1 - p_c + \frac{p_c}{l-1} \sum_{k=1}^{l-1} \eta^{-\Delta_{i,j,k}}) \\
&\quad + \eta^{|j|}(1 - p_c + \frac{p_c}{l-1} \sum_{k=1}^{l-1} \eta^{\Delta_{i,j,k}})\} \qquad (5)
\end{aligned}
$$

where $\eta = p_m/(1-p_m)$ and $\Delta_{i,j,k} = |(2^k - 1) \otimes i| - |(2^k - 1) \otimes j|$. Define permutations σ_j on R^r by

$$\sigma_j < x_0, \cdots, x_{r-1} >^T = < x_{j\oplus 0}, \cdots, x_{j \oplus r-1} >^T .$$

Let M be the matrix having (i,j)th entry $m_{ij}(0)$, and define an operator $\mathcal{U}$ by

$$\mathcal{U}(\phi) = < (\sigma_0 \phi)^T M \sigma_0 \phi, \cdots, (\sigma_{r-1}\phi)^T M \sigma_{r-1} \phi >^T,$$

with yth component in the above vector denoting the probability that string y $(y = 0, 1, \ldots, r-1)$ is generated by reproduction in a population in state ϕ. Let f be the positive objective function, and $\mathcal{F}$ be a linear operator having diagonal matrix with (i,i)th entry $f(i)$. As we assume that proportional selection is used, the probability of a string i, being selected from the population in state ϕ_i, is given by $f_i \phi_i / |\mathcal{F}\phi_i|$. Therefore, we can get

$$p_i(y) = \mathcal{U}\left(\frac{\mathcal{F}\phi_i}{|\mathcal{F}\phi_i|}\right)_y .$$

Based on (4), the transition matrix is given by

$$
\begin{aligned}
q_{ij} &= n! \prod_{y=0}^{r-1} \frac{\{\mathcal{U}(\frac{\mathcal{F}\phi_i}{|\mathcal{F}\phi_i|})_y\}^{z_{y,j}}}{z_{y,j}!} \\
&= g_{ij}(p_c, p_m), \qquad (6)
\end{aligned}
$$

where $g_{ij}(p_c, p_m)$ is the continuous function of p_c and p_m.

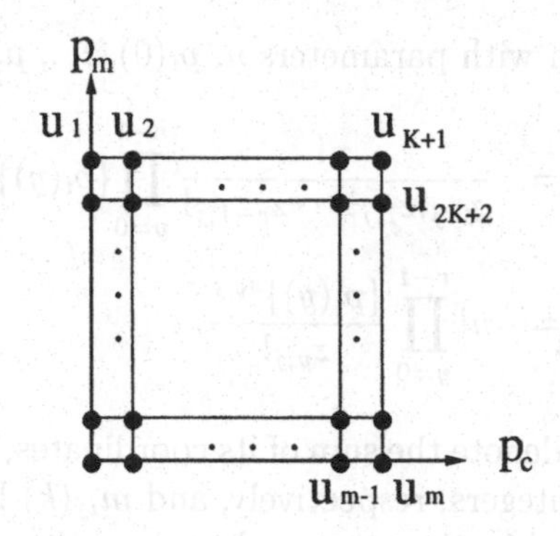

Figure 1: Control parameters inputs in genetic algorithms

3 Controlled Markov Model

It can be seen that the transition matrix (6) depends on the probabilities of crossover and mutation, and even more, the population size of GA. This means a GA can be modelled as a controlled Markov model. If we divide the control parameters p_c and p_m into K subintervals, which is shown in Fig.1. So at each time t, there are $m = (K+1) \times (K+1)$ control input options, named as u_1, u_2, $\cdots$, u_m, to control the genetic search behavior. By introducing an control input variable $u(t)$, which ranges over the finite set $U = \{u_1, \cdots, u_m\}$, then GA is equivalent to the stochastic finite systems Σ, with the following characteristics:

- $u(t)$, the control input to the system at time t, ranges over the finite set $U = \{u_1, \cdots, u_m\}$.
- $\phi(t)$, the state of the system at time t, ranges over the finite set $\Omega = \{\phi_1, \cdots, \phi_N\}$.
- The probabilistic state transitions are formally represented by the relation (7)

$$\phi(t+1) = f(\phi(t), u(t), \xi(t)) \tag{7}$$

  where the functions $f(\cdot, \cdot)$ are continuous, and $\xi(t)$ is random variable whose sample values range over finite sets $\{\xi_1, \xi_2, \cdots, \xi_r\}$.

The above system Σ possesses the Markov property, i.e.

$$\text{Prob}\{\phi(t+1)|\phi(t),\ u(t),\ \text{earlier inputs and states}\} = \text{Prob}\{\phi(t+1)|\phi(t),\ u(t)\}.$$

Based on eqn(6), the state transitions are completely described by the m $N \times N$ transition probability matrices $Q(u_1, t)$, $Q(u_2, t)$, $\cdots$, $Q(u_m, t)$, where their elements are

$$q_{ij}(u_k, t) = \text{Prob}\{\phi(t+1) = \phi_j | \phi(t) = \phi_i,\ u(t) = u_k\}.$$

At each time step t, the system is completely described probabilistically by specifying the N-vector of state probabilities $\alpha(t)$, where:

$$\alpha_i(t) = \text{Prob}\{\phi(t) = \phi_i\}, \quad i = 1, \cdots, N$$

$$\sum_{i=1}^{N} \alpha_i(t) = 1, \quad \alpha_i(t) \geq 0, \ \forall\, i, \ t.$$

A set of difference equations will be obtained for describing the transition of vector $\alpha(t)$ in terms of the input $u(t)$. Consider the $m \times N$ matrix $V(t)$, where

$$V_{ij}(t) = \text{Prob}\{u(t) = u_i \mid \phi(t) = \phi_i\} \ i = 1 \cdots, m, \ j = 1, \cdots, N \tag{8}$$

$$0 \leq V_{ij}(t) \leq 1, \ \sum_{i=1}^{m} V_{ij}(t) = 1$$

As will be seen later, the variables $V_{ij}(t)$ play the role of control. Let $v_{(l)}^T$ represent the lth row of the matrix $V(t)$

$$v_{(l)}^T(t) = (V_{l1}(t), \cdots, V_{lN}(t)).$$

Then it can be easily be seen that

$$\begin{aligned} \alpha_i(t+1) &= \sum_{j=1}^{N}\sum_{k=1}^{m} \text{Prob}\{\phi(t+1) = \phi_i, \ \phi(t) = \phi_j, \ u(t) = u_k\} \\ &= \sum_{j=1}^{N}\sum_{k=1}^{m} q_{ji}(u_k, t)\alpha_j(t)V_{kj}(t) \end{aligned} \tag{9}$$

or in matrix form

$$\alpha(t+1) = \sum_{k=1}^{m} Q^T(u_k, t) < \alpha(t) \cdot v_{(k)}(t) > . \tag{10}$$

Equation (10) is the basic dynamic equation of the system.

The input $u(t)$ that is to be used at each time t is sought so as to maximize a suitable criterion function. The optimal value of the $m \times N$ matrix of probabilities $V(t)$, which plays the role of control, has to be determined. For genetic search, it means the choice of the input over the probability distribution of the controlled parameters. If the optimal value of these variables is either 1 or 0, then the choice of the input is unambiguous. This, indeed, is the case for genetic search.

4 Stochastic Optimization Problem

In general, the stochastic optimization problem is to find the optimal control strategy to optimize the performance of GAs. Let $H(\phi)$ and ν_x denote the performance of state ϕ and the probability distribution of the controlled variables x respectively, then the problem is to find a solution of

$$F(x) = \int H(\phi)\, d\nu_x(\phi) = \max, \quad x \in S \tag{11}$$

where S is a set of feasible controls. It should be stressed that one and only one decision about x can be made. If the probability distribution ν_x is known and the integral (11) can be calculated analytically, then our problem reduces to a nonlinear deterministic constrained optimization problem. We assume however that the probability distribution ν_x is unknown. Then the stochastic optimization problem is formed as to find optimal control variables x to control the state transition in the GA so that the optimal performance is achieved.

To be more specific, let the criterion function relating to the above equivalent to the stochastic system Σ of GA be

$$J = E[W_f(\phi(t_f)) + \sum_{i=1}^{t_f} W_i(\phi(t), \phi(t+1), u(t))] \tag{12}$$

where $W_f(\phi(t_f))$ denotes the cost of terminal state, the function $W_i(\cdot,\cdot,\cdot)$ denote the cost of the state transitions, and E denotes the expectation. Also it is usually possible to rewrite it in terms of $\alpha(t)$, $V(t)$, using the transition matrices $Q(u_k, t)$ as

$$J = \Phi(\alpha(t_f)) + \sum_{i=1}^{t_f} L_i(\alpha(t), V(t)) \tag{13}$$

where the function $\Phi(\cdot)$ and $L_t(\cdot,\cdot)$ are non-negative functions of their arguments. We can restate the stochastic optimization problem as follows.

Given the system (14) and the performance criterion J in (13), choose the $m \times N$ control matrix $V(t)$, $t = 1, 2, \cdots, (t_f - 1)$ so as to maximize J satisfy the constraints (15):

$$\begin{aligned} \alpha(t+1) &= \sum_{k=1}^{m} Q^T(u_k, t) < \alpha(t) \cdot v_{(k)}(t) > . \\ v_{(l)}^T(t) &= (V_{l1}(t), \cdots, V_{lN}(t)). \end{aligned} \tag{14}$$

$$0 \le V_{ki}(t) \le 1 \quad \sum_{k=1}^{m} V_{ki}(t) = 1 \quad \forall\, i,\ t \tag{15}$$

5 Existence Theorem

The stochastic optimization problem defined in the above section can be solved by a version of the discrete maximum principle in dynamic programming [14, 15]. In general, the optimal solution is obtained by solving a 2-point boundary value problem with the constraints (15). Needless to say, the general computational procedure is not easy. However, an important observation is given in the following result.

Theorem: The elements of the optimal control matrices $V^*(t)$, $t = 1, \cdots, (t_f - 1)$, which maximize the criterion function (13) for system (14) with constraints (15), can assume only one of two values, 0 or 1.

This theorem shows that if a proper criterion function to be optimized in GA is given, then at each time step t, the choice of input $u(t)$ is deterministic, i.e., $u(t)$ can be chosen unambiguously as one of $\{ u_1, u_2, \cdots, u_m\}$. There is no need to select a randomized strategy, i.e. 30 percent of u_1, 25 percent of u_2, and 45 percent of others, etc. This shows the existence of the optimal parameters in genetic search, which may be detected using simulated approach. However, since there are so many states in the Markov model, and the numerical solution of the equilibrium distribution cannot be obtained by available computers, so it is difficult to apply this theorem directly.

Proof: To prove the theorem, the variational approach in [14, 15] is used. Similarly, let us introduce a new variable $\alpha_0(t)$ so that the criterion function J will be a function of the terminal state only. The state transition equations and the criterion function J can be rewritten, as follows:

$$\begin{pmatrix} \alpha_0(t+1) \\ \alpha(t+1) \end{pmatrix} = \left[A + \sum_{i=1}^{m} \sum_{j=1}^{N} B_{ij} V_{ij}(t) \right] \begin{pmatrix} \alpha_0(t) \\ \alpha(t) \end{pmatrix} \tag{16}$$

$$J = \sum_{i=0}^{N} \beta_i \alpha_i(t_f) \tag{17}$$

where A, B_{ij}, $i = 1, \cdots, m$; $j = 1, \cdots, N$ are all $(N+1) \times (N+1)$ matrices.

Let $V_{ijt} \triangleq V_{ij}(t)$, $i = 1, \cdots, m$, $j = 1, \cdots, N$, $t = 1, \cdots, t_f - 1$.
Let $T = t_f - 1$

$$0 \leq V_{ijt} \leq 1, \qquad \sum_{i=1}^{m} V_{ijt} = 1 \quad \forall\, j,\ t \tag{18}$$

Every tensor V whose elements satisfy (18) is a feasible solution and represents a point inside or on a $m \times N \times T$ dimensional unit hypercube.

An expression will be obtained for J by using (16) recursively

$$\begin{aligned} J(V) &= \text{Value of the criterion function at point V} \\ &= \beta^T \prod_{k=1}^{T} \left[A + \sum_{i=1}^{m} \sum_{j=1}^{N} B_{ij} V_{ijt} \right] \begin{pmatrix} \alpha_0(t) \\ \alpha(t) \end{pmatrix} \\ &= c_0 + \sum_{i=1}^{m} \sum_{j=1}^{N} \sum_{k=1}^{T} c_{ijk} V_{ijk} + \sum_{i_1,\ i_2=1}^{m} \sum_{j_1, j_2=1}^{N} \sum_{k_1,\ k_2=1,\ k_1 \neq k_2}^{T} c_{i_1 j_1 k_1 i_2 j_2 k_2} \end{aligned}$$

$$V_{i_1j_1k_1}V_{i_2j_2k_2} + \cdots + \sum_{i_1,\ i_2,\cdots,i_T=1}^{m} \sum_{j_1,j_2=1,\cdots,j_T=1}^{N} \sum_{k_1,\ j_2,\cdots,k_T=1\ k_i\neq k_j,\forall i\neq j}^{T}$$

$$c_{i_1j_1k_1i_2j_2k_2\cdots i_Tj_Tk_T}\left(\prod_{r=1}^{T} V_{i_rj_rk_r}\right). \tag{19}$$

First of all, it is desired to show that the function $J(V)$ cannot assume an extreme value in the interior of the $m \times N \times T$ cube. Consider a neighboring point $(V^0 + \delta v)$ satisfying (18),

$$\sum_{i=1}^{m} \delta V_{ijk} = 0 \qquad \forall\ j,\ k$$

or

$$\delta V_{mjk} = -\sum_{r=1}^{m-1} \delta V_{rjk} \tag{20}$$

$$J(V^0 + \delta V) \;=\; [J(V^0) + \sum_{i,\ j,\ k} d(V^0)_{ijk}\delta V_{ijk} + \sum_{i_1,\ i_2,j_1,\ j_2,k_1,k_2,k_1\neq k_2} d(V^0)_{i_1j_1k_1i_2j_2k_2} \cdot \delta V_{i_1j_1k_1}\delta V_{i_2j_2k_2} + \cdots] \tag{21}$$

where

$$d(V^0)_{ijk} \;\triangleq\; \frac{\partial J}{\partial V_{ijk}}|_{V=V^0}$$

$$d(V^0)_{i_1j_1k_1,i_2j_2k_2} \;\triangleq\; \frac{\partial^2 J}{\partial V_{i_1j_1k_1}\partial V_{i_2j_2k_2}}|_{V=V^0}, \qquad \text{etc.}$$

Substituting (20) in (21), (22) is obtained:

$$J(V^0 + \delta V) = J(V^0) + \sum_{i=1}^{T} \Delta_i \tag{22}$$

where

$$\Delta_1 \;=\; \sum_{i=1}^{m-1}\sum_{j=1}^{N}\sum_{k=1}^{T} f_{ijk}(V^0)\delta V_{ijk}$$

$$\Delta_2 \;=\; \sum_{i_1,\ i_2=1}^{m-1}\ \sum_{j_1,\ j_2=1}^{N}\ \sum_{k_1,\ k_2=1,k_1\neq k_2}^{T} f(V^0)_{i_1j_1k_1i_2j_2k_2} \cdot \delta V_{i_1j_1k_1}\delta V_{i_2j_2k_2}$$

The set of $(m-1)\times N\times T$ equations $f_{ijk}(V^0) = 0$ has either no solution V inside the cube or it has a solution, say $V = V^0$, inside the cube. In the former case, the function J has no extremum inside the cube. In the latter case, $V = V^0$ is a candidate for the extremum. Let $\Delta_2(V = V_0) \not\equiv 0$, i.e., there exists at least one coefficient $f(V^0)_{abc,\ deg}$, which is nonzero. It will be shown presently that Δ_2 can assume both positive and negative values, showing that $V = V^0$ cannot be a point of extremum.

Let

$$\begin{aligned} \delta V_{abc} &= \epsilon_1 \operatorname{sgn}[f(V^0)_{abc,deg})] \\ \delta V_{deg} &= \epsilon_2 \operatorname{sgn}[f(V^0)_{abc,deg})] \qquad \epsilon_1,\ \epsilon_2 > 0 \end{aligned}$$

and

$$\delta V_{ijk} = 0, \quad \text{otherwise}$$

Then

$$\Delta_2(V^0) > 0.$$

On the other hand, let

$$\begin{aligned} \delta V_{abc} &= \epsilon_1 \operatorname{sgn}[f(V^0)_{abc,deg})] \\ \delta V_{deg} &= -\epsilon_2 \operatorname{sgn}[f(V^0)_{abc,deg})] \qquad \epsilon_1,\ \epsilon_2 > 0 \end{aligned}$$

and

$$\delta V_{ijk} = 0, \quad \text{otherwise}$$

Then $\Delta_2(V^0) < 0$. If $\Delta_2(V^0) \equiv 0$ and $\Delta_3(V^0) \not\equiv 0$, it can be proved that $\Delta_3(V^0)$ is an indefinite form, etc. The proof is complete since all the forms Δ_i, $i = 1, 2, \cdots, T$ cannot be zero identically if $J(V)$ is not a constant.

Similarly, $J(V)$ cannot have an extremum in the interior of the faces or edges of the $m \times N \times T$ cube. Hence the extremum can occur only at the corners of the cube, thus proving the elements of the optimal control matrices $V^*(t)$, $t = 1, \cdots, (t_f - 1)$ take only one of two values, 0 or 1.

6 Conclusion

In this paper, we proved theoretically the existence of the optimal control parameters in the genetic search. The GA is modelled as a controlled Markov chain, whose transition depends on control parameters (probabilities of crossover and mutation). The basic dynamic equation for GA search behavior is derived. A stochastic optimization problem is formed based on the performance index of populations during the genetic search, in order to find the optimal values of control parameters so that the performance index is maximized. We have proved that, for the stochastic optimization problem, there exists a pure deterministic strategy which is at least as good as any other pure or mixed (randomized) strategy. It implies that if a proper criterion function to be optimized in GA is given, then at each generation t, the choice of optimal control parameters is deterministic. However, since there are so many states in the Markov model, and the numerical solution of the equilibrium distribution cannot be obtained by available computers, so it is difficult to apply this theorem directly at present. This work shows that the optimal parameters in genetic search may be detected using stochastic simulated approach [16, 17].

References

[1] Goldberg, D. E.: Genetic algorithms in search, optimization and machine learning. Reading, Massachusetts, USA: Addison-Wesley. (1989)

[2] Bäck, T.: Evolutionary Algorithms in Theory and Practice. New York: Oxford University Press. (1996)

[3] Bäck, T., Fogel, D. B. and Michalewicz, Z: Handbook of Evolutionary Computation. IOP Publishing Ltd and Oxford University Press, (1997)

[4] Grefenstette, J. J.: Optimization of control parameters for genetic algorithms. IEEE Trans. on System, Man and Cybernetics, **16**, (1986) 122–128

[5] Schaffer, D. and Morishma, A.: An adaptive crossover mechanism for genetic algorithms. Proc. Second Int. Conf. Genetic Algorithms, (1987) 36–40

[6] Davis, L.: Adapting operator probabilities in genetic algorithms. Proc. Third Int. Conf. on Genetic Algorithms. (1989) 61–69

[7] Davis, T. E.: Toward an extrapolation of the simulated annealing convergence theory onto a simple genetic algorithm. Ph.D. thesis, University of Florida, USA (1991)

[8] Fogarty, T. C.: Varying the probability of mutation in genetic algorithms. Proc. Third Int. Conf. on Genetic Algorithms, (1989) 104–109

[9] Srinivas, M. and Patnaik, L. M.: Adaptive probabilities of crossover and mutation in genetic algorithms. IEEE Trans. on System, Man and Cybernetics. **24**, (1994) 656–667

[10] Bäck, T.: Self-adaptation in genetic algorithms. Proc. of the First European Conference on Artificial Life. MIT Press, (1992) 263–271

[11] Hinterding, R.: Gaussian mutation and self-adaptation in numeric genetic algorithms. Proc. IEEE International Conference on Evolutionary Computation, IEEE Press, (1994) 384–389

[12] Vose, M. D., Modeling of genetic algorithms. In Whitley L. D. (ed), Foundations of Genetic Algorithms 2, (1992) 63–73

[13] Isaacson, D. L. and Madsen, R. W.: Markov Chains Theory and Applications. New York, USA: John Wiley and Sons. (1976)

[14] Bellman, R. and Dreyfus, S.: Applied Dynamic Programming. Princeton, N. J.: Princeton University Press. (1962)

[15] Feldbaum, A.: Optimal Control Systems. New York: Academic Press. (1965)

[16] Cao, X. R. and Wan, Y: Algorithms for sensitivity analysis of Markov systems through potentials and perturbation realization. IEEE Trans. on Contr. Syst. Tech., 6, (1998) 482–494

[17] Ho, Y. C. and Cao, X. R.: Perturbation Analysis of Discrete Event Dynamic Systems. Boston, MA: Kluwer (1991)

Tackling Epistatic Problems Using Dynastically Optimal Recombination

Carlos Cotta and José M. Troya

Dept. of Lenguajes y Ciencias de la Computación, University of Málaga,
Complejo Tecnológico (2.2.A.6), Campus de Teatinos,
E-29071, Málaga, Spain
{ccottap, troya}@lcc.uma.es

Abstract. The application of a heuristic recombination operator to epistatic problems is discussed in this paper. This operator uses an A*-like mechanism to intelligently explore the dynastic potential (set of possible offspring) of recombined solutions, exploiting partial knowledge about the epistatic relation of variables. Two case studies (the design of a brachystochrone and the Rosenbrock function) are presented, providing experimental results. The scalability of the operator for increasing problem sizes is also considered.

1 Introduction

One of the most distinctive features of genetic algorithms [4] with respect to other related techniques such as evolution strategies [10, 11] or evolutionary programming [2] is the emphasis put on the use of recombination operators. From the perspective of genetic-algorithm researchers, recombination is the fundamental search engine, while mutation plays an important but nevertheless secondary rôle. According to these considerations, the classical interpretation of the search that a genetic algorithm performs is the combination of the exploratory effects of mutation with the exploitative effects of recombination, intended to combine valuable parts of solutions independently discovered.

This interpretation is questionanle though. Consider that, in the presence of epistasis, the value of a piece of information is very dependent on the context within such information is immersed. In order words, it is necessary to transmit high-order schemata (or, more generally, high-order *formae* [6]) to allow the interchange of information valuable per se. On the one hand, this implies that algorithms using operators transmitting low-order formae can be more easily misled. On the other hand, performing a blind interchange of information without considering the available knowledge about the epistatic relations among variables may be very inappropriate.

These considerations motivate that recombination be considered in many situations just as a kind of macromutation. Nevertheless, the rationale behind recombination, i.e., combining valuable information taken from the parents, could still be useful in these situations if problem-dependent knowledge were used to determine which features of the parents must be combined and how to do this. In this sense, Cotta *et al.* [1] have proposed a framework for such a operator, namely Dynastically Optimal Recombination (DOR). This framework results in a family of operators adequate for different problem domains. This work studies the application of DOR to epistatic functions.

The remainder of the article is organised as follows. First, a brief overview of the DOR framework is presented in Sect. 2. Then, the details of the construction of particular DOR operators are given for two test problems in Sect. 3. Subsequently, experimental results are shown and discussed in Sect. 4. The scalability of the operator is considered in Sect. 5. Finally, some conclusions are outlined in Sect. 6.

2 Dynastically Optimal Forma Recombination

Before defining the Dynastically Optimal Recombination operator, some previous concepts must be stated. First of all, let $x = \{\eta_1, \cdots, \eta_n\}$ and $y = \{\zeta_1, \cdots, \zeta_n\}$ be two feasible solutions[1]. A recombination operator X can be defined as a function

$$X : \mathcal{S} \times \mathcal{S} \times \mathcal{S} \to [0, 1], \tag{1}$$

where $X(x, y, z)$ is the probability of generating z when recombining x and y using X. Clearly, the following condition must hold

$$\forall x \in \mathcal{S}, \forall y \in \mathcal{S} \;:\; \sum_{z \in \mathcal{S}} X(x, y, z) = 1 \tag{2}$$

Now, the *Immediate Dynastic Span* [8] of x and y with respect to a recombination operator X is

$$\Gamma_X^1(\{x, y\}) = \{z \mid X(x, y, z) > 0\}, \tag{3}$$

i.e., the set of solutions that can be obtained when X is applied on x and y. On the other hand, the *Dynastic Potential* $\Gamma(\{x, y\})$ of x and y is defined as

$$\Gamma(\{x, y\}) = \{z \mid \forall \xi : z \in \xi \Rightarrow (x \in \xi) \vee (y \in \xi)\}, \tag{4}$$

i.e., the set of solutions that can be constructed using nothing but the information contained in the parents. If $\Gamma_X^1(\{x, y\}) \subseteq \Gamma(\{x, y\})$, the operator X is said to be *transmitting* [9]. An example of such an operator is Radcliffe's *Random Transmitting Recombination* (RTR), defined as follows:

$$\mathrm{RTR}(x, y, z) = \begin{cases} \frac{1}{|\Gamma(\{x,y\})|} & z \in \Gamma(\{x, y\}) \\ 0 & \text{otherwise} \end{cases} \tag{5}$$

Thus, RTR returns a random individual entirely composed of its parents' formae. As stated in Sect. 1, this random selection may be inappropriate if it is

[1] The notation $z = \{\xi_1, \cdots, \xi_n\}$ is used to denote that $z \in \bigcap_{i=1}^n \xi_i$.

done on the basis of low-order formae, whose fitness information is noisy due to epistasis. Thus, problem-dependent knowledge must be used to identify higher-order formae that can be satisfactorily linked. This use of problem-dependent knowledge is formalised within the framework of *Dynastically Optimal Recombination* (DOR). To be precise, let $\phi : \mathcal{S} \to \mathbb{R}^+$ be the fitness function[2]. Then, DOR is defined as a recombination operator for which the following condition holds:

$$\mathrm{DOR}(x, y, z) = 0 \Leftrightarrow \{[z \notin \Gamma(\{x, y\})] \vee [\exists w \in \Gamma(\{x, y\}) : \phi(w) < \phi(z)]\} \quad (6)$$

Thus, DOR is a transmitting operator such that no other solution in the dynastic potential is better than any solution it generates. According to this definition, the use of DOR implies performing an exhaustive search in a small subset of the solution space. Such an exhaustive search can be efficiently done by means of the use of a subordinate A*-like algorithm as described below.

DOR uses optimistic estimations $\hat{\phi}(\Psi)$ of the fitness of partially specified solutions Ψ (i.e., $\forall z \in \mathcal{S} : \hat{\phi}(\Psi) < \phi(z)$) in order to direct the search to promising regions, pruning dynastically suboptimal solutions. More precisely, $\hat{\phi}(\Psi)$ is decomposed as $\gamma(\Psi) + \varphi(\Psi)$, where $\gamma(\Psi)$ is the known final contribution of the formae included in Ψ to the fitness of any $z \in \Psi$, and $\varphi(\Psi)$ is an optimistic estimation of the fitness contribution for the remaining underspecified epistatic relations. Although it is possible to set $\varphi(\Psi) = 0$, it is clear that the more accurate fitness estimation, the more efficient the search will be.

Thus, solutions are incrementally constructed using the formae to which any of the parents belong. In the problems considered in this work, these formae are orthogonal. In other words, let $x = \{\eta_1, \cdots, \eta_n\}$ and $y = \{\zeta_1, \cdots, \zeta_n\}$ be the solutions to be recombined. Then, any $z = \{\xi_1, \cdots, \xi_n\}$, where $\xi_i \in \{\eta_i, \zeta_i\}$, is a valid solution included in $\Gamma(\{x, y\})$. Initially, $\Psi_0^1 = \mathcal{S}$. Subsequently,

$$\Psi_{i+1}^{2\cdot j} = \Psi_i^j \cap \eta_{i+1} \quad (7)$$

$$\Psi_{i+1}^{2\cdot j+1} = \Psi_i^j \cap \zeta_{i+1} \quad (8)$$

are considered. Whenever $\bar{\phi} < \hat{\phi}(\Psi)$ (where $\bar{\phi}$ is the fitness of the best-so-far solution generated during this process), the macro-forma Ψ is closed (i.e., discarded). Otherwise, the process is repeated for open macro-formae. Obviously, this mechanism is computationally more expensive than any classical blind recombination operator but, as it will be shown in the next sections, the resulting algorithm performs better when compared to algorithms using blind recombination and executed for a computationally equivalent number of iterations.

3 Two Case Studies

Two case studies are presented in this section. First, the design of a brachystochrone is considered. Subsequently, the Rosenbrock function is approached.

[2] Minimisation is assumed without loss of generality.

3.1 Design of a Brachystochrone

The design of a brachystochrone is a classical problem of the calculus of variations. This problem involves determining the shape of a frictionless track along which a cart slides down by means of its own gravity, so as to minimise the time required to reach a destination point from a motionless state at a given starting point. To approach this problem by means of genetic algorithms, the track is divided into a number of equally spaced pillars [3]. Subsequently, the algorithm determines the height of each of these pillars.

As stated above, the objective is to minimise the time the cart requires to traverse the track. This time can be calculated as the sum of the times required to move between any two consecutive pillars, i.e.,

$$t = \sum_{i=1}^{n+1} t_i = \sum_{i=1}^{n+1} t(h_{i-1}, h_i, \lambda, v_i) \tag{9}$$

where n is the number of pillars, h_i is the height of the ith pillar (h_0 and h_{n+1} are data of the problem), λ is the distance between consecutive pillars (a problem parameter as well), and $v_i = v(v_{i-1}, h_{i-1}, h_i)$ is the velocity at the ith pillar ($v_0 = 0$). As it can be seen, this is an epistatic problem: the contribution of each variable (i.e., pillar height) depends on the values of previous variables.

According to the definition of DOR, the new solution will be generated using only the pillar heights provided by the parents. To be precise, DOR will consider partially specified solutions $\{h_1, \cdots, h_i\}, i < n$. For such a solution, times t_j and velocities v_j, $j \in \{1, \cdots, i\}$, can be calculated and hence

$$\gamma(\{h_1, \cdots, h_i\}) = \sum_{j=1}^{i} t_j, \tag{10}$$

$$\varphi(\{h_1, \cdots, h_i\}) = \frac{d_i}{v^*}, \tag{11}$$

where d_i is the straight distance from the ith pillar to the destination point and $v^* = \max(v_i, v_{n+1})$, v_{n+1} being the velocity at the end of the track (known in advance as a function of v_0, h_0 and h_{n+1}).

The values $h_0 = 2$, $h_{n+1} = 0$, $\lambda = 4 \cdot (n+1)^{-1}$, and $(2h_{n+1} - h_0) \leq h_i < h_0$, $i \in \{1, \cdots, n\}$, have been considered in all the experiments realised in this work.

3.2 The Rosenbrock Function

The Rosenbrock function is a multidimensional epistatic minimisation problem defined as:

$$f(x) = \sum_{i=1}^{n-1} [100 \cdot (x_{i+1} - x_i^2)^2 + (1 - x_i)^2] \tag{12}$$

In this problem, there exist epistatic relations between any pair of adjacent variables. Additionally, there exist non-epistatic terms as well. However, the latter have a much lower weight, and hence the search is usually dominated by the former. As a matter of fact, there exists a strong attractor located at $\boldsymbol{x} = \mathbf{0}$, where these terms become zero. The further evolution towards the global optimum ($\boldsymbol{x} = \mathbf{1}$) is usually very slow.

When considering a macro-forma $\{z_1, \cdots, z_i\}, i < n$, it is possible to calculate $\gamma(\{z_1, \cdots, z_i\})$ as

$$\gamma(\{z_1, \cdots, z_i\}) = f(z_1, \cdots, z_{i-1}) + (1 - z_i)^2. \tag{13}$$

As to $\varphi(\{z_1, \cdots, z_i\})$, it can be estimated as follows. First, a $(n-1) \times 4$ table τ is generated containing the four possible values of $(z_{i+1} - z_i^2)^2$, i.e., the four combinations in the Cartesian product $\{x_i, y_i\} \times \{x_{i+1}, y_{i+1}\}$. Now, let $\tau_j^* = \min\{\tau_{jk} : 1 \leq k \leq 4\}$. Then, $\varphi(\{z_1, \cdots, z_i\})$ is

$$\varphi(\{z_1, \cdots, z_i\}) = 100 \cdot \sum_{j=i}^{n-1} \tau_j^* + \sum_{j=i+1}^{n-1} \min\{(1 - z_j)^2 : z_j \in \{x_j, y_j\}\}. \tag{14}$$

The range $-5.12 \leq x_i \leq 5.12$, $i \in \{1, \cdots, n\}$, has been considered in this work.

4 Experimental Results

Experiments have been done using an elitist generational genetic algorithm (*popsize*=100, p_c=.9, p_m=$1/n$, where n is the dimensionality of the problem) using ranking selection [12] and binary bit-flip mutation. For the brachystochrone problem (resp. the Rosenbrock function), variables are encoded with 16 bits and the algorithm is stopped after $5 \cdot 10^5$ evaluations (resp. 64 bits and 10^5 evaluations). No fine-tuning of these parameters has been attempted. When using the DOR operator, the algorithm is stopped after having calculated an equivalent number of epistatic relations. To be precise, each time a macro-forma is extended by adding an additional forma, this is accounted as $1/n$ evaluations. Additionally, mutation is applied before recombination in the case of the DOR operator. Thus, the heuristic recombination performed by DOR is not perturbed and new genetic material is still introduced in the population.

First of all, the ability of DOR for transmitting high-order formae has been tested on the two test problems. To be precise, the order of the formae transmitted by DOR has been compared to an unbiased RTR operator. This is done by running a genetic algorithm using DOR, and measuring the length of the contiguous portions transferred from any of the parents. Subsequently, the frequency of each length is divided by the corresponding frequency in a blind operator to obtain a relative ratio. As shown in Fig. 1 for the brachystochrone design problem, and Fig. 2 for the Rosenbrock function, while low-order formae are more seldom transmitted, high-order formae are more frequently processed.

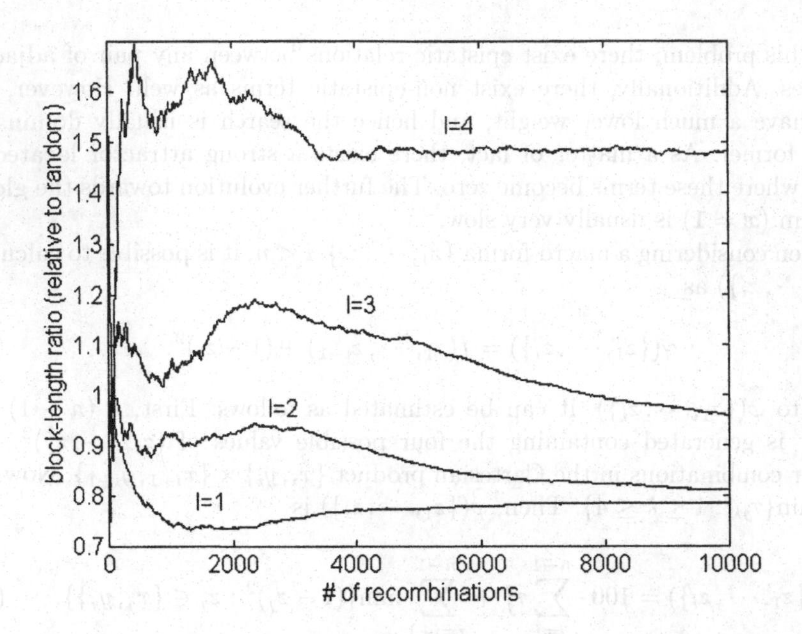

Fig. 1. Evolution of the relative frequency of the order of formae transmitted by DOR with respect to RTR. The results correspond to the brachystochrone design problem ($n = 10$).

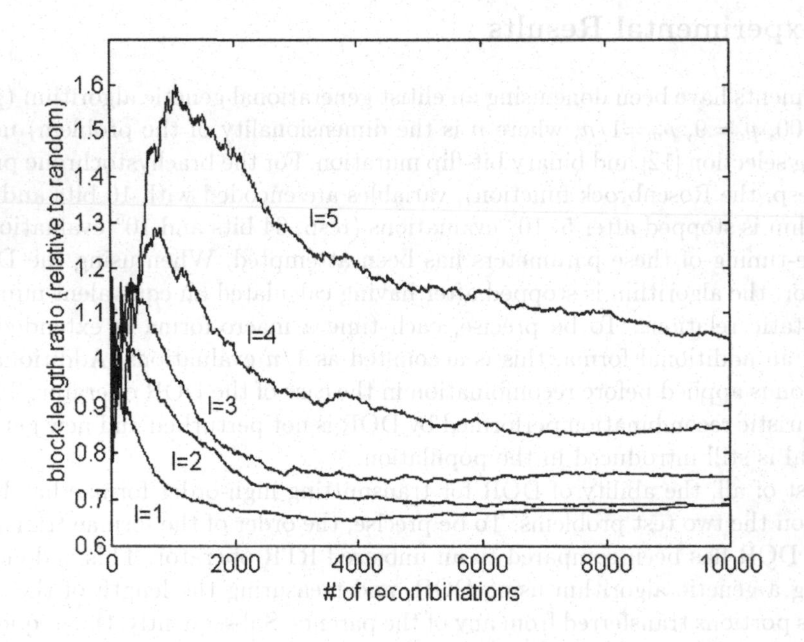

Fig. 2. Evolution of the relative frequency of the order of formae transmitted by DOR with respect to RTR. The results correspond to the Rosenbrock function ($n = 10$).

Table 1. Results (averaged for 20 runs) of a genetic algorithm using different recombination operators. The best result for each dimensionality is shown in boldface.

	Brachystochrone design (# of pillars)			Rosenbrock function (# of pillars)		
Operator	8	12	16	8	12	16
SPX	1.1577	1.1788	1.2156	13.7903	14.8529	27.7406
UX	1.1757	1.2300	1.2862	14.0393	18.3642	49.0327
AX	1.1560	1.1668	1.1828	6.2408	14.4236	35.8189
R^3	1.1540	1.1627	1.1783	6.1000	11.8783	40.2665
DOR	**1.1530**	**1.1549**	**1.1779**	**2.4963**	**5.8600**	**9.7588**

Table 1 shows the results for DOR and other classical operators such as single-point crossover (SPX), uniform crossover (UX), arithmetical crossover [5] (AX) and random respectful recombination for real genes [7] (R3). As it can be seen, the improvement is very good for the Rosenbrock function and only moderate for the brachystochrone problem. This fact admits several explanations. On the one hand, the latter problem may be harder for DOR than the Rosenbrock function and thus the allocated number of epistatic calculations may be quickly comsumed. On the other hand, the dimensionality of the instances considered may be too low to be challenging for classical operators. For this reason, experiments have been done with larger instances. The results are shown in the next section.

5 Scaling Issues

The A*-like nature of DOR makes it very sensitive to increments in the problem size and the subsequent combinatorial explosion. However, it is possible to overcome this drawback by modifying the representation *granularity*. Recall from Eqs. (7) and (8) that DOR incrementally constructs macro-formae by adding formae to which any of the parents belong. Then, it is possible to adjust the order of the formae added in each step (i.e., considering macro-formae instead of just basic formae), so as to make DOR combining larger portions of the ancestors. Let $g \geq 1$ be the order of these incremental macro-formae. Then,

$$\Psi_{i+1}^{2\cdot j} = \Psi_i^j \cap \bigcap_{j=1}^{\min(g,n-i\cdot g)} \eta_{i\cdot g+j} \tag{15}$$

$$\Psi_{i+1}^{2\cdot j+1} = \Psi_i^j \cap \bigcap_{j=1}^{\min(g,n-i\cdot g)} \zeta_{i\cdot g+j} \tag{16}$$

Notice that, for large dimensionality and small granularity, DOR is ineffective due to its computational cost, while for very large granularity, the capability for combining information from the parents is very limited. Hence, intermediate values seem to be appropriate. This fact has been empirically corroborated, suggesting a dimensionality-to-granularity ratio $\rho \approx 10$. Table 2 shows the results for larger problem instances.

Table 2. Results (averaged for 20 runs) of a genetic algorithm for larger problem instances of the brachystochrone design problem and the Rosenbrock function. The best result for each dimensionality is shown in boldface.

	Brachystochrone design (# of pillars)					
Operator	24	32	40	48	56	64
SPX	1.2713	1.3908	1.6262	1.7787	2.1049	2.5971
UX	1.3499	1.5305	1.6771	1.9379	2.2172	2.6658
AX	1.2219	1.2722	1.3534	1.4841	1.6327	1.8845
R^3	1.2149	1.2803	1.3683	1.5295	1.7357	1.9061
DOR	**1.2110**	**1.2660**	**1.3413**	**1.3886**	**1.4720**	**1.5894**
	Rosenbrock function (# of variables)					
Operator	24	32	40	48	56	64
SPX	1.2051e+2	3.7101e+2	1.1603e+3	2.7489e+3	5.9497e+3	1.1453e+4
UX	2.1405e+2	8.0481e+2	2.6152e+3	5.9924e+3	1.3691e+4	2.5647e+4
AX	1.3697e+2	3.0272e+2	5.5235e+2	1.0254e+3	1.5020e+3	2.2645e+3
R^3	1.5063e+2	4.0126e+2	8.0559e+2	1.3028e+3	2.0523e+3	2.9633e+3
DOR	**2.0453e+1**	**3.3916e+1**	**7.0529e+1**	**1.3165e+2**	**3.4566e+2**	**5.4129e+2**

Again, DOR provides the best results for both problems. Moreover, it seems to scale much better than other classical operators. It must be noted that, although the value $\rho \approx 10$ seems to be robust, an adjustment of this parameter (usually in the range $8 \leq \rho \leq 12$) may provide better results. Notice also that this ratio is problem-dependent and may take different values for other test problems.

6 Conclusions and Future Work

This work has studied the application of the DOR operator to two orthogonal epistatic problems. The experimental results are satisfactory: DOR always performs better than classical blind operators at the same computational cost. In this sense, a comment must be done regarding the criterion chosen for measuring the computational cost of DOR, i.e., accounting each extension of a forma as $1/n$ evaluations. By following this criterion, it is implicitly assumed that extending a forma n times is equivalent to a full evaluation. While this may be true in many situations, it is also frequently possible to create lookup tables of linear size, thus greatly reducing the cost of extending a forma. In this sense, the chosen measure is clearly detrimental to DOR. Despite this, DOR has provided the best results. Furthermore, it has been shown that it is possible to tackle larger problem instances with the same successful results by adjusting the granularity of the representation.

As mentioned in Sect. 1, the use of DOR is conditioned upon the availability of problem-dependent knowledge. Nevertheless, it must be noted that the use of such knowledge is in general necessary according to the *No Free Lunch Theorem* [13]. Moreover, it is the amount of problem-knowledge (included in the optimistic evaluation function $\hat{\phi}$) what determines the highest granularity-to-dimensionality ratio computationally affordable.

Future work will try to confirm these results on other test problems, focusing on the choice of the dimensionality-to-granularity ratio. Extending these results to non-orthogonal representations is a line of future work too.

References

1. C. Cotta, E. Alba, and J.M. Troya. Utilising dynastically optimal forma recombination in hybrid genetic algorithms. In A.E. Eiben, Th. Bäck, M. Schoenauer, and H.-P. Schwefel, editors, *Parallel Problem Solving From Nature - PPSN V*, volume 1498 of *Lecture Notes in Computer Science*, pages 305–314. Springer-Verlag, Berlin Heidelberg, 1998.
2. L.J. Fogel, A.J. Owens, and M.J. Walsh. *Artificial Intelligence Through Simulated Evolution.* Wiley, New York NY, 1966.
3. M. Herdy and G. Patone. Evolution strategy in action: 10 es-demonstrations. Technical Report TR-94-05, Technische Universität Berlin, 1994.
4. J.H. Holland. *Adaptation in Natural and Artificial Systems.* University of Michigan Press, Ann Harbor MI, 1975.
5. Z. Michalewicz. *Genetic Algorithms + Data Structures = Evolution Programs.* Springer-Verlag, Berlin, 1992.
6. N.J. Radcliffe. Equivalence class analysis of genetic algorithms. *Complex Systems*, 5:183–205, 1991.
7. N.J. Radcliffe. Forma analysis and random respectful recombination. In R.K. Belew and L.B. Booker, editors, *Proceedings of the Fourth International Conference on Genetic Algorithms*, pages 222–229, San Mateo CA, 1991. Morgan Kaufmann.
8. N.J. Radcliffe. The algebra of genetic algorithms. *Annals of Maths and Artificial Intelligence*, 10:339–384, 1994.
9. N.J. Radcliffe and P.D. Surry. Fundamental limitations of search algorithms: Evolutionary computing in perspective. In J. Van Leeuwen, editor, *Computer Science Today: Recent Trends and Developments*, volume 1000 of *Lecture Notes in Computer Science*, pages 275–291. Springer-Verlag, 1995.
10. I. Rechenberg. *Evolutionsstrategie: Optimierung technischer Systeme nach Prinzipien der biologischen Evolution.* Frommann-Holzboog Verlag, Stuttgart, 1973.
11. H.-P. Schwefel. Evolution strategies: A family of non-linear optimization techniques based on imitating some principles of natural evolution. *Annals of Operations Research*, 1:165–167, 1984.
12. L.D. Whitley. Using reproductive evaluation to improve genetic search and heuristic discovery. In J.J. Grefenstette, editor, *Proceedings of the Second International Conference on Genetic Algorithms*, pages 116–121, Hillsdale NJ, 1987. Lawrence Erlbaum Associates.
13. D.H. Wolpert and W.G. Macready. No free lunch theorems for search. Technical Report SFI-TR-95-02-010, Santa Fe Institute, 1995.

Extended Methods for Classification of Remotely Sensed Images Based on ARTMAP Neural Networks

Norbert Kopčo[1,2], Peter Sinčák[2], and Howard Veregin[3]

[1] Department of Cognitive and Neural Systems, Boston University
677 Beacon St., Boston, MA 02215, USA
[2] Laboratory of Artificial Intelligence, Department of Cybernetics and AI
Faculty of Electrical Engineering and Informatics, Technical University Košice
Letná 9, 040 01 Košice, Slovak Republic
[3] Department of Geography, University of Minnesota
267-19th Ave S., Minneapolis, MN 55455, USA
e-mail: cig@neuron-ai.tuke.sk

Abstract. This paper deals with two aspects of the application of ARTMAP neural networks for classification of satellite images obtained by remote sensing of the Earth. The first part contains an analysis of the influence of data representation and cluster determination method on classification accuracy. Three types of representation/determination are analyzed. Best results are obtained for Gaussian ARTMAP, an ARTMAP neural network using gaussian distributions for identification of clusters in feature space. In the second part, a method for evaluation of the classification quality is described. This method introduces a confidence index which is assigned to each individual pixel of the image, thus allowing generation of a confidence map for the classified image. The confidence index is computed conveniently exploiting features of the Gaussian ARTMAP learning algorithm. Using a threshold determining the minimal required confidence, this method allows one to generate a map which shows only pixels with prescribed minimal confidence, effectively creating an additional class containing pixels classified with subthreshold confidence.

Introduction

Satellite remote sensing is a progressive way of collecting data used for generation of various kinds of maps (vegetation maps, land-use maps, etc.). Usually, the most important part of processing of these data is classification of the multispectral imagery obtained by remote sensing. The standard approach to this task is the application of statistical classification methods, e.g., Maximum Likelihood classifier [13], or methods based on crude expert-defined rules [11]. These methods have the problem of high complexity and large data volume. It would also be desirable to develop a more autonomously functioning classifier so that subjective human participation in the task could be minimized.

A relatively new and potentionally useful method of classification of remotely sensed data is the method of classification using neural networks. In the last years, several types of neural networks have been used for classification of these data, most of them using the Backpropagation algorithm [15]. Another class of neural networks used in this domain is *Adaptive Resonance Theory* (ART) neural networks (see, e.g., [1],[14]). For an extensive discussion of the neural network versus statistical approaches to image processing see [12]. An interesting approach is described also in [10].

This study concentrates on two aspects of applying ARTMAP neural networks for classification of remotely sensed data. First, the study gives a brief intuitive description of the ARTMAP neural networks (Section 1). In Section 2, performacne of three types of ARTMAP networks, differing in the cluster determination method, is compared. Section 3 then describes a method for evaluation of classification quality for Gaussian ARTMAP neural networks.

The data for this study consists of an Landsat Thematic Mapper (TM) image of the city of Košice (located in Eastern Slovakia) and its environs. The whole image consists of 368,125 7-dimensional pixels, out of which 6,331 pixels were classified by an expert into seven thematic categories. Figure 1 shows the original image along with the seven categories identified by expert. The following classes

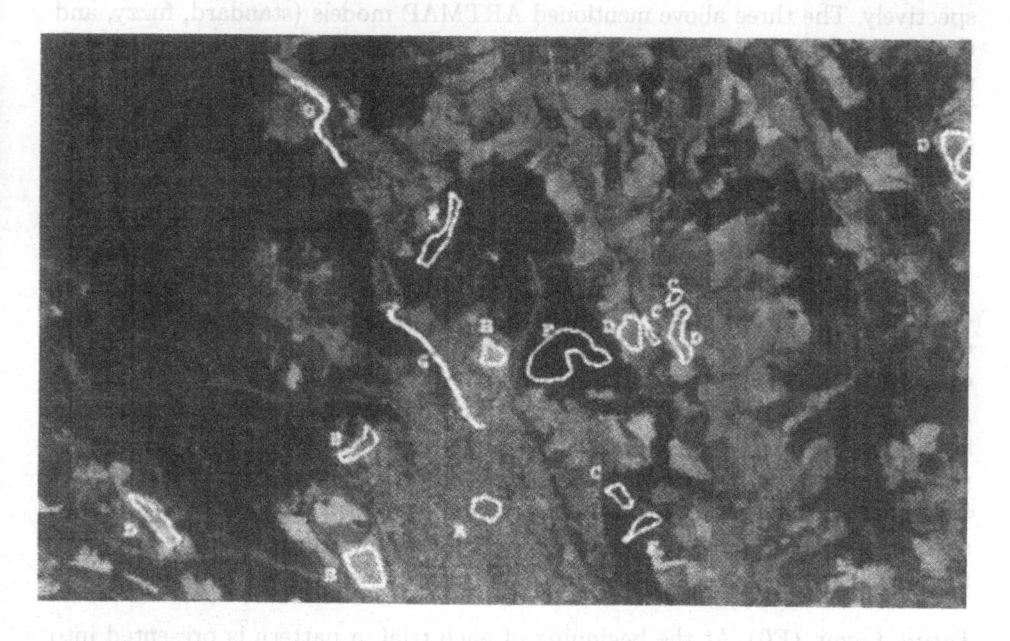

Fig. 1. Original image. Highlighted areas were classified by expert (see text).

are defined in the figure: A - urban area, B - barren fields, C - bushes, D - agricultural fields, E - meadows, F - woods, G - water.

1 ARTMAP neural networks

ARTMAP neural networks belong to a group of neural networks based on Adaptive Resonance Theory (ART), a theory of cognitive information processing in human brain [9]. The ART neural networks, as this group of neural networks is collectively known, are especially suitable for pattern recognition and classification applications using supervised as well as unsupervised learning. The first neural network model of this category, ART 1 [2], was a system for unsupervised classification of binary input data. This model was later extended to ART 2 [4] and fuzzy ART [5], ART systems for unsupervised classification of analog data. Also other modifications of the ART 1 model were introduced, for example Gaussian ART [16], which uses Gaussian distributions to define individual categories.

The next step in the development of ART models was the introduction of a new ART architecture, called ARTMAP [6], which was designed for supervised learning of arbitrary mappings of input vectors to output vectors. The ARTMAP neural network, which is the basic model of ARTMAP architecture, is in its basic version a modular system consisting of two ART 1 modules and a controlling module called Map-field. If the ART 1 modules in the ARTMAP architecture are replaced by fuzzy ART modules or Gaussian ART modules, a new ARTMAP model is obtained, called fuzzy ARTMAP [3] or Gaussian ARTMAP [16], respectively. The three above mentioned ARTMAP models (standard, fuzzy, and Gaussian ARTMAP) represent basic models of the ARTMAP architecture. The main difference among them is in the way they identify clusters in feature space. And the first part of the present paper deals with analysis of this difference in identification method, and its influence on computational properties of the three systems.

1.1 ARTMAP systems dynamics

A detailed description of the dynamics of the ART and ARTMAP systems, relevant to our study, can be found, e.g., in [6], [3], and [16], which contain a description of the standard ARTMAP architecture. A simplified version of the ARTMAP architecture was implemented for the present study. A detailed description of it can be found in [14]. In this section, an outline of the learning process in an ARTMAP system is presented with the goal of developing an intuitive understanding of the learning process. The basic structure of an ARTMAP system is shown in Figure 2. Individual blocks in the figure have the following function:

Input Layer (F0) At the beginning of each trial, a pattern is presented into this layer. The size of this layer (number of neurons) is equal to the dimensionality of input patterns (N).

Comparison Layer (F1) In the second step of each trial (see description of the algorithm below), the pattern from the Input Layer is copied into this layer. In the fourth step, the Input pattern and the optimal pattern represented

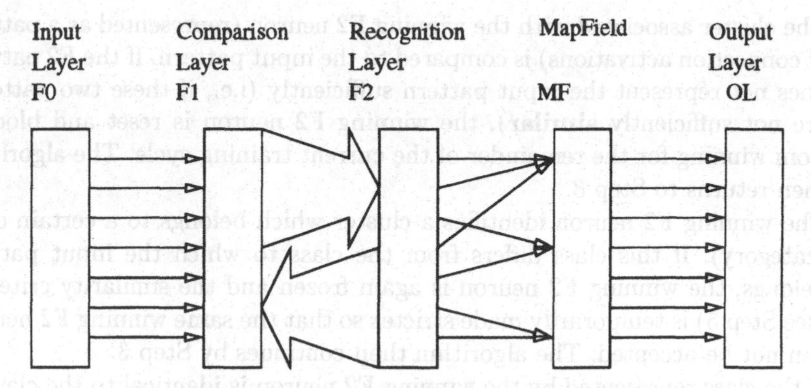

Fig. 2. Topology of ARTMAP neural networks

by the winning neuron in the Recognition Layer are compared here. If they are significantly different, a search for a new winning F2 neuron is initiated. The size of this layer is *N*.

Recognition Layer (F2) Each neuron in this layer defines one cluster in feature space (the identification method differs for different types of ARTMAP models). The size of this layer is variable. It grows during training as new clusters in feature space are identified.

MapField (MF) Each neuron in this layer represents one class and it receives input by a non-zero link from all the F2 neurons (i.e., all the clusters) which belong to the same class represented by this neuron. During training, the content of the MF layer is identical with the content of the Output Layer.

Output Layer (OL) Each neuron in this layer represents one class defined in the data set. At the beginning of each trial, a code of the class into which the input pattern should be classified is copied into this layer. Each class is represented by one neuron. The size of this layer is equal to the number of classes (M).

The ARTMAP neural network training procedure can be described as follows:

1. The input pattern is presented to the Input Layer. The corresponding class code is presented to the Output Layer.
2. The pattern from the Input Layer is identically copied into the Comparison Layer.
3. If there are no neurons in the Recognition Layer (F2) or if all the F2 neurons are reset, a new neuron is added to this layer. The connections from the F1 layer to the new neuron, which define the cluster identified by that neuron, are initialized to values identical with the input pattern. The new F2 neuron is also connected with the MapField neuron representing the class to which the input pattern belongs. The algorithm then continues by Step 1 (i.e., by a new training cycle).
4. If there are non-reset neurons in the F2 layer, the neuron in the Recognition Layer (F2) defining a cluster most suitable to include the input pattern is chosen as the winner.

5. The cluster associated with the winning F2 neuron (represented as a pattern of connection activations) is compared to the input pattern. If the F2 pattern does not represent the input pattern sufficiently (i.e., if these two patterns are not sufficiently **similar**), the winning F2 neuron is reset and blocked from winning for the remainder of the current training cycle. The algorithm then returns to Step 3.
6. The winning F2 neuron identifies a cluster which belongs to a certain class (category). If this class differs from the class to which the input pattern belongs, the winning F2 neuron is again frozen and the similarity criterion (see Step 5) is temporarily made stricter so that the same winning F2 neuron can not be accepted. The algorithm then continues by Step 3.
7. If the class represented by the winning F2 neuron is identical to the class to which the current pattern belongs, the winner is accepted and the weights of links connecting the winning F2 neuron with the F1 layer are updated so that the cluster associated with the winning F2 neuron is more similar to the current input pattern.

This procedure is repeated until the network classifies all the input patterns correctly.

In the testing/application phase, the procedure is as follows:

1. The unknown input pattern is presented into the Input Layer and copied into the Comparison Layer.
2. A winner in the Recognition Layer is found. If the pattern represented by the winner is not sufficiently similar to the input pattern, the system is not able to classify the input pattern. Otherwise, the input pattern is classified into the MapField class associated with the winning F2 neuron.

There are two adjustable parameters in the ARTMAP networks. The first of them is the baseline similarity threshold (or so-called vigilance parameter) used in the Comparison Layer computations. This parameter can range from zero to one (one meaning that identity of the compared patterns is required). The second parameter is the learning rate and again it can range from zero to one.

It has been shown before that results obtained by ARTMAP networks are dependant on the order in which the input patterns are presented during training. To suppress this presentation-order dependence, a method of *voting* is usually applied. In this method, several (usually five) independent ARTMAP networks are trained on the same data-set presented in different order. In the testing phase a new pattern is classified into the class voted for by majority of the networks.

2 Influence of data representation and cluster identification method on classification accuracy

The description given in the previous section holds, up to some minor details, for all three modifications of the ARTMAP model compared in the present study (standard, fuzzy, and Gaussian ARTMAP). The main difference among the algorithms is in the way the Recognition Layer (F2) neurons define the clusters they represent. This is closely related to the way in which the input patterns have to be pre-processed before training or testing. Individual properties of the examined models can be summarized as follows.

ARTMAP This model requires the data to be in a binary format. So, to be able to apply it on the RS data used in this study the data has to be transformed. We chose to use a mapping which transforms data into binary code by direct conversion of a decimal value into a binary value. Each digit of this value is then processed individually. This transformation has several consequences. First, the dimensionality of the patterns (and consequently the size of the network and the computation time) is considerably larger (eight-fold increase for decimal numbers ranging from 0 to 255). Second, this transformation dramatically changes relations among patterns. For example, values of 0 (00000000 binary) and 128 (10000000 binary) have in binary space the same Euclidian distance as values of 0 (00000000 binary) and 1 (00000001 binary). In ARTMAP, each cluster is defined by a hyperrectangle in feature space (e.g., a cube in the three-dimensional space). In our case it means that each cluster will be represented by a 56-dimensional hyperrectangle. It follows that this transformation will lead to creation of different clusters for binary data compared to analog data.

fuzzy ARTMAP This model accepts analog (continuous) data in the range [0, 1]. So the only transformation of data that needs to be done is rescaling. Each cluster in the Recognition Layer is again defined by a hyper-rectangle of the same dimensionality as is the dimensionality of the input data (in the present study, seven dimensions).

Gaussian ARTMAP In this model, each cluster in the Recognition Layer is defined as a Gaussian probability distribution with a mean and variance in each dimension, and ***a priori*** probability of a given cluster. This method does not require any transformation of data. But, to obtain better comparability with other methods, we rescaled the data in the same way as in the fuzzy ARTMAP experiments (range [0, 1]).

2.1 Definition of task

Although all three analyzed methods are very similar, as can be seen in the above discussion, the difference in the way they determine individual clusters should lead to significantly different results when used for the same classification task. The goal of this experiment is to compare the three classification methods in terms of classification accuracy achieved. The results should suggest the most suitable method of cluster identification for the remote sensing data used here. Also, the goal is to present Gaussian ARTMAP as a new method for classification of data in remote sensing. The expected results are as follows.

- *ARTMAP:* Because of the high dimensionality which resulting from the transformation required by this algorithm, it is expected that the algorithm will be very sensitive to small changes in data and its ability to generalize will be poor.

- *fuzzy ARTMAP:* Based on previous studies and authors' experiences it is expected that fuzzy ARTMAP will perform very well in this task.
- *Gaussian ARTMAP:* This neural network has been never before used for RS data classification. But the experiments performed by the authors previously on benchmark classification tasks suggest that it should be very powerful.

2.2 Methods

The data set consists of 6331 seven-dimensional patterns. The set was randomly divided into a training set (3166 patterns) and a test set (3165 patterns). Five copies of the training set were created, each containing the training patterns in a different random order. The training and testing sets were then transformed for each individual method (binarization for ARTMAP, scaling for fuzzy and Gaussian ARTMAP). For each method, optimal values of parameters were estimated using a validation technique in which a sequence of trainings and tests was repeated for different combinations of parameter values, training on a subset containing 90% of the training set and testing on the remaining 10% of the training set. After optimal values of parameters were found, five independent networks of each type were trained on the training sets. All the networks were then tested on the test set using the *voting* method. The parameters used in training of each network were as follows (ρ-baseline vigilance/similarity, β-learning rate, and γ-initial variance in Gaussian ARTMAP): ARTMAP ($\rho = 0.3, \beta = 1$), fuzzy ARTMAP ($\rho = 0.8, \beta = 1$), Gaussian ARTMAP ($\rho = 0.0, \beta = 1, \gamma = 0.5$).

2.3 Results and discussion

The performance of the methods was evaluated in terms of per cent of correctly classified test set patterns weighted by the size of each class (weighted PCC). Also, a confusion matrix for each of the methods was computed. Table 1 gives

Table 1. Performance (in weighted % of correctly classified test patterns) of the three methods on individual training sets and for voting

	Set #1	Set #2	Set #3	Set #4	Set #5	Voting
ARTMAP	90.76	88.95	90.27	88.47	89.51	92.20
fuzzy ARTMAP	93.72	91.48	91.66	90.82	92..16	93.95
Gaussian ARTMAP	93.90	93.57	93.49	94.24	93.09	94.04

the performance of the three methods expressed in weighted PCC. For each network type, individual performance (after training on a single training set) as well as overall performance (obtained using *voting*) is shown. The results show that the highest classification accuracy is obtained by the Gaussian ARTMAP neural network, both on individual training sets and with voting. The classification accuracy is slightly worse for fuzzy ARTMAP, and considerably worse for ARTMAP algorithm. The poor performance of the ARTMAP network can be assigned, as expected, to the poor ability of the network to generalize. This is suggested also by the fact that this network used in each training approximately 200 F2 neurons (clusters) whereas the other two methods needed only 80 to 90 F2 neurons. Another important result shown in the table is that Gaussian ARTMAP, compared to fuzzy ARTMAP, is much less sensitive to the training set ordering. This observation is supported by the smaller variance in PCC obtained when Gaussian ARTMAP was trained on individual training sets, as well as by the fact that the improvement obtained by application of *voting* on this method is much smaller than that for fuzzy ARTMAP. The difference in the performance is even stronger if non-weighted PCC is used. In this case (data not shown), Gaussian ARTMAP without voting performed better than fuzzy ARTMAP with voting.

Tables 2, 3, and 4 show the confusion matrices for each neural network model with voting. The tables show that, although there are significant differences in

Table 2. Confusion matrix for ARTMAP network with voting (weighted PCC = 92.20). Each item in the table gives the per cent of pixels from a given Actual Class (column) classified into given Predicted Class (row). The Total for each Actual Class (bottom row) gives per cent of patterns in the test set belonging to the corresponding Actual Class. The Total for each Predicted Class has analogous meaning

	Actual Class							
Predicted Class	A	B	C	D	E	F	G	Total (%)
A'	**84.82**	0.35	0.52	0.00	0.46	0.00	2.64	2.21
B'	4.02	**99.40**	1.57	0.00	0.00	0.00	0.00	36.84
C'	2.68	0.08	**75.17**	0.46	0.46	4.55	3.25	5.40
D'	0.00	0.00	0.52	**94.11**	0.00	13.13	0.61	28.78
E'	1.34	0.16	2.80	0.00	**98.92**	0.00	0.00	6.67
F'	5.80	0.00	17.13	5.46	0.00	**82.13**	1.22	15.36
G'	1.34	0.00	2.27	0.00	0.00	0.19	**92.29**	4.74
Total (%)	2.24	36.87	5.72	28.37	6.48	15.39	4.93	100.00

size of individual classes, all three methods classify almost evenly well patterns from all the categories. Also, there is no pair of categories which would be systematically confused by any of the networks. This suggests that there is no significant overlap in the data set and that the obtained performance reflects almost exclusively each network's capability to classify the data and to generalize information contained in the training data set. The color-encoded classification map obtained by the Gaussian ARTMAP network is shown in Figure 3.

All these results show that Gaussian ARTMAP is the best method for the chosen task. There is only one exception to this observation. The classification accuracy (PCC) of this network on the training set after the training was finished

Table 3. Confusion matrix for fuzzy ARTMAP network with voting (PCC = 93.95). Format as described in Table 2

Predicted Class	Actual Class A	B	C	D	E	F	G	Total (%)
A'	**88.84**	0.87	1.57	0.00	0.00	0.00	2.64	2.53
B'	2.68	**98.97**	0.00	0.00	0.00	0.00	0.00	36.56
C'	5.80	0.16	**79.55**	0.11	0.00	2.47	3.25	5.31
D'	0.00	0.00	0.52	**96.33**	0.00	8.45	0.00	28.66
E'	0.00	0.00	2.27	0.00	**100.00**	0.00	0.00	6.60
F'	1.34	0.00	12.76	3.56	0.00	**88.30**	0.61	15.39
G'	1.34	0.00	3.32	0.00	0.00	0.84	**93.51**	4.96
Total (%)	2.24	36.87	5.72	28.37	6.48	15.39	4.93	100.00

Table 4. Confusion matrix for Gaussian ARTMAP network with voting (PCC = 94.04). Format as described in Table 2

Predicted Class	Actual Class A	B	C	D	E	F	G	Total (%)
A'	**91.52**	0.35	1.05	0.11	0.00	0.00	1.22	2.34
B'	0.00	**99.48**	0.00	0.00	0.00	0.00	0.00	36.68
C'	5.80	0.00	**87.24**	0.11	0.00	4.55	4.46	6.07
D'	1.34	0.00	0.00	**97.22**	0.00	7.21	0.00	28.72
E'	0.00	0.16	0.00	0.00	**100.00**	0.00	0.00	6.54
F'	0.00	0.00	8.92	2.57	0.00	**88.04**	1.22	14.85
G'	1.34	0.00	2.80	0.00	0.00	0.19	**92.90**	4.80
Total(%)	2.24	36.87	5.72	28.37	6.48	15.39	4.93	100.00

Fig. 3. Classification map obtained by the Gaussian ARTMAP classifier

was always less than 100% (usually 97.7%), whereas for ARTMAP and fuzzy ARTMAP this parameter always reached 100%. This result can be interpreted in two ways. On the one hand it means that there may be tasks for which fuzzy ARTMAP or ARTMAP are more suitable, e.g., in case when all the data are available during the training. On the other hand this result underscores the generalization capabilities of Gaussian ARTMAP because it means that the network is able to identify and ignore noisy data in the training set.

3 Classification quality evaluation using confidence index

It is often very useful to have a measure of confidence of classification, especially in the remote sensing area where the differences in spectral content of patterns belonging to the same class can be quite considerable. There are several statistical approaches to this problem (see, e.g., [8]). But these methods are usually very computationally intensive. Therefore it would be very useful to find a method for assigning confidence to each classified pixel exploiting computations done as part of the classification process itself. In this section, a method is developed for evaluation of the classification confidence in Gaussian ARTMAP, the system which showed the best performance in classification of the remotely sensed data (see previous section).

3.1 Description of the method and results

In Gaussian ARTMAP, as mentioned in Section 2, each category comprises of a set of clusters. Each of these clusters contains patterns from a Gaussian probability distribution defined by a mean and variance in each dimension of the input space. When a new pattern is presented, the probability of that pattern belonging to each of the existing clusters is computed using a Bayes discrimination function (see [7], p. 24)

$$g_j(I) = \log\left(p(I|\omega_j)\right) + \log P(\omega_j) \tag{1}$$

where I is the input pattern, $p(I|\omega_j)$ is the conditional density of I given cluster j, and $P(\omega_j)$ is the ***a priori*** probability of cluster j. In Gaussian ARTMAP the pattern is classified into the most probable category. The probability measure of the new pattern belonging to a given category is defined as a sum of probabilities of the new pattern belonging to any of the clusters associated with that category, i.e.,

$$R_k(I) = \sum_{j \in \Omega(k)} \exp(g_j(I)) \tag{2}$$

where $\Omega(k)$ represents the set of clusters associated with category k. The new pattern is then assigned to the category k with the highest probability measure $R_k(I)$. If the ***voting*** method is used to suppress the influence of training pattern ordering (see Section 1.1), the probability measure defined by equation 2 is extended over all the networks used in *voting*

$$R_k(I) = \sum_{l=1}^{V} \left(\sum_{j \in \Omega(k)} \exp(g_{j,l}(I)) \right) \tag{3}$$

where V is the number of networks used for *voting*.

The probability measure of input I belonging to category k, $R_k(I)$, is evaluated for every category. And this fact offers a straightforward way to define a confidence of the decision made by the system when category K was chosen for input I. We define this confidence index as

$$c(I) = \frac{R_K(I)}{\sum_{k=1}^{M} R_k(I)}. \tag{4}$$

This index is evaluated for each individual pixel of the image and its value can range from 0 to 1. Using a simple encoding which assigns a shade of gray to each value of index $c(I)$ a confidence map for the classified image can be obtained. As an example, Figure 4 shows the confidence map for the image classified by

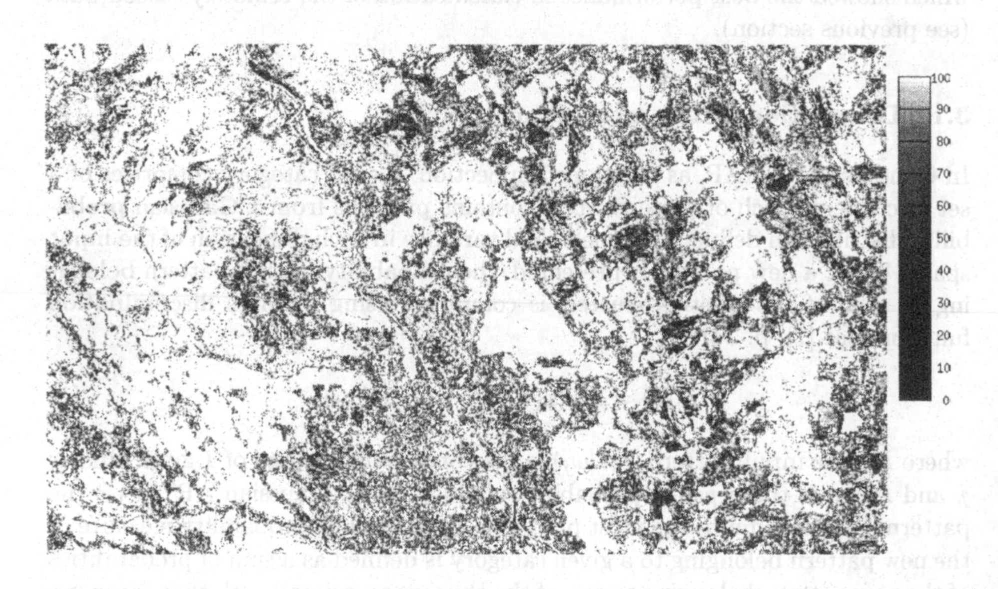

Fig. 4. Confidence map of the image classified by Gaussian ARTMAP using *voting* (confidence expressed in %)

the Gaussian ARTMAP model with *voting* (shown in Figure 3).

Next, a confidence threshold, Θ, can be introduced, which defines the minimal confidence required for a pixel to be assigned into the proposed category. Then, if the confidence for the pixel is subthreshold, the pixel can be classified as belonging to an unknown category. The choice of the confidence threshold value influences two counteracting aspects of the classification. First, the higher the value of Θ, the higher is the classification accuracy of the suprathreshold pixels. But, a higher value of Θ also means that fewer pixels will be classified into any of the known categories. To analyze these two aspects, a pair of graphs is shown in Figure 5. The dashed lines in the Figure show the threshold value ($\Theta = 0.921$)

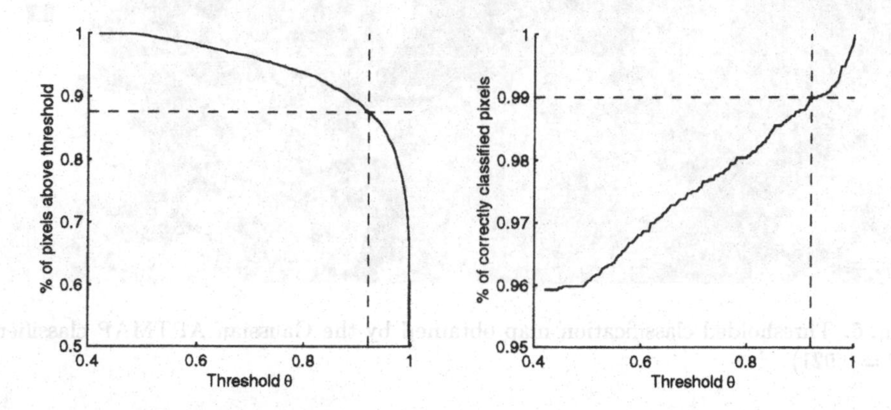

Fig. 5. Per cent of suprathreshold pixels of the image (on the left) and non-weighted per cent of correctly classified pixels (on the right) as a function of value of the confidence threshold Θ. See discussion in text.

which assures that 99% of the suprathreshold pixels will be classified correctly (non-weighted). With this threshold, the system will be able to classify 87.5% of the image. The resulting image is shown in Figure 6. In this image, the white color represents a new category of pixels unknown to the system.

4 Conclusion

This paper gives a description of the application of three different types of ARTMAP neural network for classification of images obtained from remote sensing. First, a description of the application of the ARTMAP methods for remote sensed data is given. Second, a comparison of performance of three ARTMAP neural networks differing in the input data representation method and the cluster identification method is presented. This analysis shows that Gaussian ARTMAP, which identifies clusters as data from Gaussian probability distributions, is the best classifier for this kind of data. Third, a method based on computational properties of the Gaussian ARTMAP neural network is described which allows assignment of a confidence measure, called confidence index, to the classification of each pixel. This makes it possible to create confidence maps and thresholded classification maps with prescribed classification accuracy. These maps allow a deeper insight into performance of the classifier.

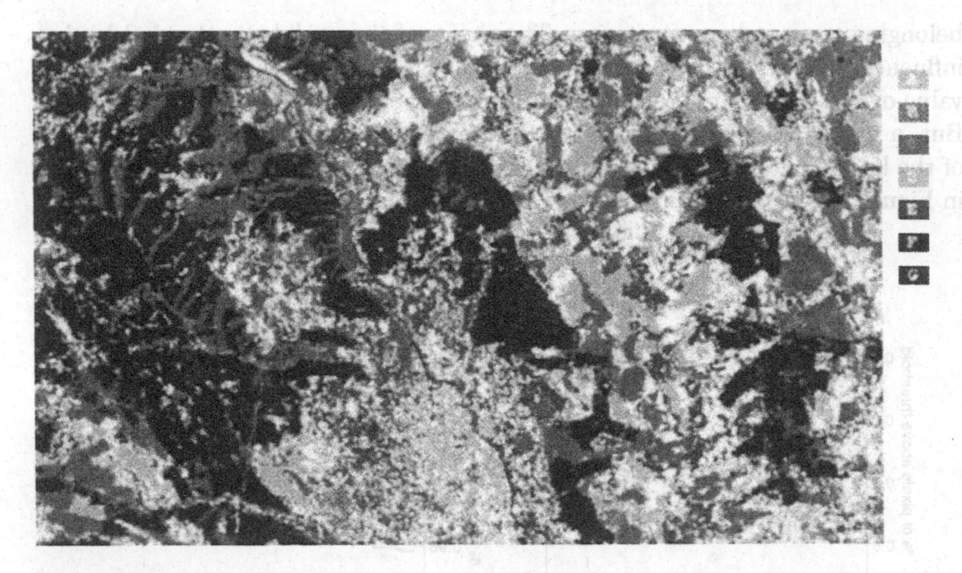

Fig. 6. Thresholded classification map obtained by the Gaussian ARTMAP classifier ($\Theta = 0.921$)

References

1. G. A. Carpenter, M. N. Gjaja, S. Gopal, and C. E. Woodcock. Art neural networks for remote sensing: Vegetation classification from landsat tm and terrain data. *IEEE Transactions on Geoscience and Remote Sensing*, 35(2):308–325, 1999.
2. G.A. Carpenter and S. Grossberg. A massively parallel architecture for a self-organizing neural pattern recognition machine. *Computer Vision, Graphics, and Image Processing*, 37:54–115, 1987.
3. G.A. Carpenter, S. Grossberg, N. Markuzon, J.H. Reynolds, and D.B. Rosen. Fuzzy ARTMAP: A neural network architecture for incremental supervised learning of analog multidimensional maps. *IEEE Transactions on Neural Networks*, 3(5):698–713, 1992.
4. Gail Carpenter and Stephen Grossberg. Art 2: Self-organization of stable category recognition codes for analog input patterns. *Applied Optics*, 1987. Special issue on neural networks.
5. Gail A. Carpenter, Stephen Grossberg, and B. Rosen David. FUZZY ART: an adaptive resonance algorithm for rapid, stable classification of analog patterns. Technical Report CAS/CNS-TR-91-006, Boston University, 1991.
6. Gail A. Carpenter, Stephen Grossberg, and John Reynolds. ARTMAP: Supervised real-time learning and classification of nonstationary data by a self-organizing neural network. *Neural Networks*, 4:565–588, 1991.
7. R.O. Duda and P.E. Hart. *Pattern Classification and Scene Analysis*. Wiley, New York, 1973.
8. K. Fukunaga. *Introduction to Statistical Pattern Recognition*. Academic Press, San Diego, CA, 1972.
9. Stephen Grossberg. Adaptive pattern classification and universal recoding, I: Feedback, expectation, olfaction, and illusions. *Biological Cybernetics*, 23:187–202, 1976.

10. International Neural Network Society. *Interactive Segmentation of Biomedical Images*, Brno, Czech Republic, May 1998. Vitium Press. ISSN 1211-421X, ISBN-80-241-1169-4.
11. B.G. Lees and K. Ritman. Decision-tree and rule-induction approach to integration of remotely sensed and gis data in mapping vegetation in disturbed or hilly environments. *Environmental Management*, 15:823–831, 1991.
12. Robert J. Mokken. Remote sensing, statistics and artificial neural networks. Some R and D perspectives., 1995.
13. J. Richards. *Remote sensing digital image analysis: An introduction.* Springer-Verlag: Berlin, 1993.
14. P. Sinčák, H. Veregin, and N. Kopčo. Conflation techniques in multispectral image processing. *Submitted to IJRS in February 1997*, 1997.
15. P. Werbos. *Beyond Regression: New Tools for Prediction and Analysis in the Behavioral Sciences.* PhD thesis, Harvard University, 1974.
16. James R. Williamson. Gaussian ARTMAP: A neural network for fast incremental learning of noisy multidimensional maps. *Neural Networks*, pages 881–897, 9 1996.

Application of Artificial Neural Network in Control of Vector Pulse-Width Modulation Inverter

Pavel Brandštetter, Martin Skotnica

Dept. of Power Electronics and Electric Drives
Technical University of Ostrava, 17. listopadu 15, 708 33 Ostrava–Poruba
pavel.brandstetter@vsb.cz, martin.skotnica@vsb.cz

Abstract. This paper describes neural controller that performs optimal switching strategy of VSI-PWM inverter. The strategy is based on rearranging the order of inverter output vectors in order to reduce the number of switched tranzistors during one time period, which leads to reduced power losses in switch tranzistors. Controller consists of one feedforward and one recursive neural network interconnected with computational blocks. The weights and thresholds of both neural nets are constructed rather than trained and require no adaption during the control process, which allows for very efficient implementation of the controller on digital signal processor.

1 Operational Principles of VSI-PWM Inverter

A voltage sourced inverter (VSI) generates three phase output voltage by connecting its phase outputs to the internal source of constant voltage (see Fig. 1). Due to the symmetry of three phase loads, as described in [1], the output voltage can be expressed as complex space vector

$$\underline{v}_o = u_1 + u_2 e^{-j2\pi/3} + u_3 e^{-j4\pi/3} \tag{1}$$

where u_1, u_2, u_3 are the phase voltages. All outputs are always connected to either positive or negative voltage. For three outputs, this gives eight possible combinations of output voltage vector (Fig. 1). Two of these combinations, $[1,1,1]$ and $[0,0,0]$, are called zero vectors because the outputs form a short circuit. Note that logical 1 means that positive voltage is connected to the phase output. Industrial applications, however, require the output voltage to be arbitrary vector. This is accomplished by fast switching of two vectors, the so called vector pulse-width modulation.

Let the output be consecutively set to three vectors $\underline{v}_a$, $\underline{v}_b$ and $\underline{v}_z$ for the time duration t_a, t_b and t_z, respectively. Then the resulting mean output voltage $\underline{v}_o$ is given by

$$\underline{v}_o = f_s(\underline{v}_a t_a + \underline{v}_b t_b + \underline{v}_z t_z) \tag{2}$$

where f_s is the inverter switching frequency and $\underline{v}_z$ is one of the zero vectors. The $\underline{v}_a$ and $\underline{v}_b$ are adjacent to the $\underline{v}_o$ vector, i.e. their switch combinations $A = [a_1, a_2, a_3]$ and $B = [b_1, b_2, b_3]$ differ only in one bit (see Fig. 1).

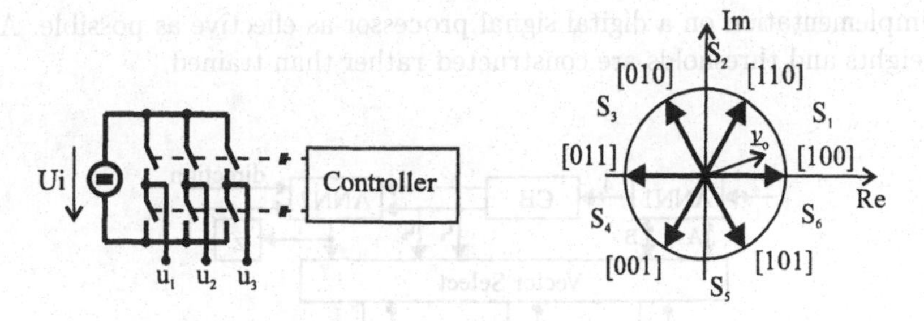

Fig. 1. Structure of voltage sourced inverter and possible output voltage vectors and their respective switch combinations

2 Optimal Switching Strategy

Equation (2) shows that the output voltage depends only on the selected vectors and the timing parameters. The order of vectors in which they are sent to the output and the selection of the zero vector ($[0, 0, 0]$ or $[1, 1, 1]$) have no influence on both the mean and effective output voltage value. This gives the controller an opportunity to optimize the vector sequence. Thermal power losses of inverter output tranzistors depend linearly on the switching rate and therefore the optimal switching strategy must choose such sequence which minimizes the switching rate. In general, the sequence of combinations $\{A, B, Z\}$ is output during one switching period. The switch combinations A and B represent the adjacent vectors $\underline{v}_a$ and $\underline{v}_b$ and always differ in at most one bit. The Z stands for zero vector and it can be chosen so that it differs from B only in one bit (Z_b). This minimizes the number of changes within the sequence to two. As a second optimization step, the number of changed bits for the two consecutive sequences has to be considered. The output vector space is divided into six sectors S_1 to S_6. Most applications require that the output vector slowly rotates ($f_o << f_s$) which means that the sequence $\{A, B, Z\}$ is mostly followed by identical sequence (with different delay times), i.e. $\{A, B, Z_b, A, B, Z_b, \ldots\}$. Because Z_b differs from A in two bits, the average switch rate for such sequence is 1.33 changes per vector. Such sequences are produced by unoptimized controllers, such as in [3]. The usual optimization, as presented for example in [2], is to use $\{A, B, Z_b, B, A, Z_a\}$ sequence order, which gives 1.0 switch change for every output vector. If we consider additional parameters, such as direction of the $\underline{v}_o$ rotation, the switching strategy can be enhanced to produce average 0.66 switch changes per output vector.

3 Controller Design

In order to implement the optimal switching strategy, a controller with two artificial neural networks was designed (see Fig. 2). It combines one feedforward and one recursive neural net with computational blocks in order to make the final implementation on a digital signal processor as effective as possible. Also, the weights and thresholds are constructed rather than trained.

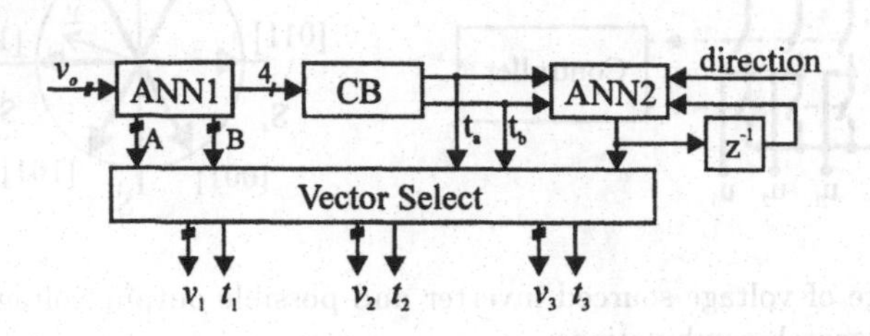

Fig. 2. Neural controller structure

The ANN1 neural net calculates two switching combinations needed for vector pulse-width modulation, $A = [a_1, a_2, a_3]$ and $B = [b_1, b_2, b_3]$. It also produces the real and imaginary part values of the respective output vectors $\underline{v}_a$ and $\underline{v}_b$. The structure of ANN1 consists of one hidden layer with six neurons and of the output layer with ten neurons. The hidden layer neurons detect the input vector angle position (sector S_1 to S_6 – see see Fig. 2). All neurons have simple step output function that is easily implemented on digital processor

$$f_o(x) = \begin{cases} 1 & \text{for } \overline{x} \cdot \overline{w} > \theta \\ 0 & \text{otherwise} \end{cases}, \tag{3}$$

where $\overline{x}$ is vector of neuron inputs, $\overline{w}$ is vector of weights and θ is activation function threshold.

The computational block CB produces delay times t_a and t_b of the $\underline{v}_a$ and $\underline{v}_b$ vectors, respectively, according to the following equations, obtained from equation (2)

$$t_a f_s = \frac{\mathrm{Im}[\underline{v}_o] \cdot \mathrm{Re}[\underline{v}_b] - \mathrm{Re}[\underline{v}_o] \cdot \mathrm{Im}[\underline{v}_b]}{\mathrm{Im}[\underline{v}_a] \cdot \mathrm{Re}[\underline{v}_b] - \mathrm{Re}[\underline{v}_a] \cdot \mathrm{Im}[\underline{v}_b]} \tag{4}$$

$$t_b f_s = \frac{\mathrm{Im}[\underline{v}_o] \cdot \mathrm{Re}[\underline{v}_a] - \mathrm{Re}[\underline{v}_o] \cdot \mathrm{Im}[\underline{v}_a]}{\mathrm{Im}[\underline{v}_a] \cdot \mathrm{Re}[\underline{v}_b] - \mathrm{Re}[\underline{v}_a] \cdot \mathrm{Im}[\underline{v}_b]}. \tag{5}$$

These delay times, together with the desired output vector $\underline{v}_o$ and direction of rotation, form input to the second neural net ANN2. The ANN2 is key part of the controller. Output of this net determines order of the output vectors that fulfils the optimal switching strategy. It contains five neurons in hidden layer and has one output neuron. It is implemented as a recurrent neural net, i.e. it takes its previous ouput as one of its inputs. Finally, the Vector Select (VS) block rearranges the order of vectors according to the output from ANN2. It also selects appropriate zero vector.

4 Results

Several simulations have been performed to test the controller. In the tests, the required output voltage was chosen as a vector of constant magnitude, rotating at constant speed f_o. The inverter switching frequency was 10 kHz. Following table presents switching statistics of one period ($T = 1/f_o$) for various output signals. The magnitude of required output voltage $|\underline{v}_o^*|$ and the output error ε are relative to the maximum inverter output value, the largest error is presented. The total number of vectors N_T is the number of generated output vectors including the zero vectors. The S_i is number of transitions of the ith switch – also given as percentage of the total number of vectors.

Table 1. Summary results of controller tests

f_o [Hz]	$\lvert\underline{v}_o^*\rvert$ [-]	$\varepsilon_{\max}$ [%]	N_T [-]	S_a [-]	S_b [-]	S_c [-]	R_S [%]
5.0	0.8	0.98	5978	1326 (22.2%)	1330 (22.3%)	1330 (22.3%)	66.4
5.0	0.1	0.98	5810	1270 (21.9%)	1274 (21.9%)	1274 (21.9%)	63.6
25.0	0.8	0.94	1194	267 (22.4%)	269 (22.6%)	266 (22.3%)	66.8
25.0	0.1	0.98	1162	258 (22.2%)	254 (21.9%)	258 (22.2%)	64.2
50.0	0.8	0.76	598	136 (22.7%)	134 (22.4%)	134 (22.4%)	67.3
50.0	0.1	0.86	582	130 (22.3%)	130 (22.3%)	130 (22.3%)	65.0

The switching strategy commonly used for inverters also optimizes the switching by rearranging order of output vectors to obtain one switch change for every output vector. This means that the switching rate of one switch is about equal to the inverter switching frequency. Presented results show that the neural controller decreases the switching rate to about 66%, which brings 33% less power losses.

5 Conclusion

Presented neural net based controller for voltage sourced inverter implements an optimal switching strategy which minimizes power losses in output tranzistors. Performed tests proved that this controller decreases the switching power losses by 33% when compared to the usual optimized switching strategy. The decreased power losses can reduce size of heatsinks and/or allow cheaper switching devices to be used which altogether leads to better price/performance index of the plant. Due to the chosen controller structure and simple neural net design it is possible to implement the controller very efficiently on digital signal processor.

References

1. Leonhard, W.: Control of Electrical Drives. Springer Verlag, Berlin, 1985
2. Brandštetter, P., Bubela, T.: Induction Motor Drive with Voltage Space Vector Modulation of Frequency Inverter. Conf. proc. VI Sympozjum Podstawowe problemy energoelektroniki i elektromechaniki, Gliwice-Ustron, Poland, 1995
3. Dinu, A. et al.: Neural Network for Control of Pulse-Width Modulation Inverters. Conf. proc. PEMC '98, Prague, Czech Republic, 1998

Modeling Multiple Microstructure Transformations in Steels with a Boltzmann Neural Net

Ernst Dieter Schmitter

Fachhochschule Osnabrueck, Fb. Werkstoffe und Verfahren
Albrechtstr.30, D-49076 Osnabrueck, Germany
Tel: +49 541 9692150 Fax: +49 541 9692999 email: schmittere@acm.org

Abstract. A generalized Boltzmann neural net is used to quantitatively model mainly diffusion controlled microstructure transformations in steels, especially in the case of multiple transformations occuring partly at the same time. Fitting the net parameters to a class of experimentally measured transformations, the prediction of other transformations, that play an important role in industrial processes, is possible with good accuracy. In contrast to other approaches (nonlinear function fitting [1], [2] or differential equation systems [3]) localized transformation development can be simulated with the possibility to include in a natural way the effects of varying grain sizes and dislocation distributions occuring for example with forming processes. The identification of the cubically (3D) arranged neurons with volume cells in the material allows a phenomenological interpretation of model parameters in terms of physical process parameters.

1 The model

A cubical volume of steel is divided it in $M = N^3$ cubical cells (typically $N = 32 \ldots 128$). Each cell has en edge length of the order $1\mu m$. The local resolution is coarse enough to allow the definition of thermodynamical quantities and fine enough to study localized transformation details. Each cell has 6 neighbours influencing the transition behaviour (periodic boundary conditions are assumed). In contrast to a standard cell automaton we identify each cell with the neuron of a Boltzmann net and perform the update of the cell state according to a statistical rule minimizing model free energy. In the following we have to distinguish strictly between model and physical thermodynamic quantities. The terms neuron and cell are used synonymously from now on. Simulating the microstructure transformations of austenite (A) to ferrite (F) (or carbide (C) in the case of hypereutectoid steels), of austenite to perlite (P) and of austenite to bainite (B) we represent the output of neuron α by a vector $z_\alpha^\dagger = (z_\alpha^F, z_\alpha^P, z_\alpha^B)$. Each component can be 0 or 1, maximally one component can be different from 0. The 0-vector represents the austenitic state.

The model energy E is given by

$$E = -\frac{1}{2} \sum_{\alpha=1,\beta\in N_\alpha}^{M} z_\alpha^\dagger \mathbf{W} z_\beta - \sum_{\alpha=1}^{M} \mathrm{q}\, z_\alpha^\dagger z_\alpha - \sum_{\alpha=1}^{M} in_\alpha^\dagger z_\alpha + \sum_{\alpha=1}^{M} b^\dagger z_\alpha$$

The first sum with Matrix $\mathbf{W}$ enforces clustering of cells with identical microstructure and via offdiagonal elements with related microstructures. For single phase transitions this sum is known from Ising models [4]. The second sum stabilizes cells against further transformations. The third sum enforces transformation according to external inputs. In our model transformations can start at grain boundaries and dislocations (heterogenuous nucleation) or randomly within grains (homogenuous nucleation) or both. Grain boundaries and dislocation lines are defined by cells α with external inputs $in_\alpha^{F,P,B} \neq 0$. The nucleation activity in cell α is given by in as a functional of the physical temperature history $T(t)$. The last sum contains the neuron thresholds (biases). The bias vector components delay the transformations according to $b^{F,P,B} = T_m ln(g\, t_{F,P,B}(T)/\Delta t - 1)$, where $t_{F,P,B}(T)$ are the characteristic transformation time scales (T_m see below). Via the counteracting principles of driving force (increasing physical free energy with undercooling) and diffusion mobility (decreasing diffusion coefficient with undercooling) the $t_{F,P,B}$ are strongly dependent on the physical temperature T and can be fitted to experimental data. Δt is the time step per iteration. g is a constant depending on the total number of cells. The delay relation for the bias $b^{F,P,B}$ leads to transition probabilities $p_{F,P,B} \propto \Delta t / t_{F,P,B}$ as can be shown using the probabilistic update rule discussed next. For the derivation of a suitable neuron update rule the total model average free energy G of the net is minimized with respect to the numbers m_F, m_P, m_B of cells in F, P, B states. The quantity G is given by

$$G(m_F, m_P, m_B) := E - T_m S = m_F u_F + m_P u_P + m_B u_B - T_m ln(\frac{M!}{m_A! m_F! m_P! m_B!})$$

where $u_{F,P,B}$ are the mean model energies, that are needed to transform a cell to F, P, B microstructures respectively. The ln-term is the mixing or configuration entropy S. S is the ln of the number of possibilities to have m_A cells in the austenitic state, m_F cells in the ferritic state and so on. We have $m_A = M - m_F - m_P - m_B$. G takes a minimum for the relative cell numbers

$$\frac{m_F}{M} = \frac{1}{1 + e^{u_F/T_m} + e^{(u_F - u_P)/T_m} + e^{(u_F - u_B)/T_m}}$$

and so on for $\frac{m_P}{M}$ and $\frac{m_B}{M}$. T_m is the model temperature governing the relative influence of E and S on the dynamic development.

We adopt the update rule, that cell α is transformed to one of F, P, B with a probability distribution defined by these frequency relations, where for $u_{F,P,B}$ the components $u_\alpha^{F,P,B}$ from expanding $E = \sum_\alpha (u_\alpha^F z_\alpha^F + u_\alpha^P z_\alpha^P + u_\alpha^B z_\alpha^B)$ are used.

2 Results and Conclusion

The model parameters have been adapted to isothermal transformation experimental data (dilatometer based). With these parameters the model is capable of predicting with satisfying accuracy continuous cooling transformation data. The neuron biases and external inputs are directly related to the physical transformation time scale and nucleation geometry respectively. Fig. 1 shows a typical cross section through the cubical volume at a certain time with clusters of F (white), P (light gray), B (dark grey) and yet untransformed A. The clusters relax to patterns as they are observed in reality. Transformations in this example nucleated at the boundaries of 4 grains. Additionally fig. 1 shows the time development of microstructure transformation for a continuous exponential cooling process of a steel Ck45. The yet untransformed volume at martensite start temperature is assumed to be transformed to martensite. In this way realistic continuous cooling transformation (CCT) diagrams can be calculated. Fig. 2 shows the time and physical temperature dependence of model thermodynamic quanitities exhibiting the typical features known from phase transitions. Research is continued by modeling the transformation behaviour during thermomechanical processes (forming) by introducing variing dislocation geometries and densities.

References

1. Vermeulen, van der Zwaag, S, Morris, P., de Weijer, T. 1997. Prediction of the continuous cooling transformation diagram of some selected steels using artificial neural networks. steel research Vol. 68, No.2, pp.72-79
2. Hougardy, H.P., 1978. Optimierung von Waermebehandlungen durch Berechnung des Umwandlungsverhaltens von Staehlen. Haerterei-Technische Mitteilungen, Vol. 33, No.3, pp.115-178
3. Leblond, J.B., Devaux, J. 1984. A new kinetic model for anisothermal metallurgical transformations in steels including effect of austenite grain size. Acta Metallurgica Vol. 32, No. 1, pp.137-146
4. Hopfield, J.J., 1982. Neural networks and physical systems with emergent collective computational abilities. Proc. of the Nat. Academy of Sciences, USA, Vol. 79, pp.2554-2558

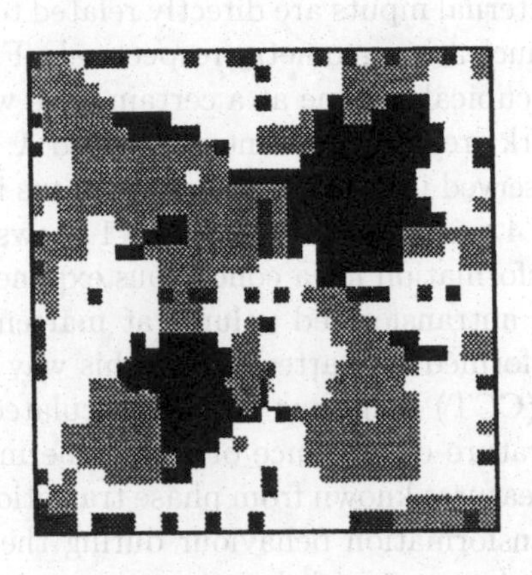

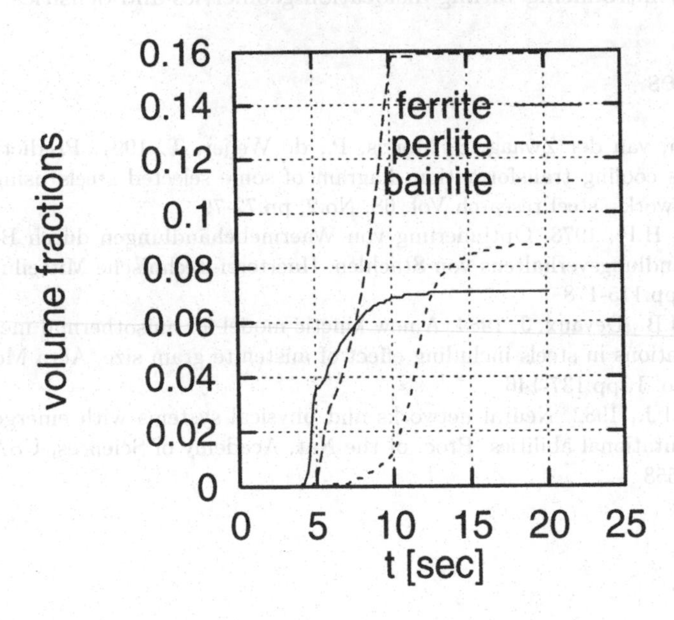

Fig. 1. Cross section through the 3D cubical model with clusters of different microstructures and time development of volume fractions for steel Ck45, exponential cooling with t(800-500 Cel.)=9s.

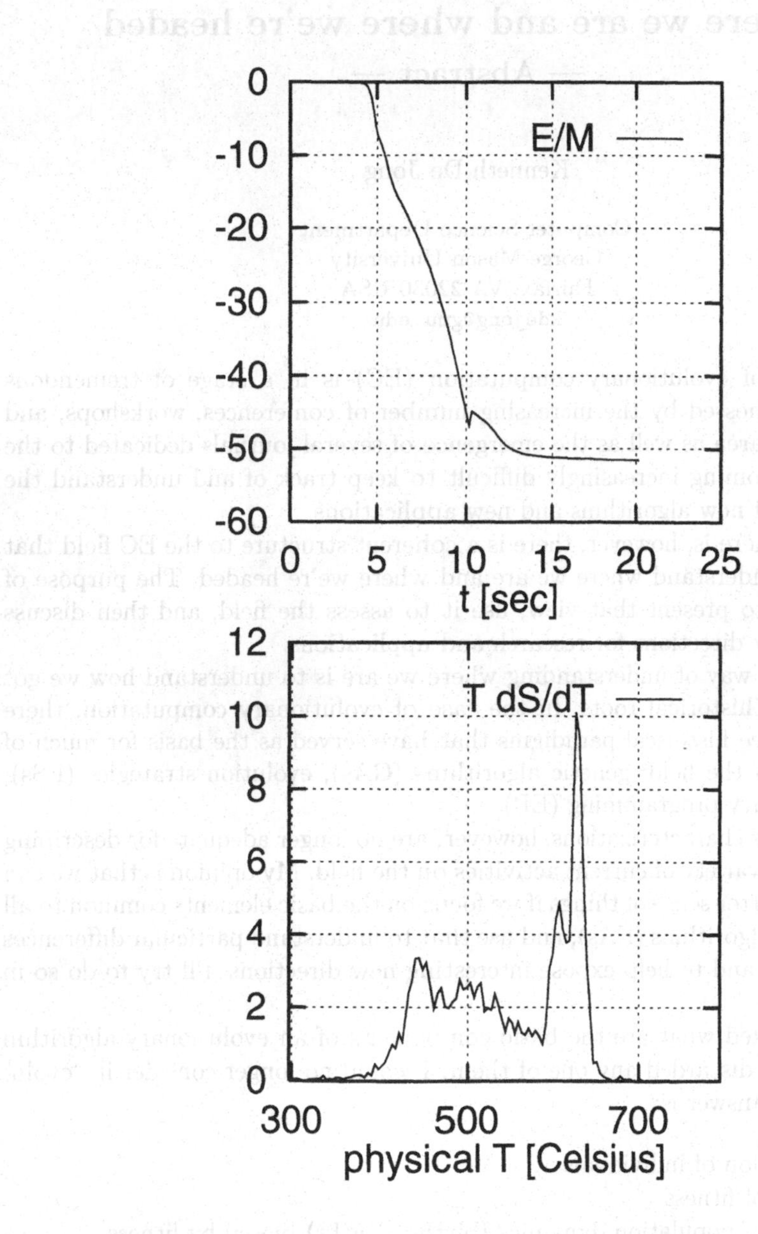

Fig. 2. Model energy per neuron for steel Ck45, exponential cooling with t(800-500 Cel.)=9s and model specific heat C=T dS/dT (with physical temperature T) showing maxima in the T-ranges of the microstructure transitions to F,P,B.

Evolutionary Computation: Where we are and where we're headed
— Abstract —

Kenneth De Jong

Computer Science Department
George Mason University
Fairfax, VA 22030 USA
kdejong@gmu.edu

The field of evolutionary computation (EC) is in a stage of tremendous growth as witnessed by the increasing number of conferences, workshops, and papers in the area as well as the emergence of several journals dedicated to the field. It is becoming increasingly difficult to keep track of and understand the wide variety of new algorithms and new applications.

I believe there is, however, there is a coherent structure to the EC field that can help us understand where we are and where we're headed. The purpose of this paper is to present that view, use it to assess the field, and then discuss important new directions for research and applications.

One useful way of understanding where we are is to understand how we got here, i.e., our historical roots. In the case of evolutionary computation, there have been three historical paradigms that have served as the basis for much of the activity in the field: genetic algorithms (GAs), evolution strategies (ESs), and evolutionary programming (EP).

These early characterizations, however, are no longer adequate for describing the enormous variety of current activities on the field. My opinion is that we can make much better sense of things if we focus on the basic elements common to all evolutionary algorithms (EAs), and use that to understand particular differences in approaches and to help expose interesting new directions. I'll try to do so in this paper.

If I am asked what are the basic components of an evolutionary algorithm such that, if I discarded any one of them, I would no longer consider it "evolutionary", my answer is:

- A population of individuals
- A notion of fitness
- A notion of population dynamics (births, deaths) biased by fitness
- A notion of inheritance of properties from parent to child

This, of course, could just as well describe any evolutionary system. In the case of *evolutionary computation*, our goal is to use such algorithms to solve problems, and so there are additional aspects involving the problem domain and how one chooses to use an EA to solve such problems. The key design decisions involve choices as to how to:

- Model the dynamics of population evolution.
- Represent individuals in the population.
- Model reproduction and inheritance.
- Exploit characteristics of fitness landscapes.

When we look at the specific details as to how these decisions have been made historically and how they are been made today, we see a surprising amount of commonality and coherence.

At the same time this perspective sheds light on a number of important open issues and directions for future research. I will briefly summarize a number of these including:

- Representation and Morphogenesis
- Inclusion of Lamarckian Properties
- Non-random Mating and Speciation
- Decentralized, Highly Parallel Models
- Self-adapting Systems
- Coevolutionary Systems
- Theoretical Analyses

This is an exciting time for the EC field. The increased level of EC activity has resulted in an infusion of new ideas and applications which are challenging old tenets and moving the field well beyond its historical roots. As a result of this rapidly changing EC landscape, a new characterization of the field is required based on core issues and important new directions to be explored.

I have attempted to articulate this new view by summarizing the current state of the field, and also pointing out important open issues which need further research. I believe that a view of this sort is an important and necessary part of the continued growth of the field.

Fuzzy Controllers by Unconventional Technologies for Tentacle Arms

Mircea Ivanescu[1] , Viorel Stoian[2]

[1]Automatic Control and Computer Dept., University of Craiova
ivanescu@robotics.ucv.ro;
[2]Automatic Control and Computer Dept., University of Craiova
stoian@robotics.ucv.ro

Abstract. A robust control system is proposed to solve the local control for a multi-chain robotic system formed by tentacle manipulators grasping a common object with hard point contacts. The two-level hierarchical control is adopted. The upper level coordinator gathers all the necessary information to resolve the distribution force. Then, the lower-level local control problem is treated as an open-chain redundant manipulator control problem.The stability and robustness of a class of the fuzzy logic control (FLC) are investigated and a robust FLC is proposed with uncertainties of the load. The fuzzy rules are established. Simulation results are presented and discussed.

1 Introduction

In the past few years, the research in coordinating robotic systems with multiple chains has received considerable attention. The problem of controlling this system in real time is more complicated. A multiple chain tentacle robotic system is more complicated. A tentacle manipulator is a manipulator with a great flexibility, with a distributed mass and torque that can take any arbitrary shape. Technologically, such systems can be obtained by using a cellular structure for each element of the arm. The control can be produced using an electrohydraulic or pneumatic action that determines the contraction or dilatation the peripheral cells. The first problem is the global coordination problem that involves coordination of several tentacles in order to assure a desired trajectory of a load. The second problem is the local control problem, which involves the control of the individual elements of the arm to achieve the desired position. The force distribution is a subproblem in which the motion is completely specified and the internal forces/torques to effect this motion are to be determined. To resolve this large-scale control problem, a two-level hierarchical control scheme [4] is used. The upper-level system collects all the necessary information and solves the interchain coordination problem, the force distribution problem. Then, the problem is decoupled into j lower-level subsystems, for every arm. The local fuzzy controllers are assigned to solve the local control. In order to obtain the fuzzy control an approximate model of the tentacle arm is used. The stability and robustness of the FLC are investigated and a robust FLC is proposed with respect to the robustness of the load uncertainties. The control strategy is based on the Direct Sliding Mode

Control (DSMC) which controls the trajectory towards the switching line and then the motion is forced directly to the origin, on the switching line. A fuzzy controller is proposed and the fuzzy rules are established by using the DSMC procedures. Efficiently considerations of the method are discussed. Numerical simulations for several control problems are presented.

2 Model for Cooperative Tentacle Robots

A multiple-chain tentacle robotics system is presented in Figure 1.

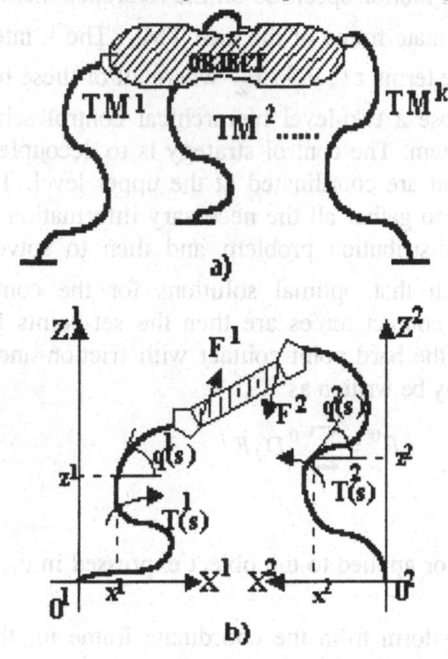

Fig. 1. A multiple-chain tentacle robotics system.

With the chains of the system forming closed-kinematics loops, the responses of individual chains are tightly coupled with one another through the reference member (object or load). The complexity of the problem is considerable increased by the presence of the tentacle manipulators, $\left(TM^j, j=1 \ldots k\right)$, the systems with, theoretically, a great mobility, which can take any position and orientation in space [1, 2]. In the Fig. 1.b is presented a plane model, a simplified structure which can take any shape in X0Z plane. The dynamic equations for each chain of the system are [1]:

$$TM^j : \rho_j A^j \int_0^s \left[\sin\left(q^j - q'^j\right)\dot{q}'^{j2} + \cos\left(q^j - q'^j\right)\ddot{q}'^j\right] ds' +$$

$$+ \rho A g \int_0^s \cos q^j ds' + \tau^j = T^j, \; j = 1,2 \qquad \textbf{(2.1)}$$

$$\int_0^{L^j} \tau^j ds = F_x^j \int_0^{L^j} \left(-\sin q^j\right) ds + F_x^j \int_0^{L^j} \cos q^j ds, \; j = 1,2 \tag{2.2}$$

where we assume that each manipulator (TM^j) has a uniform distributed mass, with a linear density ρ^j and a section A^j. We denote by s the spatial variable upon the length of the arm, $s \in [0,L^j]$. We also use the notations: q^j - Lagrange generalized coordinate for TM^j (the absolute angle), $q^j = q^j(s,t)$, $s \in [0,L^j]$, $t \in [0,t_f]$, $q'^j = q^j(s',t)$, $s' \in [0,s]$, $t \in [0,t_f]$, $T^j = T^j(s,t)$ - the distributed torque over the arm; $\tau^j = \tau^j(s,t)$- the distributed moment to give the desired motion specified on the reference member. All these sizes are expressed in the coordinate frame of the arm TM^j . The k integral equations are tightly coupled through the terms τ^j , F_x^j , F_z^j where all of these terms determine the desired motion. We propose a two-level hierarchical control scheme [4,5] for this multiple-chain robotic system. The control strategy is to decouple the system into k lower-level subsystems that are coordinated at the upper level. The function of the upper-level coordinator is to gather all the necessary information so as to formulate the corresponding force distribution problem and then to solve this constrained, optimization problem such that optimal solutions for the contact forces F^j are generated. These optimal contact forces are then the set-points for the lower-level subsystems. We consider the hard point contact with friction and the force balance equations on the object may be written as:

$$F^0 = \sum_j {}^0D_j F^j \tag{2.3}$$

where:

F^0- the resultant force vector applied to the object expressed in the inertial coordinate frame (0),

${}^0D_j$-the partial spatial transform from the coordinate frame for the arm TM^j to the inertial coordinate frame (0).

The object dynamic equations are obtained by the form:

$$M_0 \ddot{r} = GF^0 \tag{2.4}$$

where M_0 is inertial matrix of the object and r defines the object coordinate vector

$$r = (x,z,\varphi)^T \tag{2.5}$$

and r(t) represents the desired trajectory of the motion. The inequality constraints which include the friction constraints and the maximum force constraints may be associated to (2.3),

$$\sum_i A^i F^i \le B \tag{2.6}$$

where A^j is a coefficient matrix of inequality constraints and B is a boundary-value vector of inequality constraints. The problem of the contact forces F^j can be treated as an optimal control problem if we associate to the relations (2.3) - (2.6) an optimal index

$$\Psi = \sum_j C^j F^j \qquad (2.7)$$

This problem is solved in several papers [4,5,6] by the general methods of the optimization or by the specific procedures [7]. After all of the contact forces F^j are determinate, the dynamics of each arm TM^j are decoupled. Now, the equations (2.1), (2.2) can be interpreted as same decoupled equations with a given $\tau^j(s)$, $s \in [0,L^j]$ acting on the tip of the arm.

3 Approximate Model

A discrete and simplified model of (2.1), (2.2) can be obtained by using a spatial discretization.

$$s_1, s_2, \ldots s_N; \quad s_i - s_{i-1} = \Delta; \qquad (3.1)$$

$$|q^j(s_i) - q^j(s_k)| < \varepsilon; \; i, k = 1, 2, \ldots n^j \qquad (3.2)$$

where ε, Δ are constants and ε is sufficiently small.
We denote $s_i = i\Delta$, $L^j = n^j\Delta$, $q_j(s_i) = q_i^j$,

$$T^j(s_i) = T_i^j, \tau^j(s_i) = \tau_i^j \qquad (3.3)$$

and considering the arm as a lightweight arm, from (2.1), (2.2) it results [2]:

$$M^J \ddot{q}^J + C^J \dot{q}^J + D^J\left(q^J\right)F^J = T^J \qquad (3.4)$$

where M^J, C^J are $(n^J x n^J)$ contact diagonal matrixes, D is $(n^J x2)$ nonlinear matrix [2,13]

$$F^J = col\left(F_x^J, F_z^J\right) \qquad (3.5)$$

$$q^J = col\left(q_1^J \ldots q_{n^J}^J\right) \qquad (3.6)$$

$$T = col\left(T_1^J \ldots T_{n^J}^J\right) \qquad (3.7)$$

In the equation (3.4), F^J assures the load transfer on the trajectory. The uncertainty of the load m defines an uncertainty of the force F^J. We assume that

$$\left|F^{MJ} - F^J\right|_i \le \rho_i; i = 1,2 \qquad (3.8)$$

Where F^{MJ} is an estimation of the upper bound of the force.

4 Fuzzy Control

The control problem asks for determining the manipulatable torques (control variables) T^J_I such that the trajectory of the overall system (object and manipulators) will correspond as closely as possible to the desired behavior. In order to obtain the control law for a prescribed motion, we shall use the closed-loop control system from Figure 2. Let q_d^J the desired parameters of the trajectory, F_d^J the desired force applied to the j - contact point of the object, and q^J, F^J the same sizes measured on the real system. For a bounded smooth trajectory, a tracking error is:

$$e = q - q_d \tag{4.1}$$

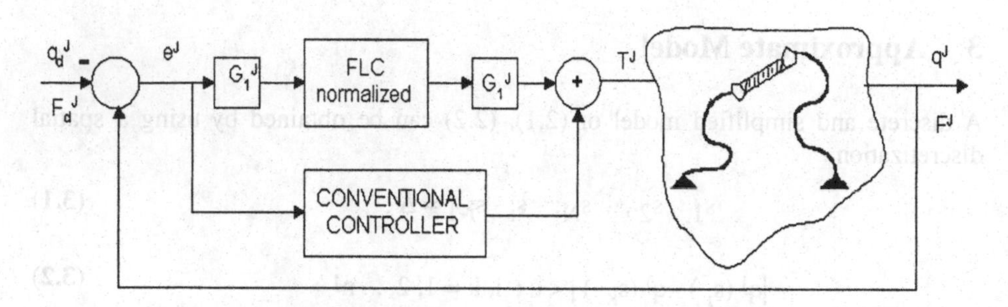

Fig. 2. The closed-loop control system.

The control system contains two parts: the first component is a conventional controller which implement a classic strategy of the motion control based on the Lyapunov stability and the second is a FLC. A fuzzy control is proposed by using the control law in the neighbourhood of the switching line (Table 1 and Figure 3) as a variable structure controller. The physical meaning of the rules is as follows: the output is zero near the switching line (s), the output is negative above the switching line, the output is positive below the diagonal line, the magnitude of the output tends to increase in accordance with magnitude of the distance between the switching line and the state $\left(e^J, \dot{e}^J\right)$. We consider that all input/output fuzzy sets are assumed to be designed on the normalized space:

$$e_{i,N}^J, \dot{e}_{i,N}^J, T_{i,N}^J \in [-1,1], i = 1,2 \ldots n^J \tag{4.2}$$

Table 1. The fuzzy rules in the neighbourhood of the switching line.

$\dot{e}^J \backslash e^J$	NBE	NSE	ZRE	PSE	PBE
PBDE	ZRC	NBC	NBC	NBC	NBC
PSDE	PBC	ZRC	NSC	NSC	NBC
ZRDE	PBC	PSC	ZRC	NSC	NBC

NSDE	PBC	PSC	PSC	ZRC	NBC
NBDE	PBC	PBC	PBC	PBC	ZRC

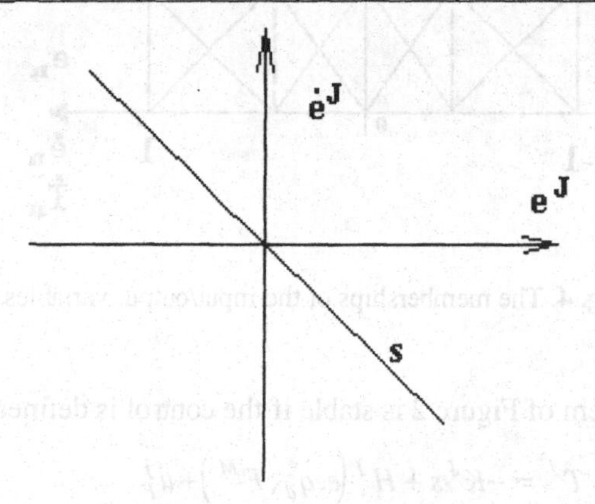

Fig. 3. The switching line s.

and the input/output gains $G_1^J = \left[G_e^J, G_{\dot{e}}^J\right]$ and G_C serve as scale factors between the normalized space and the corresponding real space of the process

$$e_{N,i}^J = G_{ei}^J e_i^J \tag{4.3}$$

$$\dot{e}_{N,i}^J = G_{\dot{e}i}^J \dot{e}_i^J \tag{4.4}$$

$$T_{N,i}^J = G_{ci} T_i^J \tag{4.5}$$

If we consider that the switching line s in the space of normalized values is defined by the diagonal:

$$\dot{e}_{N,i}^j + e_{N,i}^j = 0 \tag{4.6}$$

it corresponds to the following diagonal line in the real error phase plane [15],

$$s_i^J = \dot{e}_i^J + \sigma_i^J e_i^J = 0, \qquad \sigma_i^J = \frac{G_{ci}^J}{G_{ei}^J} > 0 \tag{4.7}$$

$$\dot{e}^J + \sigma e^J = 0 \tag{4.8}$$

where $\sigma = \text{diag}\ (\ \sigma_1, \sigma_2, \ldots \sigma_{nj}\ )$. The memberships of the input/output variables are represented in Figure 4, where NB, NS, ZR, PS, PB define the linguistic variables: NEGATIVE BIG, NEGATIVE SMALL, ZERO, POSITIVE SMALL, POSITIVE BIG, respectively.

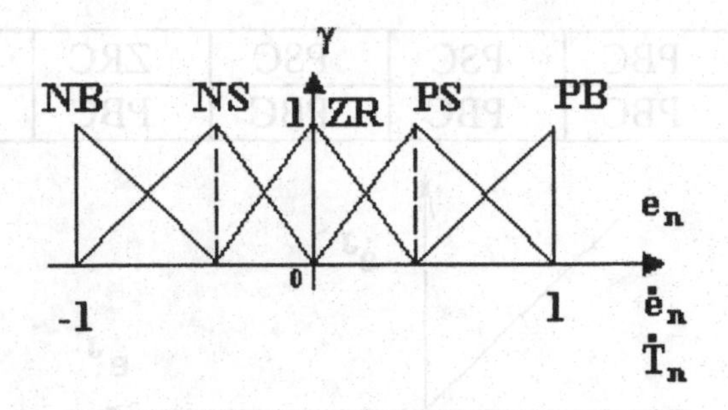

Fig. 4. The memberships of the input/output variables.

Theorem 1.
The closed-loop system of Figure 2 is stable if the control is defined by

$$T^J = -K^J s + H^J\left(e, q_d, F^M\right) + u_F^j \tag{4.9}$$

where K^J is a (n^J x n^J) symmetric positive definite matrix, which satisfies condition

$$K^j - M^j \sigma^j + C^j \text{ is a positive definite matrix} \tag{4.10}$$

and u_F is the output vector of the fuzzy controller

$$u_{Fi}^j = -k_{Fi}^j \operatorname{sgn} s_i^j \tag{4.11}$$

$$k_{Fi}^J \geq \left|H\left(e, q_d, F\right) - H^J\left(e, q_d, F^M\right)\right|_i \tag{4.12}$$

Proof. See Appendix 1.

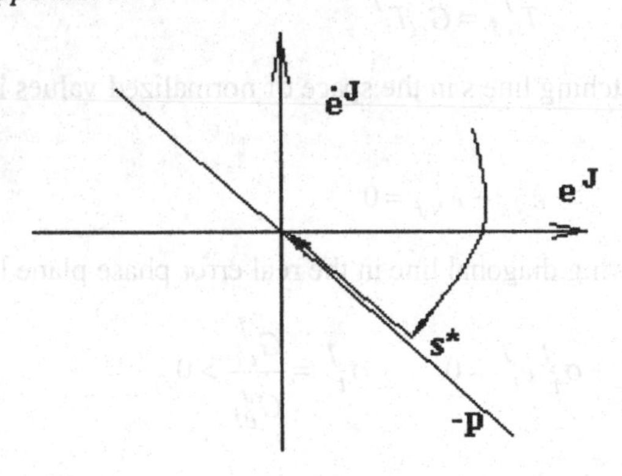

Fig. 5. The Direct Sliding Mode Control.

The Theorem 1 determines the conditions which assure the motion control in the neighborhood of the switching line. In order to accelerate the motion on the switching line, we can use the DSMC (Direct Sliding Mode Control). The DSMC was presented in [14] and it establishes conditions which force the trajectory along the switching line, directly toward the origin (Figure 5). In the first part of the motion defined by Theorem 1, we considered that the switching line has the slope - σ_i. The DSMC can require a new switching line,

$$s_i^* = \dot{e}_i^J - p_i^J e_i^J = 0 \quad \textbf{(4.13)}$$

Proposition:
The DSMC control is assured if the coefficient k_i of the controller verifies the condition,

$$(c_i + k_i)^2 \geq 4m_i(\sigma_i(-m_i\sigma_i + c_i + k_i) + h_i) \quad \textbf{(4.14)}$$

Proof. See Appendix 2.
The DSMC control can introduce a new fuzzy output variable, the coefficients k_i. In the first part of the motion k_i verifies only the condition (4.10). When the trajectory penetrates the switching line s^*_i, the k_i are increased in order to verify (4.14)

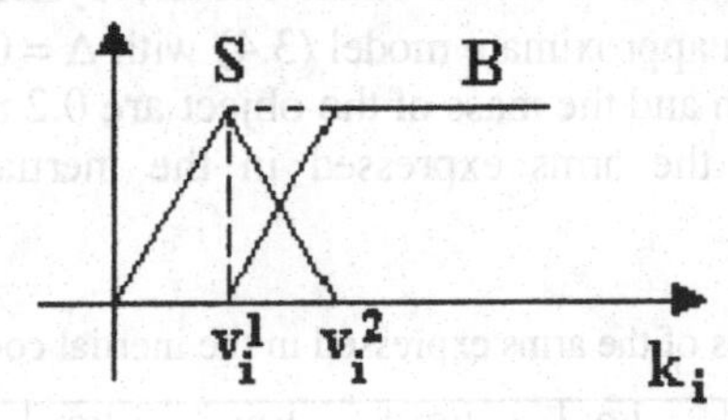

Fig. 6. The memberships functions for k_i

In Figure 6 and Table 2 are presented the memberships functions for k_i and the fuzzy rules, where we defined as v^1_i, v^2_i the average values which verify the condition (4.10) and (4.14), respectively.

Table 2. The fuzzy rules for the coefficients k_i.

$\dot{e}^J \backslash e^J$	NBE	NSE	ZRE	PSE	PBE
PBDE	B	S	S	S	S
PSDE	S	B	S	S	S
ZRDE	S	S	B	S	S
NSDE	S	S	S	B	S
NBDE	S	S	S	S	B

5 Numerical Results

The purpose of this section is to demonstrate the effectiveness of the method. This is illustrated by solving a fuzzy control problem for a two tentacle manipulator system which operates in XOZ plane (Figure 7).

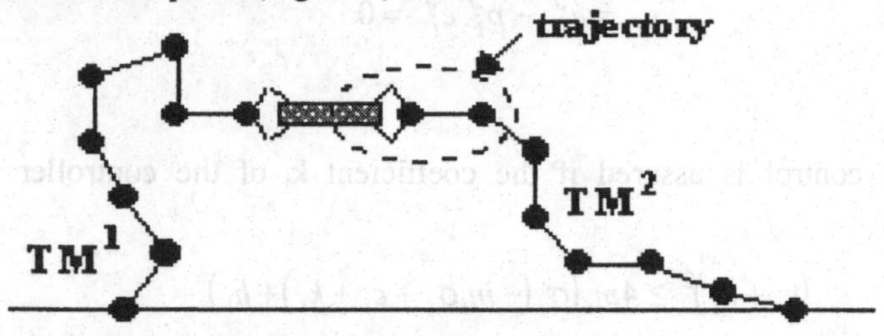

Fig. 7. The two tentacle manipulator system which operates in XOZ plane.

These two manipulators form a closed-chain robotic system by a common object which is manipulated. An approximate model (3.4) with $\Delta = 0.06$ m and $n^J = 7$ is used (J = 1, 2). Also, the length and the mass of the object are 0.2 m and 1 kg, respectively. The initial positions of the arms expressed in the inertial coordinate frame are presented in Table 3.

Table 3. The initial positions of the arms expressed in the inertial coordinate frame.

TM^j	$q_1^j(0)$	$q_2^j(0)$	$q_3^j(0)$	$q_4^j(0)$	$q_5^j(0)$	$q_6^j(0)$	$q_7^j(0)$
TM^1	$\pi/6$	$\pi/3$	$7\pi/12$	$2\pi/3$	$\pi/15$	$15\pi/8$	0
TM^2	$5\pi/6$	$4\pi/5$	$4\pi/5$	$3\pi/4$	$3\pi/4$	$2\pi/3$	π

The desired trajectory of the terminal points is defined by:

$$x = x_0 + a \sin \omega t \qquad (5.1)$$
$$z = z_0 + b \cos \omega t$$

With $x_0 = 0.2$ m, $z_0 = 0.1$ m, a = 0.3m, b = 0.1 m, $\omega = 0.8$ rad/s.
The trajectory lies the work envelopes of the both arm and does not go through any workspace singularities. The maximum force constraints are defined by:

$$F_X^J \le F_{MAX} = 50N \qquad F_Z^J \le F_{MAX} = 50N \qquad (5.2)$$

and the optimal index are used:

$$\min\left(\sum_J F_X^{J^2}\right), \min\left(\sum_J F_Z^{J^2}\right) \qquad (5.3)$$

The uncertainty domain of the mass is defined as $0.8kg \le m \le 1.4kg$. The solution of the desired trajectory for the elements of the arms is given by solving the nonlinear differential equation [13]

$$\dot{q}_d^J(t) = \left[J^{JT}(q) \quad J^J(q)\right]^{-1} J^{JT}(q)\dot{w}(t) \tag{5.4}$$

where $w=(x,z)^T$ and $J^J(q)$ is the Jacobian matrix of the arms (J=1,2).

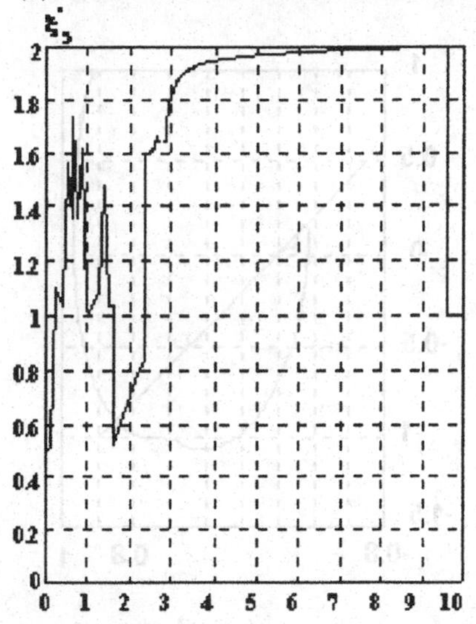

Fig. 8. The evolution of k_5^1 for a DSMC procedure.

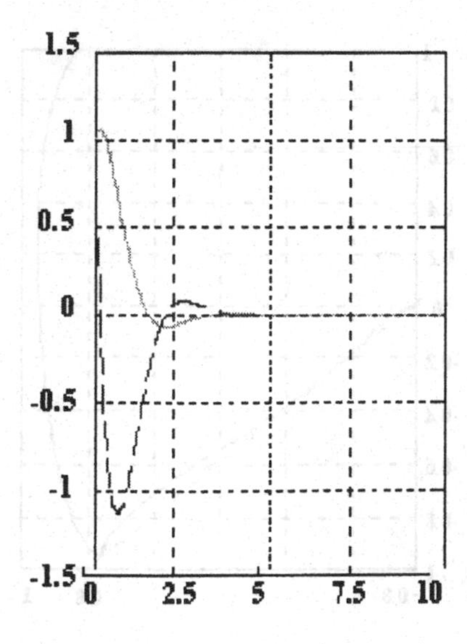

Fig. 9. The evolution of the position error e_5^1 and the position error rate $\dot{e}_5^1$.

A conventional controller with k_i^J=0.5 (i =1,..7, j=1,2) is determined. A FLC is used with the scale factors selected as $G_{e_i}^J = G_{\dot{e}_i}^J = 10$, I = 1, ... 7, J = 1, 2. The conventional and DSMC procedures are used and new switching line is computed. The condition (4.14) is verified and the new switching line is defined for $p_i^J = 1.03$: I = 1, ... 7, J = 1, 2. In Figure 8 is presented the evolution of k_5^1 for a DSMC procedure and the evolution of the position error e_5^1 and the position error rate $\dot{e}_5^1$ are presented in Figure 9. Figure 10 represents the trajectory in the plane $\left(e_5^1, \dot{e}_5^1\right)$ for conventional procedure

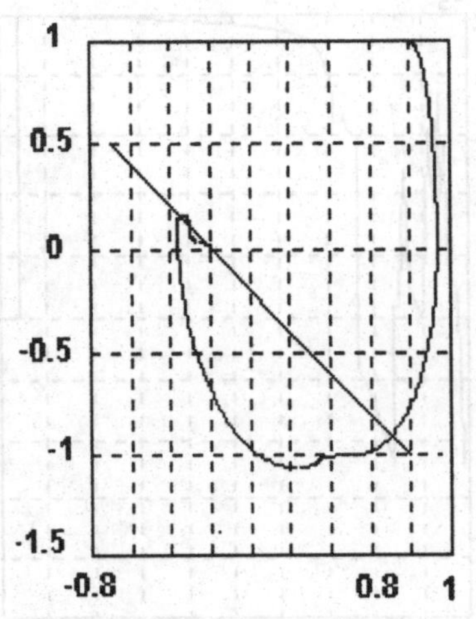

Fig. 10. The trajectory in the plane $\left(e_5^1, \dot{e}_5^1\right)$ for conventional procedure.

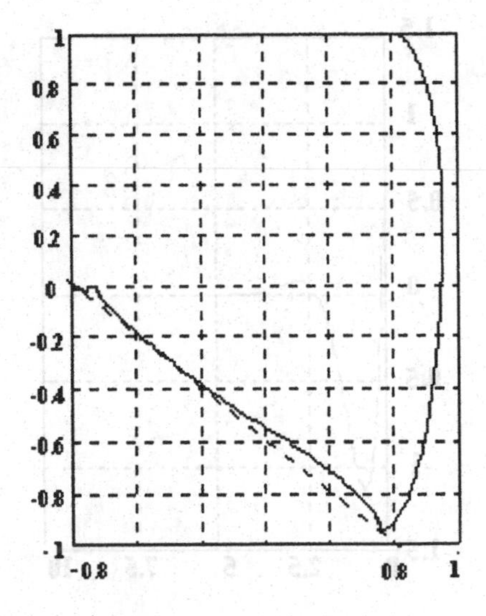

Fig. 11. The trajectory for a DSMC procedure for a new switching line.

Fig 11 represents the same trajectory for a DSMC procedure for a new switching line. Figure 12 presents the final trajectory. We can remark the error during the 1th cycle and the convergence to the desired trajectory during the 2nd cycle.

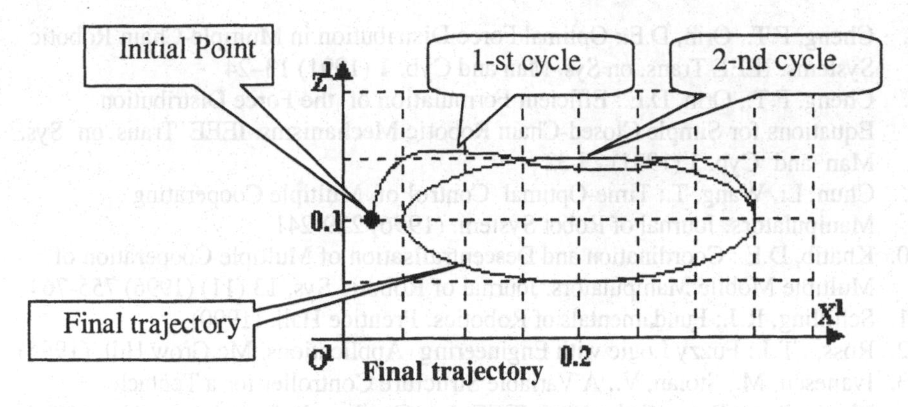

Fig. 12. The final trajectory of the load.

6 Conclusions.

The two level hierarchical control procedure is constructed in this paper to solve the large-scale control problem of a chain robotic system formed by tentacle manipulators grasping a common object. The upper-level coordinator collects all the necessary information to solve the inter-chain coordination problem, the force distribution. Then, the problem is decoupled into j lower-level subsystems, for every arm. The local fuzzy controllers are assigned to solve the local control. The stability and robustness of a class of the fuzzy logic control (FLC) are investigated and a robust FLC is proposed in order to cancel the uncertainties of the load. A DSMC procedure is used and the fuzzy rules are established. The simulation problem for two closed-chain tentacle robotic systems has also been studied.

References

1. Ivanescu, M.: Dynamic Control for a Tentacle Manipulator. Proc. of Int. Conf. Charlotte. USA (1984)
2. Ivanescu, M.: A New Manipulator Arm: A Tentacle Model. Recent Trends in Robotics. **11**(1986) 51-57
3. Silverman, L. M.: Inversion of Multivariable Linear Systems. IEEE Trans. Aut. Contr., Vol AC–14 (1969)
4. Cheng, F. T.: Control and Simulation for a Closed Chain Dual Redundant Manipulator System. Journal of Robotic Systems. (1995) 119–133
5. Zheng, Y.F., Luh, J.Y.S.: Optimal Load Distribution for Two Industrial Robots Handling a Single Object. Proc. IEEE Int. Conf. Rob. Autom. (1988) 344-349
6. Mason, M.T.: Compliance and Force Control. IEEE Trans. Sys. Man Cyb. **6** (1981) 418-432

7. Cheng, F.T., Orin, D.E.: Optimal Force Distribution in Multiple-Chain Robotic Systems. IEEE Trans. on Sys. Man and Cyb. **1** (1991) 13–24
8. Cheng, F.T., Orin, D.E.: Efficient Formulation of the Force Distribution Equations for Simple Closed-Chain Robotic Mechanisms. IEEE Trans on Sys. Man and Cyb. **1** (1991) 25-32
9. Chun, Li, Wang, T.: Time-Optimal Control of Multiple Cooperating Manipulators. Journal of Robot System. (1996) 229-241
10. Khatib, D.E.: Coordination and Descentralisation of Multiple Cooperation of Multiple Mobile Manipulators. Journal of Robotic Sys. 13 (11) (1996) 755-764
11. Schilling, R.J.: Fundamentals of Robotics. Prentice Hall. (1990)
12. Ross, . T.J.: Fuzzy Logic with Engineering Applications. Mc.Grow Hill. (1995)
13. Ivanescu, M., Stoian, V.: A Variable Structure Controller for a Tentacle Manipulator. Proc. of the 1995 IEEE Int. Conf. on Rob. and Aut. vol. 3 (1995) 3155-3160
14. Ivanescu, M., Stoian, V.: A Sequential Distributed Variable Structure Controller for a Tentacle Arm. Proc. of the 1996 IEEE Intern. Conf. on Robotics and Aut. vol. 4 (1996) 3701-3706
15. Yeong Y.S.: A Robust Fuzzy Logic Controller for Robot Manipulators. IEEE Trans. on Systems. Man and Cybernetics, vol 27, **4** (1997) 706-713

Appendix 1

We consider the dynamic model of the manipulator defined by (3.9). The parameter s^J from (4.7) represents an error measure of the closed-loop control.

$$s = \dot{e} + \sigma e \quad \textbf{(A1.1)}$$

where we cancelled the upperscript j, for the simplicity. From (3.9) and (4.1) we obtain:

$$M(\ddot{e} + \ddot{q}_d) + C(\dot{e} + \dot{q}_d) + D(e + q_d)F = T \quad \textbf{(A1.2)}$$

$$M\dot{s} - M\sigma s + M\sigma^2 e + Cs - C\sigma e + D(e + q_d)F = T$$

we separate the linear part of s, it results,

$$M\dot{s} - (M\sigma - C)s + H(e, q_d, F) = T \quad \textbf{(A1.3)}$$

where H is a (n^Jx1) nonlinear vector defined on the trajectory parameters q_d,e,F. In order to prove the stability of the closed-loop system, we use a Liapunov function by the form

(A1.4)

Differentiating (A1.4) and substituing (A1.3) we obtain:

$$\dot{V} = s^T M\dot{s} \quad \textbf{(A1.5)}$$

$$\dot{V} = s^T M\left[M^{-1}(M\sigma - C)s - M^{-1}H(e,q_d,F) + M^{-1}T\right] \quad \textbf{(A1.6)}$$

By using the control law of T , (A1.6) becomes

$$\dot{V} = s^T(k - M\sigma + C)s + s^T\left[H(e,q_d,F^M) - H(e,q_d,F)\right] + s^T u_F \quad \textbf{(A1.7)}$$

And using (4.11), [15],

$$\dot{V} \le -s^T(k - M\sigma + C)s + \sum_i \left|s_i\right|\left[\left|H(e,q_d,F^M) - H(e,q_d,F)\right|_i - k_{Fi}\right] \quad \textbf{(A1.8)}$$

If we introduce the condition (4.12) and we denote by λ_{min} the minimum eigenvalue of (K-Mσ+C), it results,

$$V \le -\lambda_{\min}\|s\| \quad \textbf{(A1.9)}$$

Appendix 2

We consider the dynamic model (A1.3) in the area around switching line ($u_{Fi} = 0$),

$$M\dot{s} - (M\sigma - C)s + H(e,q_d,F) = -Ks + H(e,q_d,F^M) \quad \textbf{(A2.1)}$$

From (A1.1) and using the properties of the matrices M,C.K,S (diagonal matrices) we obtain,

$$m_i\ddot{e}_i + (c_i + k_i)\dot{e}_i + (-m_i\sigma_i^2 + c_i\sigma_i + k_i\sigma_i)e_i + \left(H(e,q_d,F) - H(e,q_d,F^M)\right)_i = 0 \quad \textbf{(A2.2)}$$

$$\left(H(e,q_d,F) - H(e,q_d,F^M)\right)_i = (\Delta H)_i \cong \frac{\partial H}{\partial F}\frac{\partial F}{\partial q_i}e_i \quad \textbf{(A2.3)}$$

We denote this term as

$$(\Delta H)_i = h_i(q_d,e)e_i \quad \textbf{(A2.4)}$$

From (A2.2), (A2.4) we obtain the switching line,

$$\frac{\dot{e}_i}{e_i} = -p_i = \frac{1}{m_i e_i}\left[-(c_i + k_i)e_i - \sigma_i(-m_i\sigma_i + c_i + k_i)e_i - h_i(q_d,e)e_i\right] \quad \textbf{(A2.5)}$$

This equation determines the slope p_i if the following condition is verified [14],

$$(c_i + k_i)^2 \ge 4\left(\sigma_i(-m_i\sigma_i + c_i + k_i) + h_i\right)m_i \quad \textbf{(A2.6)}$$

Control of Robot Arm Approach by Fuzzy Pattern Comparison Technique

Mirjana Bonkovic, Darko Stipanicev, Maja Stula

LaRIS-Laboratory for Robotics and Intelligent Systems
Faculty of Electrical Engineering, Mechanical Engineering and Naval Architecture
UNIVERSITY OF SPLIT
R.Boskovica bb, 21000 SPLIT, C r o a t i a
Web: http://zel.fesb.hr/laris
e-mail: mirjana@fesb.hr , dstip@fesb.hr, kiki@fesb.hr

Abstract. The control of robot hand while reaching an object is a complex task involving the interaction of sensory inputs and motion commands. In this paper the system based on visual feedback is described. The whole system is shortly explained and the special emphasis is given to the target final approach control based on Fuzzy Pattern Comparison (FPC) technique.

1 Introduction

The real time control of the robot arm while reaching an object using visual feedback is quite a complex task which involves a lot of image processing and analysis [1,2,3] .

The process of a robot arm approach can be divided into two phases:

- a target approach phase, and
- an executive phase (contact with target, for example inserting a pin into the hole or grasping)

The target approach phase is based entirely on the interpretation of visual inputs, whereas the execution phase relies usually on tactical sensing.

The target approach can be further divided into two phases:

- the rough target approach and
- the final (and fine) target approach, which usually includes precise positioning of the robot arm according to the target position.

The research described in this paper is a part of noncalibrated image based robot arm approach control system. The system consists of three cameras, two of them responsible for the rough target approach and the third one for the final target approach.

The rough target approach control is based on an expert system which does not need any information about camera position [4]. The third camera becomes active when the robot arm enters its visual field. The task of its control system is to precisely position the robot arm depending on the target. This paper is concerned with the final target approach phase and describes both the main idea and its implementation on the Fuzzy Pattern Comparison (FPC) board.

2 Final target approach – problem description

The system is shown in Fig. 1.

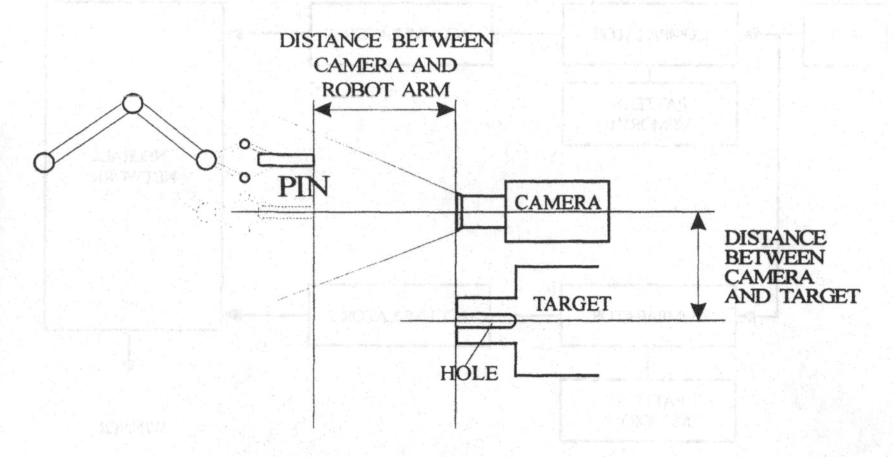

Fig. 1. Robot arm, camera for final position control and target

The final approach control system becomes active when the robot gripper enters camera visual field and reaches the predefined distance from the camera. Now the control task is to position the robot arm in the center of the camera visual field. After that, knowing the distance between the optical axe of the camera and the target, it is quite easy to finish the task and to reach the target. This control task could be interpreted as a 2D control of a robot arm around the image plane, as Fig. 2. shows.

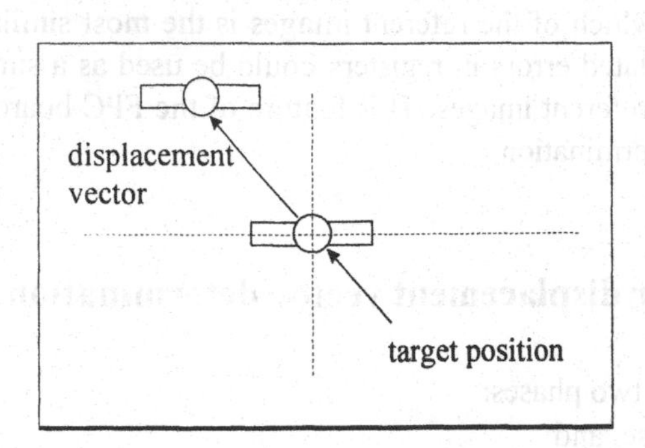

Fig. 2. The situation seen by the final approach camera

The most important part of this procedure is determination of the robot gripper displacement vector. It was calculated using the fuzzy set based algorithm and Fuzzy Pattern Comparator (FPC) board [5].

FPC is s specialized device particularly designed to provide very fast comparison of frames of data. The FPC utilizes fuzzy logic method of comparing data in order to determine the similarities or differences in groups of data. Its simplified block diagram is shown in Fig. 3.

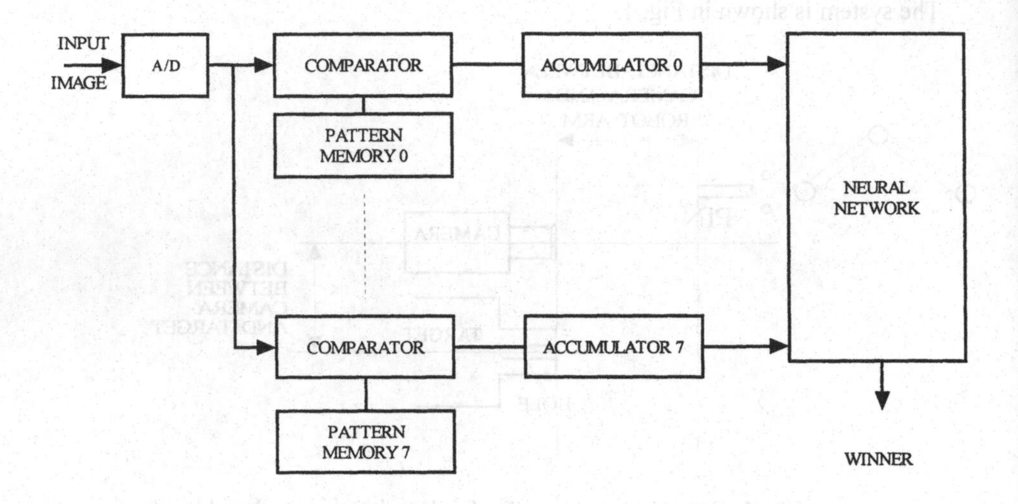

Fig. 3. The Fuzzy Pattern Comparator

Up to 8 patterns (referent images) could be stored in board pattern memories. Input image data enters serially into the device, where it is formatted into bit fields and in a parallel way compared with referent images in pattern memories. The results of the comparison are stored in registers associated to the pattern memories. The neural net gate is used to find which of the referent images is the most similar to the input image, and the accumulated errors in registers could be used as a similarity measure of the input image and referent images. This feature of the FPC board was used for displacement vector determination.

3 Robot gripper displacement vector determination

The procedure had two phases:

- the learning phase, and
- the working phase.

In the learning phase, the robot arm is moved manually in front of the camera and 6 characteristic positions were captured and stored in pattern memories: up, right-up, right-down, down, left-down, left-up.

Fig. 4. shows one of them.

Fig. 4. The referent image *"up"* from the pattern memory

In the working phase the similarity measures (accumulated errors in registers) between the input and stored reference images were used for calculation of the 6-element vector of fulfillment

$$\vec{\lambda} = [\lambda_1\ \lambda_2\ \lambda_3\ \lambda_4\ \lambda_5\ \lambda_6],\ \lambda_i \in [0,1]$$

λ_i corespondents to the i-th referent image. The input image and the pattern images are identical if $\lambda_i=1$.
$\vec{\lambda}$ is used as an input to fuzzy algorithm shown in Fig. 5, which gives the direction and magnitude of the displacement vector.

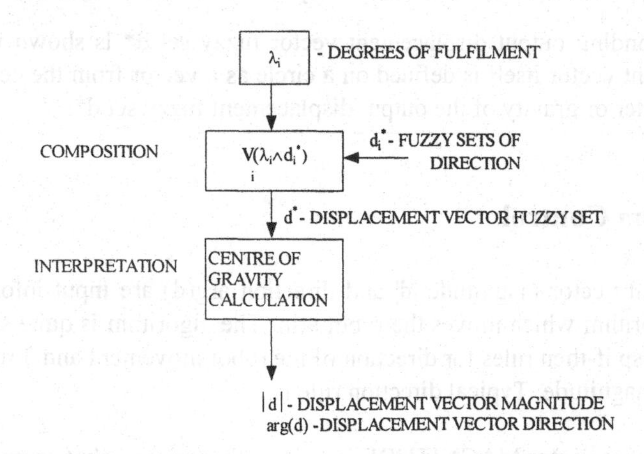

Fig. 5. Displacement vector calculation

Fig. 6. shows fuzzy sets of directions defined on the displacement vector circle. The shape of the displacement vector fuzzy set calculated by classical max-min composition carries information about relative position of the robot gripper according to the optical axe of the camera. For example for the starting position of the robot gripper, seen

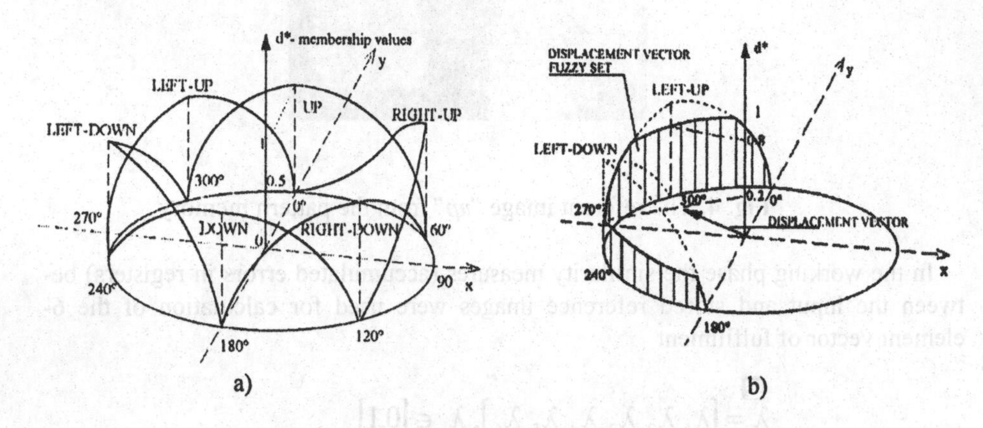

Fig. 6. a) Fuzzy sets of direction defined on a circle
b) Example of the displacement vector fuzzy set

by the camera somewhere between left-down and left-up as Fig. 9a) shows, the vector of fulfillment will be

$$\vec{\lambda} = [0 \quad 0 \quad 0 \quad 0 \quad 0.2 \quad 0.8]$$

and its corresponding output displacement vector fuzzy set d* is shown in Fig. 6b. The displacement vector itself is defined on a circle as a vector from the center of the circle to the center of gravity of the output displacement fuzzy set d*.

4 Robot Arm Control

Displacement vector magnitude |d| and direction arg(d) are input information to the control algorithm which moves the robot arm. The algorithm is quite simple one, defined by 8 crisp if-then rules for direction of the robot movement and 3 rules for the control action magnitude. Typical direction rule is:

"If displacement vector MAGNITUDE is *left-up*, then move robot segment BODY to *right*, and robot segment ELBOW to *down*."

And control action magnitude rules were:

"If MAGNITUDE is *central*, then STOP, else
If MAGNITUDE is *middle*, then STEP is 1, else
If MAGNITUDE is *big*, then STEP is 4."

Algorithm is adapted for RRR robot configuration as Fig. 8 shows, so terms like BODY and ELBOW for robot segments were used. Fig. 7 shows the definition of linguistic terms *left-up*, *down*, *central* ... In our experiments boundaries were crisp and the results were quite satisfactory, but fuzzy boundaries and fuzzy control algorithm could be applied, too.

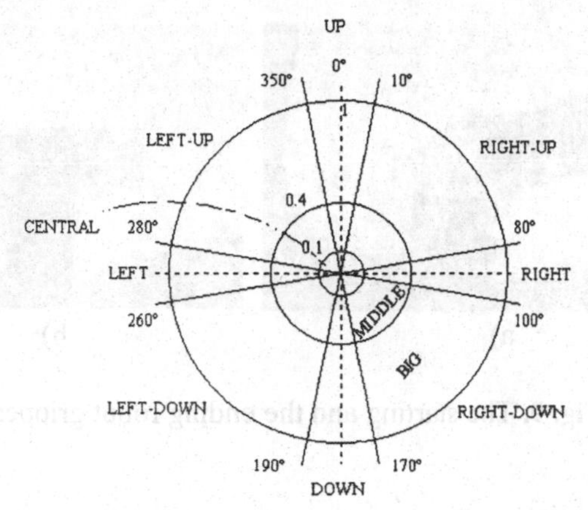

Fig. 7. Definition of linguistic terms used for robot arm control

For displacement vector shown in Fig. 6b) the robot BODY will be moved to the right 4 steps (looking from the camera side) and ELBOW down 4 steps. The value of one step was determined experimentally according to the robot speed.

5 Experiments

In experiments the educational robot MICROROBOT TechMower shown in Fig. 8. was used .

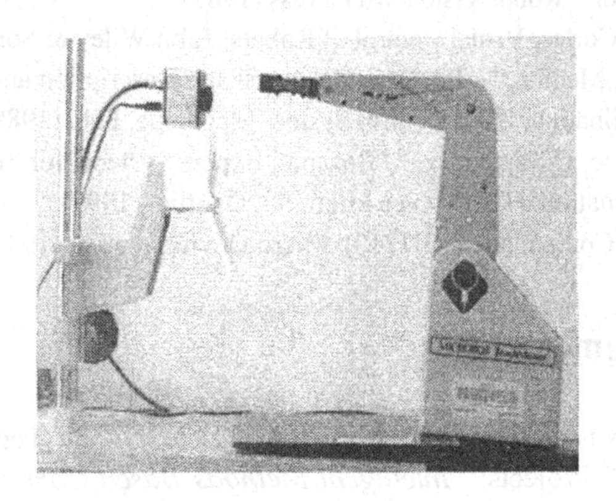

Fig. 8. Photo of the experimental system

The control algorithm was implemented on standard PC/486 computer and the real time robot movement control was achieved. Fig. 9 shows the starting and ending positions of the robot gripper seen by final approach camera. The processed image was 192x132 pixels digitalised in 4 bits.

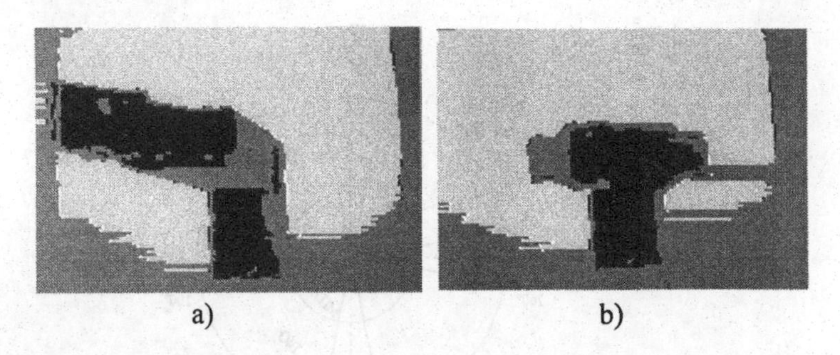

a) b)

Fig. 9. The starting and the ending robot gripper image

6 Conclusion

The simplicity is an advantage, specially in technical systems. Experiment described in this paper shows that a rather complicated control task could be solved using quite inexpensive equipment and fuzzy control principles.

Future work will be oriented toward 3D vector displacement determination and inclusion of gripper orientation control.

References

1. B. K. P. Horn: Robot Vision, MIT Press (1981)
2. 2. Peter I. Corlee, Visual Control of Robots, John Wiley & Sons Inc (1996)
3. K. Rao, G. Medion, H. Lin & G. Bekey: Shape Description and Grasping for Robot Hand-Eye Coordination, IEEE Control System Magazine, Feb. (1989) pp.22-29
4. M.Bonkovic, D.Stipanicev, M.Bozanic, Expert Systems for 3D Control of Robotics Arm, XXII International Conference Mipro 99, Opatija, (1999)
5. Fuzzy Set Comparator (MD1210), Micro Devices, USA, (1989)

Acknowledgment

This work was supported by Ministry of Science and Technology of Republic of Croatia through Projects: *"Intelligent Methods Based Complex System Control"* and *"Fuzzy Pattern Comparator Applications"*.

A Fuzzy Shapes Characterization for Robotics

Nicolae Varachiu

National Institute of Microtechnology - IMT Bucharest,
PO BOX: 38-160, Bucharest, Romania
E-mail: nichi@imt.pub.ro; nichi@imt.ro

Abstract We proposed [9] an original membership function building for fuzzy pattern recognition. This method uses only one shape feature – the compactness – leading to a simple and fast determination of membership function. We improved this method adding the possibility of control for the slopes of functions and unifying their representation [10,11]. We implemented this method in machine vision area, where for the robot's eye it must process a continuos image data series in real time. In the reason to reduce the need of preliminary human inspection, we proposed to add a new feature – the elongatedness – to be used in conjunction with the membership function built from compactness [12]. In this paper we propose a unified membership function, built by using the compactness and the elongatedness features of shapes together.

1 Introduction

An important application of pattern recognition is in the machine vision area. A robot, with various sensors and other features that allow them to have sensory functions for perceiving the circumstances and the environment around them, with intelligent information processing function became an "intelligent robot". Machine vision capability – the robot eye – is one of the most important sensory function of an intelligent robot. To obtain this function we need a camera (or more) together with an image processing and a pattern recognition method embedded in the robot controller. In general, the amount of data to be processed for pattern (shape) recognition is large scale. Since the computer that can be used with robots are in general small ones centered on microprocessors, the amount of memory is also small in most cases. However, if it is wanted to use image processing and pattern (shape) recognition for a robot's eye, it must process a continuous image data series in *real time* [8].

In the last decades, the scientific community agrees the idea that ambiguity, the uncertainty is a property of real world. We try to build some examples to support the position that *fuzziness* is an alternative to *randomness* in describing uncertainty.

2 Image representation

The largest utilization of robots is in industrial applications. Thus, the objects that we deal with are mechanical components in different shapes. In general they are "normal" geometric shapes, and they have practically uniform brightness in an usual illumination ambient. In applications, it is possible to provide a different (opposite) brightness for background (e.g. belt conveyor), thus being simple to obtain a binary image (two levels, only black and white pixels). The binary image contains all the necessary information for pattern (shape) representation, and also it is not memory consuming – specially if we use further coding methods (e.g. Run Length Coding – RLC).

3 About uncertainty

There are several sources of uncertainty and we identify uncertainty due to [1], [6]: inaccurate measurements (Resolutional Uncertainty), random occurrences (Probabilistic Uncertainty) and vague descriptions (Fuzzy Uncertainty).

We agree with idea to consider fuzziness as an alternative to randomness for describing uncertainty, and *not* to consider fuzziness as a clever disguise for randomness. Two models, one "fuzzy" and other "random" possess different kind of information: fuzzy memberships – which represent *similarities* of objects to imprecisely defined properties, and respectively probabilities – which convey information about *relative frequencies* [2].

Ambiguity is a property of natural environment and of physical phenomena, and fuzziness is a type of non-statistical uncertainty. For instance, if we consider an inexact oval [4], it is a non-statistic case, all the facts are in, but the uncertainty in saying that the shape is an ellipse remains. It is more natural to say that *it is a fuzzy* ellipse, *not* that the oval *is probably* an ellipse.

4 Compactness and elongatedness features

As features for geometric shapes – figures – the simplest metric properties are *area* and *length of perimeter*. Used in combination they defined the **compactness** (named in [3] thinness ratio) as follows:

Let X be the universe of discourse of geometric shapes.

By definition, compactness c of an object $x \in X$ is $c(x) = 4\pi A / P^2$ where A is the area and P is the perimeter of x.

It is obvious that $0 \leq c(x) \leq 1$. A famous theorem of antiquity is that compactness has a maximum value of 1, which it achieves if the geometric shape is a circle. For likewise objects (from Thales Theorem point of view), c is the same regardless the dimensions of its. This property recommends compactness feature to be used when we classify objects by only "pure" shape considerations, regardless dimensions.

Compactness c is easy to be computed in practical applications, if the image is represented in binary coded way – as it is presented in above 2 paragraph.

The compactness c is the same for likewise objects, but it is possible that two different figures to have the same "c". For instance, different geometric shapes like three unit square arranged in "line", and after in "corner" has the same c:

Fig. 1. Two different shapes with the same "c"

The **elongatedness** e(x) is defined similar as aspect ratio in [Du]. The aspect ratio of an rectangle is the ratio of its length to its width. We consider the elongatedness as the ratio of minimum dimension to the maximum dimension of the rectangle. We obtain the maximum value of one for a square, and it is obvious that $0 \le e(x) \le 1$. For an arbitrary figure, we enclose it in some rectangle and consider the elongatedness of the figure the elongatedness of circumscribed rectangle. For calculating the elongatedness we used rectangles with the sides parallel to the eigenvectors of the scatter matrix of the figure points. The eigenvectors correspond physically to the directions about which the figure has maximum and minimum moments of inertia, and represent our intuitive notion of the direction in which the figure is fattest and thinnest.

Even compactness is a simple way of characterizing shapes, it has to pay attention in using it alone. As we shown in Fig. 1, two different shapes can have the same c, but their elongatedness are respectively 0.33 and 0.75, it means enough different. In this example one object is convex and the other one is concave. Even two different shapes are convex, it is possible to have the same c (see Fig. 2 for x_1 and x_2)

Let us now examine the following example:

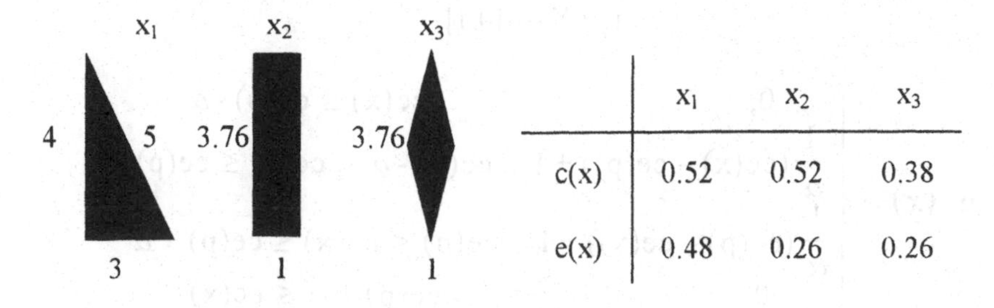

	x_1	x_2	x_3
c(x)	0.52	0.52	0.38
e(x)	0.48	0.26	0.26

Fig. 2. Three different convex shapes and their corresponding values of c(x) and e(x)

We have three different convex shapes, having the first two the same compactness, $c(x_1) = c(x_2) = 0.52$ and the last two the same elongatedness, $e(x_1) = e(x_2) = 0.26$. As we can observe in the table, when we have the same value for one feature, the second is different enough. For this reason, we propose to use both this features for membership function building.

5 Membership function building

In one class problem, we measure by membership function the degree of *similarity* of different objects to a prototype-object, or the belongness degree of the current object to the fuzzy prototype class.

We introduced in [9] a simple triangular membership function with three different forms u_{circ}, u_{pa} and u_{pb} for the cases when prototype p is a circle ($c(p)=1$), it has $c(p)<0.5$ and it has $c(p)\geq 0.5$ respectively. This manner of introducing the membership functions is a simple one, but for a part of applications it is rigid, being not able to control the slopes of function. In [10] we proposed a unified form of membership function u_p with variable slopes, and in [12] we added the elongatedness feature to be used together with u_p to simplify a thorough (and time consuming) preliminary inspection by human operator.

We propose now a unified membership function, built by using the compactness and the elongatedness features of shapes together.

Let $c(x)$ and $e(x)$ be the compactness and the elongatedness of a shape x.

We define a new compound feature $ce(x)$ as:

$$ce(x) = \lambda c(x) + (1-\lambda)e(x), \text{ where } \lambda \in [0,1]$$

Because $c(x), e(x) \in [0,1]$ as we already showed, and this closed interval is a convex set, the above definition preserve $ce(x)$ in the same set, i.e. $0 \leq ce(x) \leq 1$

Also, if the prototype p is a circle, $ce(p)=1$, because $c(p)=1$ and $e(p)=1$.

In practical applications, we used $\lambda=0.5$.

The membership function u_p is:

$$u_p : X \rightarrow [0,1]$$

$$u_p(x) = \begin{cases} 0, & ce(x) \leq ce(p) - \alpha \\ \frac{1}{\alpha}(ce(x) - ce(p)) + 1, & ce(p) - \alpha < ce(x) \leq ce(p) \\ \frac{1}{\alpha}(ce(p) - ce(x)) + 1, & ce(p) < ce(x) \leq ce(p) + \alpha \\ 0, & ce(p) + \alpha \leq ce(x) \end{cases}$$

where α is a factor controlling the slopes of membership function.

The factor α has three different ranges for the respectively three cases:
i) If the prototype p is a circle, ce(p)=1 and $\alpha \in (0, 1]$

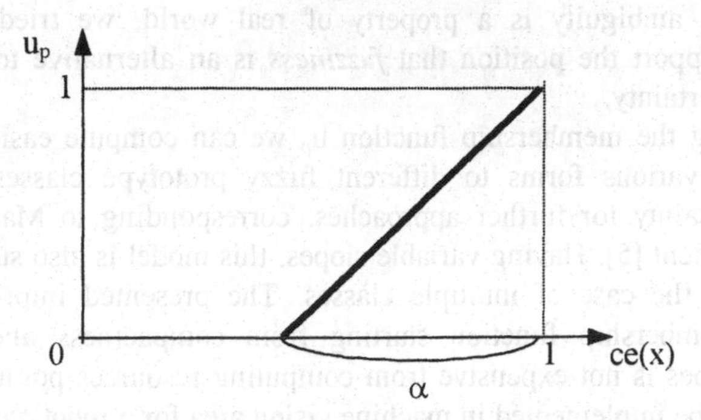

Fig. 3. Membership function form for a circle prototype

If the prototype p differ from a circle, i.e. ce(p) < 1, the range for α will be selected for two cases:
ii) For 0 < ce(p) < 0.5, $\alpha \in (0, c(p)]$

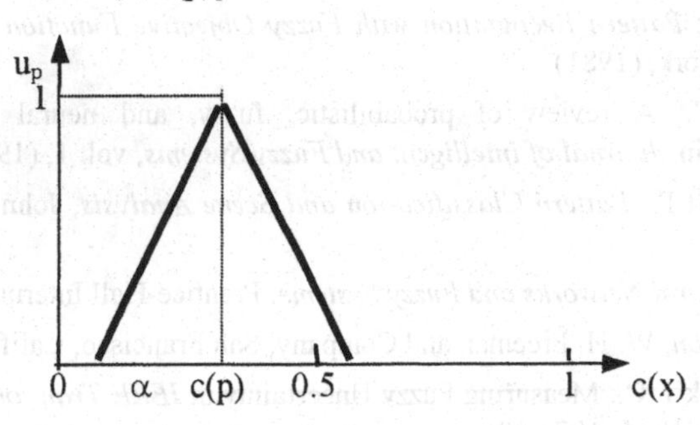

Fig. 4. Membership function form in case ii)

iii) For $0.5 \leq c(p) < 1$, $\alpha \in (0, 1\text{-}c(p)]$

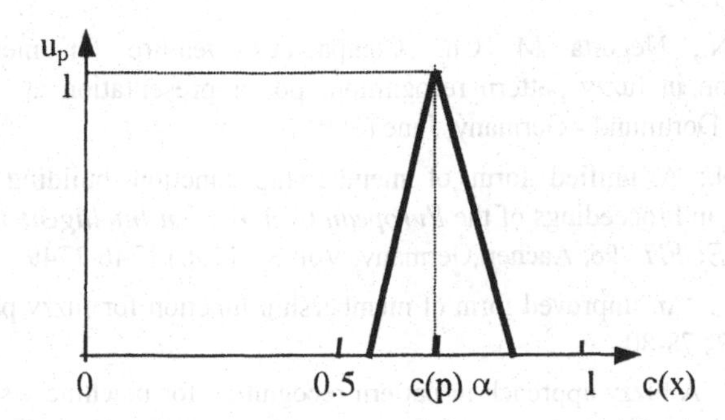

Fig. 5.. Membership function form in case iii)

6 Concluding remarks

Observing that ambiguity is a property of real world, we tried to build some examples to support the position that *fuzziness* is an alternative to *randomness* in describing uncertainty.

By proposing the membership function u_p we can compute easier the degree of belongness of various forms to different fuzzy prototype classes. Also we can preserve uncertainty for further approaches, corresponding to Marr's Principle of Least Commitment [5]. Having variable slopes, this model is also suitable to control the overlap in the case of multiple classes. The presented improved method of building a membership function starting from compactness and elongatedness features of shapes is not expensive from computing resources point of view, and it was suitable to be implemented in machine vision area for a robot's eye.

References

1. Bezdek J. C.: *Pattern Recognition with Fuzzy Objective Function Algorithms*, Plenum Press, New York, (1981)
2. Bezdek J. C.: A review of probabilistic, fuzzy, and neural models for Pattern Recognition, in *Journal of Intelligent and Fuzzy Systems*, vol. 1, (1993) 1-25
3. Duda R., Hart P.: *Pattern Classification and Scene Analysis*, John Wiley & Sons Inc. (1973)
4. Kosko B.: *Neural Networks and Fuzzy Systems*, Prentice-Hall International Inc., (1992)
5. Marr D.: *Vision*, W. H. Freeman and Company, San Francisco, California, (1982)
6. Pal S., Bezdek J. C.: Measuring Fuzzy Uncertainty in *IEEE Tran. on Fuzzy Systems*, vol. 2, no. 2, May (1994) 107-118
7. Reusch B., Dascalu D. (editors): *Real world applications of intelligent technologies*, IMT-Bucharest, (1997)
8. Terano T., Asai K., Sugeno M.: *Fuzzy Systems Theory and its Applications.*, Academic Press Inc., (1992)
9. Varachiu N., Negoita M. Gh.: Compactness feature for membership function determination in fuzzy pattern recognition, poster presentation at *4th Fuzzy Days of Dortmund,* Dortmund - Germany, June (1994)
10. Varachiu N.: A unified form of membership function building for fuzzy pattern recognition, in Proceedings of the *European Congress on Intelligent Techniques and Soft Computing EUFIT '96,* Aachen,Germany, Vol. 3, (1996) 1746-1749
11. Varachiu N.: "An improved form of membership function for fuzzy pattern recognition", in [7], (1997) 78-80
12. Varachiu N.: A fuzzy approach in pattern recognition for machine vision, in Proceedings of the *3rd Joint Conference of Information Sciences,* Research Triangle Park, North Carolina, USA, Vol. 3, (1997) 235-237

ART-based Automatic Generation of Membership Functions for Fuzzy Controllers in Robotics

G. Attolico and A. Itta and G. Cicirelli and T. D'Orazio

I.E.S.I. - C.N.R., Via Amendola 166/5, 70126 Bari, ITALY
attolico@iesi.ba.cnr.it,
Phone: +39-080-5481969,
Fax.: +39-080-5484311

Abstract. Designing fuzzy controllers involves both identifying suitable fuzzy dicretizations of the input and output spaces (number of fuzzy terms and related membership functions) and drawing effective rule bases (hopefully compact and complete). Learning from examples is a very efficient way for acquiring knowledge. Moreover, providing reliable examples is easier than directly designing and coding a control strategy. Supervised learning is therefore appealing for automatically building control systems. It may also be used as quick start-up for applications requiring further on-line learning (through reinforce-based techniques or processing of further appropriate examples). Supervised techniques for building rule bases from suitable examples generally rely on pre-defined fuzzy discretizations of the input and output spaces. This paper proposes the use of an ART-based neural architecture for identifying, starting from examples, an appropriate set of fuzzy terms and associated membership functions. These data are then used by an ID3-based machine learning algorithm for building fuzzy control rules. The ART framework provides fast convergence and incremental building of classes, gracefully accounting for the integration of new sample data. The whole chain (the neural architecture for building fuzzy discretizations and the machine learning algorithm for drawing fuzzy rules) has been proved on examples provided by several operators, with different skills, driving a real vehicle along the right-hand wall in an indoor environment. The obtained results are shown, discussed and compared with the performance of controllers using human defined fuzzy partitions.

1 Introduction

Basic reactive behaviors (straight associations of control commands to sensory data) are becoming popular bricks for building general and robust control systems in robotics. There is a strong interest in building robot with high levels of autonomy, moving from the dependency on the knowledge directly provided by the designer toward the capacity of self-organizing classes for interpreting sensory data and of self-designing strategies for accomplishing the assigned tasks.

Developing tools for deriving a control system from suitable training examples provided by a human operator executing the task at hand can both make easier and less subjective the design phase and, at the same time, provide the base for a control system evolving on the base of its own experience. Unfortunately, the automatic indentification of the function (often non-linear and complex) describing even a simple reactive behavior in a multidimensional domain is not so easy. Fuzzy rules, besides providing smooth control (a very desired property), have the advantage of being derivable through appropriate learning algorithms while remaining easily understandable by humans for direct examination and, eventually, modification. Fuzzy logic is therefore a suitable way for representing knowledge in a system where engineers and algorithms are expected to intermix their work.

Identifying relevant input and output variables, defining their fuzzy discretizations (terms and related membership functions) and drawing a suitable rule base expressing the control strategy involve increasing levels of subjectivity. Input and output variables naturally arise from the problem, but their expressions in fuzzy terms is less immediate and can eavily impact the effectiveness of automatically generated fuzzy rules. Finally, identifying and coding a control strategy into working rule bases can be difficult for a task less than trivial and strongly affected by the designer skills. Supervised learning may successfully extract rules from appropriate and meaningful training examples [1–4] but generally requires previously defined input and output fuzzy spaces.

The paper describes a part of a larger system for supporting the development of fuzzy reactive controllers from examples recorded by observing a human operator accomplishing the task at hand. The whole system supports the fuzzy discretization of system variables and the subsequent derivation of effective fuzzy rules. It allows an operator (having no knowledge about fuzzy logic and its application to robot control) to drive a vehicle, using different modalities, on the base of both external (sonar, CCD camera, ...) and internal sensory data that are stored with the corresponding control commands. A wide support is available during the acquisition (both on-line and off-line evaluation of driving, incremental extension of the training set for critical spatial configuration, vocal and graphical representations of the sensory data acquired by the vehicle, ...) in order to increase the meaningfulness and consistency of training examples.

We propose an ART-based schema for extracting suitable fuzzy terms and membership functions for each system variable. The results are naturally tuned to the driving style of the teaching operator. Then, a previously developed supervised learning algorithm is used for identifying the sensory data really relevant for the task at hand and the appropriate actions reproducing the strategy followed by the operator.

The system has been verified on the wall-following basic behavior (moving a vehicle at a constant distance from the wall on its right hand, avoiding collisions and reducing linear and rotational accelerations). Linear and steering velocities are set on the base of data provided by a sonar ring.

Fuzzy discretizations obtained by the ART-based schema have been compared with manually designed fuzzy terms and membership functions. The results show that the neural architecture is flexible and can adapt to different driving styles.

The resulting system can derive working fuzzy controller from examples provided by operators just skilled in driving the vehicle. No knowledge about the system architecture, fuzzy theory or control applications is required. Even the designer is only expected to tune a few parameters in the algorithms involved in the process.

2 Membership Functions from an ART-based Neural Architecture

The Adaptive Resonance Theory claims to provide a tool for solving the *plasticity-stability* dilemma. It provides a framework for building neural architectures that can acquire new knowledge (either adding new prototypes or modifying the existing ones) without destroying the learned classes and without human supervision [5], [6], [7]. The ART models offer a suitable tool for on-line incremental unsupervised learning of sensory classes. They require only few presentations of the examples: this property is important for providing as soon as possible the control system with meaningful sensory classes for organizing the input data. Moreover, their incremental learning (acquisition of new knowledge with controlled impact on previously defined classes) reduces the risk of sudden losses of performance due to the system evolution.

The ART-based architecture described in [7] has been choosen for the task of autonomously selecting suitable fuzzy terms and membership functions for the task at hand. This member of the ART family is able to deal with real-valued input vectors (as the data provided by the sonar ring) and can use a fast-learning modality. Its architecture is shown in the Fig. 1.

The equation of the generic F_1 sublayer unit is:

$$\epsilon \frac{d}{dt} x_k = -Ax_k + (1 - Bx_k)J_k^+ - (C + Dx_k)J_k^- \tag{1}$$

where A, B, C and D are costant. In our implementation, B e C have been set equal to zero. The Table 1 specifies the values required by the equation for each sublayer and for the orienting subsystem.

The non-linear function is:

$$f(x) = \begin{cases} 0 & 0 \leq x \leq \theta \\ x & x > \theta \end{cases} \tag{2}$$

The F_2 output function, $g(y)$, is:

$$g(y_j) = \begin{cases} d & T_j = \max_k\{T_k\} \forall x \\ 0 & \text{else} \end{cases} \tag{3}$$

The equations of top-down and bottom-up weights are, respectively:

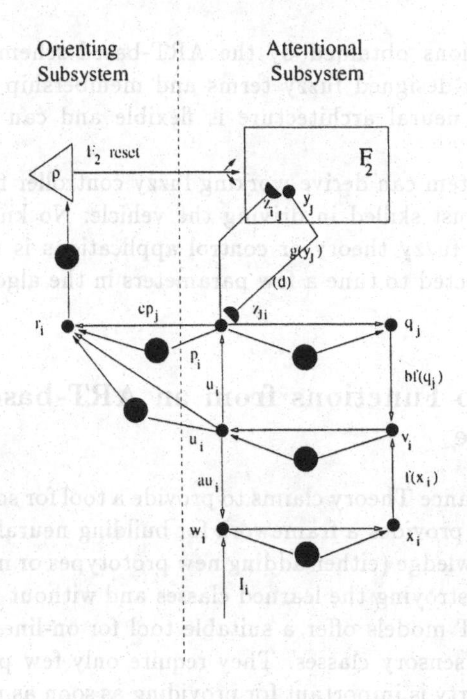

Fig. 1. The architecture of the ART2 network used in our experiments

$$\frac{d}{dt}z_{ji} = g(y_j)(p_i - z_{ji}) \tag{4}$$

$$\frac{d}{dt}z_{ij} = g(y_j)(p_i - z_{ij}) \tag{5}$$

The fast-learning approach does calculates the asymptotic solutions of differential equations.

The use of ART neural architectures requires the identification of suitable values for a few parameters. The a parameter is used only in the ART2 first layer. It balances the relative importance of the top-down with respect to the input vector I. The ART2 behavior is not really sensitive to changes in this parameter: therefore, the value 10, suggested by Carpenter and Grossberg [6],[8], has been used.

The parameters c and d, according to the suggestions of Grossberg, have to be chosen in order to make $cd/(1-d)$ close to 1. That means increasing the sensitivity of the system to mismatches. We have experimentally verified an increase of stability of the network associated to this condition. The final values used in our experiments have been: $d = 0.999999999$ and $c = 0.000000001$.

The same global effects have been observed on the b parameter. Its increase produced greater sensitivity to mismatches and greater stability [6]. The value $b = 999999999$ has been used in our experiments.

Table 1. Factors for each F_1 layer and for the **r** layer. a, b, c and e are costant. y_j is the activity of the jth unit on the F_2 layer and $g(y)$ is the output function of F_2. The non-linear function, $f(x)$, is described below.

layers	A	D	J_i^+	J_i^-
w	1	1	$I_i + au_i$	0
x	e	1	w_i	$\lVert \mathbf{w} \rVert$
u	e	1	v_i	$\lVert \mathbf{v} \rVert$
v	1	1	$f(x_i) + bf(q_i)$	0
p	1	1	$u_i + \sum_j g(y_j)z_{ij}$	0
q	e	1	p_i	$\lVert \mathbf{p} \rVert$
r	e	1	$u_i + cp_i$	$\lVert \mathbf{u} \rVert + \lVert c\mathbf{p} \rVert$

The θ parameter helps in suppressing the error in the input patterns by flattening useless details lower than its value. We have verified in our experiments that even very low values of θ could result in loss of useful data. This parameter has therefore be set to 0 leaving to the subsequent processing the task of coping with the noise present in the data.

The ART network suggest the fuzzy partition for each input set. The same schema can eventually provide suggestions for the fuzzy partition of output variables, even if terms and membership functions of control commands can usually be set more easily on the base of the required behavior.

The designer is only expected to evaluate the results and eventually tune the vigilance parameter of the ART network. A few guidelines for choosing this value have been identified: generally the evolution of the classes, associated to changes in this parameter over a small range, provides enough information for picking up the right value without much effort. More detailes will be given during the description of experimental results.

3 ID3-based Technique for Building Fuzzy Rule Bases

The fuzzy discretizations identified at the previous stage of the process are used by a supervised learning algorithm for building the rule base of the system [9].

The algorithm used is inspired to a decision-tree building algorithm, described in [10] that extends the ID3 technique proposed by Quinlan. This recursive method builds a decision tree for classifying all the provided examples with respect to an associated tag (the control commands in our case). Each leaf of the tree is expected to collect samples all belonging to the same class (that is having the same tag). Each path from the root to a leaf can be associated to a rule. Branch selection depends on the value of variables checked along the way.

At each step, an input variable is choosen on the base of its information contents (related to its capacity of separating the classes of control commands). The choosen variable becomes the root of a sub-tree having a branch for each possible value of the variable.

An appropriate condition can be posed for avoiding unnecessary expansion of the tree: a new sub-tree is generated only if it significantly reduces the error rate in classification (that is in control command selection). A nice property of the algorithm is that a complete coverage of the input space (that is the association of a control action to each possible sensory situation) can be obtained without generating all the possible rules (that is a rule for each single spatial configuration). In fact, paths from the root to a leaf shorter than the maximum depth of the tree (the number of input variables) generate rules, with less fuzzy predicates in the antecedent, that cover larger parts of the input space.

Modifications have been introduced in the method in order to fully exploit the fuzzy representation of input data. All the expressions involved in the evaluation of the information content of variables and the predicates associated to the paths in the tree have been converted in their fuzzy counterpart. Moreover, each time a sample activates two adjacent output fuzzy terms, it is allowed to contribute to more than a single rule. Eventually two rules can be generated for the same sensory configuration (and weighted on the base of their relevance) in order to obtain a finer control of the vehicle. Detailed description of this other module of the system is beyond the scope of this paper, focusing on the determination of membership functions. The method for the generation of the rule base has been kept unchanged in order to make meaningful the comparison of results.

4 Experimental Results

A basic navigation behavior, the wall-following, has been used as test-bed for the proposed method. It consists in driving along the wall on the right hand at a constant distance from it. It is useful for exploring an unknown environment (keeping a fixed hand does imply the visit of all the part of the environment) while providing a simple but effective obstacle avoidance (each obstacle is treated as a part of the wall). Moreover, a direct comparison with performance obtained by fuzzy controllers executing the same task but obtained by manually designed membership functions has been possible.

A real vehicle, based on a Nomad 200 platform, has been used for all the experiments. Its sonar ring, composed by 16 sensors, has been used as sensor by the controllers. Only 11 sensors have been used by the controllers: the rear sensors have been disabled (the vehicle was not allowed to move backward and the operator, during the collection of examples, followed the vehicle making meaningless these readings).

They have been grouped into sets: this allows lower dimensionality of the input space and, at the same time, increases the robustness of sensory data (comparison of adjacent sensors can reduce the risk of wrong measures due to reflection of the sonar beam) using the minimum reading for each group. The controllers obtained by manually designed membership functions have used five groups (shown in the left part of Fig. 2).

For the controllers using automatically extracted membership functions different groups have been used (see the right part of Fig. 2). In order to avoid

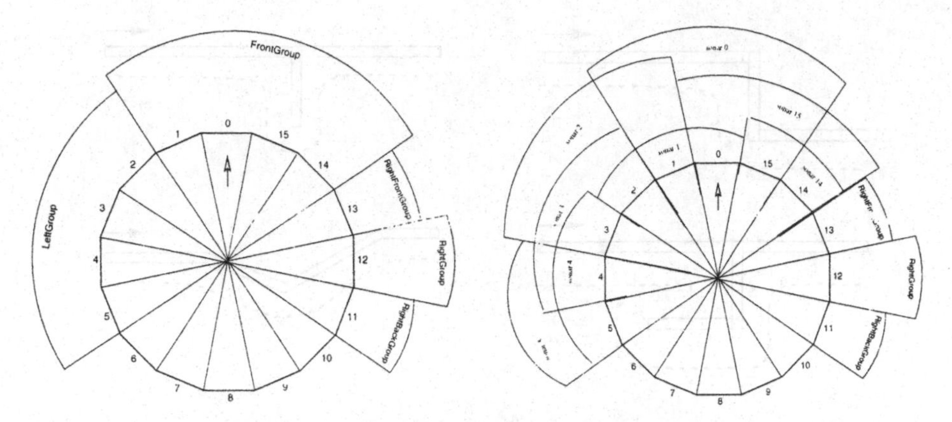

Fig. 2. The sonar groups used by the controllers using manually designed (on the left) and automatically extracted (on the right) membership functions

any constraint on the system, each sensor individually contribute to the input vector. All the sensors, but the three on the right, are filtered by taking the minimum value in a group including the adjacent two: due to their different orientation this policy provides greater chances of having a sensor with acceptable incidence angle of the sonar beam with respect to the sensed surface. Moreover, the choosen value is less probably affected by reflection errors and represents a conservative choice. The three sonar sensors on the right side of the vehicle have been kept unfiltered because they are used for estimating the attitude of the vehicle with respect to the wall: the same filtering would make their readings completely useless for this task.

The acquisition of training sets (as long as the final tests of the fuzzy controllers) has been done using the vehicle in a real environment (our lab, with the use of moveable wood panels for increasing the complexity of spatial configurations). Training data have been acquired using several operators, in order to confront the method with different driving styles and check its capability of adapting to the characteristics of each driver. Fig. 3 shows the five spatial configurations used for acquiring the training data.

A previously developed system has been used as the interface between the operator and the vehicle. It allows different modalities for this interaction and records all the pairs sensory-data/control-commands that provide the examples for the training phase. The Fig. 4 describes the characteristics of each operator: knowledge about fuzzy logic, about its application to robot control and about the algorithms used for generating membership functions and rules, skill in driving the vehicle using the software or hardware joystick and finally their use of visual or audio support during training sessions.

The experiments demonstrated that only skill in the execution of the task are needed: operator with deeper knowledge about fuzzy logic and about the algorithms used by the system provided poorer examples than operator with less knowledge but superior skill in driving.

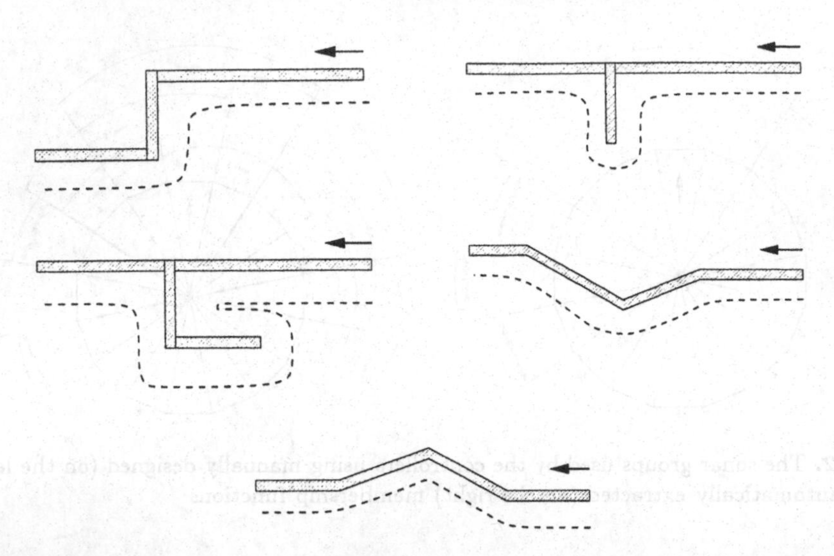

Fig. 3. The five paths used for acquiring training data. Each operator did two or three runs for each situation. The arrow shows the direction of navigation. The dotted lines represent the desired navigation path.

A numeric vector with 13 value (the 11 readings of the sonar groups plus the linear and rotational speeds) has been provided in input to the ART2 network.

The ART2, with the parameter settings described in the previous section, divided the training examples into groups. For each group a representative vector has been evaluated as mean of the vectors belonging to the group. We have kept only the vector elements smaller than 1 meter, in order to fix the horizon of the region of interest around the vehicle (the desired distance from the wall was 60 cm). Each representative vector is projected onto the axes of the input space providing mid-points for the triangular membership functions. On each axis (fuzzy variable) the first and last membership functions become trapezoidal (extended to the extremes of the acceptable range).

The Fig. 5 show the membership obtained by the system for two sonar groups (the right-front and the right ones) starting from the examples provided by the operator #1. They can be compared with the fixed functions used by the previously developed controllers (Fig. 6).

A critical parameter in the ART-based architecture is the so called vigilance parameter ρ. It controls the degree of granularity expected from the network. The number of sensory classes identified by the network increases with the value of ρ. Because the other parameters can be kept fixed, the only task of the designer is the identification, for example by trial and error, of the correct value for ρ, instead of fixing the parameters for all the membership functions. We have found that even this fine tuning is often unnecessary. We slowly increase the ρ value and track the evolution of the produced classes (in terms of both number

Operator #1	Deep knowledge about fuzzy logic applied to robotics. Good understanding of the desired behavior, of driving style effects on quality of examples and, therefore, on derived fuzzy rules. High manual skill using hardware joystick. Vocal messages supporting training sessions.
Operator #2	Deep knowledge about fuzzy logic applied to robotics. Good understanding of the desired behavior, of driving style effects on quality of examples and on derived fuzzy rules. Lower manual skill using hardware joystick. Vocal messages and landmarks on floor supporting training sessions.
Operator #3	Limited knowledge about fuzzy logic applied to robotics. Lower understanding of the desired behavior, of driving style effects on quality of examples and on derived fuzzy rules. Low manual skill using hardware joystick. Vocal messages supporting training sessions.
Operator #4	No knowledge about fuzzy logic applied to robotics. Good understanding of the desired behavior, none of driving style effects on quality of examples and on derived fuzzy rules. Good manual skill using hardware joystick. Vocal messages and landmarks on floor supporting training sessions.
Operator #5	No knowledge about fuzzy logic applied to robotics. Limited understanding of the desired behavior, none of driving style effects on quality of examples and on derived fuzzy rules. Low manual skill using hardware joystick. No support during training sessions.

Fig. 4. The different capacities of the five operators used for testing the approach

and prototypes). In the beginning of this process, the classes change slowly and stay constant for significative changes of the ρ parameter. The greater of these intervals of constant classes usually produces effective membership functions. In this way we do not need to test each set of membership functions, produced by different ρ values, by generating the corresponding rule base and running the controller on the real vehicle. Therefore, to the advantage of reducing the task of the engineer to simply look for the right value of ρ, in most cases the system could automatically find all the required settings, reducing the trial-and-error costs to the minimum.

The obtained membership functions are expected to better adapt to the driving style of each operator. This generally results in better working controller (the system receive higher scores or build-up working controller in cases when the fixed membership functions were unable to do so). On the other side, this adaptation can be useless or dangerous when training examples are not very effective. In the case of the operators #2 and #5 the resulting controllers were unable to avoid collisions. While in the latter case the fixed membership too were uneffective, in the former one the old method provided a working controller. We have done a coarse verification of these training data by building controllers for each specific situation: in most cases these sub-controllers were wrong. We suspect that inconsistencies in the data can sometimes be reduced by using all the five situations together but, in other cases, they can affect the controllers. The automatic extracted functions, that follows more closely the data, can be more sensitive to this problem with respect to the superimposed ones.

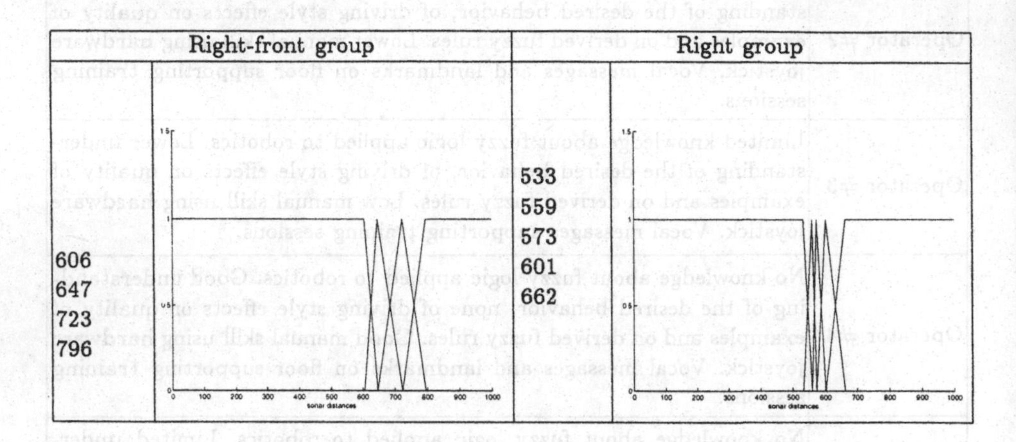

Fig. 5. Membership functions identified by the ART-based neural architecture for two sonar groups (right-front and right). The numerical values represents the vertices of membership functions (in mm).

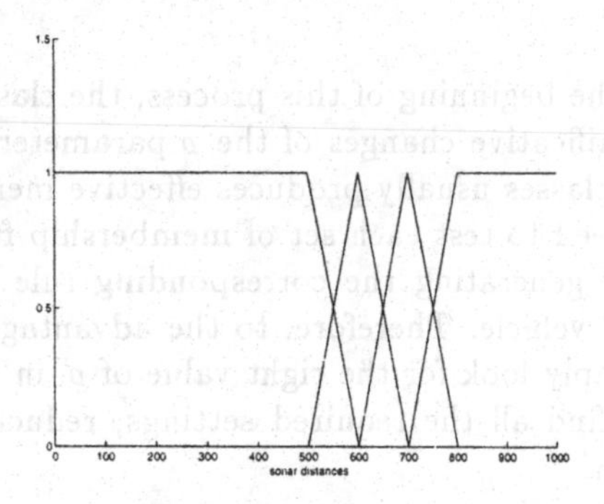

Fig. 6. Membership functions manually designed for the three groups on the right of the vehicle. They are the only groups that are kept unchanged in the ART-based fuzzy controllers. They can therefore be directly compared with the functions provided by the neural architecture

The Fig. 7 shows the results obtained by the controllers, using manually designed and automatically extracted membership functions, on a test environment. When both versions of membership functions provide working controllers, the ART-based version obtains a better score. The ART-based controller provides a working controller for the 3° operator for wich the manually designed functions was unable to avoid collisions. For the 2° operator the ART-based controller was unable to avoid collision while the manually designed version did it. Further investigations are in progress for verify the previously exposed possible explanation for this failure. For the 5° operator none of the two controllers were able to avoid collisions with the wall, probably due to intrinsic deficiencies in the training data.

5 Conclusions

The paper addresses the problem of supporting the design of reactive fuzzy controllers in robotics. The main goal is to simplify the design process and to reduce the subjectivity in identifying the different components of the fuzzy system.

The support tools starts from collected training examples (that is sensory-data/control-command pairs) of the desired control strategy. Obtaining effective examples, while not always a trivial task, seems in most cases much easier than explicitly state the control strategy.

The two following phases, that is defining suitable membership functions for fuzzifying the problem and drawing an appropriate rule base, are critical for the effectiveness of the final solution.

The paper proposes an ART-based neural architecture for extracting membership functions from training data. The neural architectures belonging to the ART family share the desired properties of dealing with the *stability/plasticity* dilemma: they can incrementally learn, in an unsupervised fashion, new knowledge (both in terms of refinement of available classes or creation of new ones) while preserving what already acquired. This is important for dinamically building sensory classes in a system without compromising the reached performance.

A general set-up has been identified for the parameters of the ART model. The vigilance parameter, the only needing a change for adapting to different training sets, can be found by tracking the effects of its changes on the behavior of the generated classes. It has been experimentally verified that this analysis provide effective values with small efforts and without requiring the time involved by the trial-and-error processes. In any case, the approach requires the set of a single parameter instead of the large number of values required for fully specify the fuzzification and defuzzification phases of a control system.

Encouraging results have been obtained on a task, the wall-following, useful as a basic brick for more complex navigation behaviors. In most cases the fuzzy controllers using membership functions identified by the ART architecture outperform controllers derived by manually drawn fuzzy terms. In a few cases, a failure has been observed, probably due to the data-driven nature of the approach, sensitive to inconsistency in the training sets. Further investigations are in progress for coping with this problem, by working on the support during collection of the training examples and on the following processing. Wider test-

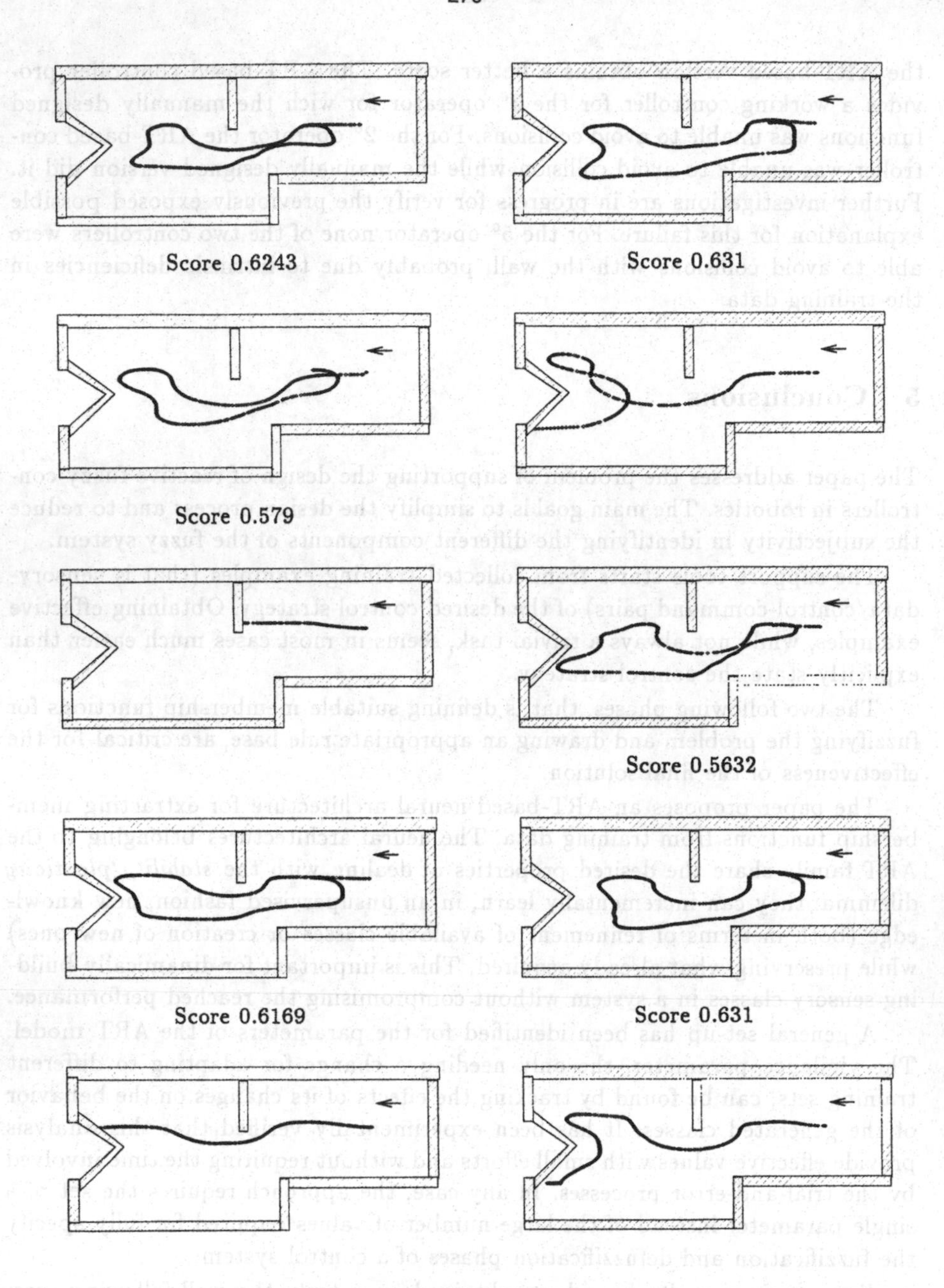

Fig. 7. Each row of the figure shows the performance obtained by controllers derived from the same training set using membership functions manually designed (on the left column) and extracted by the ART-based approach (on the right column). The operators runs from top (the first) to bottom (the fifth). A score is provided for working controllers in order to quantify the quality of driving: it is obtained by suitable evaluation functions taking into account distances, from frontal obstacles and from the wall, and amount of linear and rotational accelerations.

ing is also in progress for verifying the whole system on different basic reactive behavior useful for autonomous navigation.

References

1. S. Abe and M. Lan. Fuzzy rules extraction directly from numerical data for function approximation. *IEEE Transaction on Systems, Man and Cybernetics*, 25(1):119–129, January 1995.
2. L.X. Wang. Stable adaptive fuzzy control of nonlinear systems. *IEEE Transaction on Fuzzy Systems*, 1(2):146–155, May 1993.
3. C. Baroglio, A. Giordana, M. Kaiser, M. Nuttin, and R. Piola. Learning controllers for industrial robots. *Machine Learning*, 1996.
4. P. Reignier. Supervised incremental learning of fuzzy rules. *Int. Journal on Robotics and Autonomous Systems*, 16:57–71, 1995.
5. Gail A. Carpenter and Stephen Grossberg. A massively parallel architecture for a self-organizing neural pattern recognition machine. *Computer Vision, Graphics and Image Processing*, 37:54–155, 1987.
6. Gail A. Carpenter and Stephen Grossberg. Art2: self-organization of stable category recognition codes for analog input patterns. *Applied Optics*, 26(23), 1987.
7. James Freeman and David Skapura. *Neural networks. Algorithms, applications and programming techniques.* Addison-Wesley, 1991.
8. Gail A. Carpenter, Stephen Grossberg, and David B. Rosen. Art2-a: an adaptive resonance algorithm for rapid category learning and recognition. *Neural Networks*, 4:493–504, 1991.
9. G. Castellano, G. Attolico, and A. Distante. Automatic generation of fuzzy rules for reactive robot controllers. *Robotics and Autonomous Systems*, 22:133–149, 1997.
10. S.C. Hsu, J.Y.J Hsu, and I.J. Chiang. Automatic generation of fuzzy control rules by machine learning methods. In *Proceedings of the IEEE International Conference on Robotics and Automations*, pages 287–292, Nagoya, Japan, 1995.

On Two Types of L^M Fuzzy Lattices

Goran Trajkovski and Bojan Čukić

Department of Computer Science and Electrical Engineering
West Virginia University
PO Box 6109
Morgantown, WV 265006-6109, USA
E-mail: {goran,cukic}@csee.wvu.edu

Abstract. The features of fuzzy lattices valued by lattices can be observed in the light of more general results from fuzzy algebraic and fuzzy relational structures. By approaching directly the problem of defining the notion of a fuzzy lattice, several directions can be taken. In this paper we present an idea and two of its variations to define what is referred to as L^M fuzzy lattices. The idea is to fuzzify the membership functions of elements of the carrier of an ordinary, crisp, lattice. L_1^M fuzzy lattices require for the cuts of the structure to be sublattices of the lattice whose carrier's membership function has been the subject of fuzzification. More generally, L_2^M fuzzy lattices require that the cuts are lattices themselves, not insisting on being substructures of the crisp lattice.
The structure of the families of cuts in both cases are presented, as well as an algorithm to construct L^M fuzzy lattice with a given set of cuts. Alternative approaches to defining fuzzy lattices are discussed at the end of the paper.

1 Introduction

Lattices are among the most extensively used and discussed structures in (crisp) mathematics. They are considered as relational, ordered structures on one hand, and as algebras, on the other [3,4].

From the historical perspective, three paradigmatical shifts in mathematics that give the foundations of this paper need to be mentioned. First, Zadeh defined the $[0, 1]$–valued fuzzy sets [5], [7], [14], then Goguen generalized them to the notion of L–fuzzy sets [6]. Taking another track, few years after the inception of the notion of fuzzy set, Rosenfield started the pioneering work in the domain of fuzzification of the algebraic objects in his work on fuzzy groups [8]. Our paper is a contribution to the theory founded on the ideas of these authors and their followers.

The problem discussed in the paper is the following, [13]: How to define lattice–valued (L) fuzzy lattices? Inspired by the work of Ajmal [1-2], Yuan and Wu [15] and of Šešelja and Tepavčević, [9-11], we present several possible definitions.

The motivation for carrying out this work is to provide stronger results and algorithms for fuzzy lattices than initially obtained for more general structures like fuzzy algebras and fuzzy relational structures. The fuzzy lattices valued by lattices have interesting properties, ranging from the diversibility in defining, to providing simpler and fundamentally different algorithms for certain constructions which are much more complex in the general case of, say, fuzzy algebras.

One type of L–fuzzy lattices is defined in [12]. These lattices are referred to as L^O fuzzy lattices, since they are the result of fuzzifying the order in a crisp lattice (the superscript ^{O} stands for "ordering"–the subject to fuzzification). In our approach, a new direction is taken. Two definitions of L–fuzzy lattices, obtained by fuzzification of the membership of the elements of the lattice structure carrier are given. They are referred to as L^M fuzzy lattices in a general context (the superscript ^{M} stands for the "membership function"), or as L_1^M and L_2^M fuzzy lattices when the particular subtype needs to be emphasized.

The paper is organized as follows. Section 2 provides some basic definitions and notations that are used throughout the paper. Sections 3 and 4 give definitions and examples of L_1^M and L_2^M fuzzy lattices and discuss the structure of their families of cuts. Section 5 describes ideas for further research in the field.

2 Basic Definitions and Notations

In this section notations and definitions are given to ease the reading of the paper.

A poset $(M, \leq)$ is a lattice if for every two-element subset $\{x, y\}$ of the carrier M, its supremum (denoted as $\sup_M\{x, y\}$ or $x \vee_M y$) and infimum ($\inf_M\{x, y\}$ or $x \wedge_M y$) are also elements of the carrier M. The subscripts are omitted whenever the carrier is known in a given context.

The subset A of the lattice M is a sublattice of M if the following implication holds:

$$a, b \in A \Rightarrow a \wedge_M b, a \vee_M b \in A. \tag{1}$$

The poset M is said to be a complete lattice if for every subset A of M, there exist $\inf_M A$ and $\sup_M A$.

In the paper only finite lattices are considered (the carrier is of finite cardinality).

An L–fuzzy set $\overline{M}$ is defined as a mapping

$$\overline{M} : M \to L \tag{2}$$

form the universe of discourse M to the lattice L.

Let $\overline{M}$ be an L - fuzzy set. Then

$$\overline{M}_p : M \to \{0, 1\}, \tag{3}$$

such that

$$\overline{M}_p(x) = 1 \Leftrightarrow \overline{M}(x) \geq p, \quad x \in M, \tag{4}$$

is the p–cut (or p-level) of $\overline{M}$. M_p usually denotes the (crisp) set whose characteristic function is given by (3) and (4).

3 L_1^M Fuzzy Lattices

The L–fuzzy lattices defined in this section are lattice valued fuzzy lattices obtained by fuzzification of the membership of elements of the carrier of the crisp lattice M, and will be referred to as L_1^M fuzzy lattices.

3.1 The Kick-Off Definition and an Example

One of the easiest and most natural ways to define an L–fuzzy lattice is via cuts.

Definition 1. Let $(M, \wedge, \vee)$ be a lattice and L a complete lattice with bottom element 0_L and top element 1_L. The mapping

$$\overline{M} : M \to L, \tag{5}$$

is called L_1^M fuzzy lattice if all the p–cuts ($p \in L$) of $\overline{M}$ are sublattices of M.

This definition covers the crisp lattice as its trivial case ($L = \{0_L, 1_L\}$).

This approach in defining L–fuzzy lattice is primarily motivated by the definitions of fuzzy algebras [9]. It is preferred to the generalization of the definitions in [1] and [15], because it is directly generalized when poset and relational structure valued structures are concerned (see [10]).

An example of L_1^M fuzzy lattice to illustrate the above definition is given in the sequel.

Example 1. The carrier of the lattice $(M, \wedge_M, \vee_M)$ is the set $M = \{0_M, a, b, c, d, e, f, g, h, 1_M\}$, where the carrier of the valuating lattice is $L = \{0_L, p, q, r, 1_L\}$ (see Fig. 1).

The mapping

$$\overline{M} = \begin{pmatrix} 0_M & a & b & c & d & e & f & g & h & 1_M \\ q & q & r & q & q & q & 1_L & r & p & 1_L \end{pmatrix}, \tag{6}$$

is an L_1^M fuzzy lattice, since all its p–cuts ($p \in L$) are sublattices of the lattice M, as can be seen in Fig. 2.

3.2 Alternative Definitions

In this subsection we state two propositions that give us alternate ways of defining L_1^M fuzzy lattices.

Knowning the fact that the cuts of L–fuzzy algebras are ordinary subalgebras of the fuzzified carrier [9], the following proposition can be stated.

Proposition 2. *Let $(M, \wedge_M, \vee_M)$ be a lattice and $(L, \wedge_L, \vee_L)$ be a complete lattice with bottom 0_L and top 1_L. Then the mapping $\overline{M} : M \to L$ is an L_1^M fuzzy lattice iff both of the following relations hold for all $x, y \in M$:*

$$\overline{M}(x \wedge_M y) \geq \overline{M}(x) \wedge_L \overline{M}(y), \tag{7}$$

and

$$\overline{M}(x \vee_M y) \geq \overline{M}(x) \wedge_L \overline{M}(y). \tag{8}$$

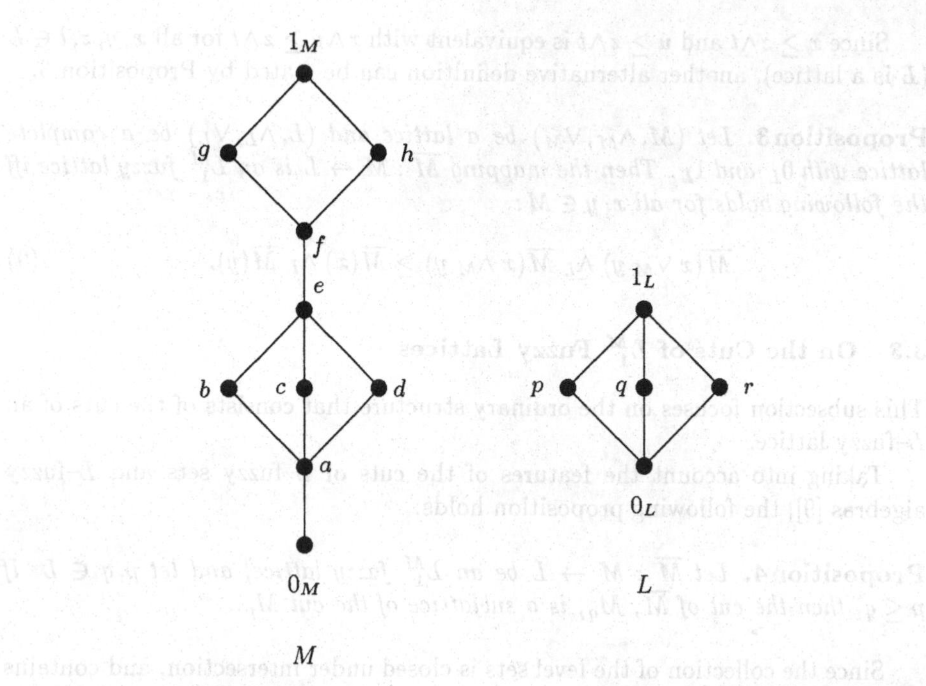

Fig. 1. The valuated lattice M and the valuating lattice L in Example 1.

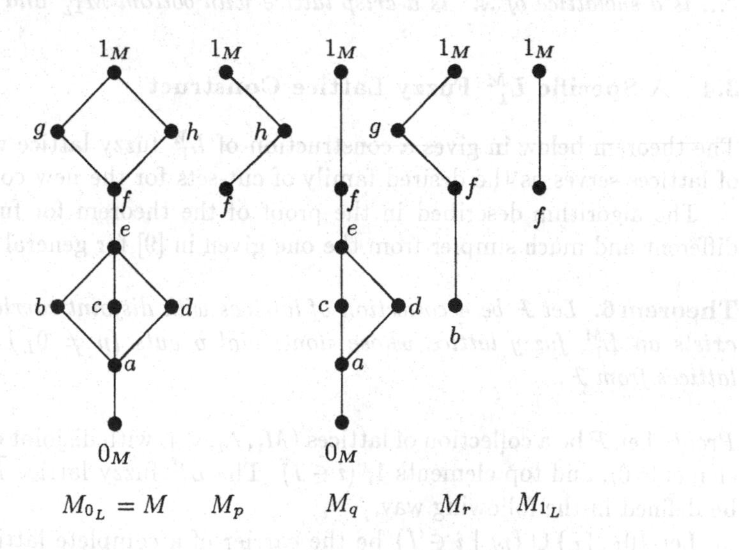

Fig. 2. The cuts of $\overline{M}$ from Example 1.

Since $x \geq z \wedge t$ and $u \geq z \wedge t$ is equivalent with $x \wedge u \geq z \wedge t$ for all $x, y, z, t \in L$ (L is a lattice), another alternative definition can be stated by Proposition 3.

Proposition 3. *Let $(M, \wedge_M, \vee_M)$ be a lattice and $(L, \wedge_L, \vee_L)$ be a complete lattice with 0_L and 1_L. Then the mapping $\overline{M} : M \to L$ is an L_1^M fuzzy lattice iff the following holds for all $x, y \in M$:*

$$\overline{M}(x \vee_M y) \wedge_L \overline{M}(x \wedge_M y) \geq \overline{M}(x) \wedge_L \overline{M}(y). \tag{9}$$

3.3 On the Cuts of L_1^M Fuzzy Lattices

This subsection focuses on the ordinary structure that consists of the cuts of an L–fuzzy lattice.

Taking into account the features of the cuts of L–fuzzy sets and L–fuzzy algebras [9], the following proposition holds.

Proposition 4. *Let $\overline{M} : M \to L$ be an L_1^M fuzzy lattice, and let $p, q \in L$. If $p \leq q$, then the cut of $\overline{M}$, M_q, is a sublattice of the cut M_p.*

Since the collection of the level sets is closed under intersection, and contains the greatest element, the following feature of this structure can be stated.

Proposition 5. *The family of cuts of an L_1^M fuzzy lattice, ordered by the relation "... is a sublattice of ..." is a crisp lattice with bottom M_{1_L} and top $M_{0_L} = M$.*

3.4 A Specific L_1^M Fuzzy Lattice Construct

The theorem below in gives a construction of L_1^M fuzzy lattice when the family of lattices serves as the desired family of cut-sets for the new construct.

The algorithm described in the proof of the theorem for fuzzy lattices are different and much simpler from the one given in [9] for general fuzzy algebras.

Theorem 6. *Let $\mathcal{F}$ be a collection of lattices with disjoint carriers. Then, there exists an L_1^M fuzzy lattice whose nontrivial p–cuts ($p \neq 0_L$) are exactly the lattices from $\mathcal{F}$.*

Proof. Let $\mathcal{F}$ be a collection of lattices $(M_i, \wedge_i, \vee_i)$, with disjoint carriers, bottom elements 0_i, and top elements 1_i ($i \in I$). The L_1^M fuzzy lattice $\overline{M} : M \to L$ can be defined in the following way.

Let $\{0_L, 1_L\} \cup \{p_i \mid i \in I\}$ be the carrier of a complete lattice L, such that O_L is the bottom element, 1_L is the top element and p_i are at the same time the atoms and the co–atoms of L (Fig. 3).

Let M be the poset

$$0_M \oplus \overset{\circ}{\bigcup_{i \in I}} M_i \oplus 1_M,$$

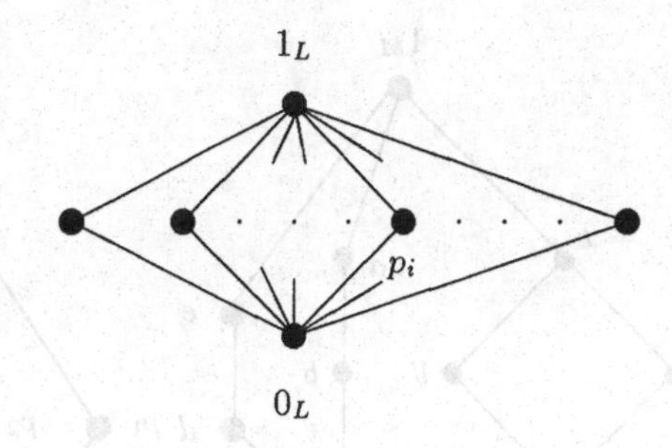

Fig. 3. The valuating lattice L from the proof of Theorem 6.

where 0_L and 1_L are one-element lattices, $\oplus$ is a linear sum and $\overset{\circ}{\bigcup}$ is a disjoint union of lattices [4]. It is easy to see that M is a lattice.

We define the mapping $\overline{M} : M \rightarrow L$ to be given by:

$$\overline{M}(x) = p_i \Leftrightarrow x \in M_i \text{ for } i \in I, \tag{10}$$

$$\overline{M}(0_M) = \overline{M}(1_M) = 0_L. \tag{11}$$

Clearly that all the nontrivial p–cuts of $\overline{M}$ are exactly the lattices from $\mathcal{F}$. □

The following example illustrates the theorem.

Example 2. Let $\mathcal{F}$ consists of M_1, M_2 and M_3, as shown in Fig. 4. The lattices M and L, constructed according to the algorithm in the proof of Theorem 6, are shown in Fig. 5.

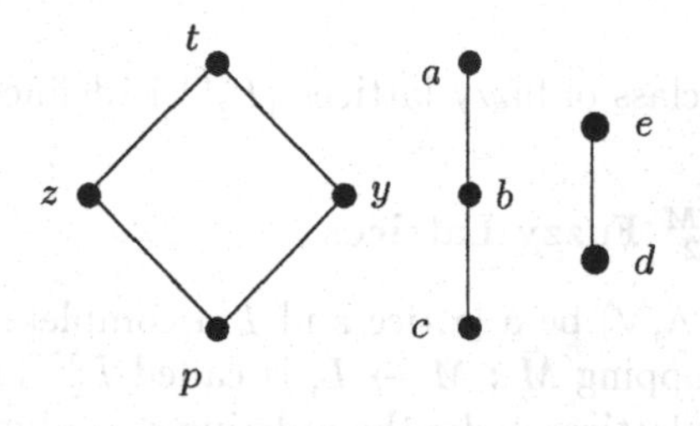

Fig. 4. The family $\mathcal{F}$ in Example 2.

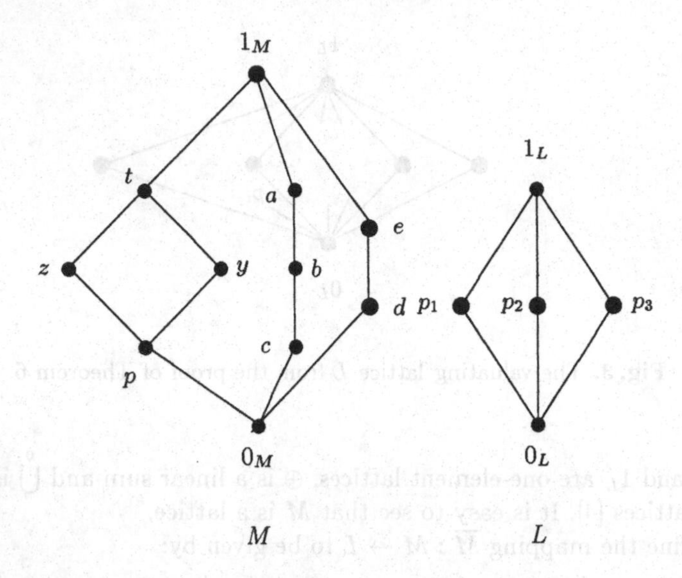

Fig. 5. The resulting L_1^M fuzzy lattice in Example 2.

Then, the required fuzzy lattice is given by the mapping $\overline{M}$:

$$\overline{M}: \begin{pmatrix} a & b & c & d & e & p & t & y & z & 0_M & 1_M \\ p_2 & p_2 & p_2 & p_3 & p_3 & p_1 & p_1 & p_1 & p_1 & 0_L & 0_L \end{pmatrix} \tag{12}$$

There are other ways to construct a lattice which fulfills the conditions of Theorem 6. For instance, it is sufficient to take the linear sum of the collection of lattices in $\mathcal{F}$. Setting then $\overline{M}$ to be the same as in the proof of Theorem 6, we obtain another fuzzy lattice that satisfies the required conditions.

The final remark is that the same construction can be used without further restrictions in the case of L_2^M fuzzy lattices discussed next.

4 L_2^M Fuzzy Lattices

In this section a wider class of fuzzy lattices (L_2^M) is defined and discussed.

4.1 Definition of L_2^M Fuzzy Lattices

Definition 7. Let $(M, \wedge, \vee)$ be a lattice and L a complete lattice with bottom 0_L and top 1_L. The mapping $\overline{M} : M \to L$, is called L_2^M fuzzy lattice if all the p–cuts ($p \in L$) of $\overline{M}$ are lattices under the ordering $\leq$ (induced by $\wedge$ and $\vee$ of M. In other words, it is insisted only that for every $p \in L$ the structure $(M_p, \leq_{/M_p})$ is a lattice per sé.

Every L_1^M fuzzy lattice is an L_2^M fuzzy lattice, because every sublattice of a given lattice is a lattice by definition. The reverse, however, does not hold. Example 3 gives arguments to support the claim.

Example 3. Let M be the lattice from Example 1. The mapping $\overline{M} : M \to L$ is given as follows:

$$\overline{M}(x) = 0_L,\ x \in \{b, e\}, \tag{13}$$

$$\overline{M}(x) = 1_L,\ \text{otherwise}. \tag{14}$$

By careful observation it can be seen that all the cuts are lattices, but not necessarily sublattices of M. For example, the supremum of the elements b and d in M is e, but in the cut M_{1_L} the supremum is different (f).

Equivalent expressions can be used to state the requirement for a mapping to be an L_2^M fuzzy set, similarly to Section 2.

Proposition 8. *Let $(M, \wedge_M, \vee_M)$ be a lattice and $(L, \wedge_L, \vee_L)$ be a complete lattice with 0_L and 1_L. Then the mapping $\overline{M} : M \to L$ is an L_2^M fuzzy lattice iff the following implication holds for all cuts $(M_p, \wedge_p, \vee_p)$ $(p \in L)$ of $\overline{M}$ and all $x, y \in M$:*

$$\overline{M}(x) \geq q, \overline{M}(y) \geq q \Rightarrow \tag{15}$$

$$\Rightarrow \overline{M}(x \wedge_p y) \geq q, \overline{M}(x \vee_p y) \geq q, \quad x, y \in M_p, q \in L. \tag{16}$$

□

4.2 The Cuts of the L_2^M Fuzzy Lattices

Here we give two propositions on the features of the cuts L_2^M fuzzy lattices.

Proposition 9. *Let $\overline{M} : M \to L$ be an L_2^M fuzzy lattice, and $p, q \in L$. If $p \leq q$, then the restriction*

$$\leq_{M/M_p} \subseteq M_p \times M_p \subseteq M \times M \tag{17}$$

is a superset of the restriction $\leq_M / M_q$.

Proof. Let $p \leq q$, and let $(x, y) \in \leq_{M/M_q}$, for some $x, y \in M$. This means that $x \leq_M y$ and $x, y \in M_q$, i.e. $x \leq_M y$, and $\overline{M}(x), \overline{M}(y) \geq q \geq q$. Therefore, $(x, y) \in \leq_{M/M_p}$. □

Theorem 10 follows directly form Proposition 9.

Theorem 10. *The cuts of the L_2^M fuzzy lattices, ordered by (set) inclusion, is a crisp lattice with bottom M_{1_L} and top M_{0_L}.*

5 Concluding Remarks

In this paper, we defined two types of fuzzy lattices valued by lattices. Results that state alternate definitions of the fuzzy structures are also presented. Further on, the features of the cut-sets families of the fuzzy lattices are discussed. A theorem for a specific construction of a L_1^M fuzzy lattice with predefined cuts is also stated and proved. Since every L_M^1 lattice is an L_2^M lattice, the same result (Theorem 6) holds for L_2^M fuzzy lattices.

The extension of this work is stating the interconnectedness of the L^M and L^O fuzzy lattices [12]. Instead of fuzzifying the membership of the elements in the carrier of a lattice, the L^O lattices are obtained by fuzzification of the ordering relation of the carrier. In other words, if M is a lattice and $\leq$ is its ordering relation, every weakly-reflexive L fuzzy ordering relation $\overline{\rho} : M^2 \to L$ is called L_1^O (L_2^O) fuzzy lattice if all the cut structures $(M, \overline{\rho}_p)$, $p \in L$ are sublattices of M (lattices per sé for the definition of L_2^O lattices). L needs to have a unique atom in order to formulate any interconnectedness between the L^M and L^O lattices [13].

Moreover, all the ideas can be generalized in order to obtain poset and relational–structure valued fuzzy lattices.

It is also expected that the extension of the work considering ideals and filters, as well as lattices that are not finite, would bring new and interesting results.

References

1. Ajmal, N., Thomas, K.V.: Fuzzy Lattices. Information Sciences, **79** (1994), 271–291.
2. Ajmal, N., Thomas, K.V.: The Lattices of Fuzzy Ideals of a Ring. Fuzzy Sets and Systems, **74** (1995), 371–379.
3. Birkhoff, G.: Lattice Theory, Amer. Math. Soc, Providence R.I., 1984.
4. Davey, B.A., Priestley, H.A.: Introduction to Lattices and Order, Cambridge University Press, 1990.
5. Dubois, D., Prade, H.: Fuzzy Sets and Systems: Theory and Applications, Academic Press, New York, 1980.
6. Goguen, J.A.: L–Fuzzy Sets. J. Math. Appli. **18** (1967), 145–174.
7. Klir, G.J., Yuan, B.: Fuzzy Sets and Fuzzy Logic, Prentice Hall, Upper Saddle River, NJ, 1995.
8. Rosenfield, A.: Fuzzy Groups. J. Math. Anal. Appl., **35** (1971), 512–517.
9. Šešelja, B., Tepavčević, A.: Partially Ordered and Relational Valued Algebras and Congruences. Review of Research, Fac. Sci., Univ. Novi Sad, Math. Ser. **23** (1993), 273–287.
10. Šešelja, B., Tepavčević, A.: Partially Ordered and Relational Valued Fuzzy Relations I. Fuzzy Sets and Systems, **72** (1995), 205–213.
11. Šešelja, B., Tepavčević, A.: Fuzzy Boolean Algebras. Automated Reasoning, IFIP Trans. **A–19** (1992), 83–88.
12. Trajkovski, G.: An Approach Toward Defining L fuzzy lattices, Proc. NAFIPS'99, Pensacola Beach, FL, USA (1998), 221–225.

13. Trajkovski G., Fuzzy Relations and Fuzzy Lattices, M.Sc. Thesis, University "St. Cyril and Methodius" - Skopje, 1997.
14. Yager, R.R., Ovchinikov, S., Tong, R.M., Nguyen, H.T. (eds.): Fuzzy Sets and Applications: Selected papers by L.A. Zadeh, John Willey & Sons, New York, 1987.
15. Yuan, B., Wu, W.: Fuzzy ideals on a distributive lattice. Fuzzy Sets and Systems, **35** (1990), 231- 240.

Generated Connectives in Many Valued Logic

Radko Mesiar
STU, Radlinského 11, 813 68 Bratislava, Slovakia
and
ÚTIA AV ČR, P. O. Box 18, 182 08 Praha, Czech Republic

1 Introduction

Many valued logic is based on the truth values range and the corresponding connectives of conjunction, disjunction, negation, implication, etc. Depending on the corresponding required properties, several relationships between MV connectives occur. So, e.g., de Morgan rules brings together conjunction, disjunction and negation; residuation relates conjunction and implication; negation is often required to be consistent with the corresponding implication; etc. On the other hand, up to the negation, all other mentioned connectives are binary operations and their real evaluation may be rather time consuming. Therefore a representation by means of (one argument) functions which are called generators is considered. Recall such a wellknown representation of Archimedean continuous t-norms (as conjunctions on [0,1]) or t-conorms (as disjunctions on [0,1]) which is due to Ling [7] and in its general form was shown already by Mostert and Shields [9]. Note that the choice of an additive generator (which is strictly decreasing) of a continuous Archimedean t-norm T is unique up to a positive multiplicative constant. Further, if a continuous Archimedean t-norm T is generated by an additive generator f, then the dual t-conorm S, $S(x,y) = 1 - T(1-x, 1-y)$ is generated by an additive generator $g = f(1-x)$. More, the corresponding residual implicator I is also defined by means of f, $I(x,y) = f^{(-1)}(\max(0, f(y) - f(x)))$. See [2].

Another known case when the generators play a key role is the case of smooth discrete t-norms [4, 8, 3]. Recall that if the truth values range is a finite subset $\mathbf{H}$ of any linearly ordered space (e.g., finite set $\mathbf{H}$ of some real values $x_0 < x_1 < \ldots < x_n$), then a smooth discrete t-norm D is a conjunction on $\mathbf{H}$ which is associative, commutative, non-decreasing, whose neutral element is the maximal element x_n of $\mathbf{H}$ and such that the output decrease caused by the decrease of an input cannot exceed the relevant input decrease (thinking on the index of mentioned values). It is easy to show, see e.g. [8], that there is unique Archimedean smooth discrete t-norm D (for a fixed n), $D(x_i, x_j) = x_r$ where $r = \max(0, i + j - n)$.

Now, it is easy to see that D is generated by any strictly decreasing (continuous) real function f such that $f(x_i) = (n-i)\frac{f(x_0)}{n}$. The corresponding residual implicator $J : \boldsymbol{H}^2 \to \boldsymbol{H}$ is then defined by

$$J(x_i, x_j) = f^{-1}\left(\max\left(0, f(x_j) - f(x_i)\right)\right) = f^{-1}\left(f(0)\max(0, i-j)\right) = x_{\min(n, n-i+j)}.$$

What concerns negations, the involutive case (on real truth values ranges) was completely solved by Trillas [10] and the generators here play the role of isomorphic transformations (from the algebraic point of view).

In this contribution, we will continue in dealing with real truth values from the unit interval $[0,1]$ (a general case is a matter of rescaling only). We will distinguish the discrete scales (and then $x_0 = 0$ and $x_n = 1$) and the continuous scale $[0,1]$. We will focus on the case of conjunctions (the disjunctions can be treated in a similar style by the duality) and the relevant residual implications.

2 Generated Conjunctors on [0,1]

Definition 1. Let **H** be a real scale. A binary operation $C : \boldsymbol{H}^2 \to \boldsymbol{H}$ is called a conjunctor if it is commutative, nondecreasing and 1 is its neutral element, i.e., $C(x,1) = x$ for all $x \in \boldsymbol{H}$. If C is also associative, then it is called a t-norm (a discrete t-norm).

Note that the class of all conjunctors on a given real scale **H** is closed under idempotent aggregations (what is not true in the case of the class of all t-norms). So, e.g., for any two conjunctors C_1 and C_2 on **H**, also the operators $\min(C_1, C_2)$, $\max(C_1, C_2)$, $\lambda C_1 + (1-\lambda)C_2$, $\lambda \in [0,1]$, $(C_1 C_2)^{\frac{1}{2}}$, etc., are conjunctors on **H**.

Definition 2. Let $f : [0,1] \to [0,\infty]$ be a strictly decreasing mapping with $f(1) = 0$. Then f will be called a conjunctive additive generator.

Recall that if a conjunctive additive generator f is continuous then the mapping $T : [0,1]^2 \to [0,1]$ defined by

$$T(x,y) = f^{-1}\left(\min\left(f(0), f(x) + f(y)\right)\right)$$

is a continuous Archimedean t-norm T, see [6,7] and vice-versa, each continuous Archimedean t-norm T is derived by means of a continuous conjunctive additive generator f which is uniquelly determined by T up to a positive multiplicative constant. In general, the Archimedean property of a conjunctor means that the power sequences $\{x^{(n)}\}$, $x^{(1)} = x$, $x^{(n)} = C\left(x, x^{(n-1)}\right)$ for $n \geq 2$, $x \in \boldsymbol{H}$, which are always decreasing, have a limit 0 for any $x \in \boldsymbol{H} \setminus \{0,1\}$. Equivalently, a conjunctor C is Archimedean if and only if for any $x, y \in \boldsymbol{H} \setminus \{0,1\}$ there is some $n \in N$ so that $x^{(n)} < y$.

Each continuous Archimedean t-norm T fulfills the cancellation law,

$$T(x,y) = T(x,z) \quad \text{if and only if} \quad T(x,y) = 0 \quad \text{or} \quad y = z.$$

However, there are Archimedean t-norms (on $[0,1]$ see Smutná [11]) as well as the discrete Archimedean t-norms violating the above cancellation law. Consequently, they cannot be generated by means of continuous additive generators. Therefore, the case of non-continuous additive generators is worth a deeper investigation. Note that in the context of t-norms the non-continuous additive generators are studied e.g. in [5,12,13].

Theorem 1. Let $f : [0, 1] \to [0, \infty]$ be a conjunctive additive generator and let $f^{(-1)} : [0, \infty] \to [0, 1]$ be the corresponding pseudo-inverse defined by $f^{(-1)}(t) = \sup(x \in [0, 1]; f(x) > t)$. Then the binary operation $C : [0, 1]^2 \to [0, 1]$ defined by $C(x, y) = f^{(-1)}(f(x) + f(y))$ is a conjunctor on $[0, 1]$.

The proof is straigthforward and therefore omited.

The function f in the above theorem is called an additive generator of the conjunctor C. As far as the value $f(0)$ has no influence on the resulting conjunctor C, see [6], we will always suppose that f is right-continuous in the point 0. Further, the multiplication of f by a positive constant again has no influence on the resulting conjunctor C, but there may be two different additive generators of the same conjunctor C which are not linearly dependent - take e.g. the weakest t-norm T_D, i.e., the drastic product, see [5].

Note that the pseudo-inverse $f^{(-1)}$ of a strictly decreasing function f is always a continuous function and consequently the left-continuity (the right-continuity) of f results to the left-continuity (the right-continuity) of the generated conjunctor C. In the case of a left-continuous conjunctor C generated by means of an additive generator f, we can represent the corresponding residual implication I.

Theorem 2. Let C be a left-continuous conjuctor on $[0, 1]$ generated by a left-continuous additive generator f. Then the corresponding residual implication I is given by

$$I(x, y) = f^{(-1)}\left(\max\left(0, f(y^+) - f(x)\right)\right),$$

where $f^{(-1)}$ is the corresponding pseudo-inverse and $f(y^+) = \lim_{x \to y^+} f(z)$ whenever $y < 1$ and $f(1^+) = 0$.

Proof: For a given left-continuous conjunctor C, the corresponding residual operator is defined by, see e.g. [2],

$$I(x, y) = \sup\left(z \in [0, 1]; C(x, z) \leq y\right).$$

If $x \leq y$ then $I(x, y) = 1$ and the theorem obviously holds. Suppose that $x > y$. Then

$$\begin{aligned}
I(x, y) &= \sup\left(z \in [0, 1];\ f^{(-1)}\left(f(x) + f(z)\right) \leq y\right) \\
&= \sup\left(z \in [0, 1];\ \sup\left(t \in [0, 1];\ f(t) > f(x) + f(z)\right) \leq y\right) \\
&= \sup\left(z \in [0, 1];\ \text{ if } f(t) > f(x) + f(z) \text{ then } t \leq y\right) \\
&= \sup\left(z \in [0, 1];\ f(x) + f(z) \geq f(t) \text{ whenever } 1 \geq t > y\right) \\
&= \sup\left(z \in [0, 1];\ f(z) \geq f(y^+) - f(x)\right) \\
&= f^{(-1)}\left(f(y^+) - f(x)\right),
\end{aligned}$$

where the last equality follows from the left-continuity of the conjunctive additive generator f.

Note that the negation n derived from I, $n(x) = I(x,0) = f^{(-1)}(f(0) - f(x))$, fulfills
$n o n o n = n$ (for a general proof, see e.g. [1]). The standart negation $n_S(x) = 1-x$ corresponds to any left-continuous bounded conjunctive additive generator f whose graph is symmetric with respect to the point $\left(\frac{1}{2}, \frac{f(0)}{2}\right)$ (possibly up to discontinuity points). Further, the left-continuity of f ensures the Archimedianity of the corresponding generated conjunctor C. However, a conjunctor C generated by a left-continuous additive generator f is a t-norm (i.e., C is associative) only if f is continuous. On the other hand, there are right-continuous additive generators generating (non-Archimedean) t-norms with non-trivial idempotent elements, see [13]. For further investigations, we address the following problem.

Open problem. Let C be a left-continuous Archimedean conjunctor on $[0,1]$. Is then C necessarily generated by a (left-continuous) additive generator f?

3 Discrete Generated Conjunctors

Similarly we can discuss the case of discrete conjunctors. Note that several interesting results concerning discrete t-norms can be found in [1, 3].

Theorem 3. Let **H** be a given finite scale (in $[0,1]$) and let $f : \boldsymbol{H} \to [0,1]$ be a conjunctive additive generator. Let $f^{(-1)} : [0,\infty] \to \boldsymbol{H}$ be the corresponding discrete pseudo-inverse defined by $f^{(-1)}(t) = \sup(x \in \boldsymbol{H};\ f(x) \geq t)$. Then the binary operation $C : \boldsymbol{H}^2 \to \boldsymbol{H}$ defined by $C(x,y) = f^{(-1)}(f(x) + f(y))$ is a discrete conjunctor.

Now, the boundedness of $f(x_{i-1}) - f(x_{n-1}),\ \ i = 1, \ldots, n$, ensures the Archimedean property of the corresponding generated discrete conjunctor C, independently of the underlying conjunctive additive generator f, and the corresponding residual discrete implicator I is given by

$$I(x,y) = f^{(-1)}(f(y) - f(x)).$$

Again, generated discrete conjunctors offers several open problems. For example, it is not known yet when a generated discrete t-norm can be extended to a generated t-norm on [0,1](observe that there is always an extension of any generated discrete conjunctor to a generated conjunctor on [0,1]).

ACKNOWLEDGEMENTS The work on this paper was suported by action COST 15 and grants VEGA 1/4064/97 and GA ČR 402/99/0032.

References

1. B. De Baets, *Oplossen van vaagrelationele vergelijkingen: een ordetheoretisch benadering,* Ph.D. Thesis, University of Gent, 1995, 389.
2. B. De Baets, R. Mesiar, *Residual implicators of continuous t-norms,* Proc. EUFIT'96, Aachen, 1996,27–31.
3. B. De Baets, R. Mesiar, *Triangular norm on product latices,* Fuzzy Sets and Systems, to appear.
4. L. Godo, C. Sierra, *A new approach to connective generation in the framework of exepert systmes using fuzzy logic,* Proc. 18th Int. Symposium on MV Logic (Palma de Mallorca, Spain), IEEE Computer Press, 1988, 157–162.
5. E. P. Klement, R. Mesiar, E. Pap, *Additive generators of t-norms which are not necessarily continuous,* Proc. EUFIT'96, Aachen, 1996, 70–73.
6. E. P. Klement, R. Mesiar, E. Pap, Triangular Norms; monograph in preparation.
7. C. M. Ling, *Representation of associative functions,* Publ. Math. Debrecen **12** (1965), 189–212.
8. G. Mayor, J. Torrens, *On a class of operators for expert systems,* Int. J. Intell. Systems **8** (1993) 771–778.
9. P. S. Mostert, A. L. Shields, *On the structure of semigroups on a compact manifold with boundary.* Ann. of Math. **65** (1957), 117–143.
10. E.Trillas, *Sobres funciones de negación en la teoríja de conjuntas difusos,* Stochastica **3** (1979) 47–60.
11. D. Smutná, *A note on non-continuous t-norms,* Busefal, to appear.
12. P.Viceník, *A note to a construction of t-norms based on pseudo-inverses of monotone functions,* Fuzzy Sets and Systems, to appear.
13. P. Viceník, *A note on generators of t-norms,* Busefal **75** (1998) 33–38.

Conjugacy Classes of Fuzzy Implications

Michał Baczyński, Józef Drewniak

Department of Mathematics, Silesian University,
40-007 Katowice, ul. Bankowa 14, Poland.
{michal,drewniak}@ux2.math.us.edu.pl

Abstract. We discuss the conjugacy problem in the family of fuzzy implications. Particularly we examine a compatibility of conjugacy classes with induced order and induced convergence in the family of fuzzy implications. Conjugacy classes can be indexed by elements of adequate groups.

1 Conjugacy Problem

Many authors describe similarity relations between binary operations in the unit interval (e.g. characterization of triangular norms in [15], Chapter 5; characterization of continuous fuzzy implications in [6], Chapter 1, or characterization of generalized means in [1], p.279). We consider in detail a similarity relation in the family of all fuzzy implications.

Definition 1 ([2]). *A function $I\colon [0,1]^2 \to [0,1]$ is called fuzzy implication if it is monotonic with respect to both variables and fulfils the binary implication truth table:*

$$I(0,0) = I(0,1) = I(1,1) = 1, \quad I(1,0) = 0 \ . \tag{1}$$

Set of all fuzzy implications is denoted by FI.

By monotonicity

$$I(0,x) = I(x,1) = 1, \qquad \text{for } x \in [0,1] \ , \tag{2}$$

and I is decreasing with respect to the first variable and increasing with respect to the second.

Example 1. The most important multivalued implications (cf. [5]) fulfil the above definition:

$$I_{LK}(x,y) = \min(1-x+y,1) \ , \qquad \text{(Łukasiewicz [12])}$$

$$I_{GD}(x,y) = \begin{cases} 1, & x \le y \\ y, & x > y \end{cases} \ , \qquad \text{(Gödel [8])}$$

$$I'_{GD}(x,y) = \begin{cases} 1, & x \le y \\ 1-x, & x > y \end{cases} \ ,$$

$$I_{RC}(x,y) = 1 - x + xy \;, \qquad \text{(Reichenbach [13])}$$
$$I_{DN}(x,y) = \max(1-x,y) \;, \qquad \text{(Dienes [3])}$$
$$I_{GG}(x,y) = \begin{cases} 1, & x \leq y \\ \frac{y}{x}, & x > y \end{cases} , \qquad \text{(Goguen [7])}$$
$$I'_{GG}(x,y) = \begin{cases} 1, & x \leq y \\ \frac{1-x}{1-y}, & x > y \end{cases} ,$$
$$I_{RS}(x,y) = \begin{cases} 1, & x \leq y \\ 0, & x > y \end{cases} , \qquad \text{(Rescher [14])}$$

for $x, y \in [0,1]$.

After [11], Chapter 8 we put

Definition 2. *Fuzzy implications $I, J \in FI$ are conjugate if there exists a bijection φ: $[0,1] \to [0,1]$ such that $J = I^*_\varphi$, where*

$$I^*(x,y) = I^*_\varphi(x,y) = \varphi^{-1}(I(\varphi(x),\varphi(y))), \qquad \textit{for } x, y \in [0,1] \; . \tag{3}$$

At first, we ask about suitable assumptions on a bijection φ.

Problem 1. Does formula (3) give fuzzy implication?

Remark 1. An arbitrary bijection is not suitable. E.g. if

$$\varphi(0) = 0.5, \qquad \varphi(0.5) = 1, \qquad \varphi(1) = 0 \;,$$

then

$$I^*(1,1) = \varphi^{-1}(I(\varphi(1),\varphi(1))) = \varphi^{-1}(I(0,0)) = \varphi^{-1}(1) = 0.5 \neq 1 \; .$$

Remark 2. A discontinuous bijection is not suitable. E.g. if φ increases in intervals:

$$\varphi\colon \left[0,\frac{1}{3}\right] \to \left[\frac{2}{3},1\right], \; \varphi\colon \left(\frac{1}{3},\frac{2}{3}\right) \to \left(\frac{1}{3},\frac{2}{3}\right), \quad \varphi\colon \left[\frac{2}{3},1\right] \to \left[0,\frac{1}{3}\right] \;,$$

then (cf. [2])

$$I^*\left(0,\frac{1}{3}\right) = \varphi^{-1}\left(I\left(\varphi(0),\varphi\left(\frac{1}{3}\right)\right)\right) = \varphi^{-1}\left(I\left(\frac{2}{3},1\right)\right) = \varphi^{-1}(1) = \frac{1}{3} \neq 1 \; .$$

Remark 3. A decreasing bijection is not suitable. E.g. if $\varphi(x) = 1 - x$ for $x \in [0,1]$, then

$$I^*(0,0) = 1 - I(1-0, 1-0) = 1 - I(1,1) = 1 - 1 = 0 \neq 1 \; .$$

This explains reasonable assumptions for the positive answer of Problem 1. To this effect we mention that (cf. [9] §28.5)

Lemma 1. *A bijection on an interval is monotonic iff it is continuous.*

Lemma 2. *Any increasing bijection* $\varphi: [a,b] \to [a,b]$ *has at least two fixed points*

$$\varphi(a) = a, \qquad \varphi(b) = b \ . \tag{4}$$

Proof. Let $\varphi^{-1}(a) = c \geq a$. Thus

$$a \leq \varphi(a) \leq \varphi(c) = a$$

and we obtain $\varphi(a) = a$. The proof of the second part in (4) is similar. □

Solution of Problem 1 gives the following

Theorem 1. *A bijection* $\varphi: [0,1] \to [0,1]$ *is increasing iff*

$$\forall_{I \in FI} \ (I^*_\varphi \in FI) \ . \tag{5}$$

Proof. If φ is increasing, then by Lemma 1, it is continuous and φ^{-1} is also increasing and continuous. Therefore, I^*_φ is monotonic with respect to both variables as a composition of monotonic functions. By Lemma 2, φ has at least two fixed points

$$\varphi(0) = 0, \qquad \varphi(1) = 1 \ , \tag{6}$$

which used in (3) gives analogue of (1) for I^*.

Conversely, assume (5) and suppose that

$$\exists_{x,y \in [0,1]} \ (x < y, \ \varphi(x) > \varphi(y)) \ . \tag{7}$$

Putting e.g. $I = I_{GD}$ in (3) we see that

$$\begin{aligned} I^*_{GD}(x,y) &= \varphi^{-1}(I_{GD}(\varphi(x), \varphi(y))) = \varphi^{-1}(\varphi(y)) = y \ , \\ \varphi(1) &= \varphi(I^*_{GD}(1,1)) = I_{GD}(\varphi(1), \varphi(1)) = 1 \ , \\ I^*_{GD}(x,x) &= \varphi^{-1}(I_{GD}(\varphi(x), \varphi(x))) = \varphi^{-1}(1)) = 1 \ . \end{aligned}$$

Since $I^*_{GD} \in FI$, then

$$1 = I^*_{GD}(x,x) \leq I^*_{GD}(x,y) = y \ .$$

Thus $\varphi(y) = \varphi(1) = 1$ contradictory to $\varphi(y) < \varphi(x) \leq 1$. Therefore, our supposition (7) is false, which proves that φ is increasing. □

Usually, we consider families of bijections in (3).

Definition 3. *Let* Φ *denote a certain family of increasing bijections on* $[0,1]$. *Fuzzy implication* $J \in FI$ *is* Φ*-conjugate with* $I \in FI$ *if*

$$\exists_{\varphi \in \Phi} \ (J = I^*_\varphi) \ . \tag{8}$$

We say that $I \in FI$ *is* Φ*-selfconjugate if*

$$\exists_{\varphi \in \Phi} \ (I^*_\varphi = I) \ . \tag{9}$$

Problem 2. Does condition (8) describe an equivalence relation in FI?

Later we shall prove that (cf. Theorem 5)

Lemma 3. *Let* $\varphi, \psi \colon [0,1] \to [0,1]$ *be increasing bijections and* $I = I_{LK}$. *If* $I^*_\varphi = I^*_\psi$, *then* $\varphi = \psi$.

The positive answer of Problem 2 depends on the algebraic structure $(\Phi, \circ)$ with composition operation.

Theorem 2. *Relation* (8) *is an equivalence iff* $(\Phi, \circ)$ *is a group of bijections.*

Proof. Let Φ be a fixed family of increasing bijections on $[0,1]$. Assume that relation (8) is an equivalence and denote it by $J \sim I$. Let $I = I_{LK}$ and $\varphi, \psi \in \Phi$. First, putting $J = I^*_\varphi$, $K = J^*_\psi$ we see that $K \sim J$ and $J \sim I$. Thus, by transitivity $K \sim I$, or $\underset{\chi \in \Phi}{\exists} \left(K = I^*_\chi\right)$. So, $K = J^*_\psi = (I^*_\varphi)^*_\psi = I^*_{\varphi \circ \psi} = I^*_\chi$ and from Lemma 3 we get $\varphi \circ \psi = \chi \in \Phi$. Therefore, the composition is an interior operation in Φ, i.e. $(\Phi, \circ)$ is a semigroup. Next, by reflexivity, $I \sim I$, or $\underset{\varphi \in \Phi}{\exists} \left(I = I^*_\varphi\right)$. But $I = I^*_{id}$, and from Lemma 3 we get $id = \varphi \in \Phi$, i.e semigroup $(\Phi, \circ)$ has the identity element. Finally, putting $J = I^*_\varphi$, since $J \sim I$, then by symmetry $I \sim J$, or $\underset{\psi \in \Phi}{\exists} \left(I = J^*_\psi\right)$. Thus, $I = J^*_\psi = (I^*_\varphi)^*_\psi = I^*_{\varphi \circ \psi} = I^*_{id}$. From Lemma 3 we get $\varphi \circ \psi = id$ and, similarly, $\psi \circ \varphi = id$. Therefore, $\varphi^{-1} = \psi \in \Phi$, and all elements of semigroup $(\Phi, \circ)$ are invertible. This proves that $(\Phi, \circ)$ is a group.

Conversely, let $(\Phi, \circ)$ be group. First, since $\varphi = id \in \Phi$, then $I^*_{id} = I$, so $I \sim I$, i.e. relation (8) is reflexive. Next, since $\varphi \in \Phi$ implies $\varphi^{-1} \in \Phi$, then $J \sim I$ implies $I \sim J$, i.e. relation (8) is symmetric. Finally, if $K \sim J$ and $J \sim I$ then $\underset{\varphi, \psi \in \Phi}{\exists} \left(K = J^*_\psi,\ J = I^*_\varphi\right)$ and $K = J^*_\psi = (I^*_\varphi)^*_\psi = I^*_{\varphi \circ \psi} = I^*_\chi$, where $\chi = \varphi \circ \psi \in \Phi$. Therefore, $K \sim I$, and relation (8) is transitive. This proves that relation (8) is an equivalence and finishes the proof. □

Later we will consider diverse groups of increasing bijections but from now, by Φ, we denote the family of all increasing bijections on $[0,1]$.

2 Partial Ordering

In FI we can consider partial order induced from $[0,1]$:

$$(I \le J) \Longleftrightarrow \underset{x,y \in [0,1]}{\forall} \left(I(x,y) \le J(x,y)\right) , \tag{10}$$

for $I, J \in FI$.

Problem 3. Is relation (8) compatible with order (10)?

First, we prove that transformation (3) saves partial order (10) and lattice order (It was proved in [2] that $(FI, \min, \max)$ is a lattice).

Theorem 3 **(cf. [2]).** *For any $I, J \in FI$, $\varphi \in \Phi$, we have*

$$I \leq J \Longleftrightarrow I^* \leq J^* , \tag{11}$$

$$\min(I,J)^* = \min(I^*,J^*), \qquad \max(I,J)^* = \max(I^*,J^*) \; . \tag{12}$$

Proof. Let $x, y \in [0,1]$. Since φ is an increasing bijection, then

$$\begin{aligned} I(x,y) \leq J(x,y) &\Longleftrightarrow I(\varphi(x),\varphi(y)) \leq J(\varphi(x),\varphi(y)) \\ &\Longleftrightarrow \varphi^{-1}(I(\varphi(x),\varphi(y))) \leq \varphi^{-1}(J(\varphi(x),\varphi(y))) \\ &\Longleftrightarrow I^*(x,y) \leq J^*(x,y) \; , \end{aligned}$$

which proves (11).
Now, we consider the left side of (12). Using (11) we get

$$\min(I,J)^* \leq I^*, \qquad \min(I,J)^* \leq J^*$$

and therefore

$$\min(I,J)^* \leq \min(I^*,J^*) \; . \tag{13}$$

Now let us observe that $(I^*_\varphi)^*_{\varphi^{-1}} = I$. Since

$$\min(I^*_\varphi, J^*_\varphi) \leq I^*_\varphi, \qquad \min(I^*_\varphi, J^*_\varphi) \leq J^*_\varphi \; ,$$

then (cf. (11))

$$\begin{aligned} \min(I^*_\varphi, J^*_\varphi)^*_{\varphi^{-1}} \leq I, \qquad &\min(I^*_\varphi, J^*_\varphi)^*_{\varphi^{-1}} \leq J \; , \\ \min(I^*_\varphi, J^*_\varphi)^*_{\varphi^{-1}} &\leq \min(I,J) \; , \\ \min(I^*_\varphi, J^*_\varphi) &\leq \min(I,J)^*_\varphi \; , \end{aligned}$$

which, together with (13), gives the left side of (12), and the right side we obtain in a similar way. □

If J is selfconjugate, i.e. $J^* = J$, then directly from (11) we get

Corollary 1. *For any $I, J \in FI$ if J is selfconjugate, then*

$$I \leq J \Longrightarrow \underset{\varphi \in \Phi}{\forall} \; (I^*_\varphi \leq J) \; , \tag{14}$$

$$I \geq J \Longrightarrow \underset{\varphi \in \Phi}{\forall} \; (I^*_\varphi \geq J) \; . \tag{15}$$

In general, conjugacy classes are not comparable.

Example 2. Let us observe that $I_{DN} \leq I_{LK}$, and let $\varphi(x) = \sqrt{x}$ for $x \in [0,1]$. Putting $x = \frac{1}{4}$, $y = \frac{1}{16}$ we see that

$$I^*_{LK}\left(\frac{1}{4}, \frac{1}{16}\right) = \left(\frac{3}{4}\right)^2 < \frac{3}{4} = I_{DN}\left(\frac{1}{4}, \frac{1}{16}\right) \; .$$

Simultaneously,

$$I^*_{LK}\left(\frac{1}{16},\frac{1}{4}\right) = 1 > \frac{15}{16} = I_{DN}\left(\frac{1}{16},\frac{1}{4}\right) \ ,$$

which shows that I_{DN} and I^*_{LK} are non-comparable.

This example brings the negative answer to Problem 3. Therefore, partial order (10) does not induce a partial order between equivalency classes of relation (8). In this situation, we look for particular cases of comparability between elements of conjugacy classes.

Theorem 4. *Let $I, J \in FI$, $\varphi, \psi \in \Phi$. If I^*_φ and J^*_ψ are comparable, then I^*_χ and J are comparable, where $\chi = \varphi \circ \psi^{-1}$. Particularly*

$$J^*_\psi \le I^*_\varphi \Longleftrightarrow J \le I^*_\chi \ , \tag{16}$$

$$J^*_\psi = I^*_\varphi \Longleftrightarrow J = I^*_\chi \ . \tag{17}$$

Proof. Let $x, y \in [0,1]$ and $u = \psi(x)$, $v = \psi(y)$. We have the following equivalences

$$\begin{aligned} J^*_\psi(x,y) \le I^*_\varphi(x,y) &\Longleftrightarrow \psi^{-1}(J(\psi(x),\psi(y))) \le \varphi^{-1}(I(\varphi(x),\varphi(y))) \\ &\Longleftrightarrow J(\psi(x),\psi(y)) \le \psi(\varphi^{-1}(I(\varphi(x),\varphi(y)))) \\ &\Longleftrightarrow J(u,v) \le \psi(\varphi^{-1}(I(\varphi(\psi^{-1}(u)),\varphi(\psi^{-1}(v))))) \\ &\Longleftrightarrow J(u,v) \le \chi^{-1}(I(\chi(u)),\chi(v))) \\ &\Longleftrightarrow J(u,v) \le I^*_\chi(u,v) \ . \end{aligned}$$

This proves (16) and implies (17). □

3 Implication Classes

Using operation (3) for $\varphi \in \Phi$, and fuzzy implications listed in Example 1, we obtain

$$I^*_{LK}(x,y) = \min(\varphi^{-1}(1 - \varphi(x) + \varphi(y)), 1) \ , \tag{18}$$

$$I^*_{GD}{}'(x,y) = \begin{cases} 1, & x \le y \\ \varphi^{-1}(1 - \varphi(x)) & x > y. \end{cases} \ , \tag{19}$$

$$I^*_{RC}(x,y) = \varphi^{-1}(1 - \varphi(x) + \varphi(x)\varphi(y)) \ , \tag{20}$$

$$I^*_{DN}(x,y) = \max(\varphi^{-1}(1 - \varphi(x)), y) \ , \tag{21}$$

$$I^*_{GG}(x,y) = \begin{cases} 1, & x \le y \\ \varphi^{-1}\left(\frac{\varphi(y)}{\varphi(x)}\right), & x > y \end{cases} \ , \tag{22}$$

$$I^*_{GG'}(x,y) = \begin{cases} 1, & x \le y \\ \varphi^{-1}\left(\frac{1-\varphi(x)}{1-\varphi(y)}\right), & x > y \end{cases} \tag{23}$$

for $x, y \in [0,1]$. Moreover,

$$I^*_{GD} = I_{GD}, \qquad I^*_{RS} = I_{RS} \tag{24}$$

(one element conjugacy classes).

From (24) we see that a conjugacy class can contain one element (selfconjugate implications). This leads to the following

Problem 4. How many elements have conjugacy classes?

The answer depends on a generator of a class. Therefore, we will examine in detail, formulas (18)–(23).

Lemma 4. *Let $h \in \Phi$. Functional equation*

$$h(\min(1, 1-x+y)) = \min(1, 1-h(x)+h(y)), \qquad x, y \in [0,1] \tag{25}$$

has exactly one solution $h = id$.

Proof. For $y = 0$ in (25) we get

$$h(1-x) = 1-h(x), \qquad x \in [0,1] \ . \tag{26}$$

If $x \geq y$ in (25), then

$$h(1-x+y) = 1-h(x)+h(y) \ ,$$

and for $x = 1-z$ we obtain (cf. (26))

$$h(z+y) = h(z)+h(y), \qquad z, y \in [0,1], \ z+y \leq 1 \ , \tag{27}$$

which is the additive Cauchy functional equation. By [1], p.48, there exists $c \geq 0$ such that

$$h(x) = c \cdot x, \qquad x \in [0,1] \ , \tag{28}$$

but $c = 1$ because $h(1) = 1$ and therefore $h = id$. □

Lemma 5. *Let $h \in \Phi$. Functional equation*

$$h(1-x+xy) = 1-h(x)+h(x)h(y), \qquad x, y \in [0,1] \ , \tag{29}$$

has exactly one solution $h = id$.

Proof. Putting $y = 0$ in (29) we get (26) and for $y = 1-z$ in (29) we have

$$h(1-xz) = 1-h(x)h(z), \qquad x, z \in [0,1] \ .$$

This connected with (26) gives the multiplicative Cauchy functional equation

$$h(xz) = h(x)h(z), \qquad x, z \in [0,1] \ . \tag{30}$$

By [10], p.311, there exists $c > 0$ such that

$$h(x) = x^c, \qquad x \in [0,1] \ . \tag{31}$$

Usage of this solution in (26) for $x = \frac{1}{2}$ gives $(\frac{1}{2})^c = \frac{1}{2}$, which proves that $c = 1$ and $h = id$. □

Lemma 6. *Let $h \in \Phi$. Functional equation*

$$h\left(\frac{1-x}{1-y}\right) = \frac{1-h(x)}{1-h(y)}, \qquad x, y \in [0,1],\ x > y\ , \tag{32}$$

has exactly one solution $h = id$.

Proof. Putting $y = 0$ in (32) we get (26), and for $x = 1 - z$, $y = 1 - v$ in (32), we have

$$h\left(\frac{u}{v}\right) = \frac{h(u)}{h(v)}, \qquad u, v \in [0,1],\ v > u\ . \tag{33}$$

Now, for $u = zv$, $z < 1$, we have

$$h(z)h(v) = h(zv), \qquad v, z \in [0,1],\ v > 0,\ z < 1\ ,$$

which is the main part of functional equation (30) and the argumentation from the proof of Lemma 5 leads us to $h = id$. □

As a direct consequence of Lemmas 4 – 6 we have

Theorem 5. *Formulas* (18), (20) *and* (23) *provide one-to-one correspondences* $*\colon \Phi \to FI$, *i.e. all implications from these formulas are different.*

Proof. Let $\varphi, \psi \in \Phi$, $I = I_{LK}$. If $I^*_\varphi = I^*_\psi$ then, by Theorem 4, we see that $I^*_h = I$ for $h = \varphi \circ \psi^{-1}$. Since bijection h fulfils functional equation (25), by Lemma 4 we get $h = id$, i.e. $\varphi = \psi$. This proves theorem for formula (18), and proof of the remaining part is similar with applications of Lemmas 5 and 6.

Directly from this theorem we get

Corollary 2. *Conjugacy classes of I_{LK}, I_{RC} and I'_{GG} are equipotent to Φ.*

After proofs of Lemmas 6 and 5 we may write

Lemma 7. *Function $h \in \Phi$ fulfils functional equation* (33) *iff there exists $c > 0$ such that h is given by formula* (31).

Applying this Lemma in the way of the proof of Theorem 5 we obtain

Theorem 6. *Implication* (22) *reduces to I_{GG} iff $\varphi \in \Phi_0$, where*

$$\Phi_0 = \left\{\varphi \in \Phi : \underset{c>0}{\exists}\ (\varphi(x) = x^c,\ x \in [0,1])\right\}\ . \tag{34}$$

Corollary 3. *Let $\varphi, \psi \in \Phi$, $I = I_{GG}$. We have*

$$I^*_\varphi = I^*_\psi \iff \varphi \circ \psi^{-1} \in \Phi_0\ . \tag{35}$$

Lemma 8. *Function $\varphi \in \Phi$ fulfils functional equation*

$$\varphi(\max(1-x,y)) = \max(1-\varphi(x),\varphi(y)), \qquad x,y \in [0,1] \tag{36}$$

iff there exists an increasing bijection $h\colon [0,\frac{1}{2}] \to [0,\frac{1}{2}]$ such that

$$\varphi(x) = \begin{cases} h(x), & x \in [0,\frac{1}{2}] \\ 1-h(1-x) & x \in [\frac{1}{2},1] \end{cases} . \tag{37}$$

Proof. Putting $y = 0$ in (36) we get

$$\varphi(x) = 1 - \varphi(1-x), \qquad x \in [0,1] \tag{38}$$

and $x = \frac{1}{2}$ leads to $\varphi(\frac{1}{2}) = \frac{1}{2}$, which proves that φ has an additional fixed point $x = \frac{1}{2}$ (cf. (6)) and φ is a bijection in $[0,\frac{1}{2}]$. Let $h(x) = \varphi(x)$ for $x \in [0,\frac{1}{2}]$. If $x \in [\frac{1}{2},1]$, then $1-x \in [0,\frac{1}{2}]$ and, from (28), we get $\varphi(x) = 1-h(1-x)$, i.e. φ fulfils (37).

Conversely, if φ has form (37) with an increasing bijection $h\colon [0,\frac{1}{2}] \to [0,\frac{1}{2}]$, then $\varphi \in \Phi$ and fulfils (38). Indeed, if $x \in [0,\frac{1}{2}]$, then $1-x \in [\frac{1}{2},1]$ and $\varphi(1-x) = 1-h(x) = 1-\varphi(x)$. Similarly, if $x \in [\frac{1}{2},1]$, then $1-x \in [0,\frac{1}{2}]$ and $\varphi(x) = 1-h(1-x) = 1-\varphi(1-x)$, which proves (38). Moreover, for $x,y \in [0,1]$, we have

$$\max(1-\varphi(x),\varphi(y)) = \max(\varphi(1-x),\varphi(y)) = \varphi(\max(1-x,y)) ,$$

which proves (36). □

Directly from the above proof we can write

Lemma 9. *Function $\varphi \in \Phi$ fulfils functional equation* (38) *iff it has the form* (37) *with an increasing bijection $h\colon [0,\frac{1}{2}] \to [\frac{1}{2},1]$.*

Applying Lemmas 8 and 9 in the way of the proof of Theorem 5 we obtain

Theorem 7. *Implication* (21) *reduces to I_{DN} (implication* (19) *reduces to I'_{GD}) iff $\varphi \in \Phi_1$, where*

$$\Phi_1 = \{\varphi \in \Phi : \varphi \textit{ is given by } (37) \textit{ for increasing bijection } h\colon [0,\tfrac{1}{2}] \to [0,\tfrac{1}{2}]\} . \tag{39}$$

Corollary 4. *Let $\varphi,\psi \in \Phi$. If $I = I_{DN}$ or $I = I'_{GD}$ then*

$$I^*_\varphi = I^*_\psi \iff \varphi \circ \psi^{-1} \in \Phi_1 . \tag{40}$$

From Theorems 5 – 7 we see that conjugacy classes (18) – (23) may have diverse cardinalities, but the full answer to Problem 4 is not known.

4 Bounds of Conjugacy Classes

Because of negative answer to Problem 3, we look for characteristic implications, which can separate equivalence classes of relation (8).

Problem 5. Does conjugacy classes have the smallest and the greatest elements?

We say, that a sequence of fuzzy implications is convergent, if it is convergent as the sequence of real functions. Since limit saves inequalities and constant values, then we have

Lemma 10. *If (I_n) is a convergent sequence of fuzzy implications then its limit is also a fuzzy implication.*

Theorem 8. *Conjugacy classes of fuzzy implications have greatest lower bounds and least upper bounds in FI. Particularly for classes* (18) – (23) *we have*

$$\sup I^*_{\mathrm{LK}} = \sup I^*_{RC} = \sup I^*_{DN} = J_1 \ , \tag{41}$$

$$\sup I^*_{GG} = J_2, \qquad \sup {I^*_{GG}}' = \sup {I^*_{GD}}' = J_3 \ , \tag{42}$$

$$\inf I^*_{\mathrm{LK}} = \inf I^*_{GG} = I_{GD}, \qquad \inf I^*_{RC} = \inf I^*_{DN} = J_4 \ , \tag{43}$$

$$\inf {I^*_{GG}}' = \inf {I^*_{GD}}' = I_{RS} \ , \tag{44}$$

where

$$J_1(x,y) = \begin{cases} 1, & x<1 \\ y, & x=1 \end{cases}, \qquad J_2(x,y) = \begin{cases} 1, & x<1 \wedge y>0 \vee x=0 \\ y, & x=1 \\ 0, & x>0 \wedge y=0 \end{cases}, \tag{45}$$

$$J_3(x,y) = \begin{cases} 1, & x<1 \vee y=0 \\ y, & x=1 \wedge y<1 \end{cases}, \qquad J_4(x,y) = \begin{cases} 1, & x=0 \\ y, & x>0 \end{cases}. \tag{46}$$

Proof. Let $I \in FI$, $\psi \in \Phi$, $x, y \in [0,1]$ and

$$J = \sup_{\varphi \in \Phi} I^*_{\varphi}, \qquad K = \inf_{\varphi \in \Phi} I^*_{\varphi} \ . \tag{47}$$

After Lemma 10 we see that $J, K \in FI$. Since $\psi \in \Phi$ is continuous (cf. Lemma 1), then

$$\begin{aligned} J^*_{\psi}(x,y) &= \psi^{-1}(J(\psi(x),\psi(y))) = \psi^{-1}(\sup_{\varphi\in\Phi} I^*_{\varphi}(\psi(x),\psi(y))) \\ &= \sup_{\varphi\in\Phi} \psi^{-1}(\varphi^{-1}(I(\varphi(\psi(x)),\varphi(\psi(y))))) \\ &= \sup_{\chi\in\Phi\circ\psi} I^*_{\chi}(x,y) = \sup_{\chi\in\Phi} I^*_{\chi}(x,y) = J(x,y) \ , \end{aligned}$$

because $\Phi \circ \psi = \Phi$ in a group $(\Phi, \circ)$. Therefore, $J^*_{\psi} = J$ for $\psi \in \Phi$, which proves that fuzzy implication J is selfconjugate. Similar arguments prove that fuzzy implication K is selfconjugate.

By a direct verification we also see that functions $J_1 - J_4$ are fuzzy implications and are selfconjugate.
Particular results (41) – (44) can be proved in three steps. First, we observe that

$$K \le I \le J \ , \tag{48}$$

for suitable fuzzy implications I, J, K (e.g. $I_{GD} \le I_{LK} \le J_1$). Next, (48) implies

$$K \le I_\varphi^* \le J, \qquad \varphi \in \Phi \ , \tag{49}$$

in virtue of Corollary 1. Finally, we put $\varphi_n, \psi_n \in \Phi$, such that

$$\lim_{n\to\infty} I_{\varphi_n}^* = J, \qquad \lim_{n\to\infty} I_{\psi_n}^* = K \ , \tag{50}$$

which proves (47) for fuzzy implications I listed in Example 1.
We complete this proof by presenting suitable sequences of bijections in (50):

$$\begin{array}{lll} \varphi_n(x) = x^n, & \psi_n(x) = 1-(1-x)^n, & \text{for } I_{LK} \ , \\ \varphi_n(x) = x^n, & \psi_n(x) = \sqrt[n]{x}, & \text{for } I'_{GD}, I_{RC}, I_{DN}, I'_{GG} \ , \\ \varphi_n(x) = 1-\sqrt[n]{1-x}, & \psi_n(x) = 1-(1-x)^n, & \text{for } I_{GG} \ , \end{array}$$

where $n \in \mathbb{N}$, $x \in [0,1]$.

The above proved theorem gives positive answer to Problem 5. However many questions concerning conjugacy classes remain unsolved.

5 Open Problems

Problem 6. Is quotient space $FI/_\sim$ finite?

Problem 7. Which implications generate equipotent classes?

Problem 8. Is family of equipotent classes finite?

Problem 9. Is family of cardinal numbers of conjugacy classes finite?

Problem 10. Does exist conjugacy class of finite cardinality greater than 1?

Remark 4. Recently it was proved that the family of all singleton classes is finite. There exist exactly 18 different selfconjugate fuzzy implications (cf. [4]).

References

1. J. Aczél, Lectures on functional equations and their applications, Acad. Press, New York 1966.
2. M. Baczyński, J. Drewniak, Monotonic fuzzy implications, BUSEFAL **76** (1998).
3. Z.P. Dienes, On an implication function in many-valued systems of logic, *J. Symb. Logic* **14** (1949) 95-97.

4. J. Drewniak, Selfconjugate fuzzy implications, BUSEFAL (to appear).
5. D. Dubois, H. Prade, Fuzzy sets in approximate reasoning. Part 1: Inference with possibility distributions, *Fuzzy Sets Syst.* **40** (1991) 143-202.
6. J.C. Fodor, M. Roubens, Fuzzy preference modelling and multicriteria decision support, Kluwer, Dordrecht 1994.
7. J.A. Goguen, The logic of inexact concepts, *Synthese* **19** (1969) 325-373.
8. K. Gödel, Zum intuitionistischen Aussagenkalkül, *Auzeiger der Akademie der Wissenschaften in Wien, Mathematisch, naturwissenschaftliche Klasse* **69** (1932) 65-66.
9. H.B. Griffiths, P.J. Hilton, Classical mathematics, Van Nostrand, London 1970.
10. M. Kuczma, An introduction to the theory of functional equations and inequalities, Silesian University, Katowice 1985.
11. M. Kuczma, B. Choczewski, R. Ger, Iterative functional equations, Cambridge Univerity Press, Cambridge 1990.
12. J. Łukasiewicz, A numerical interpretation of the theory of propositions (Polish), *Ruch Filozoficzny* **7** (1923) 92-93 (translated in: L. Borkowski (ed.), Jan Łukasiewicz selected works, North Holland - Amsterdam, PWN - Warszawa 1970, pp.129-130).
13. H. Reichenbach, Wahrscheinlichtkeitslogik, *Erkenntnis* **5** (1935-36) 37-43.
14. N. Rescher, Many-valued logic, McGraw-Hill, New York 1969.
15. B. Schweizer, A. Sklar, Probablistic metric spaces, North-Holland, Amsterdam 1983.

Characterization of Dienes Implication

Michał Baczyński

Department of Mathematics, Silesian University,
40-007 Katowice, ul. Bankowa 14, Poland.
michal@ux2.math.us.edu.pl

Abstract. Our main goal in this paper is to present a characterization of implications which are similar to Dienes implication. Our investigations are inspired by the paper of Smets, Magrez [10], where they proved the characterization of implications similar to Łukasiewicz implication.

1 Introduction

In many papers we can find characterization theorems for logical connectives, which are used in fuzzy logic, e.g characterization of triangular norms in [9, 5], characterization of strong negation in [11, 4], or characterization of fuzzy implications in [10, 5]. This paper presents the proof of the characterization theorem for implications, which are similar to Dienes implication. We use here the notation presented by Fodor, Roubens [5].

Definition 1 ([5]). *Any function* $I\colon [0,1]^2 \to [0,1]$ *is called fuzzy implication if it fulfils the following conditions* $(x, y, z \in [0,1])$*:*

- ***I1.*** $(x \le z) \Longrightarrow (I(x,y) \ge I(z,y))$,
- ***I2.*** $(y \le z) \Longrightarrow (I(x,y) \le I(x,z))$,
- ***I3.*** $I(0,y) = 1$,
- ***I4.*** $I(x,1) = 1$,
- ***I5.*** $I(1,0) = 0$.

The set of all fuzzy implications will be denoted by FI *and the set of all continuous fuzzy implications is denoted by* CFI.

Example 1. We list here four implication functions completed e.g. by Fodor, Roubens [5]. All of them belong to FI.

$$I_{LK}(x,y) = \min(1-x+y, 1) \; ; \qquad \text{(Łukasiewicz [7])}$$

$$I_{GD}(x,y) = \begin{cases} 1, & \text{if } x \le y \\ y, & \text{if } x > y \end{cases} \; ; \qquad \text{(Gödel [6])}$$

$$I_{DN}(x,y) = \max(1-x, y) \; ; \qquad \text{(Dienes [3])}$$

$$I_{RS}(x,y) = \begin{cases} 1, & \text{if } x \le y \\ 0, & \text{if } x > y \end{cases} \; ; \qquad \text{(Rescher [8])}$$

for $x, y \in [0,1]$.

2 Conjugate Implications

Definition 2. *Let $\varphi\colon [0,1] \to [0,1]$ be an increasing bijection, $I \in FI$. We say that the function*

$$I^*(x,y) = I^*_\varphi(x,y) = \varphi^{-1}(I(\varphi(x),\varphi(y))), \qquad x,y \in [0,1] \tag{1}$$

is φ-conjugate to I. Implication $I \in FI$ is called selfconjugate if $I^*_\varphi = I$ for all φ.

Function φ from the above definition is continuous, since every monotonic bijection must be continuous. We prove that transformation (1) has values in FI.

Theorem 1 (cf. [1]). *Let $\varphi\colon [0,1] \to [0,1]$ be an increasing bijection. For any $I \in FI$ ($I \in CFI$)*

$$I^*_\varphi \in FI \;\; (I^*_\varphi \in CFI) \; . \tag{2}$$

Proof. Let $x,y \in [0,1]$. From assumptions on function φ, we get

$$x \le y \iff \varphi(x) \le \varphi(y) \iff \varphi^{-1}(x) \le \varphi^{-1}(y) \; , \tag{3}$$

$$\varphi(0) = 0 = \varphi^{-1}(0), \quad \varphi(1) = 1 = \varphi^{-1}(1) \; . \tag{4}$$

Let $I \in FI$, $x,y,z \in [0,1]$. If $x \le z$ then, by (3),

$$I^*_\varphi(x,y) = \varphi^{-1}(I(\varphi(x),\varphi(y))) \ge \varphi^{-1}(I(\varphi(z),\varphi(y))) = I^*(z,y) \; .$$

If $y \le z$ then, similarly,

$$I^*_\varphi(x,y) = \varphi^{-1}(I(\varphi(x),\varphi(y))) \le \varphi^{-1}(I(\varphi(x),\varphi(z))) = I^*(x,z) \; .$$

Therefore, I^*_φ fulfils axioms I1, I2. Moreover, from (4), we get

$$I^*_\varphi(0,y) = \varphi^{-1}(I(\varphi(0),\varphi(y))) = \varphi^{-1}(I(0,\varphi(y))) = \varphi^{-1}(1) = 1 \; ,$$
$$I^*_\varphi(x,1) = \varphi^{-1}(I(\varphi(x),\varphi(1))) = \varphi^{-1}(I(\varphi(x),1)) = \varphi^{-1}(1) = 1 \; ,$$
$$I^*_\varphi(1,0) = \varphi^{-1}(I(\varphi(1),\varphi(0))) = \varphi^{-1}(I(1,0)) = \varphi^{-1}(0) = 0 \; ,$$

so I^*_φ fulfils axioms I3, I4, and I5.

If, additionally, $I \in CFI$, then I^*_φ is a composition of continuous functions, and we obtain the second assertion in (2). □

Example 2. For implications from Example 1, we have

$$I^*_{RS} = I_{RS}, \qquad I^*_{GD} = I_{GD} \; .$$

For example,

$$I^*_{GD}(x,y) = \varphi^{-1}(I_{GD}(\varphi(x),\varphi(y))) = \begin{cases} \varphi^{-1}(1), & \text{if } \varphi(x) \le \varphi(y) \\ \varphi^{-1}(\varphi(y)), & \text{if } \varphi(x) > \varphi(y) \end{cases}$$

$$= \begin{cases} 1, & \text{if } x \le y \\ y, & \text{if } x > y \end{cases} = I_{GD}(x,y) \; ,$$

so these implications are selfconjugate. For the next two implications, we get new fuzzy implications:

$$I^*_{LK}(x,y) = \min(\varphi^{-1}(1-\varphi(x)+\varphi(y)),1), \qquad x,y \in [0,1] , \qquad (5)$$

$$I^*_{DN}(x,y) = \max(\varphi^{-1}(1-\varphi(x)),y), \qquad x,y \in [0,1] . \qquad (6)$$

The characterization of implications (5) was obtained by Smets and Magrez in [10].

Theorem 2 (Smets, Magrez, [10]). *A function $I \in CFI$ satisfies*

(i) $I(x,I(y,z)) = I(y,I(x,z))$, *for all* $x,y,z \in [0,1]$,
(ii) $x \leq y \Longleftrightarrow I(x,y) = 1$, *for all* $x,y \in [0,1]$

iff there exists an increasing bijection $\varphi\colon [0,1] \to [0,1]$ *such that* $I = I^*_{\text{LK}}$.

Proof of this theorem can be found in [10], or in the book of Fodor, Roubens [5].

In this paper, we prove the characterization of implications (6).

3 Characterization of Strong Negation

To obtain the main result, first we consider the properties of strong negation.

Definition 3 ([5]). *Any function* $n\colon [0,1] \to [0,1]$ *is called strong negation if it fulfils the following conditions:*

N1. $n(0) = 1, \qquad n(1) = 0$,
N2. *n is strictly increasing* ,
N3. *n is continuous* ,
N4. $n(n(x)) = x$, *for all* $x \in [0,1]$.

One of the most important properties of strong negation n is that there exists a unique value $s \in (0,1)$, such that $n(s) = s$; such s is called a symmetry point of n. Using this property, we prove the representation theorem of strong negations.

Theorem 3 (cf. [11]). *A function* $n\colon [0,1] \to [0,1]$ *is a strong negation iff there exists an increasing bijection* $\varphi\colon [0,1] \to [0,1]$ *such that*

$$n(x) = \varphi^{-1}(1-\varphi(x)), \qquad x \in [0,1] . \qquad (7)$$

Proof. Let $h\colon [0,s] \to [0,\frac{1}{2}]$ will be an increasing bijection, where s is a symmetry point of n. We consider the function

$$\varphi(x) = \begin{cases} h(x), & \text{if } x \leq s \\ 1-h(n(x)), & \text{if } x > s \end{cases}, \qquad x \in [0,1] .$$

Such defined function φ is an increasing bijection. We show (7). Let $x \in [0,1]$. We consider two cases. If $x \leq s$, then

$$\varphi(n(x)) = 1-h(n(n(x))) = 1-h(x) = 1-\varphi(x) .$$

If $x > s$, then

$$\varphi(n(x)) = h(n(x)) = 1 - (1 - h(n(x))) = 1 - \varphi(x) .$$

Since φ is a bijection, then we get (7).

Conversely, let us assume (7), where φ: $[0,1] \to [0,1]$ is an increasing bijection. We show that such defined function fulfils N1 – N4. First, we can see N3, since function φ is a continuous. From properties of function φ, we get N1:

$$n(0) = \varphi^{-1}(1 - \varphi(0)) = \varphi^{-1}(1) = 1 ,$$
$$n(1) = \varphi^{-1}(1 - \varphi(1)) = \varphi^{-1}(0) = 0 .$$

Moreover, from (7), we get N4,

$$n(n(x)) = \varphi^{-1}(1 - \varphi(\varphi^{-1}(1 - \varphi(x))))$$
$$= \varphi^{-1}(1 - (1 - \varphi(x))) = \varphi^{-1}(\varphi(x)) = x .$$

Monotonicity of function n is a consequence of monotonicity of function φ, which proves N2. □

Additionally, we prove the characterization theorem of function maximum.

Theorem 4 (cf. [2]). *A function* $S\colon [0,1]^2 \to [0,1]$ *satisfies*

S1. *S is increasing with respect to both variables* ,
S2. $S(x,0) = S(0,x) = x,$ *for all* $x \in [0,1]$,
S3. $S(x,x) = x,$ *for all* $x \in [0,1]$

iff $S = \max$.

Proof. Let $x, y \in [0,1]$. If $x \le y$, then $y = S(0,y) \le S(x,y) \le S(y,y) = y$, so $S(x,y) = y$. If $y \le x$, then $x = S(x,0) \le S(x,y) \le S(x,x) = x$, so $S(x,y) = x$. Hence, $S(x,y) = \max(x,y)$.

The converse implication is obvious. □

4 Main Result

First, we show how strong negation can be defined by fuzzy implication.

Lemma 1. *If* $I \in CFI$ *satisfies*

$$I(I(x,0),0) = x, \qquad \textit{for all } x \in [0,1] , \tag{8}$$

then function $n\colon [0,1] \to [0,1]$,

$$n(x) = I(x,0), \qquad x \in [0,1] , \tag{9}$$

is a strong negation.

Proof. Function (9) is decreasing by I1. We shall verify axioms from Definition 2. From I3 and I5 we get N1:

$$n(0) = I(0,0) = 1, \quad n(1) = I(1,0) = 0 .$$

Function n fulfils N3, since I is continuous. Moreover, N4 follows from (8),

$$n(n(x)) = I(I(x,0),0) = x .$$

Hence, n is a bijection, so n must be strictly decreasing (fulfils N2). □

Theorem 5. *A function $I \in CFI$ satisfies*

$$I(I(x,0),0) = x , \tag{10}$$

$$I(I(x,0),0) = x , \tag{11}$$

$$I(1,x) = x , \tag{12}$$

for all $x \in [0,1]$, iff there exists an increasing bijection φ: $[0,1] \to [0,1]$ such that $I = I^*_{DN}$, i.e.

$$I(x,y) = \max(\varphi^{-1}(1-\varphi(x)),y), \qquad x,y \in [0,1] . \tag{13}$$

Proof. Let $I \in CFI$ fulfils (10) – (12). We prove that function $S\colon [0,1]^2 \to [0,1]$ defined by

$$S(x,y) = I(I(x,0),y), \qquad x,y \in [0,1] , \tag{14}$$

satisfies S1 – S3. First, we show that function S is increasing with respect to both variables. Let $x,y,z \in [0,1]$. If $x \leq z$ then, from I1,

$$S(x,y) = I(I(x,0),y) \leq I(I(z,0),y) = S(z,y) .$$

If $y \leq z$ then, from I2,

$$S(x,y) = I(I(x,0),y) \leq I(I(x,0),z) = S(x,z) ,$$

which proves S1. Moreover, for any $x \in [0,1]$, we have

$$S(0,x) = I(I(0,0),x) = I(1,x) = x ,$$
$$S(x,0) = I(I(x,0),0) = x ,$$
$$S(x,x) = I(I(x,0),x) = x ,$$

which give S2 – S3. Hence, from Theorem 4, we obtain that $S = \max$. Next, by (14) and (11) we have

$$S(I(x,0),y) = I(I(I(x,0),0),y) = I(x,y) .$$

Therefore,

$$I(x,y) = \max(I(x,0),y), \qquad x,y \in [0,1] . \tag{15}$$

From Lemma 1, function (9) is a strong negation. Thus, Theorem 3 implies that there exists an increasing bijection $\varphi\colon [0,1] \to [0,1]$, such that

$$I(x,0) = \varphi^{-1}(1-\varphi(x)), \qquad x \in [0,1] \; . \tag{16}$$

From (15) and (16), we obtain that

$$I(x,y) = \max(\varphi^{-1}(1-\varphi(x)), y) = I^*_{DN}(x,y), \qquad x,y \in [0,1] \; . \tag{17}$$

Conversely, let us assume (17), where $\varphi\colon [0,1] \to [0,1]$ is an increasing bijection. First, we show that $I \in CFI$. Let $x,y,z \in [0,1]$. If $x \leq z$ then, from (3),

$$I(x,y) = \max(\varphi^{-1}(1-\varphi(x)), y) \geq \max(\varphi^{-1}(1-\varphi(z)), y) = I(z,y) \; .$$

If $y \leq z$, then

$$I(x,y) = \max(\varphi^{-1}(1-\varphi(x)), y) \leq \max(\varphi^{-1}(1-\varphi(x)), z) = I(x,z) \; .$$

Moreover,

$$\begin{aligned} I(0,y) &= \max(\varphi^{-1}(1-\varphi(0)), y) = \max(1,y) = y \; , \\ I(x,1) &= \max(\varphi^{-1}(1-\varphi(x)), 1) = 1 \; , \\ I(1,0) &= \max(\varphi^{-1}(1-\varphi(1)), 0) = \max(0,0) = 0 \; . \end{aligned}$$

Additionally, function I is continuous, since functions φ and max are continuous; so we proved that $I \in CFI$. Now, we show that I satisfies conditions (10) – (12). Let $x \in [0,1]$. We have

$$\begin{aligned} I(I(x,0),0) &= \max(\varphi^{-1}(1-\varphi(\max(\varphi^{-1}(1-\varphi(x)),0))),0) \\ &= \max(\varphi^{-1}(1-\varphi(\varphi^{-1}(1-\varphi(x)))),0) \\ &= \max(\varphi^{-1}(\varphi(x)),0) = \max(x,0) = x \; , \\ I(I(x,0),x) &= \max(\varphi^{-1}(1-\varphi(\max(\varphi^{-1}(1-\varphi(x)),0))),x) \\ &= \max(\varphi^{-1}(1-\varphi(\varphi^{-1}(1-\varphi(x)))),x) \\ &= \max(\varphi^{-1}(\varphi(x)),x) = \max(x,x) = x \; , \\ I(1,x) &= \max(\varphi^{-1}(1-\varphi(1)),x) = \max(0,x) = x \; , \end{aligned}$$

which ends the proof. □

References

1. M. Baczyński, J. Drewniak, Monotonic fuzzy implications, BUSEFAL **76** (1998).
2. E. Czogała, J. Drewniak, Associative monotonic operations in fuzzy set theory, *Fuzzy Sets Syst.* **12** (1984) 249–269.
3. Z.P. Dienes, On an implication function in many-valued systems of logic, *J. Symb. Logic* **14** (1949) 95–97.

4. J.C. Fodor, A new look at fuzzy connectives, *Fuzzy Sets Syst.* **57** (1993) 141–148.
5. J.C. Fodor, M. Roubens, Fuzzy preference modelling and multicriteria decision support, Kluwer, Dordrecht 1994.
6. K. Gödel, Zum intuitionistischen Aussagenkalkül, *Auzeiger der Akademie der Wissenschaften in Wien, Mathematisch, naturwissenschaftliche Klasse* **69** (1932) 65-66.
7. J. Łukasiewicz, A numerical interpretation of the theory of propositions (Polish), *Ruch Filozoficzny* **7** (1923) 92–93 (translated in: L. Borkowski (ed.), Jan Łukasiewicz selected works, North Holland - Amsterdam, PWN - Warszawa 1970, pp.129–130).
8. N. Rescher, Many-valued logic, McGraw-Hill, New York 1969.
9. B. Schweizer, A. Sklar, Probablistic metric spaces, North-Holland, Amsterdam 1983.
10. P. Smets, P. Magrez, Implication in fuzzy logic, *Int. J. Approx. Reasoning* **1** (1987) 327–347.
11. E. Trillas, C. Alsina, L. Valverde, Do we neeed max, min and $1 - j$ in fuzzy set theory? (in: R. Yager (ed.), Fuzzy set and possibility theory, Pergamon Press, New York 1982, pp. 275–297).

Neural Networks Based on Multi-valued and Universal Binary Neurons: Theory, Application to Image Processing and Recognition

Igor N. Aizenberg

Grushevskogo 27, kv. 32, Uzhgorod, 294015, Ukraine
ina@karpaty.uzhgorod.ua

Abstract. Multi-valued and universal binary neurons (MVN and UBN) are the neural processing elements with complex-valued weights and high functionality. It is possible to implement an arbitrary mapping described by partial-defined multiple-valued function on the single MVN and an arbitrary mapping described by partial-defined or fully-defined Boolean function (which can be not threshold) on the single UBN. The fast-converged learning algorithms are existing for both types of neurons. Such features of the MVN and UBN may be used for solution of the different kinds of problems. One of the most successful applications of the MVN and UBN is their usage as basic neurons in the Cellular Neural Networks (CNN) for solution of the image processing and image analysis problems. Another effective application of the MVN is their use as the basic neurons in the neural networks oriented to the image recognition.

1 Introduction

An intensive developing of the neural networks as a high-speed parallel processing systems during last years makes their application to image processing, analysis and recognition very attractive. E.g., many of image processing algorithms in spatial domain may be reduced to the same operation, which it is necessary to perform over all the pixels of image. For example, many of the algorithms of linear and nonlinear filtering in spatial domain are reduced exactly to the processing within the local window around each pixel of the image. We will concentrate here exactly on the nonlinear filtering algorithms that are reduced to the local processing within the window around the pixel and based on the nonlinear transformation of the result of linear convolution with the weighting kernel (template).

Since a local processing within the local window around the pixel is not recursive, and may be organized simultaneously for all the pixels, and independently each of other, it is natural to organize this process using some appropriate kind of neural network. The most appropriate neural network for solution of the considered problems is the Cellular Neural Network (CNN). CNN has been introduced in the pioneer paper of Chua and Yang [1] as special high-speed parallel neural structure for image processing and recognition, and then intensively developed [2-9]. Many results from simple filters for binary images [1-2] to algorithms for processing of the color images [7] and even to design of the CNN universal machine (CNNUM) [8], were carried out during the nine-years period of development of CNN theory and its applications. CNN is a locally connected network (each neuron is connected with an only limited number of other ones - only with neurons from its nearest neighborhood). Depending on the type of neurons that are basic elements of the network it is possible to distinguish classical (or continuous-time) CNN (CCNN) [1], discrete-time CNN (DTCNN) [9] (oriented especially on binary image processing), CNN based on multi-valued neurons (CNN-MVN) [10] and CNN based on universal binary neurons (CNN-UBN) [11]. CNN-MVN makes possible processing defined by multiple-valued threshold functions, and CNN-UBN makes possible processing defined not only by threshold, but also by arbitrary Boolean function. These properties of the CNN-MVN and CNN-UBN will be used here for developing the conception of the nonlinear cellular neural filtering (NCNF).

Multi-valued neural element (MVN), which is based on the ideas of multiple-valued threshold logic [12], has been introduced in [10]. Different kinds of networks, which are based on MVN have been proposed [10, 13-15]. Successful application of these networks to simulation of the associative memory [10, 13-15], image recognition and segmentation [14, 15], time-series prediction [13, 14] confirms their high efficiency. Solution of the image recognition problems using neural networks became very popular during last years. Many corresponding examples are available (e.g., [16, 18]). On the other hand many authors consider image recognition reduced to the analysis of the some number of orthogonal spectrum coefficients on the different neural networks [18, 19]. Here we would like to propose new approach to image recognition based on the following background: 1) high functionality of the MVN and quick convergence of the learning algorithm for them; 2) well-known fact about concentration of the signal energy in the small number of low-frequency spectral coefficients [19]. Some types of the MVN-based neural networks were proposed during last years for solution of the problem of image recognition in associative memory: cellular network [10], network with random connections [13, 14] and Hopfield-like network [15]. A disadvantage of the both networks is impossibility of the recognition of shifted or rotated image, also as image with changed dynamic range. To break these disadvantages and to more effective using of multi-valued neurons features, we would like to propose here new type of the network, learning strategy and data representation (frequency domain will be used instead of spatial one).

2 Multi-Valued and Universal Binary Neurons

Universal Binary Neuron (UBN) performs the mappings described by arbitrary Boolean function of n variables and Multi-Valued Neuron (MVN) performs the mappings described by full-defined threshold or partial-defined k-valued function (function of k-valued logic), where k is in general arbitrary integer>0.

Common basic mathematical background of the UBN and MVN is the following. An arbitrary Boolean function of n variables or k-valued function of n variables is represented by n+1 complex-valued weights $w_0, w_1, ..., w_n$:

$$f(x_1,...,x_n) = P(w_0 + w_1x_1 + ... + w_nx_n), \tag{1}$$

where $x_1,..., x_n$ are the variables, of which performed function depends (neuron inputs) and P is the activation function, which is defined by the following way.

1) For Multi-Valued Neuron:

$$P(z) = \exp(i2\pi j/k), \text{ if } 2\pi\,(j+1)/k > \arg(z) \geq 2\pi j/k, \tag{2a}$$

or with integer-valued output:

$$P(z) = j, \text{ if } 2\pi(j+1)/k > \arg(z) \geq 2\pi j/k, \tag{2b}$$

where j=0, 1, ..., k-1 are the values of the k-valued logic, i is imaginary unity, $z = w_0 + w_1x_1 + w_nx_n$ is the weighted sum, arg(z) is the argument of the complex number z. (values of the function and of the variables are coded by the complex numbers, which are k-th power roots of unity: $e^j = \exp(i2p\,j/k)$, $j \in \{0,1,...,k\text{-}1\}$, i is imaginary unity, in other words values of the k-valued logic are represented by k-th power roots of unity: $j \rightarrow \varepsilon^j$);

2) For Universal Binary Neuron

$$P(z) = (-1)^j, \text{ if } (2\pi\,(j+1)/m) > \arg(z) \geq (2\pi j/m) \tag{2c}$$

where m is the positive integer, j is non-negative integer $0 \leq j < m$, $z = w_0 + w_1x_1 + w_nx_n$ is the weighted sum, arg(z) is the argument of the complex number z. Evidently, functions (2a)–(2c) separate the complex plane on k sectors (2a, 2b) and m sectors (2c), respectively.

One of the quickly converged learning algorithms for MVN, which is based on the error-correction learning rule, has been considered in [13-14]:

$$W_{l+1} = W_l + \frac{1}{(n+1)} C_l (\varepsilon^q \text{-} \varepsilon^s) \overline{X} \tag{3}$$

where W_l and W_{l+1} are current and next weighting vectors, $\overline{X}$ is the vector of the neuron's input signals (complex-conjugated), ε is the primitive k-th power root of unity (k is chosen from (2ab)), C_l is the scale coefficient, q is the number of the desired sector on the complex plane, s is the number of the sector, to which actual value of the weighted sum is fallen, n is the number of neuron inputs. Learning for the UBN may be also reduced to the rule (3). For UBN q has to be chosen on each step based on the following rule: if weighted sum Z gets into "incorrect" sector number s, both of the neighborhood sectors will be "correct":

$$q = s\text{-}1 \pmod m), \text{ if } Z \text{ is closer to } (s\text{-}1)\text{-th sector} \tag{4a}$$

$$q = s+1 \pmod m), \text{ if } Z \text{ is closer to } (s+1)\text{-th sector.} \tag{4b}$$

3 MVN-UBN based Neural Networks and their Applications

3.1 MVN-UBN based CNN and Nonlinear Cellular Neural Filtering

Nonlinear cellular neural filters (NCNF) are based on the nonlinearity of the activation functions (2) of the universal binary neuron (UBN) and multi-valued neuron (MVN). NCNF include the multi-valued non-linear filters (MVF) and cellular Boolean filters (CBF). Idea of cellular neural filtering has been initiated by idea of cellular neural network (CNN) [1]. CNN is a locally connected network (see Fig. 1).

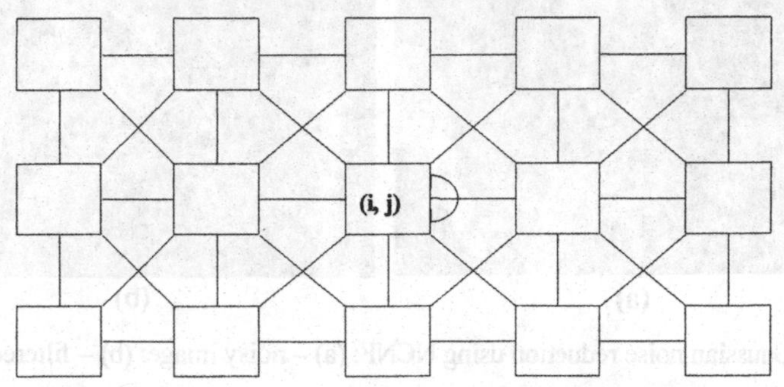

Fig. 1. Structure of the CNN with 3 x 3 local connections

Such a local connection property is very convenient for implementation of the different 2-D spatial domain filters. CNBF with n x m window may be presented by following way:

$$\hat{B}_{ij} = F(w_0 + \sum_{\substack{i-n \le k \le i+n \\ j-m \le l \le j+m}} w_{kl} Y_{kl}), \tag{5}$$

where F is the function (2b) for multi-valued filters, or function (2c) for cellular Boolean filters. Some explanations concerning Y and B should be given. If F is the function (2b), and (5) presents multi-valued filter then we have the following. Let $0 \le B \le k-1$ is a dynamic range of 2-D signal. Let consider a set of the k-th power roots of unity. We can put the root of unity

$$e^B = \exp(i2\pi B/k) = Y \tag{6}$$

to the correspondence with the real value B. Thus, we have the univalent mapping $B \leftrightarrow \varepsilon^B$, where ε is a primitive k-th root of a unity. So Y in (5) is a complex-valued value of a signal obtained according to (6). If F is the function (2c), and (5) presents cellular Boolean filter then we have the following: (5) describes processing of a single binary plane of the image. All binary planes have to be obtained directly, without thresholding. The resulting binary planes have to be integrated into resulting image. An idea of the cellular Boolean filtering has been proposed in [20], and an idea of the multi-valued filtering has been proposed in [21]. NCNF is a result of integration both of these nonlinear neural filters. NCNF may be effectively used either for noise reduction (Fig. 2), high and meadle frequency amplification (Fig. 3) and edge detection (Fig. 3-4) depending on weighting template and type of the function F in the equation (5).

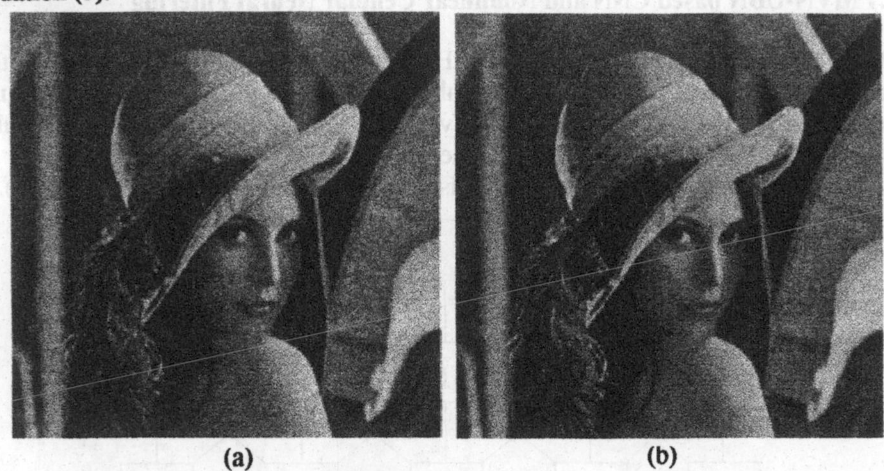

Fig. 2. Gaussian noise reduction using NCNF: (a) – noisy image; (b) – filtered image

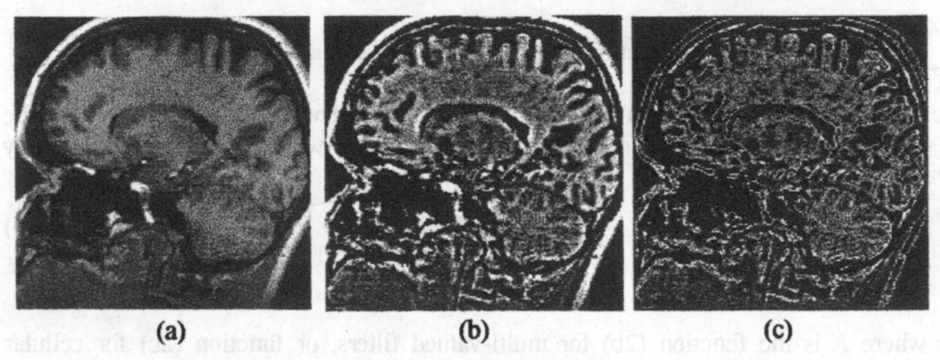

Fig. 3. Frequency correction and edge detection using NCNF: (a) – input image; (b) – high and meddle frequency correction within a 9x9 window; (c) – edge detection on image (b)

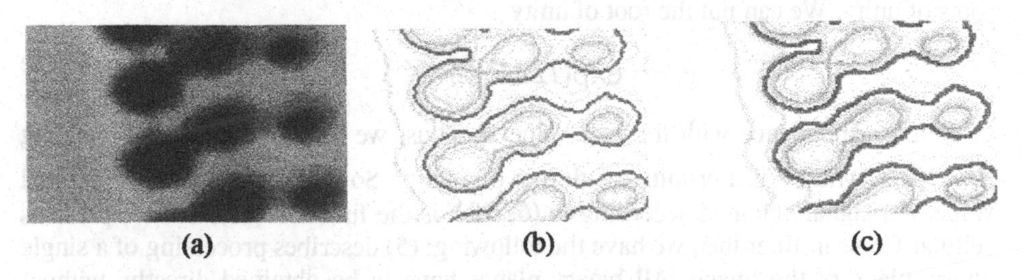

Fig. 4. Edge detection using NCNF: (a) – input image; (b) – detection of upward brightness jumps; (c) – detection of downward brightness jumps

3.2 MVN based neural network for image recognition

As it was mentioned above a disadvantage of the networks used as associative memories is impossibility to recognize shifted or rotated image, also as image with changed dynamic range. MVN based neural networks that have been proposed as associative memories in [13-15] have the same disadvantage.

To use more effectively the MVN features, and to break the disadvantages mentioned above we would like to consider here a new type of the network, learning strategy and data representation (frequency domain will be used instead of spatial one).

Consider N classes of objects, which are presented by images of n x m pixels. The problem is formulated into the following way: we have to create recognition system based on a neural network, which makes possible successful identification of the objects by fast learning on the minimal number of representatives from all classes.

To make our method invariant to the rotations, shifting, and to make possible recognition of other images of the same objects we will move to frequency domain representation of objects. It has been observed (see e.g., [18-19]) that objects belonging to the same class must have similar coefficients corresponding to low spectral coefficients. For different classes of discrete signals (with different nature and length from 64 until 512) sets of the lowest (quarter to half) coefficients are very close each other for signals from the same class from the point of view of learning and analysis on the neural network [11]. This observation is true for different orthogonal transformations. In the terms of neural networks to classify object we have to train a neural network with the learning set contained the spectra of representatives of our classes. Then the weights obtained by learning will be used for classification of unknown objects.

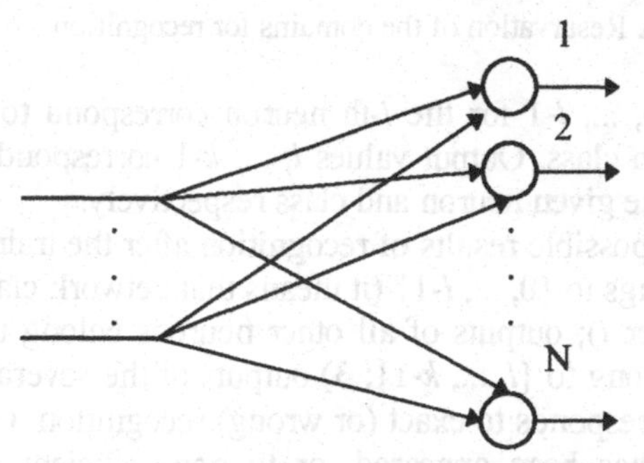

Fig. 5. MVN based neural network for image recognition

We propose the following structure of the MVN based neural network for the solution of our problem. It is single-layer network, which has to contain the same number of neurons as the number of classes we have to identify (Fig. 5). Each neuron has to recognize pattern belongency to its class and to reject any pattern from any other class. Taking into account that single MVN could perform arbitrary mapping, it is easy to conclude that exactly such a structure of the network that was just chosen is optimal and the most effective.

To ensure more precise representation of the spectral coefficients in the neural network they have to be normalized, and their new dynamic range after normalization will be [0, 511]. More exactly, they will take discrete values from the set {0, 1, 2, ..., 511}. We will use phases of a lower part of the Fourier transformation coefficients for the frequency domain representation of our data. We are basing here on such a property of Fourier transformation that phase contains more information about the signal than amplitude (this fact is investigated e.g., in [22].

The best results have been obtained experimentally, when for classification of the pattern as belongs to the given class we reserved the first $l=k/2$ (from the $k=512$) sectors on the complex plane (see (2ab)). Other $k/2$ sectors correspond to the rejected patterns (Fig. 6).

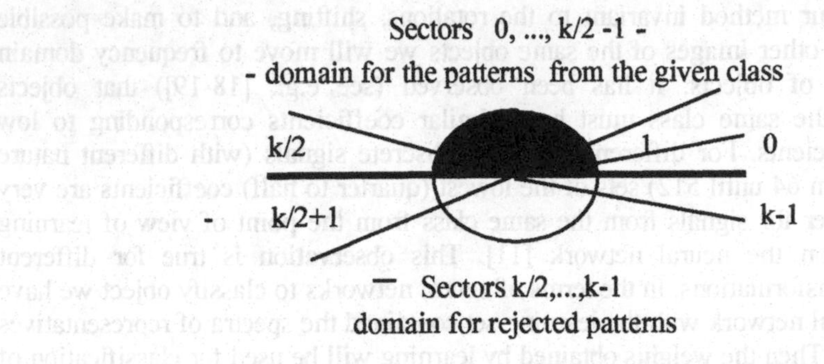

Fig. 6. Reservation of the domains for recognition

Thus, output values 0, ..., l-1 for the i-th neuron correspond to classification of object as belonging to i-th class. Output values l, ..., k-1 correspond to classification of object as rejected for the given neuron and class respectively.

Hence there are three possible results of recognition after the training: 1) output of the neuron number i belongs to {0, ..., l-1} (it means that network classified pattern as belonging to class number i); outputs of all other neurons belong to {l, ..., k-1}; 2) outputs of all neurons belong to {l, ..., k-1}; 3) outputs of the several neurons belong to {0, ..., l-1}. Case 1 corresponds to exact (or wrong) recognition. Case 2 means that a new class of objects has been appeared, or to non-sufficient learning, or not-representative learning set. Case 3 means that number of neuron inputs is small or inverse is large, or that learning has been provided on the not-representative learning set.

The proposed structure of the MVN based network and approach to solve of the recognition problem has been evaluated on the example of face recognition. Experiments have been performed on the software simulator of the neural network. We used MIT faces data base [23], which was supplemented by some images from the data base used in our previous work on associative memories (see [13-14]). So our testing data base contained 64 x 64 portraits of 20 people (27 images per person with different dynamic range, conditions of light, situation in field). Our task was training of the neural network to recognize twenty classes. Fragment of the data base is presented in Fig.7 (each class is presented by single image within this fragment).

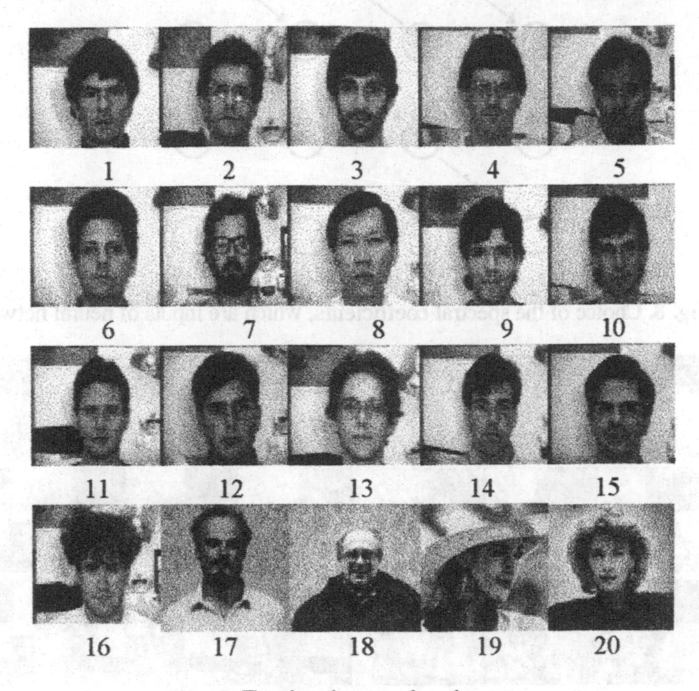

Fig. 7. Testing image data base

According to the structure proposed above, our single-layer network contains twenty MVNs (the same number, as number of classes). For each neuron we have the following learning set: 16 images from the class corresponding to given neuron, and 2 images for each other class (so 38 images from other classes). Let describe the results.

According to the scheme presented in Fig. 6 sectors 0, ..., 255 have been reserved for classification of the image, as belonging to the current class, and sectors 256, ..., 511 have been reserved for classification of the images from other classes. The learning algorithm with the learning rule (3) has been used. So for each neuron q=127 for patterns from the current class, and q=383 for other patterns in the learning rule (3). The results of recognition sequentially improved with increasing of number of network inputs.

The best results have been obtained for 405 inputs of the network, or for 405 spectral coefficients, which are inputs of the network, and beginning from this number the results stabilized. Phase of spectral coefficients has been chosen according to "zigzag" rule (Fig. 8).

For all classes 100% successful recognition has been gotten. For classes "2" and "13" 2 images from another class ("13" for "2", and "2" for "13") also have been identified as "its", but this mistake has been easy corrected by additional learning. A reason of this mistake is evidently, again the same background of images, and very similar glasses of both persons whose portraits establish the corresponding classes.

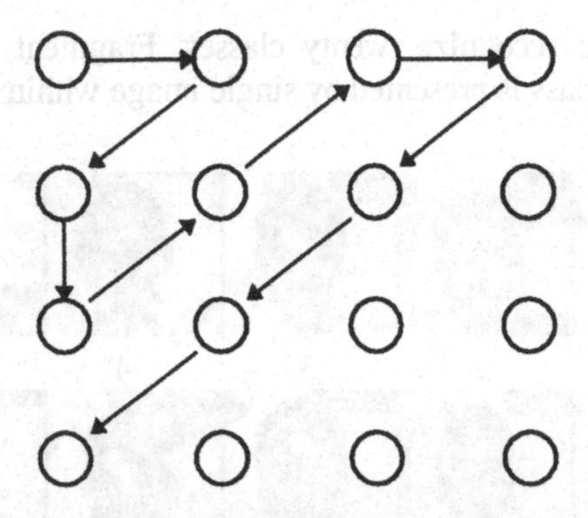

Fig. 8. Choice of the spectral coefficients, which are inputs of neural network

Fig. 9. Class 1: 100% successful recognition. **Fig. 10.** Class 2: 100% successful recognition

Fig. 11. Class "17" - 100% successful recognition

To estimate a store of the precision ensured by the learning Table 1 contains numbers of sectors (from 512), to which the weighted sum has been fallen for images from the class No 17 (see Fig. 11).

It should be mentioned that using frequency domain data representation it is very easy to recognize the noisy objects (see Fig. 11, Table 1). Indeed, we use low frequency spectral coefficients for the data representation. At the same time noise is concentrated in the high frequency part of the spectral coefficients, which is not used.

We hope that considered examples are convinced, and show either efficiency of proposed solution for image recognition problem, and high possibilities of MVN or neural networks based on them.

Table 1. Number of sectors, which to weighted sum has fallen during recognition of the images presented in Fig. 11

Image	1	2	3	4	5	6	7	8	9	10	11	12
Sector	126	122	130	129	120	135	118	134	151	126	107	119

4 Conclusions

A main conclusion from this paper is high efficiency of multi-valued and universal binary neurons, also as different networks from them. Some effective solutions for image processing and recognition have been found. A neural network from MVN for solution of pattern recognition problems based on analysis of the low frequency Fourier spectra coefficients has been proposed, and the simulation results on the example of face recognition confirm efficiency of the considered approach. Nonlinear activation functions of MVN and UBN have been used for developing of the class of effective nonlinear, so-called cellular neural filters (NCNF), which may be successfully used either for noise removing, frequency correction (extraction of image details), and precise edge detection.

References

1. Chua L.O. and Yang L., "Cellular neural networks: Theory", *IEEE Transactions on Circuits and Systems.* Vol. 35, No 10 (1988), 1257-1290
2. "Proc. of the First *IEEE International Workshop on Cellular neural networks and their applications (CNNA-90)* " Budapest (1990).
3. "Proc. of the *Second IEEE International Workshop on Cellular Neural Networks and their Applications (CNNA-92)"*, Munich (1992).
4. "Proc. *of the Third IEEE International Workshop on Cellular Neural Networks and their Applications" (CNNA-94)"*, Rome (1994).
5. "*Proc. of the Fourth IEEE International Workshop on Cellular Neural Networks and their Applications" (CNNA-96)*, Seville, Spain (1996).
6. Roska T. and Vandewalle J. (ed.) *"Cellular Neural Networks"*, John Wiley & Sons, New York (1993).
7. Lee C.-C. and.Pineda de Gyvez J "Color Image Processing in a Cellular Neural -Network Environment", IEEE *Trans. On Neural Networks,* vol. 7, No 5, (1996), 1086-1088.
8. Roska T. and Chua L.O. "The CNN Universal Machine: An Analogic Array Computer", *IEEE Transactions on Circuits and Syst. - II,* vol. 40. (1993), 163-173.
9. Harrer H. and Nossek J.A. "Discrete-time cellular neural networks", *International Journal of Circuit Theory and Applications,* Vol.20, (1992), 453-467.

10. Aizenberg N.N., Aizenberg I.N "CNN based on multi-valued neuron as a model of associative memory for gray-scale images", *Proc. of the 2-d IEEE International Workshop on Cellular Neural Networks and their Applications, Munich, October 12-14* (1992), 36-41.
11. Aizenberg N.N., Aizenberg I.N. "Fast Convergenced Learning Algorithms for Multi-Level and Universal Binary Neurons and Solving of the some Image Processing Problems", *Lecture Notes in Computer Science*, Ed.-J.Mira, J.Cabestany, A.Prieto, Vol. 686, Springer-Verlag, Berlin-Heidelberg (1993), 230-236.
12. Aizenberg N.N., Ivaskiv Yu.L *Multiple-Valued Threshold Logic*. Naukova Dumka, Kiev, (1977) (in Russian)
13. Aizenberg N.N., Aizenberg I.N.. Krivosheev G.A. "Multi-Valued Neurons: Learning, Networks, Application to Image Recognition and Extrapolation of Temporal Series", *Lecture Notes in Computer Science*, Vol. 930, (J.Mira, F.Sandoval - Eds.), Springer-Verlag, (1995), 389-395.
14. Aizenberg N.N., Aizenberg I.N., Krivosheev G.A. "Multi-Valued Neurons: Mathematical model, Networks, Application to Pattern Recognition", *Proc. of the 13 Int.Conf. on Pattern Recognition, Vienna, August 25-30, 1996, Track D*, IEEE Computer Soc. Press, (1996), 185-189.
15. Jankowski S., Lozowski A., Zurada M. "Complex-Valued Multistate Neural Associative Memory", *IEEE Trans. on Neural Networks*, Vol. 7, (1996), 1491-1496.
16. Petkov N., Kruizinga P., Lourens T. "Motivated Approach to Face Recognition", *Lecture Notes in Computer Science*, Vol. 686, (J.Mira, F.Sandoval - Eds.), Springer-Verlag, (1993), 68-77.
17. Lawrence S., Lee Giles C., Ah Chung Tsoi and Back A.D. "Face Recognition: A Convolutional Neural-Network Approach", *IEEE Trans. on Neural Networks*, Vol. 8, (1997), pp. 98-113.
18. Foltyniewicz R. "Automatic Face Recognition via Wavelets and Mathematical Morphology", *Proc. of the 13 Int. Conf. on Pattern Recognition, Vienna, August 25-30, 1996, Track B,* IEEE Computer Soc. Press, (1996), 13-17.
19. N.Ahmed, K.R.Rao "Orthogonal Transforms for Digital Signal Processing", Springer-Verlag (1975).
20. Aizenberg I.N. "Processing of noisy and small-detailed gray-scale images using cellular neural networks" *Journal of Electronic Imaging*, vol. 6, No 3, (1997), 272-285.
21. Aizenberg I.N. "Multi-valued non-linear filters and their implementation on cellular neural networks", *Frontiers in Artificial Intelligence and Applications.* Vol. 41 *"Advances in Intelligent Systems"* (F.C.Morabito - Ed.) IOS Press, Amsterdam-Berlin-Oxford-Tokyo-Washington DC (1997), 135-140.
22. Oppenheim A.V. and. Lim S.J "The importance of phase in signals", *Proc. IEEE,* Vol. 69, (1981), pp. 529-541.
23. Turk M. and Petland A. "Eigenfaces for Recognition", Journal of Cognitive Neuroscience, Vol. 3, (1991).

Modeling of Thermal Two Dimensional Free Turbulent Jet by a Three Layer Two Time Scale Cellular Neural Network

Ariobarzan Shabani, Mohammad Bagher Menhaj, Hassan Basirat Tabrizi

Mechanical Engineering Department of Amirkabir University of Technology ,
Hafez Ave. No.424 Tehran 15914 Iran
menhaj@cic.aku.ac.ir

Abstract. A three layer Cellular Neural Network (CCN) is used to model the velocity and temperature profiles in a two dimensional turbulent free jet. CNNs are a class of neural networks which are adapted for Integrated Circuits (ICs) and can process data in parallel asynchronously at very high speeds. To implement the CNN model together with an appropriate discretization scheme variable mesh size in vertical direction was also considered. As CNN operates in continuous-time and the propagation speeds of the velocity and temperature fronts are in general unequal, two different time scales were accordingly used. These time scales were set such that the ratio of the time constants of the corresponding momentum to thermal layers in the CNN matched the turbulent apparent eddy to temperature diffusivity ratio, or namely the turbulent Prandtl number. The results were compared with computational fluid dynamics similarity based solutions and indicate acceptable agreement. From the results one can justify the use of neural networks, as a powerful new tool in the fluid dynamics problems, to tackle such numerically extensive and expensive phenomena as turbulence.

1 Introduction

Many complex computational problems can be reformulated as well-defined tasks in which the signal values are placed on a regular geometric 2-D or 3-D grid. Then, the direct interactions between signal values are limited within a finite local neighborhood [1,2]. This fact together with the need for parallel processing of data has led to an invention called Cellular Neural Networks (CNNs) [3,4]. Although they are based on some aspects of neurobiology and their distinguished features are asynchronous parallel processing and continuous-time dynamics [3]. Cellular neural networks, which are analog circuits adapted to integrated circuits [2-4], represent a class of neural networks possessing the aforementioned key features and are used to solve various kinds of partial differential equations (PDEs) [5-9].

In studying anisotropic turbulent flow fields it is desirable for practical reasons to confine oneself at first to rather simple but important elementary flow patterns [10,11]. These elementary flow patterns may be classified into two main groups: (1) Free jets, and (2) wake flows behind obstacles. The features under investigation here are velocity and temperature profiles in a free turbulent jet [11]. In this flow pattern it is possible to distinguish one main flow direction with velocity substantially greater than in any other direction [1,10,11].

A suitable solution for turbulent free jet velocity and temperature equations requires the use of variable mesh size and an appropriate discretization scheme. Here a solution of turbulent free jet velocity and temperature profiles is considered.

In section two PDE approximation with CNN and architecture of CNN are considered. In section three the relevant flow field PDE's are introduced and the cloning templates are obtained. And finally in section four the simulation results together with conclusions are presented.

2. PDE Approximation with CNN

Cellular neural networks can be characterized by a large system of ordinary differential equations [3,4]. Cellular neural networks like PDEs have the property that their dynamic behaviors depend only on their spatial local interactions [3,4].
The CNN, spatially discrete dynamical system, approximates the PDE describing the desired spatially continuous system. When the PDE is discretized some error is introduced and this is the only difference between the two systems [3-6]. Although the CNN operates in continuous-time, its transitory behavior does not correspond to transient of the original system. Its stationary state, however, corresponds to the steady state solution of the PDE [3,4]. Many important PDEs can be accurately solved by a CNN [7-9].

2.1 CNN Structure

Before starting to solve the turbulent two-dimensional thermal free jet equations, we have to gain an insight in the CNN structure and its governing equations. As neural networks it is a large-scale nonlinear analog circuit which processes signals in real time. The basic circuit unit of a cellular neural network is called a "Cell". It contains linear and nonlinear circuit elements which are typically linear capacitors, linear resistors; linear and nonlinear controlled sources and independent sources [3,4]. The structure of CNN is similar to that found in cellular automata [3,4,12]; namely, any cell in a CNN is connected directly only to its neighbor cells. The ith row and jth column cell of a particular cell arrangement is denoted by C (i, j). The r-neighborhood N_r of radius r of a cell, C (i,j), in a M N cell CNN is defined by [3,4]

$$N_r(i,j) = \{C(k,l) \mid \max[|k-i|,|l-j|] \le r, \quad 1 \le k \le M, \quad 1 \le l \le N\} \tag{1}$$

Where r is a positive integer. Using u, x, and y to denote input, state and output respectively, the governing equations are the followings [5-7]:
State equation:

$$C\frac{dv_{xij}(t)}{dt} = -\frac{1}{R_x}v_{xij}(t) + I + \sum_{C(k,l)\in N_r(i,j)} A(i,j;k,l)v_{ukl}(t) + \sum_{C(k,l)\in N_r(i,j)} B(i,j;k,l)v_{ykl}(t) +$$

$$\sum_{C(k,l)\in N_r(i,j)} C(v_{uij}(t),v_{ukl}(t)) + \sum_{C(k,l)\in N_r(i,j)} D(v_{yij}(t),v_{ykl}(t)) \qquad 1 \le i \le M\,; 1 \le j \le N \tag{2a}$$

Output equation:

$$v_{yij}(t) = Sigmoid\,(v_{xij}(t)), \quad 1 \le i \le M\,; 1 \le j \le N \tag{2b}$$

Input equation:

$$v_{uij}(t) = E_{ij}\ , \quad 1 \le i \le M\ ; 1 \le j \le N \tag{2c}$$

Constraint conditions:

$$\left|v_{xij}(0)\right| \le 1\ \ , \quad 1 \le i \le M\ ; 1 \le j \le N \tag{2d}$$

$$\left|v_{uij}(t)\right| \le 1\ \ , \quad 1 \le i \le M\ ; 1 \le j \le N \tag{2e}$$

Where C is a linear capacitor; R_x is a linear resistor; I is an independent current source; I_{xy} (i, j; k, l) and I_{xu} (i, j; k, l) are linear voltage-controlled current sources with characteristics I_{xy} (i, j; k, l) = A (i, j; k, l) V_{ykl} and I_{xu} (i, j; k, l) = B (i, j; k, l) V_{ukl} for all C (k, l) N_r (i, j); (1/ R_x) V_{xij} is a peicewise-linear voltage controlled current source [5,6]; E_{ij} is a time-invariant independent voltage source,

$$D(v_{yij}(t), v_{ykl}(t)) = \begin{bmatrix} f_{11} & \cdots & f_{1N} \\ \cdot & & \cdot \\ f_{M1} & \cdots & f_{MN} \end{bmatrix}, \tag{3a}$$

and

$$C(v_{uij}(t), v_{ukl}(t)) = \begin{bmatrix} g_{11} & \cdots & g_{1N} \\ \cdot & & \cdot \\ g_{M1} & \cdots & g_{MN} \end{bmatrix} \tag{3b}$$

In the above,

$$f_{kl} = d_{kl} v_{yij}(t) v^{\alpha}{}_{ykl}(t) \tag{3c}$$

$$g_{kl} = c_{kl} v_{uij}(t) v^{\beta}{}_{ukl}(t) \tag{3d}$$

and d_{kl}, c_{kl}, α, β are constants.

3. The Thermal Two Dimensional Turbulent Free Jet Equatons

Such constant property flow is described mathematically by the following set of nonlinear partial equations [1,10,11]:

$$\frac{\partial V}{\partial t} + (V.\nabla)V = -\frac{1}{\rho}\nabla P + \nu\,\nabla^2 V \tag{4a}$$

$$\nabla . V = 0 \tag{4b}$$

$$C_p \rho \left(\frac{\partial T}{\partial t} + (V.\nabla)T \right) = k \nabla^2 T + \mu \Phi + V.\nabla P \tag{4c}$$

In which V is the velocity vector, T is the temperature, t is time and here the rest of the variables are constant coefficients and as a result some of the terms are deleted.
To obtain a spatially discrete-time system, spatial derivatives are replaced with appropriate difference terms. Therefore, for example, for the u-velocity component we have

$$\frac{du_{ij}}{d\ t} = kh_x \left(\frac{u_{i-1j} - 2u_{ij} + u_{i+1j}}{h_y^2} \right) - \left(\frac{-(u_{i-1j})^2 + (u_{i+1j})^2}{2h_x} \right) - \left(\frac{-(u_{ij-1}v_{ij-1}) + (u_{ij+1}v_{ij+1})}{2h_y} \right) \tag{5}$$

Here h_x and h_y are the fixed and the variable step sizes in the x and y directions, respectively. The descritized form of the other equations is derived similarly.
As each CNN cell has one real output, three layers are needed to represent u and v velocity components and the temperature. We map the three variables u, v and T into respective CNN layers such that the state voltage v_{xij} (t) of respective CNN cells at mesh point (i, j) is associated with u (ih_x, jh_y, t), v (ih_x, jh_y, t) and T (ih_x, jh_y, t), respectively [1,2]. Also, the propagation speed of the temperature and the velocity layers are different since in the actual flow field the diffusivity ratios of momentum to thermal fronts, namely the Prandtl number, are different. Here, the ratio of the time scales of the velocity to temperature layers is set equal to the Prandtl number of the flow.
Now by using the central linear segment of the unity-gain sigmoid output function, the CNN templates acting on and between the layers are obtained from the complete set of equations (5). The first term in (5) including u is implemented by

$$A_{ij,kl}^{u,u} = \frac{h_x}{h_y^2} \begin{bmatrix} 0 & k & 0 \\ 0 & -2k & 0 \\ 0 & k & 0 \end{bmatrix} [u_{kl}]$$

, the nonlinear terms in (5) are implemented by

$$C_{ij,kl}^{uu,u} = \frac{1}{2h_x} \begin{bmatrix} 0 & 0 & 0 \\ -1 & 0 & 1 \\ 0 & 0 & 0 \end{bmatrix} [u_i . u_k]$$

$$C_{ij,kl}^{uv,u} = \frac{1}{2h_y} \begin{bmatrix} 0 & -1 & 0 \\ 0 & 0 & 0 \\ 0 & 1 & 0 \end{bmatrix} [u_j . v_l]$$

and for the v-component of the velocity and the temperature layers we obtain the templates similarly. Here the upper right indices indicate from which layer and to which layer they connect. The boundary conditions are those presented in reference 11, and the initial condition is

$t = 0;\ u = v = 0,\ T=0$

3.1 Treatment of the Initial and Boundary Conditions

Initial values are easily incorporated into the CNN model. The initial state of the PDE is directly associated with the initial state of the CNN's state equation which is a finite system of ordinary differential equations [3,4]. Imposing prescribed boundary conditions is done by using some fixed state cells at appropriate locations at the edges.

4. Simulation Results and Conlusions

The three layer CNN each consisting of 40×40 cells, corresponding to 40×40 grid points in the turbulent two dimensional thermal free jet, has been tested through numerical simulations, and it's performance was compared with the results obtained from that of the computational fluid dynamics solution [1,11].
Velocity and temperature distributions at various sections for a free stream velocity of U = 10 (m/s) and free stream temperature of 100 'C were calculated.
Figure 1 shows the calculated CNN velocity profiles at a distance of 0.01 (m) from the orifice, and figures 2 and 3 show the comparison between the calculated temperatures of the CNN and CFD models for a distance of 0.02 (m) and 0.04 (m), respectively. From figures 2 and 3 a close agreement between the CNN approximation and the CFD solution is established.

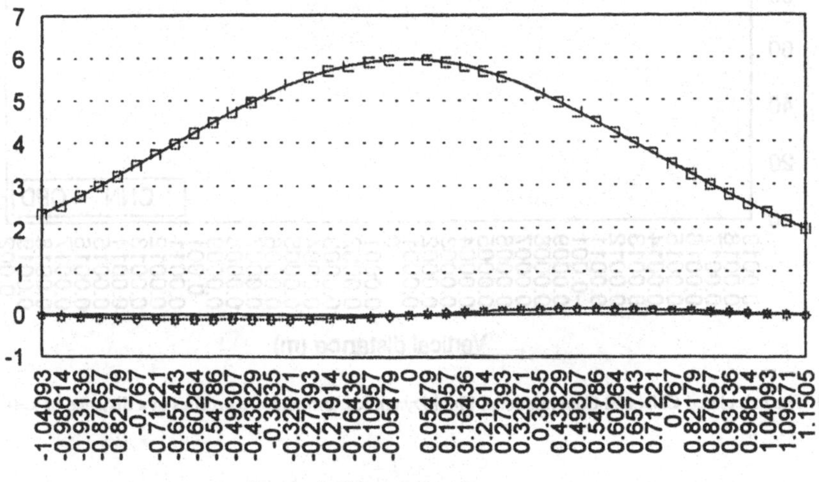

Fig. 1. The calculated CNN velocity profile at a distance of x=0.01 m

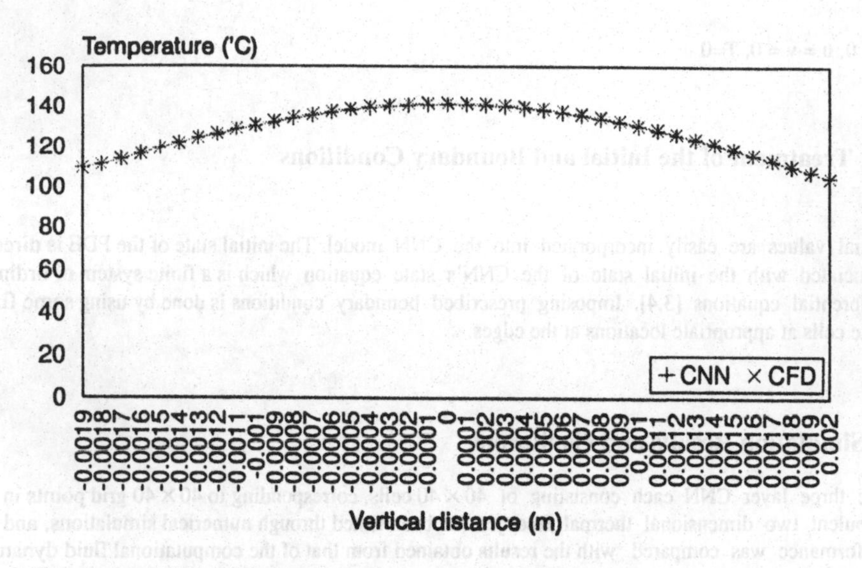

Fig. 2. The calculated CNN and Computational Fluid Dynamics temperature profiles at a distance of x=0.02 m.

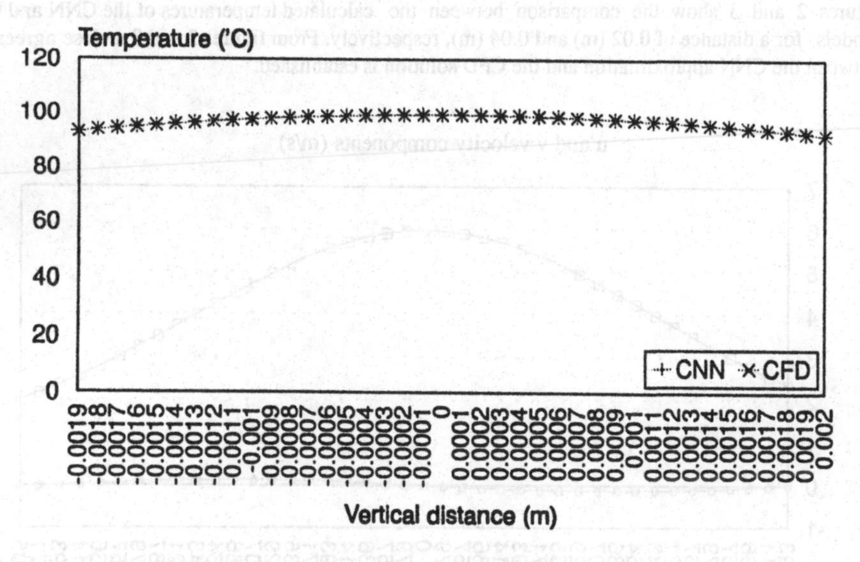

Fig. 3. The calculated CNN and Computational Fluid Dynamics temperature profiles at a distance of x=0.04 m.

A better formulation is the one which makes use of adaptive permissible x and y grid steps. Also, it is worth noting that the size of these steps affects the stability of such networks, i.e. they cannot be selected arbitrarily [3].
As the CFD solution provides a good approximation for the flow field, the results indicate the capability of the cellular neural network to obtain an acceptable velocity profile with the same practical accuracy of the similarity solution at a much shorter time.
Also, the large number of grid points and the long time required to accurately model turbulent flows in general fits the Very Large Scale Integration (VLSI) implementation of cellular neural networks and its supercomputer processing power [12], and therefore one can justify the use of such neural networks to deal with more complex turbulent flows, frequently occurring in natural and industrial applications [10,11], at high speeds with sufficient accuracy.

References

1. S.V. Patankar and D.B. Spalding,"Heat and mass transfer in boundary layers", Int. Textbook Company Ltd., 1970.
2. L.O. Chua and T. Roska , "The CNN Paradigm", IEEE Trans. Circuits and Systems, CAS-40, 147-156 (1993)
3. L.O. Chua and L.Yang, "Cellular Neural Networks: theory", IEEE Trans. Circuits and Systems, CAS-35, 1257-1272 (1988)
4. L.O. Chua and L.Yang, "Cellular Neural Networks: theory", IEEE Trans. Circuits and Systems, CAS-35, 1273-1290 (1988)
5. T.Roska, D. Wolf, T. Kozek, R. Tetzlaff, L.O. Chua, and L. Yang,"Solving partial differential equations by CNN", Proc. 11th Eur. Conf. on circuit theory and design, Davos, August 1993, pp. 1477-1482.
6. M.W.M.G. Dissananyake and N.Phan-Thien ,"Neural-network-based approximation for solving partial differential equations", Communications in Numerical Methods in Engineering, Vol. 10, No. 3, March 1993.
7. T.Roska, L.O. Chua, D. Wolf, T. Kozek, R. Tetzlaff and F. Puffer," Simulating nonlinear waves and partial differential equations via. CNN", IEEE Trans. on Circuits and Systems I: Fundamental theory and applications, Vol. 42, No. 10, Oct. 1995, pp.805-815.
8. T. Kozek, L.O. Chua, T.Roska, D. Wolf, R. Tetzlaff, F. Puffer and K. Lotz," Simulating nonlinear waves and partial differential equations via. CNN", IEEE Trans. on Circuits and Systems I: Fundamental theory and applications, Vol. 42, No. 10, Oct. 1995, pp. 816-820.
9. D. Gobovic and M.E Zaghloul,"Analog cellular neural network with application to partial differential equations", Proc. IEEE Int. Symp. on Circuits and Systems, June 1994, Vol. 6, pp. 359-362.
10. J.O. Hinze,"Turbulence", McGraw Hill, 1959
11. H. Schlichting,"Boundary layer theory", McGraw Hill, 1968.
12. T.Roska, and L.O. Chua,"The CNN universal machine: An analogic array computer", IEEE Trans. Circuits and Systems, CAS-40, 163-173 (1993).

A Neural Approach for Detection of Road Direction in Autonomous Navigation

Victor Neagoe[1], Marius Valcu[1], and Bogdan Sabac[1]

[1] Polytechnic University of Bucharest, Department of Applied Electronics and Information Engineering, Splaiul Independentei 313, Bucharest 16, Romania 77206
Vic@gitprai.pub.ro

Abstract. The paper presents an original approach for visual identification of road direction of an autonomous vehicle using an improved Radial Basis Function (RBF) neural network. We present the results of designing, software implementation, training, and testing of our RBF model for automatic road direction detection as a function of the input image. The path to be identified was quantified in 5 output directions. For training and testing the neural model, we used two lots of real road scenes: 50 images for training and other 50 images for test. The score of correct road recognition was of 100% both for estimation lot and also for the test lot. We have also designed a driving simulator to evaluate the performances of the neural road follower.

1 Introduction

The automatic detection of the path to be followed by the vehicle in autonomous navigation has proven to be a formidable task when dealing with outdoor scenes. Many of the model-based road following systems perform in very limited domains, and have a tendency to break down when environmental variables such as road width and lightening conditions change, and in the presence of external noise. Recently, connectionist approaches have shown promise in autonomous navigation, and specifically in autonomous road following. One of the most successful implementation architecture for visual road following is ALVINN (Autonomous Land Vehicle in a Neural Network) that was developed at Carnegie Mellon University, Pittsburgh, USA [7]. ALVINN is based on a feed-forward network (multilayer perceptron), where the network is fed directly with image data at a low-resolution level. Del Bimbo et al [2] used a highly parallel image processing system as a preprocessing stage followed by a feedback neural network for road direction identification. At the Polytechnic University of Bucharest, V. Neagoe and his team proposed and simulated several connectionist models for road following (some *with* and other *without* an image preprocessing system); the models have been trained and tested using real input images [5], [6].

Within this paper we present an original approach of Radial Basis Function (RBF) network for visual autonomous road following. it is based on the special RBF training procedure given in [4].

2 The Radial Basis Function Network for Road Detection

The RBF network architecture consists of an input layer with N units, a hidden layer with L neurons, and an output layer with M neurons (Fig. 1). The input layer contains a number of neurons equal to the number of pixels in the input image of the road. Units called receptive fields driven by radial basis activation functions compose the hidden layer. The input neurons are fully connected by weighted links to the receptive fields of the hidden layer. The hidden layer is described by the set of parameters {w(j,h), r(j,h); h=1,...,N; j=1,...,L}, defining the corresponding RBF activation functions of the neurons. The receptive fields are connected by the weights {λ(k, j), k=1,...,M; j=1,...,L} to the neurons of the output layer. The activation function of the hidden layer neurons is a radial basis function (called also "potential" function); we use an N-dimensional Gaussian function, namely

$$\sigma_{1i}(j)=\sum_{h=1}^{N}\left[\frac{w(j,h)-x_i(h)}{r(j,h)}\right]^2 \quad (j=1,\ldots,L;i=1,\ldots,Q) \qquad (1)$$

where the parameters {w (j,h), r(j,h); h=1,...N; j=1,...L} correspond to the connection between the input "h" and the hidden neuron of index "j"; the parameter "i" is the index of the input vector (image); Q is the number of input vectors .

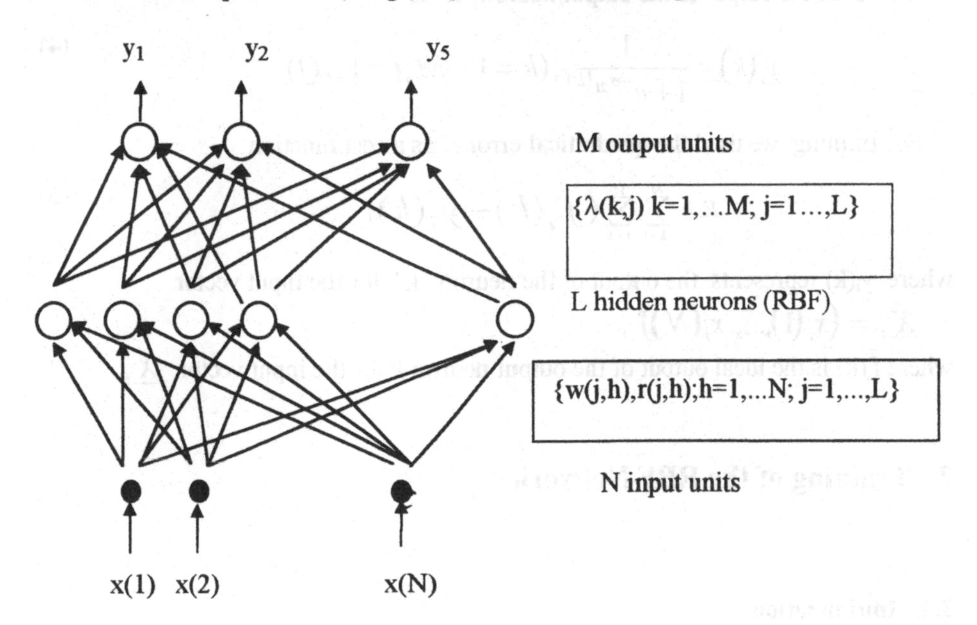

Fig. 1. RBF for visual road detection

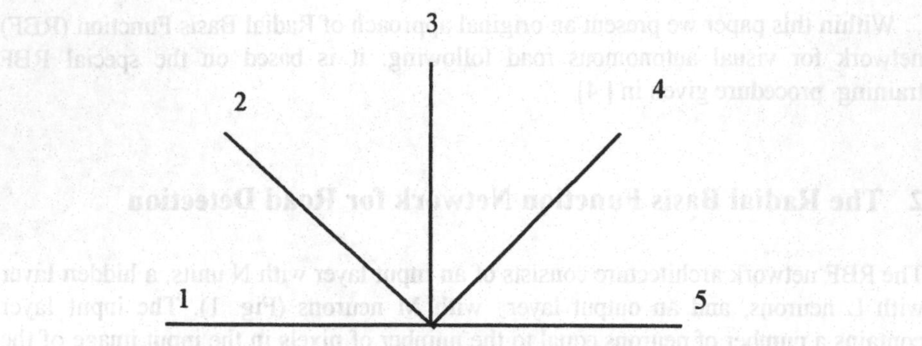

Fig. 2. Quantization of the path to be followed in 5 directions.

The nonlinear output function of the hidden neuron "j" is sigmoidal

$$u_i(j) = e^{-\sigma_U(j)}, (j = 1, \ldots, N; i = 1, \ldots Q) \tag{2}$$

The activation function of an output neuron "k" is

$$\sigma_{2i}(k) = \sum_{j=1}^{M} u_i(j) \lambda(k, j), (k = 1, \ldots, M; i = 1, \ldots, Q) \tag{3}$$

where the weight {λ(k,j); k=1,...,M; j=1,...,L} characterize the connections between the hidden neuron "j" and the output neuron "k".

The nonlinear output of the output neuron "k" is

$$y_i(k) = \frac{1}{1 + e^{-\sigma_{2i}(k)}}, (k = 1, \ldots M; i = 1, \ldots Q) \tag{4}$$

For training, we used the quadratical error E as a cost function

$$E = \sum_{k=1}^{M} \sum_{i=1}^{Q} (f_i(k) - y_i(k))^2 \tag{5}$$

where $y_i(k)$ represents the output of the neuron "k" for the input vector

$$\underline{X}_i = (x_i(1), \ldots, x_i(N))^t,$$

where $f_i(k)$ is the ideal output of the output neuron k for the input vector $\underline{X}_i$.

3 Training of the RBF Network

3.1 Initialization

The initial weight values have been chosen according to the following relations

$$w(j,h)=\sum_{k=1}^{M}\frac{1}{Q_k}\sum_{i/\underline{X}_i\in cls_k}x_i(h),(j=1,...L,h=1,...N) \quad (6)$$

$$r(j,h)=\frac{1}{Q}\sum_{i/\underline{X}_i\in cls_k}(x_i(h)-w(j,h))^2,(j=1,..L;h=1,..N) \quad (7)$$

$$\lambda(k,j)=0,\ (k=1,\ldots,M;\ j=1,\ldots,L) \quad (8)$$

where Q_k is the number of input vectors belonging to the class cls_k.

3.2 Refining of the weight set {λ(k,j); k=1,...M; j=1,...L}

$$\lambda^{t+1}(k,j)=\lambda^t(k,j)-\Delta^t\lambda(k,j) \quad (9)$$

where t represents the step index of the training algorithm and

$$\Delta^{t+1}\lambda(k,j)=\eta\left(\frac{\partial E}{\partial\lambda(k,j)}\right)_{t+1}+\alpha\Delta^t\lambda(k,j),(\alpha\in(0,1)) \quad (10)$$

the parameter η is called the learning rate;

$$\frac{\partial E}{\partial\lambda(k,j)}=(-2)\sum_{i=1}^{Q}u_i(j)y_i(k)(1-y_i(k))(f_i(k)-y_i(k)) \quad (11)$$

3.3 Refining of the weight set {r(j,h); j=1,...L; h=1,...N}

$$r^{t+1}(j,h)=r^t(j,h)-\Delta^t r(j,h) \quad (12)$$

where

$$\Delta^{t+1}r(j,h)=\eta\left(\frac{\partial E}{\partial r(j,h)}\right)_{t+1}+\alpha\Delta^t r(j,h),(\alpha\in(0,1)) \quad (13)$$

and

$$\frac{\partial E}{\partial r(j,h)}=(-4)\sum_{k=1}^{M}\sum_{i=1}^{Q}u_i(j)\lambda(k,j)y_i(k)(1-y_i(k))(f_i(k)-y_i(k))\frac{[w(j,h)-x_i(h)]^2}{[r(j,h)]^3} \quad (14)$$

3.4 Refining of the weight set {w(j,h); j=1,...L; h=1,...N}

$$w^{t+1}(j,h) = w^{t}(j,h) - \Delta^{t} w(j,h) \tag{15}$$

where

$$\Delta^{t+1} w(j,h) = \eta \left(\frac{\partial E}{\partial w(j,h)} \right)_{t+1} + \alpha \Delta^{t} w(j,h), (\alpha \in (0,1)) \tag{16}$$

and

$$\frac{\partial E}{\partial w(j,h)} = 4 \sum_{k=1}^{M} \sum_{i=1}^{Q} u_i(j) \lambda(k,j) y_i(k)(1 - y_i(k))(f_i(k) - y_i(k)) \frac{[w(j,h) - x_i(h)]}{[r(j,h)]^2} \tag{17}$$

4 Experimental Results

Our processing sequence for road identification contains the following stages:

(α) reduction of the input image resolution (of the initial 256 level blue component of the real color picture) from 256 x 256 pixels to 32 x 32 pixels (by moving an averaging window of 8 x 8 pixels on the original 256 x 256 real image).

(β) neural processing using the above presented model of RBF neural network. The network architecture was characterized by: N= 1024 input neurons (corresponding to the processed picture of 64 x 64 pixels), L=5 or L=10 hidden units, and M=5 output units (corresponding to the number of 5 quantified output directions shown in Fig. 2: *sharp left, wide left, straight ahead, wide right, sharp right*).

4.1 Real road scenes

For training and testing the proposed neural model, we used two lots of real road scenes of 256 x 256 pixels: 50 images for training (10 pictures for each direction) and 50 images (different of the previous ones) for test. A few of the road images used in our experiment are shown in Fig. 3.

Using the values of training parameters $\alpha = 0.5$ and $\eta = 0.01$ for the RBF architecture 1024:5:5, one obtains a score of correct recognition of 100 % both for the estimation lot as well as for the test lot (Fig. 4).

The test phase (whose results are shown in Figs. 4, 5, and 6) consists of: (a) applying all the road images belonging to the training lot as well as those of the test lot to the input of the RBF net; (b) computing the maximum output (meaning to identify the road direction); (c) comparing the identified direction to the ideal one (meaning to evaluate the classification error).

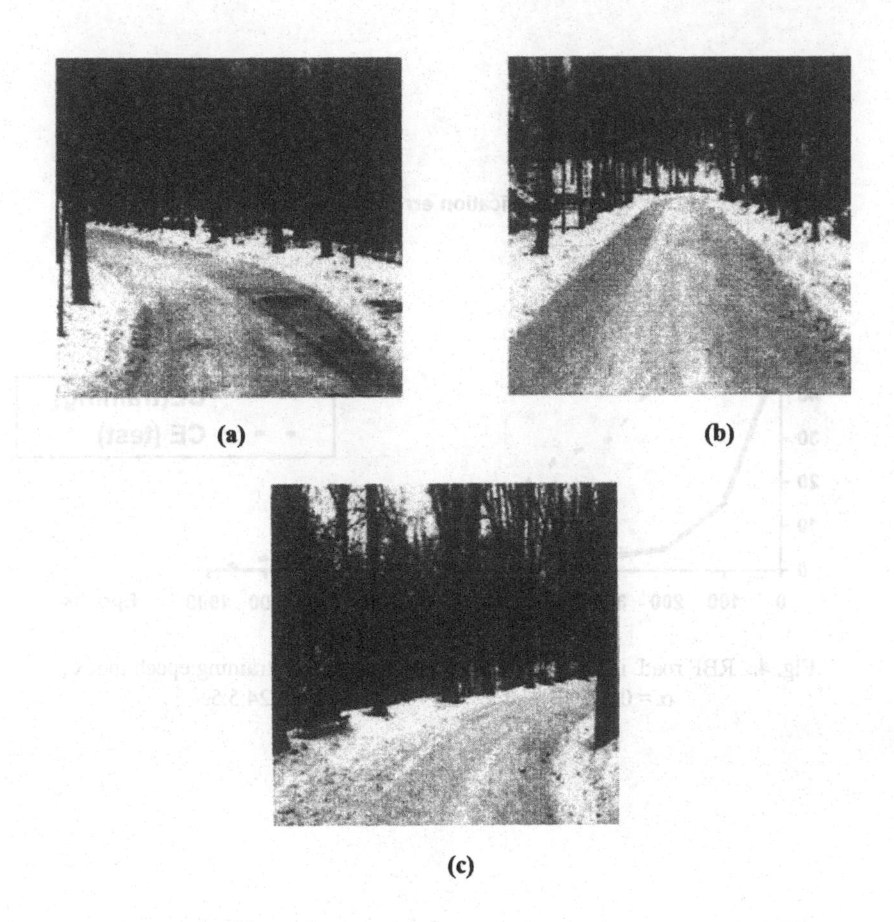

Fig. 3. Road images used for experimentation of the presented RBF model. (winter). (a) "sharp left"; (b) "straight ahead"; (c)"sharp right".

In Fig. 4, we can see that all the images of the training lot are correctly classified after 400 epochs , while the images of the test lot are classified without errors after 1000 epochs only. In Fig. 5, the classification error is shown for the training parametrs $\alpha = 0.1$ and $\eta = 0.1$; we remark that the results given in Fig. 4 are better than those given in Fig. 5, therefore, the values of parameters $\alpha = 0.5$ and $\eta = 0.01$ are optimal for our experiment. In Fig. 6, we can see the influence of the number of hidden neurons on classification performances (by choosing L= 10). The conclusion is the best results for our experiment correspond to the RBF architecture design used in Fig. 4 (RBF 1024:5:5).

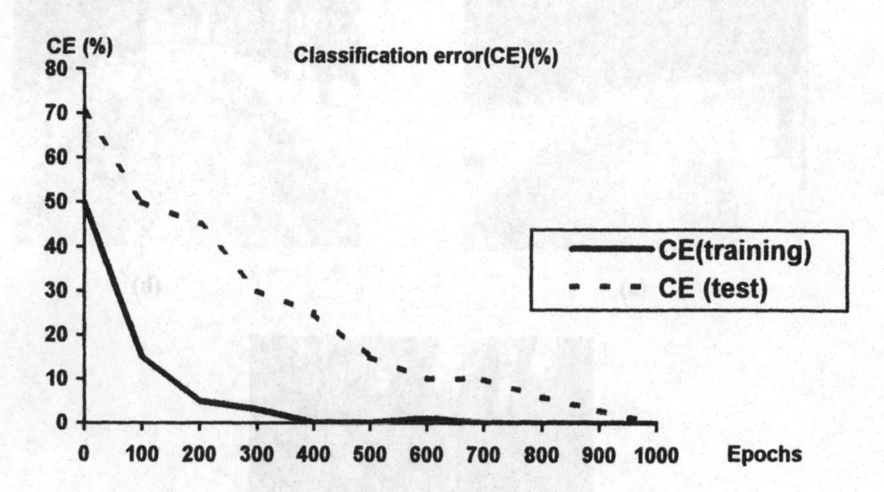

Fig. 4. RBF road identification error as a function of training epoch index ; $\alpha = 0.5$ and $\eta = 0.01$; RBF 1024:5:5.

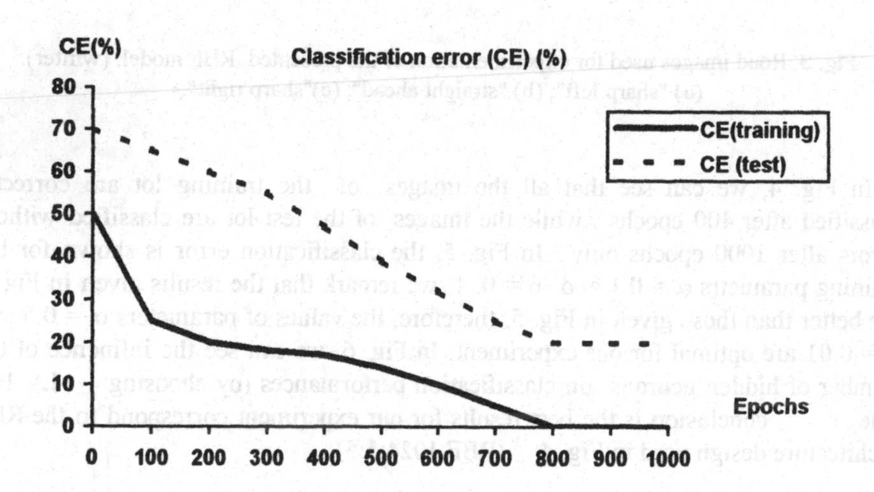

Fig. 5 . RBF road identification error as a function of training epoch index ; $\alpha = 0.1$ and $\eta = 0.1$; RBF 1024:10:5.

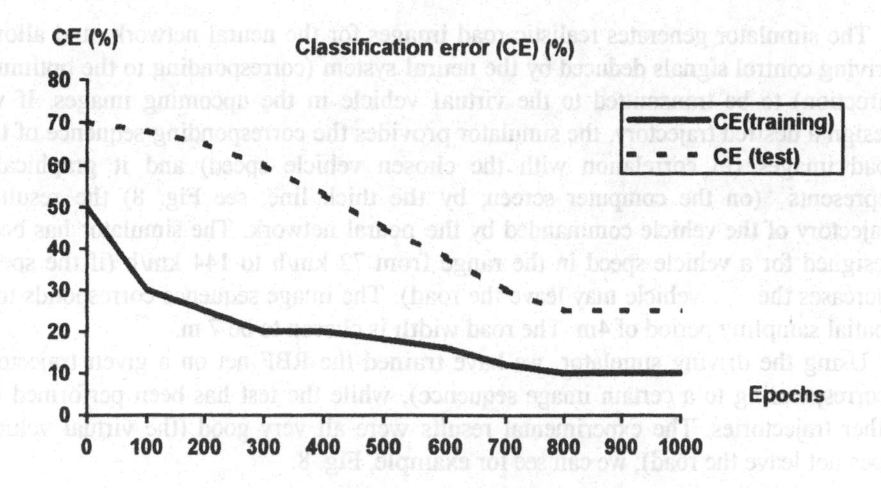

Fig. 6. RBF road identification error as a function of training epoch index ; $\alpha = 0.5$ and $\eta = 0.01$; RBF 1024:10:5.

4.2 Driving simulator

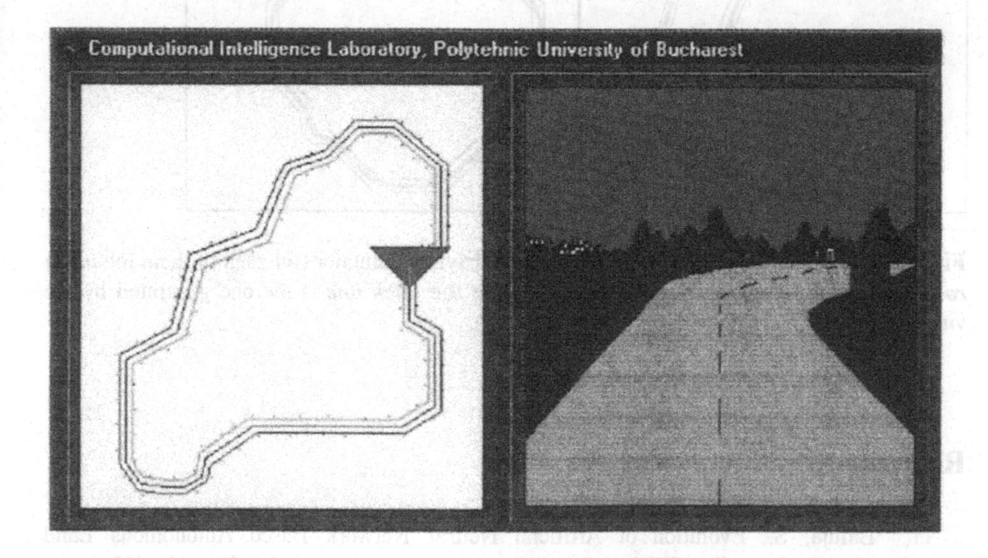

Fig. 7. Driving simulator for RBF net training and testing

We have also designed a driving simulator (performed in our Computational Intelligence Laboratory, Department of Applied Elećtronics and Information Engineering of the Polytechnic University of Bucharest) to evaluate the performances of the neural road follower (Fig. 7). The purposes of designing the simulator were to make its input and output interfaces similar to the real road following application.

The simulator generates realistic road images for the neural network, and allows driving control signals deduced by the neural system (corresponding to the optimum direction) to be transmitted to the virtual vehicle in the upcoming images. If we design a desired trajectory, the simulator provides the corresponding sequence of the road images (in correlation with the chosen vehicle speed) and it graphically represents (on the computer screen, by the thick line, see Fig. 8) the resulted trajectory of the vehicle commanded by the neural network. The simulator has been designed for a vehicle speed in the range from 72 km/h to 144 km/h (if the speed increases the vehicle may leave the road). The image sequence corresponds to a spatial sampling period of 4m. The road width is chosen to be 7 m.

Using the driving simulator, we have trained the RBF net on a given trajectory (corresponding to a certain image sequence), while the test has been performed on other trajectories. The experimental results were all very good (the virtual vehicle does not leave the road); we can see for example, Fig. 8.

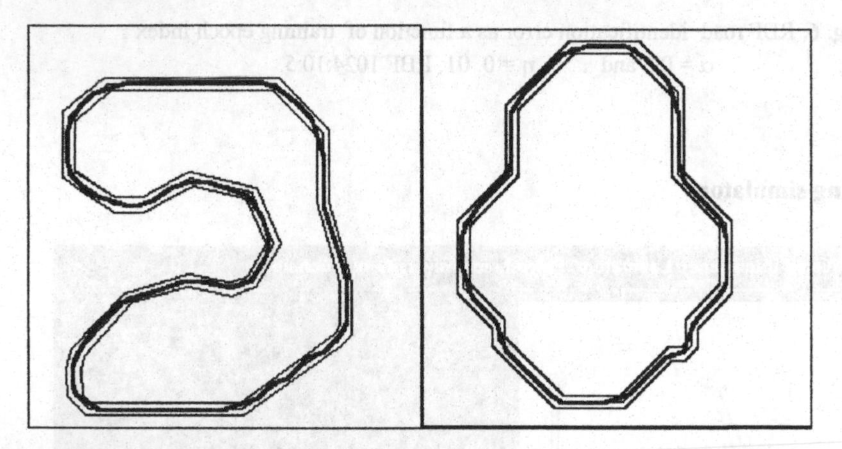

Fig. 8. Example of two trajectories designed on driving simulator (for each of them the *mean road line* is the desired trajectory while the *thick line* is the one generated by the virtual vehicle driven with the neural network).

References

1. Baluja, S.: Evolution of Artificial Neural Network Based Autonomous Land Vehicle Controller. IEEE Trans. Syst., Man, and Cybern. 3 (1996) 450-463
2. Del Bimbo, A., Landi, L., Santini, S.: Determination of Road Directions Using Feedback Neural Nets. Signal Processing. 1-2 (1993) 147-160.
3. Kuan, D., Philipps, G., Hsueh, A.- C. Autonomous Robotic Vehicle Road Following. IEEE Trans. Patt. Anal. Mach. Intell. 5 (1988) 648-658
4. Neagoe, V.: Intelligent Detection of FSK Signals Using a Refined Radial Basis Function. In: Fazekas, K. (ed.): Intelligent Terminals and Source and Channel Coding. HTE- Scientific Society for Communications, Budapest (1993) 201-209.

5. Neagoe, V., Lemahieu, I., Cristureanu, D., Hargalas, D., Constantinescu, B., Popescu, C.: A Parallel Image Processing for Optimum Direction Finding in Autonomous Vehicle Navigation. In: D'Hollander, E.H., Joubert, G. R., Peters, F. J.,. Trystam, T. (eds.): Parallel Computing: State of the Art and Perspectives. Advances in Parallel Computing, Vol. 11. Elsevier, Amsterdam New York Tokyo (1996). 681-684
6. Neagoe, V., Sabac, B.: A Connectionist Model For Road Identification in an Autonomous Vehicle Controller. In: Proceedings of the 27th International Conference of the Romanian Technical Military Academy . Section X. Technical Military Academy Publishing House, Bucharest (1997) 62-69
7. Pomerlau, D. A.: Efficient Training of Artificial Neural Networks for Autonomous Navigation. Neural Comp. 1 (1991) 88-97
8. Rosenblum, M., Davis, L. S.,: An Improved Radial Basis Function Network for Visual Autonomous Road Folowing. IEEE Trans. Neural Networks. 5 (1996) 1111-1120
9. Thorpe, C.,. Hebert, M. -H , Kanade, T., Shaffer S.-A: Vision and Navigation for the Carnegie-Mellon NAVLAB ", IEEE Trans. Patt. Anal. Mach. Intell. 3 (1988) 362-273

Appendix: Training Algorithm for the Proposed Refined RBF

Step 0: Let $\{X_i; i=1,\ldots,Q\}$ be the set of training input vectors and $\{f_i(k); i=1,\ldots,Q\}$ be the ideal outputs;
Step 1: Perform topology initialization: choose N, L, and M.
Step 2: Initialize neuron weights $\{\lambda(k,j), w(j,h), r(j,h); k=1,\ldots,M, j=1,\ldots,L, h=1,\ldots,N\}$ using relations (6), (7), and (8);
Step 3: Initialize training rate η;
Step 4: Compute the network outputs $\{y_i(k); k=1,\ldots,M, i=1,\ldots,Q\}$ by applying relation (4);
Step 5: Compute the cost function E (relation (5));
Step 6: Update the weights $\{\lambda(k,j); k=1,\ldots,M, j=1,\ldots,L\}$ using relations (9), (10), and (11));
Step 7: Compute the network outputs $\{y_i(k); k=1,\ldots,M, i=1,\ldots,Q\}$ (relation (4));
Step 8: Update the weights $\{r(j,h); j=1,\ldots,L, h=1,\ldots,N\}$ using relations (12), (13), and (14);
Step 9: Compute the network outputs $\{y_i(k); k=1,\ldots,M, i=1,\ldots,Q\}$ with relation (4);
Step 10: Update the weights $\{w(j,h); j=1,\ldots,5, h=1,\ldots,N\}$ (relations (15), (16), and (17)) ;
Step 11: Compute the network outputs $\{y_i(k); k=1,\ldots,M, i=1,\ldots,Q\}$ with relation (4);
Step 12: Compute the cost function E (relation (5));
Step 13: If $|E^{t+1}-E^t|<\varepsilon$ (where $t\in\mathbf{N}$, where t is the index of iteration and ε is a previously established threshold) and all the input vectors have been correctly classified, then STOP (**N** is the set of natural numbers; E^t is the cost function evaluated at the iteration t);
Step 14: If $|E^{t+1}-E^t|\geq\varepsilon$ and $t<t_{max}$, then go to Step 5.

A Neural Segmentation of Multispectral Satellite Images

Victor Neagoe[1] and Ionel Fratila[1]

[1]Polytechnic University of Bucharest, Department of Applied Electronics and Information Engineering, Splaiul Independentei 313, Bucharest 16, Romania 77206
Vic@gitprai.pub.ro

Abstract. A neural model for segmentation of multispectral satellite images is presented, implying the following processing stages: (a) feature extraction using the 2-dimensional discrete cosine transform (2-d DCT) applied on the image segment centered in the current pixel, for each frame of the spectral sequence; (b) a neural self-organizing map having as input the concatenation of the feature vectors assigned to the projections of the current pixel in all the image bands (computed in stage (a)). The software implementation of the model for multispectral satellite images SPOT leads to interesting experimental results.

1 Introduction

Satellite image segmentation is the process by which individual image pixels are grouped into partitions, according to some intrinsic properties of the image (gray levels, contrast, spectral values or textural properties). Purpose is to provide segmented images from remotely sensed multispectral data. One useful application is *vegetation identification and classification*. A second application is the *cloud segmentation*, representing a fundamental for analysis of satellite data. A third application is to use segmentation as a *preprocessing stage for image compression*.

By comparison with a previous neural model for satellite image segmentation performed by the same authors [3], the present paper brings the following news: (α) it uses the 2-d discrete cosine transform (DCT) (instead of the discrete Fourier transform (DFT)) of the elementary image segment centered in the current pixel and it applies the zonal filtering of the signal energy in the corresponding 2-d DCT space to extract the features for each pixel; (β) extension of the model to *multispectral satellite images*, namely the vector of characteristics of each pixel is obtained by concatenation of the corresponding features considered in all the spectral frames of the image.

2 Model Description

2.1 Feature Selection

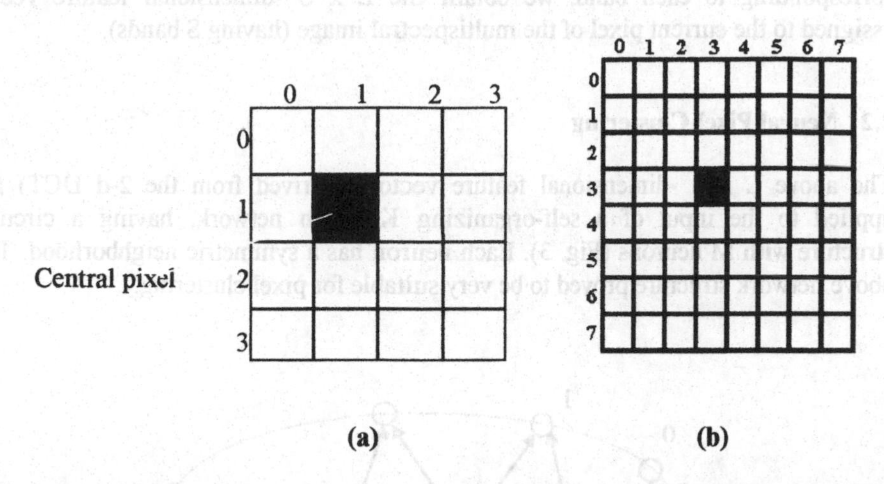

Fig. 1. Elementary image segment (of n x n pixels) centered in the current pixel to compute the 2-d DCT for the j-th band (j = 1,..., S). (a) n=4; (b) n=8.

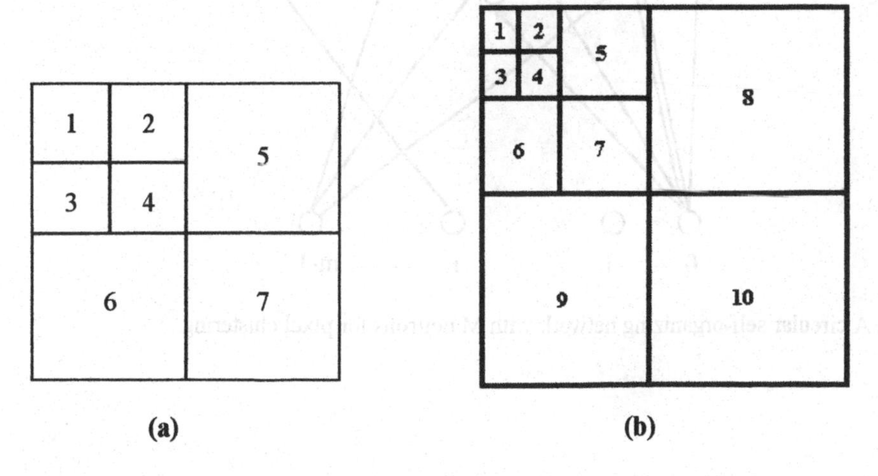

Fig. 2. Illustration of zonal filtering (L regions) in the 2-d DCT space. (a) N=4, L= 7; (b) n=8, L=10.

Consider the elementary square of n x n pixels having as a center the current pixel projected in the j-th spectral band (j=1,..., S) (see Fig. 1) and compute the 2-d discrete cosine transform (2-d DCT) of the corresponding image segment. Then, we compute the L zonal energies in the 2-d DCT space, by averaging the 2-d DCT coefficient absolute values (Fig. 2); thus, we obtain L features characterizing the current pixel in the j-th spectral band. By the concatenation of the features corresponding to each band, we obtain the L x S -dimensional feature vector assigned to the current pixel of the multispectral image (having S bands).

2.2 Neural Pixel Clustering

The above L x S -dimensional feature vectors (derived from the 2-d DCT) are applied to the input of a self-organizing Kohonen network, having a circular structure with M neurons (Fig. 3). Each neuron has a symmetric neighborhood. The above network structure proved to be very suitable for pixel clustering.

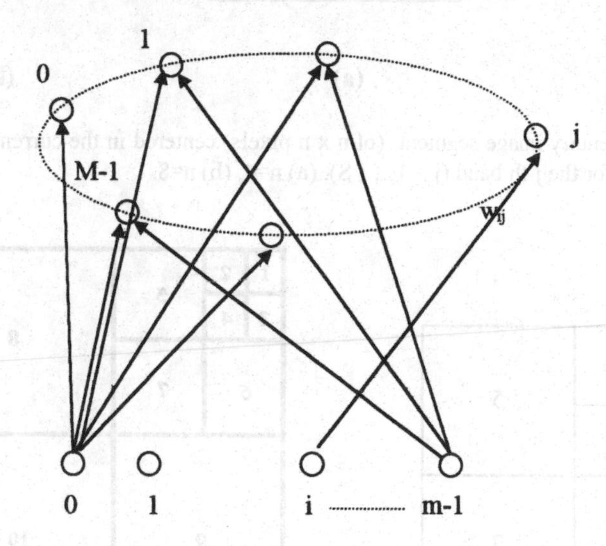

Fig. 3. A circular self-organizing network with M neurons for pixel clustering.

3 Experimental Results

3.1 Brodatz textures

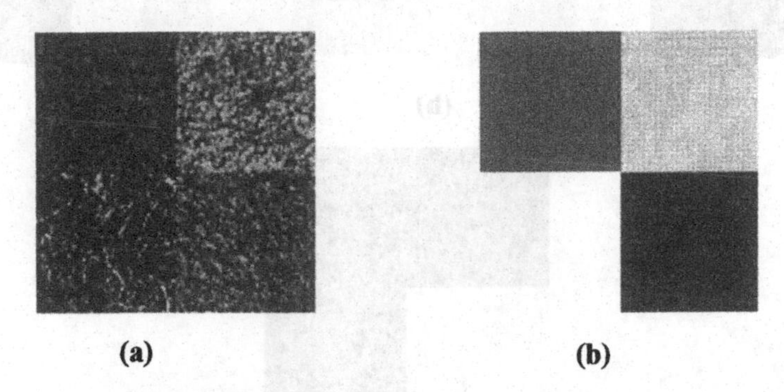

Fig. 4. (a) Original image of 256 x 256 pixels generated by concatenation of 4 natural Brodatz textures; (b) Segmentation of the image (a). The 2-d DCT has been applied on adjacent squares with the size of n=64 pixels. The model uses L=10 spectral features and M=4 neurons of the ring neural network.

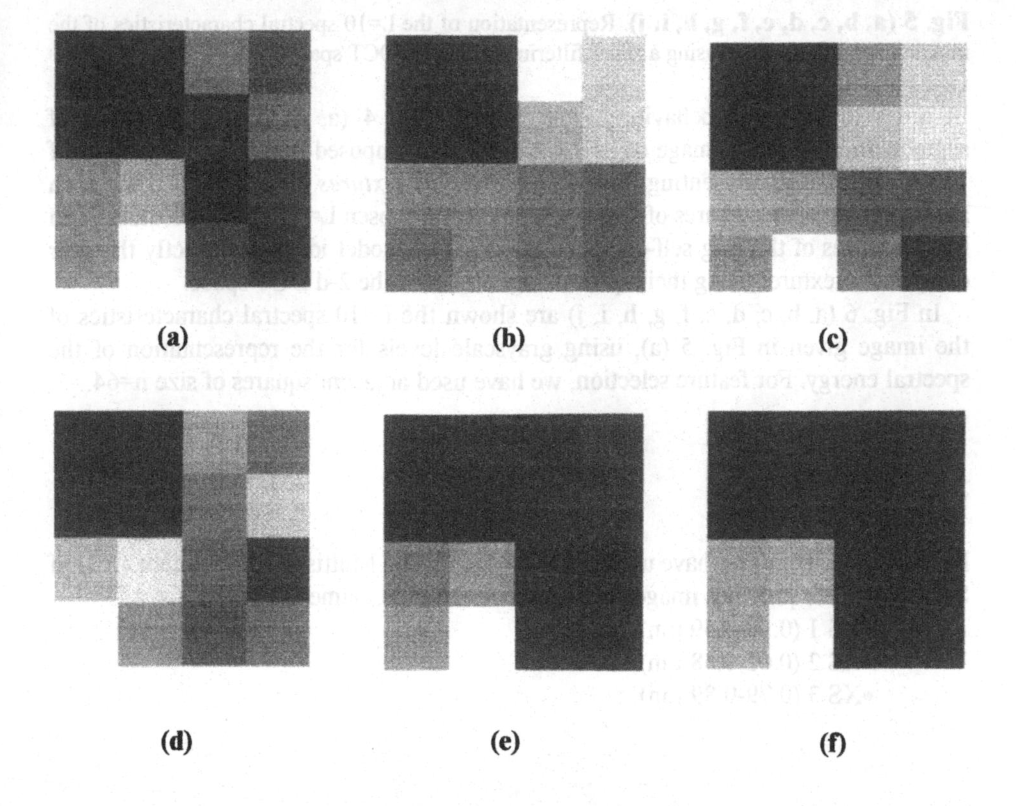

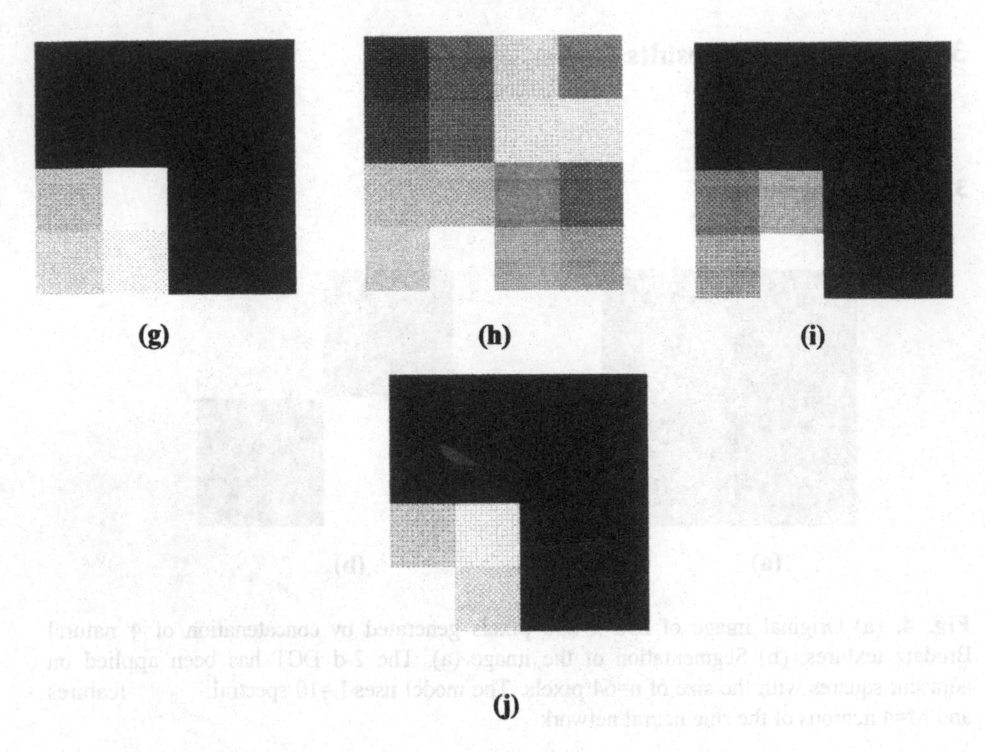

Fig. 5 (a, b, c, d, e, f, g, h, i, j). Representation of the L=10 spectral characteristics of the image given in Fig. 4 (a), using a zonal filtering in the 2-d DCT space.

In order to point out behavior of our model, Fig. 4 (a, b) shows the result of segmentation of a test image of 256 x 256 pixels composed of 4 image segments of 128 x 128 pixels representing four natural *Brodatz textures*. The 2-d DCT has been applied on adjacent squares of size n=64. We have chosen L=10 spectral features and M=4 neurons of the ring self-organizing map. The model identifies exactly the four component textures using their spectral signatures in the 2-d DCT space.

In Fig. 6 (a, b, c, d, e, f, g, h, i, j) are shown the L=10 spectral characteristics of the image given in Fig. 5 (a), using grayscale levels for the representation of the spectral energy. For feature selection, we have used adjacent squares of size n=64.

3.2 SPOT images

For our experiment, we have used SPOT images . The Multispectral Scanner (XS) of SPOT satellites provides images in S=3 spectral bands, namely:

- •XS 1 (0.50-0.59 μm);
- •XS 2 (0.61-0.68 μm);
- •XS 3 (0.79-0.89 μm).

Each pixel (corresponding to a zone of 20m x 20m) is represented with 8 bits/pixel/band.

Particularly, we used a SPOT image of 256 x 256 pixels projected in S=3 bands; our experimental frames are called XS0325_1.raw, XS0325_2.raw, and XS0325_3.raw (Fig. 6 (a, b, c)), and they correspond to a mountain area in the Switzerland.

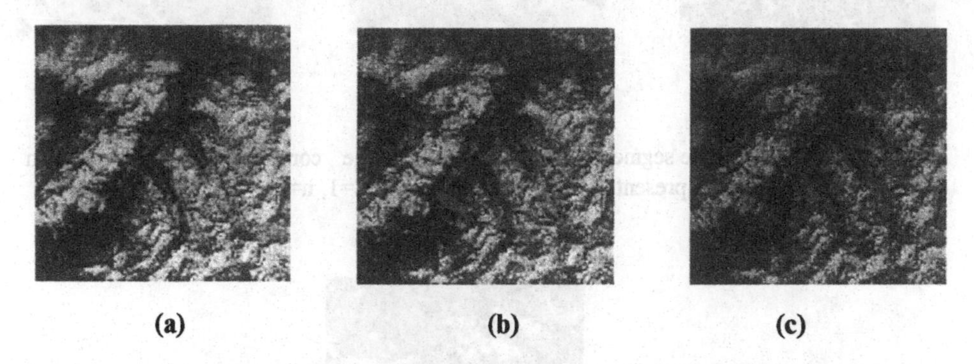

(a) (b) (c)

Fig. 6. Original multispectral SPOT image (S = 3 bands) . (a) XS0325_1.raw; (b) XS0325_2.raw; (c) XS0325_3.raw.

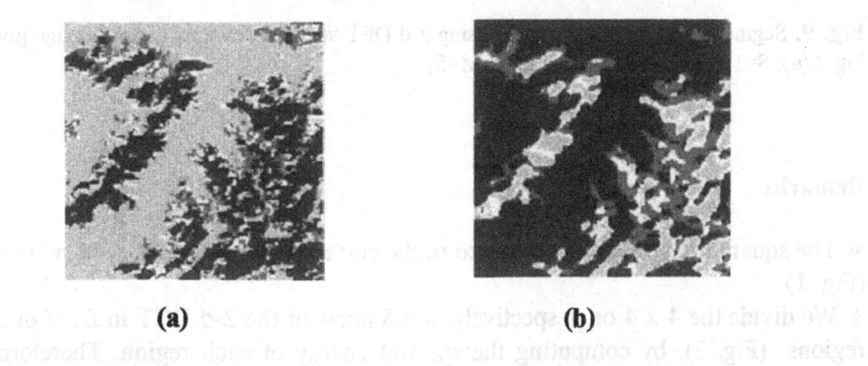

(a) (b)

Fig. 7. Segmentation of the multispectral satellite image given in Fig.6 (a,b,c) using the proposed 2-d DCT neural model. (a) (S=3, n=4, L=7, M=5); (b) (S=3, n=8, L=10, M=5).

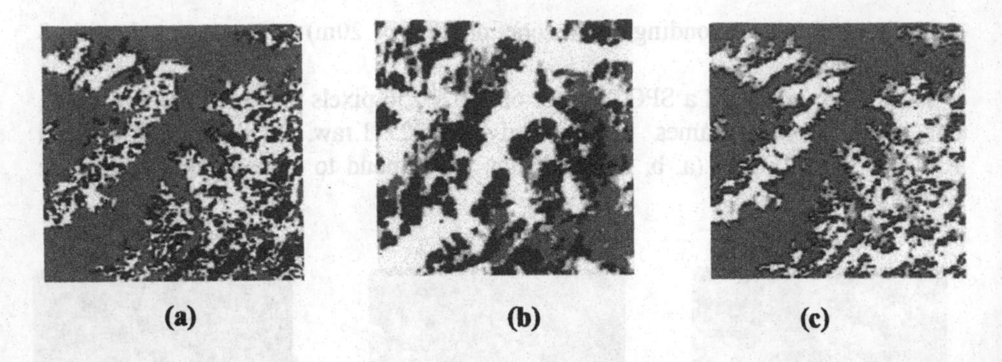

(a) (b) (c)

Fig. 8. (a, b, c). Separate segmentation of each band of the considered SPOT image given in Fig. 6(a, b, c) using the present 2-d DCT- neural model (S=1, n=4, L=7, M=5) .

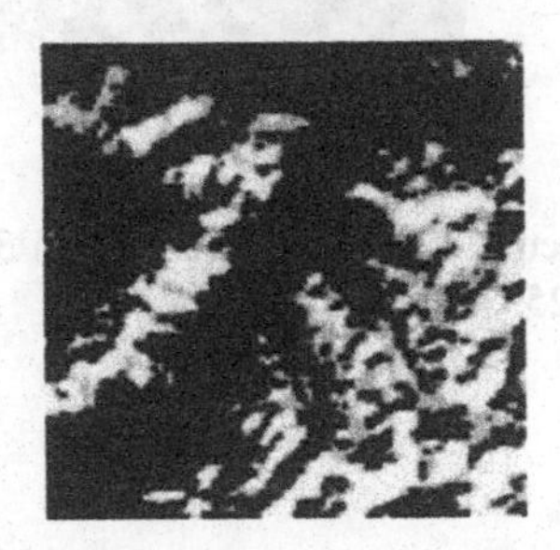

Fig. 9. Segmentation of the 1-st band using 2-d DFT spectral features (input image given in Fig. 6(a); S=1, n=5, L=4 DFT features, M=5).

Remarks

♦ The square image segment centered in the current pixel has the size of n=4 or n=8 (Fig. 1).

♦ We divide the 4 x 4 or, respectively, 8 x 8 areas of the 2-d DCT in *L =7* or *L=10* regions (Fig. 2), by computing the spectral energy of each region. Therefore, the feature vector assigned to each pixel has L x S = 21 components, or, respectively L x S=30 ones.

♦ The circular Kohonen network has *M=5* neurons, corresponding to the maximum number of clusters. According to the previous statements, the net has 21 or 30 inputs (Fig. 3).

♦ After segmentation, all the pixel belonging to the same cluster are represented by the same gray level (thus, the maximum number of gray levels is five). We can associate a pseudo-color to each of the above-mentioned levels of gray, in order to improve the visibility of the segmentation.

♦ During the training of the self-organizing map, the initial neighborhood of each neuron is the whole circular network (5 neurons). Then, the neighborhood radius decreases with 2 neurons (one to the left and one to the right) at every set of 10 iterations (one iteration corresponds to the processing of the whole picture). Thus, after 20 iterations, the neighborhood reduces to a single neuron.

♦ Using the input *multispectral* SPOT image , given in Fig. 6, and applying the proposed model, we obtain the segmented images given in Fig. 7 (a) (for S=3, $n=4$, $L=7$, M=5) and respectively in Fig. 7(b) (for S=3, $n=8$, $L=10$, M=5).

♦ For comparison, we give in Fig. 8 the segmented images obtained by the present method but using one band only (input frames given in Fig. 6 (a, b, c); $S=1$, n=4, L=7, M=5), namely the corresponding Kohonen network for pixel clustering has 7 inputs instead of 21. If we compare Fig. 8 with Fig. 7 (a), we remark that by processing a single band we can loose some useful information for image classification.

♦ Furthermore, we also give in Fig. 9, for comparison, the segmented image obtained by applying (for the first band, corresponding to the image of Fig. 6(a)) the previous model presented in [3], based on the 2-d DFT . We have used a window of 5 x 5 pixels, and computed L=4 DFT spectral features; for pixel clustering, we have chosen a self-organizing map with M=5 neurons. As a result of comparison of Fig. 8(a) and Fig. 9, we remark that the DCT seems to be more sensitive than the DFT for feature extraction.

♦ The advantages of the proposed model are the following: (a) the 2-d DCT has very good approximation and compression properties (*high quality feature selection*); (b) by concatenation of the L-dimensional spectral feature vectors corresponding to the projections of a given pixel in S bands, we obtain an $L \times S$-dimensional input vector for clustering; since then, the network can provide global *decisions*, taking into consideration the information regarding all the spectral bands; (c) the *ring structure* of the self-organizing map proved to be very suitable for pixel clustering.

References

1. Carpenter, G., Gjaja, M., Gopal, S., and Woodcock, C.: ART neural networks for remote sensing: vegetation classification from LANDSAT TM and terrain data. IEEE Trans. Geoscience and Remote Sensing. 2 (1997) 308-325
2. Neagoe, V.: A circular Kohonen network for image vector quantization. In: D'Hollander, E.H., Joubert, G. R., Peters, F. J.,. Trystam, T. (eds.): Parallel Computing: State of the Art and Perspectives. Advances in Parallel Computing, Vol. 11. Elsevier, Amsterdam New York Tokyo (1996) 677-680
3. Neagoe, V., Fratila, I.: Satellite image segmentation using a circular self-organizing map. In: Proceedings of the International Symposium COMMUNICATIONS'96. Technical Military Academy Publishing House, Bucharest (1996) 442-447
4. Yhann, S., Simpson, J.: Application of neural networks to AVHRR cloud segmentation. IEEE Trans. Geoscience and Remote Sensing. 3 (1995) 590-604

Design of a Traffic Junction Controller Using Classifier System and Fuzzy Logic

Y. J. Cao, N. Ireson, L. Bull and R. Miles

Intelligent Computer Systems Centre
Faculty of Computer Studies and Mathematics
University of the West of England, Bristol, BS16 1QY, UK

Abstract. Traffic control in large cities is a difficult and non-trivial optimization problem. Most of the automated urban traffic control systems are based on deterministic algorithms and have a multi-level architecture; to achieve global optimality, hierarchical control algorithms are generally employed. However, these algorithms are often slow to react to varying conditions, and it has been recognized that incorporating computational intelligence into the lower levels can remove some burdens of algorithm calculation and decision making from higher levels. An alternative approach is to use a fully distributed architecture in which there is effectively only one (low) level of control. Such systems are aimed at increasing the response time of the controller and, again, these often incorporate computational intelligence techniques. This paper presents preliminary work into designing an intelligent local controller primarily for distributed traffic control systems. The idea is to use a classifier system with a fuzzy rule representation to determine useful junction control rules within the dynamic environment.

1 Introduction

Traffic control in large cities is a difficult and non-trivial problem. The intensively increasing number of vehicles and passengers often causes delays, congestion, accidents and other unwanted events that influence negatively many social activities. These difficulties will be solved by expanding networks of roads, constructing subways, changing drivers' habits, and so on. However, a reasonable operating traffic signal system could lower expenses within a shorter time. To this end, automated urban traffic control systems (AUTCS), such as TRANSYT[1], SCATS[2], LVA[3] and SCOOT[4], have been widely applied to optimize traffic signal timing plans to facilitate vehicle movement. These systems are usually based on sophisticated equipment and possess broad functional capabilities, for instance automated data gathering, analysis and prediction of transport situations; automated decision making and control implementation, etc. [5] ~ [9].

The policies used in such systems divide broadly into two categories: the fixed-time systems where traffic plans are generated off-line and applied on-line, and on-line systems where traffic plans are generated on-line and are applied directly to the traffic. Both methods have their advantages and disadvantages[12]. However, the common feature to all methods is that the objective function of the algorithm is based on minimizing the total delay within a network, although some use a combination of delay and stops. There are a number of varying implementations which apply these methods, however, there are still many open problems, caused by the following circumstances:

- most AUTCS have a centralized structure, i.e. information gathering and processing, as well as control computations, are carried out in a centralized manner, in this case efficiency is decreased due to the large volume and the heterogeneous character of information [5].
- distributed AUTCS also possess considerable drawbacks, caused by the inefficient accounting of interactions between subsystems and the complex communication structure [6].
- most of the existing (centralized and distributed) AUTCS operate by means of entirely quantitative algorithms that do not reflect the qualitative aspects of transport process [7].

The above considerations suggest that it is desirable to use new techniques for solving transport problems in large cities. These techniques are usually based on artificial intelligence principles and sophisticated computing devices, operating with large memory, at a high speed and in parallel. They are usually associated with so-called 'distributed intelligence systems', or 'distributed intelligence control systems' (DICS) in the case of control problems [10, 11]. The general approach is based on the decomposition of the system into subsystems, which is often presented as a hierarchical multilevel control structure. To achieve global optimality, hierarchical control algorithms are generally employed. However, these algorithms have a slow speed of reaction and it has been recognized that incorporating some computational intelligence into lower levels can remove some burdens of algorithm calculation and decision making from higher levels. Alternatively, a fully distributed architecture can be used in which each subsystem is solely responsible for one aspect of the system and where a coherent global control plan emerges from the interactions of the subsystems; no hierarchical structure is included. Such an approach is aimed at increasing the speed of response of the local controller to changes in the environment. In both cases there is a need to apply intelligent algorithms to designing efficient local controllers. This paper is devoted to designing an intelligent local traffic controller primarily for a distributed architecture. The idea is to use a classifier system [13], with both evolutionary and reinforcement learning, to determine appropriate control rules within the dynamic traffic environment.

2 Urban Traffic Control and Previous Work

The overall goal of a traffic control system is to design the hardware configuration and implement a software control algorithm for the system, with the flexibility to select a suitable criterion according to traffic conditions. The four criteria are minimizing: delays, stops, fuel consumption and exhaust emission rate. Within an urban traffic network, speed is restricted, and the fuel consumption and exhaust emission rate depend greatly on traffic conditions and on the timing of the signals in the network. Computer-controlled hardware supported by a software algorithm is proposed to achieve the overall objective function in its many forms.

There is a growing body of work concerned with the use of adaptive computing techniques for the control and modelling of traffic junctions, including fuzzy logic [14] $\sim$ [17], neural networks [18], and evolutionary algorithms [19] $\sim$ [22]. Pappis and Mamdani's fuzzy logic controller (FLC) showed good performance in terms of the average delay at an intersection of two one-way roads with dynamic traffic flow rates, compared with a fixed-cycle traffic controller [14]. They used 25 control rules included in one fuzzy rule-set. In general, the fuzzy traffic controller (FTC) determines the the extension time of the green phase, with the fuzzy input variables of; the number of vehicles in the green approaches (= arrival) and the number of vehicles during the red signal period (= queue). These methods consider only the arrival and queue values of an intersection when the controllers are operating, but do not consider the total number of vehicles entering an intersection from other intersections by the second (= volume). Therefore, these controllers are not suitable for real intersections with variable flow rates. To solve this problem, Kim proposed an FLC using different control rules and different maximum extension time according to traffic volume, and consequently vehicles could flow smoothly at an intersection [15]. A similar approach was applied by Ho, who used nine fuzzy rule-sets according to arrival and queue flow rates [16]. Conventional FLCs use membership functions and control laws generated by human operators. However, this approach does not guarantee the optimal solution in fuzzy system design. Kim, *et al.* used a genetic algorithm (GA) [23] to tune the membership functions for the terms of each fuzzy variable [17].

Montana and Czerwinski [19] proposed a mechanism to control the whole network of junctions using genetic programming [25]. They evolved mobile "creatures" represented as rooted trees which return true or false, based on whether or not the creature wished to alter the traffic signal it has just examined. Mikami and Kakazu used a combination of local learning by a stochastic reinforcement learning method with a global search via a genetic algorithm. The reinforcement learning was intended to optimize the traffic flow around each crossroad, while the genetic algorithm was intended to introduce a global optimization criterion to each of the local learning processes [20, 21]. Escazut and Fogarty proposed an approach to generate a rule for each junction using classifier systems in biologically inspired configurations [22].

3 Classifier Systems and Fuzzy Logic

There are two alternative approaches to classifier systems, named as Michigan [13] and Pittsburgh [24] approaches. Some research work has combined both classifier systems with fuzzy logic. Valenzuela-Rendon proposed a fuzzy Michigan-style classifier system, which allows for inputs, outputs, and internal variables to take continuous values over given ranges. The fuzzy classifier system learns by creating fuzzy rules which relate the values of the input variables to internal or output variables. It has credit assignment and conflict resolution mechanisms which reassemble those of common classifier systems, with a fuzzy nature. The fuzzy classifier system employs a genetic algorithm to evolve adequate fuzzy rules [26]. Carse and Fogarty proposed a fuzzy classifier system using the Pittsburgh model, in which genetic operations and fitness assignment apply to complete rule sets, rather than to individual rules, thus overcoming the problem of conflicting individual and collective interests of the classifiers. The fuzzy classifier system dynamically adjusts both membership functions and fuzzy relations [27]. The work presented in this paper is different since it does not use fuzzy rules but generates the rules using a fuzzy coding strategy. In order to make our exposition self-contained, we introduce Michigan-style classifier systems and fuzzy logic briefly in this section.

3.1 Classifier systems

A classifier system is a learning system in which a set (population) of condition-action rules called *classifiers* compete to control the system and gain credit based on the system's receipt of reinforcement from the environment. A classifier's cumulative credit, termed *strength*, determines its influence in the control competition and in an evolutionary process using a genetic algorithm in which new, plausibly better, classifiers are generated from strong existing ones, and weak classifiers are discarded.

A classifier c is a condition-action pair

$$c =< \text{condition} >:< \text{action} >,$$

with the interpretation of the following decision rule: if a current observed state matches the condition, then execute the action. The condition is a string of characters from the ternary alphabet { 0, 1, # }, where # acts as a wildcard allowing generalization. The action is represented by a binary string and both conditions and actions are initialized randomly. The real-valued *strength* of a classifier is estimated in terms of rewards obtained according to a payoff function. Action selection is implemented by a competition mechanism, where a strength proportionate selection method is usually used. To modify classifier strengths, the given *credit assignment* algorithm is used, e.g. the Bucket brigade [13].

To create new classifiers a standard GA is applied, with three basic genetic operators: selection, crossover and mutation. The GA is invoked periodically and each time it replaces low strength classifiers with the offspring of the selected fitter ones (the reader is referred to [13] for full details).

3.2 Fuzzy logic

Fuzzy sets allow the possibility of degrees of membership. That is, an element might be assigned a set membership value between 0 and 1 (inclusive). For example, given the fuzzy set "long queue", we may speak of a particular queue being a member of this set to *degree* 0.8. This would be a long queue, but not the longest queue imaginable. The function which assigns this value is called the *membership function* associate with the fuzzy set. Fuzzy membership functions are the mechanism through which the fuzzy system interfaces with the outside world. A typical choice for a fuzzy membership function is a piecewise linear trapezoidal function.

A fuzzy system contains fuzzy sets defined for the input variables, output values and a set of fuzzy rules defined. A fuzzy rule is an "*if* condition *then* action" expression in which the conditions and the actions are fuzzy sets over given variables. Fuzzy rules are also called linguistic rules because they represent the way in which people usually formulate their knowledge about a given process. There are several approaches for obtaining "defuzzified" output from a fuzzy system [28]. Let us assume the output variables have four fuzzy sets associate with it: "ZE" (zero), "PS" (positive small), "PM" (positive medium), and "PL" (positive large). So we will assume that ZE, PS, PM, and PL represent specific numerical values. A fairly simple method is to calculate the nonzero activated weights, say w_1, w_2, w_3 and w_4, and then we compute the system output as the following:

$$\text{Output} = \frac{w_1 \cdot \text{NE} + w_2 \cdot \text{PS} + w_3 \cdot \text{PM} + w_4 \cdot \text{PL}}{\sum_{i=1}^{4} w_i} \tag{1}$$

With more general output fuzzy sets, determination of the defuzzified output involves computation of centroid values of regions defined by overlapping membership functions [28].

4 Traffic Signal Control Using Classifier Systems

The goal of intelligent traffic controllers is to diminish congestion caused by the stop signs of coordinated traffic lights when traffic volume is light, and pass maximum traffic flows through intersections in the case of heavy traffic volume. The classifier system based controller (CSC) designed in this paper will employ both evolutionary and reinforcement learning to determine useful control rules to be applied at a given traffic intersection where the traffic volume dynamically changes, so that vehicles can flow at an intersection smoothly. The block diagram of the proposed CSC is shown in Figure 1.

The classifier system employed is a version of Wilson's "zeroth-level" system (ZCS) [29]. ZCS is a Michigan-style classifier system, without internal memory. It is noted that a number of slight modifications have been made to the system presented in [29], including the use of a niche-based GA. The reader is referred to [30] for full details.

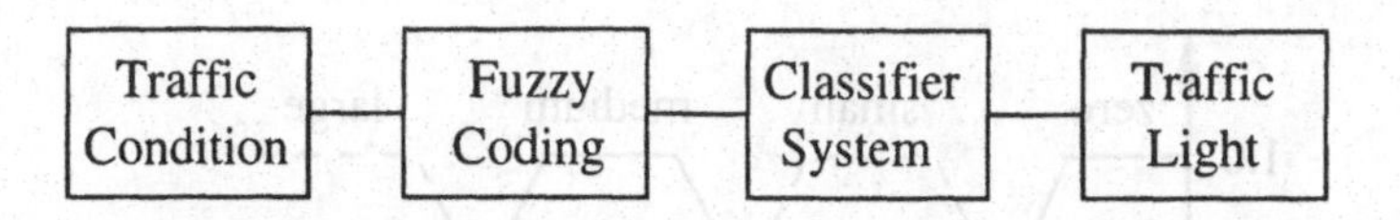

Figure 1: The block diagram of the proposed CSC

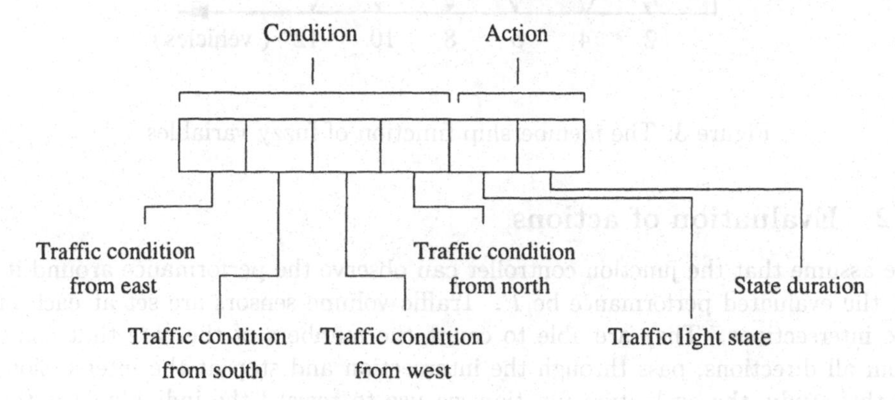

Figure 2: Structure of the classifier system rules

In order to avoid the genetic algorithm manipulating unnecessarily long rules, we extend the binary string representation in ZCS to a more general representation, which uses 0 to L ($L < 10$) for each variable (bit) position instead of the binary code. This reduces the string length significantly and appears to benefit multiple variable problems. For these hybrid strings, mutation in the GA is performed by changing an allele to a randomly determined number between 0 and L other than itself [32]. The design details of the CSC used in this paper, including the use of a Fuzzy representation, is described in the next section.

4.1 Individuals

The classifiers have the representation shown in Figure 2. The condition part of each classifier consists of four bits, which reflects the scalar level of queue length from each direction. In this application, the scalar level is set to 4, which ranges from 0 to 3, corresponding to the four linguistic variables, {zero, small, medium, large }. The action part indicates the required state of the signal and the period of duration. For instance, the rule 1302:14 says that if the queue from directions east and west are small (1) and zero (0), but the queue from directions south and north are large (3) and medium (2), then the traffic light stays green vertically (1) for 4 seconds (4). The membership function of this variable is given in Figure 3. The fuzzy output is calculated according to equation (1).

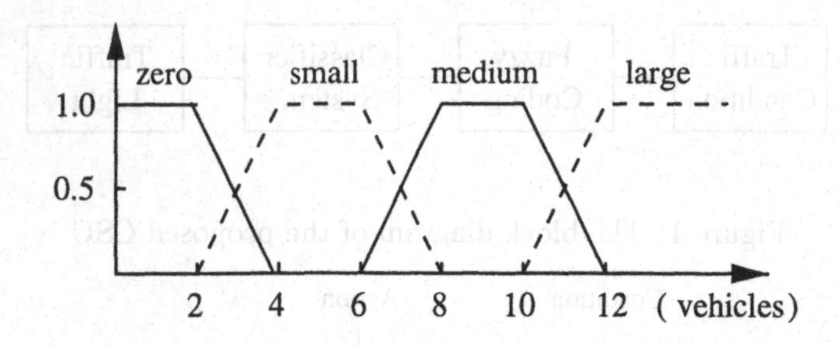

Figure 3: The membership function of fuzzy variables

4.2 Evaluation of actions

We assume that the junction controller can observe the performance around it, let the evaluated performance be P. Traffic volume sensors are set at each of the intersections. They are able to count the numbers of the cars that come from all directions, pass through the intersection and stop at the intersection. In this study, the evaluation function we use to reward the individuals is the average queue at the specific junction. Let q_i denote the queue length from direction i at the intersection ($i = 1, 2, 3, 4$), then the evaluation function is: $f = \frac{1}{4}\sum_{i=1}^{4} q_i$. We thus attempted to minimize this measure. Let us identify the k-th cycle by a subscript k, then f_k for the cycle k is calculated by observing the sensor from the beginning of the k-th cycle to the end of this cycle. Thus, the evaluated performance of the action performed at the k-th cycle is computed as $P_k = f_{k-1} - f_k$. Specifically, if $P_k > 0$, the matched classifiers containing the performed action should be rewarded, otherwise penalized.

4.3 Reinforcement learning

After the CSC has produced an action, the environment judges the output, and accordingly, gives payoff in the following manner:

- Rewards: The objective of the local signal controller is to minimize the average queue length, f_i. We have found the performance-based rewards are helpful in the environments we used in our experiments. The reward scheme we used was $r_i = \gamma P_i$, where γ is the reward constant and chosen to be 150 in the simulation.
- Punishments: We use punishments (i.e., negative rewards). We found the the use of appropriate punishments results in improved performance (in a fixed number of cycles), at least in the environments used in our experiments. We also found that large punishments could lead to instability of the classifiers and slow convergences of the rules. The appropriate punishments should be 30 percent of normal rewards.

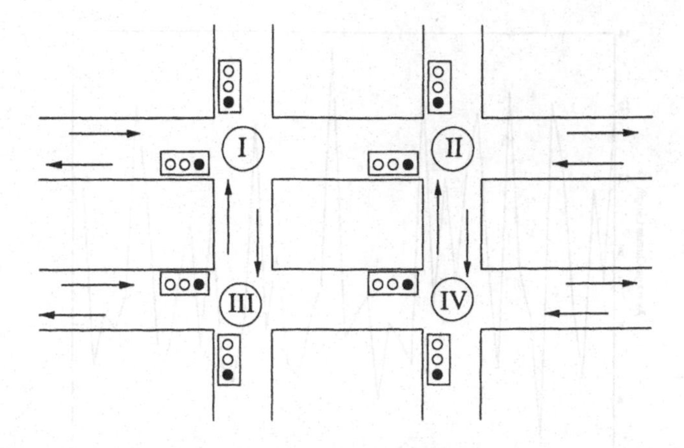

Figure 4: The simulated traffic environment

5 Simulation Results

Since developing an accurate traffic simulation is a difficult problem, as well as building an appropriate traffic controller, and the main objective of this simulation is to test whether the proposed classifier system controller works under a traffic environment or not, we developed a simplified traffic simulator, which is similar to the one used in [21]. The simulator is composed of four four-armed junctions and squared roads, which is shown in Figure 4. Each end of a road is assumed to be connected to external traffic, and cars are assumed to arrive at those roads according to a Poisson distribution. Each intersection has two "complementary" signals: when the horizontal signal is red, the vertical signal is green and *vice versa*. Each of the cars attempts to attain the same maximum speed. When a car passes an intersection, it changes its direction according to the probability associated with that intersection. Specifically, let d_i, $i = 1, 2, 3$, be the next directions for a car, that is, $\{d_i\} = \{$ right, forward, left $\}$. At each of the intersections, the probabilities $\{p_{d_i}\}$ are previously given, where p_{d_i} corresponds to the probability of selecting an action d_i for the car passing through the intersection. Roads are not endless, thus only a limited number of cars is allowed to be on the road at a given time. If a car reaches the end of the road, then the car is simply removed from the simulation, and another car is generated, entering on a randomly selected road.

Two types of junction controllers were used to control junction I in Figure 4 for the comparison: random controllers and our classifier system based controller. The random controller determines whether to change its phase or not randomly at 50% of probability and the duration of the state randomly between 1 to 6 seconds. As the major task is to test whether the proposed CSC can learn some good rules at the traffic junction, the other three junctions (II, III and IV in Figure 4) were controlled by random controllers here. The parameters used for the CSC were the same as those in [30] except that fewer rules (100) were

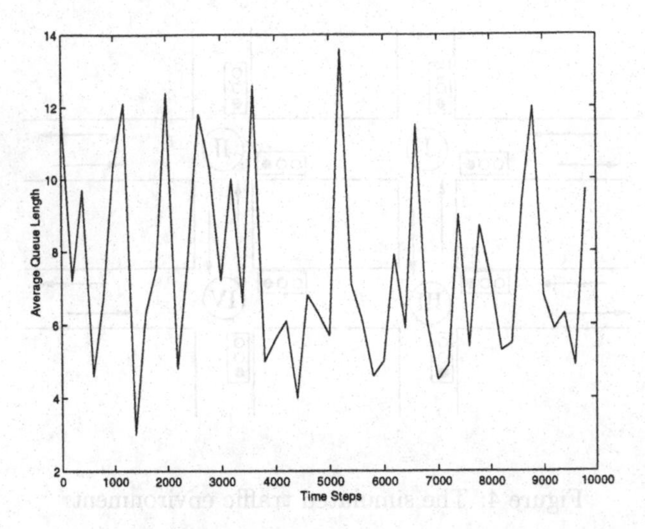

Figure 5: Performance of a random traffic junction controller

used to reduce the time taken to manipulate the rule-base and a slightly higher mutation rate was used due to the change in the number of variable values (0.05 per variable).

Experiments were carried out for three different types of traffic conditions. In these simulations, the mean arrival rates for the cars are the same but the number of cars in the area is limited to 30, 50, and 70, corresponding to a sparse, medium, and crowded traffic condition. In all cases, the CSC is found to learn how to reduce the average queue length at the specific junction. For example, Figure 5 and Figure 6 show the average performances of the random controller and CSC respectively over 10 runs in the crowded case. It can be seen that the CSC consistently reduces the average queue length over 10,000 iterations whilst the random controller's junction queue length continues to oscillate.

6 Conclusion and Future Work

In this paper we have described the application of a learning classifier system scheme to generate the effective control of a traffic junction. This preliminary work on the classifier system based controller needs, of course, a number of extensions. But these results are encouraging since we have shown that the classifier system, with both evolutionary and reinforcement learning, can determine useful control rules within a dynamic environment, and thus improve the traffic conditions.

We are currently extending this work in a number of directions, particularly examining ways of improving the use of classifier systems in distributed problem solving domains, i.e. multi-agent systems (e.g. after [31])

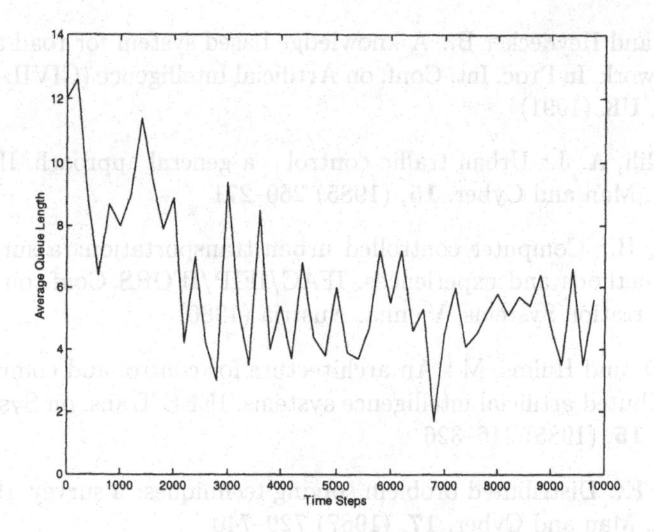

Figure 6: Performance of the classifier system based junction controller

7 Acknowledgment

This work was carried out as part of the ESPRIT Framework V Vintage project (ESPRIT 25.569).

References

[1] Robertson, D. I.: TRANSYT– A traffic network study tool. Transport and Research Laboratory, Crowthorne, England (1969)

[2] Luk, J. Y., Sims, A. G. and Lowrie, P. R.: SCATS application and field comparison with TRANSYT optimized fixed time system. In Proc. IEE Int. Conf. Road Traffic Signalling, London (1982)

[3] Lowrie, P. R.: The Sydney coordinated adaptive traffic system. In Proc. IEE Int. Conf. Road Traffic Signalling, London (1982)

[4] Hunt, P. B., Robertson, D. I., Bretherton, R. D. and Winston, R. I.: SCOOT–A traffic responsive method of co-ordinating traffic signals. Transport and Research Laboratory, Crowthorne, England (1982)

[5] Scemama, G.: Traffic control practices in urban areas. Ann. Rev. Report of the Natl Res. Inst. on Transport and Safety. Paris, France (1990)

[6] Barriere, J., Farges, J. and Henry, J.: Decentralization vs hierarchy in optimal traffic control. IFAC/IFIP/IFORS Conf. on Control in Transportation Systems, Vienna, Austria (1986)

[7] Wu, J. and Heydecker B.: A knowledge based system for road accident remedial work. In Proc. Int. Conf. on Artificial Intelligence (CIVIL-COMP91), Oxford, UK (1991)

[8] Al-Khalili, A. J.: Urban traffic control – a general approach. IEEE Trans. on Syst. Man and Cyber. **15**, (1985) 260–271

[9] Strobel, H.: Computer controlled urban transportation: a survey of concepts, methods and experiences. IFAC/IFIP/IFORS Conf. on Control in Transportation Systems, Vienna, Austria (1986)

[10] Yang, D. and Huhns, M.: An architecture for control and communications in distributed artificial intelligence systems. IEEE Trans. on Syst. Man and Cyber., **15**, (1985) 316–326

[11] Decker, K.: Distributed problem solving techniques: a survey. IEEE Trans. on Syst. Man and Cyber. **17**, (1987) 729–740

[12] Robertson, D. I.: Traffic models and optimum strategies of control: a survey. In Proc. of Int. Symp. Traffic Control Systems. University of California, (1979) 262–288

[13] Holland, J. H., Holyoak, K. J., Nisbett, R. E. and Thagard, P. R.: Induction: Processes of Inference, Learning and Discovery. MIT Press, Cambridge, MA (1986)

[14] Pappis, C. P. and Mamdani, E. H.: A fuzzy logic controller for a traffic junction. IEEE Trans. on Syst. Man and Cyber. **7**, (1977) 707–717

[15] Kim, J.: A fuzzy logic control simulator for adaptive traffic management. In Proc. IEEE Int. Conf. on Fuzzy Systems, Barcelona, (1997) 1519–1524

[16] Ho, T. K.: Fuzzy logic traffic control at a road junction with time-varying flow rates. IEE Electronics Letters, **32**, (1996) 1625–1626

[17] Kim, J. W., Kim, B. M. and Kim, J. Y.: Genetic algorithm simulation approach to determine membership functions of fuzzy traffic controller. IEE Electronics Letters, **34**, (1998) 1982–1983

[18] Ledoux, C.: Urban Traffic flow model integrating neural networks. Transportation Research, Part B: Emerging Technologies, **5**, (1997) 287–300

[19] Montana, D. J. and Czerwinski, S.: Evolving control laws for a network of traffic signals. Proc. of 1st Annual Conf. on Genetic Programming, (1996) 333–338

[20] Mikami, S. and Kakazu, K.: Self-organized control of traffic signals genetic reinforcement learning. Proceedings of the IEEE Intelligent Vehicles Symposium, (1993) 113–118

[21] Mikami, S. and Kakazu, K.: Genetic reinforcement learning for cooperative traffic signal control. Proceedings of the IEEE World Congress on Computational Intelligence, (1994) 223–229

[22] Escazut, C. and Fogarty, T. C.: Coevolving classifier systems to control traffic signals. In Koza, J. R (ed): Late breaking papers at the Genetic Programming 1997 Conference, Stanford University, (1997) 51–56

[23] Holland, J. H.: Adaptation in Natural and Artificial Systems. MIT Press, Cambridge, MA (1992)

[24] Smith, S.: A Learning System Based on Genetic Algorithms, PhD Thesis, University of Pittsburgh (1980)

[25] Koza, J. R: Genetic Programming. MIT Press, Cambridge, MA (1992)

[26] Valenzuela-Rendon, M.: The fuzzy classifier system: motivation and first results. In Schwefel, H. P. and Manner, R. (eds): Parallel Problem Solving from Nature I, Springer-Verlag, (1990) 338–342

[27] Carse B. and Fogarty, T. C.: A fuzzy classifier system using the Pittsburgh approach. In Davidor, Y., Schwefel, H. P. and Manner, R. (eds): Parallel Problem Solving from Nature III, Springer-Verlag, (1994) 260–269

[28] Kosko, B.: Neural Networks and Fuzzy Systems. Prentice-Hall International Editions (1992)

[29] Wilson, S. W.: ZCS: A zeroth level classifier system. Evolutionary Computation, 2, (1994) 1–18

[30] Bull, L: On ZCS in Multi-Agent Environments. Parallel Problem Solving From Nature - PPSN V, Springer Verlag (1998) 471–480

[31] Bull, L., Fogarty, T. C., and Snaith, M.: Evolution in Multi-Agent Systems: Evolving Communicating Classifier Systems for Gait in a Quadrupedal Robot. In Eshelman, L. J. (ed): Proceedings of the Sixth International Conference on Genetic Algorithms, Morgan Kaufmann, (1995) 382–388

[32] Cao, Y. J. and Wu, Q. H.: Mechanical design optimization by mixed-variable evolutionary programming. In Proc. of IEEE International Conf. on Evolutionary Computation, (1997) 443–446

Using Fuzzy Logic to Control Traffic Signals at Multi-phase Intersections

Jarkko Niittymäki

Helsinki University of Technology
Transportation Engineering
P.O.Box 2100, FIN-02015, HUT, Finland
jarkko.niittymaki@hut.fi

Abstract Theoretically, fuzzy control has been shown to be superior in complex problems with multi-objective decisions. Traffic signal control is a typical process, where traffic flows compete from the same time and space, and different objectives can be reached in different traffic situations. Based on recent research work, fuzzy control technology appears particularly well suited to traffic signal control situations involving multiple approaches and vehicle movements. Based on the results of our paper, we can say that the fuzzy control principles are very competitive in isolated multi-phase traffic signal control. The experiences and results of the field test and the calibration of membership functions with neural networks have been extremely promising.

1 Introduction

The control of traffic signals is one of a class of problems where a limited resource is required by a number of potential users, and the quantity of that resource is not always sufficient to meet demand. The users are competing against each other for a share of the resource and thus the control problem could be termed competitive. In the case of traffic signal control, the resource in question is green time, and the problem is made more complex by its temporal aspects and the ever-changing and stochastic nature of the demand. This means that the allocation of green time must be constantly reviewed as time passes and the traffic situation changes, in order to distribute it in the desired manner [4].

In general, traffic signal control is used to maximize the efficiency of existing traffic systems without new road construction, maintain safe traffic flows, minimize delays and hold-ups, and reduce air and noise pollution. Current traffic signal control is based on tailor-made solutions, and is poor at handling approaching traffic and cause and effect relationships. Fuzzy logic offers a number of opportunities in the field of adaptive traffic signal control. Fuzzy control has been successfully applied to a wide range of automatic control tasks, such as traffic signal control. Based on recent research work, fuzzy control technology appears particularly well suited to traffic signal control situations involving multiple approaches and vehicle movements [3]. The aim of this paper is to discuss the fuzzy traffic signal control process in general and to present the new results of fuzzy multi-phase control. The results are based on FUSICO-research project at Helsinki University of Technology in Finland.

2 Advantages of Fuzzy Control

Human decision-making and reasoning in general and in traffic and transportation in particular, are characterized by a generally good performance. Even if the decision-makers have incomplete information, and key decision attributes are imprecisely or ambigously specified, or not specified at all, and the decision-making objectives are unclear, the efficiency of human decision making is unprecedented [2]. The benefits of fuzzy logic lie in its ability to handle linguistic information by representing it as a fuzzy set. Fuzzy logic simultaneously tests this information against a variety of linguistic rules and makes a compromise decision.

For example, a police offer controlling traffic at an intersection may outperform advanced intersection controllers, even though he does not exactly know how many vehicles have been waiting for how long, or what the exact magnitude of the capacity of the intersection is. In general, fuzzy control has proved an effective way and a systematic way of solving problems with multi-objective decisions, like traffic signal control with many conflict objectives (minimizing delays, maximizing traffic safety, and minimizing fuel consumption / CO_2)

Fuzzy control also offers the benefits of a simple process structure and low maintenance cost, control adaptivity, fast evaluation time, and savings in material costs. Fuzzy signal control algorithms are also at least as good or even better in terms of effectiveness than traditional vehicle-actuated control.

3 Structure of Fuzzy Signal Controller

The design of an adaptive intersection controller naturally requires more than a suitable strategy of control. The physical and environmental design requirements have to be taken into account in the planning phase. Our fuzzy signal control process consists of seven parts (Figure 1); current traffic situation with signal status, detection and measuring part, traffic situation modeling, fuzzification interface, fuzzy inference (fuzzy decision making), defuzzification, and signal control actions. Basically, the fuzzy controller operates on a normal PC.

As much as possible the intelligence is left to the real controller, especially all safety functions. The fuzzy phase selector decides the next signal groups. Then the fuzzy green extender makes the decisions about exact timing and the length of the green phase (Figure 2). So, the main function of the fuzzy controller is to give correct orders for the signal groups. All physical and environmental design requirements are taken into account in the planning phase. One application is the use of neural networks for updating the parameters of the fuzzy membership functions.

The fuzzy controller has to collect data for the updating of the traffic situation model, which is the base of determination of signal sequence and timings. The traffic situation model contains the following elements; approaches, lanes, vehicles (size, driving characteristics), detectors, phase pictures and signal groups. The basic idea is that vehicles move along the lane according the driving dynamics and traffic models of HUTSIM-simulation program. Vehicles are generated based on detection. The additional detectors give possibilities to match vehicle movement with detector pulses. The exact traffic situation data and the good traffic model also provide an opportunity to predict the changes in the traffic flow, for example during the next 10 seconds. The fuzzy decision algorithms, which are presented later, can use the outputs of traffic situation model as input. The fuzzy timing and selection algorithms make decisions based on this data and send the decisions to the controller unit and to the signal groups.

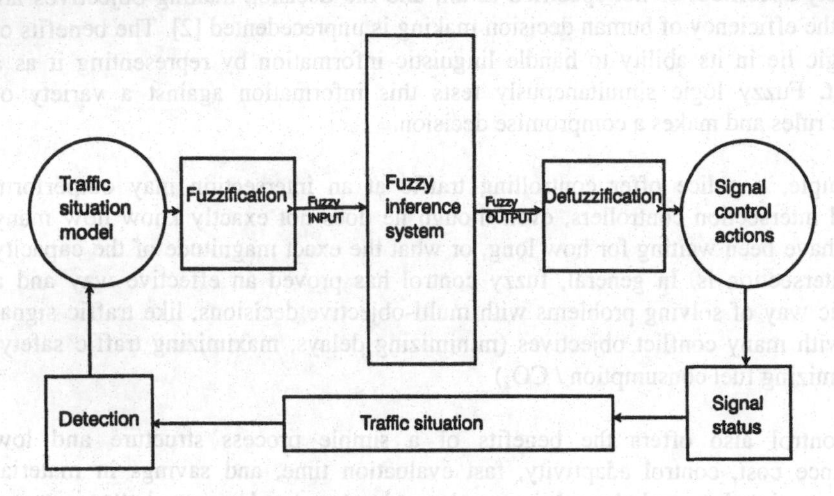

Fig.1. Fuzzy traffic signal processing.

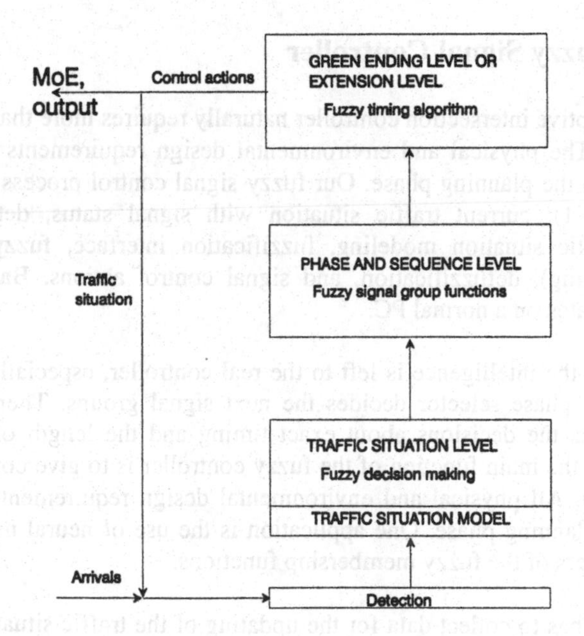

Fig.2. Structure of multi-level control algorithm.

4 Fuzzy Multi-Phase Control Algorithm

Based on Figure 2, the fuzzy rule base works at three levels:

1. *Traffic situation level*; the traffic situation is divided into three different levels (low demand, normal, oversaturated)

2. *Phase and sequence level*; the main goal of this level is to maximize the capacity by minimizing intergreen times.

3. *Green ending level or extension level*; the main goal of this level is to determine the first moment to terminate a signal group.

The fuzzy control system of the third level at the multi-phase control is shown in Figure 3.

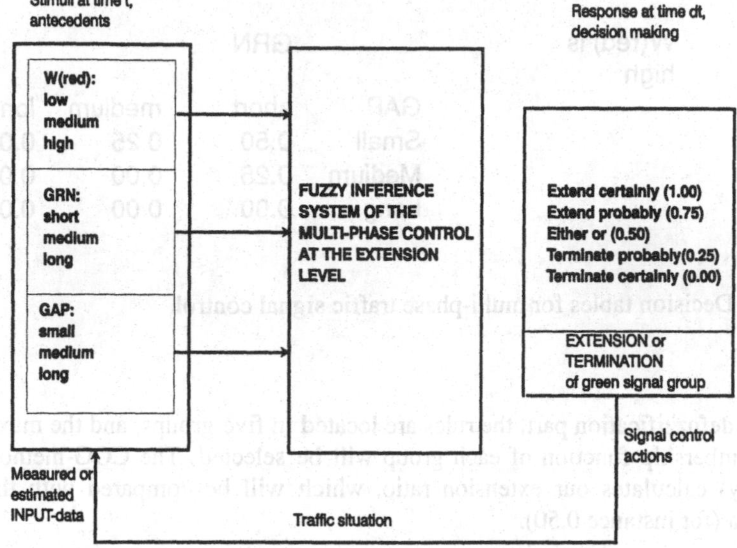

Fig.3. The fuzzy control system of the extension level.

The fuzzy rules are

If W(red) is low and GRN is short and GAP is long then extend probably (1)

...

or if W(red) is high and GRN is long and GAP is small then terminate certainly (27) .

W(red) = weight of red signal group, number of queuing vehicles,
GRN = running green time of signal group,
GAP = last gap between two approaching vehicles at the detector 80-100 meters before stop line.

W(red) is low

GRN			
GAP	Short	medium	long
Small	1.00	1.00	0.75
Medium	1.00	0.75	0.50
Long	0.75	0.50	0.25

W(red) is medium

GRN			
GAP	Short	medium	long
Small	0.75	0.50	0.25
Medium	0.50	0.50	0.25
Long	0.25	0.25	0.00

W(red) is high

GRN			
GAP	short	medium	long
Small	0.50	0.25	0.00
Medium	0.25	0.00	0.00
Long	0.00	0.00	0.00

Fig.4. Decision tables for multi-phase traffic signal control.

In the defuzzification part, the rules are located in five groups, and the maximum value of membership function of each group will be selected. The COG-method (center of gravity) calculates our extension ratio, which will be compared with the extension criteria (for instance 0.50).

5 Testing and Compared Results of Multi-Phase Control

The testing of a new control scheme such as the problem at hand requires not only the algorithm for control but also a microscopic simulation model, which allows testing of many control schemes under a realistic setting. In this respect, a sophisticated simulation model is indispensable for development and testing of an advanced signal control algorithm. Our simulation package HUTSIM gives versatile possibilities to test different traffic signal control algorithms with each other.

HUTSIM is a simulation package that has been developed at the Helsinki University of Technology. In the original design, the simulation of adaptive signal control is done by connecting a real controller to the microcomputer (PC) based simulation system. For the development of new control methods, an internal controller system has been included into HUTSIM. This system, called as HUTSIG, works so that the controller object has some measurement functions that are used to collect and analyze incoming detector data. The calculated indicators of the traffic situation are then transmitted to the control logic for timing decisions which are put into force by the group oriented signal control.

The comparison between fuzzy (FU) and vehicle-actuated (VA) control was made in real intersection in Helsinki, which was modeled for HUTSIM.

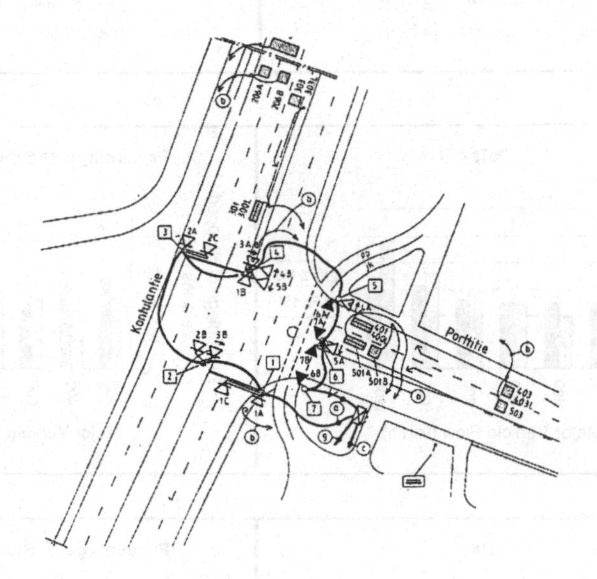

Fig.5. Layout of test intersection in Helsinki.

The simulation results proved that fuzzy control was competitive compared to traditional control methods. The compared measures of effectiveness (MoE) were average delay and percentage of stops. Totally, 18 different cases were tested :

Table 1. Simulation plan.

Major Traffic Flow	Minor/Major-ratio
200 veh/h	
400 veh/h	1:2
600 veh/h	1:5
800 veh/h	1:10
1000 veh/h	
1200 veh/h	

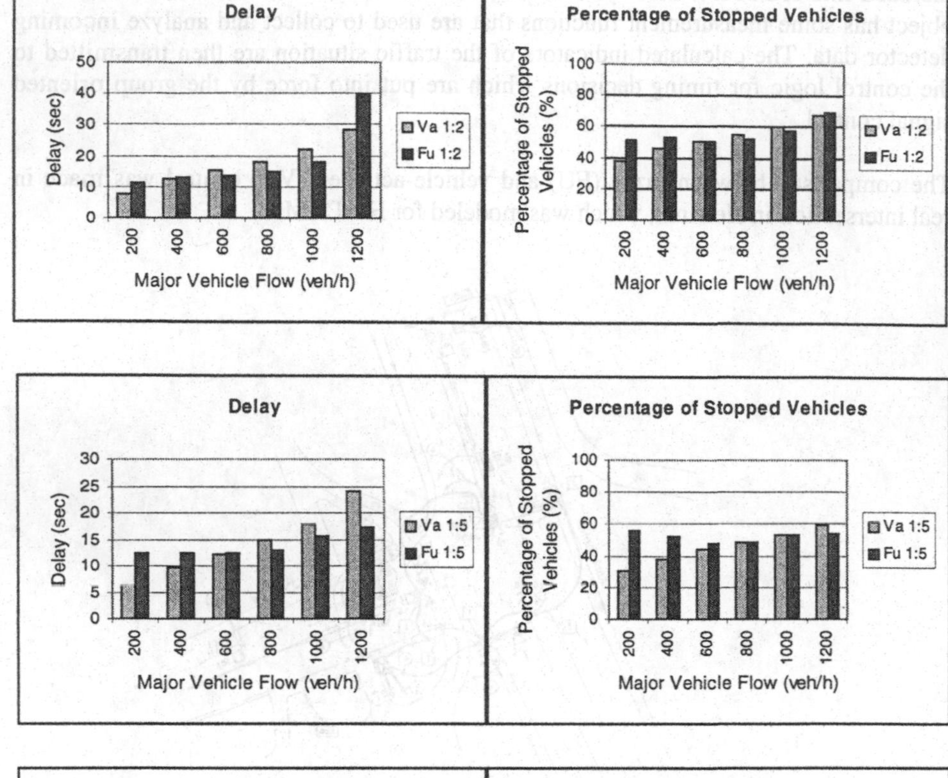

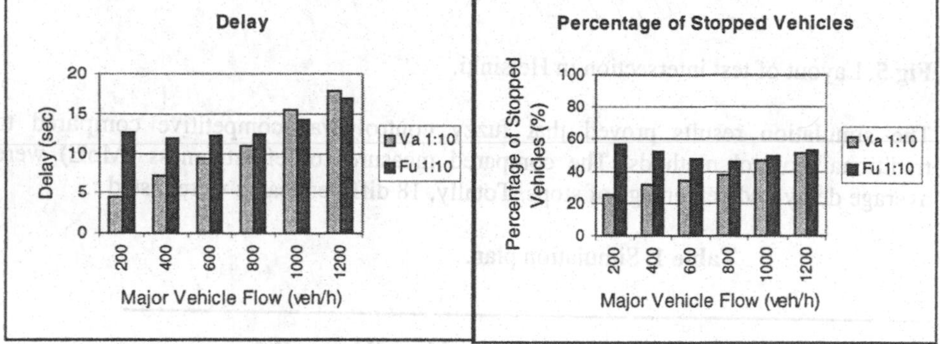

Fig.6. Compared results of VA- and Fuzzy-control.

6 Discussion of Results and Algorithm Development

The efficiency of our fuzzy algorithm was compared with the traditional vehicle-actuated control, called extension principle. The aim was that algorithm for comparison was as good as possible that is actually used in the reality. The simulation time of each case was 3600 s, and the simulated traffic situation was exactly the same in the compared cases. The results prove that the extension principle is better traffic signal control mode in the area of very low traffic volumes. However, the results indicate that the application area of fuzzy control is available. If the major traffic flow is more than 500 veh/h, the results of fuzzy control were at least as good as the results of traditional control. According to the field measurements of test intersection, the real traffic volume of major flow varies between 600 – 900 veh/h and the minor/major-ratio is approx. 1:5. Based on this, we can say that fuzzy control principles are competitive in isolated multi-phase traffic signal control.

The better results of low traffic volumes can be achieved by using the second level fuzzy decision making with the fuzzy phase selector. Our simulations are based on only the decision making of the signal group extension. The main goal of the fuzzy phase selector is to maximize the capacity by minimizing intergreen times. The basic principle is that "signal group can be kept in green while no disadvantages to other flows occur". This is also called "the method to use extra green". The main decision of this level will be the right termination moment of the green.

The second goal of this level is to determine the right phase order. The basic principle is that the phase can be skipped if there is no request or if the weight (W(p)) of this phase is low. This means that if the normal phase order is A-B-C-A the fuzzy phasing can, for example, give the orders A-B-A-C-A or A-C-A-B-C. The rules are a little bit complicated, when there are four phases, but the principle is the same as in the rules of three phases. The general principles of the rules are

- if W(p) is very high then phase p will be the next one,
- if $W(p_i)$ is high and $W(p_j)$ is zero then (i) will be the next one,
- the maximum waiting time of vehicle can not be too long.

For example, the rules of three-phase control can be

If Phase A is terminated then

if W(B) is high then next phase is B or
if W(B) is medium and W(C) is mt(high) then next phase is W(C) or
if W(B) is low and W(C) is mt(medium) then next phase is C or ***(mt=more than)***
if W(B) is lt(low) then next phase is C. ***(lt =less than)***

The fuzzy factor W(X) is the fuzzy factor for the importance of the next phase, for example the number of queuing or arriving vehicles of each phase. The exact specification will be done later.

Bingham [1] studied a way to tune our fuzzy controller by updating the parameters of the membership functions. The objective was to construct a neural learning algorithm, which modifies parameters. The results were quite good, but the learning algorithm was not found successful at stochastic traffic situations. However, we believe that neural learning or/and genetic algorithms give us some new opportunities to handle the problems of traffic signal control. Especially, the fine-tuning of membership functions as a function of detector location can be solved systematically using neural learning.

7 Conclusions

A new control method, fuzzy logic, has been presented in this paper as a possibility for the traffic signal control of future. The results of our fuzzy traffic signal applications before (signalized pedestrian crossing and two-phase control) and now (multi-phase control) have indicated that the fuzzy control offers at least equal or better performance than the traditional vehicle-actuated control. We believe that fuzzy methods are well suited to almost all kinds of control, and the biggest benefits can be achieved in complicated intersections with multi-objectives. Theoretically, fuzzy control has been shown to be superior in complex problems with multi-objective decisions. Traffic signal control is a typical process, where traffic flows compete from the same time and space, and different objectives can be reached in different traffic situations. One basic advantage of fuzzy control is that it fires many soft rules simultaneously and makes a decision, which offers the compromise. Our first real fuzzy controlled intersection was installed in June 1998. The experiences of this field test have been extremely good, and we have plans to install three fuzzy controlled intersections more in 1999.

References

1. Bingham E.: Neurofuzzy Traffic Signal Control. MOBILE 235T - Final Report. ISBN951-22-4227-3. Helsinki University of Technology. 107 p.

2. Hoogendorn S., Hoogendorn-Lanser S., Schuurman H.: Perspectives of Fuzzy Logic in Traffic Engineering. Paper presented at the 1999 Transportation Research Board 78th Annual Meeting. Transportation Research Board, Washington D.C.. 18 p.

3. Niittymäki J.: Isolated Traffic Signals – Vehicle Dynamics and Fuzzy Control. Helsinki University of Technology, Transportation Engineering, Pub. 94 (1998). 128 p.

4. Sayers T., Anderson J.: Optimisation of a Fuzzy Logic Traffic Signal Controller with Multiple Objectives. Paper presented at the 1999 Transportation Research Board 78th Annual Meeting. Transportation Research Board, Washington D.C.. 15 p.

Fuzzy Control to Non-minimal Phase Processes

W. GHARIEB, *Member IEEE*

Computer and Systems Engineering Dept.,
Faculty of Engineering - Ain Shams University,
1 El-Sarayat st., 11517 Abbassia, Cairo EGYPT
E-mail: wahied@shams.eun.eg, Fax:(202)2850617

Abstract. This paper presents a fuzzy control methodology to control non-minimal phase industrial processes. The integration between a simple fuzzy controller and Smith predictor gives a superior performance rather than using fuzzy logic controller only. The proposed fuzzy control methodology has been applied to control a fan and plate process in real time environment. The process under consideration contains a rich dynamics: time constant, transportation lag, resonant poles, non-linear characteristics and air turbulence. The obtained results showed that the proposed control methodology gives a superior performance rather than the use of fuzzy control only without dead time compensation. This paper is organized as follows: section 1 presents the motivation to use a simple fuzzy controller with a dead-time compensation in order to improve the control loop performance. Section 2 is devoted to the development of a fuzzy control with dead time compensation. In section 3, the real time application to control fan and plate process is demonstrated and the obtained results are presented. Some concluding remarks given in section 4 end the paper.

1 Introduction

Dead time or transportation lag is a common part of many industrial processes. Dead time element adds phase lag to a feedback loop and could lead to unstable response. If a standard PID controller is used, significant de-tuning is required to preserve stability and system performance will be degraded. In many cases, particularly quality loops with long dead time, it may not even be possible to use PID control at all. Dead-time compensator is now available as a standard block in many commercial digital controllers. Smith predictor is developed in 1957 and is widely used in many industrial applications. Dahlin has developed the simplest one and independently by Highan in 1968. Vogel and Edgar also developed their algorithm in 1980. Garcia and Morari developed internal model control in 1982 [1]. All algorithms require the mathematical model of industrial processes, which may be inaccurate or unavailable for complex ones.

Figure 1 shows the principle of smith predictor. Smith proposed the idea of dead-time compensation before process control computers were available to carry it out. The process model excludes the dead time of N samples. The output of the block is fed back to the controller, which then would be controlling the process without dead time. To correct for model error and unmeasured disturbances, the output of the model is delayed by N samples and subtracted from the actual controlled variable. The difference is added to the output of the model. In the absence of model error and disturbances, the difference will be zero [1].

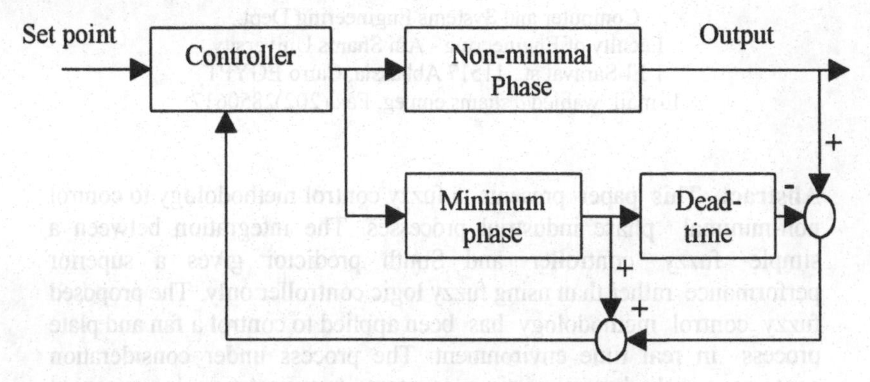

Fig. 1. Feedback control with Smith predictor

In last decade, fuzzy logic has increased attention for controlling equipment and systems to make them respond more intelligently to imprecise knowledge. Fuzzy control is a non-model base technique, it gives a good performance especially for controlling ill-defined or complex industrial processes. Fuzzy logic is developed by Lotfi Zadeh to deal with uncertainty in system representation [2]. This logic has found a variety of applications in various fields ranging from sensors, motors, steam turbines, intelligent controllers, to medical diagnosis and decision making. Mamdani and his research group developed the first attempts to control industrial processes using fuzzy logic [3] where ill-defined processes can be satisfactorily controlled using IF-THEN fuzzy rules. Fuzzy logic controller bases its decision on inputs such: (error, variation of error, …, etc) in the form of linguistic variables derived from membership functions which are used to determine the fuzzy set to which a value belongs and its degree in that set. The variables are then matched with the IF-THEN rules, and the response of each rule is obtained through fuzzy inference. The response of each rule is weighted according to the confidence or degree of membership function of its inputs, and the centroid of responses is calculated to generate the appropriate controller output [4]. Figure 2 shows a simple fuzzy controller structure. It consists of four main blocks: Fuzzification, Knowledge base, Fuzzy inference and defuzzification. Fuzzification block converts crisp measurements into linguistic variables (fuzzy labels) into the universe of discourse. Knowledge base includes input/output membership functions and If-Then fuzzy rules. Fuzzy inference computes the corresponding fuzzy decision (action). Defuzzification block converts the fuzzy action into crisp value applied to the process [5]. The simple fuzzy controller as shown in Figure 2 has a limited performance and can't respond perfectly for processes with dead time. Therefore, this paper is motivated to introduce a fuzzy control methodology for controlling industrial processes that contain a significant dead time.

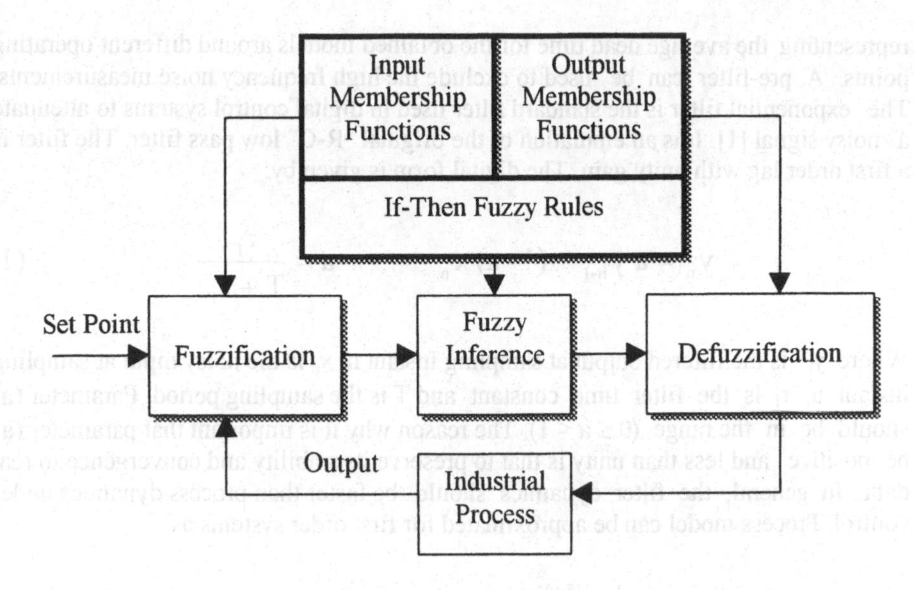

Fig. 2. Fuzzy controller structure

2 Fuzzy control with a dead-time compensation

The proposed fuzzy control methodology consists of two main parts: the former is a simple fuzzy controller structure as shown in figure 2 and the latter is a Smith predictor to compensate the effect of transportation lag. Figure 3. Presents the integration of fuzzy logic controller with smith predictor. Usually, fuzzy controller not use a mathematical model, while the Smith predictor requires it. In the proposed methodology, the required process model by Smith predictor is an approximate model, it obtained by tacking the average of process parameters around operating points.

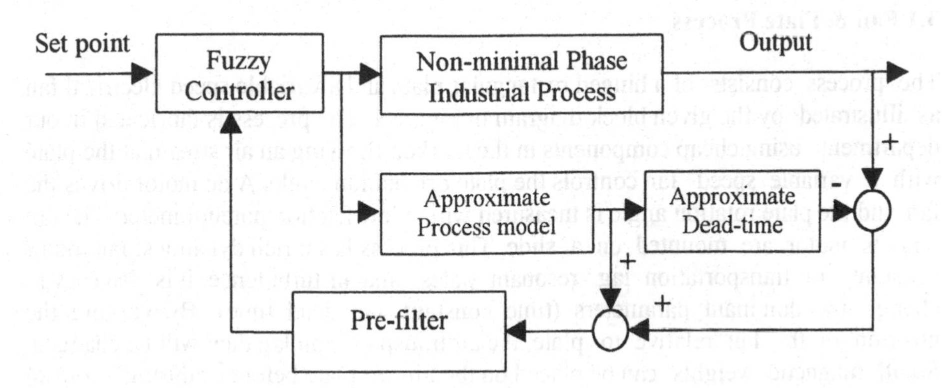

Fig. 3. Fuzzy control with dead time compensation

A set of reduced order models can be obtained by identification around different operating points to cover the region of process operation. The approximate process model can be constructed as the average values of process parameters in the form of first order or second order transfer function. The approximate dead time is N samples representing the average dead time for the obtained models around different operating points. A pre-filter can be used to exclude the high frequency noise measurements. The exponential filter is the standard filter used in digital control systems to attenuate a noisy signal [1]. It is an emulation of the original "R-C" low pass filter. The filter is a first order lag with unity gain. The digital form is given by:

$$y_n = a\, y_{n-1} + (1 - a)\, x_n \quad ; \quad a = \frac{\tau_f}{T + \tau_f} \tag{1}$$

Where y_n is the filtered output at sampling instant n, x_n is the noisy input at sampling instant n, τ_f is the filter time constant and T is the sampling period. Parameter (**a**) should be in the range $(0 \le a < 1)$. The reason why it is important that parameter (**a**) be positive and less than unity is that to preserve its stability and convergence to real data. In general, the filter dynamics should be faster than process dynamics under control. Process model can be approximated for first order systems as

$$G(s) = \frac{k\, e^{-\tau_d s}}{1 + \tau s} \tag{2}$$

Where k is the steady state gain
τ_d is the dead time
τ is the process time constant

Therefore, the parameter (**a**) should be selected in order to $\tau_f \ll \tau$. This makes the closed loop system dynamics will be dominated by process dynamics itself.

3 Application

3.1 Fan & Plate Process

The process consists of a hinged rectangular plate and a variable speed electrical fan as illustrated by the given block diagram in Figure 4. The process is fabricated in our department using cheap components in the market. Blowing an air stream at the plate with a variable speed fan controls the plate orientation angle. A dc motor drives the fan and the plate rotation angle is measured with a low friction potentiometer. The fan and its motor are mounted on a slide. This process has a rich dynamics: fan motor constant, air transportation lag, resonant poles and air turbulence. It is also easy to change the dominant parameters (time constant and dead time). By varying the position of the fan relative to plate, the air transportation lag then will be changed. Small magnetic weights can be placed on the hinged plate before or during a control experiment. These weights change the dominant time constant and also act as load disturbance. The air turbulence around the plate naturally provides the stochastic disturbance. Therefore, this process is a versatile pilot plant for studying the effects of parameter changes. Similar dynamics are also accounted in aerospace vehicles and chemical reactors.

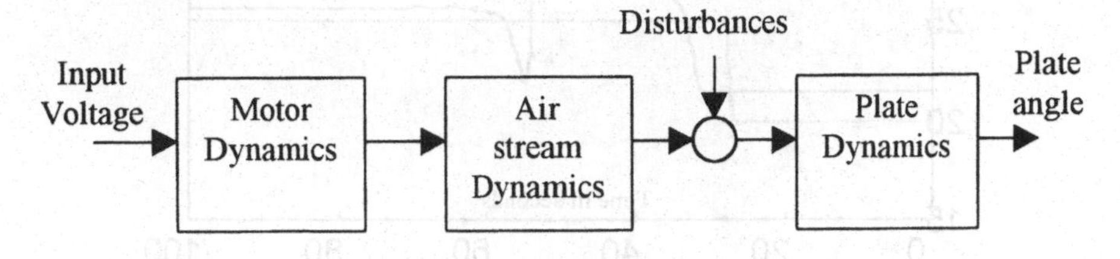

Fig. 4. Fan & Plate process

3.2 Control environment

The process can be controlled manually or by a digital controller implemented on a PC microcomputer through an interface ADDA card. Figure 5 shows the digital control environment.

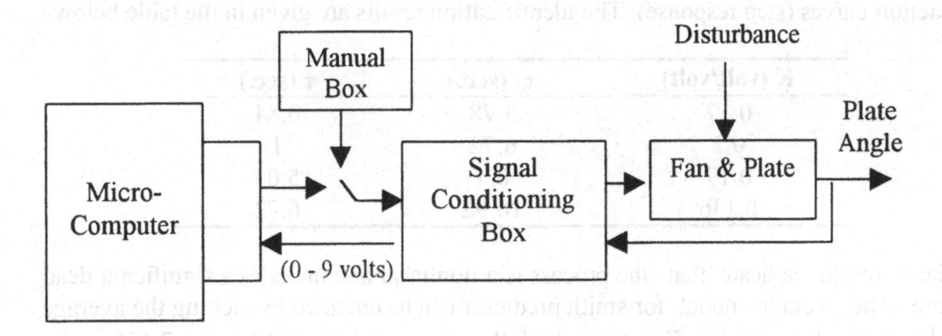

Fig. 5. Digital control environment

The microcomputer is used to implement the digital control algorithms using C++. The interface is implemented by using ADDA card. The signal conditioning box uses a voltage regulator to protect the ADDA card from the maximum measured output voltage by the potentiometer. It also used to modify the analog output signal from the ADDA card (0-10 volt) into (0-24 Volt) to be applied to the motor of the fan. The manual box has a switch to toggle between manual and computer control experiment.

3.3 Experimental results

First, we have applied fuzzy control without any numerical identification to process parameters. That means a simple fuzzy controller is used in feedback as in Figure 2. The control objective is to track the desired set point of plate angle. Seven fuzzy sets are used to fuzzify the error and five fuzzy sets are used for error variation. Five fuzzy sets are used to defuzzify the control action. Triangular membership functions are used for all fuzzy sets. The obtained result is given in figure 6.

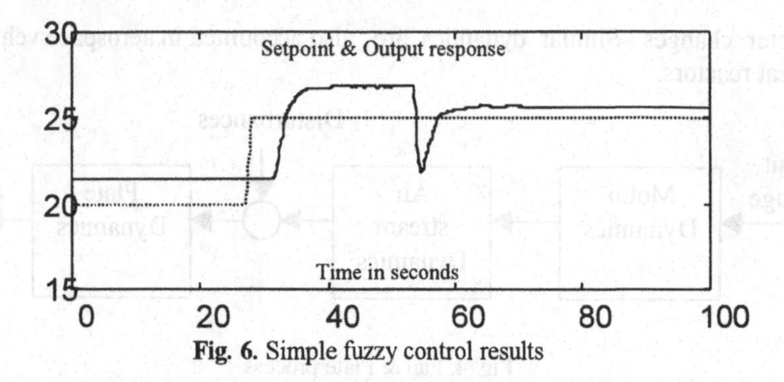

Fig. 6. Simple fuzzy control results

We observe that the tracking objective is achieved. It is clear that the response is delayed due to the dead time involved in the process behavior. An external disturbance is emulated by adding a weight on the plate at the instant (54 sec). It is shown that the objective is achieved well. A steady state error is obtained because the process is type zero and the fuzzy controller is equivalent to a nonlinear PD controller. However, adjusting the controller gain can reduce the steady state error.

Second, we identify the process model around different operating points using reaction curves (step response). The identification results are given in the table below.

K (volt/volt)	τ_d (sec.)	τ (sec.)
0.17	3.78	0.84
0.2	6.72	1
0.19	8.4	5.04
0.136	10.92	6.72

These results indicate that the process is a nonlinear and includes a significant dead time. The average model for smith predictor can be obtained by tacking the average value for each parameter. This gives the following model (K= 0.174, τ_d =7.455 and τ = 3.4). The pre-filter is designed by selecting a=0.9. The following figure represents the obtained results. We observe that the response is faster than the previous case due to compensation of the dead time. Tracking and regulation objectives are achieved.

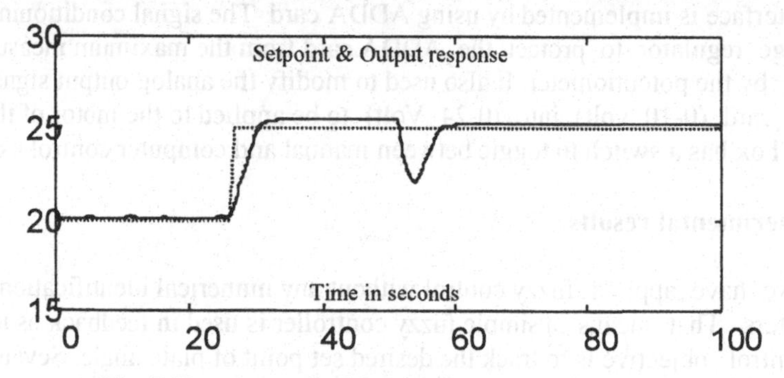

Fig. 7. Fuzzy control with dead time compensation

4 Conclusions

In this paper, a fuzzy control methodology is proposed to control non-minimal phase processes. The idea is based on the use of a Smith predictor to compensate the dead time with a simple fuzzy controller structure. This method is applied to a fan and plate process in a real time environment. The process is nonlinear with a significant dead time parameter. The obtained results affirmed that the proposed control methodology has the potential to achieve well the tracking and regulation objectives. It also speeds up the output response and decreases the effect of dead time parameter.

References

1. Carles S. Smith and Armando B. Corripio: Principles and Practice of Automatic Process Control. Ch(15), John Wiley & Sons, Inc. 2nd edition (1997)
2. L. A. Zadeh: Fuzzy Sets. Inform. Contr, Vol.8, pp. 338-353 (1965)
3. E. H. Mamdani and S. Assilian: An Experiment in Linguistic Synthesis with a Fuzzy Logic Controller. Int. J. Man-Machine Studies, Vol. 7, pp. 1-13 (1975)
4. W. Gharieb and G. Nagib: Fuzzy Control to Multivariable Systems Case Study: Helicopter Model. Proc. of 5th IEEE Int. Conf. on Fuzzy Systems, Vol. 1/3, pp. 400-405 (1996)
5. C. C. Lee: Fuzzy Logic in Control Systems: Fuzzy Logic Controller – Part I & II. IEEE Trans., Syst. Man, Cybern., vol. SMC-20, n° 2, pp. 404-435 (1990)

Robust Un-coupled Fuzzy Controller for Longitudinal and Lateral Control of an AGV

K R S Kodagoda,W S Wijesoma and E K Teoh

School of Electrical and Electronic Engineering
Nanyang Technological University
Singapore 639798.
pm1453P77@ntu.edu.sg

Abstract. The primary focus is on the development of an intelligent control scheme, which is insensitive to parametric uncertainty, load and parameter fluctuations and most importantly amenable for real time implementation. In this paper, we present an effective uncoupled direct fuzzy PD/PI control scheme for an outdoor AGV, which is a converted electrically powered golf-car. The controller performance is assessed against the required performance criteria and also against another effective nonlinear control method known as computed torque technique (CTT). It is established through simulations that the fuzzy logic controller (FLC) yields good performance even under uncertain and variable parameters in the model, unlike the CTT. And in terms of real-time implementation the availability of custom fuzzy chips and the reduced computational complexity of the fuzzy controller as against the CTT, makes the fuzzy controller, an ideal choice amongst the two schemes.

1 Introduction

The development of techniques for lateral and longitudinal control of vehicles has become an important and active research topic in the face of emerging markets for advanced AGVs and mobile robots. The particular car-like AGV, as is the case with other similar types of AGVs, is characterized by highly non-linear and complex dynamics [1]. Extraneous forces, such as those due to head winds, turning and static friction, typical of harsh outdoor environments, further complicates the modeling process and the determination of model parameters. Even if the model and the parameters are known accurately for an AGV, there are the variations in the amount of cargo in the AGV, that need be accounted for. Thus any control strategy to be useful for outdoor AGV control must able to deal with the above effectively. Linear controllers, for lateral control of AGVs, based on PD/PID, state space methods, and pole placement have been reported [2],[3],[4],[5]. However, linear controllers based on linearized models about the operating points are found to be very ineffective through simulations.

Fuzzy logic has found useful applications in control among other areas. One useful characteristic of a fuzzy controller is its applicability to nonlinear systems with model uncertainty or even unknown models. Another useful characteristic of a fuzzy logic controller is that it provides a framework for the incorporation of domain knowledge in terms of heuristic rules. Given the complexity of the AGV dynamics, the difficulty of obtaining the actual vehicle dynamic parameters, the variability of certain model parameters and the human-knowledge available on speed and steering control motivates the use of a fuzzy logic approach to control law synthesis.

Wu, et. al. [6] have developed an FLC for a vehicle to follow a lane and maintain it in the middle of the lane. It is shown that the tracking accuracy of FLC and the adaptability to parameter variations are superior to a PID controller. Hessburg and Tomizuka [7] has demonstrated through real time implementations the efficacy of an FLC for lateral control of an AGV.

In the rest of the paper we describe the particular AGV and its dynamics comprising of the driving and steering subsystems. Thereafter, a suitable intelligent control structure is proposed for drive and steering to achieve longitudinal and lateral control of the vehicle. The CTT controller is also derived for decoupled longitudinal and lateral control of the AGV. Finally we compare fuzzy un-coupled controller with CTT through simulations.

2 Vehicle Model

The vehicle considered is a *Carryall*1 golf-car. It is a front wheel steerable, rear wheel drive, electrically powered car and is suitably modified for autonomous control. A DC servomotor drives the steer system while the drive system is powered by a DC series motor (see Appendix for more details). The detailed dynamic model of the golf-car excluding the actuation motor dynamics is as follows:

$$\dot{\gamma} = \omega$$

$$\dot{\nu} = \frac{1}{(M_\nu M_\omega - N^2)} \begin{pmatrix} (M_\omega \frac{\tau_d}{r} - N\tau_s) + (NC_{s1} - M_\omega C_{d1})\,\omega^2 - M_\omega k_\nu \nu \\ + (NC_{s2} - M_\omega C_{d2})\,\nu\omega + (N\tau_s - M_\omega F_{df}) \end{pmatrix} \quad (1)$$

$$\dot{\omega} = \frac{1}{(M_\nu M_\omega - N^2)} \begin{pmatrix} (M_\nu \tau_s - N\frac{\tau_d}{r}) + (NC_{d1} - M_\nu C_{s1})\,\omega^2 + Nk_\nu \nu \\ + (NC_{d2} - M_\nu C_{s2})\,\nu\omega + ((NF_{df} - M_\nu \tau_{sf}) \end{pmatrix}$$

Where, ν is the speed of the vehicle, ω is the rate of change of steer angle (γ), τ_d and τ_s are the driving and steering torques respectively. For a detailed derivation of the above model and the definitions of the parameters (M_ν , M_ω , N , C_{s1} , C_{s2} , C_{d1} , C_{d2} , τ_{sf} and F_{df}), please refer [1]. It may be noted that the model is complex, of third order, coupled and the parameters are configuration dependent.

3 Fuzzy Longitudinal and Lateral Controller

Fuzzy logic has found useful applications in control among other areas. One useful characteristic of a fuzzy controller is its applicability to systems with model uncertainty and/or unknown models. Another useful characteristic of a fuzzy logic controller is that it provides a framework for the incorporation of domain knowledge in terms of heuristic rules. Given the complexity of the AGV dynamics, the difficulty of obtaining the actual vehicle dynamic parameters, the variability of certain model parameters and the human-knowledge available on speed and steering control motivates the use of a fuzzy logic approach to control law synthesis. The structure of the uncoupled direct fuzzy control system for longitude and lateral control is shown in Fig. 1.

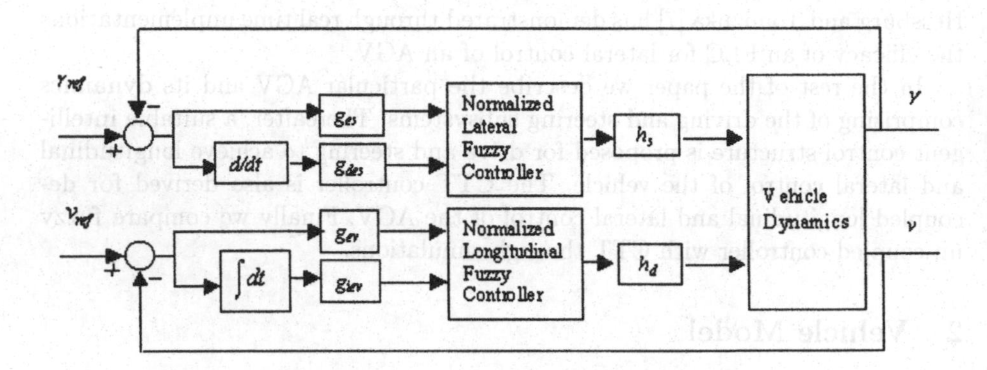

Fig. 1. Uncoupled longitudinal and lateral fuzzy controller

It may be noted that longitudinal and lateral control is achieved through the use of two separate controllers for steering and driving actuation systems. Using two separate direct fuzzy controllers instead of a single fuzzy controller can significantly reduce the complexity of the fuzzy rule base. The coupling effects of speed on steering angle (and hence angular velocity), and vice versa, are not explicitly accounted for. However, if such coupling effects are demonstrated to have an affect on the performance, these effects can be catered for in the structure proposed through explicit coupling rules. For the specific AGV used it is shown that such explicit coupling rules are not needed to achieve the desired performance.

3.1 Robust Longitudinal Fuzzy Controller

The inputs to the longitudinal fuzzy controller (LOFC) are the speed error (*ve*) and its integral (*vei*). The rule base consists of 121 rules. Triangular membership functions (Fig. 2) were chosen as it is simpler, easier to optimize and tune by using entropy equalization [10]. And also, half-overlapped triangular membership functions produce zero value in the reconstruction error. Although in general there are no precise methods to tune the fuzzy controller parameters, optimum values may be estimated which depend on the process itself and expert's control knowledge of plant. The following observations can make the tuning process more efficient [8].

- Change of scaling factors affects the entire rule base.
- Change of a membership function in the premise can only affect a row or a column in the rule base.
- Change of a rule only affects a cell in the rule base.
- Overshoot and settling time of the control system is mainly governed by the middle part of the rule base, while responsiveness is mainly governed by the outer periphery of the rule base.

The input and output gains are tuned until the system shows no further improvement. An increase of the number of rules in the rule base does not contribute to the performance significantly, but only increases the computational complexity and memory requirements of the computer.

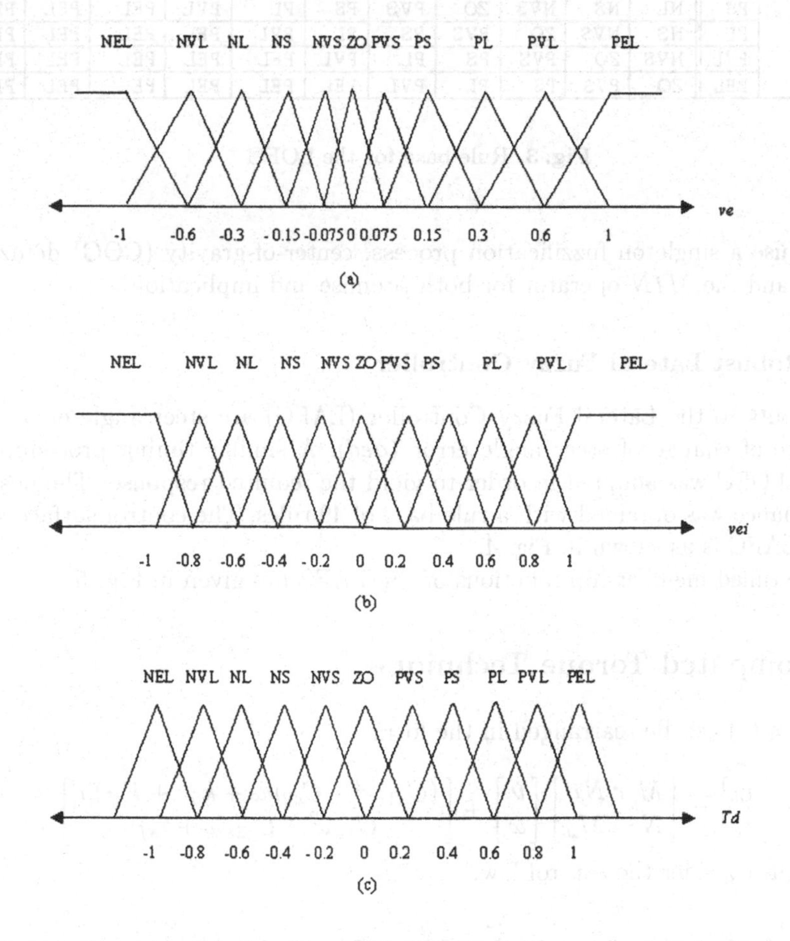

Fig. 2. Tuned membership functions of the LOFC. (a) speed error (ve), (b) integral of speed error (vei), (c) control torque (τ_d)

The body of the Fig. 3 shows fuzzy implication in the form,

IF *ve* is E **and** *vei* is D **Then** τ_d is T
where, E=D=T={NEL, NVL, NL, NS, NVS, ZO, PVS, PS, PL, PVL, PEL}

Td		*ve*										
		NEL	NVL	NL	NS	NVS	ZO	PVS	PS	PL	PVL	PEL
vei	NEL	NEL	NEL	NEL	NEL	NEL	NEL	NVL	NL	NS	NVS	ZO
	NVL	NEL	NEL	NEL	NEL	NEL	NVL	NL	NS	NVS	ZO	PVS
	NL	NEL	NEL	NEL	NEL	NVL	NL	NS	NVS	ZO	PVS	PS
	NS	NEL	NEL	NEL	NVL	NL	NS	NVS	ZO	PVS	PS	PL
	NVS	NEL	NEL	NVL	NL	NS	NVS	ZO	PVS	PS	PL	PVL
	ZO	NEL	NVL	NL	NS	NVS	ZO	PVS	PS	PL	PVL	PEL
	PVS	NVL	NL	NS	NVS	ZO	PVS	PS	PL	PVL	PEL	PEL
	PS	NL	NS	NVS	ZO	PVS	PS	PL	PVL	PEL	PEL	PEL
	PL	NS	NVS	ZO	PVS	PS	PL	PVL	PEL	PEL	PEL	PEL
	PVL	NVS	ZO	PVS	PS	PL	PVL	PEL	PEL	PEL	PEL	PEL
	PEL	ZO	PVS	PS	PL	PVL	PEL	PEL	PEL	PEL	PEL	PEL

Fig. 3. Rule base for the LOFC

We use a singleton fuzzification process, center-of-gravity (COG) defuzzification, and the MIN operator for both premise and implication.

3.2 Robust Lateral Fuzzy Controller

The inputs to the Lateral Fuzzy Controller (LAFC) are steer angle error (*ae*) and rate of change of steer angle error (*aed*). A similar tuning procedure to that of LOFC was adopted in order to yield the required response. The desired performance was obtained with a rule base of 49 rules. The control surface view of the LAFC is as shown in Fig. 4.

The tuned membership functions of the LAFC are given in Fig. 5.

4 Computed Torque Technique

Equation (1) can be rearranged in the form,

$$\begin{bmatrix} \tau_d \\ \tau_s \end{bmatrix} = \begin{bmatrix} M_\nu r & Nr \\ N & M_\omega \end{bmatrix} \begin{bmatrix} \dot{\nu} \\ \dot{\omega} \end{bmatrix} + \begin{bmatrix} \left(C_{d1}\omega^2 + C_{d2}\nu\omega + k_v v + F_{df}\right) r \\ C_{s1}\omega^2 + C_{s2}v\omega + \tau_{sf} \end{bmatrix} \tag{2}$$

Let us consider the control law:

$$\begin{bmatrix} \tau_d \\ \tau_s \end{bmatrix} = \begin{bmatrix} M_\nu r & Nr \\ N & M_\omega \end{bmatrix} \begin{bmatrix} u_d \\ u_s \end{bmatrix} + \begin{bmatrix} \left(C_{d1}\omega^2 + C_{d2}\nu\omega + k_v v + F_d f\right) r \\ C_{s1}\omega^2 + C_{s2}v\omega + \tau_{sf} \end{bmatrix} \tag{3}$$

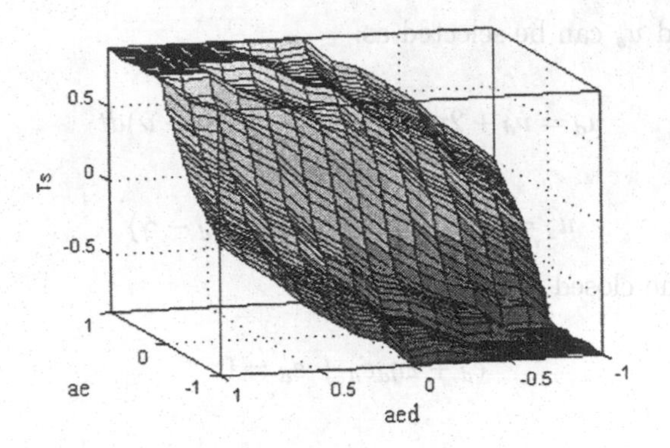

Fig. 4. Control surface view of the LAFC

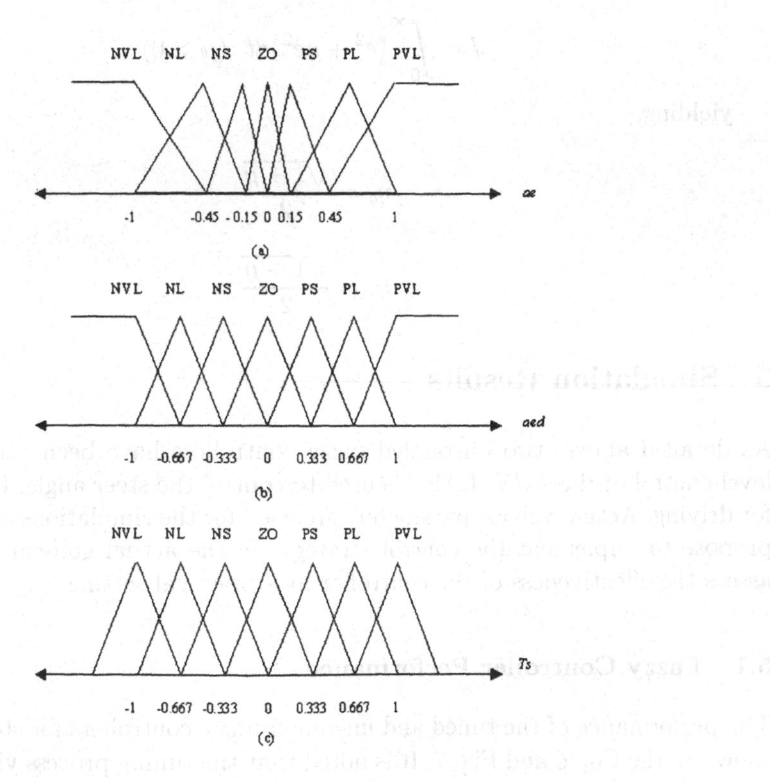

Fig. 5. Tuned membership functions of the LAFC. (a) steer angle error (ae), (b) rate of change of steer angle error (aed), (c) control torque (τ_s)

where, u_d and u_s can be selected as:

$$u_d = \dot{\nu}_d + 2\eta_d(\nu_d - \nu) + \int (\nu_d - \nu)dt \tag{4}$$

$$u_s = \ddot{\gamma}_d + 2\eta_s(\dot{\gamma}_d - \dot{\gamma}) + (\gamma_d - \gamma) \tag{5}$$

This yields the closed loop equations:

$$\ddot{e_d} + 2\eta_d \dot{e_d} + e_d = 0 \tag{6}$$

$$\ddot{e_s} + 2\eta_s \dot{e_s} + e_s = 0 \tag{7}$$

where, $e_d = (v_d - v)$ is the speed error and $e_s = (\gamma_d - \gamma)$ is the steer angle error.

Now, η_d and η_s can be chosen to minimize the performance index ,J:

$$J = \int_0^\infty (e^2 + \mu \dot{e}^2)dt \quad (\mu > 0) \tag{8}$$

yielding,

$$\eta_d = \frac{\sqrt{1+\mu}}{2} \tag{9}$$

$$\eta_s = \frac{\sqrt{1+\mu}}{2} \tag{10}$$

5 Simulation Results

As detailed above, two uncoupled fuzzy controllers have been utilized for low level control of the AGV. LAFC is used to control the steer angle. LOFC is used for driving. Actual vehicle parameters are used for the simulations as the authors propose to implement the control strategy on the actual golf-car like AGV to assess the effectiveness of the controller in a practical setting.

5.1 Fuzzy Controller Performance

The performance of the tuned and un-tuned fuzzy controllers for step inputs are shown in the Fig. 6 and Fig. 7. It is noted that the tuning process yields smother torques.

It is important to note that the same rule base is used in both the tuned and the un-tuned controllers . The normalizing gains, number of membership functions and their distribution are varied to optimize system performance. Other inferencing and defuzzification methods did not show improvements in system performance.

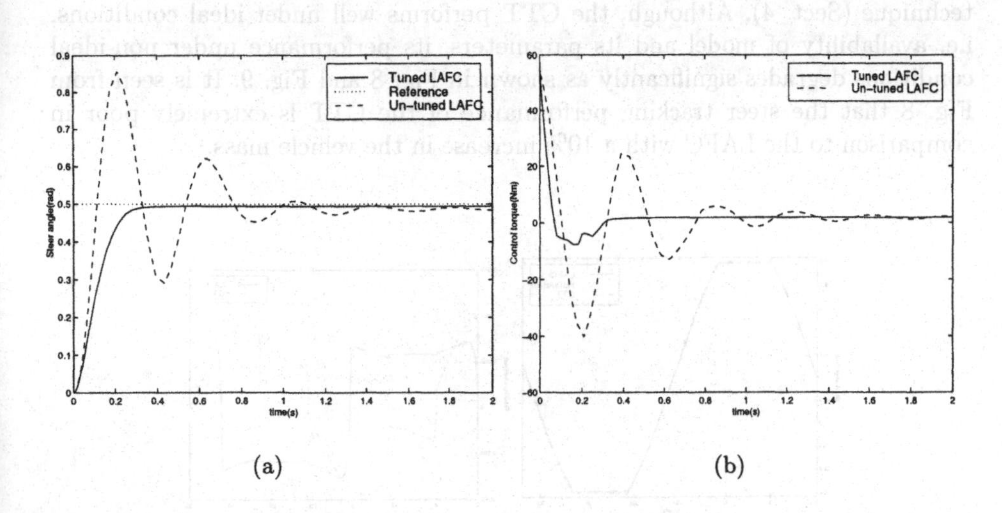

Fig. 6. Performance of LAFC before tuning (*dotted*) and after tuning (*solid*). (a) steer angle, (b) control torque

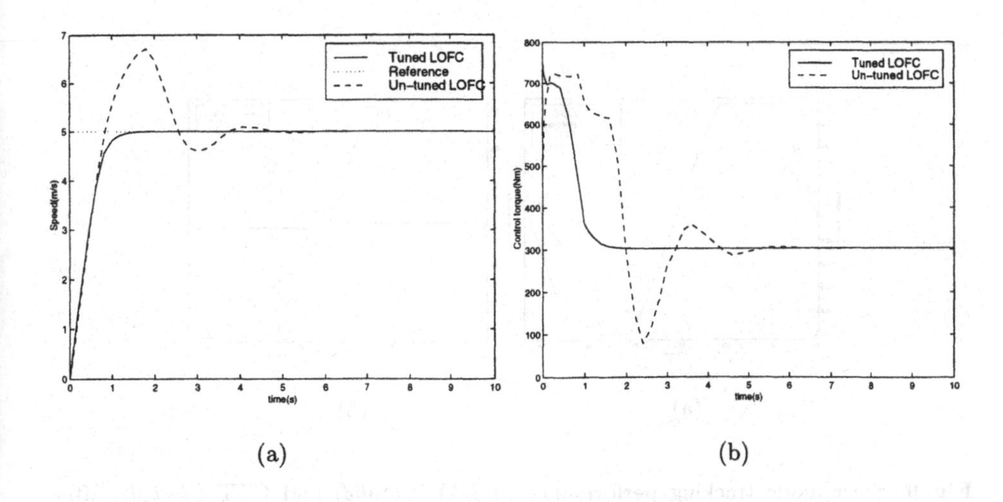

Fig. 7. Performance of LOFC before tuning (*dotted*) and after tuning (*solid*). (a) speed, (b) control torque

5.2 Robustness of LAFC

The robustness of the LAFC (Sect. 3.2) was compared with computed torque technique (Sect. 4). Although, the CTT performs well under ideal conditions, i.e. availability of model and its parameters, its performance under non-ideal conditions degrades significantly as shown in Fig. 8 and Fig. 9. It is seen from Fig. 8 that the steer tracking performance of the CTT is extremely poor in comparison to the LAFC with a 10% increase in the vehicle mass.

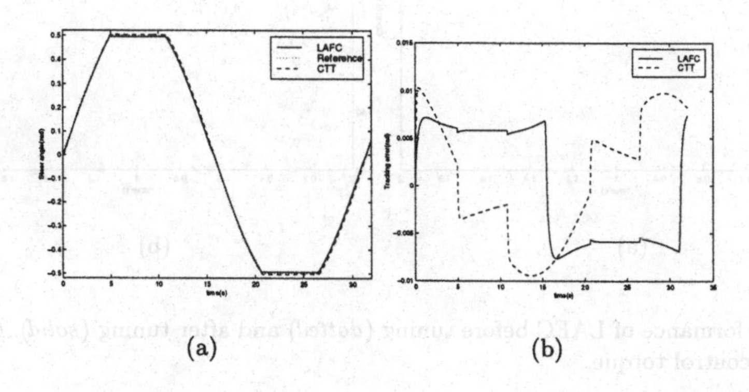

Fig. 8. Steer angle tracking performance of LAFC (*solid*) and CTT (*dotted*), with actual parameter values while the vehicle is moving at $5ms^{-1}$. (a) steer angle tracking performance, (b) associated tracking error

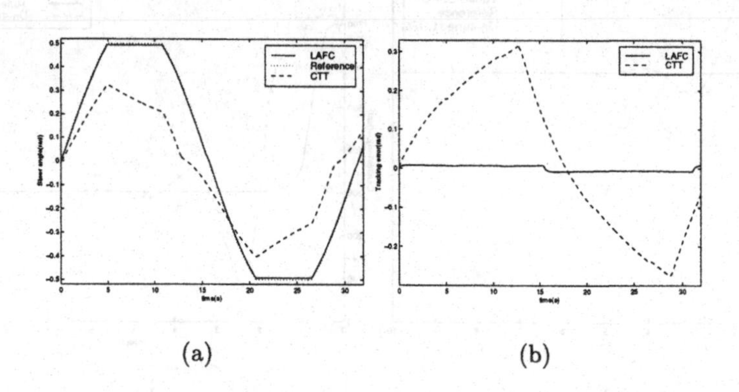

Fig. 9. Steer angle tracking performance of LAFC (*solid*) and CTT (*dotted*), after introducing a 10% increase in the vehicle load while the vehicle is moving at $5ms^{-1}$. (a) steer angle tracking performance, (b) associated tracking error

5.3 Robustness of LOFC

Fig. 10 shows the performance of both LOFC and CTT under ideal conditions. At higher speeds air drag force becomes a dominant resistive force. Therefore, we introduce a 50% increase in the coefficient of the air drag force. Fig. 11 shows the response for the CTT and LOFC under this condition. In the simulations steer angle was maintained at zero radians.

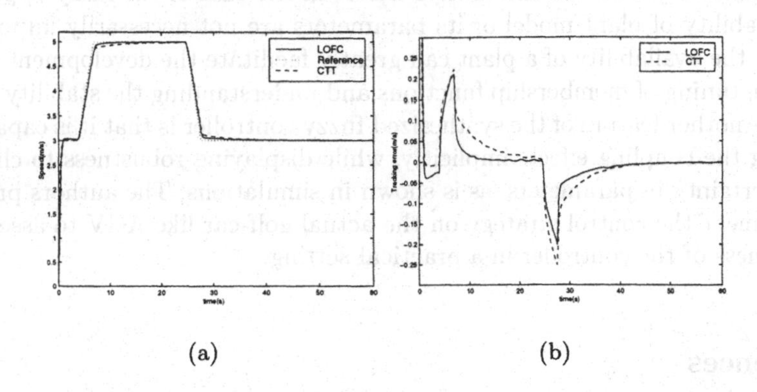

Fig. 10. Speed tracking performances of LOFC (*solid*) and CTT (*dotted*) with actual parameter values. (a) speed tracking performance, (b) associated speed error

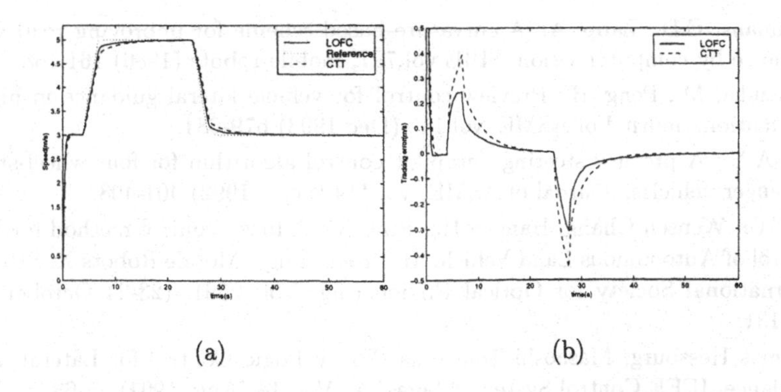

Fig. 11. Speed tracking performance of LOFC (*solid*) and CTT (*dotted*) after introducing a 50% increase in the coefficient of the air drag force. (a) speed tracking performance, (b) associated speed error

Fig. 10 and Fig. 11 shows that LOFC is less affected compared with that of CTT, due to the introduction of a higher coefficient of air drag force.

6 Conclusion

The simulations demonstrate the effectiveness of the uncoupled fuzzy controllers to parameter changes and uncertainties unlike the CTT. Also it is important to note that the CTT technique requires an exact model or near exact model of the plant to be available for de-coupling and linearization. Further, the CTT requires higher gains for achieving robustness, and at these values the controller can render the system unstable due to noise. In the case of the fuzzy approach, the availability of plant model or its parameters are not necessarily important, although the availability of a plant can greatly facilitate the development of the rule base, tuning of membership functions and understanding the stability of the system. Another feature of the synthesized fuzzy controller is that it is capable of removing the coupling effects implicitly, while displaying robustness to changes and uncertainty in parameters, as is shown in simulations. The authors propose to implement the control strategy on the actual golf-car like AGV to assess the effectiveness of the controller in a practical setting.

References

1. Xu Guangyang: Dynamic modeling and control of non-holonomic car-like vehicle, Technical report, School of Electrical and Electronic Engineering, Nanyang Technological University, (June 1998).
2. Corrier, W.H., Fenton, R.E.: On the steering of automated vehicles-a velocity adaptive controller, IEEE Trans. On Vehicular Technology, vol. VT-29, No.4, (Nov. 1980).
3. Dickmans, E.D., Zapp, A.: A curvature-based scheme for improving road vehicle guidance by computer vision, SPIE vol.727, mobile robots (1986) 161-168.
4. Tomizuka, M., Peng, H.: Preview control for vehicle lateral guidance in highway automation, Journal of ASME, vol.115,(Dec 1993) 679-684.
5. Lee, A.Y.: A preview steering autopilot control algorithm for four-wheel-steering passenger vehicles, Journal of ASME, vol.114,(Sept. 1992) 401-408.
6. Hui Wu, Wensen Chang, Hangen He, Ping Xi: A fuzzy control method for lateral control of Autonomous Land Vehicle, In Proceedings, Mobile Robots X, SPIE-The International Society for Optical Engineering, Vol. 2591, (23-24 October 1995) 125-131.
7. Thomas Hessburg, Masaoshi Tomizuka: Fuzzy Logic Control for Lateral Vehicle Guidance, IEEE Control Systems Magazine, Vol. 14,(Aug. 1994) 55-63.
8. Li Zheng: A practical guide to tune of Proportional and Integral (PI) like fuzzy controllers, IEEE Int. Conf. On Fuzzy systems, Sandiago, (1992) 633-640.
9. W. Sardha Wijesoma, K.R. Sarath Kodagoda, E.K. Teoh: Robust fuzzy PD/PI control of an AGV, submitted to ARCA'99, scheduled on (March 30-Appl. 1999), Brisbane, Australia.
10. Witold Pedrycz: Why triangular membership functions. Fuzzy Sets and Systems 64 (1994) 21-30.

Appendix: Technical data of the Golf-car

$$a = 1.650m \quad b = 0.333m \quad c = 0.097m$$
$$d = 0.49m \quad r = 0.21m \quad w = 0.21m$$
$$m_c = 450kg \quad m_w = 7.8kg \quad k_v = 40$$
$$k_t = 0.01 \quad k_r = 0.003 \quad k_b = 0.5$$

where, a , b , c , d are as defined in the Fig. 12. r and w are radius and width of the tyre respectively. m_c is mass of the chassy and m_w is mass of the tyre. k_v , k_t , k_r and k_b are coefficients of air resistance, turning friction, rolling friction and breaking force respectively.

Drive motor specifications: 48V, DC series wound, 10hp at 1125rpm.

Steer motor specification: 48V,Permanent magnet DC servo motor,0.9hp, 2750rpm.

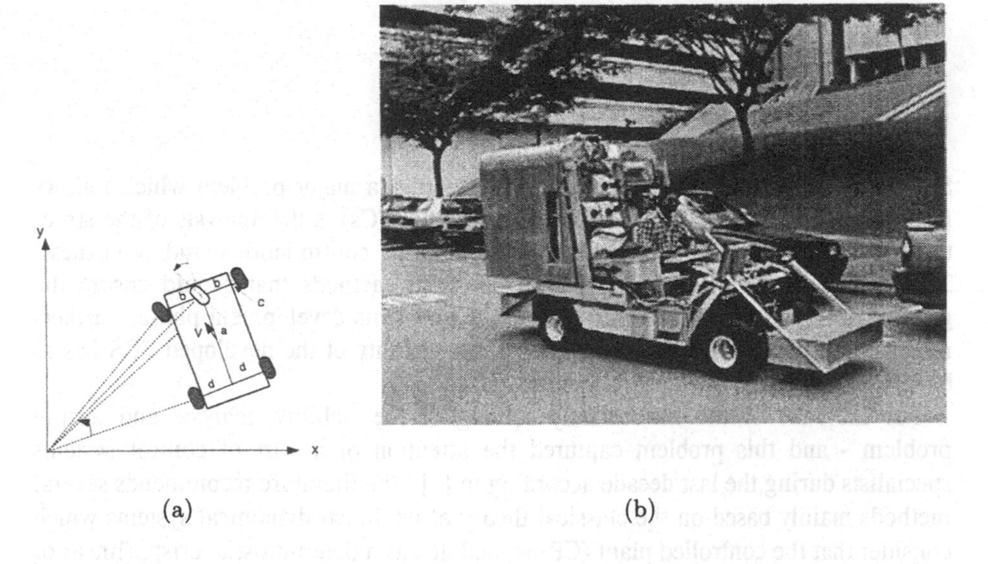

(a) (b)

Fig. 12. (a) The schematic drawing of the golf car, (b) The actual golf car

Center Manifold Theory Approach to the Stability Analysis of Fuzzy Control Systems

Radu-Emil Precup, Stefan Preitl, and Stefan Solyom

"Politehnica" University of Timisoara, Department of Automation, Bd. V. Parvan 2, RO-1900 Timisoara, Romania
{rprecup, spreitl, ssolyom}@linux1.aut.utt.ro

Abstract. The paper proposes a stability analysis method based on the application of the center manifold theory belonging to the state-space methods to the stability analysis of fuzzy control systems. The methods considers a linearized mathematical model of the second order nonlinear plant, and its only constraint is in the smooth character of the right-hand term of the state-space equations of the controlled plant. The method is exemplified by applying it to the stability analysis of a state feedback fuzzy control system meant for the position control of an electrohydraulic servosystem.

1 Introduction

Similar to the case of conventional control systems, a major problem which follows from the development of fuzzy controllers (briefly, FCs) is the analysis of the structural properties of the control system like stability, controllability and robustness. This results in the necessity of stability analysis methods that should ensure the global stability of the fuzzy control system (FCS) in its development phase. Furthermore, in the phase of FC implementation, the stability of the developed FCS has to be tested in different operating regimes.

For the sake of the immediately solving of the stability analysis and testing problem - and this problem captured the attention of a part of control systems specialists during the last decade according to [1] - the literature recommends several methods mainly based on the classical theory of nonlinear dynamical systems which consider that the controlled plant (CP) is modeled as a deterministic, crisp, (linear or nonlinear) system, and the FC is considered as particular case of nonlinear controller. These methods are developed in the time or frequency domain, and the most frequently encountered methods are based on:

- the state-space approach based on a linearized model of the nonlinear system [2], [3], [4], [5];
- the use of the hyperstability theory after Popov [6], [7], [8], [9], [10];
- the use of the stability theory after Lyapunov [11], [12], [4], [7], [13];
- the circle criterion [2], [4], [6];
- the harmonic balance method [14], [15], [16], [8].

The stability analysis method proposed in the paper belongs to the first category presented above, and it is based on the center manifold theory [17] and on the general theory of nonlinear control systems [18].

The paper is organized as follows. The following section deals with aspects concerning the mathematical models accepted in the stability analysis. Then, section 3 concerns with an outline of the proposed center manifold approach to the stability analysis of fuzzy control systems. Section 4 performs the application of the stability analysis method to the position control of an electrohydraulic servosystem by means of a state feedback fuzzy controller, and the final section highlights the conclusions.

2 Mathematical Models Involved in Stability Analysis

The dynamics of the controlled plant is considered described by the following state-space equation (in matrix expression):

$$\underline{x}^{\cdot} = \underline{f}(\underline{x}) + \underline{b}\, u , \tag{1}$$

where $\underline{x} = [x_1, x_2, \ldots, x_n]^T \in R^n$ represents the state vector, $\underline{b}$ is an [n, 1] dimensional vector of constant coefficients, u is the control signal, and $\underline{f}$: $R^n \rightarrow R^n$ stands for the process function. The only constraint imposed to $\underline{f}$ is that it must be a smooth one.

The expression (1) of the mathematical model of CP is well acknowledged and accepted for the characterization of the CP in the situations when the stability analysis of FCSs is performed [2], [5].

The relations in the sequel will be particularized for second order systems (n=2) (the generalization to higher order systems is not a difficult task). Therefore, the relation (1) transforms into (2) and (3):

$$x_1^{\cdot} = f_1(x_1, x_2) + b_1\, u , \tag{2}$$

$$x_2^{\cdot} = f_2(x_1, x_2) + b_2\, u , \tag{3}$$

where f_1, f_2: $R^2 \rightarrow R$ are smooth functions, and b_1, b_2 = const $\in R$.

Depending on the values of the constants b_1 and b_2 a coordinate transformation can be derived, and in the new coordinates the state-space mathematical model (2), (3) becomes (4), (5):

$$x_1^{\cdot} = g_1(x_1, x_2) , \tag{4}$$

$$x_2^{\cdot} = g_2(x_1, x_2) + u , \tag{5}$$

where g_1, g_2: $R^2 \rightarrow R$ are again smooth functions, and the new state variables were denoted (this is an abuse of notation, for the sake of simplicity) by x_1 and x_2.

By the application of the nonlinear state feedback control law:

$$u = - g_2(x_1, x_2) - k_1 x_1 - k_2 x_2 - h(x_1, x_2) , \tag{6}$$

with $k_1, k_2 = \text{const} \in R$ and the smooth function h: $R^2 \rightarrow$ having the features (7):

$$h(0, 0) = 0 \ , \ dh(0, 0) = 0 \ , \tag{7}$$

the closed-loop control system will be characterized by the following state-space mathematical model:

$$x_1^{\cdot} = g_1(x_1, x_2) \ , \tag{8}$$

$$x_2^{\cdot} = - k_1x_1 - k_2x_2 - h(x_1, x_2) \ . \tag{9}$$

The absence of the reference input w in (6) is justified by the fact that only the free response is analyzed, i.e. the locally asymptotically stability (LAS) of the control system around the origin (0, 0) is analyzed.

By linearizing (8) and (9) around the origin the result will be (10) and (11), respectively:

$$x_1^{\cdot} = \{[\partial g_1/\partial x_1](0, 0)\}x_1 + \{[\partial g_1/\partial x_2](0, 0)\}x_2 \ , \tag{10}$$

$$x_2^{\cdot} = - k_1x_1 - k_2x_2 \ . \tag{11}$$

For solving the problem of feedback stabilization, and the state feedback fuzzy control considered here is a particular case of the general problem from [18], it is required that [17]:

$$[\partial g_1/\partial x_1](0, 0) = [\partial g_1/\partial x_2](0, 0) = 0 \ , \tag{12}$$

and, for the sake of simplicity:

$$k_2 = 1 \ , \tag{13}$$

and the higher order terms in the Taylor series expansion of g_1 and h should depend only on x_1.

3 Outline of Center Manifold Theory Approach

By starting from the assumptions presented in the previous section and by expressing the Taylor series expansions of g_1 and h, the introduction of the coordinate transformation:

$$x_1^{\#} = x_1 \ , \tag{14}$$

$$x_2^{\#} = x_2 + k_1x_1 \ , \tag{15}$$

the center manifold (16) is derived:

$$x_2^{\#} = \varphi(x_1^{\#}) \ . \tag{16}$$

By obtaining a second order approximation of φ [17], two conditions for the LAS of the closed-loop system can be stated as (17) and (18):

$$[\partial^2 g_1/\partial x_1^{\,2}](0,0) - k_1[\partial^2 g_1/\partial x_1 \partial x_2](0,0) = 0\,, \tag{17}$$

$$[\partial^2 g_1/\partial x_1^{\,2}](0,0) - \{[\partial^2 g_1/\partial x_1 \partial x_2](0,0)\}\{[\partial^2 h/\partial x_1 \partial x_2](0,0)\} < 0\,. \tag{18}$$

The conditions (17) and (18) hold for:

$$[\partial^2 g_1/\partial x_1^{\,2}](0,0) \neq 0\,. \tag{19}$$

But, from the relations (17) and (18) it can be written down that:

$$[\partial^2 g_1/\partial x_1^{\,2}](0,0)\{k_1 - [\partial^2 h/\partial x_1 \partial x_2](0,0)\} < 0\,. \tag{20}$$

For the analysis of the LAS of the FCS it has to be taken into account the fact that $u=\varphi(\underline{i})$ from (1) is the control signal expressed through the nonlinear function φ, $\underline{i}$ is the input vector of the form $\underline{i}=\underline{T}[w;\underline{x}]$, with $\underline{T}$ being an [2, 3] dimensional constant matrix that is not necessarily to be settled, and w represents the reference input.

By summarizing the aspects presented in sections 2 and 3, the proposed stability analysis method consists in the following three steps to be proceeded:

1) Perform the necessary coordinate transformations.

2) Approximate the input-output static map (function) of the FC, φ, in order to achieve the control law (6). Three approaches can be used at least in this step:

- the approximation of the triangular and trapezoidal membership functions with exponential functions [19] (the advantage offered by such an approximation is in a continuous and differentiable input-output static map, and it was applied in [5]),
- the transposition of some approximation methods from the case of PI- or PID-fuzzy controllers to state feedback ones [20], and
- the use of genetic algorithms [21].

The first approach is considered here, and it will be applied in section 4.

3) Verify if the conditions (12), (13), (19), (20) are fulfilled. These conditions are considered as necessary and sufficient LAS conditions.

4 Application

The considered example (taken over in its simplest linearized version from [5]) is a double integrator controlled plant which is an interesting and difficult test for the fuzzy controller development and for the stability analysis of the fuzzy control system because, although it is linear, it is marginally stable having zero damping and natural frequency (with respect to other types of second order plants).

The state-space mathematical model of the CP is given in the following form:

$$x_1^{\cdot} = b'\,u\,, \tag{21}$$

$$x_2^{\cdot} = a\,x_1\,, \tag{22}$$

with the experimentally determined values of the coefficients: a = 14.05 and b' = 26.04.

A simple linear state transformation leads to the expression (4), (5) of the state-space mathematical model of CP (the state variables are again denoted by x_1 and x_2), where:

$$g_1(x_1, x_2) = a' x_1 , \; g_2(x_1, x_2) = 0 , \tag{23}$$

and the coefficient a' has the value a' = a b' = 365.862.

It can be observed that for this example the linearization is no more necessary (for the sake of simplicity of the presentation).

The FC is a state feedback nonlinear controller having two inputs ($\underline{i} = [i_1, i_2]^T$) defined as:

$$i_1 = w - x_1 , \; i_2 = x_2 , \tag{24}$$

and the state-space mathematical model (4), (5) becomes:

$$\dot{i_1} = -a' \, i_2 , \tag{25}$$

$$\dot{i_2} = u , \tag{26}$$

because w = 0 due to the fact that the LAS of the FCS around the equilibrium point placed in the origin is analyzed.

The LAS is performed in the sequel with i_1 and i_2 instead of x_1 and x_2, respectively, and according to sections 2 and 3. So, the control algorithm is expressed in terms of (6) and of the relation (24), and these two relations result in:

$$u = k_1 \, i_1 + k_2 \, i_2 - h(i_1, i_2) , \tag{27}$$

where $k_1 = 1$ and $k_2 = 1 + a' = 366.862$, and the supplementary term ($a' \, i_2$) appearing in (27) is meant for fulfilling the conditions (12).

The condition (20) transforms into:

$$k_1 > [\partial^2 h / \partial i_1 \partial i_2](0, 0) . \tag{28}$$

The condition (28) means relatively small modifications of the strictly nonlinear part of the control algorithm, and this is the reason why the function *h* has to be a smooth one. This function can be approximated by the output u^{FC} of a strictly speaking fuzzy controller, and the control algorithm of the FC is derived from (27) and is expressed as:

$$u = k_1 \, i_1 + k_2 \, i_2 + u^{FC} . \tag{29}$$

The block diagram of the FCS is presented in Fig.1, where *v* is the disturbance input, and *y* stands for the controlled output.

The input and output membership functions of the FC have the shapes presented in Fig.2, where the strictly positive parameters of the FC can be pointed out: $\{B_{i1}, B_{i2}, B_{uFC}\}$. These parameters can be determined by digital simulation in order to fulfil the condition (28). This fact can be achieved by means of a MacVicar-Whelan decision table presented in Table 1, and by using the fact that an approximation is well accepted between the triangular and trapezoidal membership functions and the exponential ones.

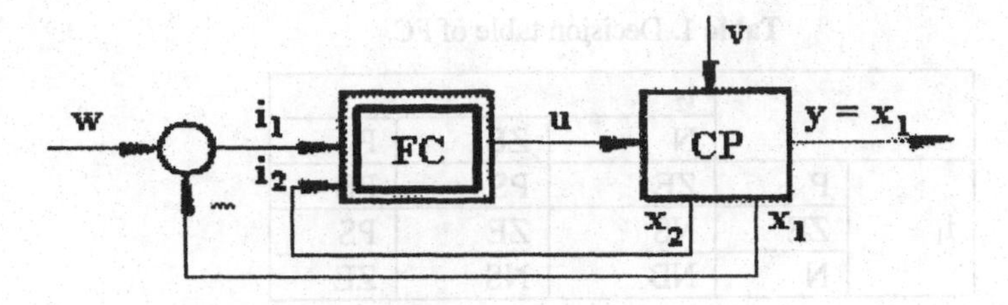

Fig.1. Structure of fuzzy control system.

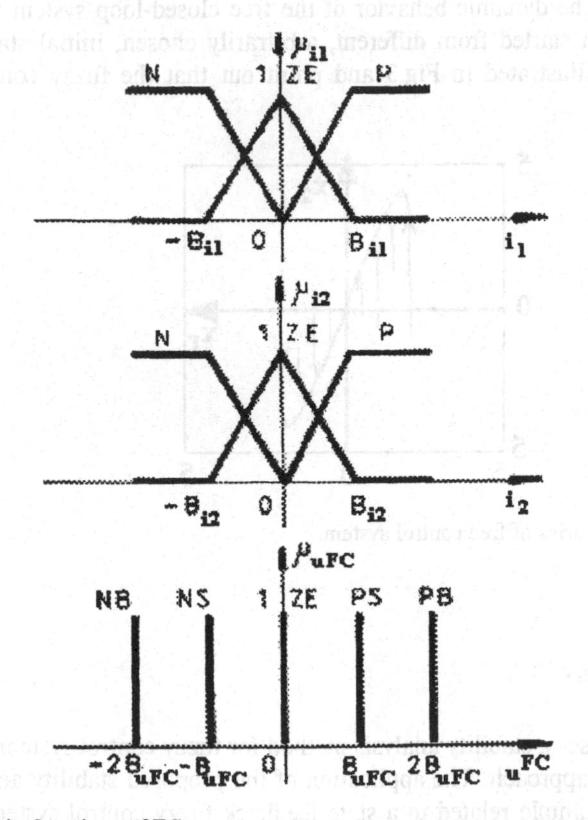

Fig.2. Membership functions of FC.

The control algorithm (29) can be seen as a state feedback one, and u^{FC} can be determined as it is proportional to the distance in the state space $<i_1, i_2, u>$ of the current state to the zero control signal plane:

$$u - k_1 i_1 - k_2 i_2 = 0 \,. \tag{30}$$

A similar approach is done in [22] in the case of a PI-fuzzy controller.

Table 1. Decision table of FC

		i_2		
		N	ZE	P
i_1	P	ZE	PS	PB
	ZE	NS	ZE	PS
	N	NB	NS	ZE

The following values of fuzzy controller parameters prove to ensure control system stability: $B_{i1} = 0.5$, $B_{i2} = 1$, $B_{uFC} = 5$.

The testing of the stability of the considered fuzzy control system is done by digital simulation. The dynamic behavior of the free closed-loop system was simulated when the system started from different, arbitrarily chosen, initial states. The state trajectories are illustrated in Fig.3 and point out that the fuzzy control system is stable.

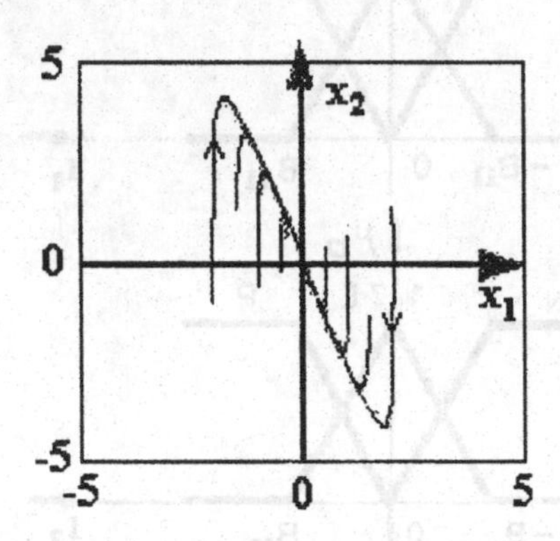

Fig.3. State trajectories of free control system.

5 Conclusions

The paper proposes a stability analysis method for fuzzy control systems based on the center manifold approach. The application of the proposed stability analysis method deals with an example related to a state feedback fuzzy control system meant for a second order, double integrator, electrohydraulic servosystem

The following remarks can be highlighted:

1) The presented stability analysis method devoted to fuzzy control system controlling a class of nonlinear systems ensures the stability of the fuzzy control system in the vicinity of the origin.

2) The method has some advantages consisting in:

- it is relatively simple;
- it is applicable to a large class of nonlinear systems; the only constraint is imposed to the controlled plant input-output function (map), which has to be a smooth one;

- it can be easily applied to Single Input-Single Output (SISO) second order systems or to higher order systems that can be reduced to second order systems (see, for example, [23]);
- it provides different guidelines for the development of fuzzy controllers being a good alternative in this direction, especially for its relative simplicity.

3) The influence of the stability analysis method on control system performance should be further investigated in order to derive connections between fuzzy controller parameters and achievable control system performance.

4) Although the presentation is focussed on second order systems, the method can be applied to higher order systems ($n > 3$); the only problem may appear in the difficult computation of the partial derivatives of FC input-output static map φ.

The method can be extended to other types of fuzzy controllers with different inference engines, fuzzification and defuzzification modules [4].

References

1. Bretthauer, G., Opitz, H.-P.: Stability of Fuzzy Systems - A Survey. Proceedings of Second EUFIT'94 European Congress, Aachen, Vol. 1 (1994) 283-290
2. Aracil, J., Ollero, A., Garcia-Cerezo, A.: Stability Indices for the Global Analysis of Expert Control Systems. IEEE Transactions on Systems, Man, and Cybernetics, Vol. 19 5 (1989) 998-1007
3. Garcia-Cerezo, A., Ollero, A.: Stability of Fuzzy Control Systems by Using Nonlinear System Theory. Proceedings of IFAC/IFIP/IMACS Symposium on Artificial Intelligence in Real-Time Control, Delft (1992) 171-176
4. Driankov, D., Hellendoorn, H., Reinfrank, M.: An Introduction to Fuzzy Control, Springer-Verlag, Berlin Heidelberg New York (1993)
5. Doboli, S., Precup, R.-E.: The Application of a Stability Analysis Method to Fuzzy Control Systems. Proceedings of Sixth IFSA'97 World Congress, Prague, Vol. 3 (1997) 452-457
6. Opitz, H.-P.: Fuzzy Control and Stability Criteria. Proceedings of First EUFIT'93 European Congress, Aachen, Vol. 1 (1993) 130-136
7. Böhm, R., Krebs, V.: Ein Ansatz zur Stabilitätsanalyse und Synthese von Fuzzy-Regelungen. Automatisierungstechnik, Vol. 41 8 (1993) 288-293
8. Bindel, T., Mikut, R.: Entwurf, Stabilitätsanalyse und Erprobung von Fuzzy-Reglern am Beispiel einer Durchflussregelung. Automatisierungstechnik, Vol. 43 5 (1995) 249-255
9. Böhm, R., Bosch, M.: Stabilitätsanalyse von Fuzzy-Mehrgrössenregelungen mit Hilfe der Hyperstabilitätstheorie. Automatisierungstechnik, Vol. 43 4 (1995) 181-186
10. Precup, R.-E., Preitl, S.: Popov-Type Stability Analysis Method for Fuzzy Control Systems. Proceedings of Fifth EUFIT'97 European Congress, Aachen, Vol. 2 (1997) 1306-1310
11. Kiendl, H., Rüger, J.: Verfahren zum Entwurf unt Stabilitätsnachweis von Fuzzy-Regelungssystemen. Workshop "Fuzzy Control" des GMA-UA 1.4.2, Dortmund, Forshungsberichte der Fakultät für ET 0392 (1992) 1-9
12. Kiendl, H., Rüger, J.: Verfahren zum Entwurf unt Stabilitätsnachweis von Regelungssystemen mit Fuzzy-Reglern. Automatisierungstechnik, Vol. 41 5 (1993) 138-144

13. Scheel, T.: Verallgemeinerte Integrale Ljapunov Funktionen und ihre Anwendung zur Stabilitätsanalyse von Fuzzy-Systeme. Workshop "Fuzzy Control" des GMA-UA 1.4.2, Dortmund, Forshungsberichte der Fakultät für ET 0295 (1995) 99-113
14. Kiendl, H.: Stabilitätsanalyse von mehrschleifigen Fuzzy-Regelungssystemen mit Hilfe der Methode der Harmonischen Balance. Workshop "Fuzzy Control" des GMA-UA 1.4.2, Dortmund, Forshungsberichte der Fakultät für ET 0392 (1992) 315-321
15. Kiendl, H.: Harmonic Balance for Fuzzy Control Systems. Proceedings of First EUFIT'93 European Congress, Aachen, Vol. 1 (1993) 127-141
16. Boll, M., Bornemann, J., Dörrscheidt, F.: Anwendung der harmonischen Balance auf Regelkreise mit unsymmetrischen Fuzzy-Komponenten und konstante Eingangsgrössen. Workshop "Fuzzy Control" des GMA-UA 1.4.2, Dortmund, Forshungsberichte der Fakultät für ET 0194 (1994) 70-84
17. Nijmeijer, H., van der Schaft, A.: Nonlinear Dynamical Control Systems. Springer-Verlag, Berlin Heidelberg New York (1990)
18. Isidori, A.: Nonlinear Control Systems. Springer-Verlag, Berlin Heidelberg New York (1989)
19. Gartner, H., Astolfi, A.: Stability Study of a Fuzzy Controlled Mobile Robot. Technical Report, Automatic Control Laboratory, ETH Zürich (1995)
20. Galichet, S., Foulloy, L.: Fuzzy Controllers: Synthesis and Equivalences. IEEE Transactions on Fuzzy Systems, Vol. 3 2 (1995) 140 - 148
21. Schröder, M., Klawonn, F., Kruse, R.: Sequential Optimization of Multidimensional Controllers Using Genetic Algorithms and Fuzzy Situations. In: Herrera, F., Verdegay, J.L. (eds.): Genetic Algorithms and Soft Computing, Springer-Verlag, Berlin Heidelberg New York (1996)
22. Precup, R.-E., Preitl, S.: Fuzzy Control of an Electrohydraulic Servosystem under Nonlinearity Constraints. . Proceedings of First EUFIT'93 European Congress, Aachen, Vol. 3 (1993) 1524-1530
23. Lohmann, B.: Order Reduction and Determination of Dominant State Variables of Nonlinear Systems. Proceedings of First MATHMOD Conference, Vienna (1994)

Stability Analysis of Fuzzy and Other Nonlinear Systems Based on the Method of Convex Decomposition

Rainer Knicker

Faculty of Electrical Engineering,
University of Dortmund,
D-44221 Dortmund, Germany,
Phone: +49 231 755 3998, Fax: +49 231 755 2752
{Knicker, Kiendl}@esr.e-technik.uni-dortmund.de
http://esr.e-technik.uni-dortmund.de

Abstract. The Convex Decomposition Stability Analysis Method (CDS method), a computer aided method for strictly proving stability, is developed further in this paper. It can now be applied to control systems consisting of a piecewise linearized plant, with a piecewise multiaffine discrete-time control law. The idea of the Convex Enclosure has been further developed and realized for applications. The efficiency of the method is demonstrated by applying it to a fuzzy control system.

1 Introduction

Rapid progress in the domain of digital process control means that very complex nonlinear controllers can now be realized. They are extremely flexible, so they can perfectly meet the desired specifications of the performance. However, an essential drawback of nonlinear control systems is that it is much more difficult or even impossible, compared with linear control systems, to establish stability. In this paper the CDS method, which allows stability to be established strictly, is extended to a more general class of nonlinear discrete-time control systems, which include fuzzy and neural network control systems.

2 Definition of $G_{H,N}$-Stability

We consider discrete-time systems $\mathbf{x}_{k+1} = \tilde{\mathbf{f}}(\mathbf{x}_k, r_k)$, where $\mathbf{x}_k$ is the n-dimensional state vector and r_k is the scalar reference variable. In applications, an important question concerning the free system $\mathbf{x}_{k+1} = \tilde{\mathbf{f}}(\mathbf{x}_k, 0) = \mathbf{f}(\mathbf{x}_k)$ is whether, for all initial states $\mathbf{x}_0$ of a given domain G of the state space, the system reaches a sufficiently small neighbourhood H of the equilibrium state $\mathbf{x}_e = \mathbf{0}$ in a given number $N > 0$ of steps and stays in H for all further steps (Fig. 1). If these conditions are met the control system is called $G_{H,N}$-stable [2].

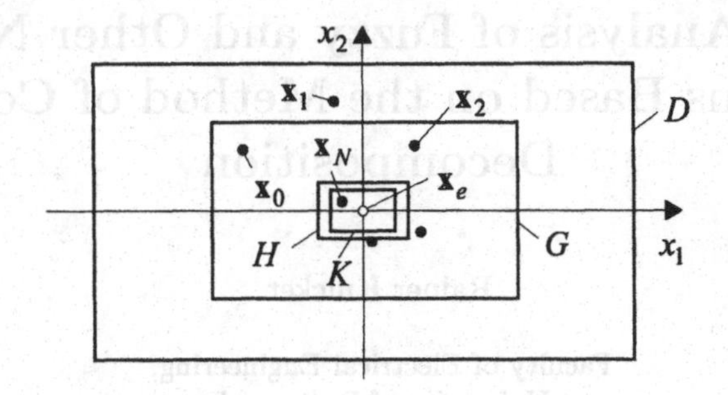

Fig. 1. Convex domains in the state space for the definition of $G_{H,N}$-stability.

Definition 1. *A system* $\mathbf{x}_{k+1} = \mathbf{f}(\mathbf{x}_k)$ *is called* $G_{H,N}$*–stable if* N *is a positive integer and the domains* G *and* H *have the properties*

$$F^n(G) \subseteq H, \quad n \geq N \tag{1}$$

with $F(P) = \{\mathbf{f}(\mathbf{x}) | \mathbf{x} \in P\}$ *and* $F^n(P) = \begin{cases} P & \text{if} \quad n = 0 \\ F(F^{n-1}(P)) & \text{if} \quad n > 0 \end{cases}$

In the case that a system meets the conditions of Def. 1 but the number of steps N is not known the system is called G_H–stable.

A sufficient criterion to check the condition (1) is to find a domain $K \subseteq H$ with the properties

$$F^M(K) \subseteq K \tag{2}$$

and

$$F^m(K) \subseteq H, m = 1, \ldots, M. \tag{3}$$

Condition (1) is established if $F^N(G) \subseteq K$ is valid.

3 The Convex Decomposition Stability Analysis Method

The method of Convex Decomposition introduced in [2] and further developed in [1, 3, 4] is a computer-aided method for establishing $G_{H,N}$-stability for a control system (Fig. 2) consisting of a linear plant

$$\mathbf{x}_{k+1} = \Phi \mathbf{x}_k + \mathbf{H} \mathbf{u}_a(\mathbf{x}_k) = \mathbf{f}_a(\mathbf{x}_k), \tag{4}$$

and a piecewise affine control law (the index *a* means *affine*)

$$\mathbf{u}_a(\mathbf{x}) = \begin{cases} \mathbf{u}_{a,1}(\mathbf{x}) = \mathbf{A}_1 \mathbf{x} + \mathbf{a}_1, \mathbf{x} \in Z_1 \\ \vdots \qquad\qquad \vdots \\ \mathbf{u}_{a,r}(\mathbf{x}) = \mathbf{A}_r \mathbf{x} + \mathbf{a}_r, \mathbf{x} \in Z_r \end{cases}, \mathbf{x} \in D. \tag{5}$$

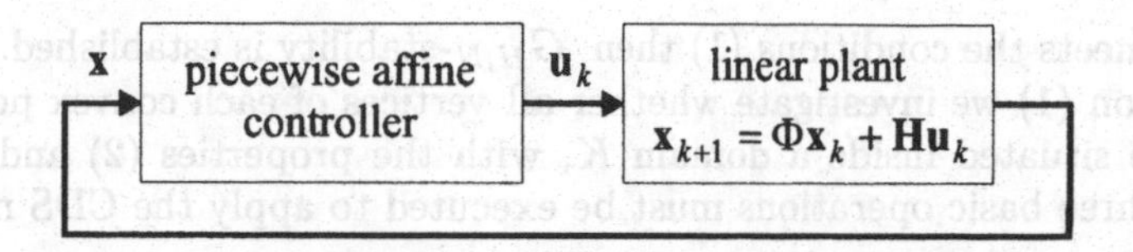

Fig. 2. Discrete-time control system with a piecewise affine controller.

The control law $\mathbf{u}_a(\mathbf{x})$, not necessarily continuous, is defined by r disjunct convex domains Z_i, bounded by hyperplanes $\mathbf{c}^T\mathbf{x} = d$ with

$$D = \bigcup_{i=1}^{r} Z_i \qquad \text{and} \qquad Z_i \cap Z_j = \emptyset, \quad \forall i \neq j, \tag{6}$$

where D is the convex domain where $\mathbf{u}_a(\mathbf{x})$ is defined in the state space (Fig. 1). For each domain Z_i an affine control law $\mathbf{u}_{a,i}(\mathbf{x})$ and consequently a piecewise affine system description

$$\mathbf{x}_{k+1} = \mathbf{f}_{a,i}(\mathbf{x}_k) = \Phi\mathbf{x}_k + \mathbf{H}\mathbf{u}_{a,i}(\mathbf{x}_k) = \underbrace{(\Phi + \mathbf{H}\mathbf{A}_i)}_{\Phi_i}\mathbf{x}_k + \underbrace{\mathbf{H}\mathbf{a}_i}_{\mathbf{h}_i}, \quad \mathbf{x}_k \in Z_i \tag{7}$$

is given. We assume that no Φ_i is singular.

A convex polytope is defined by the intersection of m half-spaces

$$l = \{\mathbf{x} | \mathbf{c}^T\mathbf{x} \leq d\}. \tag{8}$$

Combining the row vectors $\mathbf{c}^T$ of the half-spaces in a matrix $\mathbf{C}_k$ and the scalars d in a vector $\mathbf{d}_k$, the polytope P_k can be described as

$$P_k = l_{k,1} \cap \ldots \cap l_{k,m} = \{\mathbf{x} | \mathbf{C}_k\mathbf{x} \leq \mathbf{d}_k\}. \tag{9}$$

Comparison operators between vectors in this paper are defined componentwise. The vertices of the polytope are calculated by means of linear optimization [5].

The basic idea of the CDS method is that for a polytope $P_k \subseteq Z_i$ the point set

$$P_{k+1} := F_a(P_k) = \{\mathbf{x} | \underbrace{\mathbf{C}_k\Phi_i^{-1}}_{\mathbf{C}_{k+1}}\mathbf{x} \leq \underbrace{\mathbf{d}_k + \mathbf{C}_k\Phi_i^{-1}\mathbf{h}_i}_{\mathbf{d}_{k+1}}\} \tag{10}$$

can be constructed by determining the affine mappings of the vertices of P_k (7), respectively, by determining the affine mappings of the bounding hyperplanes (10). The resulting point set P_{k+1} is also a convex polytope. Therefore P_{k+1} can be decomposed into a family of convex polytopes, each situated in only one domain Z_i where a uniform affine control law is valid.

To establish $G_{H,N}$-stability we construct the point sets $G, F^1(G), F^2(G), \ldots, F^N(G)$ defined by

$$F^N(G) = \begin{cases} G & \text{if} \quad N = 0 \\ \bigcup_{i=1}^{r} F_{a,i}(F^{N-1}(G) \cap Z_i) & \text{if} \quad N > 0. \end{cases} \tag{11}$$

If $F^N(G)$ meets the conditions (1) then $G_{H,N}$-stability is established. To check the condition (1) we investigate whether all vertices of each convex polytope of $F^N(G)$ are situated inside a domain K, with the properties (2) and (3). The following three basic operations must be executed to apply the CDS method:

1. Decomposition of the polytopes into families of polytopes (subpolytopes) with each subpolytope situated in only one domain Z_i.
2. Mapping of each polytope situated in only one domain Z_i.
3. Check whether a polytope is situated inside K.

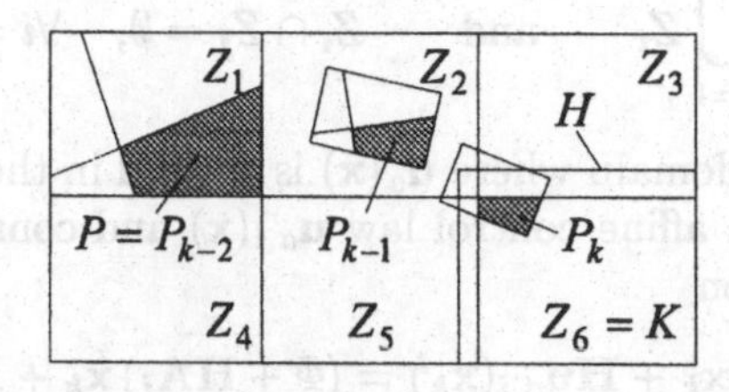

Fig. 3. Applying the mapping F_a^{-1} to P_k to determine the parts P of a domain Z_i which map to the domain K

If Z_{iH}-stability cannot be established for a domain Z_i but P_H-stability can be established for certain subpolytopes $P \subseteq Z_i$, the subpolytopes can be determined by applying the following backward mapping k-times to $P_k \subseteq K$

$$P_{k-1} := F_a^{-1}(P_k) = \{\mathbf{x} | \underbrace{\mathbf{C}_k \Phi_i}_{\mathbf{C}_{k-1}} \mathbf{x} \leq \underbrace{\mathbf{d}_k - \mathbf{C}_k \mathbf{h}_i}_{\mathbf{d}_{k-1}}\}. \tag{12}$$

This is possible without any decomposition. (Fig. 3).

4 Convex Enclosure

We consider a closed convex polytope P_k bounded by hyperplanes situated in a convex domain Z_i where a nonlinear control law $\mathbf{u}_{nl}(\mathbf{x})$ is valid (the index nl means *nonlinear*). The result of a nonlinear mapping of a hyperplane is generally not a hyperplane. Consequently, we enclose $F_{nl}(P_k)$ by a convex polytope $\hat{F}_{nl}(P_k)$ bounded by hyperplanes (Fig. 4). The nonlinear control law $\mathbf{u}_{nl}(\mathbf{x})$ can be estimated by

$$\mathbf{u}_a(\mathbf{x}) \leq \mathbf{u}_{nl}(\mathbf{x}) \leq \mathbf{u}_a(\mathbf{x}) + \mathbf{e}. \tag{13}$$

For this, we determine an affine approximation $\hat{u}_a(\mathbf{x})$ of $u_{nl}(\mathbf{x})$ and the global extrema e_{min} and e_{max} of the nonlinear function $u_{nl}(\mathbf{x}) - \hat{u}_a(\mathbf{x})$ for all $\mathbf{x} \in P_k$ for each component of $\mathbf{u}_{nl}(\mathbf{x})$. Then, the affine estimation $\mathbf{u}_a(\mathbf{x}_k) = \hat{\mathbf{u}}_a(\mathbf{x}_k) + \mathbf{e}_{min}$

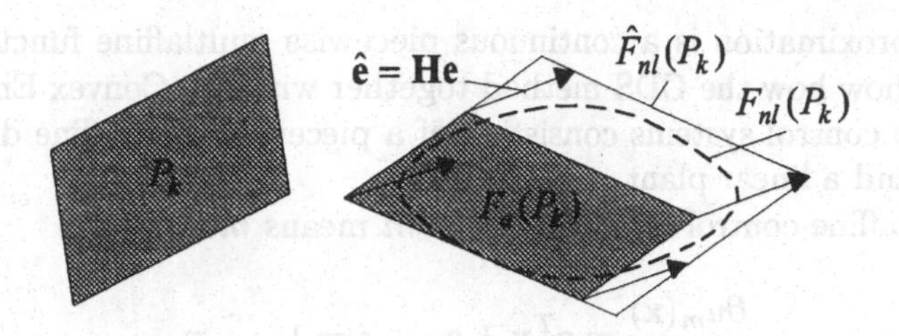

Fig. 4. Construction of the convex polytope $\hat{F}_{nl}(P_k)$ bounded by hyperplanes, which enclose $F_{nl}(P_k)$.

for eq.(13) is calculated setting $\mathbf{e} = \mathbf{e}_{max} - \mathbf{e}_{min}$. Consequently, the complete system equation is estimated by

$$\mathbf{f}_a(\mathbf{x}) \leq \mathbf{f}_{nl}(\mathbf{x}) \leq \mathbf{f}_a(\mathbf{x}) + \hat{\mathbf{e}}, \tag{14}$$

where $\hat{\mathbf{e}}$ is defined by $\hat{\mathbf{e}} = \mathbf{He}$. If the maximum approximation error $e_m = max(\mathbf{e})$ is larger than a given maximum value, P_k is divided into smaller polytopes for which a new affine estimation of the nonlinear control is determined. To construct $\hat{F}_{nl}(P_k) = \{\mathbf{x}|\mathbf{C}_{k+1}\mathbf{x} \leq \hat{\mathbf{d}}_{k+1}\}$ we calculate $F_a(P_k)$ using the estimation $\mathbf{u}_a(\mathbf{x})$ (10). For this, each hyperplane $\mathbf{c}_i^T\mathbf{x} \leq d_i$ of $F_a(P_k)$ is parallel shifted in the direction of $\hat{\mathbf{e}}$ by

$$\hat{d}_i = \begin{cases} d_i & \text{if} \quad \mathbf{c}_i^T\hat{\mathbf{e}} \leq 0 \\ d_i + \mathbf{c}_i^T\hat{\mathbf{e}} & \text{if} \quad \mathbf{c}_i^T\hat{\mathbf{e}} > 0 \end{cases}. \tag{15}$$

The main difficulty in this approach is to find an adequate affine estimation $\hat{\mathbf{u}}_a(\mathbf{x})$ and the resulting maximum estimation error $\mathbf{e}$ for each polytope P_k. The price of the Convex Enclosure is that the CDS method becomes conservative unless we decompose P_k into smaller polytopes to reduce the estimation error $\mathbf{e}$, which increases the complexity of the problem.

Due to the uncertainty of the estimation, an equivalent backward mapping according to eq. (12) for affine control laws cannot be given for polytopes constructed by the Convex Enclosure.

5 Multiaffine Control Laws

Fuzzy controllers do not produce piecewise affine control laws in general. Therefore, the CDS method cannot be applied directly. However, for a certain class of fuzzy controllers the resulting control law is piecewise multiaffine. This is valid for Takagi-Sugeno fuzzy controllers with linear conclusions [9] and for Mamdani fuzzy controllers using Sum-Prod inference, COG defuzzification and $\mu_1(e_i) + \ldots + \mu_m(e_i) = 1$ for each membership degree $\mu_j(e_i)$ corresponding to any real input value e_i [8].

Any static nonlinear controller can be approximated by an n-dimensional lookup table with a multiaffine interpolation for each cell of the table. The resulting approximation is a continuous piecewise multiaffine function. In this section, we show how the CDS method together with the Convex Enclosure can be applied to control systems consisting of a piecewise multiaffine discrete-time control law and a linear plant.

For multiaffine control laws (the index m means *multiaffine*)

$$\frac{\partial u_m(\mathbf{x})}{\partial x_i} = \mathbf{a}^T\mathbf{x} + a_0, \quad i = 1, \dots, n, \tag{16}$$

is valid. This means that a multiaffine function is linear with respect to each axis parallel to a coordinate direction. Such a function is defined by 2^n coefficients. For example a multiaffine function with a 3-dimensional vector $\mathbf{x}$ is given by

$$\begin{aligned} u_m(\mathbf{x}) = m_0 + m_1x_1 + m_2x_2 + m_4x_3 + \\ m_3x_1x_2 + m_5x_1x_3 + m_6x_2x_3 + m_7x_1x_2x_3. \end{aligned}$$

From the Box-Minimum Theorem [7], it is known that a multiaffine function defined on an n-dimensional cuboid takes its extrema in the vertices of the cuboid. When estimating a multiaffine function $u_m(\mathbf{x})$ by an affine function $u_a(\mathbf{x}) = \mathbf{a}^T\mathbf{x} + a_0$, the difference $u_e(\mathbf{x}) = u_m(\mathbf{x}) - u_a(\mathbf{x})$ is also multiaffine and hence takes its extrema in the vertices of the cuboid, so that the maximum estimation error can be calculated considering only the 2^n vertices of the cuboid. An adequate and simple way to compute the $n+1$ coefficients of the affine estimation $u_a(\mathbf{x})$ is to minimize the quadratic error $u_e(\mathbf{x})^2$, referring to the vertices of the cuboid. For this we evaluate

$$\frac{\partial Q(a_0, a_1 \dots a_n)}{\partial a_j} = \frac{\partial \sum_{i=0}^{2^n} \left(u_m - \left[a_1 \dots a_n\right]\mathbf{x} - a_0\right)^2}{\partial a_j} = 0, \quad j = 0, \dots, n, \tag{17}$$

by solving the resulting $n+1$-dimensional system of linear equations.

There are two general approaches to construct the polytope $\hat{F}_m(P_k)$, which encloses $F_m(P_k)$:

1. The domain Z_i is enclosed by a cuboid Q_Z for which an affine estimation $u_{a,j}(\mathbf{x})$ is calculated for each multiaffine function $u_{m,j}(\mathbf{x})$ of the control vector $\mathbf{u}_m(\mathbf{x})$. P_k is also enclosed by a cuboid Q_P for which the approximation error $\mathbf{e}$ is calculated with the affine approximation of the multiaffine control law for the domain Z_i (Fig. 5). This approach has the advantage that the affine approximation $\mathbf{u}_a(\mathbf{x})$ can be calculated only once for each domain Z_i. However, the resulting uncertainty of estimation for P_k may become unreasonably large.
2. P_k is enclosed by a cuboid Q_P for which a new affine estimation $\mathbf{u}_a(\mathbf{x})$ together with the estimation error $\mathbf{e}$ is calculated. This approach guarantees an optimal estimation of $\hat{F}_m(P_k)$.

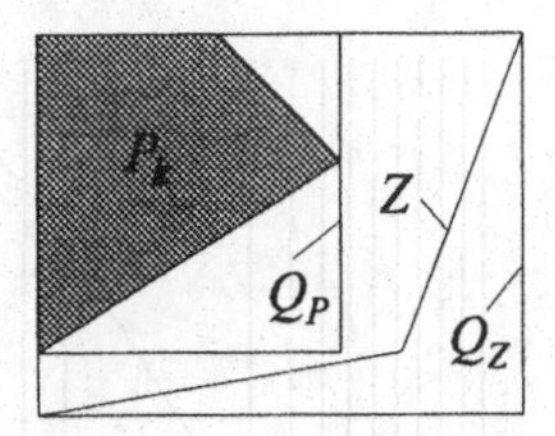

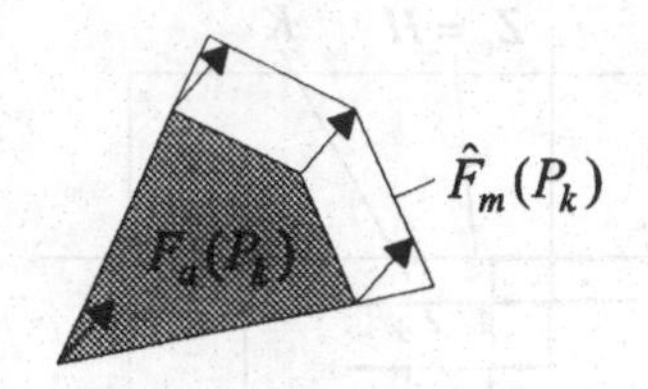

Fig. 5. Construction of the convex polytope $\hat{F}_m(P_k)$ bounded by hyperplanes that encloses $F_m(P_k)$.

6 Piecewiese Linearized Plants

Nonlinear controllers, especially fuzzy controllers and neural networks, are often applied to nonlinear dynamic plants. In many cases, the nonlinearities of the plant can be approximated by piecewise affine functions in a reasonable way, so that we obtain piecewise linearized dynamic systems of the form

$$\dot{\mathbf{x}} = \mathbf{A}_i\mathbf{x} + \mathbf{B}_i\mathbf{u}(\mathbf{x}) + \mathbf{d}_i. \tag{18}$$

The linearization must be determined so that no $\mathbf{A}_i$ is singular. With the linear transformation

$$\mathbf{x} = \hat{\mathbf{x}} - \mathbf{A}_i^{-1}\mathbf{d}_i = \hat{\mathbf{x}} + \mathbf{x}_e \tag{19}$$

the system equation (18) becomes

$$\dot{\hat{\mathbf{x}}} = \mathbf{A}_i\hat{\mathbf{x}} + \mathbf{B}_i\mathbf{u}(\hat{\mathbf{x}}), \tag{20}$$

for which the equivalent discrete-time system $\hat{\mathbf{x}}_{k+1} = \boldsymbol{\Phi}_i\hat{\mathbf{x}}_k + \mathbf{H}_i\mathbf{u}(\hat{\mathbf{x}}_k)$ with the sampling time τ can be determined. To calculate the discrete-time mapping of a polytope P_k, the linear transformation (19) is first applied to P_k to obtain $\hat{P}_k$. The resulting polytope $\hat{P}_{k+1}$ is calculated with eq. (10) and transformed back to obtain P_{k+1}. Another approach to obtaining a piecewise linearized plant is to linearize the system equation at a number of working points. This approach has the drawback that the resulting system equation may be discontinuous [6].

7 Example

We consider the nonlinear model of an inverse pendulum

$$\begin{aligned} \dot{\phi} &= \omega \\ \dot{\omega} &= 1.8sin(\phi) + u, \end{aligned}$$

with the state variables ϕ and ω and the equilibrium state $\mathbf{x}_e = (\phi = 0, \omega = 0)$. The domain of definition for ϕ is bounded by

$$D_\phi = \{\phi | -\frac{\pi}{2} \leq \phi \leq \frac{\pi}{2}\}. \tag{21}$$

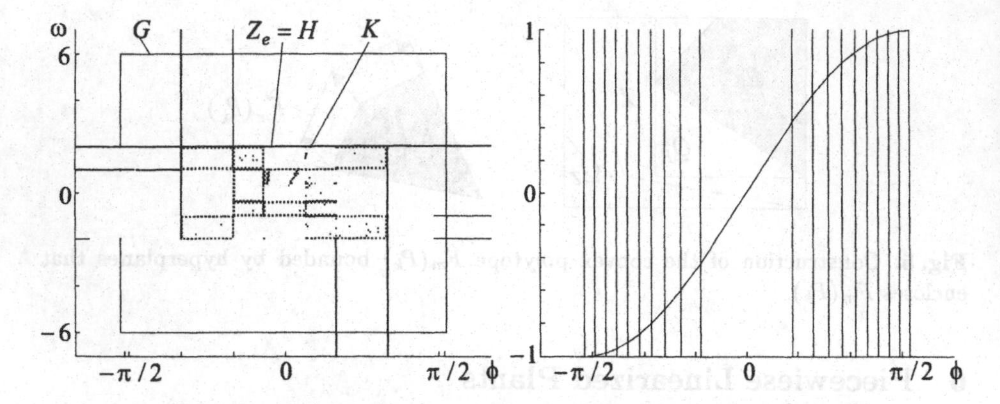

Fig. 6. Thirtyfive boxes parallel to the axes where an affine and a multiaffine, respectively, control law are valid (left). Linearization of the sine function with 20 supporting points and the resulting domains where a uniform linearized plant model is valid (right).

To control this plant we apply a piecewise multiaffine fuzzy control law $u = f_m(\phi, \omega)$ where the controller output is bounded by $-7 \leq u \leq 7$. The control law is given by 35 affine and multiaffine, respectively, partial functions, each valid in one box parallel to the axes (Fig. 6, left). For the shaded boxes, a multiaffine control law is valid, while for the other boxes an affine control law including the special cases of linear and constant control laws is valid. The fuzzy controller is designed so that it supplies a linear control law $\mathbf{a}^T\mathbf{x}$ in the neighbourhood of the equilibrium state $\mathbf{x}_e$. The sampling time for the fuzzy controller is $\tau = 0.05$ s. Here we consider the domain G defined by the cuboid

$$G = \left\{ \begin{pmatrix} \phi \\ \omega \end{pmatrix} \middle| \begin{pmatrix} -\frac{\pi}{2} \\ -6 \end{pmatrix} \leq \begin{pmatrix} \phi \\ \omega \end{pmatrix} \leq \begin{pmatrix} \frac{\pi}{2} \\ 6 \end{pmatrix} \right\}. \tag{22}$$

We investigate the system with respect to G_H-stability with the additional condition that no system trajectory leaves the domain D_ϕ (21). If G_H-stability cannot be established, we want to determine the polytopes $P \subseteq G$ for which P_H-stability is established.

We approximate the sine function by a piecewise affine function $a_{1,i}\mathbf{x} + a_{0,i}$ with 20 supporting points in D_ϕ (Fig. 6, right), so that the plant is described piecewise by the linear equation

$$\begin{pmatrix} \dot{\phi} \\ \dot{\omega} \end{pmatrix} = \begin{pmatrix} 0 & 1 \\ 1.8a_{1,i} & 0 \end{pmatrix} \begin{pmatrix} \phi \\ \omega \end{pmatrix} + \begin{pmatrix} 0 \\ 1 \end{pmatrix} \mathbf{u} + \begin{pmatrix} 0 \\ a_{0,i} \end{pmatrix}. \tag{23}$$

Altogether, there are 131 domains Z_i where an affine and a multiaffine, respectively, system equation is valid (Fig. 7, left). The equilibrium state is situated in the centre of the domain

$$Z_e \left\{ \begin{pmatrix} \phi \\ \omega \end{pmatrix} \middle| \begin{pmatrix} -0.2 \\ -0.4 \end{pmatrix} \leq \begin{pmatrix} \phi \\ \omega \end{pmatrix} \leq \begin{pmatrix} 0.2 \\ 0.4 \end{pmatrix} \right\} \tag{24}$$

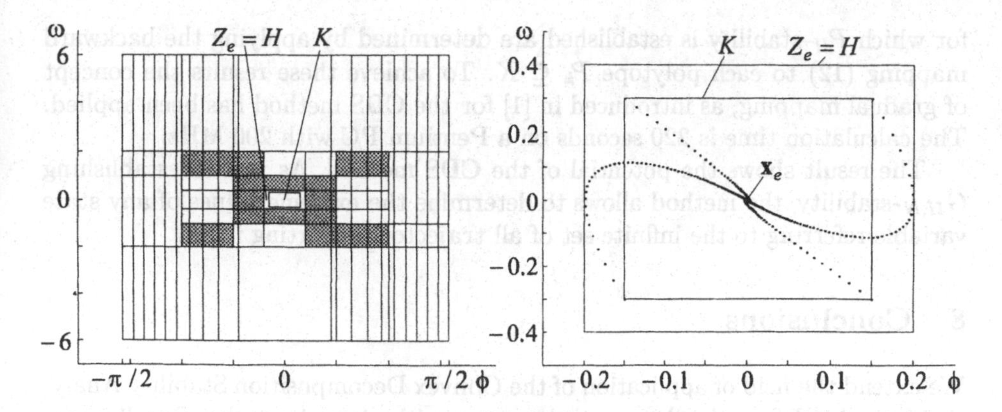

Fig. 7. Decomposition of the domain G into 131 domains $Z_1, \ldots, Z_{131}$ where an affine and a multiaffine, respectively, control law and a uniform linearized plant model are valid (left). Domains $H = Z_e$ and K with the properties $\mathbf{x}_e \in K \subseteq H$ (right).

where a linear asymptotic stable system equation is valid. It turns out that all trajectories starting in the vertices of the cuboid

$$K = \left\{ \begin{pmatrix} \phi \\ \omega \end{pmatrix} \middle| \begin{pmatrix} -0.15 \\ -0.3 \end{pmatrix} \leq \begin{pmatrix} \phi \\ \omega \end{pmatrix} \leq \begin{pmatrix} 0.15 \\ 0.3 \end{pmatrix} \right\} \tag{25}$$

reach the domain K without leaving the domain H (Fig. 7, right) in M steps. Therefore the domain K has the desired properties (2) and (3). Additionally all trajectories starting in K reach the equilibruim state asymptotically.

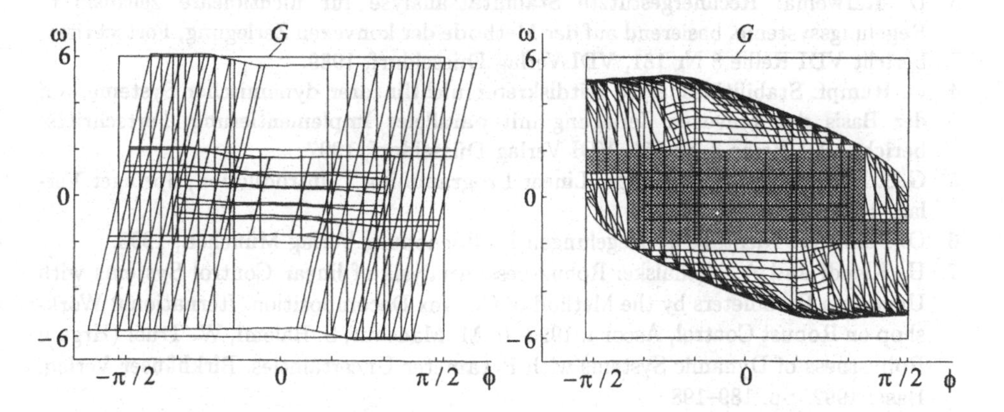

Fig. 8. The mapping $F(G)$ of G (left). The result of the CDS method with backward mapping of subpolytopes P for which P_H-stability is established (right).

The complete domain inside G for which G_H-stability is established, with the additional condition for ϕ, is shown in Fig. 8 (right). The shaded boxes show the domains Z_i for which Z_{iH}-stability is established. The subpolytopes $P \subseteq Z_i$

for which P_H-stability is established are determined by applying the backward mapping (12) to each polytope $P_k \subseteq K$. To achieve these results the concept of gradual mapping, as introduced in [1] for the CDS method has been applied. The calculation time is 220 seconds on a Pentium PC with 200 MHz.

The result shows the potential of the CDS method. As well as establishing $G_{H,N}$-stability, the method allows to determine the extreme values of any state variable referring to the infinite set of all trajectories starting in G.

8 Conclusions

We extend the field of application of the Convex Decomposition Stability Analysis Method (CDS method) to control systems with piecewise multiaffine discrete-time control laws and piecewise linearized plants. Moreover, we can establish $G_{H,N}$-stability strictly for more general controllers, those that can be approximated by piecewise multiaffine functions. Consequently, very complex nonlinear controllers like fuzzy controllers and neural networks can be considered. The efficiency of the improvements introduced has been demonstrated here on a complex 2-dimensional example. Moreover, they can be applied directly to higher-dimensional problems, limited only by the computational expenditure.

References

1. H. Kiendl: Fuzzy Control methodenorientiert, Oldenbourg Verlag München, 1997.
2. H. Kiendl: Robustheitsanalyse von Regelungssystemen mit der Methode der konvexen Zerlegung, Automatisierungstechnik 35 (1987),pp. 192–202.
3. D. Karweina: Rechnergestützte Stabilitätsanalyse für nichtlineare zeitdiskrete Regelungssysteme, basierend auf der Methode der konvexen Zerlegung, Fortschrittsbericht VDI Reihe 8 Nr 181, VDI Verlag Düsseldorf, 1989.
4. O. Rumpf: Stabilitätsanalyse zeitdiskreter nichtlinearer dynamischer Systeme, auf der Basis der konvexen Zerlegung mit paralleler Implementierung, Fortschrittsbericht VDI Reihe 8 Nr 651, VDI Verlag Düsseldorf, 1997.
5. G.B. Dantzig and M.N. Thapa: Linear Programming 1: Introduction, Springer Verlag, 1997.
6. O. Föllinger: Nichtlineare Regelungen I, Oldenbourg Verlag München, 1993.
7. H. Kiendl and A. Michalske: Robustness Analysis of linear Control Systems with Uncertain Parameters by the Method of Convex Decomposition. Iternational Workshop on Robust Control, Ascona, 1992. In M. Mansour, S. Balemi, W. Truöl (Hrgs.): Robustness of Dynamic Systems with Parameter Uncertainties. Birkhäuser Verlag, Basel 1992, pp. 189–198.
8. E.-W. Jüngst, K.D. Meyer-Gramann, Fuzzy Control-Schnell und kostengünstig implementiert mit Standard-Hardware, 2. Workshop "Fuzzy Control" des GMA-UA 1.4.2, Forschungsbericht Nr. 0295, 1995, pp. 10–23.
9. T. Takagi, M. Sugeno, Fuzzy Identification of Systems and Its Application to Modeling and Control, IEEE Transaction on Systems, Man, and Cyberbetics 15 (1), 1985, pp. 116–132.

Fuzzy Utilities Comparison in Multicriteria Analysis

Hepu Deng[1] and Chung-Hsing Yeh[2]

[1] Gippsland School of Computing and Information Technology, Monash University
Churchill, Victoria, 3842, Australia
Hepu.Deng@infotech.monash.edu.au

[2] School of Business Systems, Monash University
Clayton, Victoria, 3168, Australia
Chyeh@bs.monash.edu.au

Abstract. This paper presents a new approach for comparing fuzzy utilities in fuzzy multicriteria analysis. The approach developed combines the merit of two prominent concepts individually used in the existing methods: the fuzzy reference set and the degree of dominance. The decisive information of the fuzzy utilities being compared is sensibly used. The computation involved is simple, and the underlying concepts are logically sound and comprehensible. The comparative study conducted on benchmark cases in the literature shows that the approach compares favorably with other methods examined.

1 Introduction

Decision making often takes place in a fuzzy environment where the information available is imprecise or uncertain. For fuzzy decision problems of prioritising or evaluating a finite set of alternatives involving multiple criteria, the application of fuzzy set theory to multicriteria analysis (MA) models under the framework of utility theory has proven to be an effective approach [4]. In fuzzy MA the overall utility of an alternative with respect to all criteria is often represented by a fuzzy number, referred to as fuzzy utility. The ranking of the alternatives is thus based on the comparison of their corresponding fuzzy utilities.

Many methods for comparing fuzzy numbers have been developed in the literature. However, no single method is dominant in all situations for which the method can be used [2, 4]. With the nature of the fuzzy ranking problem each method proposed has to be judged by its own merit, since an overall evaluation of the method based on specific criteria would be subjective to some degree. Although most methods produce satisfactory results for clear-cut ranking problems they may generate an unreasonable outcome under certain circumstances, resulting from the lack of discrimination between fuzzy utilities that differ only slightly from each other [4]. The need for comparing similar fuzzy utilities is likely to grow as the problem size increases. Therefore, a fuzzy ranking method with a high degree of discrimination ability is desirable for effectively handling large-scale fuzzy decision making problems.

This paper proposes a new ranking approach that combines the two prominent concepts in fuzzy ranking: the fuzzy reference set and the degree of dominance. These two concepts are individually adopted by some existing methods and systematically summarized by this study. The approach is computationally simple and its underlying concepts are logically sound and comprehensible. In the following we first discuss the two concepts to pave the way for the methodology development. A comparative study is then followed to demonstrate the performance of the approach in its capability of distinguishing similar fuzzy numbers.

2 The Reference Fuzzy Set

Let $\varpi = \{A_i\}$ $(i \in N = \{1, 2, \ldots, n\})$ be n normal fuzzy numbers to be compared for ranking the corresponding alternatives. Each fuzzy number A_i is represented as

$$A_i = \{(x_i, \mu_{A_i}(x_i)), x_i \in X\} \tag{1}$$

where X is a subset of the universe of discourse R, and $\mu_{A_i}(x_i)$ is the membership function of x_i in A_i. To allow the n fuzzy numbers to be straightforwardly compared, a fuzzy reference set is defined in (2) to serve as the common comparison base.

$$Y = \{(y, \mu_Y(y)), y \in R\}. \tag{2}$$

Each fuzzy number A_i is then compared with the fuzzy reference set Y. As a result, a fuzzy set $Z_i = \{(z_i, \mu_{Z_i}(z_i)), z_i \in R\}$ can be obtained for each A_i to represent its relative ranking in ϖ, defined as

$$\mu_{Z_i}(z_i) \;=\; \mu_Y(y) \;*\; \mu_{A_i}(x_i) \tag{3}$$

where $*$ denotes a given operator for defining the relationship between Y and A_i. In fuzzy MA, Z_i can be considered to represent the relative utility of alternative i. As a consequence the comparison of all fuzzy numbers A_i is based on the value of Z_i.

The concept of using the fuzzy reference set for comparing fuzzy numbers has been applied by several researchers, including Jain [5, 6], Chen [3], Yager [14], Kerre [7], and Nakamura [10]. In their studies, different definitions of the fuzzy reference set, such as the fuzzy maximum and the fuzzy minimum, are used with a view to making use of the decisive information characterizing the fuzzy numbers being compared including (a) their absolute position and relative position on the real line, and (b) the shape, the spread, or the area of each fuzzy number.

Jain [5, 6] first introduces the concept of a fuzzy maximum (called the maximizing set) as the fuzzy reference set. The fuzzy maximum is defined as $Y = \{(y, \mu_{max}(y)), y \in R\}$ with the membership function given as

$$\mu_{\max}(y) = \left(\frac{y}{x_{\max}}\right)^k, k > 0, \tag{4}$$

where k is an integer that can be assigned in a given context to indicate the decision-maker's attitude towards risk, and x_{max} is the largest value of all supports $S(A_i)$ of the fuzzy numbers A_i $(i = 1, 2, \ldots, n)$, determined by

$$x_{max} = \sup\{\bigcup_{i=1}^{n} S(A_i)\}; \qquad S(A_i) = \{x_i, \mu_{A_i}(x_i) > 0\} \tag{5}$$

where sup{·} is the supremum of all $S(A_i)$. With the fuzzy maximum Y defined, a fuzzy set Z_i is obtained by

$$\mu_{Z_i}(x_i) = \mu_Y(x_i) \wedge \mu_{A_i}(x_i) \tag{6}$$

where $\wedge$ denotes the min operator. In fuzzy MA, the fuzzy set Z_i represents the grades of membership of each fuzzy utility A_i in Y. The maximum value of the grade of membership in each Z_i, i.e. $\sup_{x_i \in X} Z_i(x_i)$, is regarded as the degree of optimality of its corresponding alternative, on which the ranking is based.

This ranking method is not logically sound as only the information on the decreasing right part of the fuzzy numbers is used, that is, the increasing left part of the fuzzy numbers is not considered. This results in counter-intuitive ranking in some cases [2, 4]. In addition, the membership value of Jain's fuzzy maximum may be negative which contradicts the definition of the membership function, if negative support is contained in some fuzzy numbers and k is an odd integer [3]. If k is an even integer, and if the absolute value of the negative support x is greater than the absolute value of x_{max}, then $\mu_{max}(x) > 1$. This also conflicts with the definition of the membership function.

To avoid the problem associated with Jain's definition, Chen [3] defines the fuzzy maximum as

$$Y_{max} = \{(y, \mu_{max}(y)), \mu_{max}(y) = (\frac{y - x_{min}}{x_{max} - x_{min}})^k, y \in X\}$$

The parameter k is used in the same way as Jain's in (4). This definition uses both the largest and smallest values of all supports of the fuzzy numbers, i.e. the relative position of the fuzzy numbers is considered. To achieve a better degree of discrimination between similar fuzzy numbers, a fuzzy minimum is defined as

$$Y_{min} = \{(y, \mu_{min}(y)), \mu_{min}(y) = (\frac{x_{max} - y}{x_{max} - x_{min}})^k, y \in X\}.$$

This makes use of the information from the increasing left part of the fuzzy numbers. As a result, both right and left utility values of each fuzzy number contribute to its relative ranking.

However, Chen's approach may still produce unreasonable outcomes under certain situations [4]. This is due to using the same process as Jain [5, 6] for obtaining the ranking value to represent the degree of optimality of the corresponding alternative. In addition, the ignorance of the absolute position of the fuzzy numbers on the real line results in the same ranking value for different sets of fuzzy numbers which have the same relative position [4]. The definition of the fuzzy maximum and the fuzzy minimum by considering only the relative position of the fuzzy numbers prevents these definitions from being applicable for situations where the comparison between different sets of fuzzy numbers (i.e. different groups of alternatives) is required.

Yager [14] defines the fuzzy maximum $Y = \{(y, \mu_{max}(y)), \mu_{max}(y) = y, y \in R\}$ as the fuzzy reference set. This definition considers the absolute position of the fuzzy numbers on the real line when they are compared with the fuzzy maximum. This definition is independent of the fuzzy numbers to be compared. Therefore, the fuzzy maximum can be consistently used for all ranking cases. The relative ranking Z_i is measured by the Hamming distance between Y and each fuzzy number A_i to represent the closeness of each fuzzy number to the fuzzy maximum. The definition of the relationship between Y and A_i by the Hamming distance is solely based on area measurement that ignores the relative position of the fuzzy numbers on the real line. As a result, this method conflicts with intuition in some cases [4].

Kerre [7] uses the concept in crisp decision cases where the best alternative has the maximum gain, to define the fuzzy maximum between two fuzzy numbers A_i and A_j $(i, j \in N, i \neq j)$ as $Y = \{(y, \mu_{max}(y)), \mu_{max}(y) = \sup_{y=(x_i \vee x_j)} [\mu_{A_i}(x_i) \wedge \mu_{A_j}(x_j)], y \in X\}$. The fuzzy number whose Hamming distance to the fuzzy maximum is smaller is preferred. Following a similar concept, Nakamura [10] defines a fuzzy minimum between two fuzzy numbers A_i and A_j $(i, j \in N, i \neq j)$ as $Y = \{(y, \mu_{min}(y)), \mu_{min}(y) = \sup_{y=(x_i \wedge x_j)} [\mu_{A_i}(x_i) \wedge \mu_{A_j}(x_j)], y \in X\}$. With four Hamming distance measurements, a fuzzy number that is farther from the fuzzy minimum is considered larger. Both definitions require considerable computational effort for fuzzy numbers with a continuous membership function. In addition, due to the use of the Hamming distance for defining Z_i, both methods produce unsatisfactory results [4].

The above studies suggest that the performance of the ranking method using the concept of the fuzzy reference set is highly influenced by (a) the definition of Y and (b) in particular the way that Z_i is obtained, that is, the relationship between Y and A_i in (3). It is evident that a good fuzzy ranking method needs to logically link the definition of Y and the generation of Z_i in order to effectively use all decisive information for distinguishing between the fuzzy numbers.

In our approach, we take advantage of Yager's definition [14] (also given in [4]) to define the fuzzy maximum and the fuzzy minimum because (a) it considers the absolute position of the fuzzy numbers, (b) it involves no computation as the definition is independent of the fuzzy numbers to be compared, thus being used for all applications, and (c) it permits the comparison between different sets of fuzzy numbers (i.e. different groups of decision alternatives) as they are all compared on the same fuzzy reference sets, resulting in comparable ranking values.

To make use of the relative position of the fuzzy numbers and other information for achieving a high degree of discrimination, the dominance concept is applied. This allows us to make best use of the concept of the fuzzy reference set.

3 The Degree of Dominance

To compare two fuzzy numbers A_i and A_j $(i \neq j)$ to determine how much larger A_i is over A_j, the dominance concept is introduced. The most widely used definition is based on the maximum grade of membership of A_i in A_j. This concept is similar to Jain's definition on the degree of optimality [5, 6], although it is applied in a different context. For example, in Bass and Kwakernaak [1], this indicates the degree to which a fuzzy number A_i ($i \in N$) is ranked first when compared with the best fuzzy number A_j ($j \in N, j \neq i$), identified with the use of a conditional fuzzy set. This concept is shared by Watson *et al.* [14] who describe each of pairwise comparisons of fuzzy numbers as a fuzzy implication. However, the problem with these methods, mainly resulting from the definition of dominance, is illustrated in [4]. Tong and Bonissone [11] directly use the definition to represent the dominance of A_i over A_j, although their method produces a linguistic solution rather than a numerical ranking. Despite their intention of reflecting the separation between two fuzzy numbers by choosing this definition, the information about the overall shape of fuzzy numbers is ignored.

Departing from the above concept Tseng and Klein [12] define the degree of dominance between two fuzzy numbers by comparing their overlap and non-overlap areas, in line with the concept of Hamming distance. This method shows an advantage over some existing methods. However, the relative demerits of this method include (a) that the computation of the areas is not straightforward, and (b) that an additional, often tedious, pairwise comparison process is needed for comparing a large set of fuzzy numbers.

In our approach, we define the degree of dominance of A_i over A_j as their arithmetic difference, that is, $A_i - A_j$. This fuzzy set difference indicates the grade of membership of the crisp values representing their relative closeness on the real line. This definition is similar in concept to the comparison of crisp numbers on the real line. When pairs in a set of crisp numbers are compared, the arithmetic difference between any one pair of numbers can be regarded as an indication of how much larger (positive or negative) one is over the other. This value indicates their relative closeness in comparison with other pairs. This relative closeness can be regarded as a ranking index when comparing a large set of fuzzy numbers with a fuzzy reference set on the same universe of discourse, where the relationship between the fuzzy reference set and each fuzzy number (i.e. Z_i) is measured in relative terms.

Fuzzy arithmetic has been well developed to perform standard arithmetic operations on fuzzy numbers [4]. The fuzzy set difference D_{i-j} between A_i and A_j can be calculated by fuzzy subtraction, as

$$D_{i-j} = A_i - A_j = \{(z, \mu_{D_{i-j}}(z)), z \in R\} \tag{7}$$

where the membership function of D_{i-j} is defined as

$$\mu_{D_{i-j}}(z) = \sup_{z=x_i-x_j} (\min(\mu_{A_i}(x_i), \mu_{A_j}(x_j)), x_i, x_j \in X). \tag{8}$$

To determine how much larger A_i is over A_j, a defuzzification process is required to extract a single scalar value from D_{i-j}, which can best represent D_{i-j}. The centroid method [15] is commonly regarded as an effective defuzzification technique. As a result, the degree of dominance of A_i over A_j is determined by

$$d(A_i - A_j) = \frac{\int_{S(D_{i-j})} z\mu_{D_{i-j}}(z)dz}{\int_{S(D_{i-j})} \mu_{D_{i-j}}(z)dz} \tag{9}$$

where $S(D_{i-j}) = \{z, \mu_{D_{i-j}}(z) > 0, z \in R\}$ is the support of D_{i-j}. A_i dominates A_j if $d(A_i-A_j) > 0$, and A_i is dominated by A_j if $d(A_i-A_j) < 0$. The larger the value of $d(A_i-A_j)$, the higher the degree of dominance A_i over A_j. When a set of fuzzy numbers are compared on the same universe of discourse, the value of $d(A_i\text{-}A_j)$ indicates the degree of relative closeness between A_i and A_j.

A special indifference situation is given in Fig. 1, where $d(A_i\text{-}A_j) = 0$. Fuzzy numbers A_i and A_j share the same central tendency with different amounts of dispersion. With the perception that human intuition would favor a fuzzy number with a larger mean value and a smaller dispersion [8], A_j is preferred to A_i. To solve this problem, the standard deviation of fuzzy number A_i is used as the preference index for distinguishing between symmetrical fuzzy numbers with the same mean value, given by

$$\sigma_i = [\frac{\int_{S(A_i)} x^2 \mu_{A_i}(x)dx}{\int_{S(A_i)} \mu_{A_i}(x)dx} - (\frac{\int_{S(A_i)} x\mu_{A_i}(x)dx}{\int_{S(A_i)} \mu_{A_i}(x)dx})^2]^{1/2}, \tag{10}$$

where $S(A_i) = \{x, \mu_{A_i}(x) > 0, x \in X\}$ is the support of A_i.

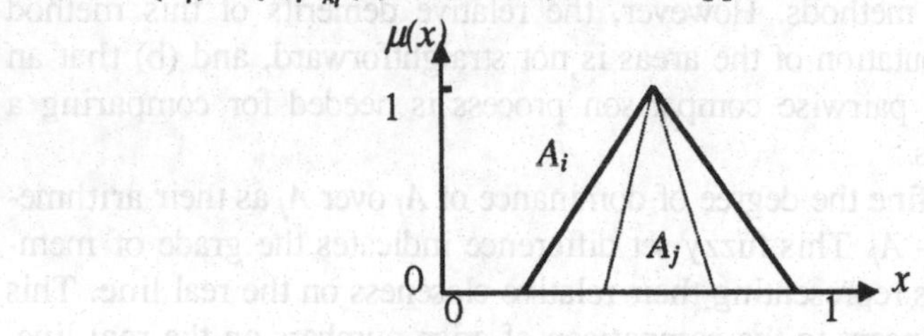

Fig. 1. Nondiscrimination case with the centroid method

It is worth noting that using the centroid of the fuzzy number alone does not provide good discrimination ability [4]. Without using a fuzzy reference set, Yager [15] directly uses the x-axis value of the centroid of the fuzzy number as the ranking value. Murakami *et al* [9] extend Yager's method by considering both the x-axis and y-axis values of the centroid of the fuzzy number. However, both methods produce the same ranking result for triangular or rectangular fuzzy numbers [2, 4]. This seems to suggest that the x-axis value of the centroid is the only rational index for comparing fuzzy numbers [4], which is used by our approach.

4 The Approach

In this section we present a new fuzzy ranking approach that incorporates the two concepts discussed above. The approach involves the use of two fuzzy reference sets: the fuzzy maximum and the fuzzy minimum. The rationale of the approach is that a fuzzy number is preferred if it is dominated by the fuzzy maximum by a smaller degree (i.e. closer to the fuzzy maximum), and at the same time dominates the fuzzy minimum by a larger degree (i.e. farther away from the fuzzy minimum). The approach makes use of all decisive information associated with the fuzzy numbers. The definition of the two fuzzy reference sets uses the absolute position of the fuzzy numbers. The use of fuzzy subtraction for defining the relationship between the fuzzy reference set and the fuzzy number considers the relative position, and the shape (increasing left part and decreasing right part) and the area of each fuzzy number.

In the approach developed, the fuzzy maximum ($Y_{max} = \{(y, \mu_{max}(y)), y \in R\}$) and the fuzzy minimum ($Y_{min} = \{(y, \mu_{min}(y)), y \in R\}$) are always given by

$$\mu_{max}(y) = \begin{cases} y, \text{if } 0 \le y \le 1 \\ 0, \text{otherwise.} \end{cases} \tag{11}$$

$$\mu_{min}(y) = \begin{cases} 1-y, \text{if } 0 \le y \le 1 \\ 0, \text{ otherwise.} \end{cases} \tag{12}$$

Based on our definition in (7)-(9), the degree to which the fuzzy maximum dominates each fuzzy number A_i ($i = 1, 2, \ldots, n$) can be expressed in general form as

$$d_i^+ = d(Y_{max} - A_i) = \frac{\int_{S(D_{max-i})} z\mu_{D_{max-i}}(z)dz}{\int_{S(D_{max-i})} \mu_{D_{max-i}}(z)dz}, \tag{13}$$

where $\quad D_{max-i} = Y_{max} - A_i, S(D_{max-i}) = \{z, \mu_{D_{max-i}}(z) > 0, z \in R\}.$ (14)

The value of d_i^+ indicates the degree of closeness of each fuzzy number A_i to the fuzzy maximum Y_{max}. A rational decision-maker would prefer a fuzzy number which is closer to the fuzzy maximum, indicated by a smaller value of d_i^+.

Similarly, the degree of dominance of each fuzzy number A_i over the fuzzy minimum Y_{min} is given as

$$d_i^- = d(A_i - Y_{min}) = \frac{\int_{S(D_{i-min})} z\mu_{D_{i-min}}(z)dz}{\int_{S(D_{i-min})} \mu_{D_{i-min}}(z)dz}, \tag{15}$$

where $\quad D_{i-min} = A_i - Y_{min}, \qquad S(D_{i-min}) = \{z, \mu_{D_{i-min}}(z) > 0, z \in R\}.$ (16)

The value of d_i^- represents the degree of closeness of each fuzzy number A_i to the fuzzy minimum. Fuzzy numbers that are farther away from the fuzzy minimum, indicated by a larger value of d_i^-, are considered to be preferred.

An overall preference index for fuzzy number A_i ($i = 1, 2, \ldots, n$) is obtained by

$$P_i = \frac{d_i^-}{d_i^+ + d_i^-} \tag{17}$$

The larger the preference index P_i, the more preferred the fuzzy number A_i. Clearly, the smaller the d_i^+ or the larger the d_i^-, the larger the P_i. This implies that a fuzzy number A_i is preferred (a larger P_i) if it is closer to the fuzzy maximum (a smaller d_i^+) and it is farther away from the fuzzy minimum (a larger d_i^-).

In practical applications, it is of utmost importance to keep the information to be processed to a minimum, in particular for large-scale problems. In situations where the type of fuzzy sets used is not the major concern, fuzzy numbers with triangular or trapezoidal membership functions are commonly used to simplify the computational process. Without loss of generality, triangular fuzzy numbers are used here to demonstrate the simplicity of the approach and to facilitate the presentation of the comparative study in the next section. A triangular fuzzy number A_i ($i \in N$) is usually characterized by three real numbers, expressed as (a_i, b_i, c_i) and given as

$$\mu_{A_i}(x) = \begin{cases} \dfrac{x-a_i}{b_i-a_i}, & \text{if } a_i \le x \le b_i \\ \dfrac{c_i-x}{c_i-b_i}, & \text{if } b_i \le x \le c_i \\ 0, & \text{otherwise.} \end{cases} \tag{18}$$

where b_i is the most possible value of A_i, and a_i and c_i are the lower and upper bounds used to reflect the fuzziness of the information.

As triangular fuzzy numbers, Y_{max} and A_i are represented as $(a_{max}, b_{max}, c_{max})$ and (a_i, b_i, c_i) respectively. In this case, d_i^+ in (13) and $D_{max\text{-}i}$ in (14) can be calculated by

$$d_i^+ = \frac{a_{\max-i} + b_{\max-i} + c_{\max-i}}{3}, \tag{19}$$

$$D_{max\text{-}i} = (a_{max\text{-}i}, b_{max\text{-}i}, c_{max\text{-}i}) = = (a_{max} - c_i, b_{max} - b_i, c_{max} - a_i). \tag{20}$$

The same formulas apply to d_i^- in (15) and $D_{i\text{-}min}$ in (16). Clearly, the computational process of the approach is simple for comparing triangular or trapezoidal fuzzy numbers.

5 Comparative Study

With the merits of simplicity and logicality, the approach developed has been applied in all cases used in the literature to examine its performance on discrimination ability. The examination result shows that the approach produces satisfactory outcomes for all cases. It is noteworthy that the approach also gives satisfactory results for all cases examined if Chen's fuzzy maximum and fuzzy minimum [3] are used as the fuzzy reference sets.

As a comparison, Table 1 shows the ranking results of representatives of the methods, using one of the concepts discussed, on seven benchmark cases. In Table 1, some results are adopted from [2] and [4], and some are calculated by this study. For easy comparison, ranking values produced by Kerre [7] and Yager [15] are transformed by subtracting the original index value from the crisp number 1, so that they are listed in descending order. The first row shows the correct ranking suggested by the literature [2, 4]. The last row shows the ranking values given by the approach developed. All methods, except for our approach, give unsatisfactory results for one or more cases.

Table 1. Comparison results

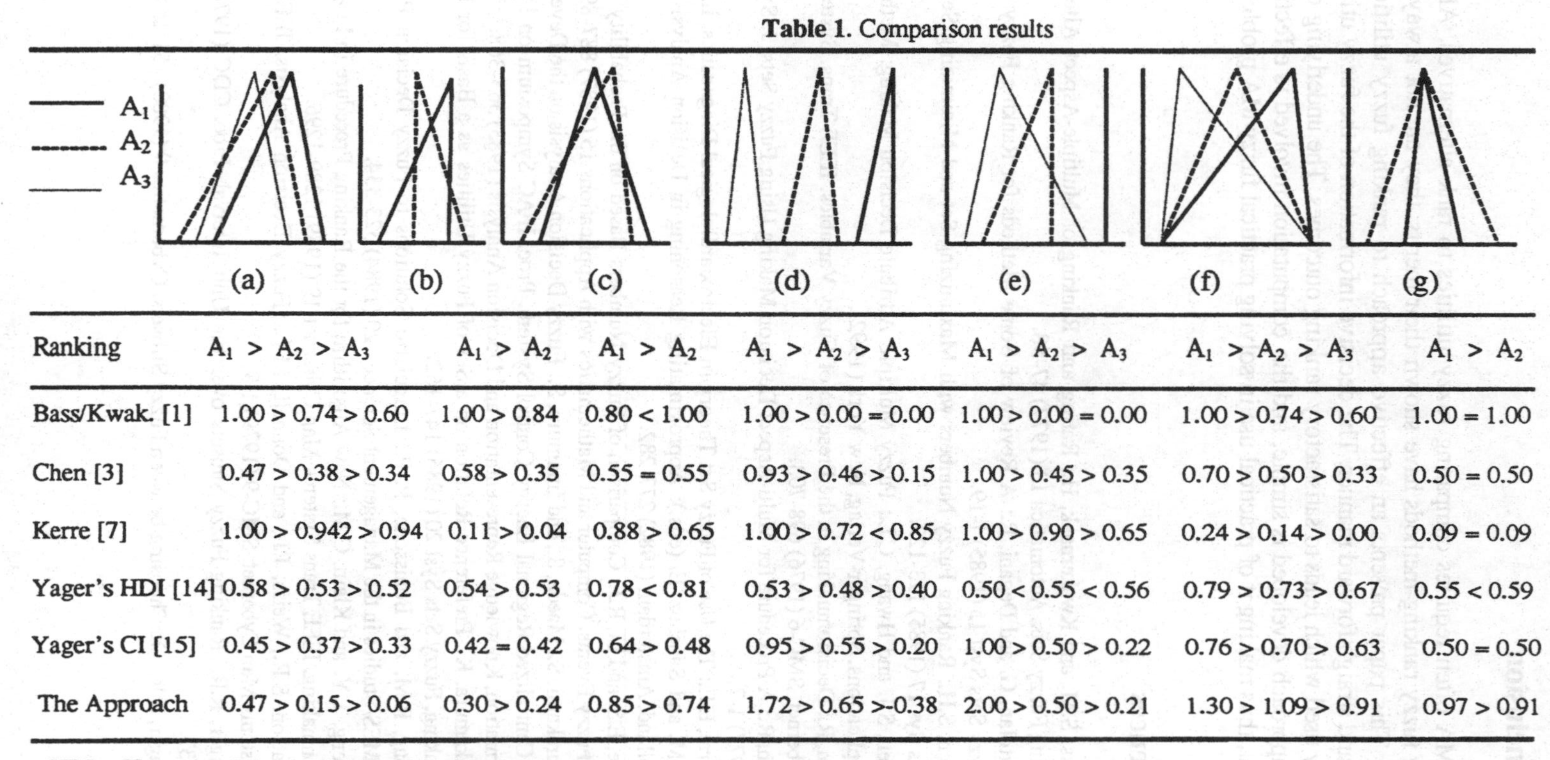

	(a)	(b)	(c)	(d)	(e)	(f)	(g)
Ranking	$A_1 > A_2 > A_3$	$A_1 > A_2$	$A_1 > A_2$	$A_1 > A_2 > A_3$	$A_1 > A_2 > A_3$	$A_1 > A_2 > A_3$	$A_1 > A_2$
Bass/Kwak. [1]	1.00 > 0.74 > 0.60	1.00 > 0.84	0.80 < 1.00	1.00 > 0.00 = 0.00	1.00 > 0.00 = 0.00	1.00 > 0.74 > 0.60	1.00 = 1.00
Chen [3]	0.47 > 0.38 > 0.34	0.58 > 0.35	0.55 = 0.55	0.93 > 0.46 > 0.15	1.00 > 0.45 > 0.35	0.70 > 0.50 > 0.33	0.50 = 0.50
Kerre [7]	1.00 > 0.942 > 0.94	0.11 > 0.04	0.88 > 0.65	1.00 > 0.72 < 0.85	1.00 > 0.90 > 0.65	0.24 > 0.14 > 0.00	0.09 = 0.09
Yager's HDI [14]	0.58 > 0.53 > 0.52	0.54 > 0.53	0.78 < 0.81	0.53 > 0.48 > 0.40	0.50 < 0.55 < 0.56	0.79 > 0.73 > 0.67	0.55 < 0.59
Yager's CI [15]	0.45 > 0.37 > 0.33	0.42 = 0.42	0.64 > 0.48	0.95 > 0.55 > 0.20	1.00 > 0.50 > 0.22	0.76 > 0.70 > 0.63	0.50 = 0.50
The Approach	0.47 > 0.15 > 0.06	0.30 > 0.24	0.85 > 0.74	1.72 > 0.65 >-0.38	2.00 > 0.50 > 0.21	1.30 > 1.09 > 0.91	0.97 > 0.91

HDI = Hamming Distance Index, CI = Centroid Index.

6 Conclusion

Fuzzy MA often requires comparing fuzzy utilities to rank alternatives. Although existing fuzzy ranking methods have shown their merits, they are not always satisfactory. This paper presents an effective approach to ranking fuzzy utilities in a simple and straightforward manner. The decisive information of the fuzzy utilities is sensibly used which leads to satisfactory ranking outcomes. The underlying concept of the approach developed is simple, and the computation involved is effective and efficient, thus making it of practical use in solving practical fuzzy MA problems.

References

1. Bass, S.M. and Kwakernaak, H.: Rating and Ranking of Multiple-Aspect Alternatives Using Fuzzy Sets. Automatica 13 (1977) 47-58.
2. Bortolan, G. and Degani, R.: A Review of Some Methods for Ranking Fuzzy Subsets. Fuzzy Sets Syst 15 (1985) 1-19.
3. Chen, S.H.: Ranking Fuzzy Numbers with Maximizing Set and Minimizing Set. Fuzzy Sets Sys 17 (1985) 113-129.
4. Chen, S.J. and Hwang, C.L.: Fuzzy Multiple Attribute Decision Making: Methods and Applications. Springer-Verlag, New York (1992).
5. Jain, R.: Decisionmaking in the Presence of Fuzzy Variables. IEEE Trans. Systems Man Cybernet. SMC-6 (1976) 698-703.
6. Jain, R.: A Procedure for Multi-Aspect Decision Making Using Fuzzy Sets. J. Syst Sci 8 (1977) 1-7.
7. Kerre, E.E.: The Use of Fuzzy Set Theory in Electrocardiological Diagnostics. In: Gupta, M.M. and Sanchez, E. (eds.): Approximating Reasoning in Decision Analysis. North-Holland, Amsterdam (1982) 277-282.
8 Lee, E.S. and Li, R.J.: Comparison of Fuzzy Numbers Based on the Probability Measure of Fuzzy Events. Computer and Mathematics with Applications 15 (1987) 887-896.
9. Murakami, S., Maeda, S., and Imamura, S.: Fuzzy Decision Analysis on the Development of Centralized Regional Energy Control System. Proc. IFAC Symposium on Fuzzy Information, Knowledge Representation and Decision Analysis (1983) 363-368.
10. Nakamura, K.: Preference Relation on a Set of Fuzzy Utilities as a Basis for Decision Making. Fuzzy Sets Syst 20 (1986) 147-162.
11. Tong, R.M. and Bonissone, P. P.: Linguistic Solutions to Fuzzy Decision Problems. TIMES/Studies in the Management Sciences 20 (1984) 323-334.
12. Tseng, T.Y. and Klein, C.M.: New Algorithm for the Ranking Procedure in Fuzzy Decisionmaking. IEEE Trans Systems Man Cybernet 19 (1989) 1289-1296.
13. Watson, S.R., Weiss, J.J., and Donnell, M.L.: Fuzzy Decision Analysis. IEEE Trans. Systems Man Cybernet. SMC-9 (1979) 1-9.
14. Yager, R.R.: Ranking Fuzzy Subsets over the Unit Interval. Proc. CDC (1978) 1435-1437.
15. Yager, R.R.: On Choosing between Fuzzy Subsets. Cybernetics 9 (1980) 151-154.

A Possibilistic Formalization of Case-Based Reasoning and Decision Making

Eyke Hüllermeier*

IRIT - Institut de Recherche
en Informatique de Toulouse
Université Paul Sabatier
eyke@irit.fr

Abstract. The idea of *case-based decision making* has recently emerged as a new paradigm for decision making under uncertainty. It combines principles from decision theory and *case-based reasoning*, a problem solving method in artificial intelligence. In this paper, we propose a formalization of case-based reasoning which is based on possibility theory and utilizes approximate reasoning techniques. The corresponding approach to case-based decision making is realized as a two-stage process. In the first stage, the decision maker applies case-based reasoning in order to quantify the uncertainty associated with different decisions in form of possibility distributions on the set of consequences. In the second stage, generalizations of expected utility theory are used for choosing among acts resp. the associated distributions.

1 Introduction

The idea of *case-based decision making* (CBDM) has recently emerged as a new paradigm for decision making under uncertainty. It combines principles from decision theory and case-based reasoning (CBR). The approach of CBR, which has now become an important problem solving method in artificial intelligence, is based on the hypothesis that "similar problems have similar solutions." More generally, the idea is to exploit the experience from similar cases in the past and to adapt then successful solutions to the current situation. The idea of basing a decision theory on principles of similarity and case-based reasoning is originally due to Gilboa and Schmeidler [6]. Alternative approaches to case-based decision making have been proposed in [5] and [9].

Utilizing CBR for problem solving requires the formalization of the CBR hypothesis and the underlying reasoning principle. In this paper, which combines ideas from [5, 6, 9], we formalize the CBR hypothesis and the corresponding reasoning method within the framework of possibility theory and approximate reasoning. Case-based decision making is then realized as a two-stage process. The decision maker, faced with a certain problem, uses his experience from precedent cases in connection with the CBR principle in order to characterize his

* This work has been partly supported by a TMR research grant funded by the European Commission.

uncertainty concerning the outcome associated with different decisions. More precisely, he derives a characterization in form of a possibility distribution over consequences based on the above mentioned formalization of CBR. The problem of choosing among decisions is then reduced to the problem of choosing among possibility distributions. Formally, this problem is closely related to decision making under risk, where the decision maker has to choose among probability distributions. As an important difference between the approach to case-based decision making proposed in this paper and previous approaches let us mention that case-based reasoning is not realized at the "decision level" directly. Rather, CBR is used for quantifying and updating the uncertain knowledge of the decision maker.

The remaining part of the paper is structured as follows: Section 2 gives a brief overview of approaches to CBDM. In Section 3, case-based reasoning is formalized as some kind of approximate reasoning, and the problem of decision making is considered. The paper concludes with a summary in Section 4.

2 Case-Based Decision Making

The approach of Gilboa and Schmeidler. In their case-based decision theory (CBDT) Gilboa and Schmeidler [6] assume a (finite) memory $\mathcal{M}$ of triples (p, a, r), where p is a problem, a is an act, and r is the (uniquely defined) consequence resulting from choosing that act for solving the corresponding problem. Faced with a new problem p_0, a decision maker estimates the utility of an act a based on its performance in similar problems in the past. The evaluation proposed in [6] is of the form

$$V(a) = V_{p_0,\mathcal{M}}(a) := \sum_{(p,a,r)\in\mathcal{M}} sim(p, p_0) \cdot u(r), \tag{1}$$

where $u(r) = u(r(p, a))$ is the utility associated with the result r and sim measures the similarity of problems. The summation over the empty set yields the "default value" 0. As in expected utility theory (EUT), the decision maker is supposed to choose an act maximizing (1). Of course, the assumption behind (1) is that the utility of an act a is similar under similar conditions, i.e., for similar problems.[1] Alternatively, a "normalized version" of the linear functional (1) has been proposed, which results from replacing sim in (1) by the "normalized similarity" $sim'(p, p_0) := sim(p, p_0)/\sum_{(p',a,r)\in\mathcal{M}} sim(p', p_0)$. Theoretical details of CBDT, including special assumptions concerning the memory $\mathcal{M}$ and an axiomatic characterization of decision principle (1), can be found in [6].

Alternative approaches. In order to avoid certain questionable properties of (1), notably its accumulative nature and the fact that it somehow "compensates"

[1] Observe that this is generally not equivalent to the assumption that acts entail similar *results* for similar problems.

between observations which support and those which suggest to refuse an act a, case-based decision making has been realized as some kind of similarity-based approximate reasoning in [5]. The evaluation

$$V(a) = V_{p_0,\mathcal{M}}(a) := \min_{(p,a,r)\in\mathcal{M}} sim(p,p_0) \rightarrow u(r) \tag{2}$$

(with $\text{rg}(sim) = \text{rg}(u) = [0,1]$) can be interpreted as a generalized truth value of the proposition that "choosing the act a for problems similar to p_0 has *always* resulted in good outcomes." As a special realization of (2) the evaluation

$$V(a) = V_{p_0,\mathcal{M}}(a) := \min_{(p,a,r)\in\mathcal{M}} \max\{1 - sim(p,p_0), u(r)\} \tag{3}$$

is proposed. Moreover, an *optimistic* counterpart of (3) is introduced, which can be seen as a generalized truth degree of the proposition that "there is at least one problem similar to p_0 for which the act a has led to a good result." Modifications which avoid some technical problems of these evaluations are also proposed.

In [9] CBDM has been based on a probabilistic formalization of the hypothesis that "the more similar two situations are, the more *likely* it is that the corresponding outcomes are similar." Here, a situation is thought of as a tuple $s = (p,a)$ consisting of a problem and an act. Instead of deriving a "point estimation" such as (1), this approach utilizes case-based information in order to obtain a probability measure characterizing the belief of a decision maker concerning an unknown utility $u(r(p_0,a))$. This idea is closely related with the *non-deterministic* formalization of CBR in [5]. The problem of choosing among decisions then appears as one of choosing among probability distributions. Thus, the approach is close to classical (statistical) decision theory, especially to decision making under risk. The possibilistic formalization of CBDM proposed in the next section has been developed with similar ideas in mind.

3 A Possibilistic Formalization of CBDM

Suppose that we are given a set $\mathcal{Q}$ of problems, a set $\mathcal{A}$ of possible acts, a set $\mathcal{R}$ of results, and a utility function $u : \mathcal{R} \rightarrow U = [0,1]$. We define $\mathcal{S} := \mathcal{Q}\times\mathcal{A}$ as the set of situations. The result r associated with a situation $s = (p,a)$ is assumed to be a function of p and a, i.e., $r = r(p,a)$. We also assume the existence of similarity functions $sim_{\mathcal{Q}} : \mathcal{Q}\times\mathcal{Q} \rightarrow [0,1]$ and $sim_{\mathcal{R}} : \mathcal{R}\times\mathcal{R} \rightarrow [0,1]$. These functions may be associated with fuzzy (similarity) relations $\widetilde{S}_{\mathcal{Q}}$ and $\widetilde{S}_{\mathcal{R}}$ modelling the fuzzy concepts of *similar problems* and *similar outcomes*, respectively. From $sim_{\mathcal{Q}}$ we may derive a similarity measure $sim_{\mathcal{S}}$ over $\mathcal{S}$ resp. a similarity relation $\widetilde{S}_{\mathcal{S}}$ simply by defining $sim_{\mathcal{S}}((p,a),(p',a')) = sim_{\mathcal{Q}}(p,p')$ if $a = a'$, and 0 otherwise, which has the same effect as the summation in (1): in order to estimate the utility of an act a given the problem p, only cases with the same act are taken into account. Observe, however, that often a more general definition of $sim_{\mathcal{S}}$ (e.g., as a function of $sim_{\mathcal{Q}}$ and a similarity $sim_{\mathcal{A}}$ on $\mathcal{A}\times\mathcal{A}$) seems appropriate. This is the case, for instance, if acts are somehow comparable and if we may assume that a small variation of an act does not influence the result considerably [7].

Suppose the agent to have a memory

$$\mathcal{M} = \{\langle s_1, r_1\rangle, \langle s_2, r_2\rangle, \ldots, \langle s_n, r_n\rangle\} \tag{4}$$

at his disposal, where $s_k = (p_k, a_k) \in \mathcal{S}$ and $r_k = r(p_k, a_k)$ $(k = 1, \ldots, n)$. Moreover, suppose that he has to choose an act for a new problem p_0. In other words, he has to select one element from the (restricted) set $\mathcal{S}_{p_0} := \{p_0\} \times \mathcal{A} \subset \mathcal{S}$ of situations associated with the problem p_0. This choice will depend on the utility expected from each of the situations $(p_0, a) \in \mathcal{S}_{p_0}$. If, for a certain act a, the situation (p_0, a) has not been encountered so far (i.e., there is no case $\langle s, r\rangle \in \mathcal{M}$ such that $s = (p_0, a)$), the agent will generally be uncertain about the result $r(p_0, a)$ and, hence, about the utility $v_0 = u(r(p_0, a))$.

3.1 Formalizing the CBR hypothesis

Consider the following formulation of the CBR principle, which is a ***possibilistic*** counterpart of the probabilistic formulation in [9]: "The more similar two situations are, the more *certain* it is that the associated outcomes are similar." This formulation is also closely related with a possibilistic formalization of the CBR principle proposed in [5]. It is possible to assume a relation between situations and utility values as well and to conclude from the similarity of situations onto the similarity of *utilities* directly: "The more similar two situations are, the more certain it is that the associated utility values are similar." Of course, these formulations are equivalent if outcomes are defined in terms of utility values (in which case u is the identity id.)

An adequate formalization of the above CBR hypothesis can be given by means of a *certainty rule*, which is a special kind of a fuzzy rule. A certainty rule corresponds to statements of the form "the more X is $\widetilde{A}$, the more certain Y lies in $\widetilde{B}$," where $\widetilde{A}$ and $\widetilde{B}$ are fuzzy sets. According to Zadeh's possibility theory-based approach to approximate reasoning [12], such a rule imposes a constraint on the possibility of tuples (x, y) in form of an upper bound:

$$(\forall\, x \in \mathcal{X})\,(\forall\, y \in \mathcal{Y}) \,:\, \pi(y|x) \leq \max\{1 - \widetilde{A}(x), \widetilde{B}(y)\}, \tag{5}$$

where $\mathcal{X}$ and $\mathcal{Y}$ are the domains of the variables X and Y, respectively. Since a certainty rule is thought of as a rule which holds true in normal cases but still allows for exceptional situations it seems well-suited as a formal model of the heuristic CBR principle. Within the context of CBDM, a certainty rule appears in the following form: "The more similar two problems p, p' and acts a, a' are, the more certain it is that the resulting consequences r and r' are similar."

3.2 CBR as approximate reasoning

Consider a new situation s_0, with unknown associated result r_0 and utility $v_0 = u(r_0)$, and a case $\langle s_1, r_1\rangle$. According to (5), this case restricts the possibility of

r_0 to be realized by some outcome $r \in \mathcal{R}$ as follows:

$$\pi(r) \leq \max\{1 - \widetilde{S}_{\mathcal{S}}(s_0, s_1), \widetilde{S}_{\mathcal{R}}(r, r_1)\} . \tag{6}$$

Given a complete memory (4), each case $\langle s_k, r_k \rangle \in \mathcal{M}$ defines a constraint of the form (6). Since the "certainty-semantics" corresponds to the implication-based approach to fuzzy rules, the rules associated with these cases should be combined conjunctively. Thus, modelling the (generalized) conjunction by means of the min-operator, and applying a minimal specificity principle,[2] we obtain the following possiblity distribution which characterizes the uncertainty concerning the unknown outcome r_0:

$$\pi(r) = \min_{k=1,\ldots,n} \max\{1 - \widetilde{S}_{\mathcal{S}}(s_0, s_k), \widetilde{S}_{\mathcal{R}}(r, r_k)\} \tag{7}$$

for all $r \in \mathcal{R}$. In order to take the degree to which the CBR hypothesis actually holds true for a certain application into account, we generalize (7) as follows:

$$\pi(r) = \min_{k=1,\ldots,n} \max\{1 - m(\widetilde{S}_{\mathcal{S}}(s_0, s_k)), \widetilde{S}_{\mathcal{R}}(r, r_k)\}, \tag{8}$$

where $m : [0,1] \to [0,1]$ is a (non-decreasing) function which allows for adapting the strength of the underlying certainty rule.[3] This generalization is closely related with the aspect of (case-based) *learning*:[4] (8) corresponds to a formalization of the *generic* "CBR knowledge" of the decision maker, and m allows for an adequate adaptation of this knowledge. For instance, $m \equiv 0$ means that the similarity of situations does not constrain the similarity of results, i.e., that the CBR hypothesis does not apply at all, whereas $m = \text{id}$ means that two results are always at least as similar (in the sense of $\widetilde{S}_{\mathcal{R}}$) as are the corresponding situations (in the sense of $\widetilde{S}_{\mathcal{S}}$.) The function m may also be utilized in order to guarantee the possibility distributions (8) to be normalized.

In order to summarize the results we have obtained so far imagine a decision maker having a memory (4) at his disposal. Moreover, suppose this decision maker to rely on the CBR hypothesis. Particularly, assume that he utilizes the possibilistic formalization of this hypothesis as outlined above. Then, faced with a new problem p_0, he will derive a possibility distribution $\pi_{a,\mathcal{M}}$ on the set $\mathcal{R}$ of outcomes for each act $a \in \mathcal{A}$ according to (8):

$$\pi_{a,\mathcal{M}}(r) = \min_{k=1,\ldots,n} \max\{1 - m(\widetilde{S}_{\mathcal{S}}((p_0, a), (p_k, a_k))), \widetilde{S}_{\mathcal{R}}(r, r_k)\} .$$

At this stage, the problem of choosing an act $a \in \mathcal{A}$ turns out to be the problem of choosing among the possibility distributions

$$\{\pi_{a,\mathcal{M}} \mid a \in \mathcal{A}\}. \tag{9}$$

[2] The constraint (6) only entails an upper bound of a possibility distribution.

[3] In connection with fuzzy rules, such as function is also called a *modifier* [2].

[4] The topic of case-based learning constitutes an essential part of the overall approach to CBDM based on (8). It is, however, not addressed in this paper.

Thus, our agent actually needs a preference relation $\preceq$ over the possibility distributions on $\mathcal{R}$ resp. a decision criterion which represents such a relation. Before considering this problem let us make a few remarks concerning the approach presented so far.

- A counterpart of (8) based on the concept of a ***possibility-rule*** has been proposed in [5]. Moreover, (8) can be seen as a special realization of

$$\mu_{a,\mathcal{M}}(r) := \min_{k=1,\ldots,n} \eta_{sim_S(s_0,s_k)}(sim_{\mathcal{R}}(r, r_k)),$$

  where $\{\eta_\alpha \mid 0 \leq \alpha \leq 1\}$ is a parametrized class of fuzzy measures. This evaluation has been derived in [10] in connection with a probabilistic formalization of CBR.
- The constraint (6) becomes trivial if $\widetilde{S}_{\mathcal{S}}(s_0, s_1) = 0$. The resulting possibility distribution $\pi \equiv 1$ is a model of *complete ignorance*: the outcome r_1 tells us nothing about the (unknown) outcome r_0 because the corresponding situations are not similar at all. This is a very reasonable property which seems to avoid problems mentioned in connection with the evaluations (1) and (3). Indeed, those problems mainly occur due to the fact that such point estimations are only partly able to reproduce certain characteristics of a memory.
- Our approach, as it has been developed so far, allows for a purely *ordinal* interpretation if $1 - m(\cdot)$ in (8) is replaced by the order-reversing function of an (ordinal) similarity scale (provided that the similarity scales for situations and outcomes are "commensurable," if not identical.)
- The method can be extended in a straightforward way such that it allows for the integration of different sources of knowledge. Suppose, for example, the decision maker to know that choosing act a_0 for problem p_0 will yield an outcome (= real number) r_0 which is certainly *very close* to 0. This knowledge, which may be independent of the decision maker's experience (in form of observed cases,) can be formalized as a constraint in form of a possibility distribution π on $\mathcal{R}$. Thus, it defines the same kind of constraint as the cases from the memory do. It is then possible to combine the distribution π conjunctively with the distributions $\pi_{a,\mathcal{M}}$ $(a \in \mathcal{A})$ derived from the memory and, hence, to constrain the unknown outcome r_0 further:

$$(\forall r \in \mathcal{R}) \,:\, \widetilde{\pi}_{a,\mathcal{M}}(r) = \min\{\pi(r), \pi_{a,\mathcal{M}}(r)\}\,.$$

  It is interesting to compare this approach of combining knowledge and data to the probabilistic one discussed in [9]. There, the combination is achieved via the Bayesian approach to belief revision: knowledge which influences the belief concerning the unknown outcome r_0 is formalized by means of a prior probability distribution. This knowledge is then revised in the light of observed data, i.e., cases from the memory.

3.3 Decision Making

The decision problem which has been left open in Section 3.2 can be characterized formally as a triple $(\mathcal{R}, \Pi, \preceq)$, where Π is the class of (normalized) possibility distributions on the set $\mathcal{R}$ of results and $\preceq$ is a relation over Π characterizing the preferences of the decision maker concerning such distributions. In this connection we are interested in decision criteria representing $\preceq$.

A qualitative approach. Dubois and Prade [4] have recently proposed a purely qualitative decision theory in which uncertainty and utility are represented by possibility measures and qualitative utility functions, respectively. The corresponding decision criteron is derived from an axiomatic framework, which can be seen as a qualitative counterpart of the axioms of Von Neumann and Morgenstern's expected utility theory.

Let $\preceq$ be a preference relation over the class Π of normalized possibility distributions on a finite set $\mathcal{R} = \{r_1, \dots, r_n\}$ of consequences. Moreover, let U and V be finite linear scales of preference and uncertainty with minimal elements 0 and maximal elements 1, respectively. The "commensuration" between these scales is achieved via an order-preserving mapping $h : V \rightarrow U$ from the plausibility scale to the preference scale which satisfies $h(0) = 0$ and $h(1) = 1$. Based on a certain set of axioms the existence of a utility function $u : \mathcal{R} \rightarrow U$ and a decision criterion QU which represents the preference relation $\preceq$ are derived:

$$\mathrm{QU}(\pi) := \min_{r \in \mathcal{R}} \max \{n(h(\pi(r))), u(r)\}. \tag{10}$$

That is, $\pi \preceq \pi' \Leftrightarrow \mathrm{QU}(\pi) \leq \mathrm{QU}(\pi')$. Here, n is the order-reversing function on U. Based on a slight modification of the underlying axiomatic system an optimistic counterpart of the (pessimistic) decision criterion (10) is obtained. In [1] a generalization of this axiomatic approach has been proposed in which the max-operator in (10) is replaced by a general t-norm.

If we interprete our approach to CBR as a purely qualitative one as outlined in Section 3.2 and if we also assure the commensuration of the plausibility scale and the preference scale, then the theory of [4] can be applied for solving the decision problem. That is, the corresponding decision criteria can be used in order to choose from (9) the most preferred distribution and, hence, the most preferred act $a^* \in \mathcal{A}$. Applying (10) to the possibility distributions (9), for instance, leads to the following evaluation of an act $a \in \mathcal{A}$:[5]

$$\mathrm{V}(a) = \mathrm{QU}(\pi_{a,\mathcal{M}}) = \min_{r \in \mathcal{R}} \max \{n(h(\pi_{a,\mathcal{M}}(r))), u(r)\}. \tag{11}$$

Thus, given a memory $\mathcal{M}$ and a new problem p_0, the decision maker can assign each act $a \in \mathcal{A}$ the utility value (11). If he relies on the CBR hypothesis (resp. our formalization thereof) and accepts the axioms underlying the decision criterion (10), he will choose an act $a^* \in \mathcal{A}$ which maximizes (11).

A probabilistic approach. The decision criteron (10) and its optimistic counterpart may appear inappropriate in some situations due to their extremely pessimistic and optimistic nature, respectively. The wish of a decision maker for a

[5] See [1] for a related approach.

compromise between an extremely pessimistic and an extremely optimistic behavior naturally leads to the idea of deriving a generalized expected utility by averaging over the utility values of possible outcomes. A reasonable candidate for such a generalized expectation is the *Choquet-expected utility* (CEU)

$$\mathrm{CEU}(\pi) = (\mathrm{C}) \int_{\mathcal{R}} (u \circ r)\, d\pi, \tag{12}$$

which is the Choquet integral of the utility $(u \circ r)$ with respect to the possibility measure $\pi : 2^{\mathcal{R}} \to [0,1]$ (induced by the corresponding possibility distribution.) Of course, for (12) to make sense we have to give up the ordinal interpretation of the scales for preference and uncertainty.

It is well-known that a possibility measure $\pi : \mathcal{A} \to [0,1]$ can be interpreted as an upper probability envelope. In this case, π is associated with a class $\mathcal{P}(\pi)$ of probability measures $\mu : \mathcal{A} \to \mathbb{R}_{\geq 0}$ such that upper probabilities coincide with possibility degrees: $\mu^*(A) := \sup\{\mu(A) \mid \mu \in \mathcal{P}(\pi)\} = \pi(A)$ for all $A \in \mathcal{A}$. According to this interpretation, a possibility measure is a characterization of some underlying but imprecisely known probability μ_0. That is, μ_0 is not known exactly but is known to be an element of $\mathcal{P}(\pi)$.[6] Now, given a class of probability measures $\mathcal{P}(\pi)$ one can show that (under some conditions [8])

$$\mathrm{CEU}(\pi) = \sup_{\mu \in \mathcal{P}(\pi)} \mathrm{EU}(\mu), \tag{13}$$

where $\mathrm{EU}(\mu)$ denotes the expected utility $\int_{\mathcal{R}} (u \circ r)\, d\mu$ associated with the measure μ. That is, $\mathrm{CEU}(\pi)$ provides an upper bound $\mathrm{EU}^*(\pi)$ for the (unknown) expected utility $\mathrm{EU}(\mu_0)$.

The evaluation (12) might still be rather optimistic. For $\pi_1 \equiv 1$ (and finite $\mathcal{R}$,) for instance, we obtain $\mathrm{CEU}(\pi_1) = \max_{r \in \mathcal{R}} u(r)$.[7] It seems, therefore, natural to look at a lower bound $\mathrm{EU}_*(\pi)$ for the expected utility $\mathrm{EU}(\mu_0)$ as well. The latter is given by

$$\mathrm{CEU}(\mathcal{N}) = (\mathrm{C}) \int_{\mathcal{R}} (u \circ r)\, d\mathcal{N},$$

where $\mathcal{N}$ is the necessity measure associated with π. A straightforward generalization of the well-known Hurwicz rule for decision making under uncertainty is to choose an act which maximizes the scalar evaluation

$$V(\pi) = \lambda \cdot \mathrm{EU}^*(\pi) + (1 - \lambda) \cdot \mathrm{EU}_*(\pi),$$

where the parameter $0 \leq \lambda \leq 1$ characterizes the optimism of the decision maker.

The interpretation of possibility measures as upper probabilities can also be adopted for the CBR part of our approach to CBDM. More precisely, it is possible to interprete the rule-based approach to approximate reasoning, which

[6] Observe that $\mathcal{P}(\pi) \neq \emptyset$ if π is normalized.

[7] The same value is assigned to π_1 by the optimistic counterpart of (10).

we have proposed as a formalization of CBR in Section 3.2, as some kind of approximate probabilistic reasoning. Particularly, the min-operator can then be justified as a fuzzy set conjunction [3]. This observation, in connection with the probabilistic interpretation of Choquet-expected utility, suggests a probabilistic interpretation of the complete approach to CBDM.

4 Summary

We have proposed an approach to case-based decision making in which the decision maker utilizes possibility theory for modelling uncertainty, approximate reasoning techniques for formalizing case-based reasoning, and generalized decision theories for solving the problem of choosing among acts. It was shown that the approach can be interpreted as a purely qualitative method if we make use of the qualitative decision theory proposed in [4]. It is, however, also possible to adopt a probabilistic point of view if preferences are represented by the Choquet integral. In this case, possibility measures are interpreted as upper probabilities, and the Choquet expected utility provides an upper bound for the expected utility of some underlying but not precisely known probability measure. Seen from this perspective, the approach realizes some kind of approximate probabilistic CBR.

As a main difference between the possibilistic approach to CBDM presented in this paper and previous proposals it should be mentioned that the former realizes a two-stage process, in which the actual decision problem is only solved in the second stage by means of (more or less) classical techniques from decision theory. In fact, CBR is not used for selecting an act directly. Rather, it is applied for reasoning about *uncertainty*, i.e., for deriving a possibility distribution characterizing the decision maker's uncertainty concerning the result of certain decisions. As opposed to this, evaluations such as (1) and (3) utilize the concept of similarity for concluding on the agent's *preference* directly.

Our approach shares some similarities with the transferable belief model of Smets [11], where knowledge representation at the "credal" (knowledge representation) level is kept separate from decision making at the "pignistic" (decision) level. Particularly, CBR takes place at the decision maker's knowledge representation level.

The possibility of making uncertainty concerning the outcome of decisions explicite before a decision is made seems to be a reasonable feature of the model which emphasizes the heuristic character of CBR. Moreover, it may avoid (technical) problems of previous formalizations of CBDM, in which CBR is used for decision making directly. Imagine, for example, a memory which does not contain any cases similar to the current problem, which means that the memory is actually empty. If, however, no cases exist, it seems strange that a *case-based* reasoning procedure can be used for estimating the utility of choosing some act for solving a problem. Rather than assigning a "default utility" it seems natural that (a formalization of) CBR suggests *complete ignorance* about the unknown utility. In our model, this is adequately reflected by a corresponding possibility distribution.

As a further advantage of the two-stage model it deserves mentioning that it may provide a flexible architecture for reasoning and decision making and, hence, may contribute to the (decision-theoretic) design of "intelligent agents." Particularly, it seems adequate for integrating different kinds of information for decision making. The integration of different information sources at the knowledge representation level, an example of which was given in Section 3.2, seems to be less problematic than doing the same at the decision level.

References

1. D. Dubois, L. Godo, H. Prade, and A. Zapico. Making decision in a qualitative setting: from decision under uncertainty to case-based decision. In A.G. Cohn, L. Schubert, and S.C. Shapiro, editors, *Proceedings of the 6th International Conference on Principles of Knowledge Represenation and Reasoning (*KR-98*)*, pages 594–605, Trento, Italy, 1998.
2. D. Dubois and H. Prade. Gradual inference rules in approximate reasoning. *Information Sciences*, 61(1,2):103–122, 1992.
3. D. Dubois and H. Prade. When upper probabilities are possibility measures. *Fuzzy Sets and Systems*, 49:65–74, 1992.
4. D. Dubois and H. Prade. Possibility theory as a basis for qualitative decision theory. In *Proceedings 14th International Joint Conference on Artificial Intelligence (IJCAI-95)*, pages 1924–1930, Montreal, 1995.
5. D. Dubois, H. Prade, F. Esteva, P. Garcia, L. Godo, and R.L. de Mantaras. Fuzzy set modelling in case-based reasoning. *International Journal of Intelligent Systems*, 13:345–373, 1998.
6. I. Gilboa and D. Schmeidler. Case-based decision theory. *Quarterly Journal of Economics*, 110(4):605–639, 1995.
7. I. Gilboa and D. Schmeidler. Act similarity in case-based decision theory. *Economic Theory*, 9:47–61, 1997.
8. M. Grabisch, H.T. Nguyen, and E.A. Walker. *Fundamentals of Uncertainty Calculi with Applications to Fuzzy Inference.* Kluwer Academic Publishers, 1995.
9. E. Hüllermeier. A Bayesian approach to case-based probabilistic reasoning. In *Proceedings IPMU-98, Seventh International Conference on Information Processing and Management of Uncertainty in Knowledge-Based Systems*, pages 1296–1303, Paris, La Sorbonne, July 1998. Editions E.D.K.
10. E. Hüllermeier. A probabilistic approach to case-based inference. Technical Report 99-02 R, IRIT - Institut de Recherche en Informatique de Toulouse, Université Paul Sabatier, January 1999.
11. P. Smets and R. Kennes. The transferable belief model. *Artificial Intelligence*, 66:191–234, 1994.
12. L.A. Zadeh. A theory of approximate reasoning. In J.E. Hayes, D. Mitchie, and L.I. Mikulich, editors, *Machine Intelligence, Vol. 9*, pages 149–194. Wiley, New York, 1979.

Parameter Determination for Nano-Scale Modeling

Marc Thomas, Christian Pacha, and Karl Goser

Universität Dortmund, D-44221 Dortmund, Germany,
thomas@luzi.e-technik.uni-dortmund.de,
WWW home page: http://www-be.e-technik.uni-dortmund.de

Abstract. Simulation model creation is an important task in electronic device development and control. Parallelizing in architecture and technology research activities increases the significance of available high quality device models. Enhanced device complexity, especially in nano-scale technology, shows difficulties in classical methods for parameter determination. Some of these problems can be solved by using evolutionary approaches. In this work an algorithm for an automated parameter determination task is presented and prospects and limitations of such an approach are studied.

1 Introduction

In semiconductor device development, a precise knowledge of the technology parameters is of fundamental relevance for circuit design and fabrication control. Devices in current technologies such as CMOS and Bipolar are described by physical models, which show quite simple analytical characteristics like linear correlation. Parameters in these models have to be determined by fitting to one or multiple measured input-output-curves. Physical based models give advantage of parameters appearing as extrema, gradients or intersections. In opposite to this analytical or semi-empirical models are more complicated. They are optimized for circuit simulation purposes and achieve a better accuracy in input-output-curve fitting. Parameters in such models have only indirect influence in the output characteristics. In these cases automatic parameter determination can lead to problems in the convergence behaviour. Additionally some parameters can support minor effects which increases optimization difficulty.

In the emerging nano-scale technology, models for simulation tasks show much more complexity, based on the influence of quantum and single charge effects. In these cases the classical methods for parameter determination are not suitable any more. They do not cope with the strong non-linear dependencies of single electron transistors and resonant tunneling devices [GPKR97]. Even experts need multiple hours to fit a model for simple devices in this area. In this situation an autonomous on-line process control seems illusory.

2 Problem

In this work an alternative way to parameter determination for a resonant-tunneling-diode (RTD) model is presented. This quite simple device got a key role in nano-scale technology and is a suitable object to support system design for nano-electronics. The model given in table 1 is proposed in [B+98]. The current-voltage-dependency $I_{RTD}(V)$ is given as summation of resonant currents I_{res} and the leakage current I_{lkg}. Parameter f is a normalizing factor. The eight variable parameters in this model are listed in table 2.

Table 1. Basic model as current-voltage-curve for resonant tunneling diode

$$I_{\mathrm{RTD}}(V) = I_{\mathrm{res}}(V) + I_{\mathrm{res}}(-V) + I_{\mathrm{lkg}}(V)$$

$$I_{\mathrm{res}}(V) = \frac{I_P}{f}(1 + \frac{2}{\pi}\arctan\left(\frac{V_N - V}{V_W} \cdot \frac{nkT}{q(V_N - V_P)}\right) \cdot \ln\left[1 + \exp\left(\frac{q(V - V_P)}{nkT}\right)\right]$$

$$I_{\mathrm{lkg}}(V) = I_V \cdot \frac{\sinh\left(\frac{qV}{n_V kT}\right)}{\sinh\left(\frac{qV_V}{n_V kT}\right)}$$

$$f = 1 - \sqrt{2V_W/\pi(V_N - V_P)}$$

Parameters are current and voltage values except the two last which are numerical. These have to be adjusted to make the model curve an appropriate fit to the measured one. An example for the RTD DU595, inspected by W. Prost at the University Duisburg, given in figure 1 expresses an essential difficulty. The size of the device causes that the measurement equipment has an enormous influence to the measurement itself. As a consequence the curves descending part is hard to record. Therefore a global fitting-criterion does not work.

As each parameter is in $\mathbb{R}$ the optimization problem is in $\mathbb{R}^8$. That means a mapping $f: \mathbb{R}^8 \to \mathbb{R}$ to express a quality measure for each parameter-set is needed. Therefore a kind of distance measure between the analytical model and the experimental curve is defined. Four different criteria form as a weighted sum the needed distance measure. For the curves rising parts, the mean of the absolute differences and of the gradients differences are calculated. Additionally the local maximum gap and the mean of negative values are summed up. This

Table 2. Parameters in rtd-model

V_W	resonant width
V_P	peak-voltage
V_N	voltage at maximum negative differential resistance
V_V	valley-voltage
I_P	peak-current
I_V	valley-current
n	resonant current parameter
n_V	thermo-ionic current parameter

combination is intuitivly gained and cannot be derived from any information. It is the most complicated part in every automated optimization process.

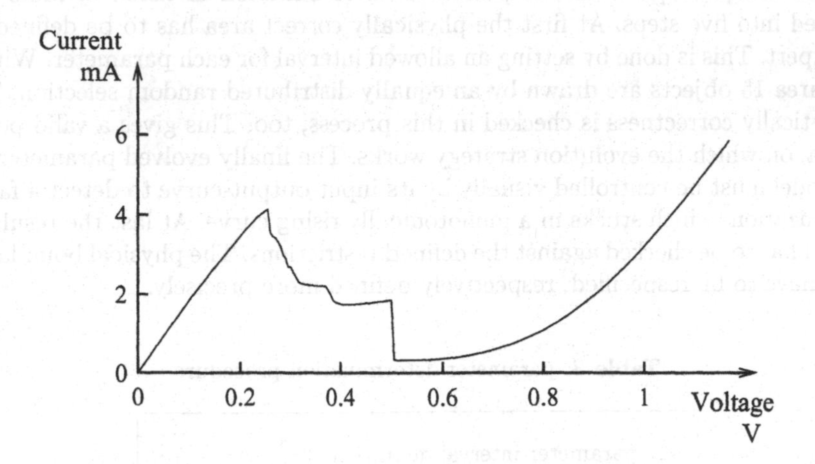

Fig. 1. IV characteristic of a resonant tunneling diode

Two cases have to be taken into account. First, this quality measure can only be calculated, if I_{RTD} is defined over the whole input space. (In our case 0V to 1.2V.) Parameter-sets performing this we call analytically correct. Consequently, the search space, defined by analytically correct items, cannot be determined a priori and may be disjointed. Second, allowed parameter intervals are restricted with regard to physical limitations. This is for example a peak-current between 1 and 6 mA. These predefinitions prevent the algorithm to follow a senseless fitting.

3 Algorithm

Parameter determination is done by utilizing evolution strategies (ES). A quality measure for each parameter-set, also called fitness-function in this context, is sufficient for its use. They are advantageous to genetic algorithms because of geno-type and pheno-type identity in this application. For a real-valued parameter space, standard methods for recombination and mutation already exists. Selection in optimization problems is easily done by taking the best individuals.

In this work a standard 15,100-ES is used [Bäc96]. Recombination is done intermediate for object variables and discrete for mutation variables. Mutation is realized by adding self-adjusted gaussian noise. By pre-filtering analytically incorrect parameter-sets, provided an appropriate starting set the creation of 100 analytically correct individuals in each epoch is guaranteed. Out of this set the best 15, regarding the fitness-function are taken for the next population. The fitness-function for physically correct parameter-sets is defined by the quality measure as named above. In case of physically incorrect parameter-sets the distance to the area of correct objects is taken as fitness value. A more stringent handling of the physical restrictions lets the ES tend to stuck in sub-optima.

The complete procedure of parameter determination as listed in table 3 is divided into five steps. At first the physically correct area has to be defined by an expert. This is done by setting an allowed interval for each parameter. Within this area 15 objects are drawn by an equally distributed random selection. The analytically correctness is checked in this process, too. This gives a valid population, on which the evolution strategy works. The finally evolved parameter-set or model must be controlled visually by its input-output-curve to detect a failed optimization, which stucks in a monotonically rising curve. At last the resulting model has to be checked against the defined restrictions. The physical boundaries may have to be respecified, respectively defined more precisely.

Table 3. parameter determination procedure

1. parameter interval definition
2. random search
3. 15,100-ES
4. visual valuing of evolved model
5. checking restrictions validity

Simulation runs showed a sure achievement of an appropriate solution. The algorithm got a better model parameter-set in every run regarding the quality measure than the expert did in multiple hours. One sample run is documented in figure 2. All simulations reached the experts quality measure level within the first 200 generations.

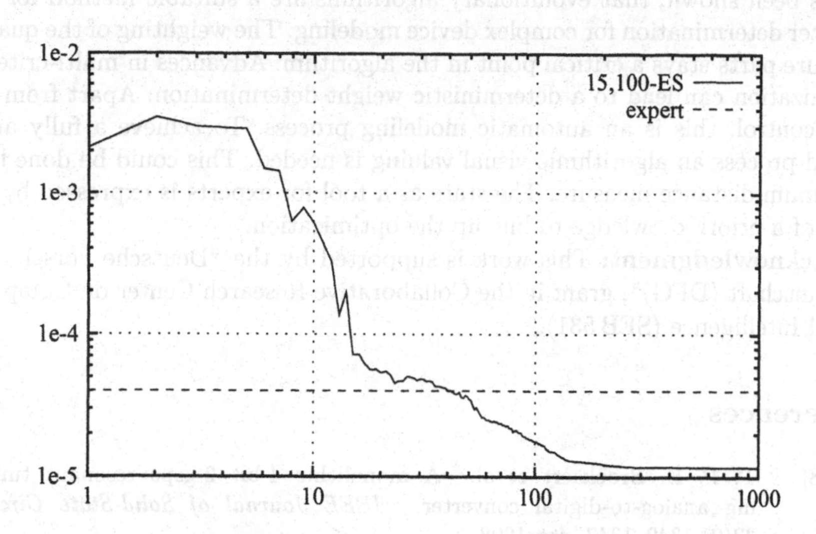

Fig. 2. Quality measure of best model in generation as solid line. The dotted line gives fitness level of the experts model.

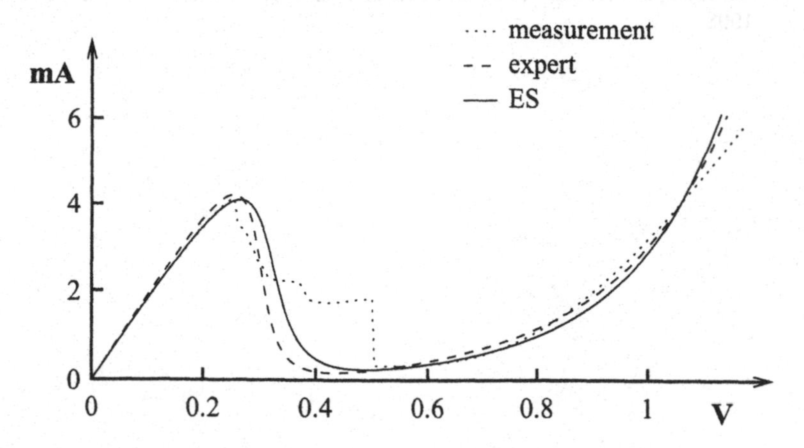

Fig. 3. Resonant tunneling diode: Measurement (dotted line) and its analytical model by an expert (dashed line) and the algorithm (solid line)

4 Conclusion

It has been shown, that evolutionary algorithms are a suitable method for parameter determination for complex device modeling. The weighting of the quality measure parts stays a critical point in the algorithm. Advances in multi-criterial optimization can lead to a deterministic weight determination. Apart from the final control, this is an automatic modeling process. To achieve a fully automated process an algorithmic visual valuing is needed. This could be done by a maximum distance measure. The state as a tool for experts is expressed by the need of a priori knowledge to line up the optimization.

Acknowledgment: This work is supported by the "Deutsche Forschungsgemeinschaft (DFG)", grant in the Collaborative Research Center on Computational Intelligence (SFB 531).

References

[B+98] T. P. E. Broekaert et al. A monolithic 4-bit 2-gsps resonant tunneling analog-to-digital converter. *IEEE Journal of Solid-State Circuits*, 33(9):1342–1347, sep 1998.

[Bäc96] T. Bäck. *Evolutionary algorithms in theory and practice: evolution strategies, evolutionary programming, genetic algorithms.* Oxford University Press, New York, 1996.

[GPKR97] K. Goser, C. Pacha, A. Kanstein, and M. L. Rossmann. Aspects of systems and circuits for nanoelectronics. *Proceedings of the IEEE, Special Issue on Namometer-Scale Science & Technology*, pages 558–573, April 1997.

[PGBP98] C. Pacha, K. Goser, A. Brennemann, and W. Prost. A threshold logic full adder based on resonant tunneling transistors. In *European Solid-State Circuits Conference (ESSCIRC'98)*, pages 427–431, The Hague, NL, sep 1998.

Optimizing Routing Algorithms in Telecommunication Networks with Neural Networks and Fuzzy Logic

Iris Gilsdorf, Wilfried Brauer

Siemens Business Services, Otto-Hahn-Ring 6, D-81739 München, Germany
iris.gilsdorf@mch.sni.de
Institut für Informatik, Technische Universität München, D-80290 München, Germany
brauer@informatik.tu-muenchen.de

Abstract. In the future the span of time between two switching actions in transport networks will become shorter and shorter. If the choice of routes is not done in the right way, the quality of the transport network will deteriorate quickly. The combination of Fuzzy Rulebases and Neural Networks offers an excellent possibility to take into account all the essential aspects for the evaluation of the possible routes. We will present three different architectures for shortest routing strategies in transport networks which try to preserve the once established network quality.

1 Introduction

The importance of telecommunications networks has increased dramatically in recent years. High availability and reliability are key issues as a result of the growing public and corporate dependence on these networks. The operators of transport networks, i.e. the telecommunications companies, are forced for competitive reasons to make optimum use of their networks. They therefore invest a lot of time and money in configuring their transport networks, with both transport-specific network dimensioning and cost optimization playing key roles [1]. After the planning phase, a transport network can be considered to be a system that is optimally configured in both the economic and technical senses. The system is changed with each switching action during operation. Switching actions are used to create link connections in order to adapt the capacity of the communication line to meet market requirements. These switching actions are also used to switch over and reroute link connections. Rerouting is necessary, for example, in order to perform maintenance activities. However, an optimal state of the system must always be guaranteed.
The route for these connections is chosen, in accordance with the task at hand, using a graph which represents the network structure. Parameters describing the current network state are stored at the edges of these graphs. Telecommunication management networks (TMN) are used today to carry out the varied tasks that arise during the operation of a transport network. Historically, the experts at the switching

centers selected the routes according to their own rules. Now, with the new technology, shortest path routing strategies are being used. These shortest path routing strategies use a metric based on one selected parameter. The traditional shortest path algorithms, like for example Dijkstra [2], use this metric in order to determine the shortest route. Descriptions of procedures that improve these traditional algorithms through use of neural networks and fuzzy logic can be found in literature. However, these also require a metric for the edges [3],[4],[5]. The changes made to the transport network as a result of the route selection only ever take account of one aspect and may therefore disrupt the optimized network configuration.

2 Constraining the task

The goal of this project is to find a method of choosing routes that best preserve the quality of the network as defined in the original network configuration.
In order to demonstrate the benefits of the models presented below, we should prove that the network quality improves through using one of the models described below in comparison with the traditional route search approach. But the network quality is not defined in the literature. It is assumed here that operators are attempting to maximize their revenue and yet still satisfy the customer. A simple cost-benefit calculation will be performed for validation purposes. This topic will not be dealt with any further in this article.
A transport network can be represented as a set of network nodes that are linked together using various media. Information must be transported in this network as data. The network nodes represent the points of receipt and dispatch for the data. The various media and also the different protocols used in these networks do not play a role in this investigation.
A transport network is structured hierarchically. Channels of lower transmission speeds are embedded in higher-speed channels through multiplexing. Each transmission speed thus gives rise to a subnetwork comprising channels for this transmission speed. If a transmission line for a particular transmission speed is to be switched from a source to a destination, a route is sought in this subnetwork. Knowledge of the routing of the channels in the higher hierarchies is not necessarily available at the level at which the route search takes place. This fact and the associated issues will not be investigated here. It is assumed for the task at hand that all data for the route search is also available in this subnetwork and that the desire is to improve the quality of this subnetwork.
An edge in this context refers to the logical combination of all channels that have the same endpoints and the same transmission speeds. The nodes of the graph are the endpoints for the channels. These can be network nodes based on synchronous digital hierarchy (SDH) [6] or network nodes based on plesiochronous digital hierarchy (PDH) but also an administrative clustering of equipment. This investigation only considers the possibility of different types of nodes.

Please be aware that what follows is a presentation of a project. Concrete end results are not yet available since not all implementations are yet concluded.

3 Solution models

Telecommunication management networks store the parameters mentioned above for the edges of the graphs in which the route search is performed. A list of parameters therefore exists for each edge, which characterize the particular edge. The experts at the switching centers used an empirical rule system for selecting the routes. It therefore seems reasonable to propose a fuzzy rule base. This rule system "calculates" a real number from the edge parameters, the so-called edge value. A similar approach was also used by H. Hellendoorn in a different problem space [7]. We aim at a rule system that can whether at least preserve the optimized configuration of the transport network if not improve it in comparison with the traditional procedure.

3.1 The fuzzy rule base

The rule system for weighting the edges incorporates all the different aspects. In order to define the route, the customer type for whom the line is to be provided must be considered. It may happen that a high cost line will be used in order to achieve high quality transmission. Different selection criteria come into play for the edges used in this case than for edges of lines which cost less but which will not guarantee high quality.

This leads to a classification of routes, whose edges are to be weighted for the selection process, into different route categories. A separate rule system is proposed for each route category. It goes without saying that the route categories and the associated fuzzy rule bases depend on the operators and their network specifications. This data is extremely confidential and therefore cannot be described in any greater detail here.

The presented solution models are implemented for a selected route category. Routes that place considerable demands on the quality of the transmission line will be considered. These are termed high-priority routes.

The weighting of the edges takes account of economic and operational requirements. These requirements are assessed differently however.

Figure 1 shows the weighting of the edges in a schematic diagram. In general, it applies for the edge weighting that the smaller the edge value, the more likely it is that the edge will be selected.

Three different proposed solutions are pursued on the basis of this fuzzy system

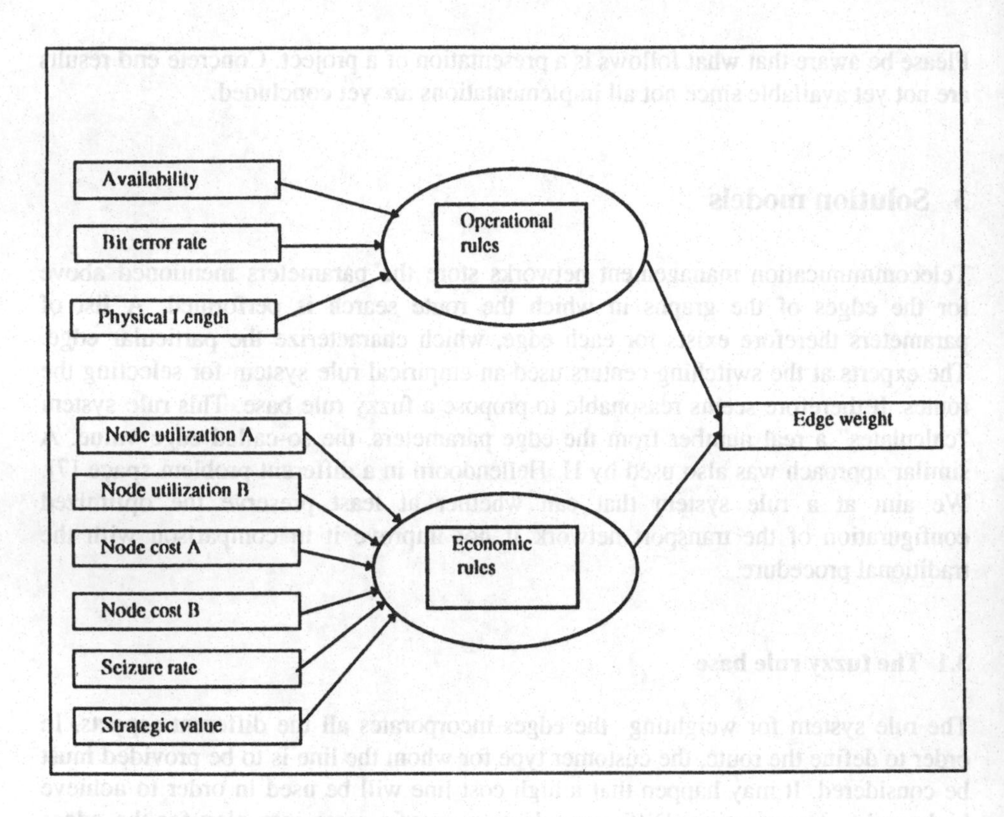

Fig. 1. Structure of the Fuzzy rule base

3.2 Fuzzy rule base for edge weighting

The fuzzy rule base defines a mapping from the multi-dimensional space of the edge parameters to the single-dimensional space of real numbers. This will be done in a manner that produces a metric for the edges upon which the Dijkstra algorithm can be used.

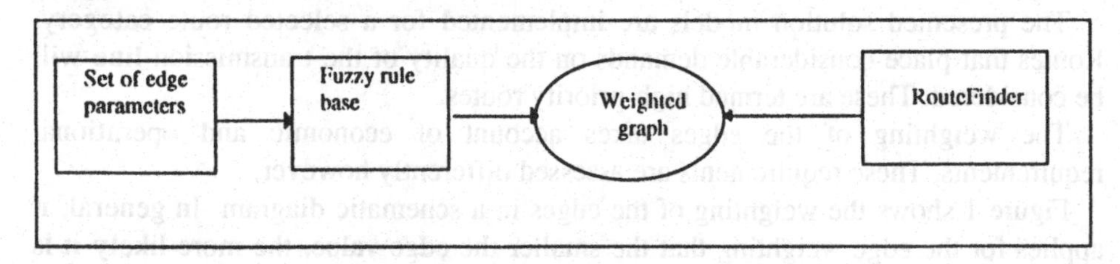

Fig. 2. Fuzzy rule base for edge weighting

Since each route found and used changes the edge parameters, the fuzzy system has to subsequently weight the graph again. It must still be investigated whether or not this is actually essential after each individual route selection.

This system only incorporates the analysis results and the recorded expert knowledge. Should the conditions governing the edge weighting change, for example because the used technologies or the economic requirements change, the rule system can be adapted accordingly. This is the greatest benefit of this technology.

This system cannot use implicit knowledge, which is available in the form of known "good routes". The second proposed solution therefore gives rise to a much more flexible system.

3.3 Fuzzy-Neural-controller for edge weighting

The fuzzy rule base is transformed reversibly into a neural network. The reversible function is necessary so that the rules can be checked and information can flow back to the experts. Knowledge that exists unconsciously but which is available in the data can be incorporated in the system in this way. Furthermore, it is possible that any incorrect rules can be recognized and removed from the system.

The trained system will be used like the fuzzy controller described in the previous section. The architecture of this system is shown in Figure 3.

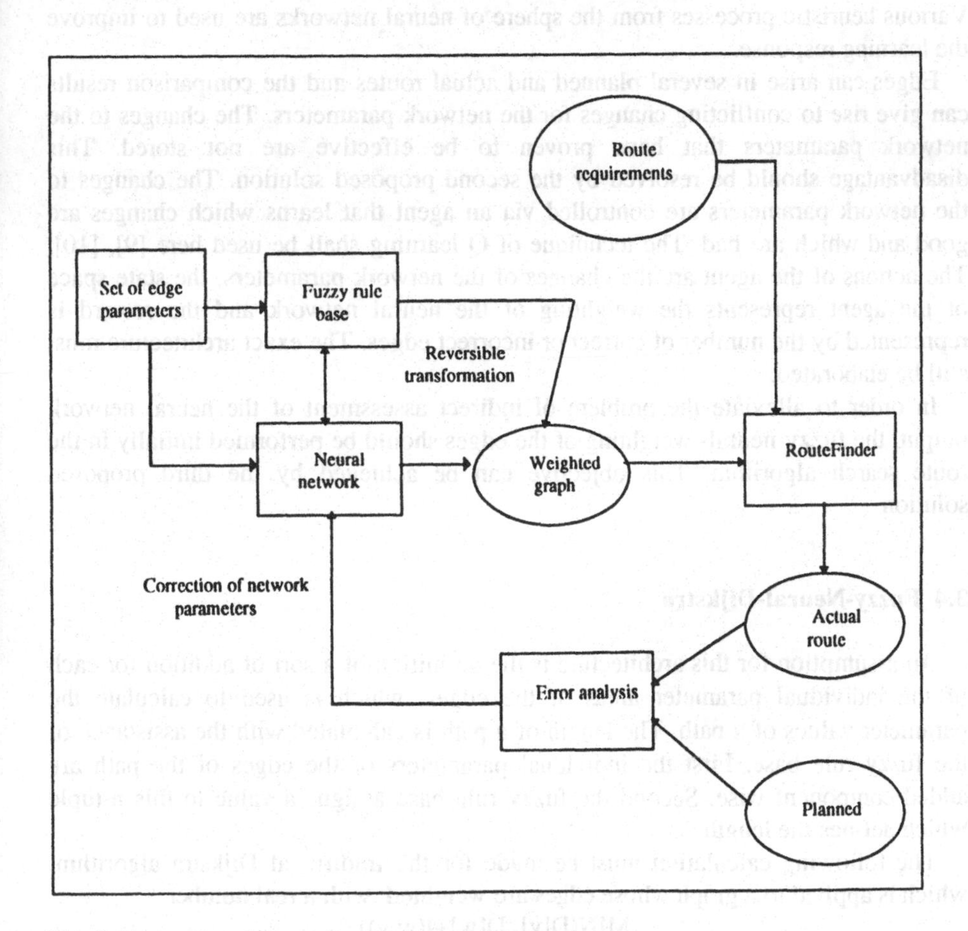

Fig. 3. Fuzzy-Neural-controller for edge weighting

The procedure developed by A. Bücherl is used and extended for the reversible transformation of the fuzzy rule base into a neural network [8].

The neural network is trained by a selection of "good" routes. It would be easy if the knowledge about the correct weights of the edges were available. However, this local information is not available. There is simply a knowledge of which routes have proven favorable. However, this is not the output of the neural network. For this reason, the output of the network must be assessed indirectly. Once all edges have been weighted by the neural network and the weighted graph is available for the route search, the weighting is regarded as good if the planned routes are found. The training is then ended. It is not yet certain how the network parameters can be changed if the results of the route search do not agree with the planned routes. Two approaches will be pursued here.

The simple solution is to compare the planned route with the actual route for a single route request. The values of the edges that agree in terms of planned and actual route are reduced by a certain adjustment value. The values of the edges that occur in the actual route, although they don't exist in the planned route, are incremented by an adjustment value. This adjustment value is then repropagated. Various heuristic processes from the sphere of neural networks are used to improve the learning response.

Edges can arise in several planned and actual routes and the comparison results can give rise to conflicting changes for the network parameters. The changes to the network parameters that have proven to be effective are not stored. This disadvantage should be resolved by the second proposed solution. The changes to the network parameters are controlled via an agent that learns which changes are good and which are bad. The technique of Q learning shall be used here [9], [10]. The actions of the agent are the changes of the network parameters, the state space of the agent represents the weighting of the neural network and the reward is represented by the number of correct or incorrect edges. The exact architecture must still be elaborated.

In order to alleviate the problem of indirect assessment of the neural network output, the fuzzy-neural- weighting of the edges should be performed initially in the route search algorithm. This objective can be achieved by the third proposed solution

3.4 Fuzzy-Neural-Dijkstra

An assumption for this architecture is the definition of a sort of addition for each of the individual parameter areas of the edges, which is used to calculate the parameter values of a path. The length of a path is calculated with the assistance of the fuzzy rule base. First the individual parameters of the edges of the path are added component wise. Second the fuzzy rule base assigns a value to this n-tuple which defines the length.

The following calculation must be made for the traditional Dijkstra algorithm, which is applied to a graph whose edges are weighted with a real number:

$$MIN(D[v], D[w]+l(w,v))$$

D[v] is the length of the shortest path from the source node to node v, D[w] is the corresponding length to node w. l(w,v) is the length of the edge from w to v. In our example, the individual edges are not weighted by real numbers but by n different parameters. The n values of the individual parameters of the shortest path are stored in D[w]. Therefore in the expression D[w]+l(w,v), two n-tuples are added component by component.

The fuzzy rule base is used to assign a real number to the two n-tuples that are compared. The calculation of the MIN is hence trivial.

When defining the addition, one can specify that for some parameters, rules are to be formulated in order to determine a third value from two specified values. A fuzzy rule base can therefore also be used for this work.

4 Further work

The architecture of the fuzzy-neural-Dijkstra, in particular, will continue to be examined and refined in the future with performance also playing an important role. It will also be examined whether the system improves if a fixed number of previously calculated edges are included in the calculation presented in section 3.4 in addition to the new edge in order to determine the value of the route.

Acknowledgement

The first author would like to express her thanks to the SBS MP 5 department for their support and to her colleagues, particularly Mr. W. Scherer, for many fruitful discussions and suggestions.

References

[1] M.Wizgall. Planung und Optimieren von Telekommunikationsnetzen, Strategisches Instrument im Wettbewerb. NET, Heft 11,1996

[2] Alfred V. Aho, John E. Hopcroft, Feffrey D. Ullman. The Design and Analysis of Computer Algorithms. Addision-Wesley Publishing Company 1974

[3] H.E.Rauch, T.Winarske. Neural Networks for Routing Communication Traffic. IEEE Control System Magazine, pp. 26 - 30, April 1988

[4] Faouzi Kamoun, M.K.Mehmet Ali. A Neural Network Shortest Path Algorithm for Optimum routing in packet-switched communications Networks. IEEE Global Telecommunications Conference, GLOBECOM 91,p. 120-4 vol.1 of 3 vol.xxxv+2150 pp.

[5] M.Collett, W.Pedrycz. Application of Neural Networks For Routing in Telecommunications Networks. IEEE 1993

[6] ITU-T, Recommendation G.803. Architectures of Transport Networks Based on the Synchronous Digital Hierarchy. ITU 1993

[7] H.Hellendoorn. Fuzzy Control in Telecommunication. 1996 Bienniel Conference of the North American Fuzzy Information Society- NAFIPS,Berkeley, CA, 1996, pp. 444-448

[8]Angela Bücherl. Reversible Übertragung von regelbasierten Systemen in adaptiver Form durch Implementierung als neuronale Netze. Diplomarbeit 1994, Technische Universität München, Prof.Dr.W.Brauer
[9] Christopher J C H Watkins, Peter Dayan. Q-Learning. Machine Learning, Nr. 8, 1990
[10] Norihiko Ono, Yoshihiro Fukuta. Learning Coordinated Behavior in a Continuous Enviroment. Distributed Artifial Intelligence Meets Machine Learning, Springer 1997, pp. 73-81

Optimum Work Roll Profile Selection in the Hot Rolling of Wide Steel Strip Using Computational Intelligence

Lars Nolle[1], Alun Armstrong[1], Adrian Hopgood[1] and Andrew Ware[2]

[1]Faculty of Technology, The Open University in Wales,
Cardiff CF1 9SA UK
lnolle@glam.ac.uk
{D.A.Armstrong, A.A.Hopgood}@open.ac.uk
[2]School of Accounting and Mathematics, University of Glamorgan,
Pontypridd CF37 1DL UK
jaware@glam.ac.uk

Abstract. The finishing train of a hot strip mill has been modelled by using a constant volume element model. The accuracy of the model has been increased by using an Artificial Neural Network (ANN). A non-linear Rank Based Genetic Algorithm has been developed for the optimization of the work roll profiles in the finishing stands of the simulated hot strip mill. It has been compared with eight other experimental optimization algorithms: Random Walk, Hill Climbing, Simulated Annealing (SA) and five different Genetic Algorithms (GA). Finally, the work roll profiles have been optimized by the non-linear Rank Based Genetic Algorithm. The quality of the strip from the simulated mill was significantly improved.

1 Introduction

There is a world-wide overcapacity for wide steel strip. In such a "buyers' market", producers need to offer a high quality product at a competitive price in order to retain existing customers and win new ones. Producers are under pressure to improve their productivity by automating as many task as possible and by optimizing process parameters to maximise efficiency and quality. One of the most critical processes is the hot rolling of the steel strip.

2 Problem Domain

In a rolling mill a steel slab is reduced in thickness by rolling between two driven work rolls in a mill stand (Fig. 1). To a first approximation, the mass flow and the width can be treated as constant. The velocity of the outgoing strip depends on the amount of reduction. A typical hot rolling mill finishing train might have as many as 7 or 8 close-coupled stands.

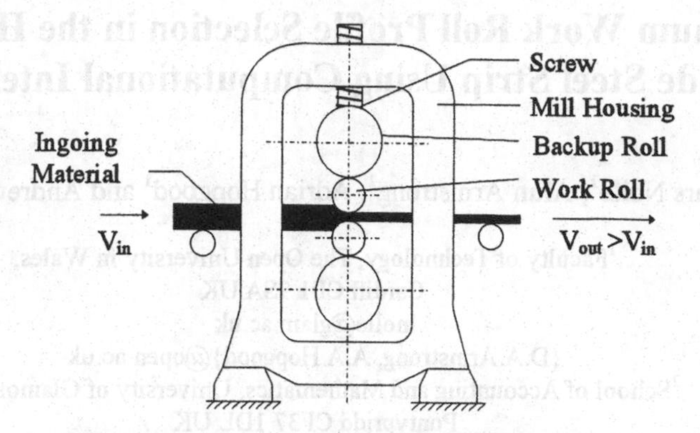

Fig. 1. Layout of a 4-high rolling mill stand.

2.1 Hot Mill Train

A hot-rolling mill train transforms steel slabs into flat strip by reducing the thickness, from some 200 millimetres to some two millimetres. Fig. 2 shows a typical hot strip mill train, consisting of a roughing mill (stands R1-R2) and finishing stands (F1-F7).

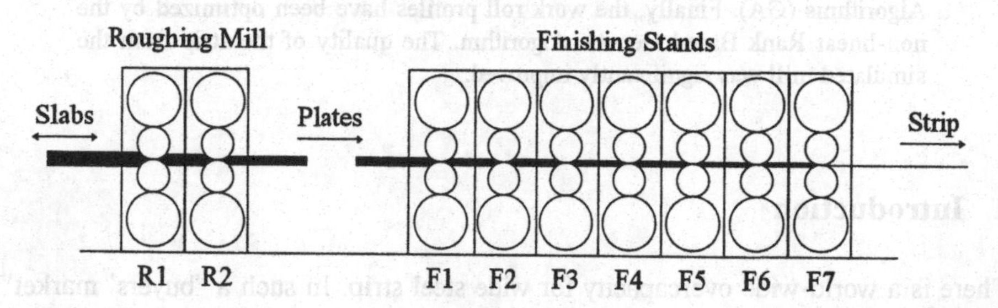

Fig. 2. Hot strip mill train.

The roughing mill usually comprises one or more stands which may operate in some plants as a reversing mill, i.e. the slabs are reduced in thickness in several passes by going through the stand(s) in both directions. When the slab or plate has reached the desired thickness of approximately 35 mm it is rolled by the "close-coupled" finishing stands in one pass. Strip dimensions, metallurgical composition, and the number of slabs to be rolled, together with other process dependent variables, are known as a *rolling program* or *rolling schedule*.

2.2 Strip Quality

Significant quality parameters of steel strip include: *dimensions*, *profile* and *flatness*. Strip profile is defined as variation in thickness across the width of the strip. It is usually quantified by a single value, the *crown*, defined as the difference in thickness between the centre line and a line at least 40 mm away from the edge of the strip (European Standard EN 10 051). Positive values represent convex strip profiles and negative values concave profiles. For satisfactory tracking during subsequent cold rolling a convex strip camber of about 0.5% - 2.5% of the final strip thickness is required [1]. Flatness - or the degree of planarity - is quantified in *I-Units*, smaller values of I-Units representing better flatness.

Modern steelmaking techniques and the subsequent working and heat treatment of the rolled strip usually afford close control of the mechanical properties and geometrical dimensions. In selecting a supplier, customers rank profile and flatness as major quality discriminators. Tolerances on dimensions and profile of continuous hot-rolled un-coated steel plate, sheet and strip are also defined in European Standard EN 10 051. Tolerances on flatness are also published [2].

Both the flatness and profile of outgoing strip depend crucially on the geometry of the loaded gap between top and bottom work rolls. As a consequence of the high forces employed, the work rolls bend during the rolling process, despite being supported by larger diameter back-up rolls. Fig. 3a shows a pair of cylindrical work rolls. In Fig. 3b the effects of the loading can be seen. Due to contact with the strip at temperatures between 800°C and 1200°C the rolls expand, despite being continuously cooled during the rolling operation. Fig. 3c shows the effect of thermal expansion of the unloaded work rolls on the roll gap.

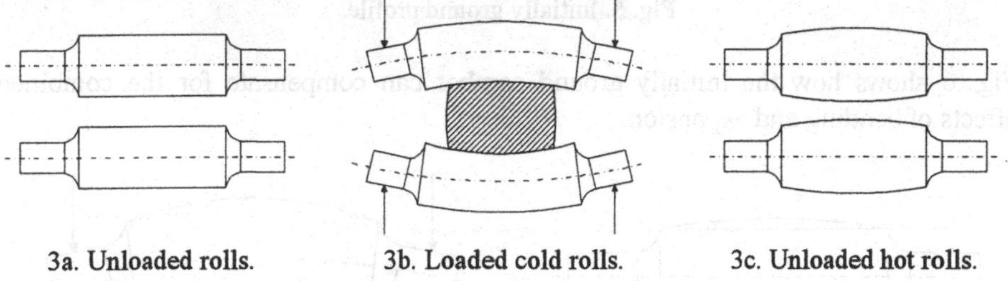

3a. Unloaded rolls. **3b. Loaded cold rolls.** **3c. Unloaded hot rolls.**

Fig. 3. Factors effecting roll gap.

If the geometry of the roll gap does not match that of the in-going strip, the extra material has to flow towards the sides. If the thickness becomes less then about 8mm, this flow across the width cannot take place any longer and will result in partial extra strip length, and therewith in a wavy surface (Fig. 4b).

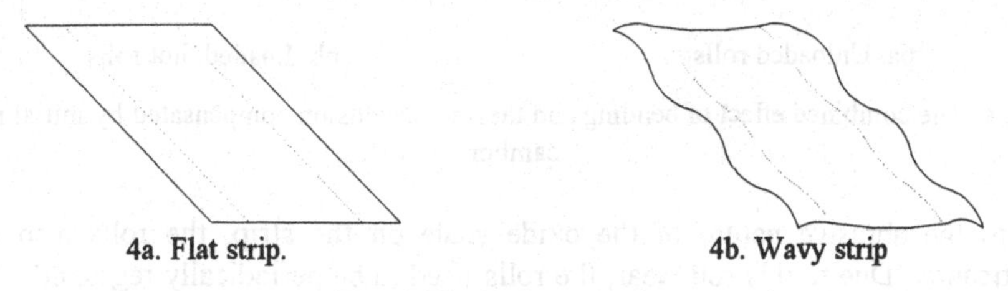

4a. Flat strip. **4b. Wavy strip**

Fig. 4. Flat and wavy strip.

2.3 Bad Shape

The effects of bending and thermal expansion on the roll gaps, and the strip tension between adjacent mill stands, results in a non-uniform distribution of the internal stress over the width of the strip. This can produce either latent or manifest bad shape, depending on the magnitude of the applied tension and the strip thickness [3]. Bad shape, latent or manifest, is unacceptable to customers, because it can cause problems in further manufacturing processes.

2.4 Initially Ground Work Roll Profiles

To compensate for the predicted bending and thermal expansion, work rolls are ground to a convex or concave camber, which is usually sinusoidal in shape (Fig. 5).

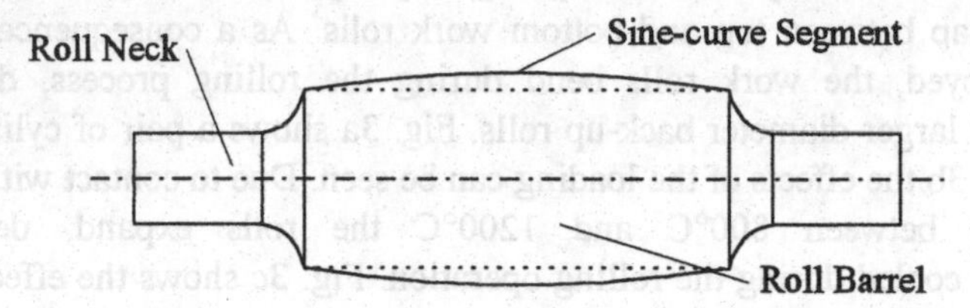

Fig. 5. Initially ground profile.

Fig. 6 shows how the initially ground camber can compensate for the combined effects of bending and expansion.

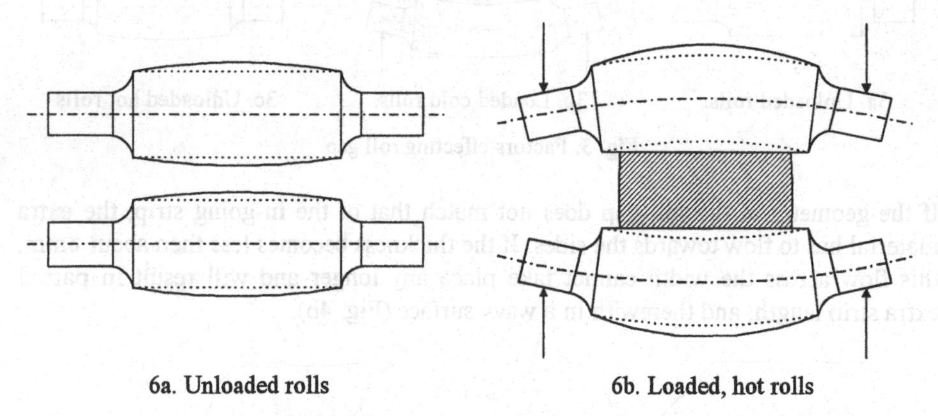

Fig. 6. The combined effect of bending and thermal expansion compensated by initial roll camber.

Due to the abrasive nature of the oxide scale on the strip, the rolls also wear significantly. Due to this roll wear, the rolls need to be periodically reground after a specified duty cycle (normally about four hours), to re-establish the specified profile.

2.5 Roll Profile Specification

The challenge is to find suitable work roll profiles - for each rolling program - capable of producing strip flatness and profile to specified tolerances. In a new mill, these profiles are initially specified individually for every single roll program. These are often later changed, e.g. by the rolling mill technical personnel in an effort to establish optimum profiles! This fine-tuning of the roll profiles is nearly always carried out empirically (Fig. 7).

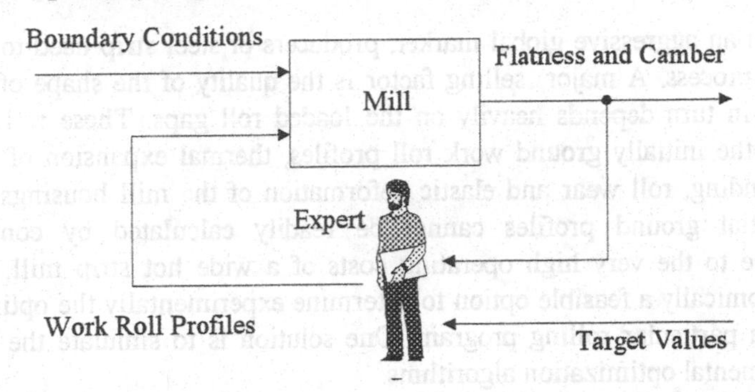

Fig. 7. Iterative optimization of work roll profiles by rolling mill expert.

Due to the lack of accurate model equations and auxiliary information, like derivatives of the transfer function of the mill train, traditional calculus-based optimization methods cannot be applied. If a new rolling program is to be introduced, it is a far from straightforward task to select the optimum work roll profiles for each of the stands involved.

2.6 Optimization of Work Roll Profiles

The seemingly obvious solution of experimenting with different profiles in an empirical way is not acceptable because of financial reasons - the earning capacity of a modern hot strip mill is thousands of pounds per minute, and the mills are usually operated 24 hours a day.

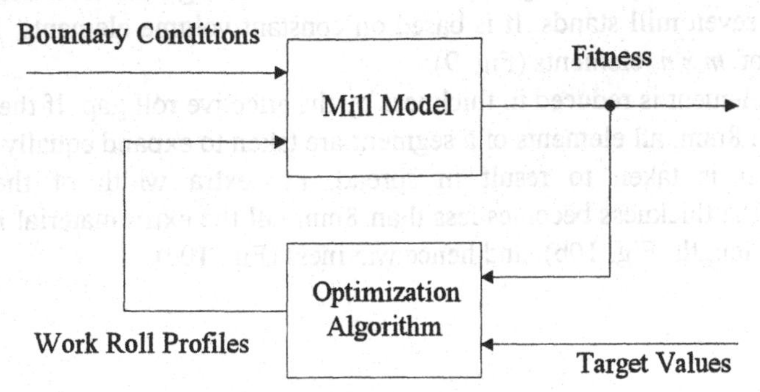

Fig. 8. The optimization loop.

Any unscheduled interruption of strip production leads to considerable financial loss. The use of unsuitable roll profiles can seriously damage the mill train. One solution is to simulate the mill and then apply experimental optimization algorithms. Fig. 8 shows the closed optimization loop, containing the mill model and an optimization algorithm.

2.7 Summary

To survive in an aggressive global market, producers of steel strip need to optimize their rolling process. A major selling factor is the quality of the shape of the steel strip, which in turn depends heavily on the loaded roll gaps. These roll gaps are functions of the initially ground work roll profiles, thermal expansion of the work rolls, roll bending, roll wear and elastic deformation of the mill housings [4]. The required initial ground profiles cannot be readily calculated by conventional methods. Due to the very high operating costs of a wide hot strip mill, it is not usually economically a feasible option to determine experimentally the optimum roll profiles for a particular rolling program. One solution is to simulate the mill and apply experimental optimization algorithms.

3 Modelling the Rolling Mill Train

In the past, the accuracy of analytical mill models has been increased by on-line adaptation to certain process parameters [5]. More recently, Artificial Neural Networks (ANN) have been used successfully to compensate for model errors [6], or for modelling the complete transfer function of a mill [7]. In this research, an ANN has been used to adapt a model to the hot strip mill, using sample data kindly supplied by Thyssen Krupp.

3.1 Constant Volume Element Model

The simulation (developed using C++) models the finishing train of a hot strip mill consisting of seven mill stands. It is based on constant volume elements. Each steel slab consists of $m \times n$ elements (Fig. 9).

A volume element is reduced in thickness by the effective roll gap. If the thickness is greater then 8mm, all elements of a segment are taken to expand equally in length. Extra materiel is taken to result in spread, i.e. extra width of the element (Fig. 10a). If the thickness becomes less than 8mm, all the extra material is taken to produce extra length (Fig. 10b), and hence waviness (Fig. 10c).

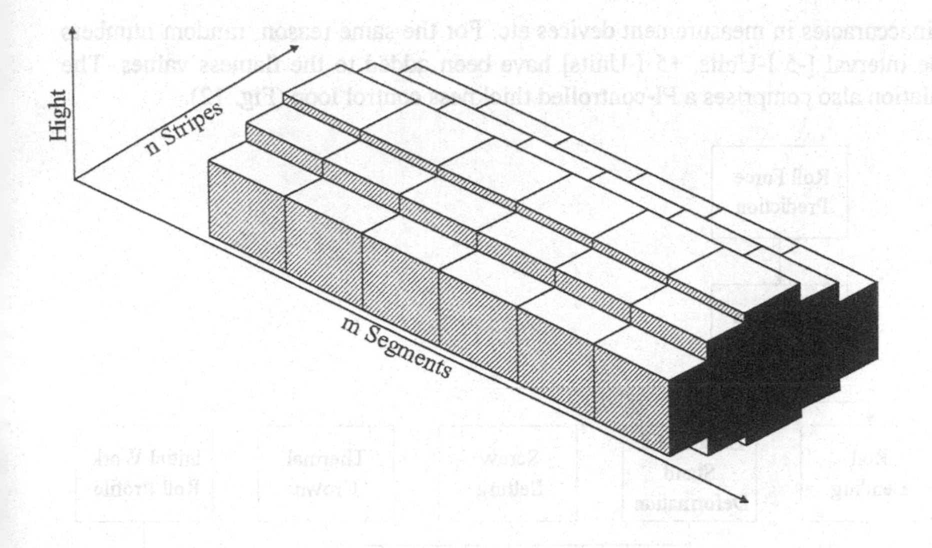

Fig. 9. Simulated steel slab.

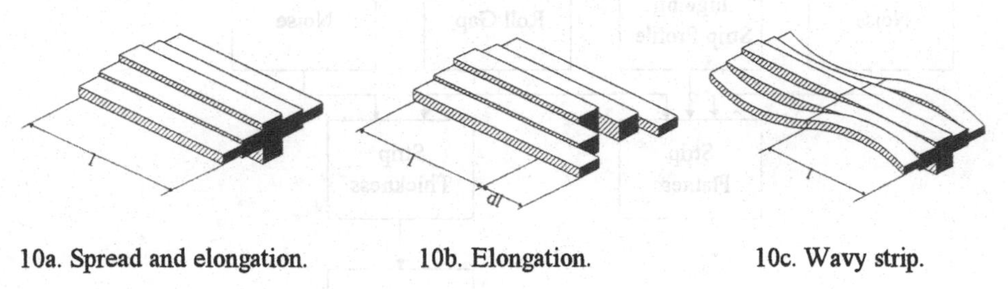

10a. Spread and elongation. 10b. Elongation. 10c. Wavy strip.

Fig. 10. Elongation of a strip segment.

The degree of flatness f of stripe i was calculated as follows:

$$f(i) = \frac{dl_i}{l} \cdot 10^5 \text{ [I-Units].} \quad (1)$$

3.2 Mill Stand Model

Fig. 11 shows the model used to calculate the effective roll gap of a single stand. The roll force has been calculated by means of a formula found empirically by *Ekelund* [8] and subsequently corrected by an Artificial Neural Network, that has been trained with actual process data derived from a commercial hot strip mill. The roll deflection due to bending stress and shear stress has been calculated by using a set of equations published by *Winkler* [9]. The thermal crown of the work rolls has been calculated by the means of the formula stated by *Wusatowski* [10]. Finally, equally distributed random numbers in the interval [-0.001mm,+0.001mm] have been added to the calculated strip thickness values in order to simulate non-deterministic influences, like inaccuracies in measurement devices etc. For the same reason, random numbers in the interval [-5 I-Units, +5 I-Units] have been added to the flatness values. The simulation also comprises a PI-controlled thickness control loop (Fig. 12).

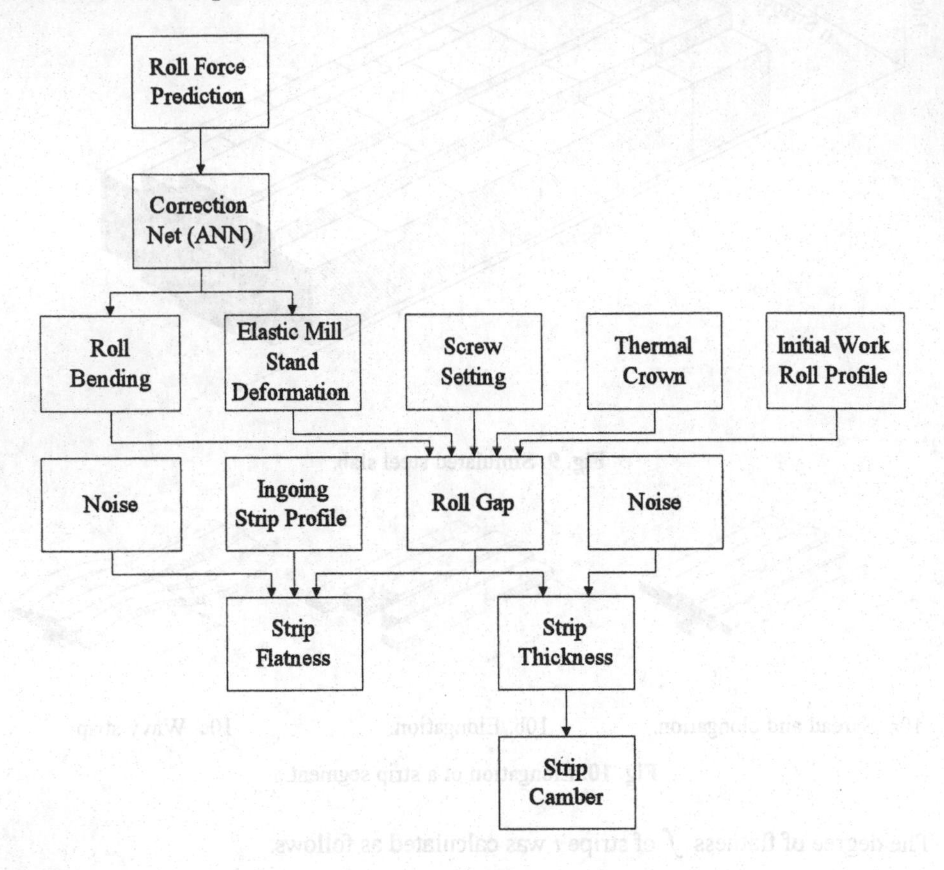

Fig. 11. Mill Stand Model.

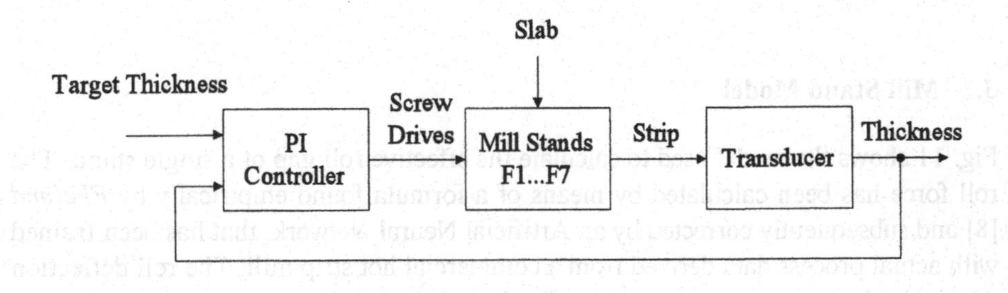

Fig. 12. Thickness control loop.

3.3 Use of an Artificial Neural Network to Correct the Predicted Roll Force

The most important parameter in the process model is "roll force". It has the greatest influence on the effective roll gap and thence on the accuracy of the model calculations. Due to a lack of satisfactory roll force models, the predicted roll forces are usually approximations of the reality. Therefore an Artificial Neural Network has been used as a synthesis network [11] to correct the predicted roll force.

The experiments were carried out on a 233 MHz Pentium PC with 64 Mb RAM using the *LINUX 2.0.29* operating system, running the well known neural network package *SNNS v4.1*, developed at the *Institute for Parallel and Distributed High Performance Systems (IPVR)* of the University of Stuttgart, Germany.

A three layer feed-forward network was used [12][13] and trained by a standard back-propagation algorithm (η=0.2, d=0.0) [14]. The number of hidden units was calculated to be 9 by the mean of *Kolmogorov's theorem* [15].

The input parameters for the network were: strip temperature, strip width, thickness reduction and the predicted roll force. The network output was the corrected roll force. The training set and the test set each contained 49 samples.

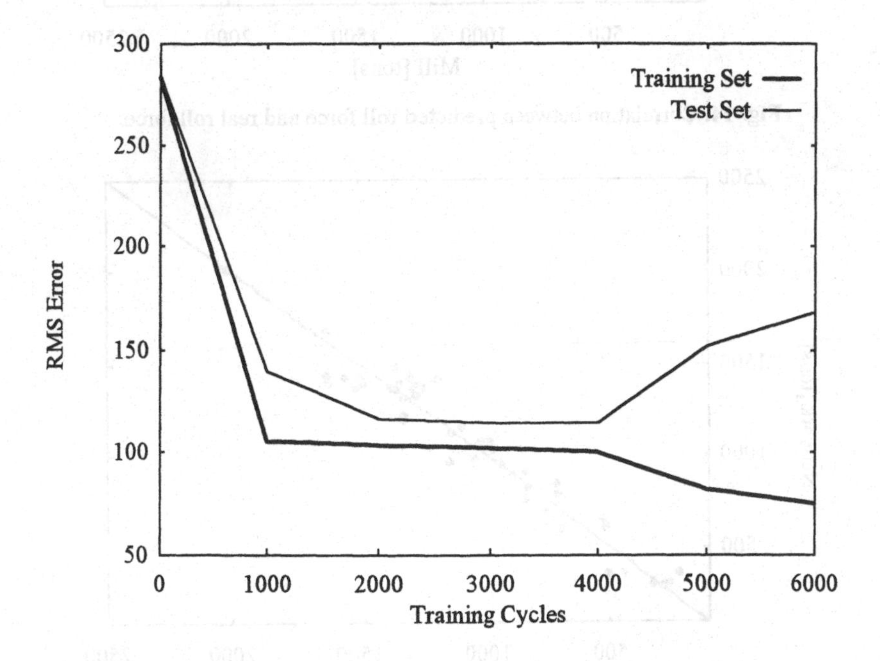

Fig. 13. **The development of the RMS error during the training phase.**

Fig. 13 shows the development of the RMS-error for the training set and the verification set (test-set) during training. It can be seen that the RMS-error for the verification set reaches a minimum after approximately 4000 training cycles. It is considered that the network has been trained optimally after 4000 cycles.

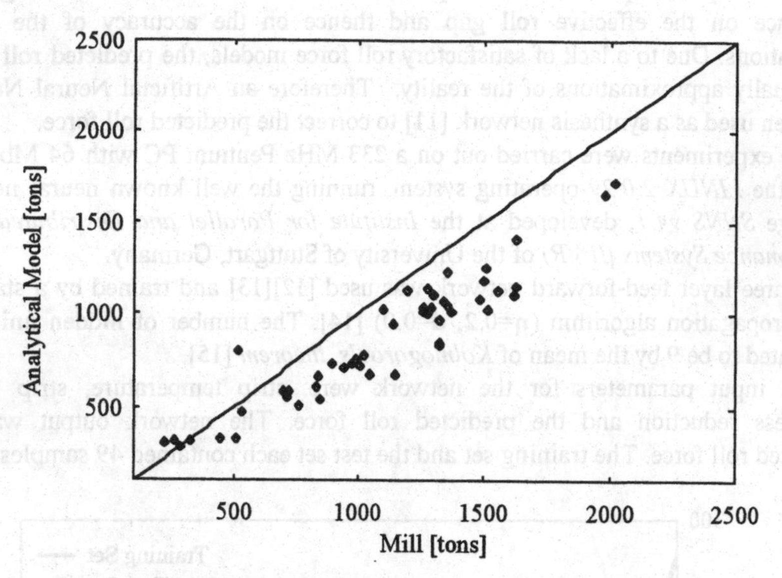

Fig. 14. Correlation between predicted roll force and real roll force.

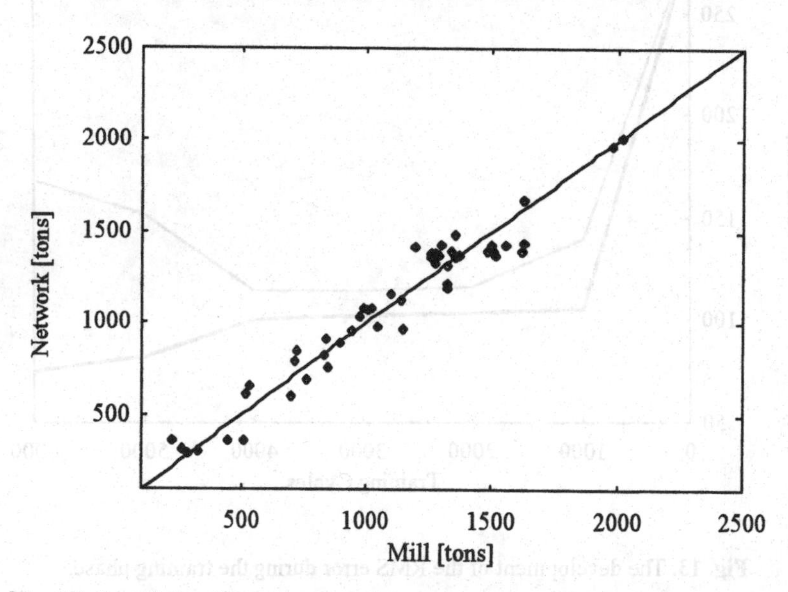

Fig. 15. Correlation between network output and real roll force for the test set.

The RMS error has been reduced from 283.15 I-Units to 100.12 I-Units and the correlation factor r has been increased from 0.9402 to 0.9723 by using the synthesis network.

3.4 Discussion

This model is a compromise between accuracy and computational costs. In order to keep the computation time within an acceptable range, for use together with Genetic Algorithms, the following assumptions were made:

- No roll flattening (roll flattening can be neglected during the hot rolling of steel strip),
- symmetrical roll gap, i.e. strip centralised,
- balanced inter-stand tension,
- constant roll force distribution over the roll barrel.

4 A Non-linear Rank Based Genetic Algorithm

For standard GAs, the selection probability of one individual depends on its absolute fitness value compared to the rest of the genepool. For Rank Based GAs [16] (Fig. 15), the selection probability of one individual depends only on its position in the ranking and is therefore disconnected from its absolute fitness value.

Baker [16] has shown, that Rank Based GAs have better performance for small genepool sizes. He used a linear selection function, but he also proposed the development of non-linear selection functions as well. During this research, such a non-linear selection function has been developed (Equation 2).

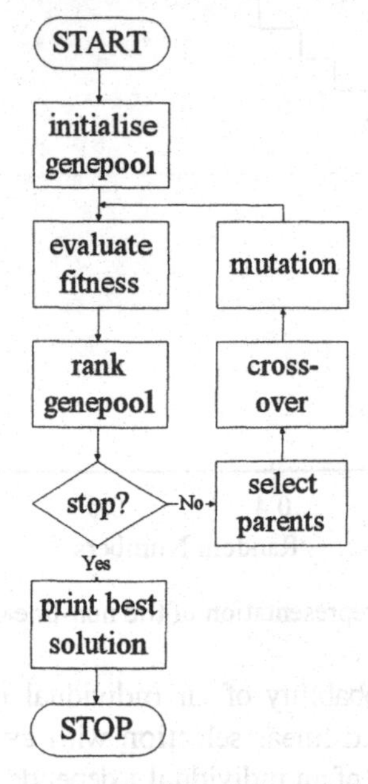

Fig. 16. Flow chart of a standard Rank Based Genetic Algorithm.

$$n(x,c) = ceil\left[\frac{N}{1-\frac{1}{e^{c}}}\cdot\left(1-\frac{1}{e^{c\cdot x}}\right)\right] \quad (2)$$

where:

n:	Individual chosen from the current genepool to join the mating pool,
N:	number of individuals in genepool,
x:	equally distributed random number [0, 1],
c:	constant which controls the non-linearity of the function,
ceil(a):	function returning the smallest integer that is not less than its argument *a*.

Note: in contrast to the original Rank Based GA, the best individual becomes the highest rank in this case. The degree of non-linearity can be controlled by changing the constant c. This allows on-line adaptation of the algorithm. For example, c could be adjusted to the ratio (on-line performance/off-line performance). Fig. 17 shows the graphical representation of the function where c=3.0.

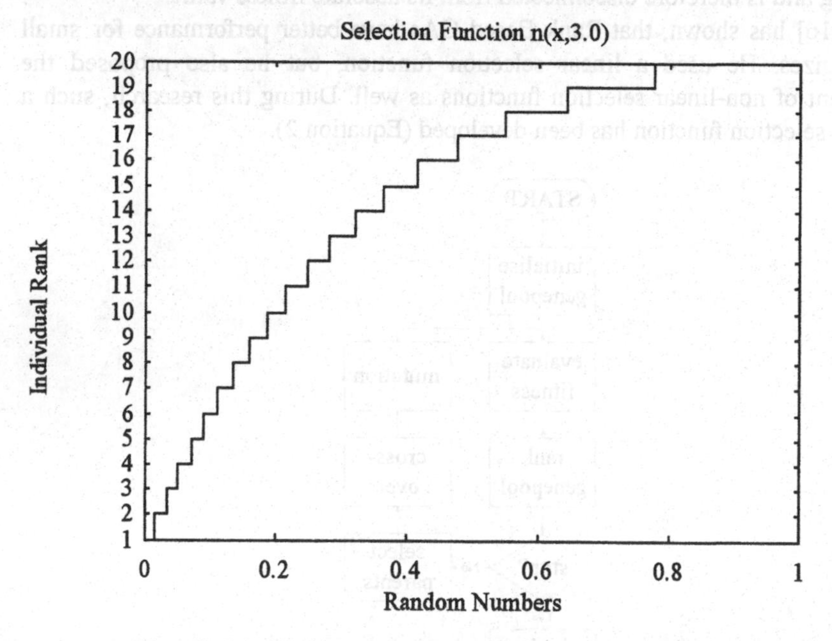

Fig. 17. Graphical representation of the non-linear selection function.

For c=0.2, the selection probability of an individual i is similar to the selection probability using the standard linear selection with e=1.1 (Fig. 18). Fig. 19 shows how the selection probability of an individual i depends on its rank for greater values of c.

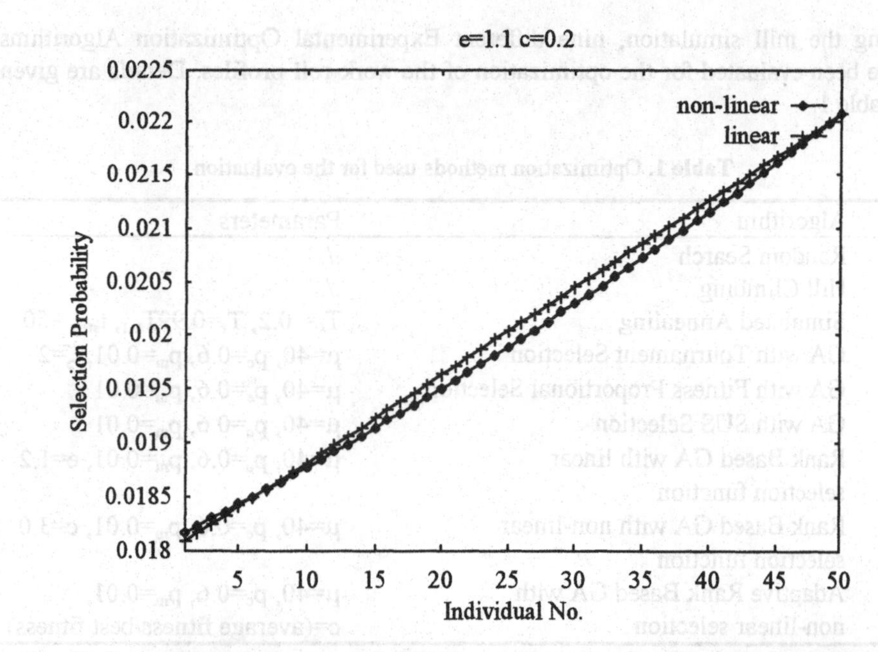

Fig. 18. For c=0.2, the selection probabilities are almost equal for linear and non-linear selection.

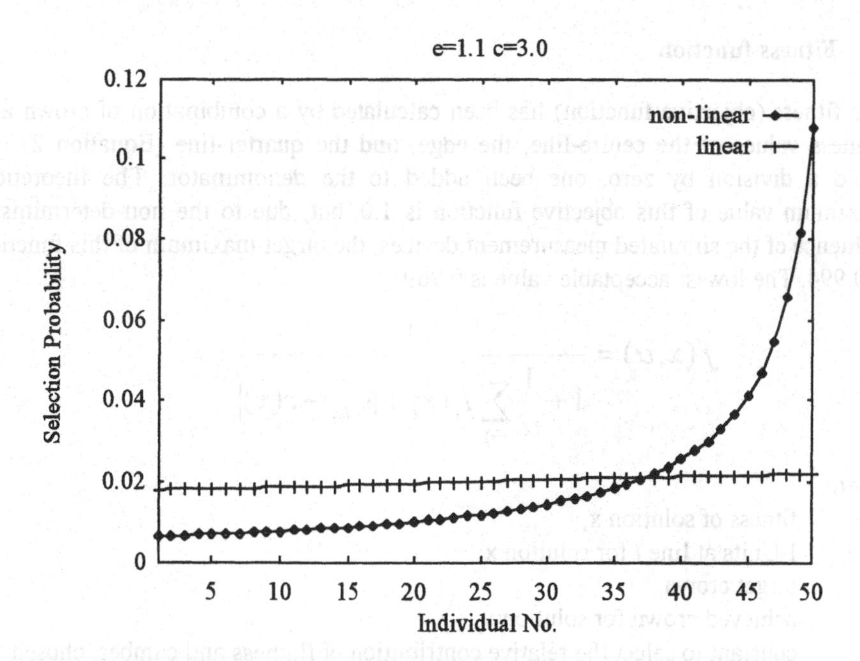

Fig. 19. Larger values of c (e.g. c=3.0) increase the selection probability of the fitter individuals in the genepool.

5 Comparison of Methods

Using the mill simulation, nine different Experimental Optimization Algorithms have been evaluated for the optimization of the work roll profiles. Details are given in table 1.

Table 1. Optimization methods used for the evaluation.

No.	Algorithm	Parameters
1	Random Search	./.
2	Hill Climbing	./.
3	Simulated Annealing	T_0= 0.2, T_t=0.99T_{t-1}, t_{step} =50
4	GA with Tournament Selection	μ=40, p_c=0.6, p_m=0.01, ζ=2
5	GA with Fitness Proportional Selection	μ=40, p_c=0.6, p_m=0.01
6	GA with SUS Selection	μ=40, p_c=0.6, p_m=0.01
7	Rank Based GA with linear selection function	μ=40, p_c=0.6, p_m=0.01, e=1.2
8	Rank Based GA with non-linear selection function	μ=40, p_c=0.6, p_m=0.01, c=3.0
9	Adaptive Rank Based GA with non-linear selection	μ=40, p_c=0.6, p_m=0.01, c=(average fitness/best fitness)

(Note: one-point cross-over has been used for all GAs.)

5.1 Fitness-function

The fitness (objective function) has been calculated by a combination of crown and flatness values of the centre-line, the edge, and the quarter-line (Equation 2). To avoid a division by zero, one been added to the denominator. The theoretical maximum value of this objective function is 1.0, but, due to the non-deterministic influence of the simulated measurement devices, the target maximum of this function is 0.994. The lowest acceptable value is 0.969

$$f(x,\alpha) = \frac{1}{1+\frac{1}{\alpha}\sum_{i=1}^{3} I_i(x) + |c_{aim} - c(x)|} \quad (3)$$

where:

- $f(x)$: fitness of solution x,
- $I_i(x)$: I-Units at line i for solution x,
- c_{aim}: target crown,
- $c(x)$: achieved crown for solution x,
- α: constant to select the relative contribution of flatness and camber, chosen to be 5000 for the experiments.

5.2 Experimental Results

Each algorithm has been applied 50 times. In order to allow for a fair comparison of the different algorithms, the measure *"guess"* has been introduced rather then using *"generation"*. One *guess* equals one model calculation, e.g. a GA with a genepool containing 50 individuals carries out 500 guesses in 10 generations. Each run has been allowed 400 guesses. Table 2 gives the results for the average fitness found by each method. Table 3 gives the average number of guesses taken by the method to find the best result during one run.

Table 2. Fitness results.

Method	Average Fitness	Standard Deviation
Random Walk	0.93261044	0.07308076
Hill Climbing	0.93091918	0.071281203
Simulated Annealing	0.9766379	0.033922189
GA Tournament	0.9608671	0.031718002
GA Roulette Wheel	0.98206098	0.012783254
GA SUS	0.98624604	0.005983528
linear Rank Based GA	0.983616	0.008452708
non-linear Rank Based GA	0.98894954	0.006316525
Adaptive Rank Based GA	0.98897732	0.006726639

Table 3. Guesses results.

Method	Average Number of Guesses	Standard Deviation
Random Walk	2097.76	1191.972736
Hill Climbing	2721	854.9758321
Simulated Annealing	2472.24	983.4050707
GA Tournament	2412	1364.96692
GA Roulette Wheel	2655.2	1146.248967
GA SUS	2809.6	941.3688687
linear Rank Based GA	2366.4	1165.528276
non-linear Rank Based GA	2683.2	939.1907937
Adaptive Rank Based GA	2760.8	899.0694509

5.3 Discussion

It can be seen, that the variation in the average number of guesses is not significant, while there are considerable differences in the achieved average fitness (Fig. 20).

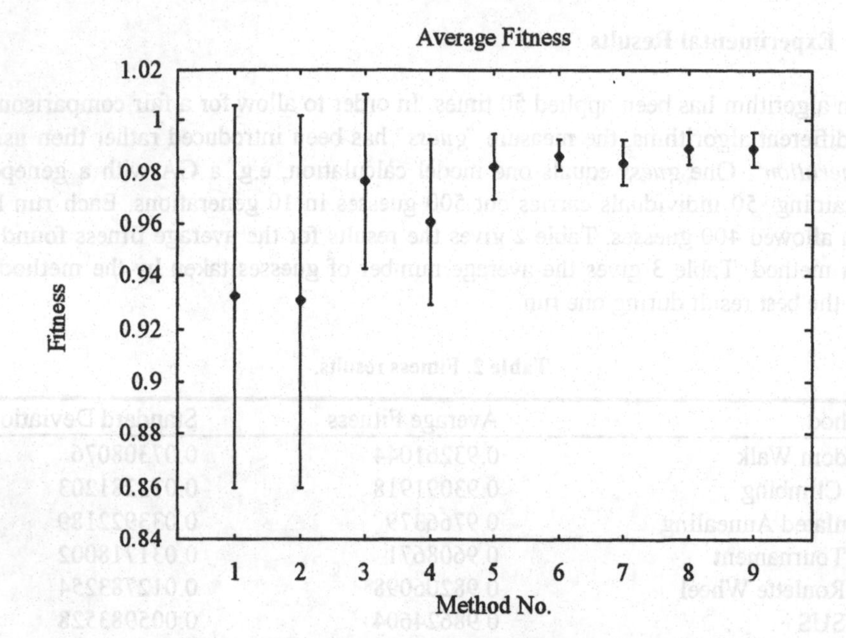

Fig. 20. Experimental Results. The dots represent the achieved average fitness, the error bars represent the standard deviation.

In this problem domain, Rank Based GAs clearly out-perform Random Walk, Simulated Annealing [17][18], and fitness proportional GAs.

The adaptive version of the non-linear Rank Based GA (method 9) shows no significant improvement compared to the non-adaptive one. Therefore, the ratio (best fitness/average fitness) is not appropriate to calculate c. It should be investigated whether other ways of calculating c lead to better results.

The second best method is SA. Interestingly, Hill Climbing has been declassified to Random Search due to the non-deterministic disturbance of the (simulated) measurement equipment.

6 Optimum Work Roll Profiles

For the final optimization, the non-linear Rank Based GA was selected. The genepool contained 40 individuals of 42 bit length. Uniform cross-over [19] was used with a cross-over probability of 0.6. The mutation probability was 0.01 and c has been chosen to be 3.0.

Table 3 shows the simulation results for a typical rolling program using the original work roll profiles and the profiles found by the GA after 63 generations. The target crown was 0.031 mm. The required flatness values were 0 I-Units, the (simulated) inaccuracy of the flatness transducer was ± 5 I-Units.

Table 4. Improvement by optimized work roll profiles.

Parameter	Target value	Old profiles	New profiles	Improvement
Strip crown [mm]	0.031	0.081	0.032	98 %
Flatness edge [I-Units]	0	5	3	40 %
Flatness quarter [I-Units]	0	205	29	86 %
Flatness middle [I-Units]	0	287	1	99 %

It will be observed, that it has prove possible to reduce the profile error from 61.73% to 3.13% and the mean flatness error from 165.67 I-Units to 11.00 I-Units.

7 Conclusions

The accuracy of an analytical mill model has been considerably improved by using Artificial Neural Networks to compensate for the model error. For this particular optimization problem, Nine Experimental Optimization Algorithms have been compared. A non-linear Rank-Based Genetic Algorithm gave the best performance. Hence, this algorithm was selected and used to optimize the work roll profiles in the simulated hot strip mill. The quality of the strip profile and flatness from the simulated mill was significantly increased.

It has been demonstrated that Genetic Algorithms are capable of solving optimization problems of physical systems, even if these systems have non-deterministic behaviour, provided adequate models of these physical systems are available.

Acknowledgements

The financial support of this research by *Metrology Systems Wales, UK* is gratefully acknowledged. The authors also wish to thank the management of *Thyssen Krupp Stahl AG, Bochum* for supplying the process data, and in particular Dr.-Ing. Ingolf Jäckel for his help and advice.

References

1. Wilms, Vogtmann, Klöckner, Beisemann, Rohde: Steuerung von Profil und Planheit in Warmbreitbandstraßen, Stahl u. Eisen 105 (1985) Nr. 22, pp 1181-1190

2. Hensel, Poluchin: Technologie der Metallformung, Deutscher Verlag für Grundstoffindustrie, Leipzig, 1990
3. Sheppard, Roberts: Shape Control and Correction in Strip and Sheet, International Metallurgical Reviews 1973, Vol. 8, pp 1-17
4. Emicke, Lucas: Einflüße auf die Walzgenauigkeit beim Warmwalzen von Blechen und Bändern, Neue Hütte, 1. Jg. Heft 5, 1956, pp 257-274
5. Takashi, Satou, Yabuta: Adaptive Technology for Thickness Control of Finisher Set-up on Hot Strip Mill, ISIJ International, Vol. 35 (1995), No. 1, pp 42-49
6. Ortmann, Burkhard: Modernisierung der Automatisierungssysteme der Warmbreitbandstraße bei Voest-Alpine Stahl Linz, Stahl u. Eisen 115 (1995) Nr.2, pp 35-40
7. Nolle, Armstrong, Ware: Optimisation of Work Roll Profiles in a 2-hight Rolling Mill Stand by Means of Computational Intelligence, Proc. of the 4th Int. Mendel Conf., Brno, 1998, pp 277-284
8. Hensel, Arnold, Thilo Spittel: Kraft- und Arbeitsbedarf bildsamer Formgebungsverfahren, VEB Deutscher Verlag für Grundstoffindustrie, Leipzig, 1978
9. Winkler W.: Theoretische Untersuchungen über die Einflüße, welche die Stärkenunterschiede beim Warmwalzen von Breitstreifen bestimmen, Dr.-Ing. Dissertation, TH Achen, 1943
10. Wusatowski, Zygmunt: Fundamentals of Rollling, Pergamon Press Ltd, Oxford, London, Edinburgh, 1969
11. Portmann, Lindhoff, Sorgel, Gramckow: Application of neural networks in rolling mill automation, Iron and Steel Engineer, Feb. 1995, pp 33-36
12. Hornik Stinchcombe, White: Multilayer Feedforward Networks are Universal Approximators, Neural Networks, Vol. 2, 1989, pp 359-366
13. Hornik: Approximation Capabilities of Multilayer Feedforward Networks, Neural Networks, Vol. 4, 1991, pp 251-257
14. Picton, Phil: Introduction to Neural Networks, The MacMillian Press LTD, 1994
15. Sprecher, D.: On the Structure of Continuous Functions of Several Variables, Transaction of the American Mathematical Society, Vol. 115, 1964, pp 340-355
16. Baker: Adaptive Selection Methods for Genetic Algorithms, Proc. on an Int. Conf. on GA's and Their Application, Hilldsdale, 1985, pp 101-111
17. Kirkpatrick, S., C.D. Gelatt, Jr., M. P. Vecchi: Optimization by Simulated Annealing, Science, 13 May 1983, Vol. 220, No. 4598
18. Metropolis, A., W. Rosenbluth, M. N. Rosenbluth, H. Teller, E. Teller: Equation of State Calculations by Fast Computing Machines, The Journal of Chemical Physics, Vol. 21, No. 6, June 1953
19. Syswerda: Uniform Crossover in Genetic Algorithms, Proceedings of International Conference on Genetic Algorithm 1989 (ICGA'89), pp 2-9

Analysing Epileptic Events On-Line by Soft-Computing-Systems

M. Reuter, C. Zemke

Department of Computer Science
Technical University of Clausthal
Julius-Albert-Str. 4
e_mail: reuter@informatik.tu-clausthal.de , czemke@informatik.tu-clausthal.de

Abstract. Normally EEG-signatures are analysed in the frequency domain while a common FFT- or wave-let-transformation transforms the time signals of the electromagnetic activities of the (wet) neurons into the frequency domain. In this paper, we will show that better and faster results can be obtained, if instead of these common algorithm the FD-operation is applied to the pure time-signal and a DAS-calculation for eliminating the noisy contributions is done next. Especially we will compare the signatures of some epilepsy events analysed by common signal analysis and by the class-oriented DAS- and FD-operations and discuss the results.

Keywords: EEG-supervision, FD-Spectra, epilepsia-recognition

1 Mathematical Background

As the mathematical theory of the DAS and FD-spectra has been discussed in many papers [see references: Reuter], we will only summarise the fundamental basic assumptions of the DAS- and FD-calculations to enable a comparison to the common FFT- and wave-let-algorithm. The basic assumptions for the DAS- and FD-algorithm are derived in the last 10 years from the researches regarding the physiology of the ear while the background mathematics of this new spectra representations is based on a more non-numerical and more symbolic way as the common signal-processing algorithm. As the experimental results show, the application of this 'physiological oriented' statistic methods like the class-related analysis for the DAS and the evaluation of the higher statistical moments by the FD-spectra leads to better results regarding the analysis of noisy and masked signatures as common signal processing algorithm do.

1.1 The Difference-Autopower-Spectrum (DAS)

The basic-idea of the DAS is to divide the x-and y-axis of a measured data set into classes to calculate the mean-value of these classes and to use the class-related mean-values as the new zero-line of the DAS. As many technological applications show, this kind of fitting curve is extremely robust and flexible for all kinds of noisy contributions of a signature and is an extremely useful pre-processing algorithm for all kinds of neural classifiers.

For a DAS algorithm the following parameters must be chosen:

- Beginning of the spectrum to be analysed (m)
- End of the spectrum to be analysed (f)
- class-width of the first DAS-Fitting (H)
- class-width of the second DAS-Fitting (Z)

The DAS-operator itself is characterised by the symbol Δ, while the DAS parameters are given in the bracket that follows. The spectra on which the DAS-operation will act are characterised by the expression Ampl (f_i). So the following expression for the DAS-operation holds:

$$\text{DAS Ampl}(f_i) = \Delta(m,f,H,Z)\,\text{Ampl}(f_i)$$

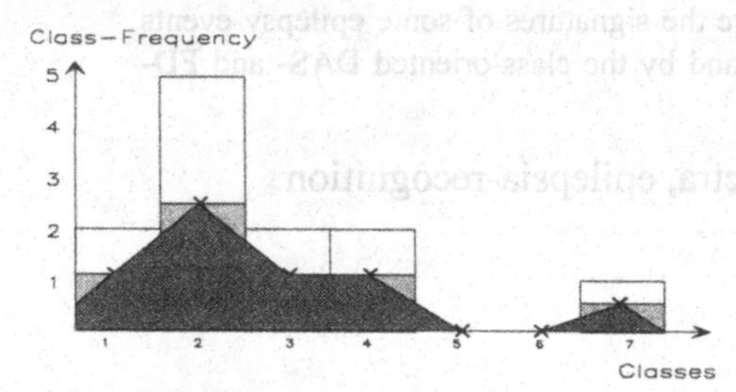

The figure left hand side shows an example of a DAS. The grey area notifies the information that will be lost, as the zero-line of the DAS will be the upper shape of this area and the positive contributions of the resulting spectrum will be taken only.

1.2 The mFD-operator

Mathematically a mFD-spectrum can be defined by the following equation:

$$\text{mFD} = \sum_{\Delta f=1}^{B} \sum_{f=m}^{M} Ampl(f) * Ampl(f+\Delta f)\delta(f-\Delta f) \tag{1}$$

where the

δ -function has the following form: $\delta = 1$ for $f = \Delta f$ $\delta = 0$ for $f \neq \Delta f$ (see Figure below).

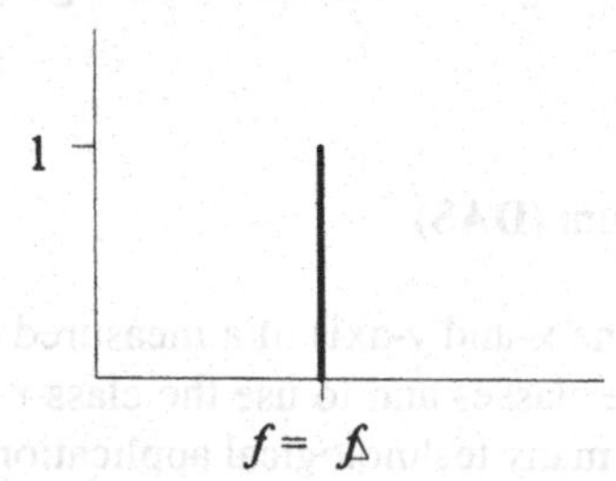

For a mFD-algorithm the following parameters must be chosen:

- **Beginning of the spectrum to be analysed** **(m)**
- **End of the spectrum to be analysed** **(M)**
- **smallest clasp-width** **(I)**
- **largest clasp-width** **(B)**
- **kind of amplitude-combination (here multiplication only)** **(*)**

The mFD-operator is characterised by the symbol Γ, while the mFD parameters are given in the bracket that follows. The spectra on which the FD-operation will act are characterised by the expression Ampl(f_i). All mFD-spectra therefore can be characterised by the following operator

$$\text{mFD Ampl}(f_i) = \Gamma^*(m,f,I,B)\ \text{Ampl}(f_i)$$

Example: **The figure on the right hand side shows a very simple frequency spectrum, which should be demodulated by the mFD-operator of the form:**

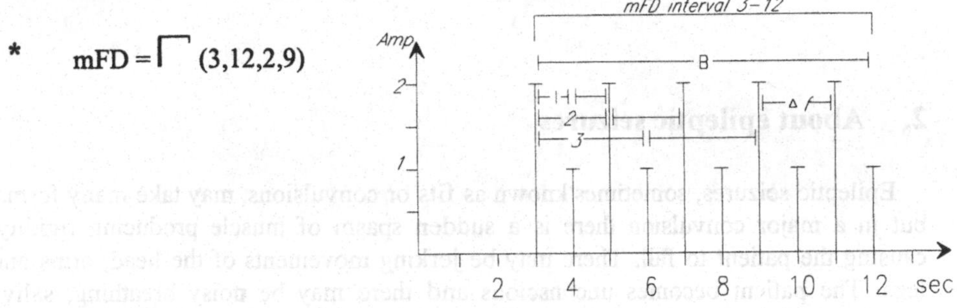

The resulting FD-components are

clasp	2	3	4	5	6	7	8	9
value	20	14	15	10	10	6	5	2

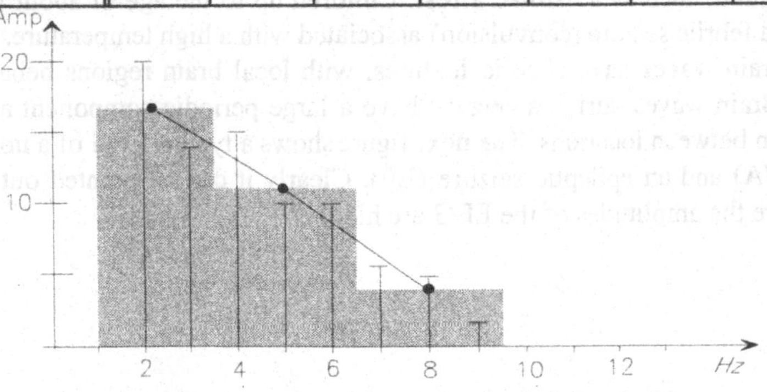

The resulting mFD-spectrum, shown on the left-hand side, seems to wear a large contribution of noise. To extract these noisy contributions the mFD-spectrum should

be fitted by a DAS operation, as the resulting spectrum will show only the amplitudes over the solid line.

The mFD-spectrum, fitted by the DAS-operation is shown on the right hand side. Clearly now it can be pointed out that the resulting spectrum has the desired ideal form, as the fundamental frequency and their harmonics occur clearly. All technical artefacts are eliminated and the harmonics occur sharply.

As the clasp-operation is scanning the spectra in the time domain, the resolution of the recorded signature will be the only indicator for the frequency-co-ordinates of the FD-spectrum. So it will be absolutely necessary to integrate the sampling rate into the clasp-width, to ensure the frequency-axis is standardised in the right way.

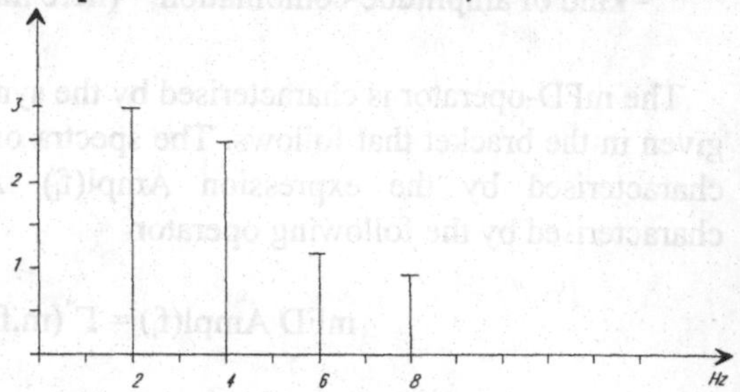

2. About epileptic seizures

Epileptic seizures, sometimes known as fits or convulsions, may take many forms, but in a major convulsion there is a sudden spasm of muscle producing rigidity, causing the patient to fall. There may be jerking movements of the head, arms and legs. The patient becomes unconscious and there may be noisy breathing, saliva bubbling from the mouth and urinary incontinence. After the seizure, which may last 2-3 minutes, the patient may sleep, be dazed or confused for some time, and will require supervision until fully conscious. The cause of seizures can be epilepsy, head injury, tetanus, some poisons, and any situation where the blood and oxygen supply to the brain is impaired, such as a cardiac arrest. Children up to the age of about three years may have a febrile seizure (convulsion) associated with a high temperature.

Nonseizure brain waves have chaotic features, with local brain regions behaving independently. Brain waves during a seizure have a large periodic component and a strong correlation between locations. The next figure shows a typical EEG of a normal brain activity (NA) and an epileptic seizure (EE). Clearly it can be pointed out that during the seizure the amplitudes of the EEG are highly.

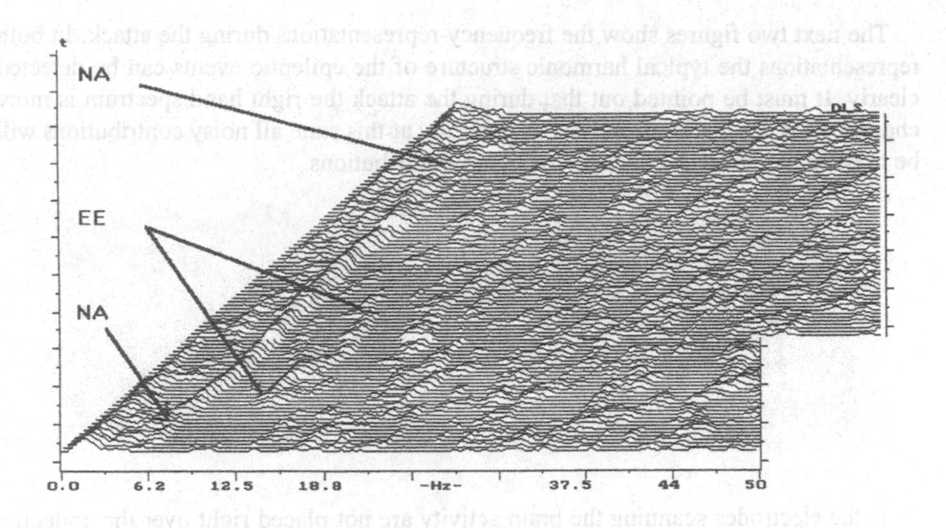

3 Experimental results

The following figures show EEG-spectra of a patient with an epileptic seizure. On the left-hand side the waterfall-diagrams are shown, which result from the operations: common FFT, calculation of the dezible-autopower-spectrum and DAS-calculation. On the right-hand side the comparing waterfall-diagrams which result from the FD- and DAS-operation are shown. The first two pictures show the beginning of the epileptic event. Clearly it can be pointed out that the disappearance of the alpha rhythm (at ca. 25 sec.) and an increasing activity of the theta-delta-band leads to the epileptic seizure (at ca. 27 sec.), pointed out by the following large and harmonic activity-structures. Please watch the deterministic change of the brain activity during the start of the attack, which clearly can be identified as a bifurcation out of the theta-delta-band into two activity-bands in the delta- and alpha-band. Also the pause before the attack is typical for all kinds of epileptic events and has been used for an early warning system [4].

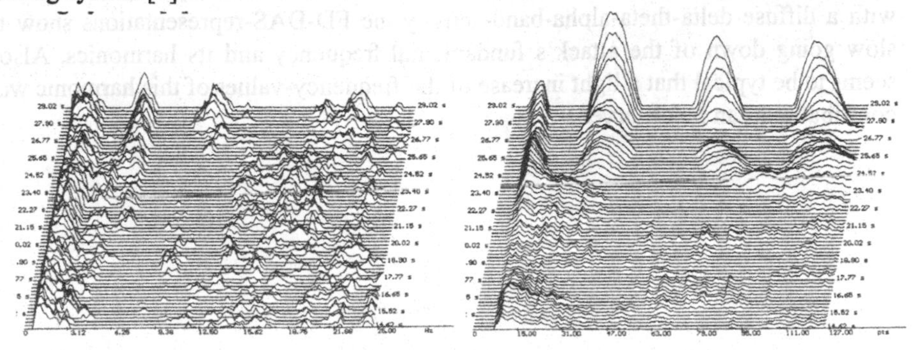

The next two figures show the frequency-representations during the attack. In both representations the typical harmonic structure of the epileptic events can be detected clearly. It must be pointed out that during the attack the right hand spectrum is more characteristic for the brain activity as normally at this state all noisy contributions will be masked by the large epileptic-frequency contributions.

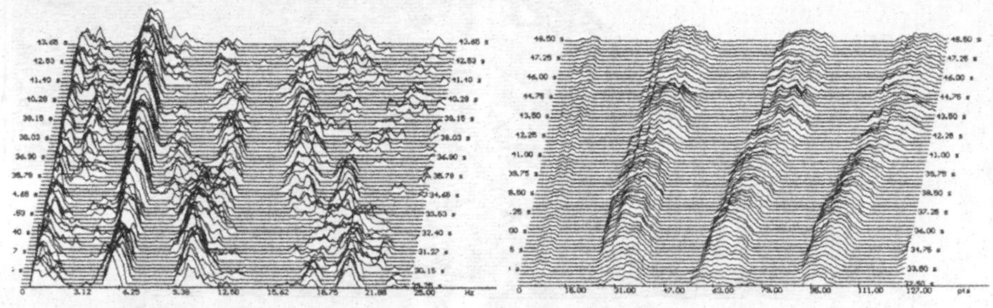

If the electrodes scanning the brain activity are not placed right over the epileptic-source, one will find the following waterfall-diagrams. It is typical for this electrode-placement that the attack-activity seems to occur suddenly. As our investigations show on the other hand, the typical pause before an epileptic seizure now can be pointed out more clearly, if the FD-DAS-signature analysis is taken into account.

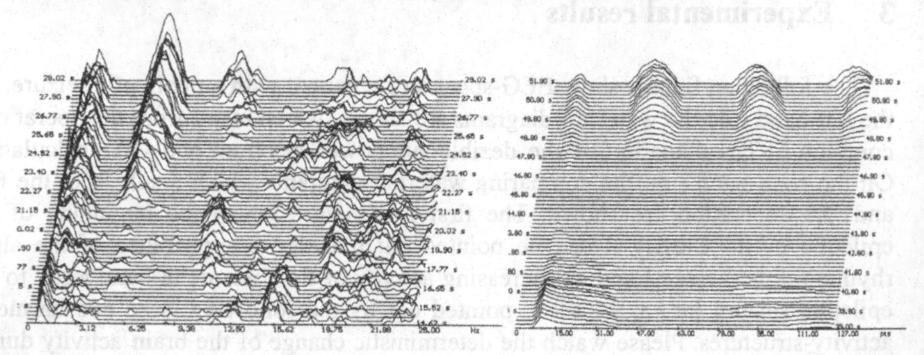

The end of the epileptic seizure is shown in the next two figures. At this stage of the attack we found sometimes that the two representations lead to different pictures. While the common signature analysis shows a more or less abrupt end of the attack with a diffuse delta-theta-alpha-band-activity the FD-DAS-representations show the slow going down of the attack's fundamental frequency and its harmonics. Also it seems to be typical that a light increase of the frequency-values of this harmonic wave contributions can be observed.

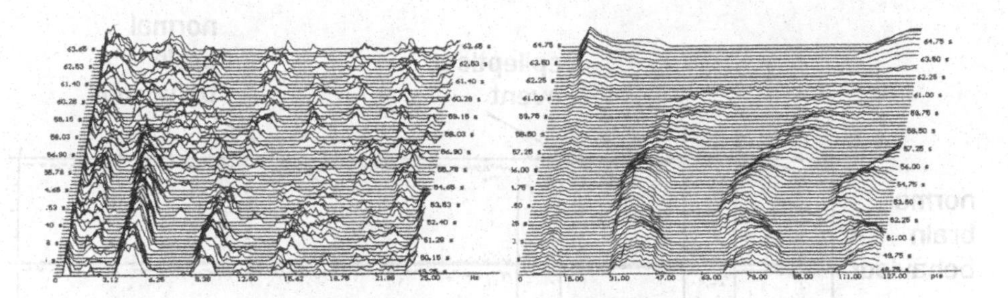

The EEG-signatures of an epileptic seizure of another patient are shown in the next figures. Qualitative the same frequency behavior as before can be observed, but only one bifurcation point occurs at the lower frequency domain. We are not quite sure at this moment, if this more or less solitary occurring restoring of the alpha-rhythm combined with a more or less complex harmonic structure can be used as characteristic feature of an epileptic seizure (which means that only one bifurcation-point can be detected) or if the beta-activity (and therefore the existence of a second bifurcation point in the higher frequency area) more reflects the character of epilepsia.

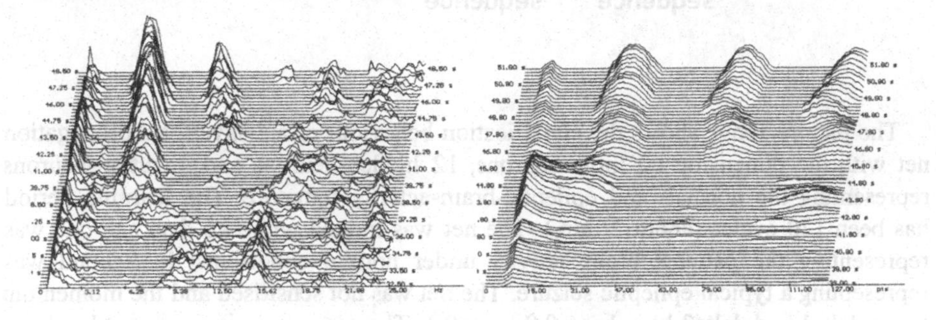

Another advantage for analysing EEG-signatures in the time domain is the possibility to detect the so-called „spikes“ of the pre-epileptic brain activity, which mostly get lost if the time signature is transferred in the frequency domain by a common FFT-algorithm [5]. But as the DAS-FD-operator scans the time spectrum itself, these spikes will be a part of the FD-spectrum also, if they have a harmonic structure or are correlated with other electrical-activities of larger frequencies. Nevertheless the FD-operator combines the FFT- and wave-let-features in a favourite form.

In addition to this special applications we have shown that the DAS-FD-operator can be used for all kinds of time signals. The only restriction, which has to be taken into account, is that the pure signal has not to be covered with an offset information, as the FD-operation is highly sensitive for such contributions. This sensitive behaviour results from the multiplication of the amplitudes. Otherwise the offset information can be eliminated easily, if a DAS-operation is used as first pre-processing step.

We have also tested both signal-forms as input pattern for neural nets and Fuzzy-classificators and found that both representations lead to the same classification behaviour.

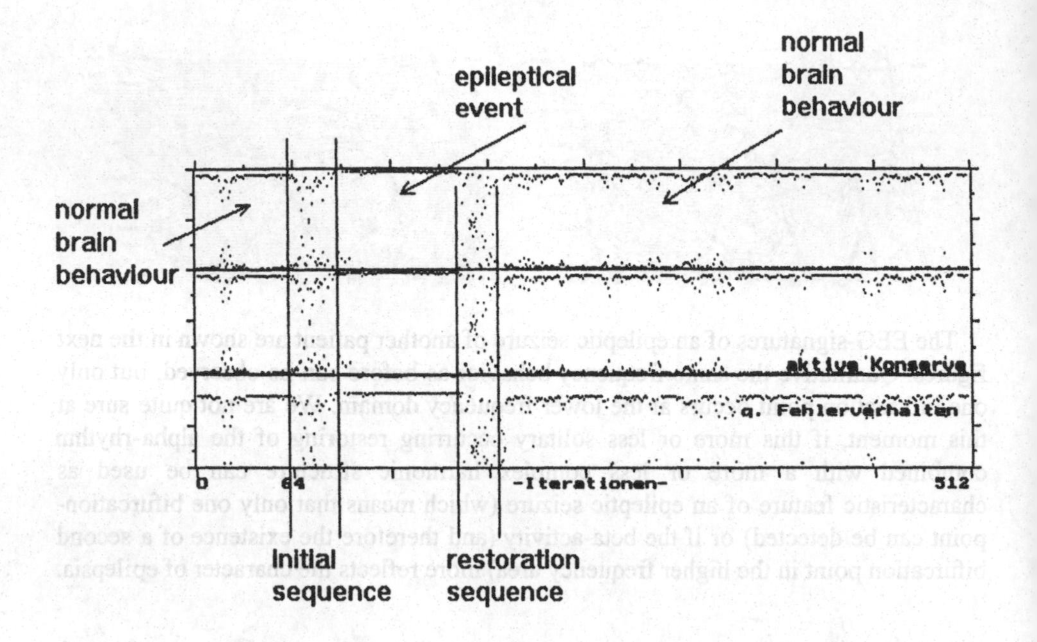

The picture above shows the classification behaviour of a 3-layer backpropagation net with the dimension 60 input-neurons, 12 hidden-neurons and 2 output-neurons representing the normal- and epileptic-brain-activity-signature. The learning period has been 120 cycles (about 2 min.). The net was trained with two data set, one was representing the patients brain activity under normal conditions the second was representing a typical epileptic seizure. The net was not sensitised and the momentum terms delta1 and delta2 have been 0.9 reps. 0.1. The pre-processing operator has been a DAS-FD-operator.

The classification results is encoded in that way, that the classification results of the two output neurons are represented in the upper stripes while the third stripeshows the RMS of an fictive learning phase. If the dots, which represent a classification-result, are near by the upper border of a stripe, the net classifies the current EEG-spectrum with a large confidence, otherwise with a lower confidence. (Normally the dots are coloured, for a quick interpretation by the user.[1]) Clearly it can be pointed out that about 30 sec. before the epileptic event occurs the neural net, which can be used to alarm the patient that an epileptic seizure will soon occur, announces an initial sequence. It is important to mention that our investigations showed that during this initial sequence the patient felt quiet well and that the typical epileptic features of behaviour's change became not evident until the end of this sequence.

From our experiments we learned that the use FD-spectra is a very simple way to calculate feature specific representations which can be used as pre-processing step for

---

[1] All experiments have been performed with the soft-ware-tool "zz2", distributed by the RT&S Ingenieurbüro, Clausthal-Zellerfeld

neural nets to minimise the necessary learning steps by a factor 10-15. But mention that it is most important for the resolution of these representations and for the study of the time behaviour of the sources of the signals, to choose the right parameters for the calculation of the DLS and FD-spectra.

4 Conclusions

The use of the FD-DAS-operations instead of the time intensive FFT- and autopower-calculations seems to open a wide range for a new quality of on-line-recording- and on-line-analysing-tools. Up now we tested our systems for some special medical and industrial applications and found that the new operators can be used instead of the common signal processing algorithm with no problems or loosing adequate information. So the results published in this paper leads to a very optimistic view to the future of the on-line-analysis done by the FD-DAS-operations instead the common FFT- or wave-let-algorithm. Nevertheless the mathematical proof that both 'pictures' of the spectra correspondent in the frequency-space must be done next, as we didn't know exactly up today, why the clasp-procedure leads to the almost same results as the common FFT-dezible-autopower-spectra-operations. But we will work on this and will come up with further results and mathematical proofs hopefully in the early next time.

5 References

[1] M. Reuter, 'Frequency Difference Spectra and Their Use as a New Pre-processing Step for Acoustic Classifiers/ Identifiers', EUFIT '93 Proceedings, September 1993, pp. 436-442.

[2] M. Reuter, D.P.F. Möller, C. Zemke, 'The Detection of Abnormal Brain Activities in their Early Stage by Using New Methods of Signal Processing', EUFIT '97, Aachen 1997

[3] H. Werner, 'Time Trajectories in Sleep Stages'; Proceedings EUFIT, September 1996 Aachen., pp. 2107-2111

[4] D.P.F. Möller, M. Reuter, 'Sensitive Neural Networks and Fuzzy-Classificators and Their Impact for Medical Applications', Proceedings on the IEEE Conference of Man, Machine and Cybernetics, Beijing, VR China 1996

On Interactive Linguistic Summarization of Databases via a Fuzzy-Logic-Based Querying Add-On to Microsoft Access®

Janusz Kacprzyk and Slawomir Zadrozny

Systems Research Institute, Polish Academy of Sciences
ul. Newelska 6, 01-447 Warsaw,Poland
E-mail: {kacprzyk,zadrozny}@ibspan.waw.pl

Abstract. We propose an interactive approach to the linguistic summarization of databases. The point of departure is that a fully automatic generation of summaries is presently impossible, and an interaction with the user is necessary; the user is to specify a class of summaries of interest or relevance. For an implementation of such an interactive approach, a human-friendly database querying interface is employed, and we propose to use our FQUERY for Access querying add-on. An implementation of two relevant types of summaries is shown.

1. Introduction

Though the availability of data is not a problem today, for all kinds of companies and organizations, this does not make by itself the use of those data more productive. More important are relevant, nontrivial dependencies that are encoded in those data. On the other hand, the limits of human perception are not wider than before. This all implies a need for an automatic generation of compact representations of what is really relevant in those data.

In this paper we propose an approach to the derivation of linguistic summaries of large sets of data. The summaries will be in the sense of Yager (1991), i.e. will be derived as linguistically quantified propositions, exemplified by " most of the employees are young and well paid" or "*most* our customers are *reliable*" (which may be more useful than, say "65% of our customers have paid at least 70% of their duties in less than 10 days"), with which a degree of validity (truth,...) is associated.

It is easy to notice that such summaries, to be useful, should concern relevant entities and relations. At the present state of information technology they cannot be derived automatically. The most natural and efficient way to proceed is to derive such linguistic summaries in an interactive way, i.e. to request from the user to define a (class of) summaries of interest, and then to check the contents of a database against such a class of summaries to pick out the best (most valid, most true, ...) one.

This formulation of a class of summaries is basically equivalent to the formulation of a (fuzzy) query to a database, and Kacprzyk and Zadrozny's (1994 – 97c) [cf. also Zadrozny and Kacprzyk (1995)], FQUERY for Access, a fuzzy querying add-on to Microsoft Access, is used for this purpose.

The approach presented in this paper may be viewed as a combination of two approaches presented in Kacprzyk and Zadrozny (1998a, b).

2. The Concept of a Linguistic Summary Using Fuzzy Logic with Linguistic Quantifiers

We mean here the linguistic summaries in the sense of Yager (1991). Basically, suppose that we have:

- V - a quality (attribute) of interest, with numeric and non-numeric (e.g. linguistic) values - e.g. salary in a database of workers,
- $Y=\{y_1,\ldots,y_n\}$ - a set of objects (records) that manifest quality V, e.g. the set of workers; $V(y_i)$ - values of quality V for object y_i,
- $D=\{V(y_1),\ldots,V(y_n)\}$ - a set of data (database)

A *summary* of data set consists of:

- a summarizer S (e.g. young),
- a quantity in agreement Q (e.g. most),
- truth (validity) degree T - e.g. 0.7,

and may exemplified by "T(*most* of employees are *young*)=0.7".

More specifically, if we have a summary, say "*most* (Q) of the employees (y_i's) are *young* (S)", where "*most*" is a fuzzy linguistic quantifier $\mu_Q(x), x\in[0,1]$, "*young*" is a fuzzy quality S, and $\mu_S(y_i), y_i\in Y$, then using the classic Zadeh's (1983) calculus of linguistically quantified propositions, we obtain:

$$T=\mu_Q\Big(\sum_{i=1}^{n}\mu_S(y_i)/n\Big) \qquad (1)$$

The above calculus may be replaced by, e.g., Yager's (1988) OWA operators [cf. Yager and Kacprzyk (1997)].

For more sophisticated summaries as, e.g., "most (Q) *young* (F) employees (y_i's) are *well paid* (S)", the reasoning is similar, and we obtain

$$T=\mu_Q\Big(\sum_{i=1}^{n}(\mu_S(y_i)\wedge\mu_F(y_i))/\sum_{i=1}^{n}\mu_F(y_i)\Big) \qquad (2)$$

where, $\sum_{i=1}^{n}\mu_F(y_i)\neq 0$, and $\wedge$ is a *t*-norm.

3. A General Scheme for Fuzzy Logic Based Data Summarization

The simple approach given above has some serious limitations. Basically, in its source version, it is meant for one-attribute simplified summarizers (concepts) - e.g., young. It can be extended to cover more sophisticated summaries involving some confluence of attribute values as, e.g., "*young* and *well paid*", but this should be done "manually", and leads to some combinatorial problems as a huge number of summaries should be generated and validated to find the most proper one.

The validity criterion is not trivial either, and various measures of specificity, informativeness, etc. may be employed. This relevant issue will not be discussed here, and we will refer the reader to, say, Yager (1991).

For instance, cf.. George and Srikanth (1996), let us consider the following set of labels corresponding to attribute-values and linguistic quantifiers relevant for description of workers in a database:

Q	Age	salary	experience
a few	young	low	low
many	ca. 35	medium	medium
most	middle aged	high	high
almost all	old	very high	very high

Then, we should generate the particular combinations:

- **almost none** workers are: **young, low salary, low experience**
- **a few** workers are: **young, low salary, low experience**
- ...
- **almost all** workers are: **old, very high salary, very high experience**,

whose number may be huge in practice, and calculate the validity of each summary.

This is a considerable task, and George and Srikanth (1996) use a genetic algorithm to find the most appropriate summary, with quite a sophisticated fitness function.

Clearly, when we try to linguistically summarize data, the most interesting are non-trivial, *human-consistent* summarizers (concepts) as, e.g.:

- *productive* workers,
- *difficult* orders, etc.

and it may easily be noticed that they involve a very complicated *combination of attributes* as with, e.g.: a hierarchy (not all attributes are of the same importance for the concept in question), the attribute values are ANDed and/or ORed, *k* out of *n*, *most*, etc. of them should be accounted for, etc.

The basic idea of fuzzy logic based data summarization (data mining) adopted in this paper consists in using a linguistically quantified proposition, as originated by Yager (1991), and here we extend it for using a fuzzy querying package.

We start with the reinterpretation of (1) and (2) for data summarization. Thus, (1) is meant as formally expressing a statement:

"*Most records match query S*" (3)

We assume a standard meaning of the query as a set of conditions on the values of fields from the database's tables, connected with AND and OR. We allow for fuzzy terms in a query (see next section), which implies a degree of matching from [0,1] rather a yes/no matching. Effectively, a query *S* defines a fuzzy subset (fuzzy property) on the set of the records, where the membership of them is determined by their matching degree with the query.

Similarly, (2) may be interpreted as expressing a statement of the following type:

"*Most records meeting conditions F match query S*" (4)

Thus, (4) says something about a subset of records, i.e. only those which satisfy conditions *F*. That is, in database terminology, *F* corresponds to a *filter* and (4) claims that *most* records passing through *F* match query *S*. Moreover, since *F* may be fuzzy, a record may pass through it to a degree from [0,1]. As it is more general, than (3), we will assume (4) as a basis.

Thus, we seek, for a given database, propositions of the type (4), which are highly valid (true) in the sense of formula (2). Basically, a proposition sought consists of three elements: a fuzzy filter *F* (optional), a query *S*, and a linguistic quantifier *Q*. There are two limit cases, where we:

- do not assume anything about the form of any of these elements,
- assume fixed forms of the fuzzy filter and query, and only seek a linguistic quantifier *Q*.

Obviously, in the first case data summarization will be extremely time-consuming, though may produce interesting results, not predictable by the user in some other way. In the second case the user has to guess a good candidate formula for summarization, but the evaluation is fairly simple - requires more or less only the same resources as answering a (fuzzy) query. Thus, the second case refers to the summarization known as *ad hoc queries*, extended with an automatic determination of a linguistic quantifier.

Table 1

A	all is given (or sought), i.e., fields, values and how simple conditions are linked using logical connectives,
A^{fc}	fields and linkage of simple conditions are given, but values are left out,
A^{v}	denotes sought left out values referred to in the above notation
A^{f}	only a set of fields is given, the other elements are sought

Between these two extremes there are different types of summaries, with various assumptions on what is given and what is sought. In case of a linguistic quantifier, it may be given or sought. In case of a fuzzy filter F and a fuzzy query S, more possibilities exist. Basically, both F and S consist of simple conditions, each stating what *value* a *field* should take on, and connected using logical connectives. Here we assume that the table(s) of interest for summarization are fixed. We will use the notation shown in Table 1 to describe what is given or what is sought in respect to the fuzzy filter F and query S (below A stands for F or S)

Using this notation we may propose a rough classification of summaries shown in Table 2.

Table 2

Type	Given	Sought	Remarks
1	S	Q	simple summarizing through ad-hoc query
1.1	S,F	Q	as above + the use of fuzzy filter, i.e., summary is related to a fuzzy subset of records
2	Q,S^{fc}	S^{v}	in the simplest case corresponds to the search for typical or exceptional values (see the comments given below this table)
2.1	Q,S^{fc},F	S^{v}	
3	Nothing	S,F,Q	fuzzy rules, extremely expensive computationally
3.1	S^{f},F^{f}	S,F,Q	much more viable version of above
3.2	S	F,Q	looking for causes of some pre-selected, interesting data features (machine learning alike)

Thus, we distinguish 3 main types of data summarization. Type 1 is a simple extension of fuzzy querying as in FQUERY for Access. Basically, the user has to conceive a query, which may be true for some population of records in the database. As the result of this type of summarization he or she receives some estimate of the cardinality of this population as a linguistic quantifier. The primary target of this type of summarization is certainly to propose such a query that a large proportion as, e.g., *most*, of the records satisfy it. On the other hand, it may be interesting to learn that only *few* records satisfy some meaningful query. Type 1.1 is a straight extension of Type 1 summaries by adding a fuzzy filter. Having a fuzzy querying engine dealing with fuzzy filters, the computational complexity is here the same as for Type 1.

Type 2 summaries require much more effort. Primary goal of this type of summary is to determine a typical (exceptional) values of a field. Then, query S consists of only one simple condition referring to the field under consideration. The summarizer tries to find a value, possibly fuzzy, such that the query (built of the field, equation relational operator and that value) is true for Q records. Depending on the category of Q used, e.g., *most* versus *few*, typical or exceptional values are sought, respectively. This type of summaries may be used with more complicated, regular queries, but it quickly becomes computationally infeasible (combinatorial explosion) and the interpretation of results becomes vague. Type 2.1 may produce typical (exceptional) values for some, possibly fuzzy, subpopulations of records. From the computational point of view, the same remarks apply as for Type 1 versus Type 1.1.

Type 3 is the most general. In its full version this type of summaries are to produce fuzzy rules describing the dependencies between values of particular fields. Here the use of the filter is essential, in contrast to the previous types, where it was optional. The very meaning of a fuzzy rule received is that if a record meets a filter's condition, then it meets also the query's conditions - this corresponds to a classical IF-THEN rule. For a general form of such a rule it is difficult to devise an effective and efficient algorithm looking for them. Full search may be acceptable only in case of restrictively limited sets of a rule (summarizer) building blocks, i.e. fields and their possible values. Type 3.1 summary may produce interesting results in a more reasonable time. It relies on the user pointing out promising fields to be used during the construction of a summarizer. For computational feasibility some limits should be also put on the complexity of query S and filter F in terms of the number of logical connectives allowed. Finally, Type 3.2 is here distinguished as a special case due to its practical value. First of all it makes the generation of a summarizer less time consuming and at the same time has a good interpretation. Here the query is known in an exact form and only the filter is sought, i.e. we look for causes of given data features. For example, we may set in a query, that profitability of a venture is *high* and look for the characterization of a subpopulation of ventures (records) of such a high profitability. Effectively, what is sought, is a, possibly fuzzy, filter F.

The summaries of type 1 and 2 have been implemented as an extension to our FQUERY for Access (Kacprzyk and Zadrozny, 1994 – 1997c).

4. FQUERY for Access

FQUERY for Access is an add-in to Microsoft Access that provides with fuzzy querying capabilities [cf. Kacprzyk and Zadrozny (1994 - 1997c), Zadrozny and Kacprzyk (1995)].

FQUERY for Access makes it possible to use fuzzy terms in regular queries, then submitted to the Microsoft Access's querying engine. The result is a set of records matching the query, but obviously to a *degree* from [0,1].

Briefly speaking, the following types of fuzzy terms are available:

- fuzzy values, exemplified by *low* in "profitability is *low*",
- fuzzy relations, exemplified by *much greater than* in "income is *much greater than* spending", and
- fuzzy linguistic quantifiers, exemplified by *most* in "*most* conditions have to be met".

The elements of the first two types are elementary building blocks of fuzzy queries in FQUERY for Access. They are meaningful in the context of numerical fields only. There are also other fuzzy constructs allowed which may be used with scalar fields.

If a field is to be used in a query in connection with a fuzzy value, it has to be defined as an *attribute*. The definition of an attribute consists of two numbers: the attribute's values lower (LL) and upper (UL) limit. They set the interval which the field's values are assumed to belong to, according to the user. This interval depends on the meaning of the given field. For example, for *age* (of a person), the reasonable interval would be, e.g., [18,65], in a particular context, i.e. for a specific group. Such a concept of an attribute makes it possible to universally define fuzzy values.

Fuzzy values are defined as fuzzy sets on [-10, +10]. Then, *the matching degree* $md(\cdot,\cdot)$ of a simple condition referring to attribute AT and fuzzy value FV against a record R is calculated by:

$$md(\mathrm{AT} = \mathrm{FV}, \mathrm{R}) = \mu_{\mathrm{FV}}(\tau(\mathrm{R}(\mathrm{AT}))) \tag{5}$$

where: R(AT) is the value of attribute AT in record R, μ_{FV} is the membership function of fuzzy value FV, τ: $[\mathrm{LL}_{\mathrm{AT}},\mathrm{UL}_{\mathrm{AT}}] \to [-10,10]$ is the mapping from the interval defining AT onto [-10,10] so that we may use the same fuzzy values for different fields. A meaningful interpretation is secured by τ which makes it possible to treat all fields domains as ranging over the unified interval [-10,10].

For simplicity, it is assumed that the membership functions of fuzzy values are trapezoidal as in Figure 1 and τ is assumed linear.

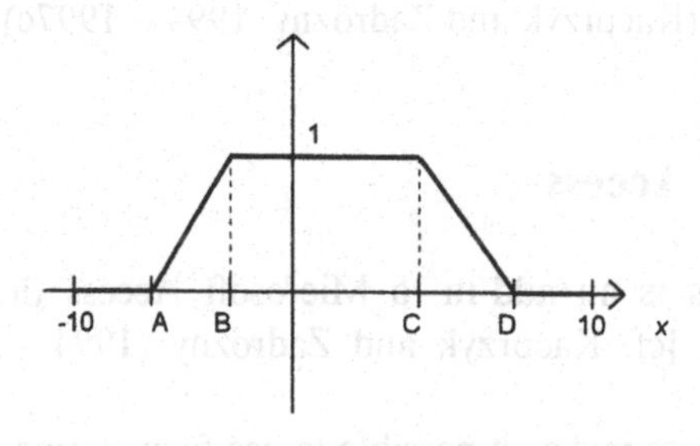

Figure 1. An example of the membership function of a fuzzy value

Linguistic quantifiers provide for a flexible aggregation of simple conditions. In FQUERY for Access the fuzzy linguistic quantifiers are defined in Zadeh's (1983) sense (see Section 2), as fuzzy set on [0, 10] interval instead of the original [0, 1]. They may be interpreted either using original Zadeh's approach or via the OWA operators (cf. Yager, 1988, Yager and Kacprzyk, 1997); Zadeh's interpretation will be used here. The membership functions of fuzzy linguistic quantifiers are assumed piece-wise linear, hence two numbers from [0,10] are needed to define a quantifier. Again, a mapping from [0,N], where N is the number of conditions aggregated, to [0,10] is employed to calculate the matching degree of a query. More precisely, the matching degree, $md(\cdot,\cdot)$, for the *query* "Q of N conditions are satisfied" for record R is equal to

$$md(\underset{i}{Q}\,\text{condition}_i, \text{R}) = \mu_Q\left(\frac{1}{\tau(N)}\tau\left(\sum_i md(\text{condition}_i, \text{R})\right)\right) \tag{6}$$

We can also assign different importance degrees for particular conditions. Then, the aggregation formula is equivalent to (2). The importance is identified with a fuzzy set on [0,1], and then treated as property F in (2).

In FQUERY for Access queries containing fuzzy terms are still syntactically correct Access's queries, through the use of parameters. Basically, Access represents the queries using SQL. Parameters, expressed as strings limited with brackets, make it possible to embed references to fuzzy terms in a query. We have assumed special naming convention for parameters corresponding to particular fuzzy terms. For example:

[FfA_FV *fuzzy value name*] will be interpreted as a fuzzy value (7)

[FfA_FQ *fuzzy quantifier name*] will be interpreted as a fuzzy quantifier

First, a fuzzy term has to be defined using a toolbar of FQUERY for Access which is stored internally. This maintenance of dictionaries of fuzzy terms defined by users, strongly supports our approach to data summarization. In fact, the package comes with a set of predefined fuzzy terms but the user may add new ones to the dictionary too.

When the user initiates the execution of a query it is automatically transformed and then run as a native query of Access. The transformation consists primarily in the replacement of parameters referring to fuzzy terms by calls to functions which secure a proper interpretation of these fuzzy terms. Then, the query is run by Access as usually.

5. Summaries via FQUERY for Access

FQUERY for Access, which extends the querying capabilities of Microsoft Access by making it possible to handle fuzzy terms, may be viewed as an interesting tool for data mining, including the generation of summaries. The simplest method of data mining through ad-hoc queries becomes much more powerful by using fuzzy terms. Nevertheless, the implementation of various types of summaries given in Section 3 seems to be worthwhile, and is fortunately relatively straightforward.

We rely on dictionaries of fuzzy terms maintained and extended by users during the subsequent sessions. The main feature supporting an easy generation of summaries is the adopted concept of context-free definitions of particular fuzzy terms. Hence, looking for a summarizer we may employ any term in the context of any attribute. Thus, we get a summarizer building blocks at hand and what is needed is an efficient procedure for their composition, compatible with the rest of the fuzzy querying system.

In case of Type 1 summaries only the list of defined linguistic quantifiers is employed. The query S is provided by the user and we are looking for a linguistic quantifier describing in a best way the proportion of records meeting this query. Hence, we are looking for a fuzzy set R in the space of linguistic quantifiers, such that:

$$\mu_R(Q) = \text{truth}(QS(X)) = \mu_Q\left(\sum_{i=1}^{n} \mu_S(x_i)/n\right) \quad (8)$$

FQUERY for Access processes the query, additionally summing up the matching degrees for all records. Thus, the sum in (8) is easily calculated. Then, the results are displayed as a list of records ordered by their matching degree. In another window, the fuzzy set of linguistic quantifiers sought is shown. We only take into account quantifiers defined by the users for querying, for efficiency. At the same time it seems quite reasonable as the quantifiers defined by the user should have a clear interpretation. Currently, FQUERY for Access does not support fuzzy filters. As soon as this capability is added, also summaries of Type 1.1 will be available. Simply, when evaluating the filter for particular records, another sum used in (2) will be calculated and the final results will be presented as in case of Type 1 summaries.

Type 2 summaries require more effort and the redesigning of the results' display. Now, we are given the quantifier and the whole query, but without some values. Thus, first of all, we have extended the syntax of the query language introducing a *placeholder* for a fuzzy value. That is, the user may leave out some values in the query's conditions and request the system to find a best fit for them. To put such a placeholder into a query, the user employs a new type of parameter. We extend the list partially shown in (7) adding the parameter [FfA_F?]. During the query processing these parameters are treated similarly as fully specified fuzzy values. However, the matching degree is calculated not just for one fuzzy value but for all fuzzy values defined in the system's dictionary. The matching degrees of the whole query against the subsequent records, calculated for different combination of fuzzy values, are summed up. Finally, it is computed for particular combinations of fuzzy values how well the query is satisfied when a given combination is put in a query. Thus, we again receive as a result a fuzzy set but this time defined in the space of fuzzy values' vectors.

Obviously, such computations are extremely time-consuming and are practically feasible only for one placeholder in a query. On the other hand, the case of one placeholder, corresponding to the search for typical or exceptional values, is a most useful form of a type 2 summary. It is again fairly easy to embed type 2 summaries in the existing fuzzy querying mechanism.

6. Concluding remarks

We proposed an interactive approach to the linguistic summarization of databases. An interaction with the user via the authors' FQUERY for Access, a fuzzy logic based querying add-on to Microsoft Access, is shown to be ab effective and efficient solution.

Literature

Bosc P. and J. Kacprzyk, Eds. (1995) Fuzziness in Database Management Systems. Physica-Verlag, Heidelberg,

George R. and R. Srikanth (1996) Data summarization using genetic algorithms and fuzzy logic, In: F. Herrera and J.L. Verdegay (Eds.): Genetic Algorithms and Soft Computing. Physica-Verlag, Heidelberg and New York, pp. 599 - 611.

Kacprzyk J. and S. Zadrozny (1994) Fuzzy querying for Microsoft Access. Proceedings of the Third IEEE Conference on Fuzzy Systems (Orlando, USA), Vol. 1, pp. 167-171.

Kacprzyk J. and S. Zadrozny (1995a) FQUERY for Access: fuzzy querying for a Windows-based DBMS. In: P. Bosc and J. Kacprzyk (Eds.) Fuzziness in Database Management Systems, Physica-Verlag, Heidelberg, pp. 415 - 433.

Kacprzyk J. and S. Zadrozny (1995b) Fuzzy queries in Microsoft Access v. 2, Proceedings of 6th IFSA World Congress (Sao Paolo, Brazil), Vol. II, pp. 341 - 344.

Kacprzyk J. and S. Zadrozny (1997a) Fuzzy queries in Microsoft Access v. 2. In: D. Dubois, H. Prade and R.R. Yager (Eds.): Fuzzy Information Engineering - A Guided Tour of Applications, Wiley, New York, 1997, pp. 223 - 232.

Kacprzyk J. and S. Zadrozny (1997b) Implementation of OWA operators in fuzzy querying for Microsoft Access. In: R.R. Yager and J. Kacprzyk (Eds.) The Ordered Weighted Averaging Operators: Theory and Applications, Kluwer, Boston 1997, pp. 293 - 306.

Kacprzyk J. and S. Zadrozny (1997c) Flexible querying using fuzzy logic: An implementation for Microsoft Access. In: T. Andreasen, H. Christiansen and H.L. Larsen (eds.): Flexible Query Answering Systems, Kluwer, Boston, 1997, pp. 247-275.

Kacprzyk J. and S. Zadrozny (1998a), "Data Mining via Linguistic Summaries of Data: An Interactive Approach", in T. Yamakawa and G. Matsumoto (eds.): Methodologies for the Conception, Design and Application of Soft Computing" (Proc. of 5th IIZUKA'98), Iizuka, Japan, 1998, pp. 668-671.

Kacprzyk J. and S. Zadrozny (1998b), "On sumarization of large datasets via a fuzzy-logic-based querying add-on to Microsoft Access", in: Intelligent Information Systems VII (Proceedings of the Workshop - Malbork, Poland), IPI PAN, Warsaw, 1998, pp.249-258.

Kacprzyk J., Zadrozny S. and Ziółkowski A. (1989) FQUERY III+: a 'human consistent' database querying system based on fuzzy logic with linguistic quantifiers. Information Systems 6, 443 - 453.

Kacprzyk J. and Ziółkowski A. (1986) Database queries with fuzzy linguistic quantifiers. IEEE Transactions on Systems, Man and Cybernetics SMC - 16, 474 - 479.

Rasmussen D. and R.R. Yager (1997) Fuzzy query language for hypothesis evaluation. In Andreasen T., H. Christiansen and H. L. Larsen (Eds.) Flexible Query Answering Systems. Kluwer, Boston/Dordrecht/London, pp. 23-43.

Yager R.R. (1988) On ordered weighted averaging aggregation operators in multi-criteria decision making. IEEE Transactions on Systems, Man and Cybernetics, 18, 183-190.

Yager R.R. (1991) On linguistic summaries of data. In: G. W. Frawley and G. Piatetsky-Shapiro (Eds.): Knowledge Discovery in Databases. AAAI/MIT Press, pp. 347 - 363.

Yager R.R. and J. Kacprzyk, Eds. (1997) The Ordered Weighted Averaging Operators: Theory and Applications, Kluwer, Boston.

Zadeh L.A. (1983) A computational approach to fuzzy quantifiers in natural languages. Computers and Maths. with Appls. 9, 149 - 184.

Zemankova M. and J. Kacprzyk (1993) The roles of fuzzy logic and management of uncertainty in building intelligent information systems, Journal of Intelligent Information Systems 2, 311-317.

A General-Purpose Fuzzy Engine for Crop Control

Moataz Ahmed[1], Ernesto Damiani[2], and Andrea G. B. Tettamanzi[2]

[1] King Fahd University of Petroleum & Minerals, Dhahran, Saudi Arabia
moataz@kfupm.edu.sa
[2] University of Milan – Polo di Crema,
Via Bramante 65, 26013 Crema (CR), Ita\ly
edamiani@crema.unimi.edu, tettaman@dsi.unimi.it

Abstract In this position paper we outline an architecture of a general-purpose *indirect fuzzy-based* crop control. Some ways of exploiting previous work on a control system for single-plant hydroponics nurseries are also discussed.

1 Introduction

Recording physical Earth phenomenon from remote locations is a time-honored technique, as first attempts at using aerial photography for environmental monitoring date back to the early 1900's, while satellite remote sensing was first introduced in the 1960's [1]. More recently, the huge amount of high-quality environmental data available from remote sensing devices has boosted the interest for applying remote sensing to critical applications such as crop control and agricultural resources optimization, increasingly perceived to be crucial for the development of agriculture in many countries.

In fact, today's remote sensors are capable of presenting data both in analog form, such as photographs that must be later digitized and line plotted, and in high resolution digital form. In the latter case, images are typically available as $N \times M \times B$ matrices, where $N \times M$ is the sensor's spatial resolution and B is the number of bands (spectral resolution). In such images, each pixel is stored with fixed number of bits (typically 8-12) called *pixel depth* or *quantization*. For example, the NASA Landsat Thematic Mapper (TM) offers a spectral resolution of 7 bands in the range 0.45 to 12.5 mm (visible blue through thermal infrared) with a spatial resolution of 30×30 meters at 8 bits per pixel quantization. Besides imaging, remote sensing can now seamlessly provide all sorts of crop related biophysical measurements, such as surface temperature, geographic location, vegetation - chlorophyll, biomass, foliar water content, and soil moisture [2].

For instance, the *ARTEMIS (Africa Real Time Environmental Monitoring Using Imaging Satellites)* system, that became operational at the FAO Remote Sensing Center in Rome in August 1988, provides real and near-real time precipitation and vegetation assessment for Africa, the Near East and South West Asia. ARTEMIS data are now, more and more, provided to users involved in early warning for food security and desert locust control at the regional and national levels in Africa [3]. The ARTEMIS products are currently distributed to its users on color hard copies, IBM-PC compatible diskettes, tapes, and as point-listing using mail, pouch and courier services.

In addition, FAO and the European Space Agency (ESA) are collaborating to develop a dedicated satellite communications system, *DIANA* (*Data and Information Available Now in Africa*), which will enable high speed digital dissemination of ARTEMIS products, and other information, to remote stations in Africa. A distinct through related approach was followed in systems such as the first European Remote Sensing Satellite [4], which was launched in July 1991 by the European Space Agency. ERS-1 uses advanced microwave techniques to acquire measurements and images regardless of clouds and sunlight conditions. In comparison to other satellite systems, the ERS-1 (and, now, ERS-2) is unique in the simultaneous measurement of many parameters, including sea state, sea surface winds, sea surface temperature, ocean circulation and sea and ice level, as well as all-weather imaging of ocean, ice and land. ERS-1 satellite is both an experimental and pre-operational system. The nature of its orbit and its complement of sensors enable a global mission providing a worldwide geographical and repetitive coverage, primarily oriented towards ocean and ice monitoring, but with an all-weather high resolution microwave imaging capability over land and coastal zones for environmental monitoring and crop control.

However, the systems described above only aim at detecting exceptional stress situations of crops, so that emergency actions can be taken. In this paper, we propose an approach we believe to be novel, relying on feeding remote sensing data, together with an evolvable model of crop behavior, to an intelligent decision support system based on a backward-chaining-based fuzzy reasoning engine. Decisions suggested by our system could be later translated into control actions to be performed by humans and/or by on-site actuators, and disseminated using techniques not unlike those used in the ARTEMIS approach. As an experimental test bed for our approach, we set up a greenhouse cell equipped with local sensors, including a video camera, and actuators.

2 Fuzzy Control Systems

It is clear that many control theorists have successfully dealt with a large class of control problems (complex, large-scale, nonlinear, non-stationary and stochastic control problems) by mathematically modeling the process and solving these analytical models to generate control. However, the analytical models tend to become complex, especially in large, intricate systems. The non-linear behavior of many practical systems and the unavailability of quantitative data regarding the input-output relations make this analytical approach even more difficult. On the other hand, it is well known that for most engineering systems, there are two important information sources:

1. Sensors which provide numerical measurements of variables and
2. Human experts who provide linguistic instructions and descriptions about the system.

The information from sensors is called numerical information and the information from human experts is called linguistic information. Numerical information is represented by numbers whereas linguistic information is represented by words such as *small*, *large*, *very large*, and so forth. Conventional engineering approaches can only make use of numerical information. Because so much human knowledge is represented in linguistic terms, incorporating it into an engineering system in a systematic and efficient manner is very important. However, linguistic information can be represented in fuzzy terms [5]. A fuzzy-logic system representation has the ability to incorporate the human knowledge, represented in linguistic terms, into the engineering system very efficiently. Fuzzy-logic systems are constructed from fuzzy *IF-THEN* rules. Representing the process to be controlled as a fuzzy logic system makes it much easier to understand its behavior, and complex requirements may be implemented in amazing simple, easily maintained, and inexpensive controllers.

Fuzzy controllers have found particular applications in industrial systems that are very complex and cannot be modeled precisely even under various assumptions and approximations. Crop control systems fall into this class of systems.

Using fuzzy logic, a crop and its surrounding environment may be described by fuzzy predicates [6]. Fuzzy predicates are predicates with associated values in a form of fuzzy sets (as opposed to the *TRUE* and *FALSE* crisp values). Each predicate describes some properties of the process and/or its surrounding environment in its present state. Each predicate may or may not have arguments. The crop surrounding environment might be represented by fuzzy predicates such as: *WIND(LOC)*, *LIGHT(LOC)*, *HUMIDITY(LOC)*, *RAIN(LOC)*, etc. Where *WIND(LOC)*, with one argument *LOC*, gives the speed of the wind at location *LOC* as a fuzzy set, and *LIGHT(LOC)* gives the level of lightening at location *LOC* as a fuzzy set. *COLOR(LEAF)* is an example of using a fuzzy predicate to represent a property, that is the color, of a given leaf, *LEAF*. The state of the crop and its surrounding environment, at any point in time, is represented by a set of fuzzy predicates.

The use of fuzzy control and classification systems for environmental monitoring is a well established subject in soft computing research [7]

In this position paper we discuss the feasibility of a complete fuzzy-based solution for crop control problems, focusing on some factors like temperature, humidity and light.

2.2 Direct Fuzzy Control vs. Indirect Fuzzy Control

Linguistic information from human experts can be classified into two categories:

1. Fuzzy control rules, which propose a suitable control action for each possible situation. For example, the following fuzzy IF-THEN rule may be used to control the level of lightening in a crop nursery:

   IF

   the growth rate is low

   THEN

   apply more lightening,

   where *low* and *more* are labels of fuzzy sets. This type of control rule is called *direct fuzzy control*.

2. Fuzzy IF-THEN rules, which describe the behavior of the plant (i.e., the crop). For example, the behavior of the crop can be described as:

   IF

   you apply more lightening

   THEN

   the growth rate will increase,

   where *more* and *increase* are labels of fuzzy sets. This type of control rules is called *indirect fuzzy control.*

3 The Approach

We propose an *indirect fuzzy control approach* for crop control; that is using fuzzy rules to describe the behavior of the crop rather than using them to specify control actions. The justification for this approach is threefold. (1) Taking an indirect control approach, a *crop model* is produced as a by-product of the control process; this crop model can then be reused for several other applications. (2) The evolvement of the crop model during the control process may be done through training with historical data only, and without a need for a real-time data; thus saving the effort and risks of experimental-based training. (3) This approach accommodates for easy change of the controller objectives (e.g., for simulation purposes).

Since, in this approach, the crop behavior is represented by *indirect* IF-THEN rules, the fuzzy controller, that is to control the crop to achieve a certain state, can be seen as a *planning* system. Planning systems are concerned with the problem of generating a sequence of actions to accomplish a goal. Normally, each action has a set of *pre-conditions* which must be satisfied before the action can be applied, and a set of *post-conditions* which will be true after the action execution. A *planning problem* is characterized by an initial state and a goal statement. A *state* is a collection of characteristics of an object that is sufficiently detailed to uniquely determine the new characteristics of the object that will result after an action. The initial state description tells the planning system the way the *world* is "right now". The *goal statement* tells the planning system the conditions which must be satisfied when the plan has been executed. The world in which the planning takes place is often called the *application domain.* We will sometimes refer to the goal statement as simply *the goal.* A plan is said to be a solution to a given problem if it is applicable in the problem's initial state, and if after plan execution, the goal is true. A plan is applicable if all the preconditions of any action, within the plan, are satisfied before applying that action.

The *indirect fuzzy control problem* as a planning problem can be described as follows. Given an initial state of a crop and its surrounding environment, a set of indirect fuzzy rules, and a desired crop goal state, find the sequence of actions which can change the state of the crop from its initial state to its desired goal state. The produced sequence of actions is a *plan.* We propose to use a backward chaining inference to generate such a plan.

4 Indirect Fuzzy Crop Control

The indirect fuzzy crop control process involves two main stages (Figure 1):

1. *Training Stage*: Off-line model identification via evolutionary computation and fine tuning, and
2. *Control Stage*: On-line control that generate control actions on the basis of the model identified during the Training Stage, as well as actual input data.

The Training Stage is concerned with building a fuzzy rule base that represents a linguistic description of the crop behavior. This rule base is to be used later during the on-line control stage in order to generate control actions.

During the Training Stage, the crop model is identified through training using data from two sources: historical data, that is readings of the environmental variables and their outcomes and supplement data generated through experiments on a small-scale crop in a laboratory setting. The objective of the supplement data is to consider extreme or unusual conditions that are not well represented in the historical data. A small-scale lab crop can be used to conduct collect such supplement data. The experiments on that small-scale lab crop are guided and constrained by expert meta-knowledge that describes feasible ranges for control variables.

Developing the crop model involves two phases: exploration, and fine tuning. *Exploration* involves searching the whole space of models to find a set of models that well represent the crop under discussion. This exploration phase can be achieved by means of suitable heuristics such as evolutionary algorithms. Such evolutionary algorithms ought to be guided and constrained by expert *meta-knowledge* in order to avoid unrealistic models. The initial population of such evolutionary algorithms can be seeded with handcrafted models based on experts' interviews.

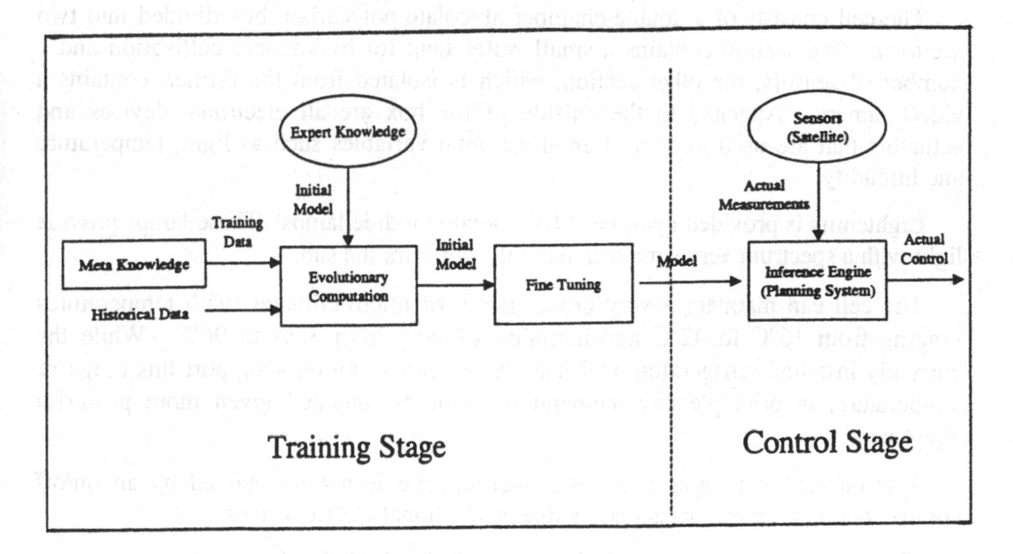

Fig. 1. The two stages of the proposed system

Once the exploration phase is complete, the fine tuning phase starts. The objective of the fine tuning phase is to conservatively adapt the set of models, that are provided as a result of the exploration phase, in order to obtain a satisfactory level of performance. The model that gives such a satisfactory level of performance is used during the on-line control stage.

At the control stage, actual measurements are read from the sensors located in the field, and the model of the system is used to generate the control action through fuzzy backward chaining.

5 Overall System Architecture

A possible architecture for implementing and deploying the above described control system consists of four blocks:

1. *Data acquisition*, involving image and sensor measurements,
2. *Data transmission* to the fuzzy engine, possibly involving long-range satellite links,
3. *On-line control* involving the generation of the control action starting from the fuzzy model of the crop behavior, and
4. *Application of control actions* that can be either automated or carried out manually following the system advice.

5.1 Using A Nursery Control System As Experimental Test-Bed

An experimental cell, that is a nursery environment, has been realized at the University of Milan – Polo di Crema [8]. This cell could be used as a test-bed for validating the control system described in the previous sections.

The cell consists of a double-chamber alveolate polycarbon box divided into two sections. One section contains a small water tank for hydroponic cultivation and a number of sensors; the other section, which is isolated from the former, contains a video camera. Attached to the outside of the box are all electronic devices and actuators that are used to control environmental variables such as light, temperature and humidity.

Lightening is provided by three 400W metallic iodide lamps. These lamps provide light with a spectrum very similar to natural light from the sun.

The cell can maintain a very broad gamut of micro-climates, with temperatures ranging from 16°C to 42°C and humidity ranging from 32% to 96%. While the currently installed refrigerating and heating machines can only support this range of temperature, in principle any temperature could be attained given more powerful equipment.

It is interesting to notice that the microclimate is not maintained by an on/off control, but by proportional actuators like proportional electro-valves.

The sensors in the cell provide measurements for the following quantities:

- *pH*, *temperature*, *conductivity*, and *oxygen* in the solution;

- Global *radiation* (via a photo diode), relative *humidity*, *temperature* and CO_2 of the environment.

All solution sensors are thermally compensated.

The video camera points to the plant. At programmable intervals, it takes snapshots, from which the values of a number of significant parameters are extracted using fuzzy techniques. These include:

- Average color of the leaves;
- Image area of the plant;
- Height and width of the plant.

All these values can serve as input to a fuzzy inference engine located on a control station, which can adjust the set points of most control variables.

Figure 2 and Figure 3 show respectively the inside and the outside of our experimental cell.

Fig. 2. The experimental cell (inside)

Possible validation of the proposed approach includes:

- Comparison between "close" and "remote" image data processing, in order to simulate satellite image acquisition and assess the controller performance when only remote sensor data are available. The objective of this comparison is to assess the sensitivity of the controller to the accuracy of the available data.
- Use of nursery control variables (such as humidity and temperature) to simulate various microclimates in order to assess the flexibility of the tuning mechanism.
- Comparisons between *coarse-grained* and *fine-grained* control actions in order to evaluate possible differences in performance caused by the use of the controller as a decision support system rather than automatically control actuators. In the latter

Fig. 3. The experimental cell (outside)

case, precise control actions are possible while in the former, control actions are to be executed by humans who always combine their own perception of the situation while applying the control action.

An experimental protocol is being designed in order to allow unbiased evaluation of results.

6 Potential Benefits

Once implemented, the system described above could greatly help rationalizing crop cultivation, especially in developing countries, where scarcity of resources is one of the major obstacles to matching agricultural production to demand.

The system is conceived to work at different scales, allowing crop control applications to wide areas of land by means of remote image and sensor data acquisition as well as small to medium intensive or high-quality cultivation by means of fixed or mobile image and data acquisition devices.

In addition to the potential benefits in the area of crop control, the crop model that is the product of the training stage is expected to be very useful for researchers that are interested in studying crop behavior. These researchers could use the model as accurate simulators during their studies.

References

1. Sabins, F.F. Jr. "Remote Sensing: Principles and Interpretation", W.H. Freeman, 1987.
2. Jensen, J.R. "Introductory Digital Image Processing: A Remote Sensing Perspective", 2nd Edition, Prentice-Hall 1996.
3. http://www.neonet.nl/ceos-idn/campaigns/ARTEMIS.html
4. http://www.asi.it/00HTL/eng/asicgs/remotesens/ers_m1.html

5. Wang, L. *Adaptive Fuzzy Systems And Control.* Prentice-Hall, Inc., 1994.
6. Ahmed, M., Damiani, E., and Rine D. Fast Recall of Reusable Fuzzy Plans Using Acyclic Directed Graph Memory, *1998 ACM Symposium on Applied Computing (ACM SAC'98), Fuzzy Applications Track,* February 27-March 1, 1998.
7. Binaghi, E., Madella, P., Montesano, M.G., and Rampini, A. "Identification of glacier snow equilibrium line using a fuzzy rule-based classifier". Proc. of International Geoscience and Remote Sensing Symposium, pp. 1768-1780, 1995.
8. Tarzia, G. *Sistema di Controllo e Ricerca per Colture Idroponiche in Serra.* Master's Thesis, University of Milan, 1998 (In Italian).
9. Chen, J., and Rine, D. Training Fuzzy Logic Based Software Components by Combining Adaptation Algorithms, Soft Computing, vol. 2, no.2, pp.48-50, 1998.
10. Mamdani, E.H. "Twenty years of fuzzy control: experiences gained and lessons learnt." In *1993 IEEE International Conference on Fuzzy Systems* (San Francisco, CA, March 28-April 1). IEEE, Piscataway, N.J., pp.339-344.

Fuzzy Control of a Physical Double Inverted Pendulum Model

Ján Vaščák

Technical University in Košice, Faculty of Electrical Engineering and Informatics,
Department of Cybernetics and AI, Computer Intelligence Group,
Letná 9, 041 20 Košice, Slovakia
vascak@tuke.sk

Abstract. This paper deals with the design of a fuzzy controller for a double inverted pendulum (DIP). The DIP model used is based on the Lagrange – Euler equation. Various design methods are described where the hierarchical approach is chosen for the controller design. The whole problem is decomposed into two hierarchical levels. The higher level serves as a supervisor which enables broader criteria to be taken into consideration, which is necessary for the control strategy of such a system. The Matlab Fuzzy Logic Toolbox was used for the algorithm implementation and the control quality evaluation of the performed experiments.

1 Introduction

The classical control theory is based on design of controllers for stable systems which are in many cases more or less nonlinear. But it is difficult to use linearization methods in systems with high degree of nonlinearities. Therefore the need of nonlinear controllers arises. However, this is a non-trivial task which, in the area of the classical state feedback control, often leads to a deadlock.

To overcome limitations of the classical control theory, other methods are also introduced into the control design process, for example neural networks and fuzzy logic. Above all, fuzzy logic is very effective and its power has been demonstrated in various fields of system theory and applications where robustness is a very welcome property which is decisive for the choice of the controller.

A typical case of an instable and nonlinear system is the inverted pendulum which has been previously described in many papers. DIP is an extension of this system. Based on their design, there are several possible kinds of DIP, most important aspect of design being the connection between the actuator and the DIP alone. The DIP are suitable means for investigation and verification of different control methods for kinetic systems with high-order nonlinearities. The goal of this paper is to show a relatively simple and effective method for design of a fuzzy logic controller for a system as difficult–to–control as the DIP is.

2 Mathematical Description of the Physical DIP

Generally, there are two basic types of DIP models. The first one is a purely mathematical model that neglects the fact that the mass of the pendulum pole (fig. 1) is distributed along its length. In this model, the whole mass is concentrated at the end of the pole, which does not correspond to reality. The second model is a physical model, taking into account the distribution of the pendulum pole mass and as well as friction. The present paper uses the latter type of model.

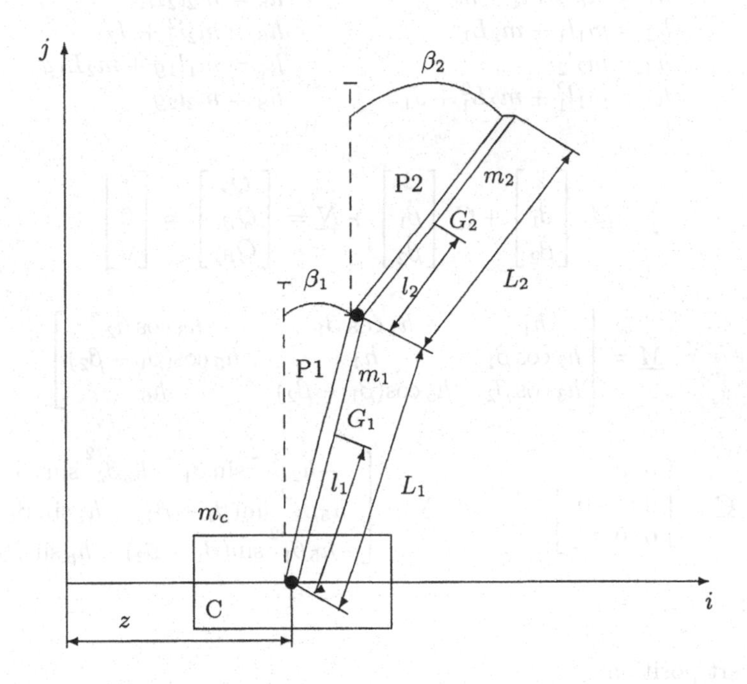

Fig. 1. Physical model of a double inverted pendulum (C - cart, P1, P2 - poles, β_1, β_2 - deviation angles of P1 and P2, respectively)

The mathematical model is derived from the Lagrange – Euler equation which describes mutual relations among kinetic, potential and external energy. All three components of energy must be balanced. If we take into account also friction, the equation is in the form:

$$\frac{d}{dt}\left(\frac{\partial L}{\partial \dot{q}}\right) - \frac{\partial L}{\partial q} + \frac{\partial D}{\partial \dot{q}} = Q_q \qquad (1)$$

where:

L – Lagrangian ($L = T - V$)
Q_q – generalized forces

q – generalized coordinates
T – kinetic energy of DIP
V – potential energy of DIP
D – Rayleigh function describing friction forces

Substituting T and V by DIP parameters we can derive a matrix form (3) of the DIP description, where (1) was decomposed to three equations by individual components of energy, with following substitutions:

$$\begin{array}{ll} h_1 = m_c + m_1 + m_2 & h_5 = m_2 l_2 L_1 \\ h_2 = m_1 l_1 + m_2 L_1 & h_6 = m_2 l_2^2 + J_2 \\ h_3 = m_2 l_2 & h_7 = m_1 l_1 g + m_2 L_1 g \\ h_4 = m_1 l_1^2 + m_2 L_1^2 + J_1 & h_8 = m_2 l_2 g \end{array} \tag{2}$$

$$\underline{M} \cdot \begin{bmatrix} \ddot{z} \\ \ddot{\beta}_1 \\ \ddot{\beta}_2 \end{bmatrix} + \underline{C} \cdot \begin{bmatrix} \dot{z} \\ \dot{\beta}_1 \\ \dot{\beta}_2 \end{bmatrix} + \underline{N} = \begin{bmatrix} Q_z \\ Q_{\beta_1} \\ Q_{\beta_2} \end{bmatrix} = \begin{bmatrix} f \\ 0 \\ 0 \end{bmatrix} \tag{3}$$

$$\underline{M} = \begin{bmatrix} h_1 & h_2 \cos\beta_1 & h_3 \cos\beta_2 \\ h_2 \cos\beta_1 & h_4 & h_5 \cos(\beta_1 - \beta_2) \\ h_3 \cos\beta_2 & h_5 \cos(\beta_1 - \beta_2) & h_6 \end{bmatrix} \tag{4}$$

$$\underline{C} = \begin{bmatrix} \nu & 0 & 0 \\ 0 & C_1 & 0 \\ 0 & 0 & C_2 \end{bmatrix} \qquad \underline{N} = \begin{bmatrix} -h_2 \dot{\beta}_1^{\,2} \sin\beta_1 - h_3 \dot{\beta}_2^{\,2} \sin\beta_2 \\ h_5 \dot{\beta}_2^{\,2} \sin(\beta_1 - \beta_2) - h_7 \sin\beta_1 \\ -h_5 \dot{\beta}_1^{\,2} \sin(\beta_1 - \beta_2) - h_8 \sin\beta_2 \end{bmatrix}$$

where:

z – cart position
β_1 – angle of the first pole P1 with respect to the vertical axis j
β_2 – angle of the second pole P2 with respect to the vertical axis j
$\dot{z}$ – the first derivative of z
$\dot{\beta}_1$ – the first derivative of β_1
$\dot{\beta}_2$ – the first derivative of β_2
J_1 – inertia momentum of the first pole
J_2 – inertia momentum of the second pole
L_1 – length of the first pole
L_2 – length of the second pole
l_1 – distance from the first pole beginning to G_1
l_2 – distance from the second pole beginning to G_2
g – acceleration of gravity
ν – friction coefficient between the cart and the track
C_1 – friction coefficient of the first pole
C_2 – friction coefficient of the second pole

3 Methods of DIP Controller Design

There are several approaches to solving the control problem of a DIP. The task is to swing up the pendulum from its lower position, which is one of two equilibrium positions of each DIP, to the upper one (the second equilibrium position). Of course, any deviation from this position causes the DIP to fall down. To prevent this a DIP control is needed. There are various control methods whose brief description is included here along with a more detailed treatment of the fuzzy logic methods.

The design of a classical (non-fuzzy) adaptive linear controller with an observer is described, e.g. in [3]. Other methods, e.g. in [6], are based on optimization of energy and limit cycles. These classical approaches are very suitable for control of swinging up of the pendulum as well as for reduction of small angle deviations. The control process can be performed smoothly in an aperiodic way with tiny oscillations. But these controllers fail for larger deviations. The main reason for this is their inability quickly to change their dynamics. The robustness is also very questionable here. To solve this problem, the use of fuzzy controllers and their considerable robustness in many cases can be the right way.

Fuzzy controllers (FC) offer several design approaches with respect to the controller structure. We can distinguish three basic structures of DIP fuzzy controllers:

- **Central FC.**
- **Cascade FC.**
- **Hierarchical FC.**

3.1 Central FC

The design of a central FC is described, e.g. in [2]. The structure of such a control loop is very simple. It consists of one controller of the MISO type (multiple input - single output). All the control tasks are solved in one module. The inputs are β_1, β_2, $\dot{\beta}_1$, $\dot{\beta}_2$ and the output is the force of the cart f. However, this causes problems with the design and tuning of the knowledge base where the search for relations among variables can be very laborious. We have to adjust many parameters without distinguishing relevant and less relevant parameters, and so the number of possible parameter combinations increases very quickly (multiplicative growth). Therefore it is hard to find a combination near the optimal solution.

3.2 Cascade FC

The approach to the design of a cascade controller is described, e.g. in [1]. This enables us partially to decompose our problem and so to decrease the number of parameters that need to be set at one stroke, i.e. in one module. The control strategy is in fig. 2.

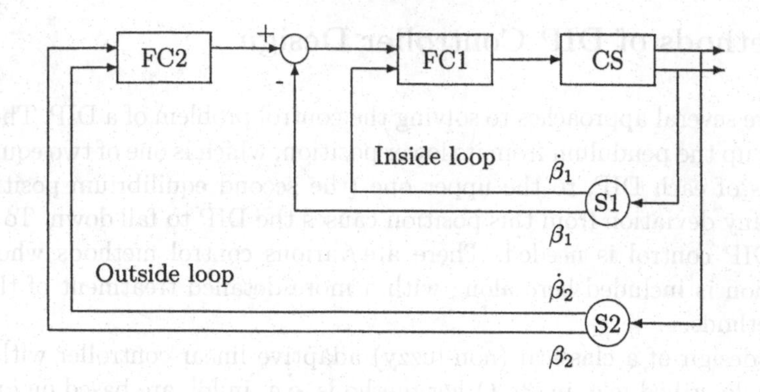

Fig. 2. Cascade control loop for DIP (S1, S2 – sensors)

The inside loop 1 controls the first pole and the outside loop 2 controls the second pole and modulates the state of the first loop. In other words, the loop 1 is dependent on the function of the second loop which is practically superior to the first loop. The control process can be described as follows:

1. The control of the first pole is carried out on base of the desired value β_{d1} for β_1 which is calculated by FC2 based on the value of β_2.
2. β_{d1} for the first pole is defined by the angle of the second pole β_2 so that $\beta_{d1} = \beta_2$.
3. The first pole is controlled with the goal to reach $\beta_{d1} = \beta_1$.
4. Because the target state for the outside loop is $\beta_{d2} = \beta_2 = 0$, the final state is $\beta_2 = \beta_1 = 0$.

The disadvantage of this cascade FC is in the fact that the control more or less depends on the needs of the second pole. If the deviation angles (β_1, β_2) have reverse signs, see fig. 5 a) or b), the control fails.

3.3 Hierarchical FC

The hierarchical control is based on decomposition of the controlled object into several subsystems which are relatively autonomous and may interact with each other [5]. The basic idea of the hierarchical control is depicted in fig. 3.

The system (in our case the whole DIP) is decomposed into N subsystems which are located at the lower level. Each subsystem solves its own control problem and communicates related information to the upper level where the coordinator is. In this way, the case is avoided in which the subsystems are subordinate to each other. This is the main difference between the cascade FC and the hierarchical approach. The coordinator summarizes the information from the lower level and sends supervisory instructions back to the lower level. In other words, it is superior to all the subsystems. The introduction of supervisor enables us to take into consideration broader criteria necessary for the control

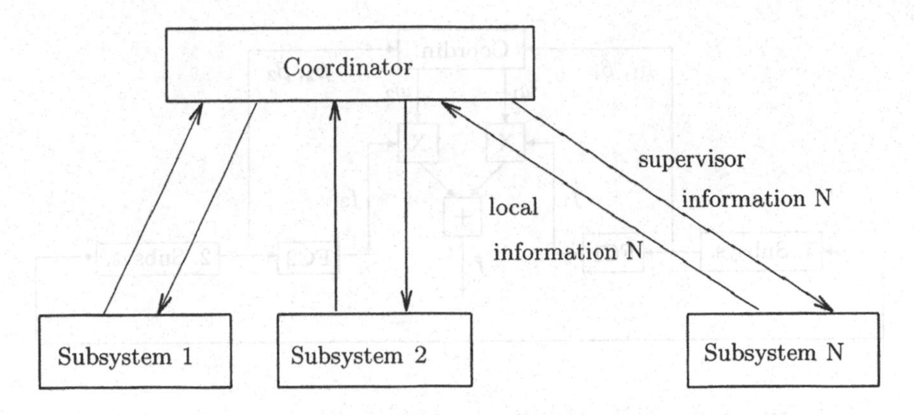

Fig. 3. Universal scheme of a Hierarchical FC

strategy of such a system. This makes the knowledge base easier to optimize. Also, partially contradictory requests on the controller output can be solved by this supervised hierarchical method, e.g. see fig. 5 a) or b). Of course, designing FC for individual subsystems is simpler, too. The set of input variables is smaller and the number of possible parameter combinations decreases proportionally which can lead to better controller design. These advantages were the reason for choosing the hierarchical FC for our problem.

4 Structure and Design of the Hierarchical FC for DIP

The hierarchical FC of DIP consists of two fuzzy controllers, FC1 and FC2, for the first and second pole, respectively (see fig. 4). The controlled system is thus divided into two subsystems – the poles are controlled separately. The inputs for FC1 are β_1, $\dot{\beta}_1$; and for FC2 are β_2, $\dot{\beta}_2$. The outputs are the forces f_1, f_2 affecting the cart. The same inputs enter the coordinator, too. This FC is a MIMO (multiple input – multiple output) system, producing on its outputs weights w_1, w_2.

Requests from FC1 and FC2 on how the cart has to move may be contradictory. Each controller tries to control its part of the pendulum regardless of needs of the other one. The most evident case of contradiction is when the poles are inclined in different directions (fig. 5 a) or b)). It means that one pole needs to move the cart, e.g. to the left, and the other one needs to perform the movement to the right. The following rule is used for solving this problem: The FC whose pole is in a "worse" state is given priority, until the situation changes. All the possible combinations of mutual positions (inclinations) of both poles are shown in fig. 5. They serve for the rule base design of the coordinator FCC.

To calculate the total force $f = g(f_1, f_2)$ or in other words, to coordinate actions of FC1 and FC2, it is needed to implement a FCC. It calculates differences $\Delta\beta = \beta_1 - \beta_2$ and $\Delta\dot{\beta} = \dot{\beta}_1 - \dot{\beta}_2$. The weights w_1, w_2 represent the deviation

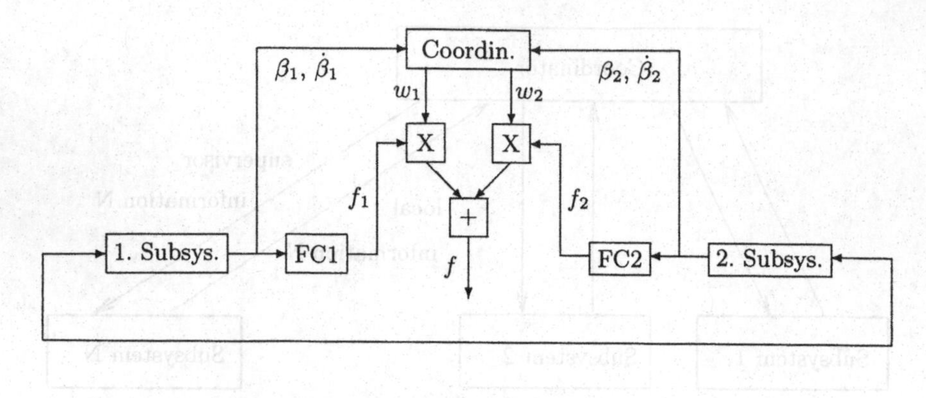

Fig. 4. Structure of the Designed Hierarchical FC for DIP

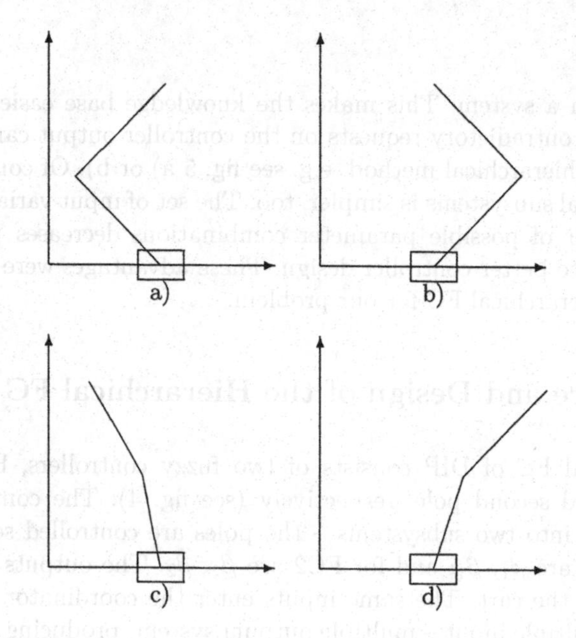

Fig. 5. Possible Combinations of Mutual Positions of P1 and P2

magnitude and its derivative for poles P1 and P2. The total force f is:

$$f = \frac{w_1 . f_1 + w_2 . f_2}{w_1 + w_2} \tag{5}$$

Both partial controllers, FC1 and FC2, and FCC are Mamdani fuzzy controllers with MIN – MAX inference and center of gravity defuzzification method. Despite of the fact that the PRODUCT – MAX controllers provide smoother transition in many cases, MIN – MAX inference is more tolerant, preferring the dominant rule able faster to respond to state changes. Especially, the area near to set point is very sensitive to any changes and we must not forget considerable inertia of a DIP. The MIN – MAX FC also provides a smooth transition between stabilisation of the one pole and the other. In such a way, partial contradictions of control requests coming from FC1 and FC2 can be partially removed, taking into account both requests with different weights.

A number of trials, where a human operator tries to control the pendulum, is necessary for creating the knowledge base. The values of inputs and outputs in these trials are used to create a control surface. Individual parts of this surface are then approximated by linear plains. Their borders represent supports on which the membership functions (MF) can be constructed. The rule base is obtained by projections of control surface into individual universes of discourse [1]. Thus we have the initial knowledge base that can be tuned by changing parameters of MF.

4.1 Knowledge Base of the DIP Hierarchical FC

The goal of the DIP control is to calculate such a value of the force f which brings the system to the upper equilibrium point where β_1, $\dot{\beta}_1$, β_2 and $\dot{\beta}_2$ are zero. No demands are put on the cart position z and its derivative $\dot{z}$. Further, we assume sufficient track length. The controller was designed for a DIP with parameters described in tab. 1.

Table 1. Physical parameters of the DIP

Parameter	Value	Parameter	Value
m_c	$2.3kg$	J_1	$0.0017kgm^2$
m_1	$0.274kg$	J_2	$0.028kgm^2$
m_2	$0.388kg$	C_1	$3.10^{-3}Nms$
L_1	$0.206m$	C_2	$5.10^{-5}Nms$
L_2	$0.71m$	ν	$5kgs^{-1}$

Each variable of FC1 and FC2 (β_1, $\dot{\beta}_1$, β_2, $\dot{\beta}_2$) can take on one of three linguistic values – Negative (N), Zero (Z) and Positive (P). N and P have trapezoidal MF, and Z is bell-shaped because of the high sensitivity in the area near the desired value. The output variables (f_1, f_2) of FC1 and FC2 have five values – Negative small (NS), Negative medium (NM), Zero (Z), Positive medium (PM) and Positive big (PB). All output MF are trapezoidal.

The inputs of FCC ($\Delta\beta$, $\Delta\dot{\beta}$) have, similarly to FC1 or FC2, three linguistic values – Negative (N), Zero (Z) and Positive (P). All MF are bell-shaped. The output variables (w_1, w_2) of FCC are singletons with the grade of membership equal to 1 for the following points: 0, 1; 0, 25; 0, 5; 0, 75 and 1.

The total number of rules in all three rule bases is 27, and they are listed in tab. 2.

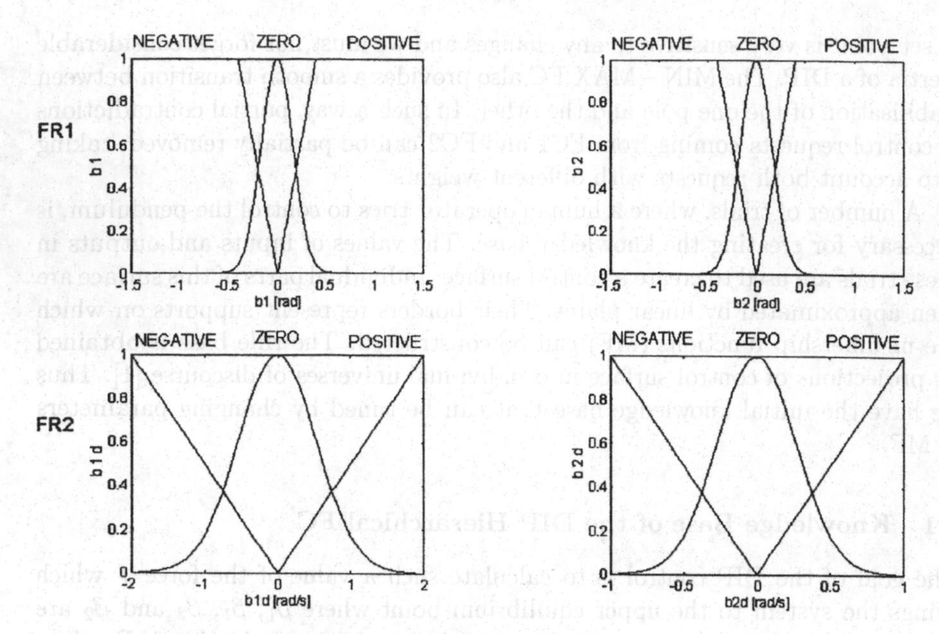

Fig. 6. MF for inputs of FC1, FC2

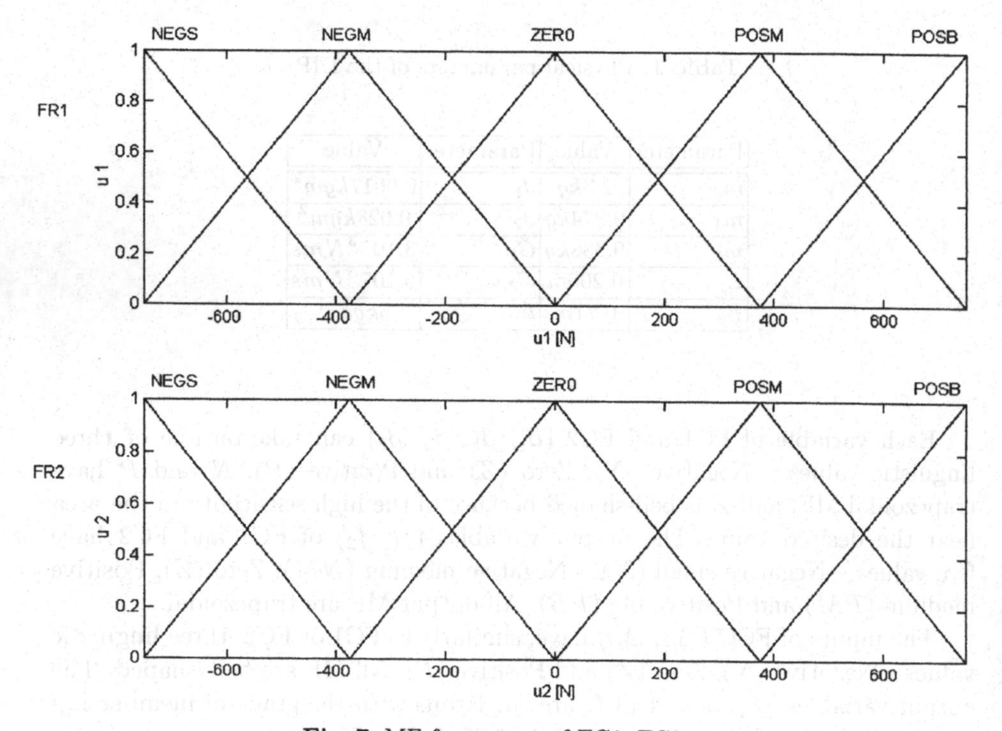

Fig. 7. MF for output of FC1, FC2

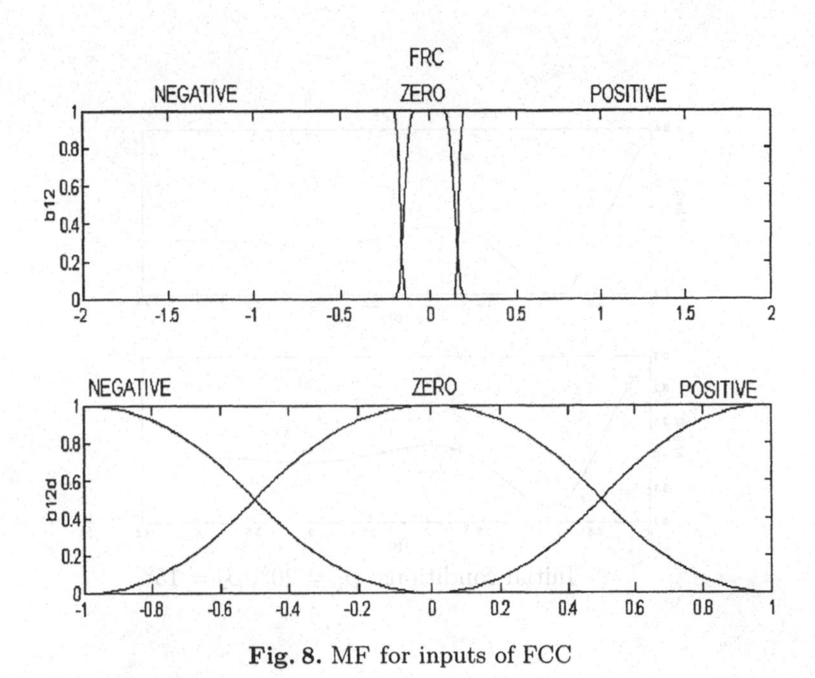

Fig. 8. MF for inputs of FCC

Table 2. Rule bases of FC1, FC2 and FCC

$\beta_1 \setminus \dot{\beta}_1$	N	Z	P
N	NS	NM	PM
Z	NS	Z	PB
P	NM	PM	PB

$\beta_2 \setminus \dot{\beta}_2$	N	Z	P
N	PB	PM	NM
Z	PB	Z	NS
P	PM	NM	NS

$\Delta\beta$	N	Z	P	N	Z	P	N	Z	P
$\Delta\dot{\beta}$	N	N	N	Z	Z	Z	P	P	P
w_1	0,75	0,75	0,1	0,1	1	0,1	0,1	0,5	0,75
w_2	0,75	0,5	0,5	0,25	1	0,25	0,5	0,75	0,75

Following graphs show responses for β_1 and β_2 of the controlled pendulum at various initial conditions and parameter changes.

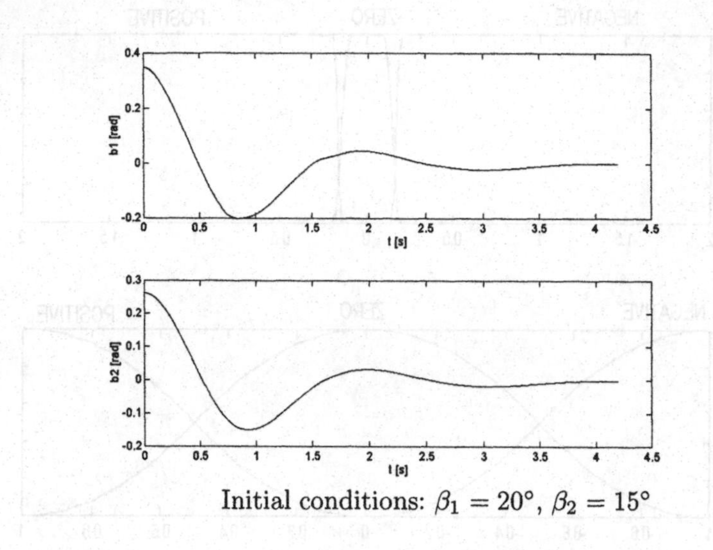

Initial conditions: $\beta_1 = 20°$, $\beta_2 = 15°$

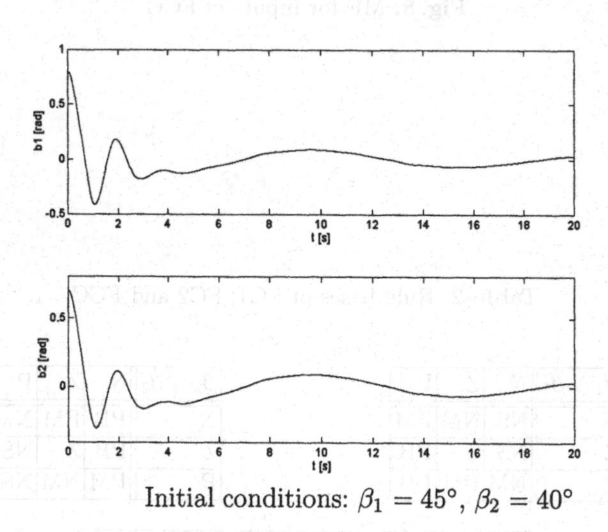

Initial conditions: $\beta_1 = 45°$, $\beta_2 = 40°$

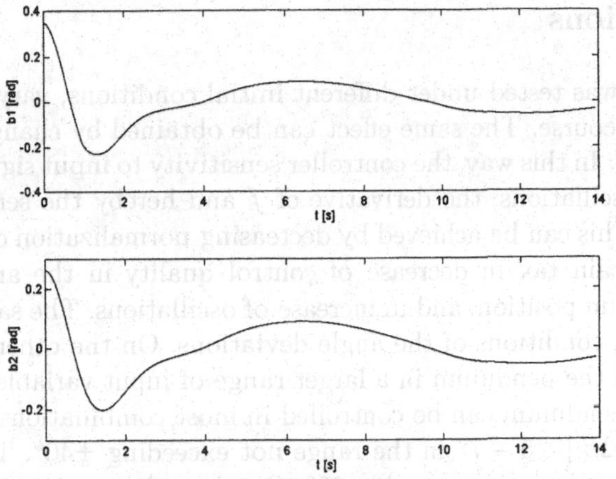

Initial conditions: $\beta_1 = 20°$, $\beta_2 = 15°$, $m1 = 1.5kg(+447\%)$

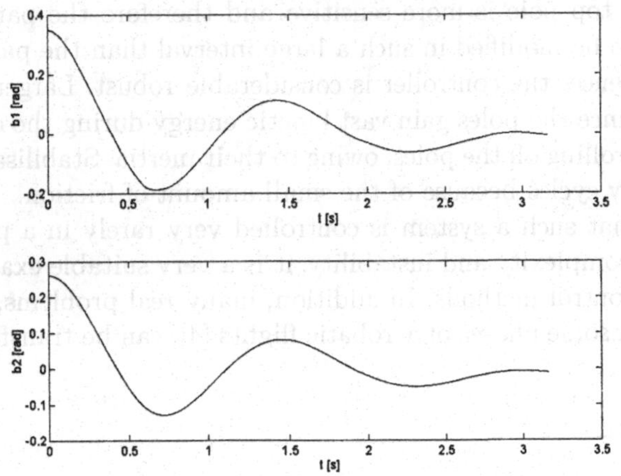

Initial conditions: $\beta_1 = 20°$, $\beta_2 = 15°$, $m2 = 1kg(+157\%)$

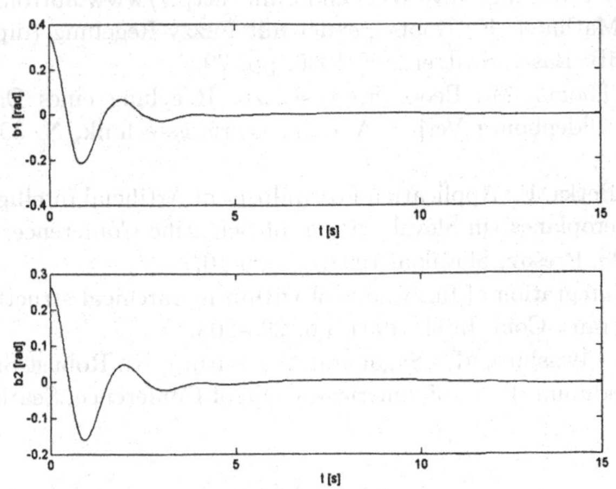

Initial conditions: $\beta_1 = 20°$, $\beta_2 = 15°$, $L1 = 1m(+385\%)$

5 Conclusions

The controller was tested under different initial conditions, varying intervals of universes of discourse. The same effect can be obtained by changing normalization coefficients. In this way, the controller sensitivity to input signals is changed. To eliminate oscillations, the derivative of f and hereby the sensitivity, has to be decreased. This can be achieved by decreasing normalization coefficients. But this brings certain tax in decrease of control quality in the area near to the erected pendulum position, and in increase of oscillations. The same applies also for small initial conditions of the angle deviations. On the other hand, this enables to control the pendulum in a larger range of input variables. Experiments show that the pendulum can be controlled in most combinations of mutual pole positions for $| \Delta\beta | < 5 - 7°$ in the range not exceeding $\pm 40°$. The best results were achieved for deviations roughly 20°. Good results are obtained for a broad interval of DIP parameter changes, too. Of course, the response on parameter changes of the top pole is more sensitive and therefore the parameters of the top pole cannot be modified in such a large interval than the parameters of the bottom one. Hence, the controller is considerable robust. Larger $| \Delta\beta |$ cannot be controlled since the poles gain vast kinetic energy during the control process, which leads to rolling off the poles, owing to their inertia. Stabilisation is possible only after many cycles because of the small amount of friction.

It is true that such a system is controlled very rarely in a pure form. But, because of its complexity and instability, it is a very suitable example for examining various control methods. In addition, many real problems, e.g. control of an aeroplane in some phases of acrobatic flights [4], can be transformed into this form.

References

1. Aptronix Inc.: Two Stage Inverted Pendulum. http://www.aptronix.com/fuzzynet
2. Bernet, F. - Mathiuet, F.: Doppelpendel mit Fuzzy-Regelung (diploma thesis - in German). ISBB, Basel, Switzerland, 1996, pp. 79.
3. Kluge, M. - Thoma, M.: Beobachtergestützte Regelung eines Doppelpendels (in German). R. Oldenbourg Verlag, Automatisierungstechnik, N. 39, 1991, pp. 304-309.
4. Kováčik, P. - Betka, F.: Application Possibilities of Artificial Intelligence in Aviation Systems of Aeroplanes (in Slovak). Proc. of Scientific Conference: Informatics and Algorithms '98, Prešov, Slovakia, 1998, pp. 99-102.
5. Langari, R.: Integration of fuzzy control within hierarchical structured control systems. IEEE Trans. Com. Intel., 1994, pp. 293-303.
6. Yamakita, M. - Iwashiro, M. - Sugahara, Y. - Furuta, K.: Robust Swing Up Control of Double Pendulum. Proc. of American Control Conference, Seatle, 1996, pp. 290-295.

Synthesis of Stable Fuzzy PD/PID Control Laws for Robotic Manipulators from a Variable Structure Systems Standpoint

W S Wijesoma and K R S Kodagoda

School of Electrical and Electronic Engineering
Nanyang Technological University
Singapore 639798.
eswwijesoma@ntu.edu.sg

Abstract. In this paper we present a systematic procedure for the synthesis of stable and robust fuzzy PD/PID controllers for robot manipulators from a variable structure systems standpoint. For linear systems and certain classes of non-linear systems variable structure systems theory has been used as a basis for synthesizing so-called fuzzy sliding mode controllers (FSMC)[1]. Variable structure systems theory has also been shown to be useful for understanding the effectiveness and stability of direct fuzzy PD controllers [2]. In this paper we will present a systematic method to exploit these two complementary roles of variable structure systems theory in order to guide the design and analysis of fuzzy PD/PID controllers for the class of non-linear systems characterized by robot dynamics.

1 Introduction

The derivation of the control methodology can be viewed as a fusion of two conventional and established hard control methodologies with a soft intelligent control method. The three control methodologies fused are the hard control methods known as the computed torque technique (CTT), and sliding mode control (SMC), and the soft method of fuzzy logic control. Each of these control methods have been applied, either individually, or some combination thereof in the literature with different degrees of success under different scenarios [3],[4]. It is well known that the method of output feedback linearization or popularly known as the computed torque technique (CTT) in the robotics community, is an effective method of control for robotic manipulators. However, in the face of uncertain and variable parameters of the model and payload variations its performance is much less to be desired. The inexact de-coupling, resulting from model parameter mismatch can be significant, especially for high performance direct drive robots. Use of high gains in the linear feedback loop of computed torque controller can alleviate the problem to an extent, but can render the robot unstable due to noise and unmodeled dynamics. However, the computed torque control methodology provides a framework for the development of non-linear robust control techniques. In particular, variable structure systems or sliding mode controllers can be synthesized within such a framework. The main advantage of the SMC technique is that systems in the sliding mode are insensitive to disturbances and parameter uncertainty, and variations. Although, sliding mode control alone can yield good tracking performance for robots, it may yield large control torques if apriori -knowledge of the dynamic model and its parameters are not exploited. This shortcoming can be overcome by developing the SMC controller within a computed torque framework [3]. However, the favorable performance of the system is achieved at the price of chattering, i.e. high speed switching of the control signal. This chattering is undesirable not only because it may excite unmodeled high frequency dynamics of the plant, but also it will result in unnecessary wear and tear of the actuator components. This shortcoming is alleviated through the introduction of a boundary layer at the expense of tracking accuracy [7]. Fuzzy control offers simple but robust control solutions that can cope with a wide operating range of non-linear systems [6]. Further, fuzzy control provides a formal methodology for representing, manipulating and implementing a human's heuristic knowledge about how to control the system. Also, unlike SMC control a fuzzy control approach, permits greater flexibility in fine tuning the controller to achieve arbitrary complex control surfaces.

There are some, but significant disadvantages of direct fuzzy control. Most importantly, it may be difficult for a human expert to effectively incorporate stability criteria in to a rule base to ensure reliable, stable, and safe operation. It is in this context that VSS theory is most significant. One direct way of exploiting VSS theory is to derive the VSS or SMC control law for the non-linear system and thereafter 'fuzzyfy' the control law using the extension principle. This yields the so-called fuzzy sliding mode controller or FSMC and is reported in the literature in many flavors for certain classes of non-linear systems [1],[5]. A fuzzy realization of an SMC controller can account for the undesirable chattering implicitly. Further, an FSMC control scheme has the added advantage of being able to solve the fuzzy control problem with a minimal realization, as the inputs required are the sliding surface (s) (and in some, $\dot{s}$) in spite of the number of state variables. However, in situations where heuristic knowledge is prevalent, the latter advantage can become a serious impediment to the successful application of this knowledge. Heuristic control knowledge is usually available in terms of the states and their changes. To facilitate the inclusion of such knowledge it is paramount that the fuzzy rules are state based. The essence of the paper is to show the development of fuzzy control laws based on fuzzy state feedback, whose stability will be guaranteed through VSS theory. This provides a framework for the inclusion of heuristics and more degrees of freedom for shaping of the control surface, whilst ensuring stability.

In the rest of the paper the overall control methodology is derived and presented for the general manipulator dynamics. The non-linear and coupled configuration dependent dynamics of the robot is de-coupled and linearized by the computed torque methodology. To overcome effects of in-exact de-coupling due to parametric model uncertainty and parameter variations and also to account for disturbances a sliding controller is synthesized. Thereafter, it is shown how to derive the rule base of a fuzzy PD controller that ensures accurate tracking while maintaining stability. Simulations involving a two link planar manipulator are presented to show that the synthesized fuzzy controller does emulate the stable VSS controller, which formed the basis for the synthesized fuzzy controller.

2 Robot Dynamics

The equation of motion of a manipulator in joint space may be written as:

$$\tau = \mathbf{M}(\theta)\ddot{\theta} + \mathbf{Q}(\theta,\dot{\theta}) + \omega \tag{1}$$

where,

$$\mathbf{Q} = \mathbf{V}(\theta,\dot{\theta}) + \mathbf{G}(\theta) + \mathbf{f}(\theta,\dot{\theta}) \tag{2}$$

$\theta, \dot{\theta}, \ddot{\theta} \in \mathbf{R}^{n\times 1}$ are the joint position, velocity and acceleration vectors, $\mathbf{M}(\theta)\in \mathbf{R}^{n\times n}$ is the inertia matrix, $\mathbf{V}(\theta,\dot{\theta}) \in \mathbf{R}^{n\times 1}$, is the centrifugal and Coriolis vector, $\mathbf{G}(\theta) \in \mathbf{R}^{n\times 1}$, is the vector of gravity terms, $\omega \in \mathbf{R}^{n\times 1}$ is the vector of disturbances, and $\tau \in \mathbf{R}^{n\times 1}$ is the actuator torque vector acting on the joints of the robot. Since the robot dynamics are linear in terms of a suitably chosen parameter vector $\Omega \in \mathbf{R}^{r\times 1}$, we may write (1) as:

$$\tau = \mathbf{W}(\theta,\dot{\theta},\ddot{\theta})\mathbf{\Omega} + \omega \tag{3}$$

where, $\mathbf{W}(\theta,\dot{\theta},\ddot{\theta})\in \mathbf{R}^{n\times r}$ is a matrix of functions.

3 Computed Torque Method (CTT)

In the computed torque method for tracking control of robots the dynamic system (1) is initially decoupled and linearized using the following control law:

$$\begin{aligned} \tau &= \hat{\mathbf{M}}\mathbf{u}(\mathbf{t}) + \hat{\mathbf{Q}}(\theta,\dot{\theta}) \\ &= \mathbf{W}(\theta,\dot{\theta},\mathbf{u})\hat{\mathbf{\Omega}} \end{aligned} \tag{4}$$

where,

$$\mathbf{u} = \ddot{\theta}_d + \mathbf{K}_\nu(\dot{\theta}_d - \dot{\theta}) + \mathbf{K}_p(\theta_d - \theta) \tag{5}$$

$\hat{\mathbf{M}}$ represents the estimate of the inertia matrix, $\mathbf{M}$, and $\hat{\mathbf{Q}}$, the estimate of centrifugal, Coriolis, gravitational and friction forces. $\mathbf{K}_\nu$, $\mathbf{K}_p \in \mathbf{R}^{n\times 1}$, are constant diagonal gain matrices with k_{vi}, k_{pi} on the diagonals, and $(\theta_d, \dot{\theta}_d, \ddot{\theta}_d)$ specifies the reference trajectory. From (1) and (4) we obtain the closed loop equation:

$$\begin{aligned} \ddot{\mathbf{E}} + \mathbf{K}_v\dot{\mathbf{E}} + \mathbf{K}_p\mathbf{E} &= \tau_{dis} \\ \tau_{dis} &= -\mathbf{M}^{-1}[(\mathbf{M} - \hat{\mathbf{M}})\mathbf{u}(\mathbf{t}) + (\mathbf{Q} - \hat{\mathbf{Q}}) + \omega] \end{aligned} \tag{6}$$

$\mathbf{E} = \theta - \theta_d$, is the tracking error. It may be noted that under perfect parameter modeling, i.e., $\mathbf{M} = \hat{\mathbf{M}}$, $\mathbf{Q} = \hat{\mathbf{Q}}$, and zero disturbances ($\omega = 0$), $\tau_{dis} = 0$, and hence tracking error converges to zero. The gains $k_{vi}, k_{pi} (i = 1, 2..., n)$ can be suitably chosen to define the error dynamics. However, in the presence of model mismatch and external disturbances, the tracking error convergence and stability associated with the CTT is a function of the model mismatch and the external disturbances as defined by (6).

4 Sliding mode control of robot manipulators with a CTT structure

Although, the computed torque technique does not guarantee robust and accurate trajectory tracking in the face of model parametric uncertainty and external disturbances, it provides a useful non-linear model based control framework for the development of robust control laws. We may synthesize a robust controller using VSS theory within a computed torque framework. We select the following overall computed torque control structure:

$$\tau = \hat{\mathbf{M}}\mathbf{u} + \hat{\mathbf{Q}} + \tau^{vss} \tag{7}$$

The choice $(\hat{\mathbf{M}}\mathbf{u} + \hat{\mathbf{Q}})$ represents the computed torque component, which is chosen to linearize and decouple the system (1). The term τ^{vss} is used to remove the effects of inexact decoupling as a result of model mismatch and bounded disturbances. From (1) and (7) we have the following closed loop equation:

$$\mathbf{u} = \ddot{\theta} + \mathbf{M}^{-1}[(\mathbf{M} - \hat{\mathbf{M}})\mathbf{u} + (\mathbf{Q} - \hat{\mathbf{Q}}) + \omega - \tau^{vss}] \tag{8}$$

Using (3) we may re-write (8) as:

$$\begin{aligned} \mathbf{u} &= \ddot{\theta} + \mathbf{M}^{-1}[\mathbf{W}(\theta, \dot{\theta}, \mathbf{u})(\mathbf{\Omega} - \hat{\mathbf{\Omega}}) + \omega - \tau^{vss}] \\ &= \ddot{\theta} + \mathbf{M}^{-1}[\mathbf{W}(\theta, \dot{\theta}, \mathbf{u})\mathbf{\Psi} + \omega - \tau^{vss}] \end{aligned} \tag{9}$$

$\mathbf{\Psi} \in \mathbf{R}^{\mathbf{r} \times 1}$ is the parameter mismatch vector representing any mismatch between the nominal ($\hat{\mathbf{\Omega}}$), and actual parameter vector $\mathbf{\Omega}$. Let us choose $\mathbf{u}$ as:

$$\mathbf{u} = \ddot{\theta}_d + \mathbf{\Lambda}(\dot{\theta}_d - \dot{\theta}) \tag{10}$$

$\mathbf{\Lambda} \in \mathbf{R}^{n \times n}$ is a diagonal matrix with elements $\lambda_i (i = 1, ..., n)$. From (9) and (10) the closed loop equation is:

$$(\ddot{\theta} - \ddot{\theta}_d) + \mathbf{\Lambda}(\dot{\theta} - \dot{\theta}_d) = \mathbf{M}^{-1}[\tau^{\mathbf{vss}} - (\mathbf{W}\mathbf{\Psi} + \omega)] \tag{11}$$

Now, let us define the switching or sliding surfaces as:

$$s = (\dot{\theta} - \dot{\theta}_d) + \mathbf{\Lambda}(\theta - \theta_d) \tag{12}$$

or, the ith switching plane is:

$$s_i = (\dot{\theta}_i - \dot{\theta}_{di}) + \lambda_i(\theta_i - \theta_{di}) \tag{13}$$

A condition for the intersection of switching planes, $s = 0$, to be attractive can be derived by defining a quasi Lyapunov function $V(t)$ as:

$$V(t) = \frac{1}{2} s^T \mathbf{M} s \tag{14}$$

Since $\mathbf{M}$ is positive definite, and $V(t) = 0$, only when $s = 0$,$V(t)$ is a positive semi-definite function. Differentiating (14) we have:

$$\dot{V(t)} = s^T \mathbf{M} \dot{s} + \frac{1}{2} s^T \dot{\mathbf{M}} s \tag{15}$$

From (12), and (11) it follows that:

$$\mathbf{M}\dot{s} = \tau^{vss} - (\mathbf{W}\mathbf{\Psi} + \omega) \tag{16}$$

Combining (16) and (15):

$$\dot{V(t)} = s^T(\tau^{vss} - (\mathbf{W}\mathbf{\Psi} + \omega)) + \frac{1}{2} s^T \dot{\mathbf{M}} s \tag{17}$$

Now, if we can choose τ^{vss} so as to ensure $\dot{V}(t)$ is negative semi-definite at all times, then it follows that the switching planes are asymptotically stable. Hence, from (12) it may be deduced that the closed loop system is stable and the tracking error converges to zero. A suitable choice for the ith component of τ^{vss} to guarantee negative semi-definiteness of $\dot{V}(t)$ is [3]:

$$\tau_i^{vss} = -sgn(s_i)[\sum_{j=1}^{r} \bar{W}_{ij}\bar{\Psi}_j + \bar{\omega}_i + \delta_i] - s_i \sum_{j=1}^{n} \frac{\bar{\dot{M}}_{ij}}{2} \tag{18}$$

where,

$$\begin{array}{ll} |\Psi_j| < \bar{\Psi}_j & j = 1, ...r \\ |\omega_i| < \bar{\omega}_i & i = 1, ...n \\ |\dot{M}_{ij}| < \bar{\dot{M}}_{ij} & i = j = 1, ...n \\ |W_{ij}| < \bar{W}_{ij} & i = j = 1, ...n \\ \delta_i > 0 & i = 1, ...n \end{array} \tag{19}$$

Therefore, the sliding mode control law with a computed torque structure for the ith link is:

$$\tau_i = \sum_{j=1}^{n} \hat{M}_{ij} u_j + \hat{Q}_i - sgn(s_i)[\sum_{j=1}^{r} \bar{W}_{ij}\bar{\Psi}_j + \bar{\omega}_i + \delta_i] - s_i \sum_{j=1}^{n} \frac{\bar{\dot{M}}_{ij}}{2} \tag{20}$$

5 Fuzzy sliding mode control of robot manipulators

From (18) it follows that the ith component of the sliding mode torque, τ^{vss} , of the overall control torque, τ can be expressed as:

$$\tau_i^{vss} = -sgn(s_i)[K_{i1}] - s_i K_{i2} \tag{21}$$

or,

$$\tau_i^{vss} = -sgn(s_i)[K_{i1} + |s_i| K_{i2}] \tag{22}$$

where,

$$\begin{aligned} K_{i1} &= \textstyle\sum_{j=1}^{r} [\bar{W}_{ij}\bar{\Psi}_j + \bar{\omega}_i + \delta_i] \\ K_{i2} &= \textstyle\sum_{j=1}^{n} \frac{\bar{M}_{ij}}{2} \end{aligned} \tag{23}$$

It may be noted that K_{i1} , and K_{i2} are positive constants whose magnitudes depend on the extent of the model's parametric uncertainty.

Now, we may choose to *'fuzzyfy'* the VSS (SMC) component, τ^{vss}, of the control law (7). It is apparent from (22) that the magnitude of the SMC torque increases with increasing magnitude of s , whilst, the sign of the VSS torque is opposite to that of s . Thus, we may choose a fuzzy system with the following rule structure to realize the VSS component, given by (22):

$$\begin{array}{l} \textbf{IF } \bar{s}_i \textit{ is PVL } \textbf{THEN } \bar{\tau}_i^{fuzzy-vss} \textit{ is PVL} \\ \textbf{IF } \bar{s}_i \textit{ is PL } \textbf{THEN } \bar{\tau}_i^{fuzzy-vss} \textit{ is PL} \\ \textbf{IF } \bar{s}_i \textit{ is PM } \textbf{THEN } \bar{\tau}_i^{fuzzy-vss} \textit{ is PM} \\ \textbf{IF } \bar{s}_i \textit{ is PS } \textbf{THEN } \bar{\tau}_i^{fuzzy-vss} \textit{ is PS} \\ \textbf{IF } \bar{s}_i \textit{ is NS } \textbf{THEN } \bar{\tau}_i^{fuzzy-vss} \textit{ is NS} \\ \textbf{IF } \bar{s}_i \textit{ is NM } \textbf{THEN } \bar{\tau}_i^{fuzzy-vss} \textit{ is NM} \\ \textbf{IF } \bar{s}_i \textit{ is NL } \textbf{THEN } \bar{\tau}_i^{fuzzy-vss} \textit{ is NL} \\ \textbf{IF } \bar{s}_i \textit{ is NVL } \textbf{THEN } \bar{\tau}_i^{fuzzy-vss} \textit{ is NVL} \end{array} \tag{24}$$

where,

$$\begin{aligned} \bar{s}_i = -s_i &= (\dot{\theta_{di}} - \dot{\theta}_i) + \lambda_i(\theta_{di} - \theta_i) \\ &= \dot{\bar{e}}_i + \lambda_i \bar{e}_i \end{aligned} \tag{25}$$

NVL (Negative, and Very Large), NL (Negative, and Large), NM(Negative, and Medium, NS(Negative and Small), PS(Positive and Small), PM (Positive and Medium), PL (Positive and Large), and PVL (Positive and Very Large), are the linguistic terms of the input $\bar{s}_i$ (=$-s_i$), and, $\bar{\tau}_i^{fuzzy-vss}$ output , defined over the universe of discourses $\bar{s}_i$, and, τ_i respectively. Fig. 1 shows an appropriate choice for the linguistic terms using triangular membership functions.

In general, we can write (24) as:

$$\textbf{IF } \bar{s}_i \textit{ is } A_i^k \textbf{ THEN } \bar{\tau}_i^{fuzzy-vss} \textit{ is } B_i^k \;\; (j = 1, ...k) \tag{26}$$

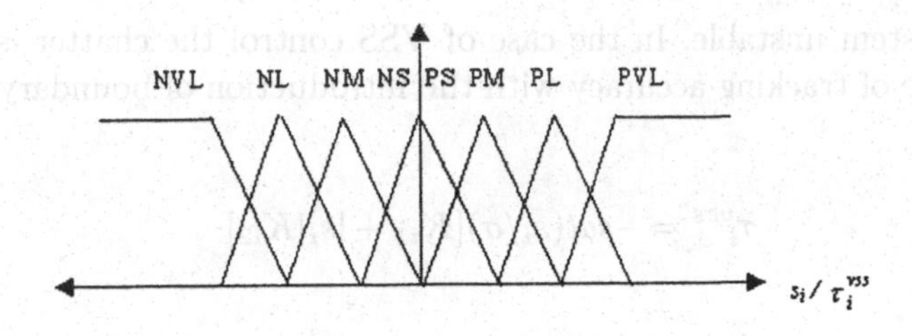

Fig. 1. Membership functions for s_i and output torque τ_i^{vss}

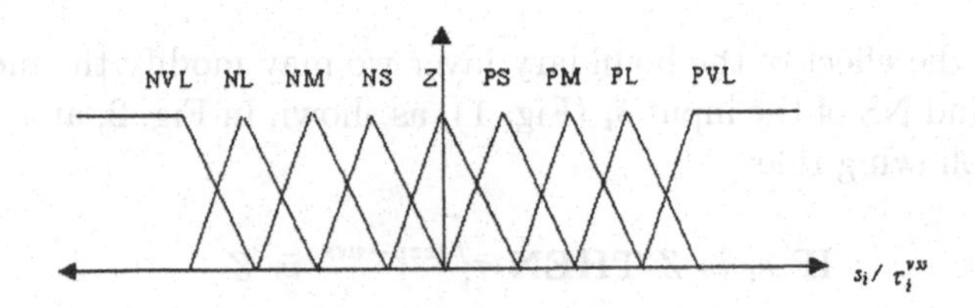

Fig. 2. Membership functions for s_i and τ_i^{vss} to avoid control chatter

A_k^i $(k = 1, ...m)$, and, B_k^i $(k = 1, ...m)$, are the linguistic values of the linguistic variables, $\bar{s}_i$ and $\bar{\tau}_i^{fuzzy-vss}$.

We, choose singleton fuzzification, product implication, and center average defuzzification. Hence, the inferred crisp VSS torque output of the fuzzy SMC is:

$$\tau_i^{fuzzy-vss}(\bar{s}_i) = \frac{\sum_{j=1}^{r} b^j \mu^j(\bar{s}_i)}{\sum \mu^j(\bar{s}_i)} \tag{27}$$

where, b^j is the center of the output fuzzy set corresponding to rule j(j=1,...k), $\mu^j (j = 1, ...k)$ is the premise of rule j, and, r, the total number of rules.

It is important to note that the universal approximation property guarantees the existence of a fuzzy system ((26), and (27), with appropriately defined membership functions) that can approximate the piece-wise continuous SMC law (22) to arbitrary accuracy [8]. However, what is really important to guarantee stability and convergence, in this case, is that the fuzzy system's output magnitude,$|\tau_i^{fuzzy-vss}|$, is greater than or equal to the VSS component's magnitude, $|\tau_i^{vss}|$. That is:

$$|\tau_i^{fuzzy-vss}(\bar{s}_i)| \geq \tau_i^{vss}(\bar{s}_i), \ \ \forall \ |\bar{s}_i| \tag{28}$$

Thus, the membership functions of the input and output given in Fig. 1 can be chosen or tuned to guarantee the above condition.

It can be easily verified that, the direct implementation of the fuzzy VSS control law (26) and (27), with membership functions appropriately chosen to guarantee condition (28), will lead to control chatter, as it's VSS counterpart. This control chatter is undesirable as it may excite unmodeled dynamics and may render the system unstable. In the case of VSS control the chatter is removed at the expense of tracking accuracy with the introduction of boundary layer (σ) as follows:

$$\tau_i^{vss} = -sat(s_i/\sigma)[K_{i1} + |s_i|K_{i2}] \tag{29}$$

where,

$$sat(x) = \begin{cases} x \text{ if } |x| < 1 \\ 1 \text{ if } |x| \geq 1 \end{cases}$$

To emulate the effect of the boundary layer we may modify the membership functions PS, and NS of the input $\bar{s}_i$ (Fig. 1), as shown in Fig. 2, and add to the rule base the following rule:

$$\textbf{IF } \bar{s}_i \ is \ Z \ \textbf{THEN } \bar{\tau}_i^{fuzzy-vss} \ is \ Z \tag{30}$$

Z (Zero) is an additional linguistic value defined for both input and output as shown in Fig. 2. It can be verified that the above modification provides for the smooth transition of the control output across the switching plane. Thus, from equations (24), (25) and (30), the resultant fuzzy sliding mode rule base with smoothing is:

$$\begin{array}{l}
\textbf{IF } \bar{s}_i \ is \ PVL \ \textbf{THEN } \bar{\tau}_i^{fuzzy-vss} \ is \ PVL \\
\textbf{IF } \bar{s}_i \ is \ PL \ \textbf{THEN } \bar{\tau}_i^{fuzzy-vss} \ is \ PL \\
\textbf{IF } \bar{s}_i \ is \ PM \ \textbf{THEN } \bar{\tau}_i^{fuzzy-vss} \ is \ PM \\
\textbf{IF } \bar{s}_i \ is \ PS \ \textbf{THEN } \bar{\tau}_i^{fuzzy-vss} \ is \ PS \\
\textbf{IF } \bar{s}_i \ is \ Z \ \textbf{THEN } \bar{\tau}_i^{fuzzy-vss} \ is \ Z \\
\textbf{IF } \bar{s}_i \ is \ NS \ \textbf{THEN } \bar{\tau}_i^{fuzzy-vss} \ is \ NS \\
\textbf{IF } \bar{s}_i \ is \ NM \ \textbf{THEN } \bar{\tau}_i^{fuzzy-vss} \ is \ NM \\
\textbf{IF } \bar{s}_i \ is \ NL \ \textbf{THEN } \bar{\tau}_i^{fuzzy-vss} \ is \ NL \\
\textbf{IF } \bar{s}_i \ is \ NVL \ \textbf{THEN } \bar{\tau}_i^{fuzzy-vss} \ is \ NVL
\end{array} \tag{31}$$

The above rule base (31) with the inference (27), defines the fuzzy SMC with smoothing.

6 Fuzzy PD control in the state space

The synthesized FSMC controller has the advantage of being able to solve the fuzzy control problem with a minimal realization, as the inputs required are the switching surface instead of the state variables. However, in situations where heuristic knowledge is prevalent, the latter advantage can be a serious impediment to the successful application of this knowledge. Heuristic control knowledge is usually available in terms of the state variables. Hence, it is desirable to implement the rule base in the state space.

Firstly, we normalize the switching plane:

$$\frac{\bar{s}_i}{N_{si}} = \frac{\bar{\dot{e}}_i}{N_{si}} + \lambda_i \frac{\bar{e}_i}{N_{si}} \tag{32}$$

or

$$\hat{s}_i = \hat{\dot{e}}_i + \hat{e}_i \tag{33}$$

where,

$$\hat{s}_i = \frac{\bar{s}_i}{N_{si}}, \hat{\dot{e}}_i = \frac{\bar{\dot{e}}_i}{N_{si}}, \hat{e}_i = \frac{\lambda_i}{N_{si}} \bar{e}_i \tag{34}$$

$\hat{\tau}_i^{fuzzy-pd}$		$\hat{\dot{e}}_i$								
		NVL	NL	NM	NS	Z	PS	PM	PL	PVL
$\hat{e}_i$	NVL	NVL	NVL	NVL	NVL		NL	NM	NS	Z
	NL	NVL	NVL	NVL		NL	NM	NS	Z	PS
	NM	NVL	NVL		NL	NM	NS	Z	PS	PM
	NS	NVL		NL	NM	NS	Z	PS	PM	PL
	Z		NL	NM	NS	Z	PS	PM	PL	
	PS	NL	NM	NS	Z	PS	PM	PL		PVL
	PM	NM	NS	Z	PS	PM	PL		PVL	PVL
	PL	NS	Z	PS	PM	PL		PVL	PVL	PVL
	PVL	Z	PS	PM	PL		PVL	PVL	PVL	PVL

Fig. 3. Equivalent state space rule base

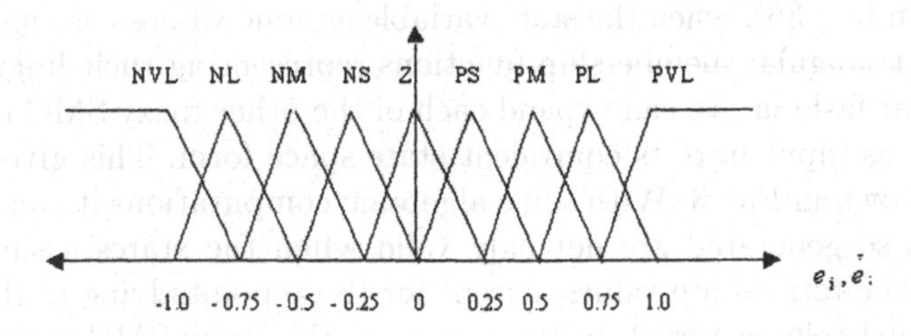

Fig. 4. Membership functions of the linguistic state variables $\hat{\dot{e}}_i$ and $\hat{e}_i$

N_{si} is a design parameter which is a positive scalar. Taking note of the fact that $\hat{s}_i$ is a linear combination of the state variables, $\hat{\dot{e}}_i$, and $\hat{e}_i$, we may re-write the rule base of the fuzzy SMC controller (31) in terms of linguistic state variables as shown in Fig. 3. The membership functions of the normalized linguistic state variables are chosen as shown in Fig. 4. It may be noted that each of the rules of the fuzzy SMC control law is now expressed as a composition of fuzzy rules in the state space. These equivalent rules in the state space are obtained as explained below.

Consider the fuzzy SMC rule:

$$\textbf{IF } \bar{s}_i \textit{ is PS } \textbf{THEN } \bar{\tau}_i^{fuzzy-vss} \textit{ is PS} \tag{35}$$

Let us for the moment, assume that the state variables take discrete values, which are the center points of the triangular membership functions corresponding to the linguistic values. Now, from (33), if $\hat{s}_i = PS_c$ (center of membership function of the linguistic value PS), it could mean that the state pair $(\hat{\dot{e}}_i, \hat{e}_i)$ can be any one of the pairs of centers:

$(PVL_c, NL_c), (PL_c, NM_c), (PM_c, NS_c), (PS_c, Z_c), (Z_c, PS_c), (NS_c, PM_c), (NM_c, PL_c), (NL_c, PVL_c)$

Thus we may write the state space equivalent of the fuzzy SMC rule (35) as:

$$\begin{array}{l}
\textbf{IF } \hat{\dot{e}} \textit{ is PVL } \textbf{AND } \hat{e}_i \textit{ is NL } \textbf{THEN } \hat{\tau}_i^{fuzzy-pd} \textit{ is PS} \\
\textbf{IF } \hat{\dot{e}} \textit{ is PL } \textbf{AND } \hat{e}_i \textit{ is NM } \textbf{THEN } \hat{\tau}_i^{fuzzy-pd} \textit{ is PS} \\
\textbf{IF } \hat{\dot{e}} \textit{ is PM } \textbf{AND } \hat{e}_i \textit{ is NS } \textbf{THEN } \hat{\tau}_i^{fuzzy-pd} \textit{ is PS} \\
\textbf{IF } \hat{\dot{e}} \textit{ is PS } \textbf{AND } \hat{e}_i \textit{ is Z } \textbf{THEN } \hat{\tau}_i^{fuzzy-pd} \textit{ is PS} \\
\textbf{IF } \hat{\dot{e}} \textit{ is Z } \textbf{AND } \hat{e}_i \textit{ is PS } \textbf{THEN } \hat{\tau}_i^{fuzzy-pd} \textit{ is PS} \\
\textbf{IF } \hat{\dot{e}} \textit{ is NS } \textbf{AND } \hat{e}_i \textit{ is PM } \textbf{THEN } \hat{\tau}_i^{fuzzy-pd} \textit{ is PS} \\
\textbf{IF } \hat{\dot{e}} \textit{ is NM } \textbf{AND } \hat{e}_i \textit{ is PL } \textbf{THEN } \hat{\tau}_i^{fuzzy-pd} \textit{ is PS} \\
\textbf{IF } \hat{\dot{e}} \textit{ is NL } \textbf{AND } \hat{e}_i \textit{ is PVL } \textbf{THEN } \hat{\tau}_i^{fuzzy-pd} \textit{ is PS}
\end{array} \tag{36}$$

It can be verified that the above rule set (36) is equivalent to the single fuzzy SMC rule given by (35), when the state variable assume values corresponding to centers of the triangular membership functions representing their linguistic values. In a similar fashion, we can expand each of the other fuzzy SMC rules (31) stated with $\hat{s}_i$ as input into its equivalent state space form. This gives the rule structure as shown in Fig. 3. With some algebraic computations it can be shown that the rules so generated are not only valid when the states assume center values, but also intermediate values, except for those points lying in the regions shown by shaded cells in Fig. 3. In these regions, the fuzzy SMC torque output magnitude is marginally higher than that produced by the fuzzy state space controller. This is due to the non-triangular saturating membership functions representing the extreme linguistic terms. However, to ensure stability and convergence it is necessary that the fuzzy state space controller produces a torque output $(\tau_i^{fuzz-pd})$ equal to or higher than that of the fuzzy SMC $(\tau_i^{fuzz-vss})$ controller at all points in the state space. This can be corrected for, either by increasing the number of membership functions, or scaling up the output gain of the fuzzy state space controller or shaping of the outer most membership functions of the state space linguistic variables or inclusion of explicit rules. The scaling up of the output gain is a convenient and practical way to correct for the deficiency, although in practical terms the difference is negligible with increasing number of membership functions. For the general case of (N+1) uniform triangular membership functions for input states, switching plane input, and the outputs, it can be shown that the output mismatch between the fuzzy SMC and the fuzzy state space controller is bounded by $\pm 1/(2N)$. Thus, for the case in point the mismatch at the shaded regions does not exceed ± 0.0625. In the limiting case ($N \rightarrow \infty$), $\tau_i^{fuzz-pd} \rightarrow \tau_i^{fuzz-vss}$. It follows that the fuzzy state space controller is equivalent to the fuzzy SMC.

7 Fuzzy PID control in state space

In order to realize equivalent fuzzy PID controllers using VSS theory, it may be noted that we can redefine the switching planes (12) to incorporate an integral term. Following the derivation given in Sect. 4 we can synthesize a CTT+VSS controller of the form given in (22). Now as was illustrated in Sect. 5 we may implement a fuzzy equivalent of the VSS control law. Thereafter, we can implement a fuzzy PID equivalent of the fuzzy SMC law as was shown in Sect. 6.

8 Simulation results

We consider a two link planar manipulator operating on a horizontal plane. For simplicity, it is assumed that the link masses m_1, and m_2 of the manipulator are concentrated at the distal ends of the links of length l_1 and l_2, and there are no static and viscous friction forces, and external disturbance torques. M and Q of the planar manipulator are:

$$M = \begin{pmatrix} m_2 l_2^2 + 2 m_2 l_1 l_2 c_2 + (m_1 + m_2) l_2^2 + I_{a1} & m_2 l_2^2 + m_2 l_1 l_2 c_2 \\ m_2 l_1 l_2 c_2 + m_2 l_2^2 & m_2 l_2^2 + I_{a2} \end{pmatrix} \tag{37}$$

$$Q = \begin{pmatrix} -m_2 l_1 l_2 s_2 \dot{\theta}_2^2 - 2 m_2 l_1 l_2 s_2 \dot{\theta}_1 \dot{\theta}_2 \\ m_2 l_1 l_2 s_2 \dot{\theta}_1^2 \end{pmatrix} \tag{38}$$

where, $s_2 = sin(\theta_2)$, $c_2 = cos(\theta_2)$, I_{a1} and I_{a2} are the actuator inertias of joint 1 and joint 2.

We choose the actual robot parameters as:

$l_1 = 1.0m$, $l_2 = 1.0m$, $m_1 = 1.0kg$, $m_2 = 3.0kg$, $I_{a1} = I_{a2} = 5kgm^2$.

For the purposes of the CTT and the VSS+CTT controller we assume that the following estimates as regards the robot's parameters are available:

$\hat{l}_1 = 1.0m$, $\hat{l}_2 = 1.0m$, $\hat{m}_1 = 0.5kg$, $\hat{m}_2 = 0.5kg$. $\hat{I}_{a1} = \hat{I}_{a2} = 5kgm^2$

It is assumed that the robot is required to track the following joint trajectories for both the links simultaneously:

$$\theta_{di} = \begin{cases} a_i + b_i(1 - cos(\frac{\pi t}{t_f})) & \text{if } t \leq t_f \\ a_i + 2b_i & \text{if } t > t_f \end{cases}$$

where, $a_1 = -90.0^o$, $b_1 = -52.5^o$, $a_2 = 170^o$, $b_2 = -60.0^o$ and $t_f = 1.5secs$.

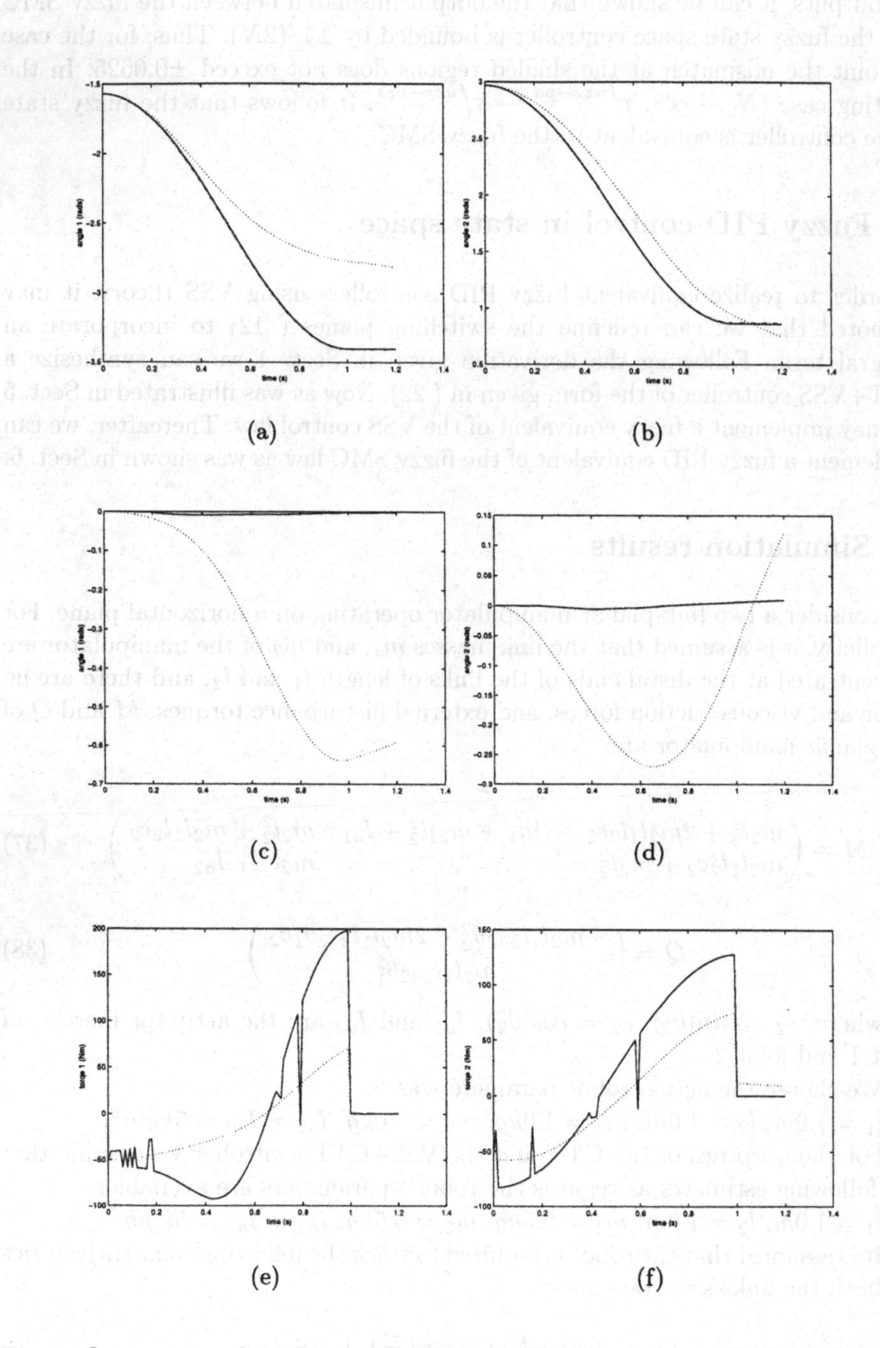

Fig. 5. Comparison of tracking performance of CTT (*dotted*) and CTT+VSS (*solid*) controllers. (a) and (b) tracking trajectories of joint 1 and joint 2, (c) and (d) tracking errors of joint 1 and joint 2, (e) and (f) control torques

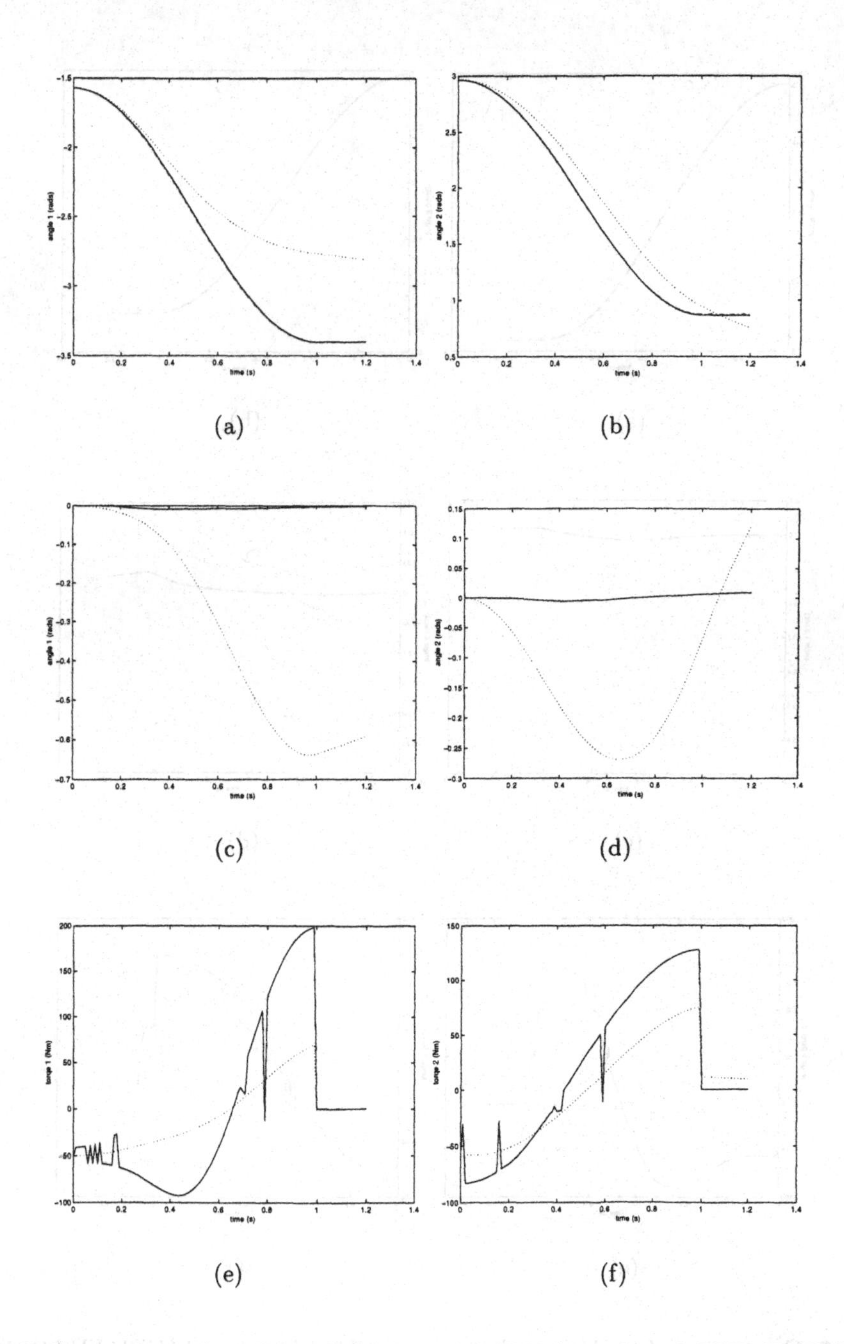

Fig. 6. Comparison of tracking performance of CTT (*dotted*) and CTT+FUZZ-VSS without smoothing (*solid*) controllers. (a) and (b) tracking trajectories of joint 1 and joint 2, (c) and (d) tracking errors of joint 1 and joint 2, (e) and (f) control torques

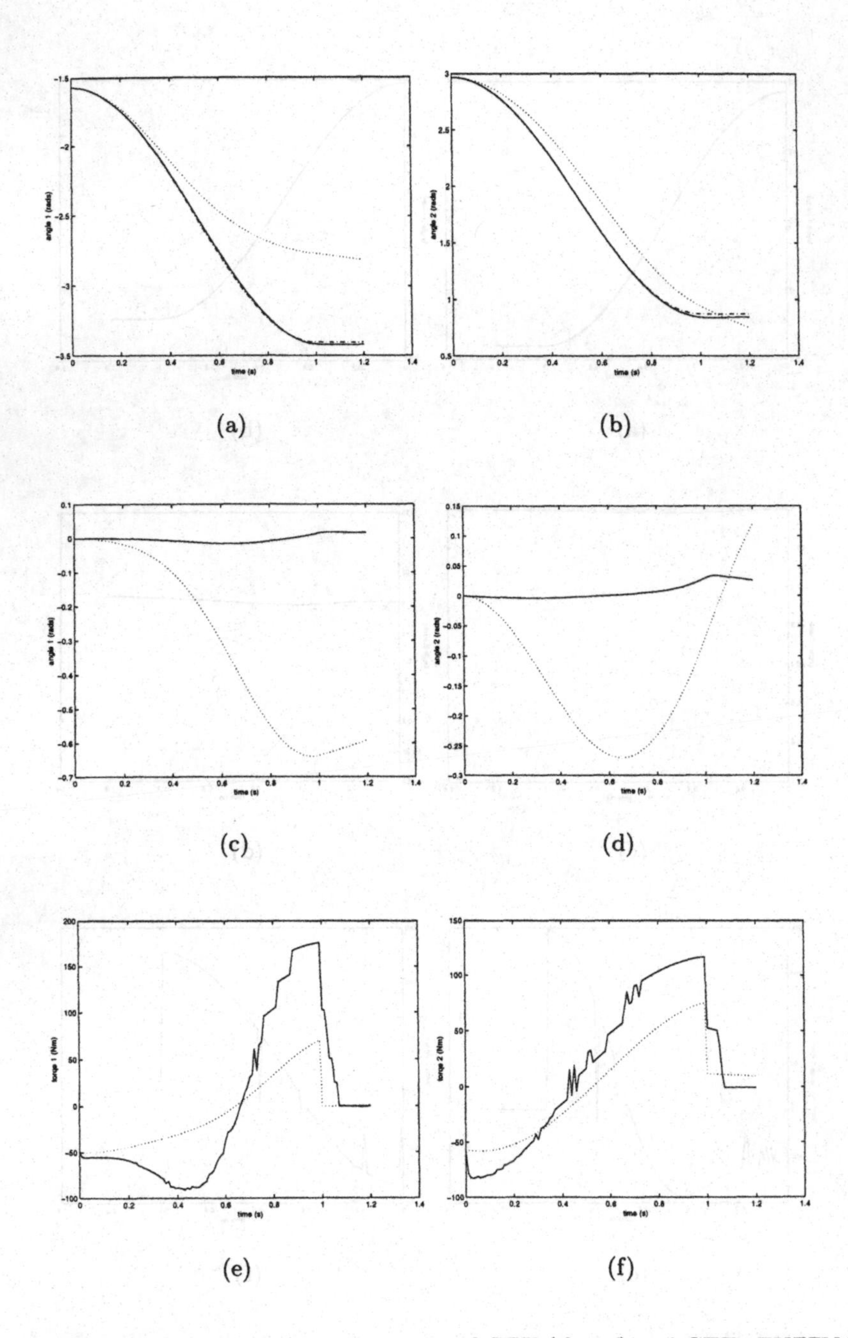

Fig. 7. Comparison of tracking performance of CTT (*dotted*) and CTT+FUZZY-VSS with smoothing (*solid*) controllers. (a) and (b) tracking trajectories of joint 1 and joint 2, (c) and (d) tracking errors of joint 1 and joint 2, (e) and (f) control torques

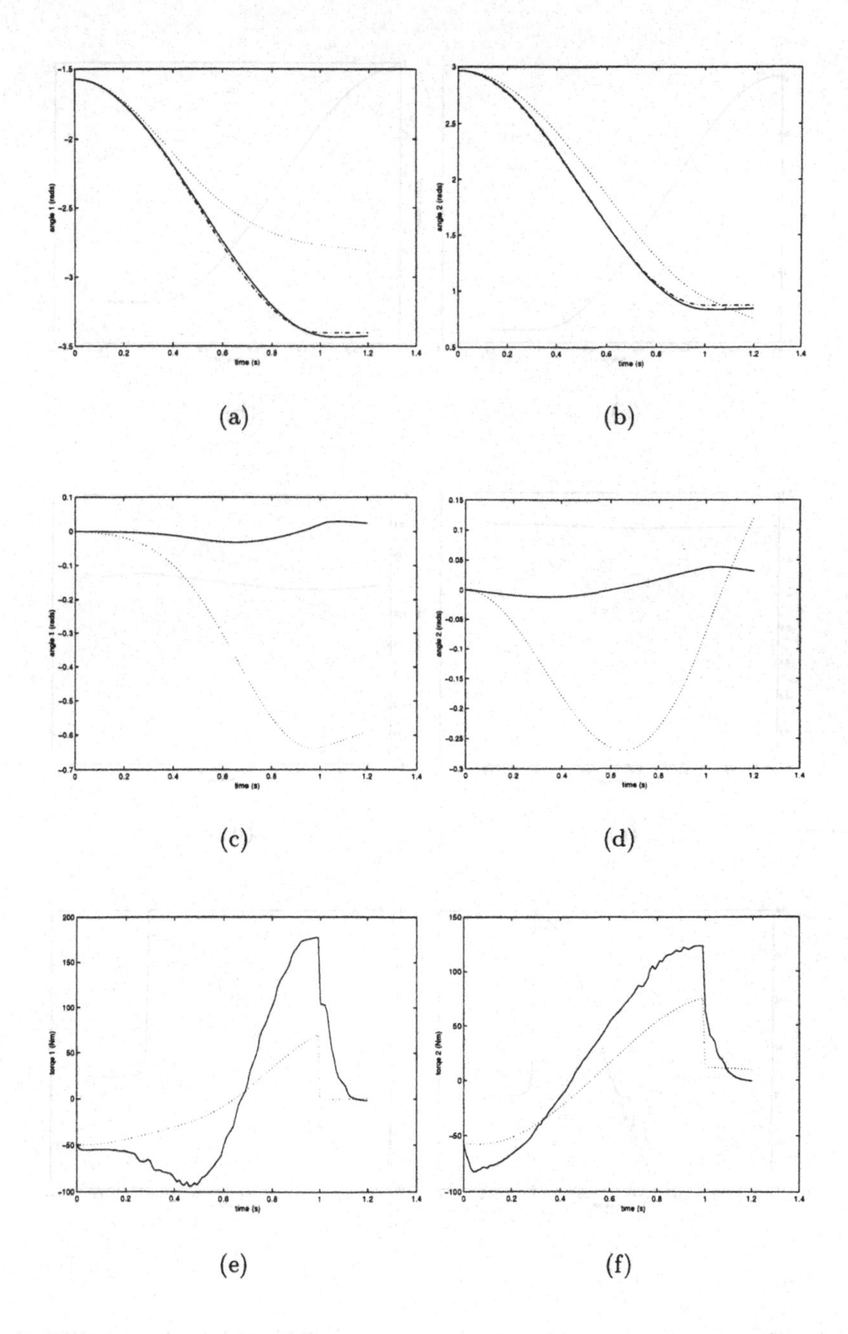

Fig. 8. Comparison of tracking performance of CTT (*dotted*) and CTT+FUZZY-PD (*solid*)controllers. (a) and (b) tracking trajectories of joint 1 and joint 2, (c) and (d) tracking errors of joint 1 and joint 2, (e) and (f) control torques

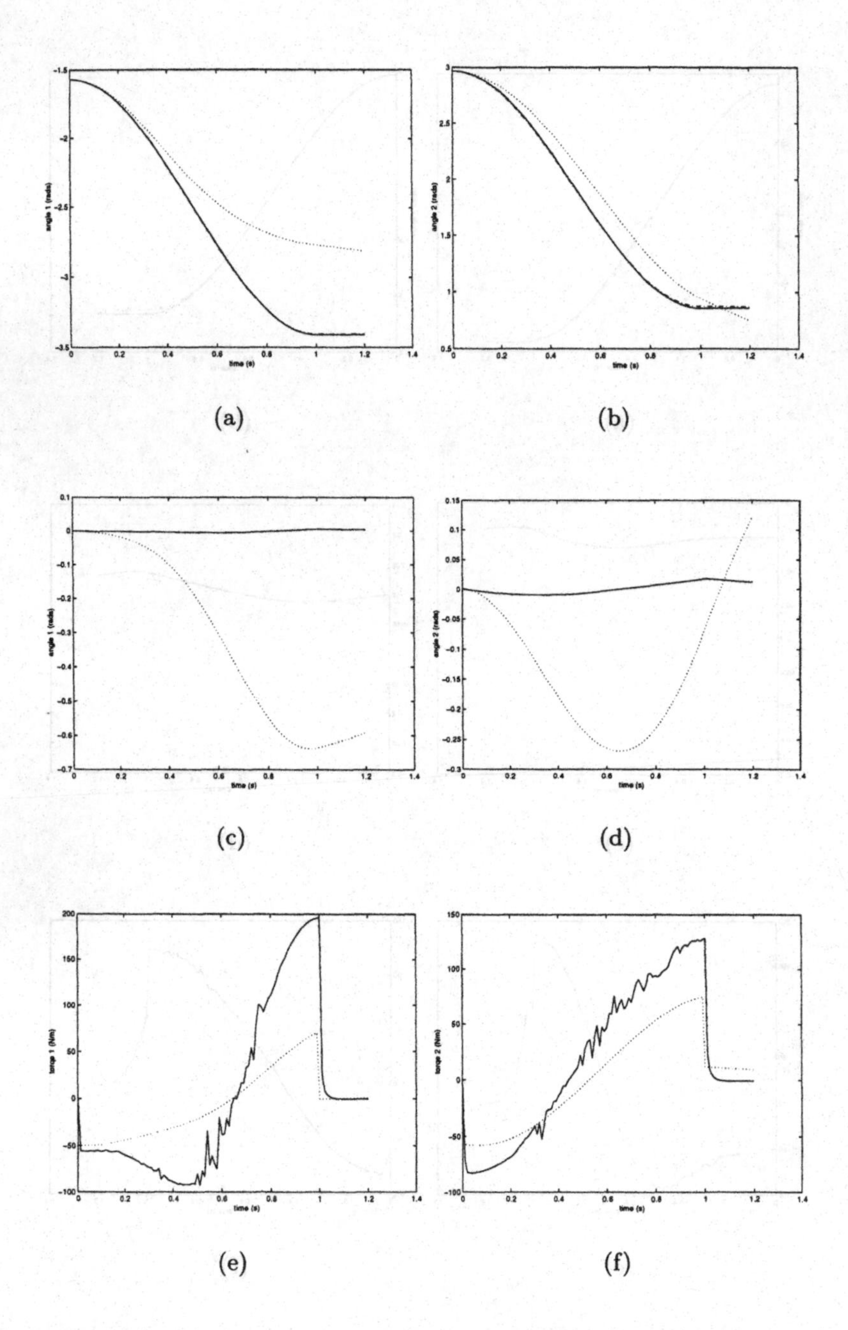

Fig. 9. Comparison of tracking performance of CTT (*dotted*) and FUZZY-PD (*solid*)controllers. (a) and (b) tracking trajectories of joint 1 and joint 2, (c) and (d) tracking errors of joint 1 and joint 2, (e) and (f) control torques

Fig. 5, compares the response of the VSS controller, with the CTT+VSS. Fig. 6 and Fig. 7 shows the response of the fuzzy implementation of the CTT+VSS controller, i.e. the fuzzy VSS/SMC without and with smoothing. The response of the equivalent fuzzy PD controller of the fuzzy VSS/SMC with smoothing is shown in Fig. 8. Fig. 9 shows the fuzzy PD controller's performance without the CTT component. However, in this case the output gains of the fuzzy PD controllers have been increased to account for the complete absence of the CTT component.

9 Conclusions

We have shown that for a class of non-linear systems characterized by robot dynamics, fuzzy PD, PID control laws can be synthesized from a variable structure systems point of view. The derivation of the control methodology can be viewed as a fusion of two conventional and established hard control methodologies with a soft intelligent control method. The three control methodologies fused are the hard control methods known as the computed torque technique (CTT), and sliding mode control (SMC), and the soft method of fuzzy logic control. The development of the control methodology has emphasized the inter play of the three control methods. The synthesized controller exhibits superior performance as a consequence of the exploitation of the best of the fused individual controlled methods. In this way we show how the good aspects of the specific techniques can be retained in the final controller, while, offsetting their disadvantages by the other methods incorporated in to the design. Simulation results have been presented on a two-link manipulator to show the effectiveness and relationship of each of the methods.

References

1. Palm, R: Robust Control by Fuzzy Sliding Mode, Automatica, Vol. 30, (1994)1429-1427,
2. Kawaji S., and Matsunaga, N.: Fuzzy Control of VSS type and its Robustness, IFSA'91 Brussels, Engineering, (July 7-12)81-88.
3. Wijesoma, W. S., and Richards, R. J.: Robust Trajectory Following of Robots using Computed Torque Structure with VSS, Int. J. Control, Vol 52, No. 4, (1990)
4. Bekit, B. W. , Seneviratne, L. D., Whidborne, J. F., and Althoefer, K.: Fuzzy PID Tuning for Robot Manipulators, Proc. 24th Conf. IECON'98, Aachen-Germany, (Aug.31-Spet.4 1998).
5. Yu, X. H., Man, Z. H., and Wu, B. L.: Design of fuzzy sliding mode control systems, Fuzzy Sets and Systems, vol. 95, No. 3, pp. 295-306.
6. Passino, K.M. and Yurkovich, S.: Fuzzy control, Addison Wesley (1998).
7. Slotine, J.J.E. and Sastry, S.S.: Tracking control of non-linear systems using sliding surfaces, with application to robot manipulators, Int. J. Control, 38(2),(1983) .
8. Wnag, X.L.: Fuzzy systems are universal approximators, In Proc. of the 1st IEEE Conf. on Fuzzy systems, San Diago, CA, (March 1992) 1163-1170.

On Data Summaries Based on Gradual Rules

Patrick Bosc[1], Olivier Pivert[1], and Laurent Ughetto[2]

[1] IRISA, ENSSAT, Technopole Anticipa, BP 447, 22305 Lannion Cedex, France
e-mail: bosc@enssat.fr, pivert@enssat.fr

[2] IRISA, IUT de Lannion, rue E. Branly, BP 150, 22302 Lannion Cedex, France
e-mail: ughetto@iut-lannion.fr

Abstract. With the increasing size of databases, the extraction of data summaries becomes more and more useful. The use of fuzzy sets seems interesting in order to extract *linguistic* summaries, i.e., statements from the natural language, containing gradual properties, which are meaningful for human operators. This paper focuses on the extraction from databases of linguistic summaries, using so-called *fuzzy gradual rules*, which encode statements of the form "the younger the employees, the smaller their bonus". The summaries considered here are more on the relations between labels of the attributes than on the data themselves. The first idea is to extract *all* the rules which are not in contradiction with tuples of a given relation. Then, the interest of these rules is questioned. For instance, some of them can reveal potential incoherence, while other are not really informative. It is then shown that in some cases, interesting information can be extracted from these rules. Last, some properties the final set of rules should verify are outlined.

Key-words: data summaries, knowledge discovery, fuzzy gradual rules, functional dependencies, coherence of rules.

1 Introduction

During the last years, the number and volume of databases have tremendously increased. As a consequence, the need for *condensed* information has also increased, and the extraction of summaries from large databases has received more and more attention. Many works in this area, known as *Knowledge Discovery*, are based on rules (as for instance *decision*, *production* or *inference* rules), which constitute a simple and natural tool for the representation of some kind of knowledge, and in particular knowledge meant for human people. However, these works are mainly in the frameworks of classical logic and/or statistics. As already stated by Yager [8], the investigation of other kinds of representation, able to capture a richer range of properties seems of interest. The use of fuzzy logic seems particularly interesting for the extraction of linguistic summaries of data [3, 6, 8]. Indeed, such summaries are statements of the natural language, as for instance: "*young* people are *well-paid*", which represent properties about the current content of a given database. These statements often link *gradual* properties of the natural language, as *young* or *well-paid*; and it is well-known that fuzzy logic allows for an accurate representation of these *gradual* properties.

The linguistic summaries discussed here are situated in the context of the relational model of data, and their interpretation is based on gradual rules [2, 4]. The idea is to capture and model summaries of the type: "the more X_1 is L_1 and ... and X_n is L_n, the more X_{n+1} is L_{n+1}", where X_i's are attributes and L_i's are linguistic labels defined by fuzzy partitions of the underlying domains of the X_i's. For instance, if the labels "*young*" and "*small*" are associated with attributes *age* and *bonus*, one might consider the summary: "the younger the employees, the smaller their bonus". Due to the form of the considered gradual rules, which have only one label in their conclusion part (L_{n+1}), what is summarized here are *simple* relations between labels (involving only one label of the output variable), rather than the attribute values of the tuples in the database.

In Section 2, some properties of gradual rules are briefly given. They constitute the basis for an algorithm aimed at the discovery of summaries that are valid on a given relation, given a fuzzy partition for each involved attribute. Then, it is shown in Section 3 that the produced rules, even if they contain valuable information which is not in contradiction with the database, can be potentially incoherent. However, this kind of incoherence allows for the extraction of additional pieces of knowledge. In addition, the informativeness of each rule can be questioned, and it is shown that some interesting information on the fuzzy partitions, and more particularly on their granularity, can be extracted in some situations.

2 Mining gradual rules

2.1 Some useful properties of gradual rules

A clear semantics of statements of the form "the younger the employees, the smaller their bonus" demands a connection between the degrees $\mu_{young}(e.age)$ and $\mu_{small}(e.bonus)$ for every employee e. This can be achieved using fuzzy implications, and more precisely with the ones known as *residuated* implications or R-implications, whose general definition is:

$$a \to b = \sup\{\delta \in [0,1],\ T(a,\delta) \leq b\}\ ,$$

where T is a triangular norm. If T is chosen as the minimum, one gets Gödel implication. Goguen and Lukasiewicz implications are obtained in a similar way, with $T(a,x) = a.x$ and $T(a,x) = \max(a + x - 1, 0)$ respectively. All these implications are completely true if the consequent is at least as true as the antecedent, and they vary in the way they penalize the occurrence of the opposite situation. The fuzzy rules modeled with such an implication are the so-called gradual rules [2, 4]. The particular case of the maximal penalty corresponds to Rescher-Gaines implication which writes:

$$a \to_{RG} b = 1 \text{ if } a \leq b, \text{ and } 0 \text{ otherwise}\ .$$

This implication takes its values only in $\{0,1\}$ instead of the unit interval, and corresponds to the *core* of any residuated implication. It corresponds to a more drastic view of the gradual rules, called *pure* gradual rules. For the sake of simplicity, only this implication is considered here. Indeed, using Rescher-Gaines implication, the relation (on the Cartesian product of the input-output spaces) modeled by a set of rules is boolean. Then, tuples of the database belong totally or not to this relation, and there is no need for the introduction of a *satisfaction threshold.*

Such a gradual rule, as for instance "the younger the employees, the smaller their bonus", modeled with Rescher-Gaines implication, means that "the closer the age of an employee is to *young*, the closer its bonus should be to *small*", where *young* and *small* are fuzzy sets (expressing gradual properties). It then means that if the membership degree of an input value (the age of an employee) to the fuzzy set *young* is α, then the membership degree of the output value (the bonus) has to be greater than α, which leads to the following inequality :

$$\mu_{young}(e.age) \leq \mu_{small}(e.bonus) \ .$$

These rules then express a closeness relation between properties (here "age = young" and "bonus = small"), more than a joint evolution of several attributes, as wrongly suggested by the statement "the younger the employees, the smaller their bonus".

The objective pursued here is to explore a database in order to discover linguistic summaries represented by gradual rules which are *valid.* Consider a relation R of the database, defined over the attributes $A_1, \ldots, A_p$, whose domains D_i are provided with fuzzy partitions $\{L_{i,1}, \ldots, L_{i,n_i}\}$, for $i = 1, \ldots, p$. The triple $\{A_i$, D_i, $P_i\}$, where $P_i = \{L_{i,1}, \ldots, L_{i,n_i}\}$ is a fuzzy partition of D_i, is called a *linguistic variable.* In most of applications, a *strict partition* P_i, also called Ruspini partition [7], is chosen. They are made of fuzzy sets with triangular or trapezoïdal functions which verify:

$$\forall x \in D_i, \quad \sum_{j=1,\ldots,n_i} \mu_{L_{i,j}}(x) = 1 \ , \qquad (1)$$

as shown on Figure 1.

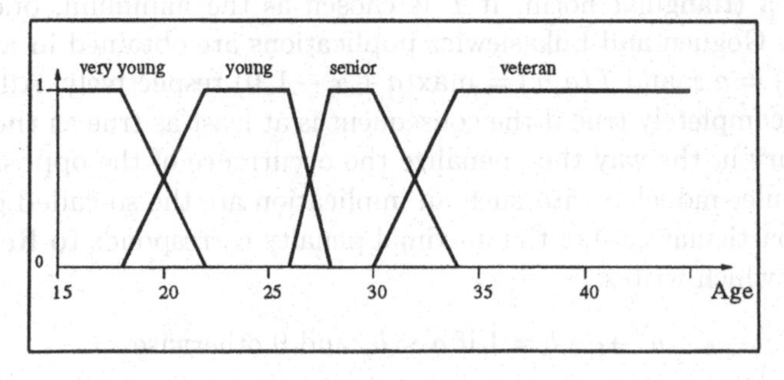

Fig. 1. Example of a linguistic partition for the attribute Age

The summary: "the more A_1 is L_{1,i_1} and ... and A_k is L_{k,i_k}, the more A_{k+1} is $L_{k+1,i_{k+1}}$ and ... and A_q is L_{q,i_q}" (with some well-chosen permutation on the A_i's, and $q \leq p$) is denoted:

$$((A_1, L_{1,i_1}), \ldots, (A_k, L_{k,i_k})) \rightarrow_{RG} ((A_{k+1}, L_{k+1,i_{k+1}}), \ldots, (A_q, L_{q,i_q})) .$$

This summary is said to be *valid* over R if and only if:

$$\forall t \in R, \min (\mu_{L_{1,i_1}}(t.A_1), \ldots, \mu_{L_{k,i_k}}(t.A_k)) \rightarrow_{RG} \min (\mu_{L_{k+1,i_{k+1}}}(t.A_{k+1}), \ldots, \mu_{L_{q,i_q}}(t.A_q)) = 1 ,$$

where t is a tuple of R. In other words, a summary is valid if it is in accordance with each tuple of R.

Such summaries satisfy some interesting properties, which are exploited in the mining algorithm. Due to some analogy between this kind of linguistic summaries and functional dependencies (see [1]), the names of the properties given here are the ones used for functional dependencies.

- **The property of augmentation** states that the condition part of a gradual rule can be extended, i.e., if one rule is valid, the addition of new conditions to its condition part leads to another valid rule.
  For instance, if the rule $(A_1, L_{1,i_1}) \rightarrow_{RG} (A_3, L_{3,i_3})$ is valid with respect to the database, then the rule $((A_1, L_{1,i_1}), (A_2, L_{2,i_2})) \rightarrow_{RG} (A_3, L_{3,i_3})$ is also valid.
- **The property of decomposition** allows to reduce the conclusion parts to singletons, i.e., if one rule is valid, the suppression of conclusions from its conclusion part leads to another valid rule.
  For instance, if the rule $(A_1, L_{1,i_1}) \rightarrow_{RG} ((A_3, L_{3,i_3}), (A_4, L_{4,i_4}))$ is valid with respect to the database, then the two rules $(A_1, L_{1,i_1}) \rightarrow_{RG} (A_3, L_{3,i_3})$ and $(A_1, L_{1,i_1}) \rightarrow_{RG} (A_4, L_{4,i_4})$ are also valid.
- **The property of union** is somewhat the opposite of the decomposition property, and states that two summaries with the same condition part can be combined into a new one whose conclusion is the conjunction (via the min operator) of the initial conclusion parts. From a practical point of view, decomposition and union allow for the search of summaries whose conclusions reduce to singletons (i.e., to one single pair $<$ *attribute*, *label* $>$).
  For instance, if the two rules $(A_1, L_{1,i_1}) \rightarrow_{RG} (A_3, L_{3,i_3})$ and $(A_1, L_{1,i_1}) \rightarrow_{RG} (A_4, L_{4,i_4})$ are valid with respect to the database, then the rule with compound conclusion $(A_1, L_{1,i_1}) \rightarrow_{RG} ((A_3, L_{3,i_3}), (A_4, L_{4,i_4}))$ is also valid.

2.2 A mining algorithm

The objective is the extraction of *linguistic* summaries from databases. Then, the produced gradual rules should be composed of fuzzy sets with a clear semantics, and as a consequence should pertain to a *linguistic variable*. In order to obtain such fuzzy sets, the partitions P_i on the domains D_i of the involved attributes A_i are supposed to be *a priori* given by an expert, instead of being automatically determined by a clustering algorithm. An example of such a partition is given on Figure 1.

Using only these fuzzy sets to construct the gradual rules is a rather strong constraint. Indeed, as a drawback, the extracted rules are probably not the most accurate ones to summarize the database. However, they present the essential advantage of using only a predetermined vocabulary, which makes sense for the involved human operators.

Then, the attributes are *a priori* partitioned into *input* and *output* variables, depending on the kind of relation to be modeled. The input (resp. output) ones are those which will appear in the condition parts (resp. conclusion parts) of the rules. Usually, the extracted rules are expected to express constraints (see [4])on the values for one attribute, depending on conditions on the other attributes, for each tuple in the database. Thus, only one attribute is generally chosen as output variable. Moreover, when more than one output attributes are involved, the property of decomposition (see the previous section) allows to decompose a rule with a compound conclusion into a set of rules with a simple one. As a consequence, all the extracted gradual rules can have a simple conclusion part, and a possibly compound condition part.

The objective is then the extraction rules of the form :

$$((A_1, L_{1,i_1}), \ldots, (A_k, L_{k,i_k})) \to_{RG} (A_{k+1}, L_{k+1,i_{k+1}}) ,$$

which are valid with respect to the tuples in the relation to be summarized. The way the extraction algorithm itself is designed is of less interest in this paper, and it is just outlined here, for the sake of brevity. Details on this algorithm can be found in [1].

The extraction algorithm is based on an iteration over tuples t of the relation R to summarize. It defines a minimal set S of summaries which are valid on R, knowing that other valid but less specific summaries could be obtained by the extension of the condition part of any summary in S, according to the augmentation property.

Roughly, the algorithm starts from an empty set S, and the tuples t of relation R are successively processed. The invariant of the loop is: "from S, it is possible to obtain any summary which is valid on already processed tuples". The progression consists in constructing the set S_t made of summaries which are valid for $\{t\}$, and to merge S and S_t to obtain the new set S.

3 Extracting additional knowledge from a set of rules

The output of the mining algorithm is a set of pure gradual rules representing linguistic summaries of relations between labels, which are valid with respect to the database. However, one may wonder about the informativeness of such summaries. Moreover, if *each* rule is valid with respect to the database, it is not impossible for the *set* of rules S to be *potentially incoherent* (or *incoherent* for short). These problems are now briefly investigated.

3.1 Coherence checking reveals empty zones in the database

The notion of coherence for gradual rules is a direct extension of the notion of coherence for a set of classical rules. A set of (classical) rules is said *coherent* if: "for any subset of rules that can be *fired* together, for a given precise value, these rules do not give contradictory conclusions".

Since attributes are partitioned into input and non-input variables for the rules, the extraction algorithm gives a set of parallel fuzzy rules, i.e., rules whose condition (resp. conclusion) parts are defined on the same universes. This set of rules then expresses a relation on the Cartesian product of the input universe (U) and output universe (V), namely $U \times V$. Then, the rules are said to be coherent if for any (allowed) value for the input variables, there is at least one value for the output variables which is totally compatible with the input value and the set of rules. Indeed, each rule is viewed as a constraint restricting the possible output values (given a condition on the input variables). Thus, it can occur that for a given input value, the triggered rules forbid all the output space. This rules are then said to be *potentially* incoherent, and not simply incoherent, since the incoherence only appears in presence of some input values.

As stated before, due to the decomposition and union property, all the summaries can have single conclusions, represented here by convex fuzzy sets. The coherence checking of the set of gradual rules is then very easy since it has been shown that it comes down to check only pairs of rules (see [5]). It is even easier when the partitions $\{L_{i,1}, \cdots, L_{i,n_i}\}$ used on each domain D_i are so-called *strict* partitions, i.e., partitions which verify: $\forall x \in D_i$, $\sum_{j=1,\dots,n_i} \mu_{L_{i,j}}(x) = 1$, as shown on Figure 1. Indeed, coherence checking only consists in verifying if the labels involved in the two rules are the same, adjacent or non-adjacent elements of the partitions.

These three cases are now illustrated on a simple example. For the sake of simplicity, two rules with only one input variable A_1 are first considered.

- The rules have *the same* condition part:
  - $(A_1, L_{1,1}) \rightarrow_{RG} (A_3, L_{3,i})$,
  - $(A_1, L_{1,1}) \rightarrow_{RG} (A_3, L_{3,j})$.

  These two rules are coherent if and only if their conclusion parts are the same, i.e., if and only if $i = j$. Otherwise, the two rules are said to be *potentially incoherent*. Indeed, the incoherence only appears in presence of a tuple t such that $\mu_{L_{1,1}}(t.A_1) > \min(\mu_{L_{3,i}}(t.A_3), \mu_{L_{3,j}}(t.A_3))$ as, for instance, a tuple t such that $\mu_{L_{1,1}}(t.A_1) = 1$.
- The rules have *adjacent* condition parts:
  - $(A_1, L_{1,1}) \rightarrow_{RG} (A_3, L_{3,i})$,
  - $(A_1, L_{1,2}) \rightarrow_{RG} (A_3, L_{3,j})$.

  Then, these two rules are coherent if and only if either their conclusion parts are the same (i.e., $i = j$), or if they are adjacent elements of the partition on D_3, i.e., if $i = j + 1$ or $i = j - 1$ (since the labels of the linguistic partition are supposed to be ordered).
- The rules have *disjoint* condition parts:
  - $(A_1, L_{1,1}) \rightarrow_{RG} (A_3, L_{3,i})$,

- $(A_1, L_{1,3}) \rightarrow_{RG} (A_3, L_{3,j})$.

Whatever the conclusion part of these rules, they are always coherent, since they cannot be triggered with the same precise input value.

All the other cases can be dealt with the same way. Consider for instance two following rules:

- $(A_1, L_{1,1}) \rightarrow_{RG} (A_3, L_{3,i})$,
- $(A_2, L_{2,1}) \rightarrow_{RG} (A_3, L_{3,j})$.

Their input condition are not on the same attribute. However, they can be considered as rules with two input conditions, replacing the missing condition with the whole input universe:

- $((A_1, L_{1,1}), (A_2, D_2)) \rightarrow_{RG} (A_3, L_{3,i})$,
- $((A_1, D_1), (A_2, L_{2,1})) \rightarrow_{RG} (A_3, L_{3,j})$.

Indeed, as conditions are combined with the min operator, the first rule means, "the rule $(A_1, L_{1,1}) \rightarrow_{RG} (A_3, L_{3,i})$ holds whatever the value for A_2 in D_2". As these two rules can be triggered to degree 1 with a precise value, the conclusions have to be the same (i.e., $i = j$) for the rules to be coherent. Indeed, a tuple t such that $\mu_{L_{1,1}}(t.A_1) = 1$ and $\mu_{L_{2,1}}(t.A_2) = 1$ corresponds to degree 1 to both conditions of the rules. Then, the coherence condition leads to the following equality: $\mu_{L_{3,i}}(t.A_3) = \mu_{L_{3,j}}(t.A_3) = 1$, which means that $i = j$, according to the linguistic partition of D_3. The detailed proofs are not given here for the sake of brevity, but can be extracted from more general propositions given in [5].

Two particular cases are especially interesting. They both stem from the fact that two *potentially* incoherent rules are incoherent only for tuples pertaining to the intersection of the condition parts of the rules. Thus, if the checking algorithm finds two potentially incoherent rules, it gives the information that the database contains no tuple corresponding to the condition parts of both rules. There are two cases :

- The conditions are on the same attribute: the two rules
  - $(size, very\ small) \rightarrow_{RG} (age, children)$
  - $(size, small) \rightarrow_{RG} (age, adult)$

  are potentially incoherent since being at the same time *children* and *adult* is impossible[1], while it is not for *very small* and *small*[2]. It then gives the information that no tuple in the database is *very small* and *small*.
- The conditions are on two different attributes: the two rules
  - $(size, very\ small) \rightarrow_{RG} (age, children)$
  - $(weight, big) \rightarrow_{RG} (age, adult)$

  are potentially incoherent for the same reason: someone can be small and heavy, but cannot be children *and* adult. However, these rules give the (more interesting) information: "no *very small* people of the database is also *big*".

[1] It is assumed here that the linguistic labels *children* and *adult* are totally disjoint.

[2] Here, the two labels *very small* and *small* are supposed to be adjacent elements of the partition, and then to have a non-empty intersection.

Checking the coherence of the gradual rules as a post-processing of the mining algorithm can then give additional information, and reveal empty zones of the database. Then, if a coherent set of rules is needed (which seems more natural), the coherence can be restored with the addition of an integrity constraint which forbid the values in the intersection of the condition parts of the two potentially incoherent rules.

3.2 Informativeness of the rules

Even if they are coherent with respect to the tuples in the database, some of the rules can be not really informative. For instance, rules with an empty condition, or the entire output space as the conclusion are totally uninformative. This is an extreme example, and these rules are generally not generated by mining algorithms. However, the informativeness of some other kinds of rules is also questionable.

A first case stems from the very definition of an implication (even the material one), which is true if the antecedent is false. In other words, a summary can be valid only because no tuple of the summarized relation fulfills its condition. Two criteria seem particularly important to determine the informativeness of a rule:

- the amount of tuples which support this rule, i.e., the tuples which more or less fulfill the condition part of this rule;
- and the degrees of support, i.e., the membership degrees of the tuples to the condition part of this rule.

Thus, rules which are supported by only a very small amount of tuples can be considered as rules modeling exceptional or unusual situations. If *summarizing* the database means removing the exceptions and focusing on the main values only, such rules are not interesting. By contrast, if it means *representing* all the database, these rules are necessary. However, in this case, this information can be valuable.

Moreover, the number of tuples which support each rule is not a sufficient information. Indeed, a rule can be supported by a large amount of tuples, but their membership degree to the condition part of this rule can be rather low. This means that there is no *typical tuple* for this rule in the database. No tuple contradicts the rule, but the relation encoded by this rule has no real representation in the database. Thus, the meaning of the rule as a linguistic summary can be questioned. Moreover, as the linguistic labels in the partitions overlap, if a rule is supported at rather low degrees (e.g., $\alpha < 0.5$), it means that another condition corresponds more accurately to these tuples (at degree $1 - \alpha > 0.5$). Then, the involved tuples should be encoded by other relations, i.e., other rules.

In conclusion, the informativeness of a rule can be measured by both the number of tuples which support this rule, and their degree of support. This double information can be encoded by means of a histogram representing the number of tuples included in each α-cut of the condition part of the rule. Depending on the expected kind of summary, an expert can use this histogram to determine the informativeness and then the interest of each of the produced rules. This can be done as an interactive post-processing added to the mining algorithm.

Further studies of the notions of informativeness and summaries could lead to an automated processing. The selection of interesting rules could then be included in the mining algorithm itself.

3.3 Informativeness of the set of rules

Some properties of the entire set of rules can also give information on the summarized database.

Summarizing a database can be considered as rewriting it in a more compact way, loosing the less information as possible. However, once the redundancy is removed, compacting the knowledge in the database means loosing information.

On the one hand, it is possible to create one rule for each tuple in the database. Then, no piece of information is loosed, but it is merely a rewriting, and not at all a summary. On the other hand, it is always possible to encode the whole database with only the rule $((A_1, D_1), \ldots, (A_k, D_k)) \rightarrow_{RG} (A_{k+1}, D_{k+1})$. However, this rule is totally uninformative. Between these two extreme cases, the choice is between the compactness of the representation, and the accuracy of the rules with respect to the original information.

Some situations are particularly noticeable. They all suppose that a majority of tuples are covered by the condition parts of the rules. This is not always the case with the kind of rules considered here. Indeed, consider a *minimal* zone of the input space, i.e., a situation encoded with one label for each attribute of the input space. A rule can be constructed using this situation for its condition part if and only if all the tuples in the database which correspond to this condition have a value for the output attribute which belong to the same label in the linguistic partition. When the values belong to at least two disjoint labels for the output attribute, no rule can be extracted using the algorithm proposed in Section 2.2. This means that there is no *simple* relation in the database, for this particular condition.

In order to extract rules for each possible condition parts of the rules, disjunctions of labels in the conclusion parts of the rules have been considered. A set of such gradual rules represent a relation on the Cartesian product of the input-output space, expressing the maximal restriction in accordance with both the tuples in the database and the granularity of the linguistic partitions. Then, these rules summarize the tuples in the database, more than the *simple* relations it contains.

When the rules are supported by a large amount of tuples in the base, some typical situations should be detected. Some examples are briefly given here.

- In case a lot of rules are obtained, each supported by only a few tuples, it means either that the tuples are evenly spread, or that the granularity of the linguistic partitions on the input attribute is too fine. Then, it corresponds more to a rewriting than to a summary of the database.
- In the opposite case, where few rules are supported by a large amount of data, it can either correspond to an ideal case of grouped tuples, or reveal that the linguistic partitions on the input attribute are too coarse and does not allow to distinguish between clusters of tuples.

– Similarly, if only one label of the output attribute appears in the conclusion part of the rules, it means either that only this label is involved in simple relations occurring in the database, or that the granularity of the output partition is too coarse.

Clearly, these situations could mean that the linguistic partitions *a priori* given by the human expert do not allow for the discovery of very interesting summaries. Thus, they could be submitted to this expert who can decide to modify the partitions or not.

In conclusion, considering the entire set of rules, it is also possible to extract additional information, either on the structure of the database, or on the granularity of the linguistic partitions.

4 Concluding remarks

This paper has shown that gradual rules are useful for the extraction of linguistics summaries from a database. If their semantics allows for an accurate representation of gradual properties under statements of the form "the more..., the more...", their use requires some care. In particular, as well as rules of the classical logic, they are aggregated conjunctively and then can be potentially incoherent. However, it has been shown here that since the coherence checking is very easy with the considered kinds of fuzzy partitions, potentially incoherent rules obtained with the extraction algorithm can be useful to extract other pieces of information. At last, it has been shown that studying the informativeness of the rules is also of interest. Then, far from being a drawback, these "problematic" rules allow to extract useful additional information. Their detection can be added to the extraction algorithm itself or, for some of them, performed as a post-processing.

References

1. P. Bosc, L. Liétard, and O. Pivert. Extended functional dependencies as a basis for linguistic summaries. In *Proc. of the 2nd Europ. Symp. on Principles of Data Mining and Knowledge Discovery (PKDD'98)*, pages 255–263, Nantes, France, 1998.
2. D. Dubois and H. Prade. Gradual inference rules in approximate reasoning. *Information Sciences*, 61(1, 2):103–122, 1992.
3. D. Dubois and H. Prade. On data summarization with fuzzy sets. In *Proc. of the 5th IFSA Congress*, pages 465–468, Seoul, Korea, 1993.
4. D. Dubois and H. Prade. What are fuzzy rules and how to use them. *Fuzzy Sets and Systems*, 84(2):169–186, 1996. Special issue in memory of Prof A. Kaufmann.
5. D. Dubois, H. Prade, and L. Ughetto. Checking the coherence and redundancy of fuzzy knowledge bases. *IEEE Trans. on Fuzzy Syst.*, 5(3):398–417, Aug. 1997.
6. D. Rasmussen and R.R. Yager. Summary SQL – A flexible fuzzy query language. In *Proc. of the Flexible Query-Answering Systems Workshop (FQAS'96)*, pages 1–18, Roskilde, Denmark, 1996.
7. E.H. Ruspini. A new approach to clustering. *Inf. and Control*, 15:22–32, 1969.
8. R.R. Yager. Linguistic summaries as a tool for database discovery. In *Proc. of the FUZZ-IEEE'95 Workshop on Database Syst. and Inf. Retrieval*, pages 79–84, 1995.

Working Towards Connectionist Modeling of Term Formation

Peter Marshall[1] and Zuhair Bandar[1]

[1] The Intelligent Systems Group, Department of Computing and Mathematics, Manchester Metropolitan University, Manchester, M15 5GD, UK.
{P.J.Marshall, Z.Bandar}@doc.mmu.ac.uk

Abstract. In recent years, there has been a conscious move away from rule based methods of term acquisition with research focusing on alternative machine learning approaches. This comes as a response to the difficulties of complete knowledge representation of term formation as a general set of rules. This paper is a continuation of our initial research into connectionist approaches to term recognition [13]. An extension to the Winner-take-all algorithm is proposed which uses exhaustive testing of weights to elucidate term and non-term forming clusters. This algorithm is applied to automatic term recognition. Initial experiments have shown improved results ranging between 1.31% and 5.86% after initial training.

1 Introduction

As technology develops, the quality of interaction between the user and the machine has also seen a steady rise [7]. Much of its success can be attributed to the availability of high quality hardware. In addition to the physical aspect, developers have started to exploit theories from Computational Linguistics [1]. Applications have been seen to incorporate more and more intelligent forms of communication e.g. textual and speech based language recognition. Many of these Natural Language Understanding applications require large scale lexicons which comprehensively model language in order to accurately process information [15]. Problems exist in the compilation of these lexicons. The lexicon must contain linguistic information not only about general purpose language but also contextual language appropriate to the application [15]. The compilation of contextual terms present a stumbling block in the way of expertise and time taken [2,3]. These problems have been seen as a catalyst for research and has influenced the creation of a number of tools which provide lexicographer with an advisory service to identification of terms [2,3,6,8,13].

1.1 Terms and Terminology

Terminology is the study of and the subject relating to the compilation, description, processing, and presentation of Terms[15]. A Term is an important notion within linguistics. It is used to represent a concept in a specialized domain [15]. Terms can be of various syntactic categories i.e. nouns, verbs or adjectives. The way terms are formed vary from sublanguage to sublanguage [2,8]. This is due to the differing patterns used in term formation [15]. Once designated, terms can consist of more than one word [3,6,8,15]. These are known as multi-word terms. The complexity exhibited in the creation of terms continues into the recognition of terms. This has influenced research towards focusing on individual sublanguages as opposed to general language [2,8]. However the problems do not end there. As language and sublanguages evolve more and more terms are being developed to represent new concepts[15]. Ideally a term should be monoreferential. Yet in practice, the problems of synonymy, homonymy and polysymy make the recognition of terms difficult [11]. This means that word forms can represent diverse concepts in differing domains and sometimes even within the same domain [2,8,15].

1.2 Automatic Term Recognition

There are several tools which have been developed to aid Automatic Term Recognition (ATR). The earliest methods used mainly linguistic information. Ananiadou used morphology to extract terms from medical corpora [2]. This approach was limited to the extraction of single word terms from medical corpora. Terms were extracted if they were derived from Greek or Latin morphemes. LEXTER was another linguistic method which extracted terms [3]. This method focused primarily on the extraction of multi-word noun phrase terms. This was performed using a syntactic grammar.

A shift in paradigm focused research away from explicit rules to the development of systems which used a combination of linguistic and statistical techniques [6,8]. The statistical measures used in these systems often evolved from the frequency of occurrence measure i.e. the number of times a term / multi-word term appears in a corpus. Terms tend to have a high frequency of occurrence in sublanguages [8]. The measure on its own does not guarantee term-hood as terms can exist within sublanguages even if they possess a low frequency [4,8]. A more accurate method of extracting word association is the association ratio [6]. This measure took into account terms which had a low frequency of occurrence.

1.3 Corpus Linguistics

Computational Linguistics has seen a rise in the use of Corpus based techniques [12]. A corpus is defined as a body of data stored in a machine readable form [12]. Corpora can be compiled in differing formats which characterize the type of data which is required by the application (e.g. text, audio, visual, etc.) [12]. Previous automatic term recognition techniques use have used text based corpora [2,3,6,8,13]. This is used in one of three ways i.e. modeling, testing or extracting. Approaches which have used machine learning techniques, corpora has been used to model term formation and then test the paradigm [13]. Where rules have been created to extract terms the quality of these rules are tested using corpora [2, 3, 6, 8]. All these methods use corpora to extract terms [2, 3, 6, 8, 13].

There is a problems with corpora. In its raw format, corpora does not provide any explicit information. However tools can be used to extract information form the text. This can be either linguistic, semantic, ontological or statistical [15].

2 Neural Networks

Neural networks exhibit characteristics which are of interest to NLU applications [14]. The most important of which are the automatic knowledge acquisition, knowledge representation and generalization features [14]. This eliminates the need to specify every aspect of a rule by allowing rules to evolve through exposure to training data vectors [5]. There are two alternative learning paradigms associated with neural networks, supervised and unsupervised learning. The main difference between the two paradigms is that supervised learning requires the expected outcome of each training vector to be incorporated with the training data.

When applying neural networks to corpus based techniques, the provision of teacher information requires an abundance of additional pre-processing [13]. As mentioned in section 1.2, the process of extracting terms requires a high level of expertise. This kind of pre-processing would be required to create the training data. The raw data contained in a corpus would have to individually tagged. Problems exist when the size of the a typical corpus can range between thousands of words and millions of words [12]. This would be a laborious and expensive [2,3].

2.1 Unsupervised Neural Networks

Unsupervised neural networks fall into three categories : Hebbian, Competitive, and Self organizing feature maps. Techniques which are derived from the Hebbian category are motivated by classic Hebbian synaptic modification hypothesis [9]. A further development evolving from this class of neural networks is the competitive learning category. These algorithms include a degree of lateral inhibition between the output neuron ensuring that only one neuron fires at anyone time [10]. The self organizing map algorithms group together techniques where significant features in the training data are discovered and used to cluster vectors together. This grouping of neural networks differ significantly in architecture from the previous two paradigms by employing a neighborhood of neurons to represent a cluster [10].

2.2 Winner-takes-all neural network

Our initial ATR paper used a competitive neural network technique, the winner take all algorithm [13]. This technique creates categories on the basis of correlations in the training data [13]. The trained network is then used to cluster test data which have similar characteristics. This network was ideal for the problem. The neural network is inhibitory by nature i.e. it prevents more than one neuron being selected at one time[10]. Thus making this approach well suited to the application of term recognition where a word form is either a term or a non term [13]. The results from the initial experiments ranged from 65% to 68%. An expansion to the algorithm was developed in the form of neuronal exploration.

2.3 Neuronal Exploration

Neuronal exploration is the empirical investigation of the weight vector through its alteration in the vector space. The purpose of neuronal exploration is to optimize the training of neural networks. The procedure starts by altering the weights in order to find the true center of vector clusters (Figure 1).

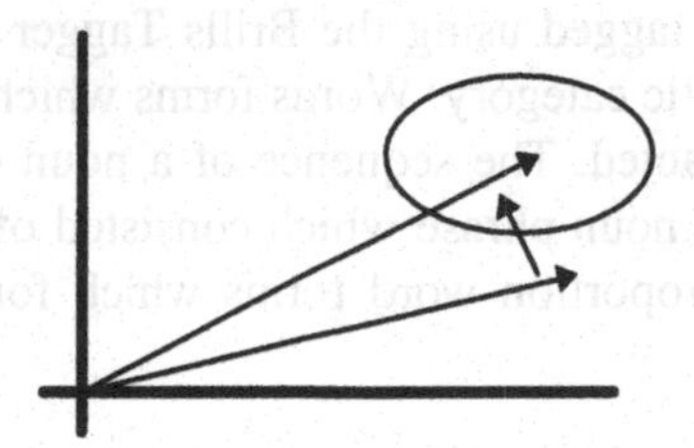

Fig. 1. Altering weight vectors

Every time the weight vector is change the result is then logged and the weights altered again. When the weight vector is comprehensively tested and the optimum values are found, it can assumed that the weights have found the center of a cluster or several important clusters. Unless the result is 100%, the complexity of the problem domain is not fully encapsulated. This requires an additional neuron to be included randomly in an attempt to find other clusters. Ideally this sequence is repeated until every data cluster is represented by a neuron, however in practice the sequence is repeated until the model has achieved an acceptable result.

3 Experiments

The following experiments use neuronal exploration in the problem area of ATR.

3.1 Experimental aim

The aim of the initial experiments were to recognize terms from medical corpora which consisted of two words and were of the syntactic type noun. The experiments form the basis for research into the extraction of terms which consist of more than one word. Medical corpora was chosen because this is an area of limited expertise in term extraction and an area which consists of many different forms of multi-word terms.

3.2 Experiment design

This section covers the sequence of events which occur in our method to automatically recognize terms.

3.2.1 Extraction of word collocations from the corpus

The corpus was prepared for syntactic categorization. Unwanted parts of text were eliminated from the corpus i.e. removing carriage returns, backslashes, full stops, and commas. The corpus was tagged using the Brills Tagger [4]. All the words in the corpus were given a syntactic category. Words forms which consisted of specific syntactic categories were extracted. The sequence of a noun or adjective followed by a Noun was used to extract noun phrase which consisted of two words. This has been seen to extract a high proportion word forms which form terms in Frantzi's ATR implementation [8].

3.2.2 Creation of the input vectors

Relevant linguistic and statistical measures were extracted for each collocation. This information was then compiled and stored as an input vector. The following is a list of measures used.

1. Frequency of the collocation including contextual cues.
2. Frequency of the collocation.
3. Frequency of the collocation including previous contextual cue.
4. Frequency of the collocation including following contextual cue.
5. Syntactic Category of the first word in the collocation.
6. Frequency of the first word in the collocation.
7. Syntactic Category of the second word in the collocation.
8. Frequency of the second word in the collocation.
9. Syntactic Category of the previous word in the collocation.
10. Frequency of the previous word as a previous word in other collocations.
11. Syntactic Category of the following word in the collocation.
12. Frequency of the following word as a following word in other collocations.
13. Association Ratio of the collocation.

3.2.3 Preparation of the data for compatibility with the neural network

Before the vector could be used by the neural network, the input vectors had to be formatted. The statistical measures which were used in the input vector were normalized between 0 and 1. Whilst the syntactic information was converted into a binary format.

3.2.4 Creation of the data sets

Once the medical corpus was processed, three data sets were extracted randomly from the corpus. The training set consisted of 2000 vectors containing 606 terms and 1304 non terms. The calibration set consisted of 10 terms and 10 non terms. The vectors in this set were formatted with its relevant teacher information. Finally a test set was created which consisted of 200 vectors containing 75 terms and 125 non terms.

3.2.5 Competitive neural network

The winner take all neural network was trained using the training data set. This neural network was initially trained with two output neurons. One neuron was used to cluster word forms which formed terms and the other one was used to cluster word forms which formed non terms.

3.2.6 Calibration of the neural network

A second neural network was used to calibrate the neurons of the competitive network. This was performed using a backpropagation neural network. The calibration set was then used to train the backpropagation network to classify the output from each neuron in their respective categories, those which classify terms and those which classify non terms.

3.2.7 Testing

Using the test data set, the vectors were propagated through the hybrid network and the results of this were then stored for comparison with other results to be extracted from the neuronal exploration.

3.2.8 Neural exploration

Each weight used in the neural network was altered slightly. After each alteration the competitive neural network was re-calibrated to account for any change in the classification of the neurons (section 3.2.6). The network was re-tested (section

3.2.7). When the vector space was explored, the network which categorized the most terms and non-terms correctly was chosen as the optimum value.

3.2.9 Additional neuron

Once the network had found the optimum set of weights for *n* number of neurons, another neuron was added and the competitive neural network was trained (section 3.2.5) without any alteration to the original neurons. The whole procedure (section 3.2.5 to 3.2.9) was repeat for six neurons.

3.3 Results

Table 1 contains the results from the experiments. The neuronal exploration was limited to a competitive network consisting of six neurons due to the computational limitations The initial training accuracy is the result which was achieved for each set of neurons after the competitive network was trained. The improvement of the results after optimization ranges from a 1.31% increase to a 5.85% increase. The results also improved though the addition of neurons to the model. This was due to the increase of vector space which allowed better representation of the problem's complexity.

Table 1. Experimental Results

No. of neurons	2	3	4	5	6
Initial training accuracy (%)	65.7	65.87	69.47	68.93	70.8
After weight optimization (%)	67.73	69.72	70.8	71.47	71.73
Improvement (%)	3.1	5.86	1.91	3.68	1.31

3.4 Conclusions

In conclusion, the improvements in results from the initial training elucidates the problems with the winner take all algorithm when it comes to finding clusters. The exhaustive exploration of the weight vector to find the optimum weight, allows us to explore the complexity of the data domain. The exploration can be applied to different paradigms which use weights to optimize results.

4 Future work

This paper summarizes a real world application and its use of unsupervised learning. Unsupervised learning has been shown to suit areas of research where there are difficulties in assembling training data with associated teacher information. The future of our work with Automatic Term Recognition lies in this area. The following points highlight our future work.

- The experiments highlighted in this paper are limited to a small subset of the entire problem. A more complete model is being developed. The model is being extended to include terms which contain varying amounts of words. This is important as there are many terms in the medical corpus which consist of more than two words. Eventually all the neural networks will be integrated into one model.

- Another issue which has to be examined is the quality of measures in the training set. The use of alternative statistical and linguistic measures is currently being explored. The use of morphology is significant especially if the model is to be extended to include single word term recognition [2].

- Once the general framework has been established the next step will be to investigate the use of other unsupervised learning techniques particularly the methods which come into the category of self organizing feature maps.

- The use of neuronal exploration has been beneficial to the winner takes all algorithm. This approach has to be refined to be less exhaustive and more intelligent in its exploration. Further research will look at its application to other neuronal paradigms.

References

1. Allen, J., Natural language understanding, 2nd edition, Benjamin Cummings (1995).
2. Ananiadou, S., A methodology for automatic term recognition, COLING 94 (1994).
3. Bourigault, D., Surface grammatical analysis for the extraction of terminological noun phrases, COLING 92 (1992).
4. Brill, E., A simple rule-based part of speech tagger, In the proc. of the Third Conference of Applied Natural Language Processing, ACL, pp 152-155 (1992).
5. Charniak, E., Statistical language learning, MIT Press (1993)
6. Church, K.W. & Hanks, P., Word Association Norms, Mutual Information and Lexicography, in Computational Linguistics, Vol.16, No. 1, pp. 22-29 (1990).
7. Dix, A., Human computer interaction, Wiley (1994).
8. Frantzi, K., Extracting Nested Collocations, COLING 96 (1996).
9. Hebb, D.O., 1949, The organization of behavior, Wiley (1949).
10. Hertz, J., Krogh, A., Palmer, R.G., Introduction to the theory of neural computation, Addison Wesley (1991).
11. Lauriston, A., Automatic recognition of complex terms : Problems and the TERMINO solution, Terminology, 1, 1, 147 - 170 (1994).
12. McEnery, T., Wilson A., Corpus linguistics, Edinburgh University Press (1996).
13. Marshall, P., Bandar, Z., Ananiadou, S., A connectionist approach to term recognition, EXPERSYS 98 (1998).
14. Reilly, R.G. and Sharkey, N.E., Connectionist approaches to natural language processing. Earlsdale (1992).
15. Sager, J.C., A practical course in terminological processing (1991).

A Quantization-Model of the Neural Nets

M. Reuter

Institute for Computer Science
TU-Clausthal
Julius-Albert-Str. 3
D-38678 Clausthal-Zellerfeld
Tel.: ++49 5323 953 115
e_mail:reuter@informatik.tu-clausthal.de

Abstract. This paper presents a new quantization model of neural nets describing the change of material configuration of neural nets by the Hamilton-Jacobi equations and the general activity of neural nets by the Schrödinger equation. Rest on the phase-space representation of the degrees of freedom of the net and on the potential-oriented formulation of the activity states of neural nets a Hamilton-operator can be defined that describes the classification- and the conditioning-state of the net as a disturbed equilibrium state. As a result of this theory it can be shown that supervised and unsupervised neural nets can be understood as the same kind of n-particle systems with only different Lagrange functions.

Keywords: quantization of neural nets, neural nets as n-particle systems, Lyapunov function of neural nets, Hamilton-function of neural nets, Schrödinger equation of neural nets.

1 Introduction of the potential oriented description of the neural nets

For every neural net a classification potential

$$E_p = E_p\left(\sum_{n=1}^{N}\left(E_p{}^{n}(y_1^n(t),\ldots,y_M^n(t)\right)\right) \quad (1)$$

and a conditioning potential

$$U_p = U_p\left(\sum_{n=1}^{N}\left(U_p{}^{n}(y_1^n(t),\ldots,y_M^n(t)\right)\right) \quad (2)$$

can be defined, where the index n indicates a neuron of the neural net with N neurons and y_m^n indicates the mth degree of freedom of the M degrees of freedom of a neuron n. If the state vector $\vec{y}(t)$ describes all neuronal and synaptic values of the network at a time t the neural network reaches steady state when

$$\dot{\vec{y}}(t) = \vec{f}(\dot{\vec{y}}(t)) = \vec{0} \tag{3}$$

holds indefinitely or until new stimuli perturb the system out of the equilibrium described by a the Lagrange function

$$L = L(f(y(t))) = E_p(y(t)) - U_p(y(t)) = 0 \tag{4}$$

If and only if the net adapts in the condition-state asymptotically to a desired vector of the net called the 'Repräsentant $RP_{-Netz} = \vec{y} = (y_1, ..., y_n)'$', this Lagrange function can be identified as a Lyapunov function for which holds

$$\dot{L} = \sum_{i=1}^{n} \frac{\partial L}{\partial x_i} \frac{\partial x_i}{\partial t} = \sum_{i=1}^{n} \frac{\partial L}{\partial x_i} \dot{x}_i \le 0 \tag{5}$$

where the x_i are the transformed co-ordinates of the state vector $\vec{y}(t)$ of the net in a co-ordination system for which holds

$$\vec{X} = \vec{0} \tag{6}$$

if the net meets its equilibrium point.

Depending on this preconditions the conditioning-state of a neural net can be understood as a non-equilibrium-state of the potential-functions E_p and U_p, which forces a change of the degrees of freedom x_i of the net to adapt the co-ordinates of the desired net structure $\vec{X} = \vec{0}$.

Furthermore a Hamilton-function H of a neural net can be defined, for which holds

$$H = 2E_p - L = 2E_p - (E_p - U_p) = E_p + U_p \tag{7}$$

If we assume that neural nets can be identified as conservative n-particle systems, for the change of the degrees of freedom is given by

$$\delta\ L = \int_{t_0}^{t_1} (E_p - U_p)dt = 0 \tag{8}$$

Following the last expression, the actual energy-function H can be separated in two Hamilton-functions

$$H = H_K + H_S \tag{9}$$

where H_S describes the energy, which forces the net to adapt to the new vector $RP_{-Netz} = \vec{Y}$ in the conditioning-phase and H_K describes the form of the classification potential of the net during the pure classification state.

If a Laurin serial develops H_S, the adaptation to the new vector $RP_{-Netz} = \vec{Y}$ will follow the formulas

$$\begin{aligned} \dot{\vec{x}} &= -(E_p - U_p)\vec{x}_0 \\ \dot{\vec{y}} &= -(E_p - U_p)\vec{y}_0 \end{aligned} \tag{10}$$

Following the last equations, the Lagrange function L is the generator of the net modifying operator

$$\Gamma(\vec{x},t) = e^{Lt} = \Gamma(\vec{y},t) = e^{Lt} \tag{11}$$

where these operators are members of a Lie-group.

If we define that a classification concept stored in the net will be called 'basic-concept' for the activity function $\Psi(\vec{x},t)$ of an out-put-neuron holds

$$\Psi(\vec{x},t) \approx 1 \tag{12}$$

and for the other neurons holds

$$\Psi(\vec{x},t) \approx 0 \tag{13}$$

if for $\Psi(\vec{x},t)$ holds [Reuter 1998]

$$\Psi(\vec{x},t) = e^{-\sqrt{\vec{E}}\,\vec{x}\ -(E_p - U_p)t} \tag{14}$$

where the expression $e^{-\sqrt{\vec{E}}\,\vec{x}}$ describes the time-independently classification behaviour of the net (represented by the configuration-vector $RP_{\_Netz} = \vec{Y}$) and the expression $e^{-(E_P - U_P)t}$ describes the disturbance of the net activity $\Psi(\vec{x}, t)$ under conditioning. If $\Delta\vec{x}(t)$ denotes the flow of the vector $\vec{x}(t)$ and if $F_i = \dot{x}_i$ stands for the conditioning-impulse of the degree of freedom x_i the conditioning of a neural net can be described as the shift of the potential of the Repräsentant $RP_{\_Netz}$ forced by the Lyapunov-Lagrange function $L = E_P - U_P \neq 0$ as shown in Figure 1.

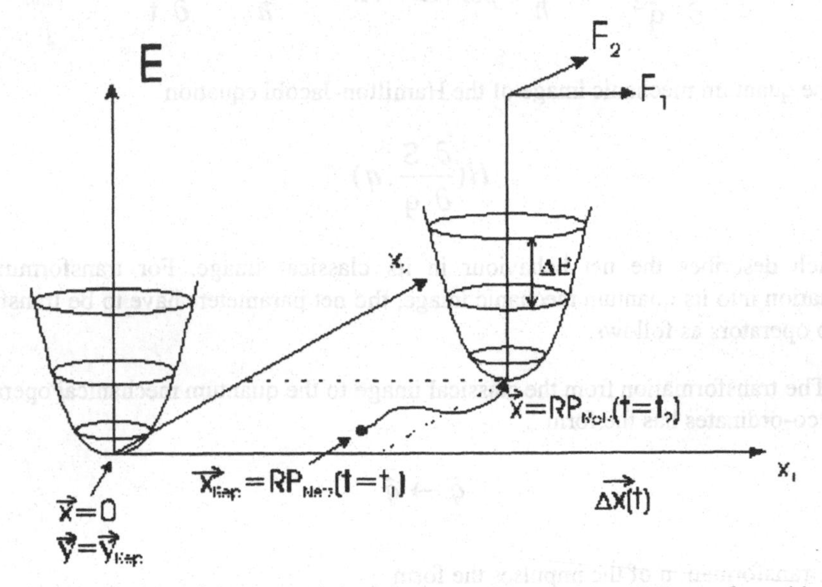

Fig. 1. The translation of the potential of the vector Repräsentant $RP_{\_Netz}$ forced by the presence of a Lyapunov-Lagrange function $L = E_P - U_P \neq 0$

2 Description of neural nets by the quantum mechanics

If the activity of a neural net is described by a context:

carrier-pattern-meaning

the material configuration of a neural net, described by the configuration of the generalised degrees of freedom q_i, p_i, gives rise to an activity pattern $\Psi(\vec{x}, t)$, which correspondents to a classification-result, whose meaning is normally carried by the operator's interpretation rules.

In this context the expression for the net activity

$$\Psi(\vec{x},t) = e^{-\sqrt{\vec{E}}\,\vec{x}\;-(E_p - U_p)t} \tag{15}$$

must follow the Hamilton equations in the same way the degrees of freedom q_i, p_i of the net do.

Next we remember that the Schrödinger equation

$$\frac{\partial^{2} \Psi(\vec{q},t)}{\partial \vec{q}^{2}} - \frac{2m}{\hbar} V_{pot}(\vec{q}) \Psi(\vec{q},t) = \frac{i2m}{\hbar} \frac{\partial \Psi(\vec{q},t)}{\partial t} \tag{16}$$

is the quantum mechanic image of the Hamilton-Jacobi equation

$$H\left(\frac{\partial S}{\partial q}, q\right) \tag{17}$$

which describes the net behaviour in its classical image. For transforming this equation into its quantum mechanic image, the net parameters have to be transformed into operators as follows:

The transformation from the classical image to the quantum mechanical operator of the co-ordinates has the form

$$\vec{q} \to \hat{\vec{q}} \tag{18}$$

the transformation of the impulses the form

$$p \to \hat{p} = \frac{\partial}{\partial q} \tag{19}$$

the transformation of the energy the form

$$E \to \frac{i}{\hbar} \frac{\partial}{\partial t} \frac{1}{\Psi(\vec{q},t)} \tag{20}$$

and the transformation of the Hamilton operator the form

$$\hat{H} = \frac{\partial^{2}}{\partial q^{2}} + \frac{2m}{\hbar^{2}} V_{pot} \tag{21}$$

Next we are remembering that

$$H \;= H_K + H_S \tag{22}$$

holds for the Hamilton-function H of the net. This leads to the following form of the Hamilton-operator $\hat{H}_K$ for the classification state of the net

$$\hat{H}_K = \frac{\partial^{\;2}}{\partial\,\vec{q}^{\,2}} + \frac{2m}{\hbar^2} U_p(\vec{q}) \tag{23}$$

and the following form for the Hamilton-operator $\hat{H}_S$ for the conditioning state of the net

$$\hat{H}_S = \frac{\partial^{\;2}}{\partial\,\vec{q}^{\,2}} - \frac{2m}{\hbar^2} U_p(\vec{q}) \tag{24}$$

These formulas lead to the following form of the Schrödinger equation

$$\left(\frac{\partial^{\;2}\Psi(\vec{q},t)}{\partial\;\vec{q}^{\,2}} + \frac{2m}{\hbar^2} U_p(\vec{q})\Psi(\vec{q},t)\right) + \left(\frac{\partial^{\;2}\Psi(\vec{q},t)}{\partial\;\vec{q}^{\,2}} - \frac{2m}{\hbar^2} U_p(\vec{q})\Psi(\vec{q},t)\right) = \frac{i}{\hbar}\frac{\partial\;\;\Psi(\vec{q},t)}{\partial\;\;t} \tag{25}$$

Separating the activity-function $\Psi(\vec{q},t)$ into a classification-related activity function $\Psi_K(\vec{q},t)$ and a conditioning-related function $\Psi_S(\vec{q},t)$ by

$$\Psi(\vec{q},t) = \Psi_K(\vec{q},t)\Psi S(\vec{q},t) \tag{26}$$

two Schrödinger equations will be given, one for the classification state

$$\left(\frac{\partial^{\;2}\Psi_K(\vec{q},t)}{\partial\;\vec{q}^{\,2}} + \frac{1}{\hbar^2} U_p(\vec{q})\Psi_K(\vec{q},t)\right) = \frac{i}{\hbar}\frac{\partial\;\;\Psi_K(\vec{q},t)}{\partial\;\;t} = \frac{1}{\hbar^2} E_K \Psi_K(\vec{q}, \tag{27}$$

and one for the conditioning state

$$\left(\frac{\partial^2 \Psi_S(\vec{q},t)}{\partial \vec{q}^2} + \frac{1}{\hbar^2} U_p(\vec{q}) \Psi_S(\vec{q},t)\right) = \frac{i}{\hbar} \frac{\partial \Psi_S(\vec{q},t)}{\partial t} = \frac{1}{\hbar^2} E_S \Psi_S(\vec{q}, \tag{28}$$

if we assume that for the mass of the net holds

$$2m = 1 \tag{29}$$

Now the activity functions $\Psi_K(\vec{q},t)$ and $\Psi_S(\vec{q},t)$ which solve these equations have to be formulated. To do so, we remember the classical activity function

$$\Psi(\vec{x},t) = e^{-\sqrt{\vec{E}_q}\vec{x} \ -(E_p - U_p)t} \tag{30}$$

and modify it the following way

$$\Psi_K(\vec{q},t) = e^{i\hbar^{-1}\left(-\sqrt{\vec{E}_q}\vec{q} \ -(E_p + U_p)t\right)} \tag{31}$$

for the classification activity and

$$\Psi_S(\vec{q},t) = e^{i\hbar^{-1}\left(-\sqrt{\vec{E}_q}\vec{q} \ -(E_p - U_p)t\right)} \tag{32}$$

for the conditioning activity. As it can be clearly pointed out, these functions solve the Schrödinger equations 27 and 28.

As the expression

$$\Psi'(\vec{q},t) = e^{i\hbar^{-1}\left(-\sqrt{\vec{E}_q}\vec{q}\right)} \tag{33}$$

is equal for both activity functions, we will get the following form for the generalised activity function $\Psi(\vec{q},t)$

$$\Psi(\vec{q},t) = e^{i\hbar^{-1}\left(-\sqrt{\vec{E}_q}\vec{q} \ -(E_p + U_p)t + (E_p - U_p)t\right)} \tag{34}$$

which can be written as

$$\Psi\ (\vec{q},t) = e^{i\hbar^{-1}\left(-\sqrt{\vec{E}_q\vec{q}}\ -(E_K+E_S)t\right)} \tag{35}$$

As for the time-independently Schrödinger equation holds

$$\left(\frac{\partial^2 \Psi_S(\vec{q})}{\partial\ \vec{q}^2} + \frac{1}{\hbar^2} U_p(\vec{q})\Psi\ (\vec{q})\right) = \frac{1}{\hbar^2} E\ \Psi\ (\vec{q},t) \tag{36}$$

in the equilibrium state E_p-U_p=0 the Schrödinger equation will be solved by the function

$$\Psi_S(\vec{q},t) = e^{i\hbar^{-1}\left(-\sqrt{\vec{E}_q\vec{q}}\ \right)} \tag{37}$$

that describes the exponential decrease of the classification potential if the components $e_i(t) = q_i - \Delta e_i'(t)$ of the input-vector $\vec{E}_i(t)$ are 'leaving' the position of the co-ordinates of the Repräsentant $RP_{\_Netz}$ by an increase of the terms $\Delta e_i'(t)$.

Next we will interpret the activity function $\Psi(\vec{q},t)$ by discussing the significance of the factor $\hbar$.

As the factor $\hbar$ describes the quantisation of the activity states $\Psi\ (\vec{q},t)$ of a neural net, it sets up the phase-room behaviour of the flow of the degrees of freedom q_i concerning the characterisations: Non-ergodic flow, ergodic mixing and ergodic non-mixing flow. Thus the value of $\hbar$ defines, if a net can be sensitised or not. Furthermore the change of $\hbar$ will define jumps of the classification behaviour and thus a kind of quick classification routine and/or a focus of attention of the net can be implemented by choosing the 'right' form of $\hbar$.

To proceed in our analysis of $\Psi\ (\vec{q},t)$, we remember that the following equation holds

$$\Psi\ (\vec{q},t) = e^{-i(\sqrt{\vec{E}_p}\vec{p}-Ht)} = \cos(\vec{E}_p\vec{p} - Ht) - i\sin(\vec{E}_p\vec{p} - Ht) \tag{38}$$

Surely the expression $(\vec{E}_p\vec{p} - Ht)$ will only make sense in the interval

$$-\frac{\pi}{2} \leq (\vec{E}_p \vec{p} - Ht) \leq \frac{\pi}{2} \tag{39}$$

as a periodical concept probability can not be observed in (common) neural nets.

On the other hand in this interval the function $\Psi(\vec{q},t)$ describes a spherical-wave and the quadratic form of $\Psi(\vec{q},t)$ describes the probability of the classification activation (classification concept) of the Repräsentant $RP_{\_Netz}$.

As for $\Psi(\vec{q},t)$ must hold

$$\Delta\Psi(\vec{q},t) = \frac{1}{\kappa^2} \frac{\partial^2 \Psi(\vec{q},t)}{\partial t^2} \tag{40}$$

where for κ holds

$$\kappa = \frac{E_p}{\sqrt{(E_p - U_p)}} \tag{41}$$

and on the other hand the formula

$$\frac{\partial \Psi(\vec{q},t)}{\partial t} = \frac{iE}{\hbar} \Psi(\vec{q},t) \tag{42}$$

holds, where

$$E = -i\hbar \frac{\partial \Psi(\vec{q},t)}{\partial t} \frac{1}{\Psi(\vec{q},t)} \tag{43}$$

it follows

$$\frac{\partial^2 \Psi(\vec{q},t)}{\partial t^2} = \frac{E^2}{\hbar^2} \Psi(\vec{q},t) \tag{44}$$

which leads with the formula

$$\Delta\Psi(\vec{q},t)=\frac{(E_p-U_p)}{U_p}\frac{E^2}{\hbar^2}\Psi(\vec{q},t) \tag{45}$$

to the following form of the time-independently Schrödinger equation

$$\Delta\Psi(\vec{q},t)+\hbar^{-2}U_p\Psi(\vec{q},t)=-i\hbar^{-2}\frac{\partial\ \Psi(\vec{q},t)}{\partial\ t} \tag{46}$$

As this equation has the same form as a diffusion-equation, for which holds

$$\Delta\Psi\ (\vec{q},t)=\frac{1}{\kappa^{\ 2}}\frac{\partial\ \Psi\ (\vec{q},t)}{\partial\ \ t} \tag{47}$$

we will introduce an information quantization by defining a smallest information quantum i (called after Weizsächer UR), for which holds

$$\frac{\partial\ \rho}{\partial\ t}=div\ \ \mathrm{i} \tag{48}$$

If κ is constant, we can separate $\Psi\ (\vec{q},t)$

$$\Psi\ (\vec{q},t)=\rho\ (\vec{q})\Psi(t) \tag{49}$$

which leads to the following form of equation 47

$$\frac{\kappa^{\ 2}}{\rho\ (\vec{q})}\Delta\rho\ \ (\vec{q})=\frac{1}{\Psi(t)}\frac{d\ \ \Psi\ (\vec{q})}{d\ \ t} \tag{50}$$

If the constant of separation can be written in the form

$$-\kappa^{\ 2}k^2 \tag{51}$$

and for $\hbar$ holds $\hbar=1$, the dynamic of $\Psi\ \ (t)$ is given by

$$\frac{d\Psi\ \ (t)}{dt}+\kappa^{\ 2}k^2\Psi(t)=0 \tag{52}$$

with the solution

$$\Psi\ (t) = \Psi(t_0)e^{-\kappa^2 k^2 (t-t_0)} \tag{53}$$

As the left part of equation 50 correspondents to the Helmholtz-equation for any kind of arbitrary wave, the general solution of equation 47 will be given by the formula

$$\Psi'(\vec{q},t) = \rho\ (\vec{q})\Psi''(t) = \sum_{k_{q2},k_{q2},k_{q3}} \left(\psi_1(\vec{k}e^{ik\vec{q}-\kappa^2 k^2 t}) + (\psi_2 \vec{k}e^{ik\vec{q}-\kappa^2 k^2 t}) \right) \tag{54}$$

As on the other hand for $\Psi\ (t)$ holds

$$\Psi_K(\vec{q},t) = e^{i\hbar^{-1}\left(-\sqrt{\vec{E}_q}\vec{q}\ -(E_p+U_p)t\right)} \tag{55}$$

we can choose an adequate functionalisation and get

$$\Psi'\ (\vec{q},t) = e^{\hbar^{-1}\left(-i\sqrt{\vec{E}_q}\vec{q}\ -(E_p+U_p)'t\right)} \tag{56}$$

Now the form of $\Psi'\ (\vec{q},t)$ can be regarded as a complex potential of the form

$$\Omega(\Psi(\vec{q},t)) = \ln(\Psi(\vec{q},t)) \tag{57}$$

where the imaginary expression $-i\sqrt{\vec{E}_q}\vec{q}$ describes the (semidefinit) streamlines of the complex potential and the expression $(E_p + U_p)'t$ describes the quantizated equi-potential-lines of the complex potential $\Omega(\Psi(\vec{q},t))$. Now the expression

$$\hbar^{-1}E_K t = const. \tag{58}$$

describes an information particle wave-front, which streams out of the source 'neuron' into the neighbourhood of the neuron and 'induces' the classification potential E_p of the Repräsentant $RP_{\_Netz}$.

These considerations shows us that our assumptions regarding the description of neural nets by the carrier-platform and by the pattern-platform lead to the same results and descriptions of the classification potential E_p, if the net is understood as a carrier of a constant energy-rate E_K=E_p+U_p.

3 Conclusions

In the context carrier-pattern-meaning a quantization model of neural nets can be derived that describes a neural classifier as an n-particle-system with a well-defined potential behaviour and quantizated energy rates. The classification-act as well as the change of the net under conditioning can be interpreted by our model as the disturbance of the equilibrium state between a classification potential E_p and a conditioning potential U_p. So every neural net can be regarded as a dynamical system, where the quantization parameter $\hbar$ rules the classification behaviour of the net. Following this model the classification act, forced by a classification potential E_p, can be identified as the reflection of an condition potential U_p, at the zero line of the Lyapunov potential L as shown in Figure 2.

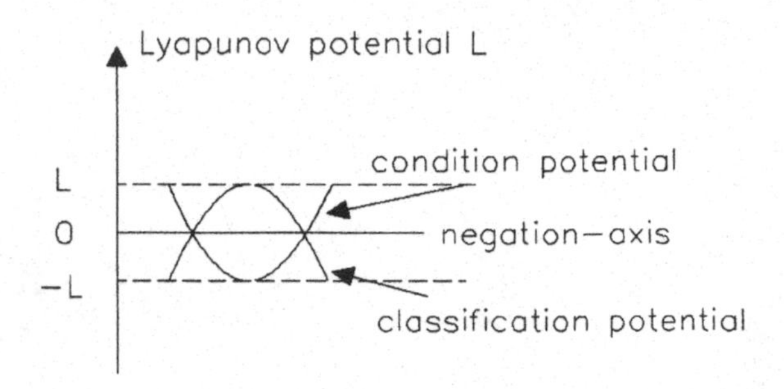

Fig. 2. Dependence between the classification potential E_p and the condition potential U_p

Similar to the theory of brain activation states the storing and the evaluation of information by dry-neural nets can now be understood as two kinds of one activation principle, which is ruled by the equilibrium state of the potentials of the degrees of freedoms of the net.

The change of these system-variables can be described in the classical image by the Hamilton-Jacobi equations. On the other hand these changes can be described in the quantum mechanical image of neural nets by the Schrödinger equation too, where the same potential structure as in the classical phase-space-image can be observed. According to this theory we can understand a classifying neuron as a source emitting classification waves in the context of a new model of quantum neurodynamics.

4 References

[Dawes] Dawes, R.L., 'Inferential Reasoning Through Soliton Properties of Quantum Neurodynamics', IEEE International Conference on System, Man and Cybernetics, Vol. 2 Chicago, II, 1992, ISBN 0-7803-0720-8/92

[Jos] Joos, G., 'Quantenmechanik', Akademische Verlagsgesellschaft, Wiesbaden, 1980, ISBN 3-400-00013-2

[Mes] Messia, A., 'Quantenmechanik', Walter de Gruyter, Berlin, 1976, ISBN 3-11-003686-X

[Prigogine] Prigogine, I., 'Vom Sein zum Werden', Piper, München, 1977, ISBN 3-492-02488-2

[Reuter] Reuter, M. 'Die potentialorientierte Beschreibung der neuronalen Netze', Clausthal, 1998

Risk Analysis Using Perceptrons and Quadratic Programming

Bernd-Jürgen Falkowski

FH Stralsund
FB Wirtschaft
Zur Schwedenschanze 15
D-18435 Deutschland

Abstract. A heuristic method (computation of weighted averages) is considered for risk analysis. It is improved upon by utilizing the Perceptron Learning Theorem and Quadratic Programming. Experimental work shows that both techniques give good results, the former one being somewhat more efficient in terms of CPU-time used. In spite of certain theoretical shortcomings it is argued that the familiar paradigm offers considerable potential for practical applications.

1 Introduction

Risk analysis may be considered as a special problem in the field of pattern recognition. Thus well-known statistical techniques (in particular Bayes' theorem, cf. [2], p. 17) are available to construct a solution. In addition the classical perceptron approach may be employed, cf. [10], p. 175. Both techniques lead to similar linear decision functions (at least under certain independence assumptions), cf. [2], p. 84, [10], p. 175. Whilst in either case the decision taken may be interpreted as a 'minimum distance' decision it is nevertheless difficult to explain the procedure to a non-specialist. This argument applies even more if Multi-Layer Neural Networks are used. Moreover all these methods have well-known inherent shortcomings. As a consequence major obstacles have to be overcome if practical applications are envisaged.

Thus we are led to consider a method in more detail which has been employed in a variety of fields (risk analysis, assessment of personnel, grading of students,...) on a heuristic basis: The computation of weighted average grades from individual grades. We improve on it by utilizing the Perceptron Learning Theorem and Quadratic Programming. Both techniques prove successful in practical tests, the former being more efficient in terms of CPU-time used.

Hence, in spite of certain theoretical shortcomings, we feel that our use of a familiar paradigm shows considerable potential for practical applications.

2 The Problem

Many large banks nowadays use so-called 'scoring systems' in order to assess the creditworthiness of their customers. Within this framework certain customer characteristics (e.g. income, property, liabilities, ...) are assigned grades (1 to 5 say, where 1 is better than 2, 2 is better than 3 ...) and a weighted average of these grades is computed. Here the weights are usually determined by ad hoc methods and it is hoped that this will lead to a reasonable decision procedure. One can, however, do rather better than this and in fact exploit the available bank records (of customers whose credit applications have been judged 'worthy'/'doubtful'/'unworthy' on the basis of past experience) in a systematic fashion.

Indeed, the problem may be formulated in abstract terms as follows: One is given m vectors $\mathbf{x}_1, \mathbf{x}_2, \ldots, \mathbf{x}_m \in \mathcal{Z}^n$ (think of these as having grades of customer characteristics as their entries) and a preference relation ' $\succ$ ' on these vectors (think of this as given by the judgement taken from the bank records, where 'worthy' is of course preferred to 'doubtful' which in turn is preferred to 'unworthy'). It is then required to find a vector $\mathbf{w} \epsilon \mathcal{R}^n$ (think of this as a weight vector) and a map $m_{\mathbf{W}} : \mathcal{Z}^n \rightarrow \mathcal{R}$ such that

$$\mathbf{x}_i \succ \mathbf{x}_j \Rightarrow \mathbf{m}_{\mathbf{W}}(\mathbf{x}_i) > \mathbf{m}_{\mathbf{W}}(\mathbf{x}_j) \tag{1}$$

Note that it is rather useful to consider such an abstract description, since the problem under consideration is thus seen to be a special case of a more general problem which arises for example in the (at first sight entirely unrelated) context of information retrieval using similarity measures in a classical vector space model, cf. [3].

Specializing the definition of $m_{\mathbf{W}}$ to $m_{\mathbf{W}}(\mathbf{x}) := - < \mathbf{x}, \mathbf{w} >$, where $< ., . >$ denotes the scalar product, it becomes clear, that condition (1) is fulfilled, if the weighted average computed using **w** accurately reflects the judgements contained within the records of the bank.

3 Scoring Systems, Perceptrons, and Quadratic Programming

For $\mathbf{x}_i \succ \mathbf{x}_j$ one can introduce the difference vectors $\mathbf{d}_{ij} := \mathbf{x}_j - \mathbf{x}_i$ and then reformulate (1) as

$$\mathbf{x}_i \succ \mathbf{x}_j \Rightarrow < \mathbf{d}_{ij}, \mathbf{w} > \quad > \quad 0 \tag{2}$$

An explicit solution to (2), if it exists, may be constructed using the perceptron learning theorem ,cf. e.g. [10]. Unfortunately very little is known about the speed of convergence in general, cf. e.g. [4], p. 89. Moreover, due to the discrete nature of the problem, it does not seem possible to apply gradient descent techniques to an error surface.

Alternatively a constructive proof of the theorem on the existence of a separating hyperplane, cf. e.g. [9], may be used to derive a necessary and sufficient condition for the existence of a solution. Indeed, one easily obtains

Theorem 1. *Suppose that S is the closed convex hull of the $\mathbf{d}_{ij}$, i.e.*

$$S := \left\{ \sum_{i,j} \lambda_{ij} \mathbf{d}_{ij} | \sum_{i,j} \lambda_{ij} = 1, \lambda_{ij} \geq 0, \mathbf{x}_i \succ \mathbf{x}_j \right\}$$

then a solution $\mathbf{w}$ *for (2) exists if and only if* $\mathbf{0} \notin S$.

Proof: ($\Rightarrow$) Suppose, to get a contradiction, that $\mathbf{0} \in S$, then for arbitrary $\mathbf{w} \in \mathcal{R}^n$ we have

$$0 = <\mathbf{0}, \mathbf{w}> = \sum_{i,j} \lambda_{ij} < \mathbf{d}_{ij}, \mathbf{w}>$$

But this implies that $< \mathbf{d}_{ij}, \mathbf{w} > \leq 0$ for some i, j.
($\Leftarrow$) This is an immediate consequence of the theorem on the existence of a separating hyperplane, cf. e.g. [9]. □

Suppose that $\mathbf{D} := [\mathbf{d}_{11}, \mathbf{d}_{12}, \ldots]$ and $\lambda := [\lambda_{11}, \lambda_{12}, \ldots]^T$ then one obtains as a further consequence [1]

Theorem 2. *Let* $\mathbf{D}' := \mathbf{D}^T\mathbf{D}$, *and suppose that the condition of theorem 1 is satisfied, then a solution vector* $\mathbf{w}$ *for (2) is given by solving the following quadratic programming problem*

minimize $<\mathbf{D}'\lambda, \lambda>$
subject to $\sum_{i,j} \lambda_{ij} = 1, \lambda_{ij} \geq 0, \mathbf{x}_i \succ \mathbf{x}_j$
and setting $\mathbf{w} = \mathbf{D}\lambda$.

Proof: This follows immediately from a constructive proof of the theorem on the existence of a separating hyperplane, cf. e.g. [3]. □

To solve this problem the well-known techniques (e.g. gradient descent) may be employed, cf. e.g. [8], however, in this case the speed of convergence heavily depends on the eigen-value-structure of $\mathbf{D}'$, which is not known in advance (note here that this approach differs essentially from using back propagation and an error surface).
Moreover, known results concerning the capacity of the perceptron indicate that the existence of a suitable weight vector is not necessarily likely, cf. e.g. [5], p.111. Since nevertheless banks seem to assume that a weighted average as described above can characterize a customer's creditworthiness it appeared desirable to conduct some experiments.

4 Experimental Results

In order to obtain a comparison between the methods sketched out above and to investigate the existence of a suitable weight vector a prototype was constructed under MS Windows using Excel and VBA (the main reason being that this is a readily available software providing suitable functionality through the so-called Solver). The Hardware used was a Pentium Laptop (166 MHz, 16 MB RAM).

[1] Note that all vectors are assumed to be column vectors and 'T denotes the transpose of a vector or a matrix.

From a bank data concerning (unfortunately only) nine (of course anonymous) customers could be obtained. In each case 14 characteristics were graded (from 1 to 5). Since cases which were representative in the eyes of the bank had been provided, it was thought that the experiments were useful in spite of the small number of customers considered. In order to provide at least a little more testing material, the entirely realistic judgements of the bank were varied slightly.

It turned out that both the quadratic programming method (QP), as implemented by using the Solver, as well as the perceptron (P), found a solution to (2) in all cases within at most 25 seconds. In all cases, however, it turned out that P performed at least as well as QP and in some cases even more than 100 percent better than QP within the bounds of error (due to measuring elapsed time and eliminating e.g. cache effects by averaging measurements only). Indeed, the longest time taken by P was 11,9 seconds.

In addition some fictitious data were constructed by extending the original bank data (36 sets in all). These data were checked for consistency of valuation by program (after all it wouldn't make sense to rate a customer non-creditworthy if his single grades were nowhere worse than those of a doubtful or creditworthy customer). Thereafter P found a solution within approximately 40 seconds, whilst the Solver Implementation of QP could not cope with the problem any more (there were too many variables).

It should be emphasized that in every case a suitable weight vector was found.

5 Discussion

The results given above provide evidence that in spite of a lack of theoretical justification the perceptron learning algorithm performs well in practice when compared with gradient descent methods. It was in any event very successfully employed for the problem at hand.

A neural network approach to risk evaluation is not new, cf. e.g. [6], [7]. Also fuzzy technology has been applied in this context, cf. e.g. [1]. However, in spite of sophisticated theoretical models (e.g. multi-layer networks), these methods apparently have not been widely used in practice. This may on the one hand be due to the fact that for those comparatively complicated models well-known shortcomings (e.g. lack of reliable procedures to find a global extremum of the error surface) put certain restrictions on their validity. On the other hand the very complexity of the results obtained (e.g. a customer may be a good risk with a certainty of 0,76) may make them unacceptable to people not entirely conversant with these techniques (e.g. managers).

Similar arguments apply to the statistical technique (and the related perceptron method) described in the introduction.

Thus the advantage of the approach outlined above seems to be twofold: Its simple theoretical 'ansatz' can be understood even by managers, and the computation of the weight vector may easily be achieved. Of course, to be held against this is the fact that a suitable weight vector may not always exist. If it does, it should be easy to compute though (that is to say that one does not need many characteristic examples), cf. [5], p.153. Moreover, since banks continue to use scoring systems in practice, it seems likely that in practical situations this difficulty will have little relevance: At worst one will choose a weight vector on the basis of (part of) past experience, which still seems preferable to entirely ad hoc methods..

References

[1] Barczewski, T.; Rust, H.J.; Weber, R.; Zygan, H.: Vorstellung eines Fuzzy Entscheidungssystems zur Bonitaetsbeurteilung von Unternehmen, in: Fuzzy Technologien bei Finanzdienstleistern, MIT-Management Intelligenter Technologien GmbH, Promenade 9, 52076 Aachen, (1997)

[2] Bishop, C.M. Neural Networks for Pattern Recognition, Oxford University Press, (1995)

[3] Falkowski, B.-J.: On Certain Generalizations of Inner Product Similarity Measures, Journal of the American Society for Information Science, Vol. 49, No. 9, (1998)

[4] Hampson, S.E.;Volper, D.J.: Feature Handling in Learning Algorithms, in: Dynamic Interactions in Neural Networks: Models and Data, Springer, (1988)

[5] Hertz, J.; Krogh, A.; Palmer, R.G.: Introduction to the Theory of Neural Computation, Addison-Wesley, (1991)

[6] Heyder, F.; Zayer, S.: Analyse von Kurszeitreihen mit Kuenstlichen Neuronalen Netzen und Competing Experts, in: Informationssysteme in der Finanzwirtschaft, Springer, (1998), pp. 489-500

[7] Locarek-Junge, H.; Prinzler, R.: Estimating Value-at-Risk Using Neural Networks, in: Informationssysteme in der Finanzwirtschaft, Springer, (1998), pp. 385-397

[8] Luenberger, D.G.: Linear and Nonlinear Programming, Addison-Wesley, (1986)

[9] McLewin, W.: Linear Programming and Applications, Manchester Student Edition, Input-Output Publishing Company, (1980)

[10] Minsky, M.L.; Papert, S.: Perceptrons, MIT Press, 3rd edition, (1988)

Finding Relevant Process Characteristics with a Method for Data-Based Complexity Reduction

J. Praczyk, H. Kiendl, and T. Slawinski

Faculty of Electrical Engineering
University of Dortmund, D-44221 Dortmund
Tel.: +49 231 755-4621 Fax: +49 231 755-2752
praczyk@esr.e-technik.uni-dortmund.de
http://esr.e-technik.uni-dortmund.de

Abstract. In this paper we present a fuzzy method that allows a reduction in the number of relevant input variables and, therefore, in the complexity of a fuzzy system. Input variables are only kept if they affect the output variable and are not correlated with any other kept input variable. The efficiency of the method is demonstrated by using it to find relevant process characteristics in the examples of a fuzzy classifier for quality control in the car industry and a load predictor used in power station management.

1 Introduction

For complex technical systems, knowledge-based modelling often fails. In such cases a data-based modelling approach using available system data may be successful. In particular, fuzzy models have proven to be useful for this, as they can be generated from available data and can be improved by adding rules generated from expert knowledge [1].

A key problem in this approach is the selection of the input variables. To reduce the number of variables and, therefore, the complexity of the resulting fuzzy model, we present a method — also based on fuzzy logic — that detects correlations between variables.

Input variables are analysed for correlation with the output variables and with other input variables. We consider two variables as correlated if there is *any* dependence between these variables. If an input variable is not correlated with the output variable, or if it is correlated with another input variable, it is not relevant for the model and is omitted. The well-known correlation [2] and transinformation [3] do not distinguish between the two possible directions of dependence between two variables. In contrast, our method for data-based complexity reduction [4] is based on the concept of mappings and does distinguish between the two directions.

Another difference between our method and existing statistical methods is that the latter are often based on an assumption about the structure of the possible correlation between the variables and analyse how well the actual correlation meets this assumption. Correlation analysis and principal component analysis [6], for example, analyse variables for *linear* correlation.

In contrast to these statistical methods, our method does not make any assumption about the nature of the possible correlation. The method tests whether a variable e_j depends on a variable e_i in such a way that a mapping $e_j = f(e_i)$ is confirmed by the available data. The method does not supply any information about the nature of this mapping, but this is not required for the selection of the input variables.

2 The Method

In the selection of input variables only those variables are essential that affect the output variable. Additionally, only those variables are essential that do not correlate with any of the considered (and kept) input variables.

To analyse potential input variables in terms of these two conditions, the concept of mappings is transferred to the case of linguistic variables:

Definition 1. *Let A be the set of linguistic values $e_{i,k}$, $k = 1, \ldots, m$, provided for the linguistic variable e_i and let B be the set of linguistic values $e_{j,l}$, $l = 1, \ldots, n$, provided for the linguistic variable e_j. A relation f is a mapping of A into B if*

$$\forall e_{i,k} \forall e_{j,l_1} \forall e_{j,l_2} ((e_{i,k}, e_{j,l_1}) \in f \wedge (e_{i,k}, e_{j,l_2}) \in f \rightarrow e_{j,l_1} = e_{j,l_2}) \quad (1)$$

is true.

To include cases in which a mapping in this strict sense cannot be established, we extend the concept of mappings by a fuzzy definition of a measure for the degree of *clearness* to which a mapping $e_j = f(e_i)$ can be established by the available data. This approach is similar to the definition of entropy on which the transinformation concept is based [3]. However, the clearness measure we define in the following does not consider data that do not contribute to establishing the mapping $e_j = f(e_i)$. For the definition of this measure we define an $n \times m$ matrix $\{a_{l,k}(e_i, e_j)\}$, the elements of which are calculated from the N data points $(x_1, y_1), \ldots, (x_N, y_N)$ in the e_i/e_j data space as given in:

$$a_{l,k}(e_i, e_j) = \sum_{d=1}^{N} \left(\mu_{i,k}(x_d) \wedge \mu_{j,l}(y_d)\right) . \quad (2)$$

Here $\mu_{i,k}$ and $\mu_{j,l}$ specify the degrees of membership of the linguistic values $e_{i,k}$ and $e_{j,l}$. The fuzzy-AND operator is chosen as the algebraic product. Fig. 1 illustrates the determination of the elements of the matrix $\{a_{l,k}(e_i, e_j)\}$ from the given data when crisp fuzzy sets are used.

The measure $\kappa(e_i, e_j)$ for the degree of clearness to which a mapping $e_j = f(e_i)$ can be established by the available data is defined as follows:

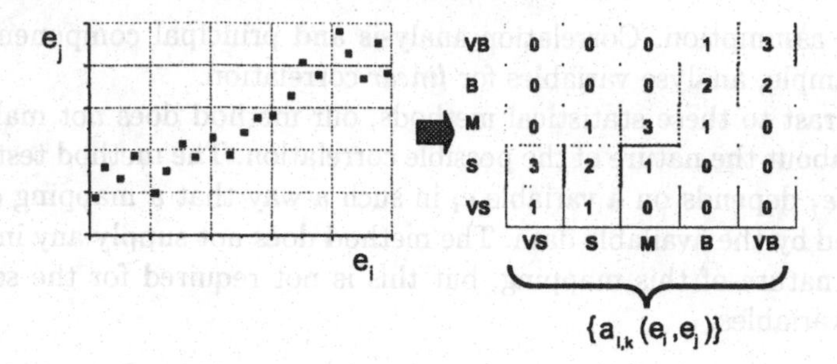

Fig. 1. Determination of the matrix $\{a_{l,k}(e_i, e_j)\}$ from given data in the e_i/e_j data space.

Definition 2. *Clearness measure* $\kappa(e_i, e_j)$

$$\kappa(e_i, e_j) := \frac{\sum_{k=1}^{m} \max_{l} a_{l,k}(e_i, e_j)}{\sum_{k=1}^{m} \sum_{l=1}^{n} a_{l,k}(e_i, e_j)} . \tag{3}$$

The values of $\kappa(e_i, e_j)$ are limited by definition to the interval [0, 1]. To illustrate the use of the clearness measure $\kappa(e_i, e_j)$ four examples of data distributions are considered in Fig. 2.

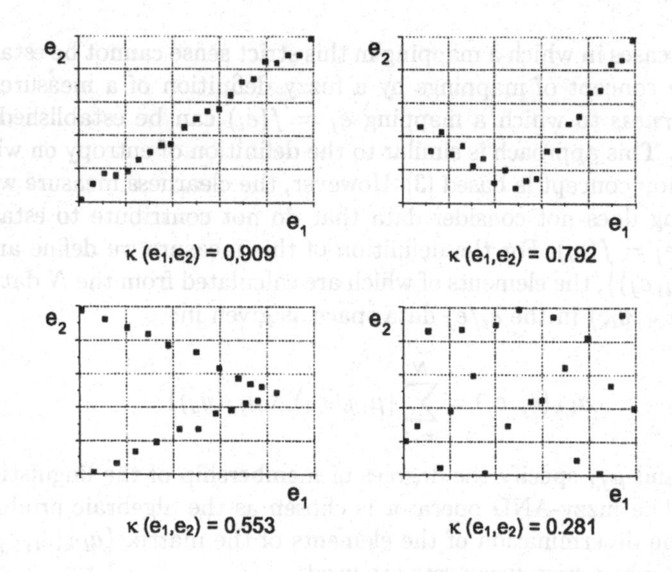

Fig. 2. Evaluation of different data distributions in the e_1/e_2 data space using the clearness measure $\kappa(e_1, e_2)$.

2.1 Non-Relevant Variables

Based on the clearness measure $\kappa(e_i, e_j)$ we define a ***relevance measure*** $\omega_u(e_i)$ that describes the relevance of the input variable e_i for the determination of the output variable u.

Definition 3. *Relevance measure* $\omega_u(e_i)$

$$\omega_u(e_i) := \kappa(e_i, u) \quad (4)$$

With this relevance measure we formulate the following

Reduction Rule 1:

If $\omega_u(e_i) < \text{Threshold}_1$, then omit e_i.

By applying this rule to all input variables, we omit all non-relevant input variables (variables that are not necessary for the determination of the output variable).

2.2 Redundant Variables

The clearness measure $\kappa(e_i, e_j)$ is also used to define a ***redundancy measure*** $\rho_u(e_i, e_j)$. It rates the degree of redundancy of an input variable e_j with respect to input variable e_i for the determination of the output variable u.

Definition 4. *Redundancy measure* $\rho_u(e_i, e_j)$

$$\rho_u(e_i, e_j) := \begin{cases} \kappa(e_i, e_j), & \textit{for } \omega_u(e_j) \le \omega_u(e_i), \\ 0 & \textit{otherwise.} \end{cases} \quad (5)$$

Based on $\rho_u(e_i, e_j)$, we formulate the

Reduction Rule 2:

If $\rho_u(e_i, e_j) > \text{Threshold}_2$, then omit e_j keeping e_i.

By applying this rule to all input variables kept after the application of reduction rule 1, we omit all redundant input variables (variables that are unnecessary given the presence of some other variable for the determination of the output variable). To apply reduction rule 2, we first calculate $\rho_u(e_i, e_j)$ for all $i, j = 1, \ldots, M$, $i \neq j$, where M is the number of potential input variables. Then, in ascending order of $\omega_u(e_j)$ the rule is applied to all variables e_j to check if e_j can be omitted while keeping e_i. While doing so, we consider chained correlations: an input variable e_j is only omitted if no other input variable e_k has been omitted by reduction rule 2 while keeping e_j. Nevertheless, if a further variable $e_l, l \neq j$, can be found, for which $\rho_u(e_l, e_k) > \text{Threshold}_2$ is true, (which means a mapping $e_k = f(e_l)$ is established by the available data to a sufficient clearness), e_j can be omitted.

3 Application 1: Classification in the Car Industry

In the car industry gearboxes must be checked to ensure they meet a specific quality standard before being installed in a new car.

Faulty gearboxes can be identified by their characteristic acoustic patterns. Therefore, human experts perform acoustic tests of gearboxes, to detect material faults, production faults and assembly faults, as well as unusual sounds during operation. The gearboxes are classified into 'o.k.' and 'not o.k.'.

To support and finally automate a human expert's decision processes, a fuzzy classifier has been developed. Due to the gearboxes' complex structure, the high number of moving parts and the influence of different operating states, 149 acoustic characteristics are derived from measured signals.

With 149 input characteristics the design of a fuzzy classifier is a very complex problem. Therefore, a knowledge-based approach is very time consuming and needs a high degree of process understanding. On the other hand, it is possible to design a fuzzy classifier. With about 200 rules and a maximum of 16 combined linguistic statements in the rule premise, it is possible to classify more than 95 per cent of the gearboxes correctly.

In a data-based approach a fuzzy classifier was generated using the fuzzy-ROSA method [1,7–9] based on 1060 data sets (1000 'o.k.' and 60 'not o.k.') using the fuzzy sets shown in Fig. 3 for all input characteristics [10,11]. With 170 rules it is possible to correctly classify 92 % (95 % for 'o.k.' and 70 % for 'not o.k') of gearboxes not among those used for generating the rule base.

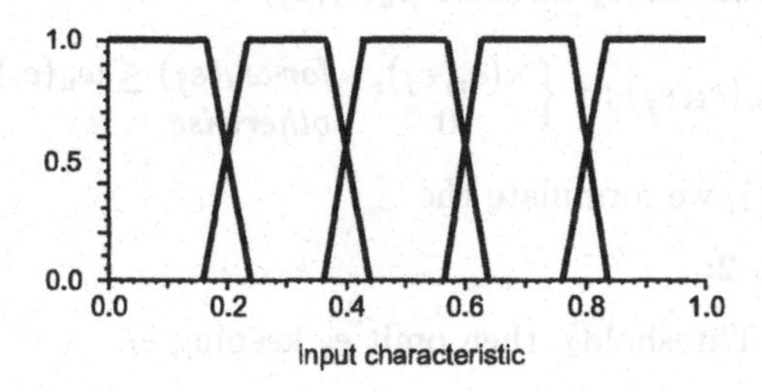

Fig. 3. Fuzzy sets used in the data-based approach for design of the fuzzy classifier.

To reduce the number of 149 potentially relevant input characteristics the method for data-based complexity reduction described above was applied. As a result it was reduced to 93 relevant non-redundant characteristics. Using only rules that combine 10 or less linguistic statements in the premise, the search space for the fuzzy-ROSA method with 149 input characteristics consists of $2.817 \cdot 10^{25}$ possible rules. With 93 characteristics the search space is reduced to $2.447 \cdot 10^{23}$ possible rules, which means a reduction by a factor of about 100.

A second fuzzy classifier was generated using only the reduced number of 93 characteristics as input variables. With 230 rules a correct classification rate of 92 per cent is achieved.

For comparison with the above method for data-based complexity reduction, we have analysed the input characteristics using the statistical correlation coefficient instead of the clearness measure κ defined in (3). With reduction rules corresponding to rules 1 and 2 defined above, 34 characteristics could be omitted. All these characteristics are part of the 56 omitted by the method for data-based complexity reduction. This result demonstrates that our method for data-based complexity reduction, which is based on the concept of mappings, has a comparatively higher potential in reducing the number of input variables.

Applying principal component analysis, the number of input characteristics can be reduced to 35. However, these variables are linear combinations of the original 149 input characteristics, so that all 149 original characteristics are needed for the determination of the 35 new input characteristics. With this method, the number of input variables for the whole system including the classifier and the determination of the new 35 input variables is still 149. Furthermore, the 35 new characteristics cannot be interpreted as process variables. The main advantages of the fuzzy classifier, its transparency and the ability to add knowledge-based rules by hand, are lost.

4 Application 2: Load Prediction in Power Station Management

For efficient power station management, prediction of the future load is indispensable. It is necessary to distinguish between short-term load prediction and long-term load prediction. While long-term load predictions are also used for administrative purposes, short-term load prediction allows anticipation of forthcoming load changes and taking appropiate action.

In [12, 13] a fuzzy system was generated using the fuzzy-ROSA method [1, 7–9] to predict the load in 15 minutes. A major problem was the selection of relevant input variables. To support the selection of input variables, 16 possibly relevant input variables were analysed using our method for complexity reduction. These are:

1. Temperature at 7.00 a.m.
2. Temperature at 3.00 p.m.
3. Long-term prediction of current load
4. Current load
5. First derivative of current load
6. Second derivative of current load
7. Load at the same time on the previous day (previous day load)
8. First derivative of previous day load
9. Second derivative of previous day load
10. Current prediction error (predicted load – current load)
11. First derivative of current prediction error
12. Prediction error at previous day (previous day prediction error)
13. First derivative of previous day prediction error

14. Time of day
15. Season (spring, summer, autumn, winter)
16. Day (workday/weekend)

The output variable chosen was the predicted difference of the load in 15 minutes to the current load. The fuzzy sets were defined as in [12, 13].

With our method for data-based complexity reduction, four relevant non-redundant input variables were extracted from the 16 variables listed above. These are variables 4 (current load), 14 (time of day), 15 (season) and 16 (day). In [12, 13], variables 14, 15 and 16 were selected using expert knowledge and verified by generating different fuzzy systems with different combinations of input variables. These three variables are included in the set of variables found with the method for data-based complexity reduction.

A fuzzy system using the input variables 4 (current load), 14 (time of day), 15 (season) and 16 (day) was generated. The mean-value of the error between the actual load and the predicted load was reduced to 40.5 MW for the year 1995 compared to 40.9 MW achieved with the fuzzy system generated with the input variables selected using expert knowledge.

With our method for data-based complexity reduction the expenditure for the selection of relevant input variables is reduced significantly. Nevertheless, the selected input variables are equivalent to those selected by a process expert except for one additional variable. The consideration of this variable in a data-based generation of a fuzzy load predictor using the fuzzy-ROSA method leads to a slightly improved prediction.

5 Conclusions

We presented a method for data-based complexity reduction, which is based on the concept of mappings and, therefore, distinguishes between the two possible directions of a correlation. It is very successful in reducing the number of input variables. In contrast to well-known statistical methods, our method does not make any assumptions about the nature of a possible correlation. It determines the degree of clearness to which a mapping $e_j = f(e_i)$, with e_i, e_j input variables, is established by the available data. The method does not supply any information about the kind of the mapping, but this is not necessary for the selection of the input variables. Dependent on the degree of clearness, input variables are omitted by applying two reduction rules.

We demonstrated the efficiency of our method in the context of data-based fuzzy modelling. Our method allowed a significant reduction in the size of the rule space that must be searched by methods for data-based rule generation such as the fuzzy-ROSA method.

Acknowledgement

This research was sponsored by the *Deutsche Forschungsgemeinschaft (DFG)*, as part of the Postgraduate Research Programme 'Modelling and Model-Based Design of Complex Technological Systems' and the Collaborative Research Center 'Computational Intelligence' (531) at the University of Dortmund.

References

1. H. Kiendl: Fuzzy Control methodenorientiert, Oldenbourg Verlag, Munich, 1997.
2. J. Hartung, B. Elpelt, K. Klosener: Statistik, 10. Aufl., Oldenbourg Verlag, Munich, 1995.
3. R. Mathar: Informationstheorie, Teubner Verlag, Stuttgart, 1996.
4. J. Praczyk, H. Kiendl, H. Jessen: Ein Verfahren zur datenbasierten Komplexitaetsreduktion, Reihe *Computational Intelligence*, Sonderforschungsbereich 531, ISSN 1433-3325, paper #47, 1998.
5. J. Praczyk, T. Slawinski, H. Kiendl: Auswahl relevanter Prozemerkmale fuer einen Fuzzy-Klassifikator durch ein Verfahren zur datenbasierten Komplexitaetsreduktion, 8. Workshop *Fuzzy Control* des GMA-FA 5.22, Forschungsberichte der Fakultaet fuer Elektrotechnik, Nr. 0298, Universitaet Dortmund, ISSN 0941-4169, S. 182-194, 1998.
6. J. Hartung, B. Elpelt: Multivariate Statistik, 2. Aufl., Oldenbourg Verlag, Munich, 1986.
7. M. Krabs: Das ROSA-Verfahren zur Modellierung dynamischer Systeme durch Regeln mit statistischer Relevanzbewertung, Dissertation, Fortschrittberichte VDI, Reihe 8, Nr. 404, VDI-Verlag, Duesseldorf, 1994.
8. M. Krabs, H. Kiendl: Automatische Generierung von Fuzzy Regeln mit dem ROSA-Verfahren, Tagungsband VDI/VDE-GMA-Aussprachetag Fuzzy Control, VDI-Berichte 1113, VDI-Verlag Duesseldorf, 1994.
9. A. Krone, H. Kiendl: An Evolutionary Concept for Generating Relevant Fuzzy Rules from Data. *Journal of Knowledge-based Intelligent engineering Systems*, Vol 1, No. 4, ISSN 1327-2314, 1997.
10. T. Slawinski, U. Schwane, J. Praczyk, A. Krone, H. Jessen, H. Kiendl, D. Lieske: Application of WINROSA for Controller Adaptation in Robotics and Classification in Quality Control, 2nd Data Analysis Symposium EUFIT '98, 1998.
11. T. Slawinski, J. Praczyk, U. Schwane, A. Krone, H. Kiendl: Data-based Generation of Fuzzy-Rules for Classification, Prediction and Control with the Fuzzy-ROSA method, accepted paper: European Control Conference ECC '99, Karlsruhe, 1999.
12. H. Jessen, T. Slawinski: Mittelwertbasierter Regeltest und Bewertung fuer das Fuzzy-ROSA-Verfahren und Anwendung zur Lastprognose, 8. Workshop *Fuzzy Control* des GMA-FA 5.22, Forschungsberichte der Fakultaet fuer Elektrotechnik, Nr. 0298, Universitaet Dortmund, ISSN 0941-4169, S. 67-81, 1998.
13. H. Jessen: A Mean-Value based Test and Rating Strategy for Automatic Fuzzy Rule Generation and Application to Load Prediction, International Conference on Computational Intelligence for Modelling, Control and Automation (CIMCA'99), Vienna (Austria), 1999.

Traffic Control in an ATM Network Using Rough Set Theory

Jerzy Martyna

Jagiellonian University, Dept. of Computer Science
ul. Nawojki 11, 30-072 Kraków, Poland

Abstract. The paper presents the concept of a rough fuzzy controller for traffic management in ATM networks. It is a very complex task due to the random nature of the traffic arrival parameters and the character of burst duration. We present the rough fuzzy controller, which contains the knowledge base described in linguistic terms. The relationship between decision rules and inference rules are interpreted from the rough set theory perspective. The presented approach may be treated as an alternative to artificial neural networks and fuzzy set theory.

1 Introduction

The Asynchronous Transfer Mode (ATM) standard is currently being designed for integration of various traffic types, ranging from digital voice to image, bulk data and video. To allocate a certain portion of bandwidth, two related control mechanisms are proposed. The first, Connection Acceptance Control (CAC) decides during the call set-up phase to assign the required bandwidth according to the required Quality of Service (QoS) parameters for both, already established connections and a new one. The second one, the Usage Parameter Control (UPC) has the role of ensuring that each source adjusts to its negotiated parameters. This policing mechanism should detect all nonconforming sources that potentially exceed traffic parameters determined for each of them. As a result, the UPC can drop or mark some cells.

Several policing mechanisms have been presented in the literature such as the leaky bucket and window mechanisms. The overview of these methods was given among others in [3], [15]. The latest methods developed are the fuzzy logic approach [7], [9], [14] and the artificial neural network approach [2], [4], [17].

The organization of this paper is as follows: In Section 2 we describe the rough set theory. In Section 3, we discuss how to use rough fuzzy policing mechanism for ATM traffic source management. Performance evaluation of the rough fuzzy controller mechanism will be presented in Section 4. Finally, the conclusion and future works are provided in Section 5.

2 Rough sets preliminaries

In this Section we recall some basic notions and definitions of rough set theory. Detailed considerations on rough set can be found in [11], [12], [8]. Some considerations concerning the relationship between decision rules and inference rules from rough set theory perspective are presented in the paper [13].

Let U be a non-empty, finite set called the universe and A be a non-empty, finite set of attributes, i.e.

$$a : U \rightarrow V_a \quad \text{for} \quad a \in A \tag{1}$$

where V_a is called the value of a. Elements of U are called objects.

With every subset of attributes $B \subseteq A$ there is an equivalence relation, denoted by $IND_A(B)$ (or $IND(B)$) called the $B-$*indiscernibility relation*. This relation is defined as follows:

$$IND(B) = \{(x, y) \in U \times U : \forall a \in B, a(x) = a(y)\} \tag{2}$$

Objects x, y satisfying relation $IND(B)$ are indiscernible by attributes from B. The sets:

$$\underline{B}(X) = \{x \in U : \quad [x]_B \subseteq X\} \tag{3}$$

$$\overline{B}(X) = \{x \in U : \quad [x]_B \cap X \neq \emptyset\} \tag{4}$$

are called $\underline{B}$-*lower* and $\overline{B}$-*upper approximation* of $X \subseteq U \in A$, respectively.

The B-boundary of X is the set

$$BN_B(X) := \underline{B}(X) - \overline{B}(X) \tag{5}$$

A set $X \subseteq U$ is definable by B if it is the union of some classes of the indiscernibility relation $IND(B)$, otherwise it is roughly definable by B.

The membership degree of an original value of the discretized attribute in a linguistic term we can calculate as follows.

We suppose that $x \in U$ is an object, Q is a set of attributes in the system; $q_i \in Q$ is an attribute. Let $q_i(x_i)$ denote an original value of quantitative attribute q_i for object x. $\mathcal{V}_i$ denotes an original domain of quantitative attribute q_i, i.e. a real interval; $\widetilde{v_{ih}}$ is a h-th fuzzy subinterval defined on $\mathcal{V}_i$, corresponding to liguistic term v_{ih}, $h = 1, \ldots, m_i$, where m_i is a number of different linguistic terms for attribute q_i. The membership function of subinterval $\widetilde{v_{ih}}$ is denoted by $\mu_{\widetilde{v_{ih}}}(q_i(x))$. It is obvious that $\forall q_i(x) \in \mathcal{V}_i \exists \widetilde{v_{ih}} \in \mathcal{F}(\widetilde{v_{ih}})$: $\mu_{\widetilde{v_{ih}}}(q_i(x)) > 0$ which means that $\mathcal{F}(\widetilde{v_{ih}})$ is a family of all fuzzy subintervals, $\widetilde{v_{ih}}$ is defined for attribute q_i and covering the whole domain $\mathcal{V}_i$.

The original value of attribute q_i may be either imprecise or crisp. We suppose that an imprecise descriptor can be modelled by a fuzzy number, e.g. $\widetilde{q_i(x)}$, which is characterized by membership function $\mu_{\widetilde{q_i}}(q_i(x))$.

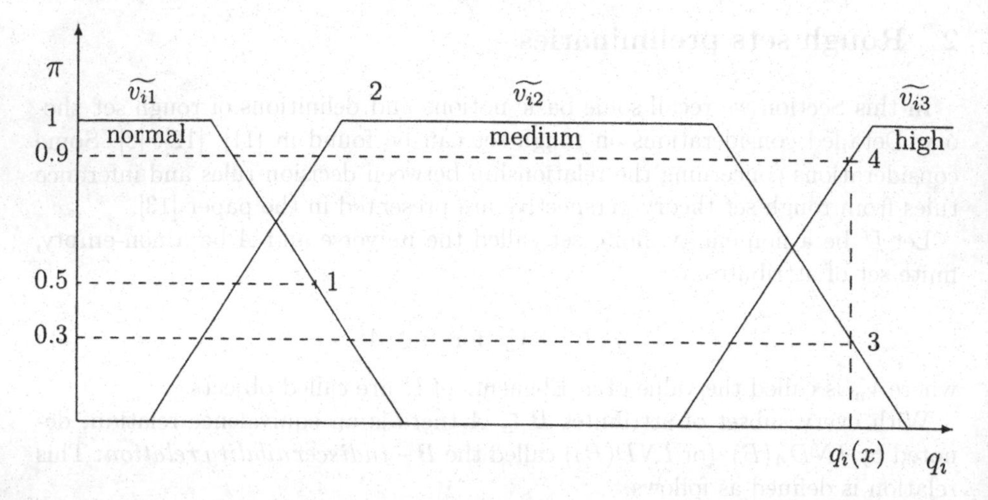

Fig. 1. Graphical illustration of computation of $\pi(\widetilde{q_i}(x), \widetilde{v_{ih}})$ and $\pi(q_i(x), \widetilde{v_{ih}})$

The membership of $\widetilde{q_i(x)}$ or $q_i(x)$ to subinterval $\widetilde{v_{ih}}$ can be evaluated by using the consistency measure for two fuzzy sets introduced by Zadeh [20]. For fuzzy number $\widetilde{q_i(x)}$ is defined as follows:

$$\pi(\widetilde{q_i(x)}, \widetilde{v_{ih}}) = \sup_{q_i(x) \in \mathcal{V}_i} \{\min[\mu_{\widetilde{q_i(x)}}(q_i(x)), \mu_{\widetilde{ih}}(q_i(x))]\} \tag{6}$$

For crisp number $q_i(x)$ is reduced to:

$$\pi(q_i(x), \widetilde{v_{ih}}) = \sup_{q_i(x) \in \mathcal{V}_i} \{\min[1, \mu_{\widetilde{v_{ih}}}(q_i(x))]\} = \mu_{\widetilde{v_{ih}}}(q_i(x)) \tag{7}$$

The calculation of mentioned above measures for two exemplary values $\widetilde{q_i(x)}$ and q_i is shown in Fig. 1. It can be seen that in case of fuzzy number $\widetilde{q_i(x)}$ degrees of membership to fuzzy subintervals $\widetilde{v_{i1}}$ and $\widetilde{v_{i2}}$ are equal to 0.5 and 1.0 (points no. 1, 2), respectively. In the case of precise value $q_i(x)$ degrees of membership to $\widetilde{v_{i2}}$ and $\widetilde{v_{i3}}$ are equal to 0.3 and 0.9 (points no. 3, 4), respectively.

Let V_i denote a finite set of linguistic terms for attribute q_i. We can transform attribute q_i into liguistic variables with fuzzy subintervals $\widetilde{v_{ih}}$. Each subinterval corresponds to a linguistic term denoted by v_{ih}. In our approach V_i is a "linguistic" domain of attribute q_i; $v_{ih} \in V_i$, $h = 1, 2, \ldots, m_i$; where m_i denotes cardinality of V_i.

Refering to [16] we introduce a multiple descriptor for object $x \in U$ and attribute q_i ($i = 1, 2, \ldots, n$ where n is a number of considered attributes.). A multiple descriptor is as follows:

$$MultDes(x, q_i) = \{(v_{ih}, \pi_{ih}(v) : \ v_{ih} \in V_i \text{ and } \pi_{ih}(x) > 0\} \tag{8}$$

where

$$\pi_{ih}(x) = \begin{cases} \pi(\widetilde{q_i(x)}, \widetilde{v_{ih}}) \text{ if descriptor } \widetilde{q_i(x)} \text{ is imprecise and } \widetilde{v_{ih}} \text{ is a fuzzy} \\ \qquad\qquad \text{subinterval corresponding to linguistic term } v_{ih} \\ \pi(q_i(x), \widetilde{v_{ih}}) \text{ if descriptor } q_i(x) \text{ is crisp and } \widetilde{v_{ih}} \text{ as above} \\ \pi(q_i(x), v_{ih}) \text{ if descriptor } q_i(x) \text{ and } v_{ih} \text{ are precise} \end{cases}$$

It can be seen that in the first two cases $\pi_{ih} \in [0,1]$ while in the first one, $\pi_{ih} = 1$ if $q_{ih}(x) = v_{ih}$ and $\pi_{ih} = 0$.

We can define an information system with multiple descriptors as 4-tuple $\widetilde{S}_{MultDes} = < U, Q, V, \widetilde{\varrho}_{MultDes} >$ where $x \in U$: $\widetilde{\varrho}_{MultDes}(x, q_i) = MultDes(x, q_i)$.

We can possibly obtain the generalized information system. By generalized information system we understand the 5-tuple $\widetilde{S} = < U', \eta, Q, V, \widetilde{\varrho} >$ where U' is a finite set of sub-objects ($x' \in U'$) coming from subdivision of object x described by attributes from set $P \subseteq Q$; η is a finite set of degrees of possibility for objects U', i.e. the set of $\eta_Q(x')$; Q is a finite set of attributes; $V = \bigcup_{i:q_i \in Q} V_i$, $\widetilde{\varrho}$ is a generalized information function such that for each object $x' \in U'$, $q_i \in Q$: $\widetilde{\varrho}(x', q_i) = (v_{ih}, \pi_{ih}(x'))$.

Each sub-object x' is associated with a degree of possibility of its realization $\pi_P(x')$, as a result of aggregation of $\pi_{ih}(x')$ for attributes $q_i \in P$. We suggest to use the Yager's parametrized t-norm operator [19] which was first given in the paper [18]. It is defined as

$$\pi_P(x') = 1 - \min\{1, [\sum_{i:q_i \in P} (1 - \pi_{ih}(x'))^w]^{\frac{1}{w}}\} \tag{9}$$

In dependence on value value w, the Yager's t-norm covers a wide range of possible conjunctive operators. For instance, if $w \to \infty$ then

$$\pi_P(x') = \min_{i:q_i \in P} \{\pi_{ih}(x')\} \tag{10}$$

To express a relative degree of possibility for sub-object x', denoted here as $\eta_P(x')$, we will take into consideration all sub-objects associated with the original x, namely

$$\eta_P(x') = \frac{\pi_P(x')}{\sum_{g=1}^{p} \pi_P(x^g)} \tag{11}$$

where p denotes the number of all possible n-elementary combinations of descriptors for the given object x, i.e. $j = 1, 2, \ldots, p$. In particular, an original object described by single descriptor corresponds to one sub-object with degree of possibility equal to 1.

Thus, the coefficient $\eta_P(x')$ can be understood as a "part" of original object x characterized by a given combination of single descriptors.

If condition and decision attributes are distinguished in the information system, it is interpreted as a decision table. Thus $Q = C \cup D$ where C is a set of condition attributes and D is a set of decision attributes. We can define a generalized decision table $\widetilde{DT}$ as 5-tuple $\widetilde{DT} = < U'', \eta, C \cup D, V, \widetilde{\varrho} >$ where U'' is a set of sub-objects x'', η is a finite set of degrees of possibilities for objects U'', i.e. it is the set of values $\eta_{C,D}(x'')$. V, $\widetilde{\varrho}$ are defined as for S assuming that $Q = C \cup D$.

3 Rough fuzzy policing mechanism for ATM traffic source management

Let us consider a simple example with multi-attribute decisions. We analyse rough sets controller which is a window-based control mechanism. Our control mechanism can accept the maximal number of cell N_i in the i-th window of size T. It was assumed that source posseses the negotiated average cell arrival rate equal to λ_n. It can transmit N cells per window with $N = T * \lambda_n$ on average.

The designed policing mechanism is based on a credit granted to a source which in the past has respected the parameters negotiated by incresing its control threshold, N_i, as long as it perseveres with non-violating behaviour. Otherwise, if the behaviour of the source is violated, the policing mechanism reduces its credit by decreasing the threshold value.

We assume that each source is characterized by the parameters: the average number of cell arrivals per window since the connection starting point, A_{0i}, and the average number of cell arrivals in the last window, A_i. The first indicates the long-term trend of the source and the second gives its current behaviour. The third parameter represents the current degree of control (degree of permissiveness) of the mechanism over the source. It is given by the value of N_i in the last window. The fourth parameter, the value of b, is the burst factor which is the quotient of the peak rate of a source R_p [cells/sec] to the average rate R_a [cell/sec]. All these parameters are the four linguistic variables which make up the rough set controller input. Burstiness is not a very accurate traffic descriptor, since two calls with similar peak and average rates can have very different traffic characteristics. Therefore, we propose here the usage of yet an other parameter, i.e. the burst length (or burst duration). It means that when burst length increases, ATM cell losses and delays increase significantly [10]. In our approach this parameter is the decision attribute, d. The data streams are divided into two classes according to attribute d taking values equal to 1 or 2. Value of d equal to 1 means that for the given the data stream the defined burst length is not exceed. On the other hand, value of d equal to 2 means that for the given data stream the determined burst length was exceeded. In our analysis we have $C = \{q_1, q_2, q_3\}$ and $D = \{d\}$.

The membership functions chosen for linquistic terms of attributes: q_1 - average number of cell arrivals per window at the beginning of connection and in the last window, q_2 - degree of permissiveness, q_3 - burst factor are given in Fig. 2, 3, 4, respectively.

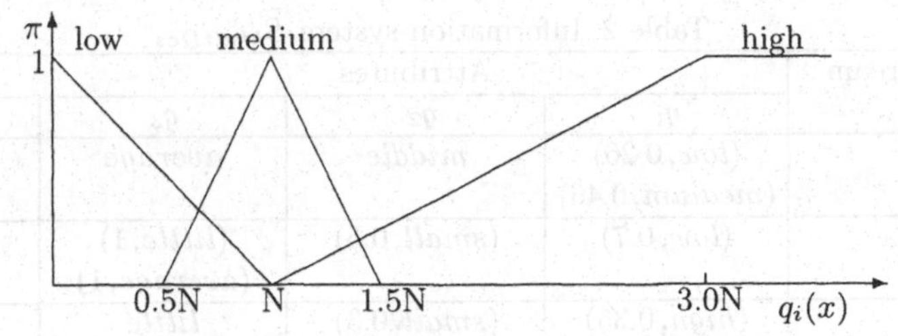

Fig. 2. Membership functions for linguistic terms of attribute A_{0i} and A_i - input variables

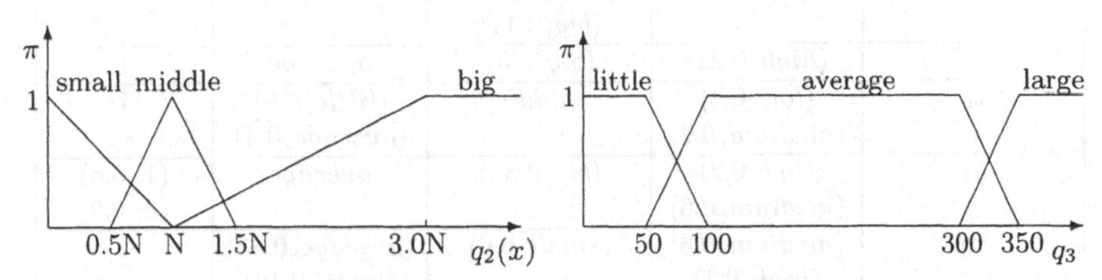

Fig. 3. Membership functions for linguistic terms of attribute N_i (degree of permissiveness) - input variable

Fig. 4. Membership functions for linguistic terms of attribute b_i (burst factor) - input variable

In our analysis we consider an ATM transmission channel through which the ATM cells were transmitted. For instance, the stream of ATM cells consists of 8 data streams (denoted here as objects x_1, x_2, $\cdots$, x_8) which are controlled by our rough set controller. This traffic can be described by three attributes q_i, $i = 1, 2, 3$. The original information system is presented in Table 1. The value of attribute d for data stream cannot be uniquely defined. For instance, we suppose that the data stream x_7 is equal to 1 with degree 0.6 and to 2 with degree 0.2.

After translation of original values of quantitative attributes to linguistic terms, we get generalized information system with multiple descriptors $\widetilde{S}_{MultDes}$, which is presented in Table 2. For object x_2 and attribute q_3 the controller defined the subset of possible values as multiple desciptor {(little, 1), (average, 1)}.

Table 1. The original information system

Data stream	Attributes			
x	q_1	q_2	q_3	d
x_1	0.75	1.0	200	1
x_2	0.3	0.5	*unknown*	1
x_3	1.7	0.7	45	1
x_4	1.0	1.2	370	2
x_5	1.5	2.2	220	1
x_6	0.6	1.0	70	1
x_7	0.8	1.7	280	1 or 2
x_8	1.2	0.4	310	2
x_9	1.8	0.9	80	1
x_{10}	0.3	1.1	250	2

Table 2. Information system $\widetilde{S}_{MultDes}$

Data stream	Attributes			
x	q_1	q_2	q_3	d
x_1	$(low, 0.26)$ $(medium, 0.46)$	$middle$	$average$	1
x_2	$(low, 0.7)$	$(small, 0.5)$	$(little, 1)$ $(average, 1)$	1
x_3	$(high, 0.35)$	$(small, 0.3)$ $(middle, 0.4)$	$little$	1
x_4	$medium$	$(middle, 0.6)$ $(big, 0.11)$	$large$	2
x_5	$(high, 0.22)$	$(big, 0.6)$	$average$	1
x_6	$(low, 0.4)$ $(medium, 0.2)$	$middle$	$(little, 0.6)$ $(average, 0.4)$	1
x_7	$(low, 0.2)$ $(medium, 0.6)$	$(big, 0.35)$	$average$	(1, 0.6) (2, 0.2)
x_8	$(medium, 0.6)$ $(high, 0.1)$	$(small, 0.6)$	$(average, 0.83)$ $(large, 0.16)$	2

We build $\widetilde{S}_{MultDes}$ in the following way. The original value of attribute $q_1(x_4)$ is equal to N. For domain $\mathcal{V}_1$ we have three fuzzy intervals $\widetilde{v_{11}}$, $\widetilde{v_{12}}$, $\widetilde{v_{13}}$ (corresponding to low, medium, high) - see Fig. 2. By using the operations in Section 2 for $q_1(x_4)$ we get: $\pi(q_1(x_4), \widetilde{v_{11}}) = 1$, $\pi(q_1(x_4), \widetilde{v_{12}}) = 0$, $\pi(q_1(x_4, \widetilde{v_{13}}) = 0$. So, the value $q_1(x_4)$ is uniquely translated to the "low" term.

Table 3. Generalized information system $\widetilde{S}$

Sub-object	Condition attributes			Decision	Degree
x'	q_1	q_2	q_3	attribute d	$\eta_C(x')$
x_1^1	$(low, 0.26)$	$middle$	$average$	1	0.36
x_1^2	$(medium, 0.46)$	$middle$	$average$	1	0.36
x_2^1	$(low, 0.7)$	$(small, 0.5)$	$little$	1	0.42
x_2^2	$(low, 0.7)$	$(small, 0.5)$	$average$	1	0.42
x_3^1	$(high, 0.35)$	$(small, 0.3)$	$little$	1	0.46
x_3^2	$(high, 0.35)$	$(middle, 0.4)$	$little$	1	0.54
x_4^1	$medium$	$(middle, 0.6)$	$large$	2	0.78
x_4^2	$medium$	$(big, 0.11)$	$large$	2	0.14
x_5	$(high, 0.22)$	$(big, 0.6)$	$average$	1	0.5
x_6^1	$(low, 0.4)$	$middle$	$(little, 0.6)$	1	0.33
x_6^2	$(low, 0.4)$	$middle$	$(average, 0.4)$	1	0.33
x_6^3	$(medium, 0.2)$	$middle$	$(little, 0.6)$	1	0.17
x_6^4	$(medium, 0.2)$	$middle$	$(average, 0.4)$	1	0.17
x_7^1	$(low, 0.2)$	$(big, 0.35)$	$average$	(1, 0.6) (2, 0.2)	0.6
x_7^2	$(medium, 0.6)$	$(big, 0.35)$	$average$	(1, 0.6) (2, 0.2)	0.2
x_8^1	$(medium, 0.6)$	$(small, 0.6)$	$(average, 0.83)$	2	0.29
x_8^2	$(medium, 0.6)$	$(small, 0.6)$	$(large, 0.16)$	2	0.12
x_8^3	$(high, 0.1)$	$(average, 0.84)$	$(average, 0.83)$	2	0.065
x_8^4	$(high, 0.1)$	$(small, 0.6)$	$(large, 0.16)$	2	0.12

Table 4. Generalized decision table $\widetilde{DT}$

Sub-object x''	Condition attributes q_1	q_2	q_3	Decision attribute d	Degree $\eta_{C,d}(x'')$
x_1^1	$(low, 0.26)$	$middle$	$average$	1	0.36
x_1^2	$(medium, 0.46)$	$middle$	$average$	1	0.36
x_2^1	$(low, 0.7)$	$(small, 0.5)$	$little$	1	0.42
x_2^2	$(low, 0.7)$	$(small, 0.5)$	$average$	1	0.42
x_3^1	$(high, 0.35)$	$(small, 0.3)$	$little$	1	0.46
x_3^2	$(high, 0.35)$	$(middle, 0.4)$	$little$	1	0.54
x_4^1	$medium$	$(middle, 0.6)$	$large$	2	0.78
x_4^2	$medium$	$(big, 0.11)$	$large$	2	0.14
x_5	$(high, 0.22)$	$(big, 0.6)$	$average$	1	0.5
x_6^1	$(low, 0.4)$	$middle$	$(little, 0.6)$	1	0.33
x_6^2	$(low, 0.4)$	$middle$	$(average, 0.4)$	1	0.33
x_6^3	$(medium, 0.2)$	$middle$	$(little, 0.6)$	1	0.17
x_6^4	$(medium, 0.2)$	$middle$	$(average, 0.4)$	1	0.17
x_7^{11}	$(low, 0.2)$	$(big, 0.35)$	$average$	1	0.45
x_7^{12}	$(low, 0.2)$	$(big, 0.35)$	$average$	2	0.15
x_7^{21}	$(medium, 0.6)$	$(big, 0.35)$	$average$	1	0.15
x_7^{22}	$(medium, 0.6)$	$(big, 0.35)$	$average$	2	0.05
x_8^1	$(medium, 0.6)$	$(small, 0.6)$	$(average, 0.83)$	2	0.29
x_8^2	$(medium, 0.6)$	$(small, 0.6)$	$(large, 0.16)$	2	0.12
x_8^3	$(high, 0.1)$	$(average, 0.84)$	$(average, 0.83)$	2	0.065
x_8^4	$(high, 0.1)$	$(small, 0.6)$	$(large, 0.16)$	2	0.12

If we consider object x_8 and attribute q_3 then $q_3(x_8) = 310$. For value $\mathcal{V}_3$ we have also three fuzzy intervals $\widetilde{v_{31}}$, $\widetilde{v_{32}}$, $\widetilde{v_{33}}$ (see Fig. 3). In this case we have: $\pi(q_3, \widetilde{v_{31}}) = 0$, $\pi(q_3, \widetilde{v_{32}}) = 0.83$, $\pi(q_3(x_8, \widetilde{v_{33}}) = 0.16$. So, the original value of $q_3(x_8)$is possibly "average" with degree 0.83 or "large" with degree 0.16. In the information system $\widetilde{S}_{MultDes}$ presented in the Table 2, we skip all degrees equal to 1 for unique descriptors for the sake of simplicity, see Table 2.

We can transform information system $\widetilde{S}_{MultDes}$ to the generalized information $\widetilde{S}$. The obtained values are presented in Table 3. For this reason we calculate the degree $\eta_C(x')$ for set C of condition attributes. To illustrate the way of calculating this parameter we consider object x_4. It will be represented by two sub-objects: x_4^1, x_4^2. If the t-norm aggregation will be implemented as *min* operation. Then, we shall have

$$\pi_C(x_4^1) = \min\ (1, 0.6, 1\ ) = 0.6$$

$$\pi_C(x_4^2) = \min\ (1, 0.11, 1) = 0.11$$

$$\eta_C(x_4^1) = \frac{0.6}{0.6+0.11} = 0.78$$

$$\eta_C(x_4^2) = \frac{0.11}{0.6+0.11} = 0.14$$

Now, we can construct the generalized decision table $\widetilde{DT}$. To illustrate this calculation we consider object x_7. It has two sub-objects x_7^1 and x_7^2. Each of them is associated with two possible decision descriptors. We can obtain sub-objects x_7^{11}, x_7^{12}, x_7^{21}, x_7^{22}. The degree $\eta_{C,d}(x')$ for these sub-objects is determined in the following way:

$$\eta_{C,d}(x_7^{11}) = \frac{0.6}{0.6+0.2} * 0.6 = 0.45$$

Table 5. Atoms obtained from generalized decision table $\widetilde{DT}$

Atom	Condition attributes			Decision	$card(X'_C)$	Sub-objects
X'_C	q_1	q_2	q_3	attribute d		in X'_C
X_C^1	$(low, 0.26)$	$middle$	$average$	1	0.69	x_1^1, x_6^2
X_C^2	$(medium, 0.46)$	$middle$	$average$	1	0.64	x_1^2
X_C^3	$(high, 0.35)$	$(small, 0.3)$	$little$	1	0.46	x_3^1
X_C^4	$(high, 0.35)$	$(middle, 0.4)$	$little$	1	0.54	x_3^2
X_C^5	$(low, 0.7)$	$(small, 0.5)$	$little$	1	0.97	x_2^1, $x6^1$
X_C^6	$(low, 0.7)$	$(small, 0.5)$	$average$	1	0.42	x_2^2
X_C^7	$medium$	$(middle, 0.6)$	$large$	2	0.78	x_4^1
X_C^8	$medium$	$(big, 0.11)$	$large$	2	0.14	x_4^2
X_C^9	$(high, 0.22)$	$(big, 0.6)$	$average$	1	0.5	x_5^1
X_C^{10}	$(medium, 0.2)$	$middle$	$(little, 0.6)$	1	0.17	x_6^3
X_C^{11}	$(medium, 0.2)$	$middle$	$(average, 0.4)$	1	0.17	x_6^4
X_C^{12}	$(low, 0.2)$	$(big, 0.35)$	$average$	1 *or* 2	0.6	x_7^{11}, x_7^{12}
X_C^{13}	$(medium, 0.6)$	$(big, 0.35)$	$average$	1 *or* 2	0.2	x_7^{21}, x_7^{22}
X_C^{14}	$(medium, 0.6)$	$(small, 0.6)$	$(average, 0.83)$	2	0.29	x_8^1
X_C^{15}	$(medium, 0.6)$	$(small, 0.6)$	$(large, 0.6)$	2	0.12	x_8^2
X_C^{16}	$(high, 0.1)$	$(small, 0.6)$	$(average, 0.83)$	2	0.13	x_8^3
X_C^{17}	$(high, 0.1)$	$(small, 0.6)$	$(large, 0.16)$	2	0.03	x_8^4

$$\eta_{C,d}(x_7^{12}) = \frac{0.2}{0.6+0.2} * 0.6 = 0.15$$

$$\eta_{C,d}(x_7^{21}) = \frac{0.6}{0.6+0.2} * 0.2 = 0.15$$

$$\eta_{C,d}(x_7^{22}) = \frac{0.2}{0.6+0.2} * 0.2 = 0.05$$

The obtained generalized decision table $\widetilde{DT}$ is given in Table 4.

Obviously, some sub-objects are possibly indiscernible. They create C-elementary sets (atoms) defined by the generalized indiscernibility relation. For instance, sub-objects x_1^1 and x_6^1 create atom X_C^1. For attribute values from the description of X_C^1 degrees of possibilities are as follows, namely

$$\pi_{11}(X_C^1) = \min(0.36, 1) = 0.36,$$

$$\pi_{21}(X_C^1) = \min(1, 1) = 1,$$

$$\pi_{31}(X_C^1) = \min(1, 1) = 1.$$

The cardinality card$(X_C^1) = \sum_{X_1^1, X_3^1} \eta_{C,d}(x'') = 0.36 + 0.33 = 0.69$. The atoms obtained from the generalized decision table $\widetilde{DT}$ are given in Table 5.
In respect of d we can classify Y= $\{Y_1, Y_2\}$ where $\{Y_1\}$ is the set of sub-objects with $d = 1$ (i.e. x_1^1, x_6^2, x_1^2, x_3^1, x_3^2, x_2^1, x_6^1, x_2^2, x_5^1, x_6^3, x_6^4, x_7^{11}, x_7^{21}) and Y_2 with $d = 2$ (i.e. x_4^1, x_4^2, x_7^{12}, x_7^{22}, x_8^1, x_8^2, x_8^3, x_8^4). We obtain $\underline{\widetilde{C}}(Y_1) = \{x_1^1, x_6^2\} \cup \{x_1^2\} \cup \{x_2^1, x_6^1\} \cup \{x_3^1\} \cup \{x_3^2\} \cup \{x_2^2\} \cup \{x_5^1\} \cup \{x_6^3\} \cup \{x_6^4\}$, $\overline{\widetilde{C}} = \{Y_1\} = \{x_1^1, x_6^2\} \cup \{x_1^2\} \cup \{x_2^1, x_6^1\} \cup \{x_2^2\} \cup \{x_5^1\} \cup \{x_3^1\} \cup \{x_3^2\} \cup \{x_6^3\} \cup \{x_6^4\} \cup \{x_7^{11}, x_7^{12}\} \cup \{x_7^{21}, x_7^{22}\}$, $BN_{\widetilde{C}}(Y_1) = BN_{\widetilde{C}}(Y_2) = \{x_7^{11}, x_7^{12}\} \cup \{x_7^{21}, x_7^{22}\}$, $\underline{\widetilde{C}}(Y_2) = \{x_4^1\} \cup \{x_4^2\} \cup \{x_8^1\} \cup \{x_8^2\} \cup \{x_8^3\} \cup \{x_8^4\}$, $\overline{\widetilde{C}}(Y_2) = \{x_4^1\} \cup \{x_4^2\} \cup \{x_8^1\} \cup \{x_8^2\} \cup \{x_8^3\} \cup \{x_8^4\} \cup \{x_7^{11}, x_7^{12}\} \cup \{x_7^{21}, x_7^{22}\}$.
$card(\underline{\widetilde{C}}(Y_1)) = 0.69 + 0.64 + 0.46 + 0.54 + 0.97 + 0.42 + 0.5 + 0.17 + 0.17 = 4.56$
$card(\overline{\widetilde{C}}(Y_1)) = 0.69 + 0.64 + 0.46 + 0.54 + 0.97 + 0.42 + 0.5 + 0.17 + 0.17 + 0.6 + 0.2 = 5.36$
$card(\underline{\widetilde{C}}(Y_2)) = 0.78 + 0.14 + 0.29 + 0.12 + 0.13 + 0.03 = 1.49$
$card(\overline{\widetilde{C}}(Y_2)) = 0.78 + 0.14 + 0.29 + 0.12 + 0.13 + 0.03 + 0.6 + 0.2 = 2.29$

The generalized accuracy of approximation regarding both classes of decision attributes is equal to: $\alpha_{\widetilde{C}}(Y_1) = \frac{4.56}{5.36} = 0.85$, $\alpha_{\widetilde{C}}(Y_2) = \frac{1.49}{2.29} = 0.65$. The generalized quality of classification is equal to $\gamma_{\widetilde{C}}(Y) = \frac{4.56+1.49}{5.36+2.29} = \frac{6.05}{7.65} = 0.79$.

In our classification process we take into consideration the following parameters: strength factor, specifity factor and matching factor [6]. We recall also that the strength factor of a rule is the ratio of all examples correctly classified by the rule to the total number of examples matched by the rule. The bigger strength is, the better. The specifity is a measure of completeness of a rule. It is the number of conditions (attribute-value) of a rule. It means that a rule with bigger number of attribute-value pairs is more specific. Specifity is here used to classify cases. The matching factor is a measure of matching of a case and a rule. It is defined as the ratio of the number of matched attribute-value pairs of a rule with a case to the total number of attribute-value pairs of the rule. The probabilistic sum of all is used to computation of matching factor.

For classification of all rules we used after [6] the following support

$$\mathcal{S} = \sum_{\text{rules } R \text{ describing } C} \text{matching\_factor}(R) * \text{strength}(R) * \text{specificity}(R)$$

Using the classification method described above and a modified version of Grzymala's LEM procedure [5], we obtained some decision rules. The obtained rules are presented below:

r_1: if $(q_1 = low) \cap (q_2 = middle) \cap (q_3 = average)$ then $(d = 1)$ $\mathcal{S}(r_1) = 1.32$
r_2: if $(q_1 = medium) \cap (q_2 = middle) \cap (q_3 = average)$ then $(d = 1)$ $\mathcal{S}(r_2) = 2.4$
r_3: if $(q_1 = high) \cap (q_2 = small) \cap (q_3 = little)$ then $(d = 1)$ $\mathcal{S}(r_3) = 0.9$

r_4: if $(q_1 = high) \cap (q_2 = middle) \cap (q_3 = little)$ then $(d = 1)$ $\mathcal{S}(r_4) = 0.9$
r_5: if $(q_1 = low) \cap (q_2 = small) \cap (q_3 = average)$ then $(d = 1)$ $\mathcal{S}(r_5) = 1.7$
r_6: if $(q_1 = low) \cap (q_2 = small) \cap (q_3 = average)$ then $(d = 1)$ $\mathcal{S}(r_6) = 0.8$
r_7: if $(q_1 = medium) \cap (q_2 = middle) \cap (q_3 = large)$ then $(d = 2)$ $\mathcal{S}(r_7) = 1.0$
r_8: if $(q_1 = high) \cap (q_2 = big) \cap (q_3 = average)$ then $(d = 1)$ $\mathcal{S}(r_8) = 0.7$
r_9: if $(q_1 = low)$ $(q_2 = big) \cap (q_3 = average)$ then $(d = 1)$ $\mathcal{S}(r_9) = 0.8$
r_{10}: if $(q_1 = low) \cap (q_2 = big) \cap (q_3 = average)$ then $(d = 2)$ $\mathcal{S}(r_{10}) = 0.6$

Only 10 rules appear in the knowledge base our policing mechanism. The remaining 7 are not included as they would never be activated.

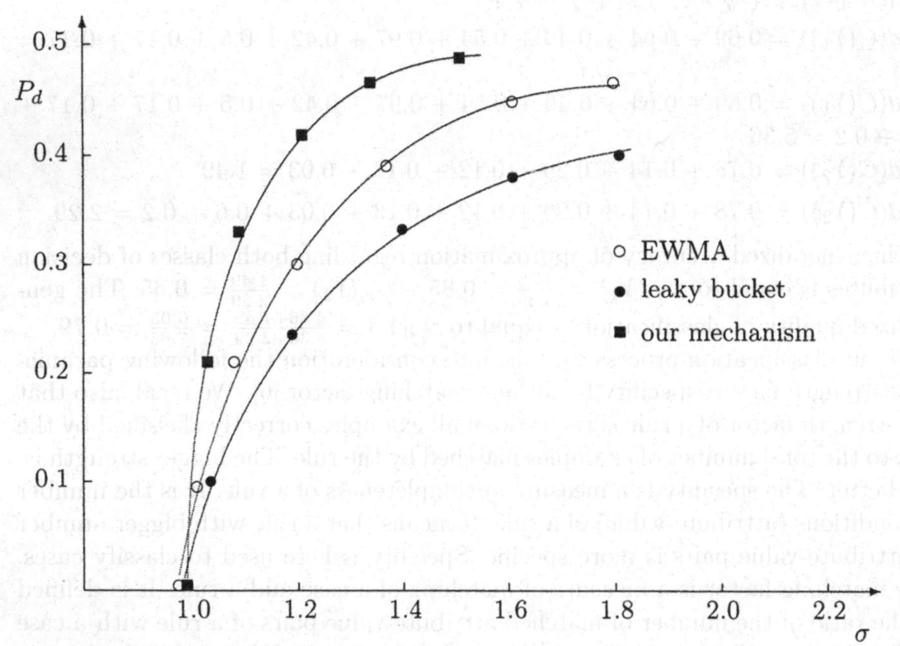

Fig. 5. Probability that the policing mechanism detects a cell as excessive versus the long-term actual mean cell rate of the source normalized to the negotiated mean cell rate

4 Performance evaluation of the rough fuzzy policing mechanism

In this section we evaluate the efficiency of proposed rough set policing mechanism. We have compared the performance of our mechanism with the Leaky Bucket (LB) flow-control and Exponentially Weighted Moving Window (EWMA) [10].

To compare all three mechanisms we used the same characteristics of ATM traffic studied by Rathgeb [15]. We assumed that the number of cells per burst has a geometric distribution with a mean of E[x] = 5 cells. The duration of the idle phase has an exponential distribution with a mean of E[s] = 0.148 sec. The intercell time during a burst is $t_c = 0.016$ sec.

We examined the probability that policing mechanism detects a cell as excessive, namely P_d. In the ideal behaviour the value of P_d is equal to zero. In the case of violating sources there is a increase of the probability P_d (see Fig. 5) where $P_d = (\sigma - 1)/\sigma$, σ is the long-term actual mean cell rate of the source normalized to the negotiated mean cell rate.

In our simulation analysis we evaluated the average number of cells emitted by a violating source before the policing mechanism takes control actions. We observed that our policing mechanism controls action after about 500 cells and after about 1000 cells it reaches an improvement over the other policing methods.

5 Conclusions

This paper outlines a conception of rough set rate-based mechanism for traffic control in an ATM network. Although we have discussed in detail only some attributes, the idea is extendible to more parameters of flow rate.

Moreover, we demonstrated that the proposed mechanism leads to results with the satisfied quality coefficient. The differences between our model and the leaky bucket flow control mechanism or EWMA scheme are generally negligible, even in the case of very irregular traffic.

Finally, a further research must lead to the design and implementation of an effective hardware implementation of rough set fuzzy controller for traffic control in an ATM network. Its architecture organization allows to parallel execution of rules or input variables.

References

1. Butto, M., Cavallero, E., Tonietti, A.: Effectiveness of the "Leaky Bucket" Policing Mechanism in ATM Networks, IEEE Journal on Selected Areas in Communications, Vol. 9, No. 3 (1991) 335 - 342
2. Cheng, R.-G., Chang, C.-J.: Neural-Network Connection-Admission Control for ATM Networks, IEE Proc.-Commun., Vol. 144, No. 2 (1997) 93 - 98
3. Dittmann, L., Jacobsen, S.B., Moth, K.: Flow Enforcement Algorithms for ATM Networks, IEEE Journal on Selected Areas in Communications, Vol. 9, No. 3 (1991) 343 - 350
4. Fan, Z., Mars, P.: Access Flow Control Scheme for ATM Networks Using Neural-Network-Based Traffic Prediction, IEE Proc.-Commun., Vol. 144, No. 5 (1997) 295 - 300
5. Grzymala-Busse, J.W.: LERS - A System Learning from Examples Based on Rough Sets, in: R. Slowinski (Ed.), *Intelligent Decision Support. Handbook of Applications and Advances of the Rough Sets Theory*, Kluwer Academic Publishers, Dordrecht, 1992, 3 - 18
6. Grzymala-Busse, J.W., Zou, X.: Classification Strategies Using Certain and Possible Rules, in: L. Polkowski, A. Skowron (Eds.), *Rough Sets and Current Trends in Computing*, LNCA, Vol. 1424, Springer-Verlag, Berlin, 1998, 37 - 44

7. Jensen, D.: B-ISDN Network Management by a Fuzzy Logic Controller, Proc. GLOBECOM'94, San Francisco, Nov. 1994
8. Mrozek, A.: Rough Sets and Dependency Analysis Among Attributes in Computer Implementations of Expert's Inference Models, Int. Journal of Man-Machine Studies, Vol. 30 (1989) 457 - 473
9. Ndousse, T.D.: Fuzzy Neural Control of Voice Cells in ATM Networks, IEEE Journal on Selected Areas in Communications, Vol. 12, No. 9 (1994) 1488 - 1494
10. Onvural,R.O.: Asynchronous Transfer Mode Networks. Performance Issues, Artech House, Boston, London, 1994
11. Pawlak, Z.: Rough Sets, Int. Journal of Computer and Information Sciences, Vol. 11, No. 5 (1982) 341 - 356
12. Pawlak, Z.: Rough Classification, Int. Journal of Man-Machine Studies, Vol. 20 (1984) 469 - 485
13. Pawlak, Z.: Reasoning About Data. A Rough Set Perspective, in: L. Polkowski, A. Skowron (Eds.), *Rough Sets and Current Trends in Computing*, LNAI, Vol. 1424, Springer-Verlag, Berlin, 1998, 25 - 34
14. Pitsillides, A., Sekercioglu, Y.A., Ramamurthy, G.: Effective Control of Traffic Flow in ATM Networks Using Fuzzy Explicit Rate Marking (*FERM*), IEEE Journal on Selected Reas in Communications, Vol. 15, No. 2 (1997) 209 - 225
15. Rathgeb, E.: Modeling and Performance Comparison of Policing Mechanism for ATM Networks, IEEE Journal on Selected Areas in Communications, Vol. 9, No. 3 (1991) 325 - 334
16. Slowinski, R., Stefanowski, J.: Handling Various Types of Uncertainty in the Rough Set Approach, Proc. of the 2nd Int. Workshop on Rough Sets and Knowledge Discovery, Banff 12 - 15. Oct. 1993, Canada, 395 - 398
17. Tarraf, A.A., Habib, I.W., Saadawi, T.N.: A Novel Neural Network Traffic Enforcement Mechanism for ATM Networks, IEEE Journal on Selected Areas in Communications, Vol. 12, No. 5 (1994) 1088 - 1095
18. Yager, R.R.: On a General Class of Fuzzy Connectives, Fuzzy Sets and Systems, Vol. 4 (1980) 235 - 242
19. Yager, R.R., Filev, D.: Essentials of Fuzzy Modeling and Control, John Wiley and Sons, Inc., New York, 1994
20. Zadeh, L.A.: Fuzzy Sets as a Basis for Theory of Possibility, Fuzzy Sets and Systems, Vol. 1 (1978) 3 - 28

Evaluation of Characteristic Temperatures of Materials Using an Approximate Reasoning Method

Bohdan S. Butkiewicz[1], and Tomasz Mroczek[1], and Marek W.Grzybek[2]

[1] Institute of Electronic Systems, Warsaw University of Technology, ul. Nowowiejska 15/19, 00-665 Warsaw, Poland
bb@ise.pw.edu.pl
http://www.ise.pw.edu.pl

[2] Industrial Institute of Electronics, ul. Dluga 44/50, 00-241 Warsaw, Poland
grzybekm@pie.edu.pl

Abstract. The characteristic temperatures of materials as softening temperature, melting temperature, etc. are defined by national or international norms. These norms give visual description of the special samples of the material at characteristic temperatures. This description is verbal, qualitative, inaccurate in mathematical sense. In the paper a system with furnace and video camera coupled with computer, developed in The Industrial Institute of Electronic in Warsaw, is presented. Fuzzy logic and approximate reasoning methods are used to denote these temperatures, basing on the series of images supplied by video camera.

1 Introduction

Fuzzy logic and approximate reasoning were at first successfully applied in the control theory and practice. Now, there are many other fields of application as artificial intelligence, pattern or voice recognition, image processing, diagnostic, economic, and some ecological problem [1]. Generally, we use the fuzzy logic approach in situations where we do not know the exact quantitative behavior of our system under external actions, but we have some qualitative information about it. In this paper we describe a method of the evaluation of the characteristic material temperature i.e. softening temperature, sintering temperature, melting temperature, etc.

2 Powder Materials Behavior at High Temperature

Many objects are made from powder materials, especially in metallurgy and ceramic or electronic industry. Very important properties of these materials are sintering, softening, melting, flowing and other effects observed while increasing the environmental temperature. These temperatures depend on the components of the powder.

There are international and national norms describing this characteristic temperatures [2]. The norms contain verbal, qualitative description of material behavior at each of

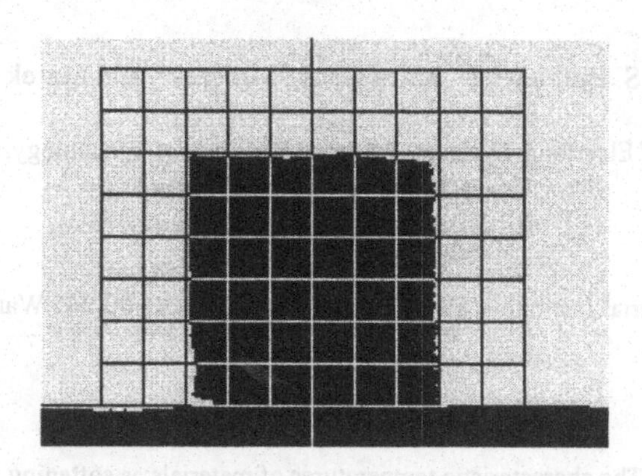

Fig. 1. Initial specimen shape before melting process, temperature 505.7°C

characteristic temperatures. Special moulded samples of the powdered material are used for investigations. The moulded sample (specimen) is prepared melting the pow-

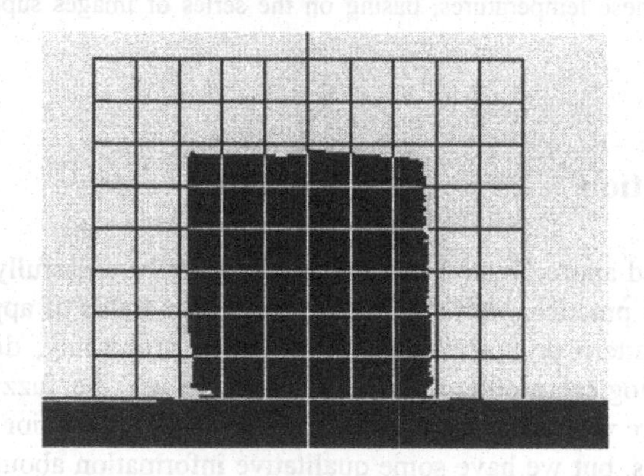

Fig. 2. . The specimen shape at sintering temperature 553.5°C

der of tested material with appropriate glue and putting into the press to obtain a briquette. The specimen has a cylindrical shape with diameter of 3 mm and height of 3 mm. Note that a glue used to join the substance grains has no influence (in chemical and physical sense) on the appointed temperatures. As it is shown in the Fig.1, the specimen has only theoretically ideal cylindrical shape. On its surface many different distortions can be seen, depending on the kind of melting material. It made difficult appointment of these temperatures. In extreme situations, in consequence of a specimen badly performed, it is not possible to determine some of desired temperatures.

The definitions of these temperatures given by Polish norm PN-82/G- 04535 are presented below.

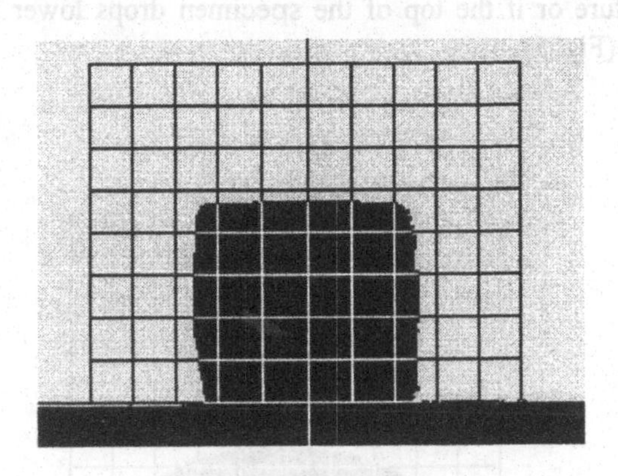

Fig. 3. Example, the specimen shape at softening temperature 710.7⁰C

Sintering temperature – **temperature, when the melting begins on the contact points of the particular grains, while simultaneously the specimen size diminish, but without changes of it initial shape (Fig.2).**
Softening temperature - **temperature, when one observes the first symptom of the specimen softening. One observes changes of the surface, round off the cylinder corners, or else the specimen begins to bulge (Fig.3).**

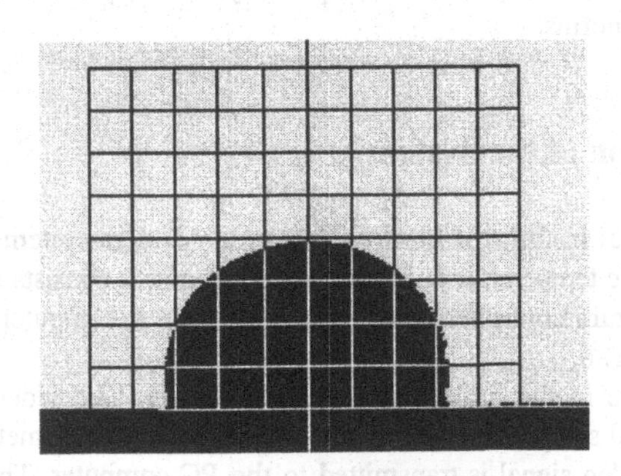

Fig. 4. . Example, the specimen shape at melting temperature 819.7⁰C

Melting temperature – **temperature, when specimen is melting, assuming hemispherical form, which height is equal to length of the specimen base, in the case of**

cubical form (not used hear), at this temperature. In the case of cylindrical form the height of hemisphere equals approximately to 2/3 of initial specimen height (Fig.4).
Flowing temperature - temperature when the specimen flow and forms a layer, whose thickness is approximately equal to 1/3 of the specimen height observed at melting temperature or if the top of the specimen drops lower than second line of measuring greed (Fig. 5).

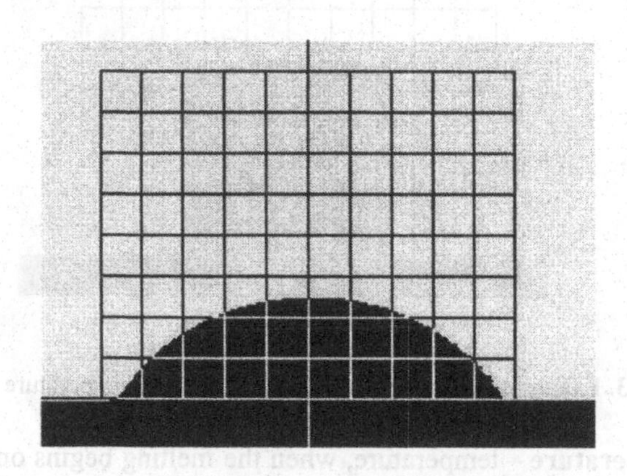

Fig. 5. The specimen shape at flowing temperature 903.1^{0}C

Some years ago, expert decided by visual inspection of the furnace, if the material attained the sintering temperature or not. If you have new component or composition of powder or environmental atmosphere you must determine this temperature. Our idea is to replace the expert and set up fuzzy rules representing the qualitative description of the norms.

3 Description of the System

In The Industrial Institute of Electronics a video camera system for the evaluation of the characteristic temperatures of materials was built. It consists of an electric furnace with a temperature controller, a video camera, and a computer. The system is presented in the Fig. 6.
The temperature in the furnace is increased slowly. The video camera observes a small cylindrical sample of material placed in the furnace. Sometimes one use cubical sample. The video signal is transmitted to the PC computer. The first version of the system was previously presented in 1995 at the International Fairs in Brussels and has obtained a gold medal. A classical algorithm, and a computer program, which evaluates these characteristic temperatures was written for the first version of the system [3], [4]. The algorithm uses and develops ideas proposed by Pitas [5] and Heijden [6].

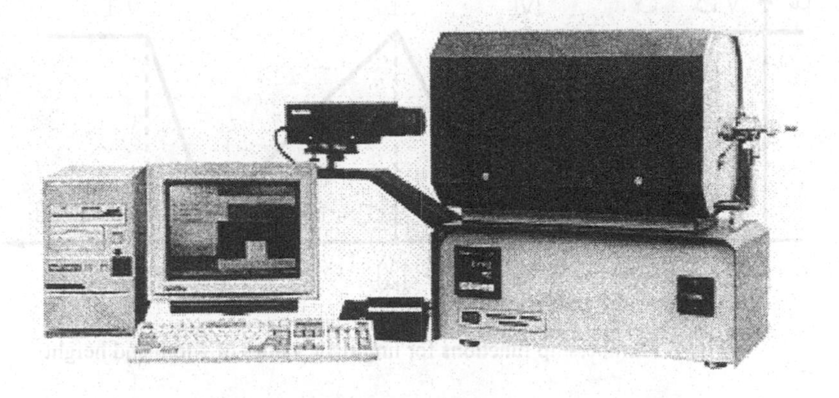

Fig. 6. The video camera system with furnace and PC computer

Now, in the second version of the system, an algorithm with fuzzy rules, based on the qualitative description of the material behavior given by the norms, was introduced. As it can be seen in the figures, a rectangular raster is introduced and the vision area of the camera is divided into small rectangular areas. All information about the material behavior is placed in rectangular areas containing a picture of the external surface of the material sample. Analyzing, the numbers of pixels in this areas, which belong to the picture of the specimen, one can decide if the characteristic temperature is attained or not. In the paper we present some details of the algorithm and rules, and results of tests performed with real powder materials.

4 Approximate Reasoning

There are many methods of approximate reasoning. For example, we have the first one compositional rule of inference proposed by Zadeh, fuzzy inferences of Mizumoto, Yager, Tsukamoto, Sugeno and Takagi, Turksen and Zhong (see ex. [7]). Up to this time, there is no general theory which reasoning is better in different situations. There are some comparisons, but rather theoretical and only good practice can help us to select the method.

In the system, standard form of reasoning rules is used. Any of the characteristic temperatures are described by set of fuzzy rules of the type if ... then. The rules use linguistic values of height, width and other parameters of the specimen picture. The height and width are expressed in percents of initial value. Thus, the range 0-160% is covered by fuzzy sets: VS-very small, S-small, M-medium, L-large, VL-very large. For height and width same shape of membership functions are used (Fig. 7). The shape is triangular or trapezoidal and it is very popular solution. However, the shapes of triangles are not symmetrical and uniformly located along the width and height spaces.

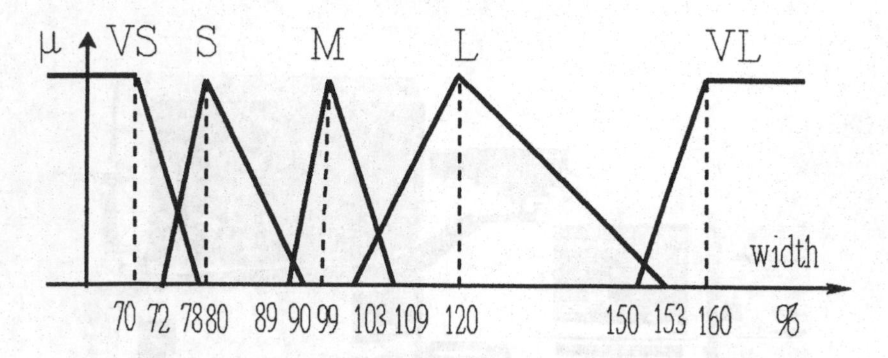

Fig. 7. Membership functions for linguistic values of width and height

The rules have typical form. Five rules are used only

if Height is L and Width is M then Tsin is True
if Height is M and Width is S then Tsof is True
if Height is M and Width is M then Tmel is True
if Height is S and Width is M then Tflo is True
if Height is S and Width is L then Tflo is True

where *L, S, M* are abbreviation of Large, Small and Medium, and *Tsin, Tsof, Tmel, Tflo* are appropriate temperatures of sintering, softening, melting, and flowing. Last two rules can be of course modified to one rule using logical *or*.

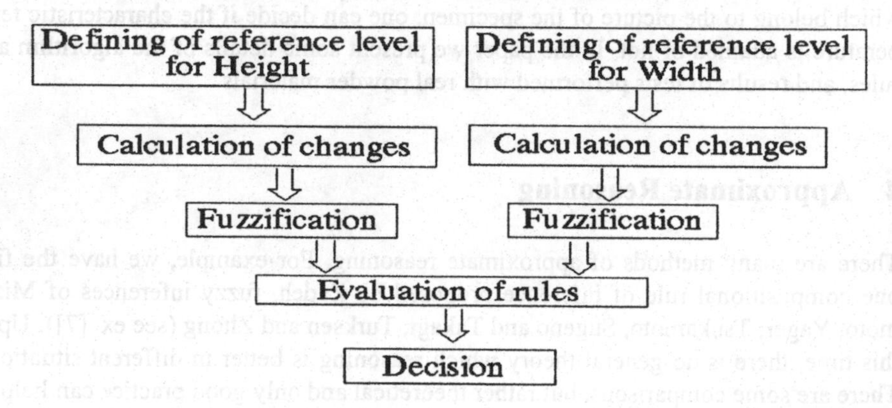

Fig. 8. General structure of approximate reasoning algorithm applied in the system

The images of the specimen are registered by a video camera. There are two options: first - image are registered every 3 seconds, second – image is registered every 1.5°C. Analyzing picture by picture registered images, weights of the rules are calculated. As the characteristic temperature of the material that temperature is chosen, when appropriate weight of the rule attains maximum. Thus, Mamdani method, using minimum and maximum as mathematical interpretation of "*and*" and "*or*", is applied. The method gives good results in the control theory, and some others fields. However, there are some differences. In the control theory decision about control signal is obtained applying defuzzification method. Usually, area, height or center of gravity methods are used. Here, the conclusions of decision rules are discrete. Using Mamdani method a fuzzy singleton is a result of each rule. It must be decided if any of characteristic temperatures is attained or not. So, the conclusion is obtained using winning rule (rule with maximal weight) as defuzzification method.. The winning rule is very simple, but may give satisfying results (see ex. [8]). New generation of controllers, fuzzy microcontrollers use winning rule as defuzzification method.

Omitting many details, general algorithm of approximate reasoning method used for evaluation of the characteristic temperatures is shown in the Fig. 8.

5 Results of Investigation

New algorithm was tested in real conditions and compared with older one. Actually, industrial orders concerned investigations of ashes of pit coal and brown coal, so the tests were performed for these materials. Computer screen images during tests are

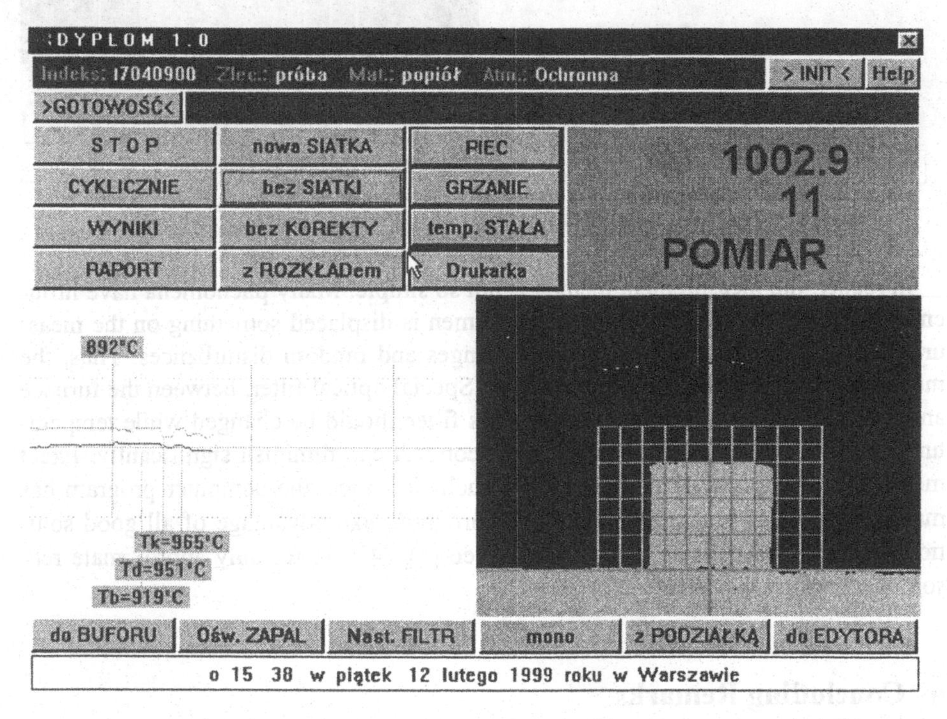

Fig. 9. Screen image during investigation of coal ashes; temperature 1002.9°C

shown in the Fig. 9 and 10 (of course for this publication beautiful colors are reduced to gray scale). The result of algorithm, i.e. the characteristic temperatures of material can be compared with appropriate shape of the specimen. For example, in the Fig. 1 –

5 the corresponding temperatures are equal to 505.7°C, 553.5°C, 710.7°C, 819.7°C, 903.1°C. Comparing successive images with obtained digital value we can to avoid a mistake.

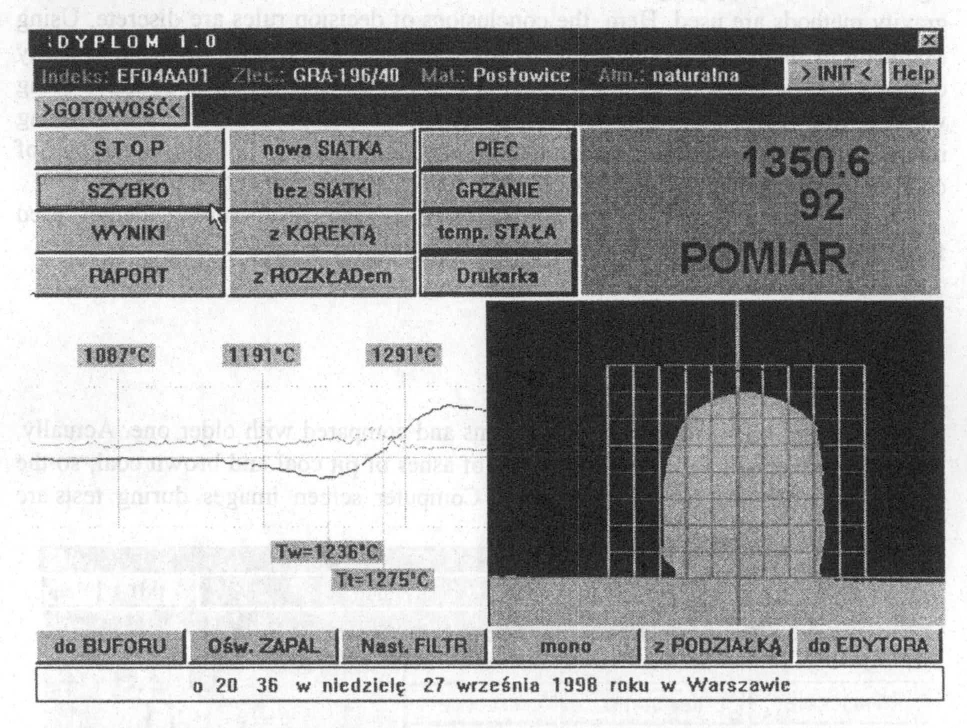

Fig. 10. Screen image during investigation of coal ashes; temperature 1350.6°C

In reality, the measurement method is not so simple. Many phenomena have influence on the result. While heating, the specimen is displaced something on the measurement table, because of dimensional changes and random disturbances. Thus, the measurement greed must be displaced also. Special optical filter, between the furnace and video camera, must be introduced. This filter should be changed while temperature increased. In counter case, the image contrast can diminish significantly. Exact measure can be impossible. To avoid all such situations, the computer program has many options and functional modes. The program take advantage of all good solutions find for first version of the system (see [3], [4]). Here, only approximate reasoning algorithm is presented.

6 Concluding Remarks

Two developed methods are compared, classic older algorithm and fuzzy algorithm. There are some differences between results obtained, but not very big. The differences attain no more than 5-10°C. It appears that the rules are very simple, may be too simple, but in practice they give good results. The method was tested on ashes of coal. The rules and membership functions at the beginning were adjusted using previously obtained images and results of older algorithm, after, using new samples of ashes. Probably, for other materials, the shapes and positions of membership triangles or trapezes in the width and height spaces should be changed. So, learning problem arises. But same problem we have with older algorithm. The algorithm could not find some of characteristic temperatures in one case, the older one in two cases, using same set of specimen images. It should be mentioned that we have always possibility of visual inspection of images using computer. All images captured during investigations are stored in a RAM disk.

The method presented is not limited to powder materials. It is used also for glasses, stones, and other materials in solid state, but general proceeding is similar.

Authors intend to develop the system. May be the rules can be modified, but first of all, looking for new measurement possibility and new physical phenomena, which can be used for measurement.

References

1. Marks II, R.J. , (ed.): Fuzzy Logic Technology and Application, IEEE Technology Update Series, The Institute of Electrical and Electronics Engineers, Inc., New York, (1994)
2. Polish Norm PN-82/G-04535, Evaluation of the characteristic temperature of melting (in Polish), PKNMiJ, (1982)
3. Grzybek, M.W.: A Method for Determining the Location of Objects and its Features in the Field of View Camera, Prace IPE, No. 117, Warszawa, (1992)
4. Grzybek, M.W.: A Method and System for Processing of Vision Image, Patent application in the Republic of Poland, No. P-313579, (1996)
5. van der Heijden, F.: Image based Measurement Systems, John Wiley & Sons, England, (1994)
6. Pitas, I.: Digital Image Processing Algorithms, Prentice Hall, (1993)
7. Lee, E.S., Zhu, Q.: Fuzzy and Evidence Reasoning, Physica-Verlag, Heidelberg (1995)
8. Butkiewicz, B.S.: Position Control System with Fuzzy Microcontroller. In: Reusch, B. (ed.): Computational Intelligence Theory and Applications, Lecture Notes in Computer Science, vol. 1226. Springer-Verlag, Berlin Heidelberg New York, (1997) 74-81

A Human Centered Architecture for Distributed Retrieval of Medical Images

R. Castelletti[1] , E. Damiani[2], G. Righini[2], and R. Khosla[3]

[1]Dipartimento Scientifico e Tecnologico, Università di Verona, Strada Le Grazie, I-47100 Verona, Italy
[2]Polo di Ricerca di Crema, Università di Milano, Via Bramante 65, I-26013 Crema, Italy
{edamiani,righini}@crema.unimi.it
[3]Dept. Computer Science and Computer Engineering, LaTrobe University, Bundoora, Melbourne, Australia
khosla@cs.latrobe.edu.au

Abstract In this paper we describe a distributed retrieval system for medical images based on a Human Centered Virtual Machine (HCVM). HCVM layered architecture allows indexing and retrieval based on the user's vision of data, seamlessly linking a nominal scale to ordinal and interval based representations suitable for intelligent search agents. Moreover, it ensures applicability of the approach to a wide range of application domains.

1 Introduction

Image indexing and retrieval by content is a time-honored problem, and many approaches have been proposed so far.
While older systems relied on the straightforward technique of providing textual descriptions to be stored together with the images, a number of more recent approaches focuses on using color, texture or shape [8] as the basis for image indexing and querying.

However, while promising from the purely technological point of view, many of these systems (see for instance [1]) explicitly renounced incorporating in the query language the naming or coding methods already familiar to user communities in specific application domains, as they turned out difficult to be mapped to mathematically satisfying definitions of similarity.

Moreover, these systems are mostly *monolithic*, i.e. they integrate indexing, search and storage facilities, requiring the whole multimedia database to be stored in the same place and available for indexing at the same time.

In this paper we follow a different line of research, describing query support to a distributed collection of medical images as an example of a *human-centered* (as against technology-centered) retrieval architecture for multimedia data.

The paper is structured as follows: in Section 2 we outline the characteristics of the problem, while in Section 3 we give a reference model for distributed image retrieval. In Section 4 the Human Centered Virtual Machine is briefly described; Section 5 and 6 describe its application to the medical images retrieval problem. Section 7 describes our current prototype, while Section 8 draws the conclusion and outlines some future work.

2 Image Collections for Medical Applications

Searching and querying image collections is a routine activity in the medical field and constitutes an important support to differential diagnosis and clinical research.

A number of medical images repertoires illustrating various pathologies are currently available over the Internet, but few of them are stored in fully-fledged image databases. Indeed, most of these collections are accessible through standard Web sites, each of them mantained by a different medical institution. Moreover, while the medical community has since long agreed on a basic textual coding for image contents (namely SNOMED, the *Systematized Nomenclature of Human and Veterinary Medicine*), no standard technique for indexing or retrieval of medical images based on color or shape is currently available or planned.

Retrieval of images tagged by a SNOMED code is easily and efficiently performed through pattern matching; Fig.1 presents some sample queries [2].

```
SNOMED (Systematized Nomenclature of Human and
Veterinary Medicine)

T: Topography

T=2*   Breathing Apparatus

T=28*   Lungs

T=281* Right Lung
```

Fig.1 Sample SNOMED patterns

Medical users are accustomed to using SNOMED patterns as search and classification codes for images depicting pathologies.

3 A Reference Model for Distributed Image Retrieval

Before describing the functionality of our agent-based architecture for medical image indexing and retrieval, we briefly outline its reference model at the highest level of abstraction, in order to clarify the use of the terminology.

Abstract Model
<u>Roles:</u> Customer Broker Supplier <u>Actions</u> Submit Search Deliver

Tab. 1 Roles and actions of the reference model

In the following sections, a technique for transparently superimposing hybrid search to SNOMED compliant Web-based collections of medical images will be presented as an alternative to storing such images in a monolithic database.

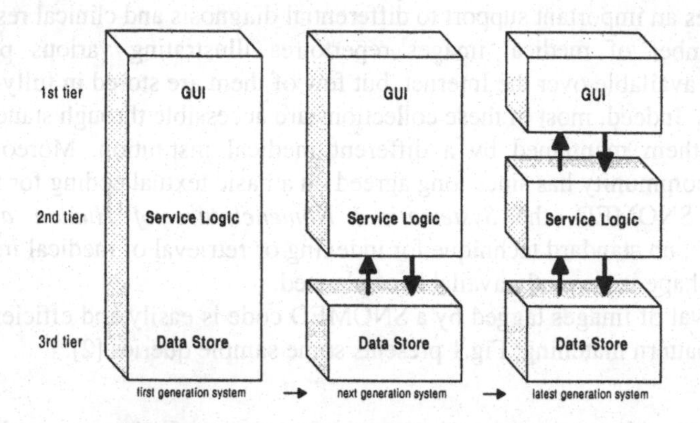

Fig. 2 An architectural view of roles

The core concepts of this reference model are the roles of *customer*, *broker* and *supplier* and the actions of *search*, *submit* and *deliver*. In Fig. 2 an architectural view of this distinction is given, outlining how service logic can be located in the broker agent, while customer and supplier respectively provide a graphical user interface and a very fast though conventional data storage.

In our system, the broker acts as a supplier of images to the customers, and as a distribution mechanism for suppliers wishing to make their images accessible over the Internet.

A basic assumption of our model is that suppliers do not index images, other perhaps than providing their standard SNOMED codes; it is left to brokers to compute and store content representations of online images, in order to be able to locate the image required by the customers. The function of the broker is therefore to provide a path whereby the customer may find and obtain from a supplier a image offering the required characteristics to the highest possible degree.

4 Human-Centered Virtual Machine (HCVM)

HCVM [9] consists of problem solving agents, intelligent agents, software agents and objects. It is based on the *Intelligent Multi-Agent Hybrid Distributed Architecture* (*IMAHDA*) described in [3].

The problem solving agents (among other aspects) employ problem solving knowledge and domain structural knowledge to realize human-centered software designs. The problem solving knowledge revolves around five information processing phases namely, pre-processing, decomposition, control, decision and post-processing. The problem solving knowledge used in these phases includes external representations of the domain (on psychological scales like nominal, ordinal and interval) and internal representations (e.g. continuous values of a neural network, rules, mathematical computation, etc.).

The problem solving knowledge including external and internal representations have been derived from various human-centered perspectives like distributed cognition [4], problem driven and activity centered systems, man-made complex systems, neurobiological control, learning, forms of knowledge, problem complexity, notion of time and reliability, and others.

HCVM has been designed to achieve three types of objectives, namely, architectural objectives, domain objectives and management objectives. Some of the architectural objectives include human-centeredness, task orientation, reliability and flexibility of methods used to accomplish tasks, and others. Some of the domain objectives include modeling of complex (sizewise and resourcewise) and time critical systems. The management objectives involve making sure the human-centered software designs/systems are scalable, maintainable, evolvable and are cost effective.

HCVM can be applied in two ways, namely domain-dependent and domain-independent. The domain-dependent application method of HCVM involves the integration of domain independent problem solving agents of HCVM with domain dependent problem solving agents (as identified through study of problem solving behaviur of the system user) of the application. More details can be found in [3]. The domain-independent application method is used whenever sufficient information is not available/provided on a application. In this method, the domain-independent problem solving agents of HCVM are used as a guide to develop human-centered software design. It may be noted that knowledge content employed by the domain-independent problem solving agents of HCVM include external and internal representations identified in the human-centered distributed-cognition approach [4].

5 HCVM and Image Retrieval

In this section we shall outline how HCVM layered architecture allows the broker design to proceed seamlessly from a human-centered representation of image content, based on a nominal scale, to the ordinal and interval-based representations that are more suitable for intelligent search agents. This will allow identification of the decision classes to be submitted to the user. As a by-product, we shall show how HCVM approach ensures generality and applicability of the retrieval system to a wide range of application domains.

5.1 A Human-Centered View of Medical Images Collections

The basic idea underlying our approach is putting the user's perception of data at the center of the knowledge organization process. The basic assumption we rely on is that the customer organizes a mental model of the information domain via a limited number of general features, and an agent-based broker should be able to fully comprehend and utilize such a model.

Following HCVM decomposition phase, we use a Decomposition Agent employing a small number of coarse-grain input features in order to identify a hierarchy of abstract classes.

In the present setting, SNOMED hierarchical encoding provides an easy and effective way to decompose the medical images domain. Moreover it has the additional advantage of being familiar to the user community.

The SNOMED-based taxonomy identified by the decomposition phase presents a model of the whole information space that is familiar to the user. However, it is not related to the solution of any particular search problem. HCVM following step, the Control Phase, uses a Control Agent to execute a further specialization on the basis of finer-grain features whose values can be drawn on an ordinal or an interval-based scale.

In the first case, values will belong to any ordinal domain such as the integers, while in the second case the feature will be associated to a fuzzy linguistic variable. In our present application, fuzzy linguistic variables will be used to represent features of image content.

The classes obtained in this phase will be directly involved in the decision support process of the broker system. It is interesting to observe that the SNOMED-based domain model developed in the previous phase allows us to easily deal at this level with the user's implicit knowledge about the (sub)domain. For instance, the user's linguistic knowledge about a pathology of the uterus may well include the fact that it involves an "abnormal" eccentricity of blobs corresponding to cells aggregates in the image. In this case, the meaning of "abnormal" depends on the image being compared with other images of the same kind and not, say, with the image of a lung.

In our model, being a class feature, the fuzzy linguistic variable `max_blob_eccentricity` may well have different definition intervals for different classes of images, thus dealing with the implicit different meanings of abnormal.

Classes developed in the Control Phase will then be used by the broker, in association with user input, to compute the Decision classes for a specific user query. In the general HCVM approach, this step involves a hybrid Decision Agent, using the search and identification technique (Rule-based decision system, Fuzzy matching, Neural Networks) which is more suited to the specific multimedia information available from the supply sites. As we shall see, user input itself must be pre-processed in order to be used by the Decision Agent.

6 A Sample Application

The user-centered design of our architecture requires the broker to compute and retain information about the available images in the format of a hierarchy. Classes at the higher levels exhibit general and structured features based on a nominal scale, while lower level classes present ordinal and interval-based features.

To decompose the chosen subdomain on the basis of the nominal scale, the Decomposition agent exploits the available body of domain knowledge, in this case the SNOMED encoding.

At the Control level, however, intervals and ordinal values must be computed on the basis of image content.

In this Section, we outline how such a computation can be carried out exploiting a broker-managed repository storing descriptors of the images available on the network. This repository is a structured collection of simplified descriptions of image properties, in the line of [5]. We consider that such a repository to be associated with each application subdomain, i.e. to each specific prefix of SNOMED coding, such as, for instance `T=3*` for the cardiac system.

In such a repository, the O-O control level classes are stored as a set of fuzzy relations, which are defined by applying an imprecise criterion through a fuzzy predicate on a crisp relation. For such a fuzzy relation, built on a set of domains Di, every t-uple is supplied with a membership degree *R*, from 0 to 1, interpreting how this t-uple satisfies a fuzzy predicate *P* applied to the relation *R*.

In the simplest case, the repository is a single fuzzy relation whose attributes are: `OID, feature, fuzzy element, weight`. Each feature corresponds to a linguistic variable, such as `max_blob_eccentricity`, having several fuzzy elements. To each fuzzy element of each feature a weight is associated, describing to which extent the corresponding property is offered by the object. From a syntactic point of view, features are expressed by nouns whereas adjectives describe fuzzy elements.

Fig. 3 shows a simplified fuzzy relation describing the properties of two images. The actual data model currently employed in our system will be discussed in detail in the next Section.

OID	Feature	Fuzzy Element	Weight
1	Max blob ellipticity	abnormal	.3
1	Max blob ellipticity	normal	.8
1	Max blob magnification	high	0.7
2	Max blob magnification	high	0.8
2	Max blob ellipticity	normal	1

Fig 3 An example of fuzzy descriptor relation

The above table was computed by the broker's Control Agent by applying to two images the definition of the `max_blob_eccentricity` and `max_blob_magnification` linguistic variables. The procedure can be outlined as follows: First, the Control Agent extracts from the images identified by the SNOMED code provided by the user the crisp values of two content-related indexes, namely the eccentricity of the main blob in the image and its relative size w.r.t the total image. The genetic algorithm used to identify blobs and compute their eccentricity is rather straightforward and will be omitted here; details can be found in [6]. It is however worthwhile to observe that in our case an approximate value of eccentricity could easily be obtained also by thresholding pixels to identify black blobs and applying to these blobs a deterministic convex-hull algorithm.

However, our current genetic approach allows us to set a maximum amount of time for index calculation per image (namely, 5 seconds). Then, indexes are fuzzified by applying linguistic variables definitions held by the Control Agent. These linguistic variables are part of the knowledge base for each value of the SNOMED code. The linguistic variables' definitions for a given subdomain are a part of the domain knowledge stored by the corresponding broker. It is important to observe that this computation takes place as the suppliers' sites are periodically polled with simple HTTP connections, allowing the broker to take updates into account.

User interaction is first used to select a broker, i.e. a part of the domain model that was built in the Decomposition phase, again via SNOMED codes.

In a second step, user input will detail the features of the desired image. User input is collected through a graphical interface in order to identify a fuzzy predicate for each interval-based or ordinal feature specified by the user.

When provided by the user, an absolute value in the definition universe of the selected predicate is also selected (for instance `max_blob_magnification` = 2).

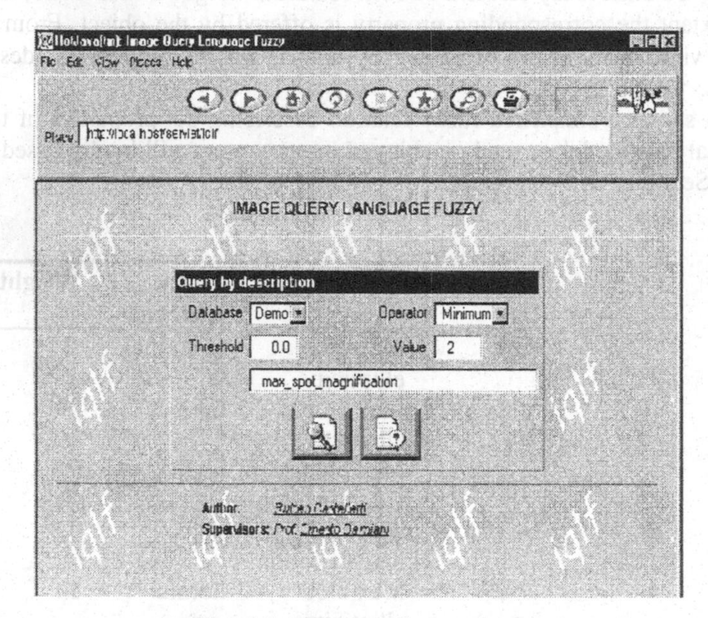

Fig. 4 Sample user interaction

Here, we assume the availability of a Thesaurus allowing features about images in a given subdomain to be uniformed through a naming discipline, in order to deal with a standard context-dependent vocabulary [5]. In the pathology application domain, such vocabulary is readily available thanks to the work of international standardization bodies.

This allows both brokers and clients to use a well-known domain-specific language to express features, without any explicit reference to the fuzzy model. Fuzzy elements and membership values i.e. the internal knowledge representation, are only computed and dealt with inside the broker.

6.1 Computation Of The Decision Classes

With reference to the previous example, a user could request to the broker an image having the following features: a `max_blob_magnification` of 2 and a `max_blob_eccentricity` of 0.8 (note that this latter value is not a fuzzy membership value, but a crisp geometrical parameter).

User input filtering computes a list of properties, each one associated with a certain fuzzy predicate and weighted by a value between 0 and 1.

These values are obtained by transformation of crisp values specified by the user according to the linguistic variable definition determined at the Control level.

The processed input defines a fuzzy request to the broker, which is nothing but another fuzzy relation, such as the one shown in Fig. 5.

max_blob_magnification	average	1
max_blob_ellipticity	abnormal	0.9

Fig. 5 A fuzzy query

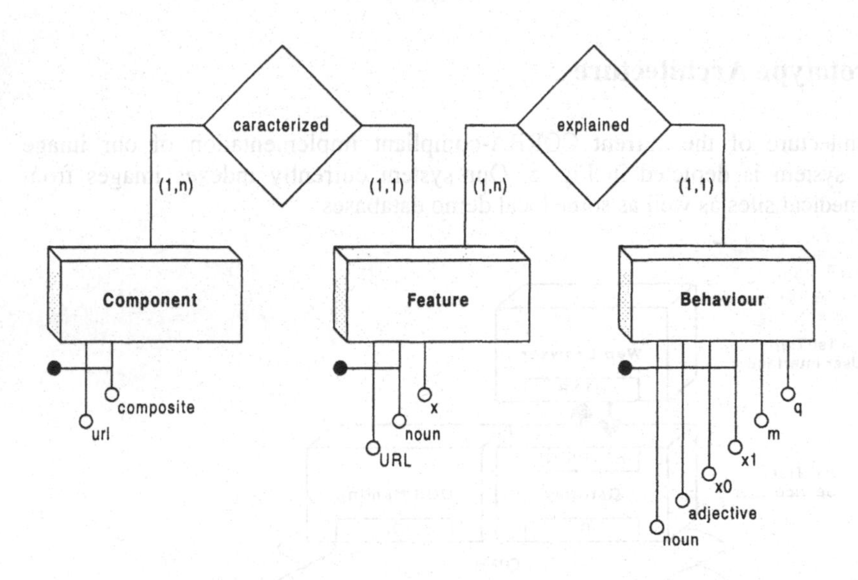

Fig. 6 Relational schema of the broker's repository

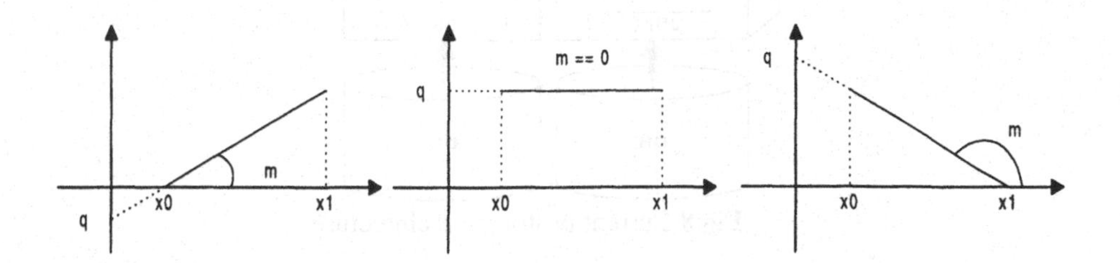

Fig.7 Membership function format

In order to compute such a list of images, our Decision Agent exploits the broker's repository, whose relational schema is depicted in Fig. 6. The entity Component has two attributes: URL is a reference to the image on the Web and Composite is the HTML document holding the image. This entity gives access to the Web pages holding the images composing a query result. Note that the URL alone would not be sufficient, as an image may well appear in multiple Web pages.

As mentioned above, image features are stored in the entity Feature, much in the same way as shown in Fig. 3. However, instead of explicitly storing fuzzy membership values, in our system we use entity `Behaviour` to store definitions of trapezoidal fuzzy elements as shown in Fig. 7, so that fuzzy membership values can be computed at run time by the broker agent as simple SQL queries.

Besides selecting the target site, the user can choose the aggregation operator the Decision Agent will use to compute the division. Available choices include fuzzy AND (min), fuzzy OR (MAX) and any mean-based aggregation operator (AVG). While the choice of aggregation operators could in principle be extended [7], it should be noted that currently supported operators are easily executable in standard SQL.

7 Prototype Architecture

The architecture of the current CORBA-compliant implementation of our image retrieval system is depicted in Fig. 8. Our system currently indexes images from several medical sites as well as some local demo databases.

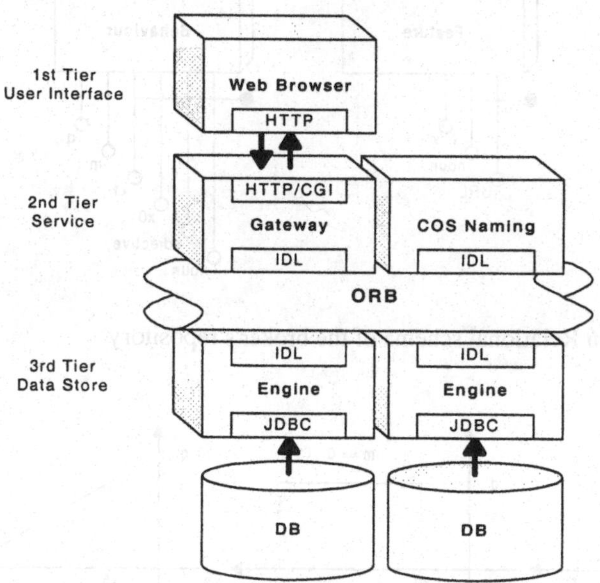

Fig.8 Current prototype architecture

The current prototype of our system was entirely implemented in Java, using CORBA support provided by JDK 1.2.

The broker agent complies to the CORBA/MASIF guidelines for the development of agent-based systems.

8 Conclusions and Future Work

The following table summarizes the proposed mapping of our Medical Images Retrieval Model to HCVM layered architecture

Medical Image Retrieval Reference Model	HCVM
Structure of a *image description* to be supplied by customer	Input filtering and pre-processing
Domain Specific Knowledge SNOMED-based classification of images in the domain of interest	Decomposition *Domain model,* hierarchy of abstract classes. Domain specific metadata.
System Functional Units Fuzzy query to the broker's database.	Control Alternative Fuzzy/NN component identification techniques, based on multimedia data and domain specific metadata
Candidate Selection Presentation of candidate images	Decision Candidate classes to be proposed to user
Transaction Network transfer of the chosen images/Request of validation of choice process	Validation Optional validation of the decision phase

Tab.2 Model mapping to HCVM layers

As we have seen, the HCVM approach allows for providing a seamless connection between a human-centred domain decomposition, based on nominal scale features, to ordinal and interval-based features suitable to be processed by intelligent agents. We are currently exploring a number of applications of this approach.

Acknowledgements The authors wish to thank Maurizio Lestani, M.D. and his colleagues at the Institute of Anatomy and Pathological Histology, Medical University of Verona, for their precious assistance on medical issues. Thanks are also due to Letizia Tanca and Patrick Bosc for their encouragement and valuable suggestions.

9 References

1. E. Binaghi, I. Gagliardi, R. Schettini Image Retrieval Using Fuzzy Evaluation of Color Similarity, Intl. Journal of Pattern Recognition and Artificial Intelligence vol.8, n.4, 1994
2. F. Mauri, A. Beltrami, V. Della Mea et al. Telepathology Using Internet Multimedia Electronic Mail, Journal of Telemedicine and Telecare, vol.6 n.3 1996
3. R. Khosla, T. Dillon, Engineering Intelligent Hybrid Multi-Agent Systems, Kluwer Academic Publishers, 1997
4. J Zhang, D. A. Norman, Distributed Cognitive Tasks, Cognitive Science, pp. 84-120, 1994
5. E. Damiani, M.G. Fugini Fuzzy Identification Of Distributed Components, Proc. of the 5th Fuzzy Days Intl. Conf., Dortmund, 1997, LNCS 1226

6. R. Castelletti, Un servizio per la gestione distribuita di immagini di interesse patologico, (in Italian) Tesi di Laurea in Scienze dell'Informazione, Dip. Scientifico e Tecnologico, Università di Verona, 1998
7. P. Bosc, D. Dubois, O. Pivert and H. Prade, Flexible Queries In Relational Databases - The Example of The Division Operator, Theoretical Computer Science, vol.171, 1997
8. C. Faloutsos, W. Equitz, M. Flickner, W. Niblack, D. Petrovic, R. Barber Efficient and Effective Querying by Image Content, Journal of Intelligent information Systems vol. 3 n. 3/4, 1994
9. R. Khosla, Human-Centered Virtual Machine of Problem Solving Agents, Intelligent Agents, Software Agents and Objects, Proceedings of the 3rd IEEE Symposium on High Assurance in Intelligent Systems, Washington, Nov. 13-14 1998

Representing the Real-Time-Behaviour of Technical Processes in Neural Nets by Using the Phase-Space-Flow of the Degrees of Freedom of the Nets

M. Reuter, D.P.F. Möller

Department of Computer Science, University of Hamburg
Chair Computer Engineering/Building F
Vogt-Kölln-St. 30
D-22527 Hamburg
E_mail: reuter@informatik.tu-clausthal.de
E_mail: dietmar.moeller@informatik.uni-hamburg.de

Abstract. The application of neural networks in supervision and control of technical processes does not only require the ability to classify states of process and to identify possible faulty or dangerous states but also the possibility to monitor the changes of the process-variables over time to predict eventually developing dangerous states. As the authors have shown [2] one way to store such real time behaviour in a neural net classification structure is the learning strategy of sensitisation. Meanwhile this method of teaching neural nets by means of sequences of sets of values in process which converge towards states of process, which is known to be faulty or dangerous, is successful in many applications. The mathematical background and the methods how to supervise a sensitisation were clarified last year. The first part of this paper introduces the theory of the sensitisation with a more complex mathematical background based on the phase-space-representation of the degrees of freedom of neural nets. The second part shows how a sensitisation can be supervised by the flow of trajectories of the net-parameters. Furthermore it will be shown by experimental results, how the classification behaviour of a sensitised and a non-sensitised neural net differ.

Keywords: Supervision of technical processes, real-time behaviour of neural nets, sensitive neural nets, learning strategies of neural nets, supervision of neural net learning strategies

1. Theory

From cognitive psychology in the context of the Chunking problem we learned of the idea to condition a neural net first by well-defined easy distinguishable data sets and to deepen and to enlarge the stored classification-information later by sensitisation in a further conditioning-step. Chunking is more or less the adaptation of a new fact using in an unknown situation 'known' facts, concepts, or models. Only from known facts or from known acting strategies it is possible to develop new strategies for solving problems. Prerequisite for a Chunking process is that all stored facts/concepts are encoded in the same way, otherwise unconnected facts/concepts cannot be connected to each other by using new concepts/facts as an integration factor. That's the way our mental model of the world successively increases from an (individual) basic concept (the first understood fact, the first model to understand how the world acts) to a very complex reaction and understanding system [1].

Transforming this model to the handling of neural nets implies that a neural net has to 'learn' a so-called basic-concept first. For instance such a concept can represent well distinguishable input patterns of the system, which have to be supervised. To prevent the net from including sensor typical output ranges in its classification behaviour and therefore 'preferring' sensors with a high output, it is necessary to normalise the input data sets, by using a proportional pre-processing [2].

If a neural classifier can separate in a sufficient way all the system-states, which are involved in a basic concept, the 'weaker' evolutionary states of the different system-states should be trained. Using these weaker states ensures that the net changes its classification structure only slightly without destroying the classification-structure representing the basic-concept. This second conditioning-phase is called 'sensitisation'.

To supervise the different learning and sensitisation periods and to ensure that the basic-concept will not be destroyed in the sensitisation-conditioning period the behaviour of the change of the net-parameters, the so-called 'flow of thetrajectories of the net-parameters' can be used. This fact is the result of a theoretical framework [3,4], which observes that neural nets, modified by sensitisation, can be described in deterministic ways by analysing their phase space behaviour.

The basic idea of this theory is that a classification potential

$$\mathrm{Ep}=\mathrm{Ep}\left(\sum_{n=1}^{N}\left(E_p{}^n(y_1^n(t),\ldots,y_M^n(t)\right)\right) \tag{1}$$

and a conditioning potential

$$\mathrm{Up}=\mathrm{Up}\left(\sum_{n=1}^{N}\left(U_p{}^n(y_1^n(t),\ldots,y_M^n(t)\right)\right) \tag{2}$$

can be defined for every neural net, where the index n indicates a neuron of the neural net with N neurons and y_m^n indicates the mth degree of freedom of the M degrees of freedom of the neuron n. If the state vector $\vec{y}(t)$ describes all neuronal and synaptic values of the network at a time t the neural network reaches steady state when

$$\dot{\vec{y}}(t)=\vec{f}(\dot{\vec{y}}(t))=\vec{0} \tag{3}$$

holds indefinitely or until new stimuli perturb the system out of the equilibrium described by a function

$$L = L(f(y(t))) = E_p(y(t)) - U_p(y(t)) = 0 \tag{4}$$

We can locally linearize $\vec{f}$ by replacing $\vec{f}$ with its Jacobian matrix of partial derivatives $\vec{J}$. The eigenvalues of $\vec{J}$ describe the system's local behaviour about an equilibrium point. If all eigenvalues have negative real parts then the local equilibrium is a fixed point and the system converges to it exponentially quickly. More abstractly, generalised eigenvalues or Lyapunov exponents describe the underlying dynamical contraction and expansion [Kosko] and L can be identified as one Lyapunov function of a neural net [Reuter 1998].

If and only if the net adapts in the conditioning state asymptotically for the Lyapunov function L holds

$$\dot{L} = \sum_{i=1}^{n} \frac{\partial L}{\partial x_i} \frac{\partial x_i}{\partial t} = \sum_{i=1}^{n} \frac{\partial L}{\partial x_i} \dot{x}_i \leq 0 \tag{5}$$

where the x_i are the transformed co-ordinates of the state vector $\vec{y}(t)$ of the net in a co-ordination system for which holds

$$\vec{X} = \vec{0} \tag{6}$$

A possible form for the Lyaponov function will be given by

$$L = \frac{1}{2} \sum_{i=1}^{n} x_i \tag{7}$$

and the dynamics of the degrees of freedom of the net, like the inter-neural weight w_{ij} or the parameters of the transferfunction f_{trans}, will be given by the formula

$$\dot{x}_i = -E_i x_i \tag{8}$$

where

$$x_i = x_i(t = 0)\, e^{-E_i t} \tag{9}$$

If we assume that the time derivation of the x_i can be written in the form

$$\dot{x}_i = -grad\ V \qquad (10)$$

the change of the x_i can be observed as trajectories, which cross the equi-potential levels of a potential, described by the values of L, orthogonal.

Depending on this preconditions the conditioning-state of a neural net can be understood as an non-equilibrium-state of the potential-functions E_p and U_p, which forces a change of the degrees of freedom x_i of the net to adapt the co-ordinates of the desired net structure $\vec{X} = \vec{0}$.

2. The neural phase space

The special form of the co-ordinates x_i and the impulses $F_i = \dot{x}_i$ makes it possible to supervise the behaviour of the net during the conditioning-state by studying the flow of the degrees of freedom in the phase-space of the net, as the expressions for the flow of the generalised co-ordinates $x_i = q_i$ and the generalised impulses $F_i = p_i$ must follow the Hamilton equations, given by

$$\frac{dq_i}{dt} = -\frac{\partial\ H(q_i, p_i)}{\partial\ p_i} \qquad (11)$$

and

$$\frac{dp_i}{dt} = \frac{\partial\ H(q_i, p_i)}{\partial\ q_i} \qquad (12)$$

As the divergence of the volume of the phase-space-fluid $\rho(t)$ (describing all degrees of freedoms of the net) will be zero, the change of the phase-space-fluid $\rho(t)$ will be described by the formula

$$-i\frac{\partial\ \rho(t)}{\partial\ t} = \hat{L}\rho(t) \qquad (13)$$

where for the operator $\hat{L}$ holds

$$\hat{L} = -i \sum_{i=1}^{n} \left(\frac{\partial\, H(q_i, p_i)}{\partial\, p_i} \frac{\partial}{\partial\, q_i} - \frac{\partial\, H(q_i, p_i)}{\partial\, q_i} \frac{\partial}{\partial\, p_i} \right) \tag{14}$$

and for $\rho(t)$ holds

$$\rho\,(t) = e^{-i\hat{L}t} \rho_0(t = 0) \tag{15}$$

By means of the spectrum of the operator $\hat{L}$ four different flows of $\rho(t)$ can be expected [Prigogine]

? non ergodic flows
? ergodic non-mixing flows
? ergodic mixing flows
? and K-flows

The behaviour of the first three flows is shown in Figure 1 (as the K-flows are following the Bäcker-transformation they cannot be visualised).

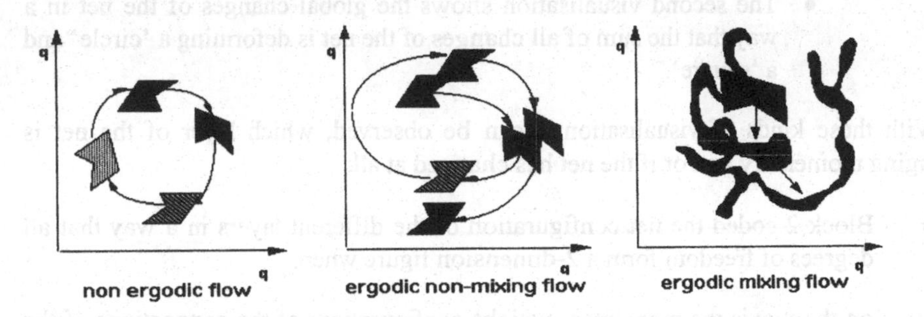

Fig. 1. The different classes of the flow of the phase-space-fluid $\rho(t)$

Guided by this classification of the flows of the phase-space-fluid $\rho(t)$ we can learn that the range of the changes of the degrees of freedom and the conditions of these changes will lead to different classifiers.

- Ergodic nets will show after a frequency of conditioning-steps the same classification behaviour as before, a neural 'clock' will be designed.
- Ergodic non-mixing nets will represent the common neural nets. After a conditioning-phase all classification criteria learned before will be gone lost.

- Ergodic mixing flows will represent nets which are not loosing all their stored classification knowledge under further conditioning; they can be 'sensitized ' and will show a kind of real-time behavior [Reuter 1996].

3 The visualisation of the phase-space-trajectories

Based on the theory discussed above an algorithm was implemented in the software-tool zz-2[1], which enables to visualise the phase-space-flow of the trajectories of the degrees of freedom of a 3-layer Backpropagation neuralnet. For supervising the different areas of the changes of the nets and the behaviour of the nets at all, the visualisation was divided in four monitoring blocks as follows:

- Block 1: To summarise the changes of the net during the conditioning state, two kinds of visualisations were implemented:
  - The first visualisation shows the sum of the changes of the weights of the connections from the input-layer to the hidden-layer during two iteration steps and the changes of the weights of the connections form the hidden-layer to the output-layer during two iteration steps.
  - The second visualisation shows the global changes of the net in a way that the sum of all changes of the net is deforming a 'circle' and a 'square'.

With these kinds of visualisation it can be observed, which layer of the net is changing momentary and/or if the net has changed at all.

- Block 2 coded the net configuration of the different layers in a way that all degrees of freedom form a 2-dimension figure where

- on the x-axis the momentary weight-configurations of the connections of the neurons are shown,
- on the y-axis the gradient of the flow of the changes of these connections is shown.

- In Block 3 and 4 the flows of individual degrees of freedom are shown. As usually the number of degrees is very large, the user can select the individual degrees of freedom to be visualised.

The phase-space-figures and the flows of the degrees of freedom of a 3-layer Backpropagation net with two out-put-neurons during a successful conditioning-phase and during a sensitisation-phase are shown in the Figures 2a-2c. It is important to mention that all data sets have been pre-processed by a DLS- and mFD-operation, so the net only has to learn a system-relevant information-structure.

---

[1] The 'zz-2' neural net work tool is a product of: RT&S Ingenieurbüro, Am Ehrenhain 1, D-38678 Clausthal-Zellerfeld, Germany

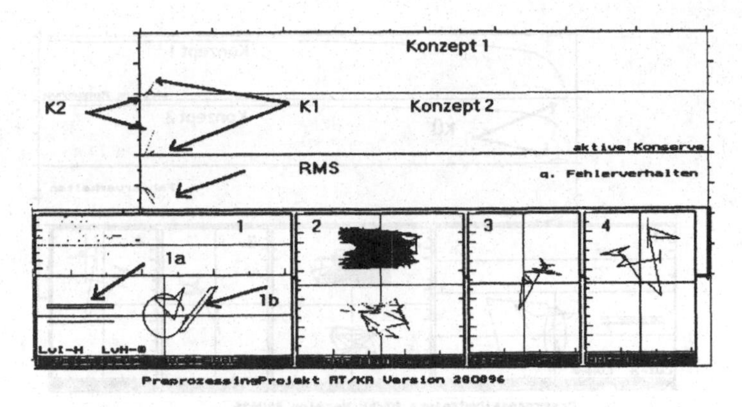

Fig. 2 a. Phase-space-behaviour of the net's degrees of freedom at the beginning of the conditioning-state

Clearly it can be pointed out of Figure 2a that the classification structure of the net at the beginning of the conditioning is far away from the desired net-structure, as the classification security indicators K1 and K2 are very low. Also the changes in both layers are very large as mentioned by the stripes 1a. The symmetry figure 'square' is extremely deformed, another indicator showing that the net is changing it's structure largely.

In Figure 2a four degrees of freedom have been selected to be supervised individually. In block 3 the phase-space-flow of two of them are shown. As this window belongs to the input-hidden-layer, the flows represent the change of the net in this area only. The corresponding block 4 shows the change of two degrees of freedom in the hidden-output-layer.

About 200 iteration steps later on the different net-parameters have totally changed (see Figure 2b). Now the classification security indicators K1 and K2 point out that the classification concept of the net has reached the desired maximum-value of '1' nearly. On the other hand the phase-space-representation of the net and the flow of degrees of freedom show that the structure of the net is still changing as the figures 'circle' and 'square' haven't reached a high degree of symmetry.

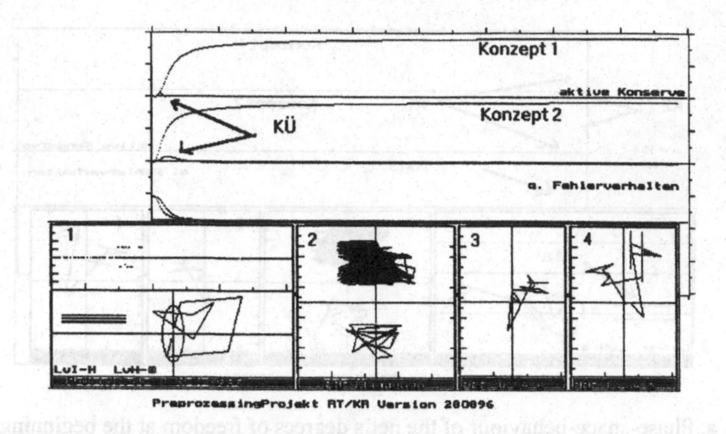

Fig. 2b. Phase-space-behaviour of the net's degrees of freedom in the middle of the conditioning-state

By the structure of block 2 we can learn that the major changes of the net's structure have been performed in the hidden-output-area, as this net-structure has been changed totally contradict to the input-hidden-structure. This result can also be verified by the changes of block 3 and 4, as a major change of the trajectories only can be detected in block 4.

As we have learned from the theory discussed above, the final structure of the net will be reached when the phase-space-flow has stopped and the geometrical figures 'circle' and 'square' have reached a high degree of symmetry. This state of the conditioning of a net is shown in Figure 2c.

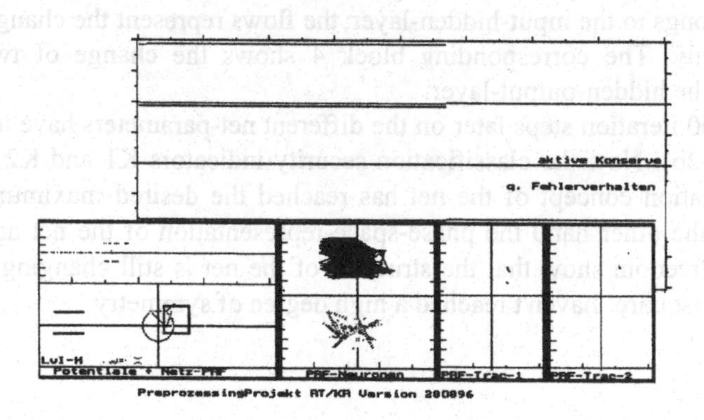

Fig. 2c. Phase-space-behaviour of the net's degrees of freedom at the end of the conditioning state

From Figure 2c it can be pointed out clearly that no major changes of the net's structure have been done during the last phase of the conditioning. In block 1 the stripes, representing the changes in the input-hidden-structure and the hidden-output-structure, are overlapping themselves and the 'circle' and the 'square' are highly symmetric. In block 3 and 4 the flow of the different degrees of freedom stopped and also the probability of the classification results, shown by the security indicators K1 and K2, did not change anymore.

Figure 3 gives an example, how the phase-space-flow-representation can be used to supervise a learning phase more effectively and how this kind of phase-space-representations enables a net-designer to control or to change the different learning parameters in accordance to a desired classification structure. In this Figure a typical momentary stoppage of the conditioning in the hidden-output-area (block 4) is shown, which is caused by a local minimum. In the beginning the conditioning-phase just started, the weight-configuration was changed largely at every iteration-step, but the Backpropagation algorithm adapted no weight-changes. Only in the input-hidden-area some smoothly changes can be detected, announced by the slowly down-ward-shift of the trajectories in block 3. The fact that the net has not reached its desired configuration can be detected also by the non-symmetry of the figures: 'circle' and 'square' in block 1.

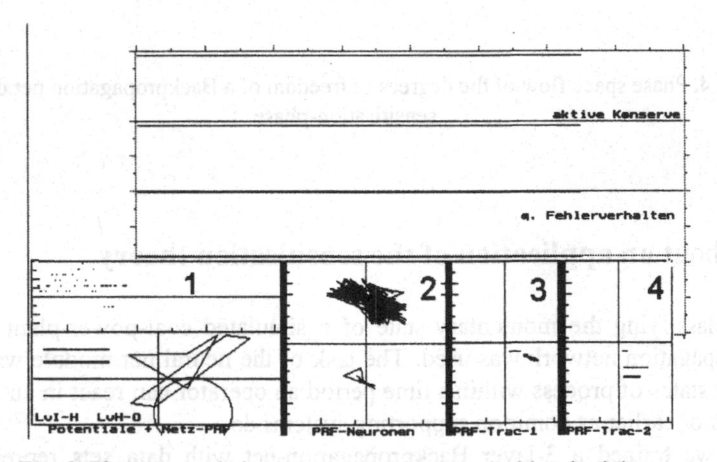

Fig. 3. Phase-space-behaviour announcing a typical local minimum

An example for a 'classical' sensitisation during a supervised conditioning-phase is given by Figure 4. Clearly it can be pointed out from Figure 4 (upper hand) that the second out-put-neuron is sensitised two times. But while the RMS-Error announce a non-discrete change of the net-structure (follow the arrows) block 3 and block 4 announce that some weights have been changed in a mixing ergodic way.

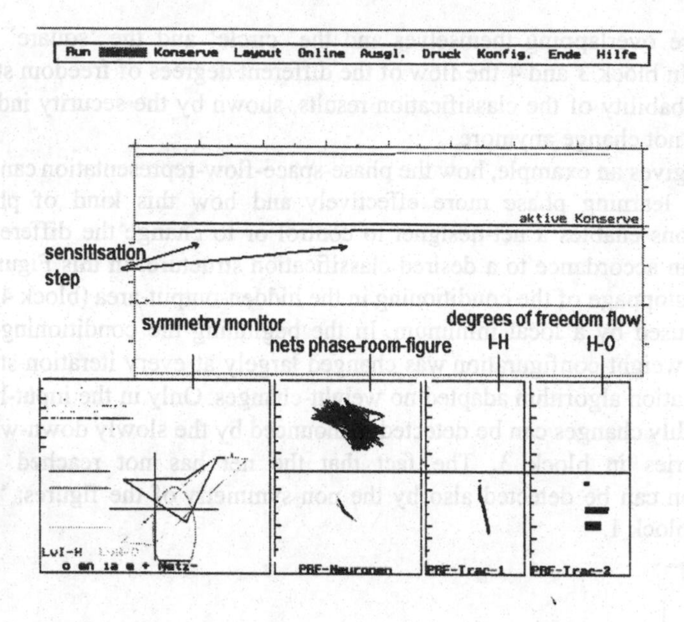

Fig. 4. Phase space flow of the degrees of freedom of a Backpropagation net during a sensitisation-phase

4. About an application of the sensitisation theory

For classifying the momentary state of a simulated coal-power-plant a common Backpropagation network was used. The task of the neural net module was to detect irregular states of process within a time period an operator can react in an appropriate way and/ or earlier as common supporting systems do.

First we trained a 3-layer Backpropagation-net with data sets representing the following situations of the simulated plant:

I) all parts of the plant are working at their operation point,
II) all faulty states are presented clearly,
III) some faulty states occurred in their evolutionary state.

The states II) and III) represent the plant after a time t_F, which begins at the state when no fault exists (at the time t_0) and ends when the different instruments indicate the typical values of the faults. The other data sets represent states before the time t_F. In this time periods faults just arise and the instruments values change their values, but normally in a so far undifferentiated manner that only well trained operatorscan detect a faulty state.

Our learning criterion was that an output neuron representing a faulty state has to have an output value higher than 0.95 while all other neurons had to have an output value lower than 0.15 when the given output range was between 0.0 and 1.0. The test results of the trained net showed that only 14 of the 20 faulty states can be distinguish in a sufficient way. Further evolutionary states were classified in anundifferentiated or wrong way.

Now this 'basic concept' was sensitised by presenting evolutionary data sets of faulty states, which have been recorded before the time t_F. Successively we used data sets recorded earlier and earlier before the time t_F. Even data sets that represented states of processes when the instruments indicated an undifferentiated evolutionary state of the plant were used. Now the test results of the net showed that the net was empowered to detect all the different faulty states even when these states were very similar to each other by their data set representation. Furthermore the tests showed that the net was empowered to classify the different evolutionary states by announcing them with a smaller probability value as when the time t_F would have been occurred. By using the learning-strategy of the sensitisation at least we were able to create neural classifier that showed real-time behaviour as now the classifier related even evolutionary states to their faulty (or non-faulty) states which will be reached clearly later on.

In Figure 5 the classification results of a non-sensitised and a sensitised net are shown.

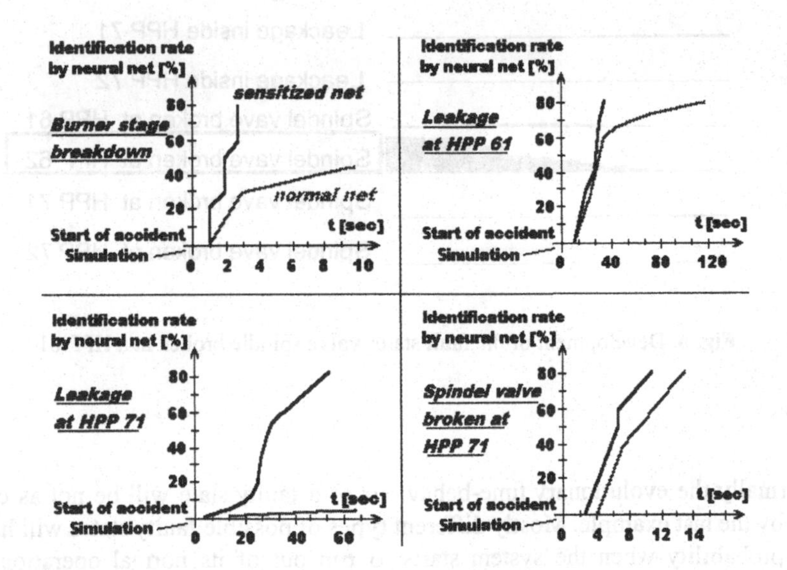

Fig. 5. Comparison of the classification-behaviour of a sensitised and a non-sensitised net

It can be pointed clearly out from Figure 5 that for the faulty states 'burner stage breakdown' and 'leakage at HPP 71' a normal net was not even able to detect the faults. For the other two faulty states shown in Figure 5 the normal net needed more time (about a factor 3) to announce the corresponding faulty states.

As an interpretation of the numerical representation of the neural classification results is more or less difficult, we used a kind of interface which enabled an operator to supervise the plant in a more sufficient way. The used interface is shown in Figure 6 and 7. The first line represents the normal state, followed by the different faulty states listened below. The left column represents the probabilities of the states of process with their special coded colours, here seen as stripes with different grey-shadings.

Figure 6 shows the evolutionary development of the state: 'valve spindle broken at a HPP 62'. The system announced quite early that the normal running state was left, but as the system tried to compensate the faulty state by the internal automatic control no exact classification result can be observed over a longer time period. After half time of the supervised time window the neural net classifier announced that the probability of the fault state 8 (valve spindle is broken) became more and more probable.

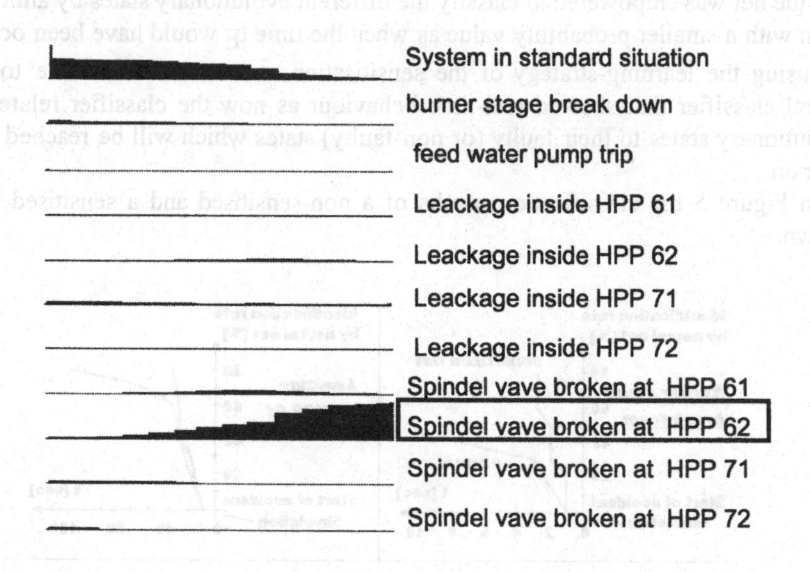

Fig. 6. Development of the fault state: valve spindle broken at a HPP 61

Normally the evolutionary time-behaviour of a faulty state will be not as clear as given by the last example. Mostly different types of possible faulty states will have the same probability when the system starts to run out of its normal operation point. That's the reason why an operator will get problems to control the system in a sufficient way by supervising the classical sensor equipment.

In Figure 7 such undifferentiated fault evolution behaviour is shown.

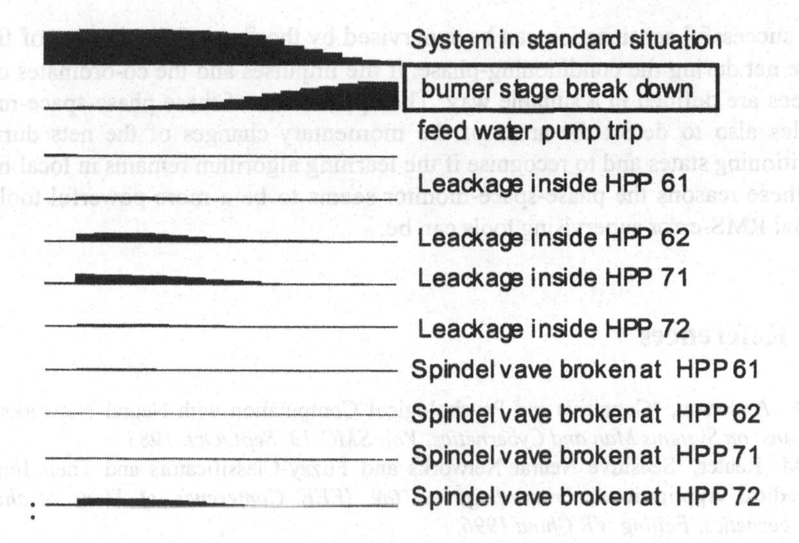

Fig. 7. Typical evolutionary behaviour of the system when a breakdown burner stage 4 as fault state occurs

Here the faulty state: 'breakdown burnerstage 4' was simulated. It's typical for this event that the net suddenly announce that something 'abnormal' had occurred without changing the probability of the normal running state but with announcing not less than 3 different faulty states concepts represented by their increasing probabilities. This is a typical effect of the automatic control that tries to compensate a faulty state. Not before the automatic control fails to compensate the faulty state the classification probability of these concepts will decrease rapidly and the probability to detect a breakdown of the burner stage 4 will increase continuously and will reach a constant and trustworthy classification level.

5.Conclusion

The application of neural nets in supervision and control of technical processes requires that the data sets have been transformed in a suitable way and that the network has been pre-learned with data sets representing well distinguishable states of the running system. The so formed network structure represents a basic concept, which can be sensitised successively by presenting the evolutionary states of faults. Such sensitised networks are able to identify even states of technical processes that 'normal' trained networks cannot classify. Furthermore these sensitised neural nets show real-time classification behaviour, as the evolutionary development of the process states can be stored in classification structure of the nets.

A successful sensitisation can be supervised by the flow of the degrees of freedom of the net during the conditioning-phase, if the impulses and the co-ordinates of these degrees are defined in a suitable way. The application of these phase-space-monitors enables also to detect the areas of the momentary changes of the nets during the conditioning states and to recognise if the learning algorithm remains in local minima. For these reasons the phase-space-monitor seems to be a more powerful tool as the normal RMS-error supervising tools can be.

6. References

[1] A. Anderson, "Cognitive and Psychological Computation with Neural Networks", *IEEE Trans. on Systems Man and Cybernetics, Vol. SMC-13, Sept./Oct. 1983*

[2] M. Reuter, 'Sensitive Neural Networks and Fuzzy-Classificators and Their Impact for Medical Applications', *Proceedings on the IEEE Conference of Man, Machine and Cybernetics, Beijing, VR China 1996*

[3] M. Reuter, A.Berger, P.E. Elzer, 'A Proposed Method for Representing the Real-Time-Behavior of Technical Processes in Neural Nets', *IEEE Congress: Man, Machine and Cybernetics, Vancover 1995*

[4] M. Reuter, 'Die potentialorientierte Beschreibung der neuronalen Netze', *Clausthal, 1998*

[5] R.L. Dawes, . 'Inferential Reasoning Through Soliton Properties of Quantum Neurodynamics', *IEEE Int. Conf. On System, Man and Cybernetics, Vol. 2, Chicago, Il, 1992*

[6] I. Prigogine, 'Vom Sein zum Werden', *Piper München 1977*

Spatial Neural Networks Based on Fractal Algorithms Biomorph Nets of Nets of ...

Thomas Kromer,Zentrum für Psychiatrie, Münsterklinik Zwiefalten , D

1 . Abstract and Introduction

Biological central nervous systems with their massive parallel structures and recurrent projections show fractal characteristics in structural and functional parameters (Babloyantz and Louren÷o 1994). Julia sets and the Mandelbrot set are the well known classical fractals with all their harmony , deterministic chaos and beauty , generated by iterated non - linear functions .The according algorithms may be transposed , based on their geometrical interpretation , directly into the massive parallel structure of neural networks working on recurrent projections. Structural organization and functional properties of those networks , their ability to process data and correspondences to biological neural networks will be discussed .

2. Fractal algorithms and their geometrical interpretation

2.1 The algorithms of Julia sets and the Mandelbrot set

Iterating the function f(1) : $z_{(n+1)} = c + z_n^2$,(c and z representing complex numbers respective points in the complex plane), will generate the beautiful fractals of the Julia sets and the Mandelbrot set (Mandelbrot 1982 , Peitgen and Richter 1986) . According to the rules of geometrical addition and multiplication of complex numbers (Pieper 1985) we can interpret function f(1) as describing a combined movement :

First , the term :" $+ z_n^2$ " in f(1) describes a movement from point z_n to the point z_n^2 . A lot of trajectories can connect these two points, one is the segment of the logarithmic spiral through z_n . (In a polar coordinate system we get a logarithmic spiral by the function f(2): $r = ae^{c*\varphi}$. Geometrical squaring of a complex number is done by doubling the angle between the vector z (from zero to the point z) and the x - axis and squaring the length of vector z (2) . Doubling the angle φ in f(2) will also cause a squaring of r . This proves point z^2 lying on the logarithmic spiral through z .)

Second , the first term of f(1), " c " (meaning the addition of complex number c), can be interpreted as describing a linear movement along vector c .

Both movements can be combined to a continuous movement along spiralic trajectories (according to Poincaré)from any point z_n to the according point $(c+z_n^2) = z_{(n+1)}$.We get two different fields of trajectories , one with segments of logarithmic spirals arising from each point z_n, the other as a field of (parallel) vectors c .We can follow the different trajectories alternately (fig 2.1c) or simultaneously (fig 2.1d , 2.1e) . Various options to visualize the developments are shown in figure (2.1 a-f) .

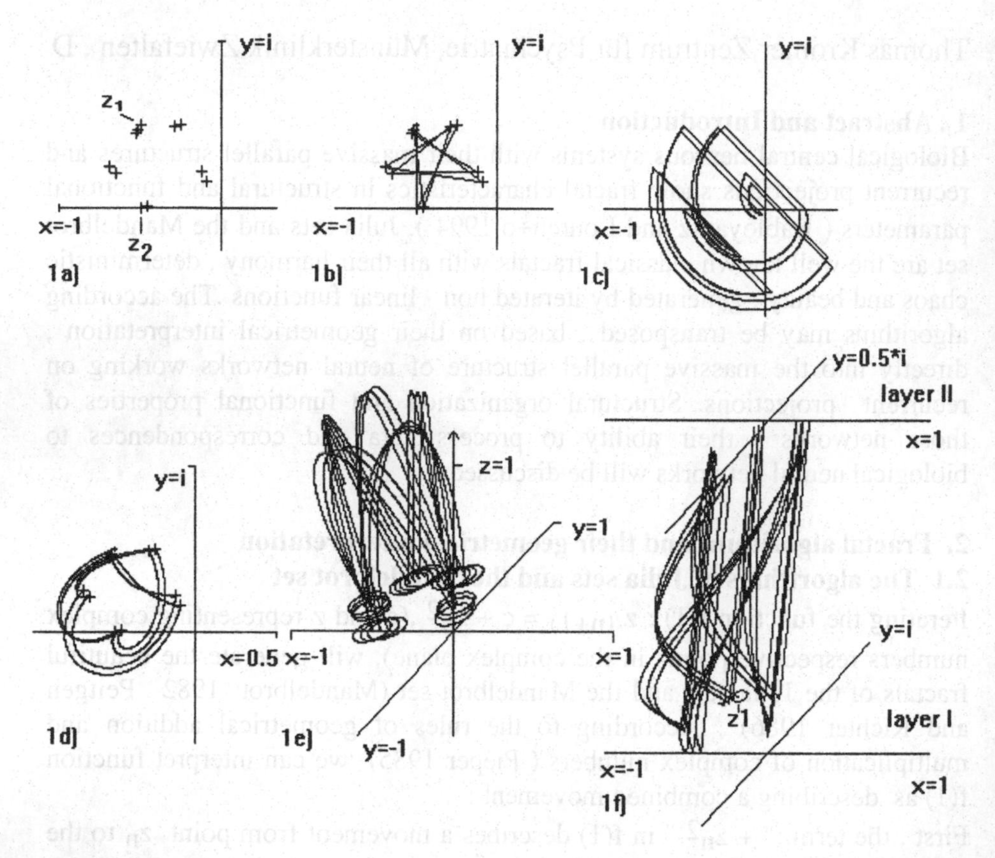

Figure 2.1 : Development of values acc. f (1) - (c = -0.5 + 0.5* i , z1 = -0.5+0.5*i; 10 iterations): a) isolated values after each iteration ; b) subsequent values connected ; c) following the trajectories of terms " +c " and " z^2 " alternately ; d) Trajectories of the combined movement acc. f(1); e) trajectories acc. Poincaré in 3 - dim. parameter - space, (intersecting points with complex plane marked by circles); f) The trajectories projecting from layer I to layer II (representing term " z^2 " and from layer II to layer I (acc. term " + c ").

2.2 Three - dimensional algorithms

We may transfer the principles of function f(1) into spatial algorithms in a three-dimensional coordinate system . Because squaring of a triple of numbers (coordinates on x - , y - and z - axis) is not defined , the direct " translation " of f(1) into a three-dimensional algorithm is impossible . The following two algorithms will transfer some of the fundamental functional properties of Julia sets to according three-dimensional fractal structures :

Algorithm I) In a three-dimensional coordinate system with x -, y -

and z-axis we can lay a plane through each point $z(x1,y1,z1)$ and the x-axis. In this oblique „ complex plane“ we square z_n. To the point " $z\,(x1,y,z1)^2$ ", which we thus find, we can easily add a three-dimensional vector $c(x,y,z)$.This addition of the constant vector c will bring us to a point $z(x2,y2,z2)$, which will be the starting point of the next iteration . This algorithm generates interesting spatial fractal structures on the basis of the two - dimensional Julia sets which are formed by the respective vector c.

Algorithm II) Before adding the vector c in algorithm I , we rotate the oblique „complex plane“ together with the point "$z\,(x1,y,z1)^2$ " around the x - axis , until ist angle with the y-axis of the three-dimensional coordinate system will be doubled . After addition of vector c the next iteration may be started . If we combine all these three partial movements (along the logarithmic spiral in the oblique "complex plane" from $z(x1,y1,z1)$ to "$z(x1,y1,z1)^2$ " , the rotation of this complex plane around its x-axis and the straight movement along the three - dimensional vector c) to one movement we get three - dimensional trajectories from every point $z(x1,y1,z1)$ to the according point $z(x2,y2,z2)$ and on this base spatial sets corresponding to two - dimensional Julia sets .

3.1 Neural networks based on fractal algorithms

In principle all trajectories can be interpreted as symbolic representations of neurons . An iteration of a fractal algorithm will move us from z_n to $z_{(n+1)}$. As well , a neuron at any point z_n in a network could send its activity along its axon (following the respective trajectory) to the point $z_{(n+1)}$ of the neural network . By this we can transpose the fields of trajectories of any analytic function into the structure of a neural network .

The inclusion - criterium of Julia sets (points belong to the Julia set , if the values we get in the following course of iterations will not leave a defined zone around zero) can be applied to the neural net too : Neurons will belong to the net (will survive) , if they can activate a certain number of neurons in a defined zone around zero .

3.2 Two - dimensional networks

We can construct neural networks with one or two layers of neurons on the basis of function f(1):

A one - layer - net we get , if the neurons send their axons from the points (z_n) directly to the points $(c + z_n^2) = z_{(n+1)}$ in the same layer (according fig.1e).

If we transpose the two terms of function f (1) into two different neural layers, we will get a two-layer network (according fig. 2.1f and fig. 3.2)

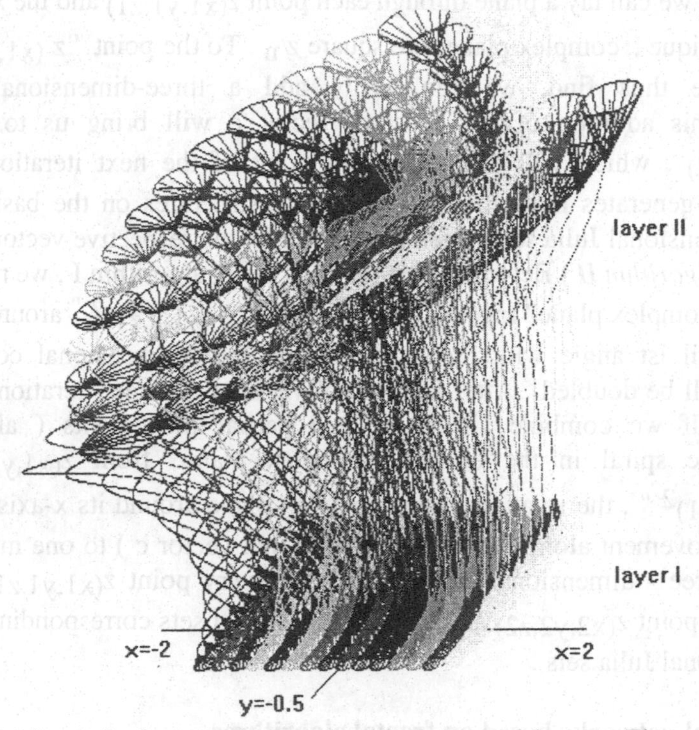

Figure 3.2 : Layer I of a neural network with two neural layers . The axons of the neurons reflect the term " z^2 " of function f(1) .

3.3 Three-dimensional neural networks

As in the two - dimensional case we interpret trajectories of three-dimensional algorithms as functional neurons . Applying the inclusion - criterium of Julia sets , we finally get three - dimensional neural networks (fig. 3.3).

4 . Structural and functional properties and features

4.1 " Julia " - a network - language?

Because these networks are based on fractal algorithms (which they perfectly perform), the patterns of activations in these networks show all properties of Julia sets like self - similarity and symmetry of generated structures and patterns, occurence of strange attractors , periodic and aperiodic orbits , Siegel discs , zones of con- and divergence, implementation of binary trees and amplification of small differences between slightly different starting patterns. Activating a neuron at any point z_n will cause a sequence of activation of neurons which will reflect the development of the values we get by iterations of function f(1) in geometrical interpretation. In many cases the activation of two different neurons will lead to an activation of the same circumscribed region in the course of the iterated activations . In this region neurons get

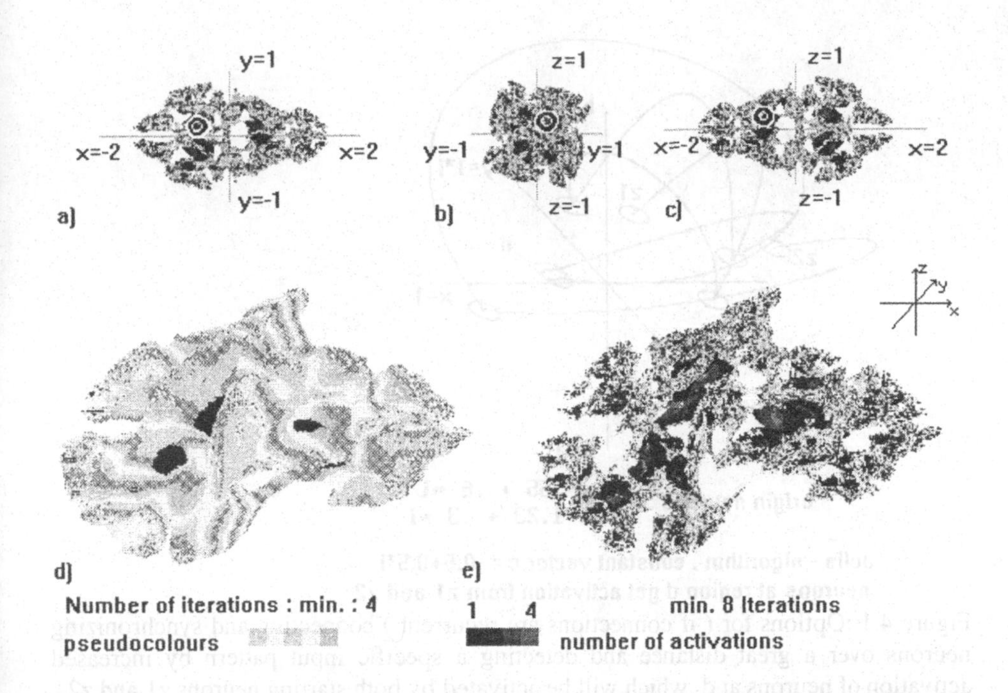

Figure 3.3 : Aspect of a three - dimensional analogon of Julia sets , produced according algorithm II , constant vector c = -0.7(x) + 0.2(y) + 0.4(z) . A)-c) : view of structure e , orthogonal to the indicated plane , d and e) 3-D impression , in d all neurons shown with at least 4 iterations before values leave a radius of 5 around zero , in e) with at least 8 iterations . Light grey colours are used to show the spatial structure , darker grey colours , to indicate neurons , which may activate in the sequence of activation neurons in the region around the point p = -0.4(x) - 0.4(y) - 0.2(z) with radius 0.09 (region marked in a-c by circle) directly or by interneurons.

activation from both starting neurons . They will be activated in a higher degree than other neurons . These more activated neurons will be able to indicate the pattern of the simultaneous activity of these two starting neurons (figure 4.1) and may work as " detector - neurons " of the specific input pattern .

4.2 Implementation of binary trees , dialectic network - structure

According to function f (1) , a neuron z as well as its negation , the neuron (-z) will activate the neuron ($c + z^2$) . Every neuron will have two predecessor - neurons and one subsequent neuron . The neurons and their connections will perfectly model the structure of a manifold binary tree . This will enable those networks to represent data with hierarchical , binary order very well (fig. 4.2). It should be very suitable for dialectic processes , each neuron z representing the thesis , its negation (- z) the antithesis , both

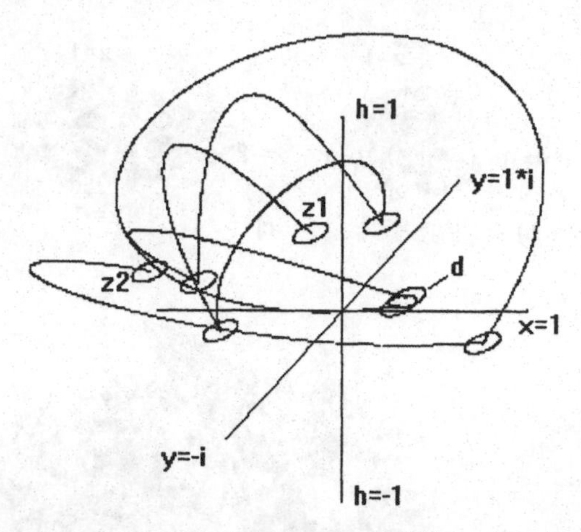

origin neurons : z1 =-.55 + .6 *i
z2 =-1.23 + .3 *i

Julia - algorithm , constant vector c = -0.5+0.5*i
neurons at region d get activation from z1 and z2

Figure 4.1 :Options for (if connections are recurrent) connecting and synchronizing neurons over a great distance and detecting a specific input pattern by increased activation of neurons at d ,which will be activated by both starting neurons z1 and z2 .

projecting their activity to the neuron ($c + z^2$) as representative of synthesis. In three- dimensional networks , based on algorithm II (2.2) , we get instead of a binary tree a quartenary structure with each neuron having four predecessor - neurons , thus increasing the connectivity of the network .

Binary tree of neuron z1

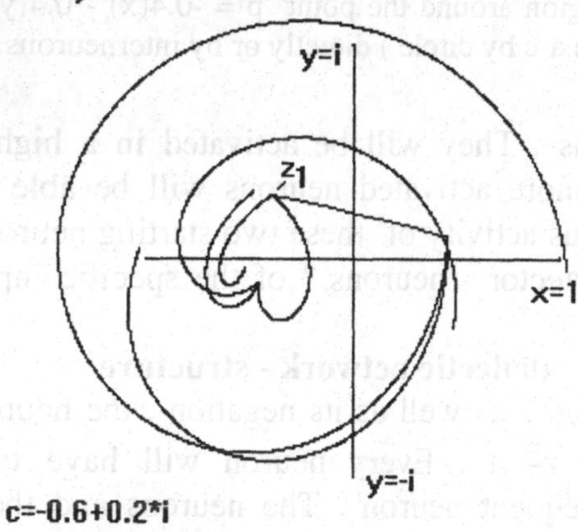

Figure 4.2 : Binary tree of the neuron z1 with two " generations " of predecessor - neurons , one subsequent neuron .

Combining excitatory and inhibitory neurons , the binary structure will allow to represent logical structures in the networks.

4.3 " Somatotopic " structure , synaptic learning , neural columns

The structure of such networks is massive parallel with neural axons preserving their relationship during their course , so the neural projections will be " somatotopic "(neighboured neurons will activate neighboured regions of the network)(fig.3.2) .

Each region will receive activation from many other neurons , due to the overlapping structures of axonal endings and due to effects of convergence in fractal functions .Many neurons will compete for activation with their overlapping arboreal dendrites. There will be effects of synaptic learning , varying the pathway of activation , which will be no longer determined only by geometrical algorithmic rules (as it will be at the very beginning) , but also by learning - history . Because every signal will have to pass on every transmission a small neural network (formed by all neurons reaching the area of transmission with their dendritic structures) , the whole net is working as net of nets of .... . (In mammalian brains the neural columns may represent these units of neural networks) .

4.4 Handling of intermediate values

Fractal functions need a "continuum" of infinitely many numbers , whereas neural nets consist of a limited quantity of neurons .Quite often an activation will not hit only one neuron exactly, but will lead to an activation - maximum between two distinct neurons . Such an intermediate result can be correctly represented by a simultaneous activity of these two neurons analogue to the degree of activation they receive from their predecessor - neurons . This transforming of intermediate values into analogue degrees of activity of the neighboured neurons could be done by the overlapping structures of dendritic and axonal endings . Using these processes , the maximum resolution of two stimuli will no longer depend on the distance between two neurons , but on the distance between two synaptic contacts. By the almost infinite number of synaptic endings in neural networks we get the " Quasi-continuum" of units , fractal processes need . This may be an important task of the arboreal structures of dendritic and axonal endings beyond their role in processes of synaptic learning .

4.5 Amplification of slight differences between input patterns

Slight differences between two starting patterns may be amplified not only by effects of learning in the individual history of the network but also because of the great sensibility of fractal structures to little changes of starting or intermediate parameters . Similar sensibility to slightly different patterns we can find in biological networks in form of hyperacuity.

4.6 Input and output

All neurons , or only neurons in specialised regions of the networks , may have connections to peripheral receptors or effectors or they themselves may represent these organs . Thus the network can communicate with the external world .

4.7 Recurrent connections

In biological nervous systems we often find recurrent projections , each neuron not only activating its subsequent neuron but also the neurons by which it has been activated itself. Such networks will not only perform one algorithm but simultaneously its reversed function , (in case of function f(1) the function f(3) $z_{(n+1)} = (z_{(n)} - c)^{1/2}$). In this case all neurons will send their activity to three others (In case of nets based on algorithm II to five other neurons because of the fourfold symmetry of these structures). Depending on the algorithm determining the net and the chosen parameters , all neurons may thereby be connected over a smaller or greater distance with almost each other neurons . The activity of one neuron may spread like a wave over the entire net .

4.8 Synchronization of distant neurons , representing complex concepts

Depending on the choice of the parameter c in function f (1) the course of activation through the neural network follows certain regularities . For instance , the activation will rotate around a fixed point steadily approaching to it , if the point is an attractor , while in case of Siegel discs the activation will rotate in an elliptical orbit around the fixed point (Peitgen and Richter 1986). In case of recurrent connections such networks could work as " phone centers" which may provide a plenitude of possibilities to connect different neurons of the network by " active axonal lines " using the functional properties of fractal structures .These possibilities to connect neurons by only few interneurons as shown in figure 4.1 may allow neurons , being activated by the same object , to synchronize their activity . Synchronous activity seems to be an important feature for representation of objects or complex concepts in central nervous systems . If one starting neuron in fig. (4.1) should represent the term " cup " , the other the term " tea " , the whole chain of active neurons in fig. (4.1) may represent the term " cup of tea " .

4.9 Dynamic temporal activation patterns

Continuous flow of sensoric perceptions will lead to specific temporal activation patterns . In the case of neural networks working with the Julia - algorithm each neuron z_1 has certain subsequent neurons z_n . For example , different temporal sequence of activation of 3 origin neurons ("t_1" , "e_1" , "n_1") will produce different clusters of active neurons (in the group of their subsequent neurons) . At any iteration (n) , " ten " would be represented by simultaneous activity of the neurons t_n , $e_{(n-1)}$, $n_{(n-2)}$, whereas the word " net " would be represented by simultaneous activity of the neurons n_n , $e_{(n-1)}$, $t_{(n-2)}$. Synaptic learning leading to different limit cycles will enable the net to differentiate between the words .

4.10 Information processing as task of the whole network

Functional behaviours stem from altered activities of large ensembles of neurons and thus they are a network property (Wu et al 1994). In fractal networks each neuron is influenced by many others and itself influencing also many other neurons . Fig. 4.10a shows an example to which degree a region can be influenced by the entire network . In reversal this region is influencing by recurrent projections wide parts of the net .

The superposition of all wave functions caused by the input neurons will result in an entire wave function . We get an information - coding related to holographic information storage (The relation becoming even closer , if we assume periodic " coherent " waves of activity, we find in brain functions (Destexhe 1994), interfering with the activity waves caused by the fractal pathways).

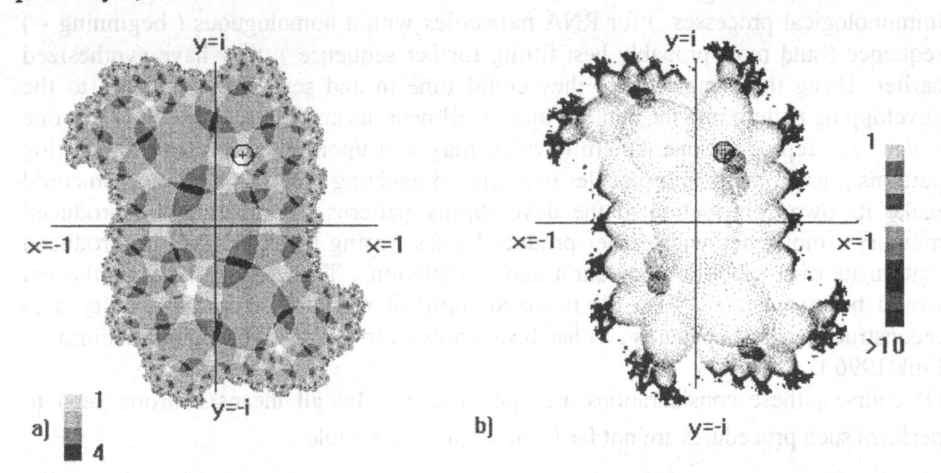

Figure 4.10 : Neural net based on a Julia - algorithm , constant vector c = 0.25 + 0.25*i . A) Indicated by pseudocolours is the degree , to which the network regions can activate neurons in a region around z = 0.17 + 0.43*i with radius 0.065 (marked by a black circle) in the course of activation sequences . B) Showing to which degree the neurons of the mentioned circumscribed region can activate other regions of the net by recurrent connections (reversal of function f(1) according function f(3)).

Each input pattern will cause a highly specific activation pattern as result of the interference of the influences of each input neuron to the whole net . Regarding Fig. 4.10b we can interpret the diagram of the strength of influence of one region or neuron to the net as diagram of some sort of a wave function .

4.11 Options for associative memory functions

Following considerations may be hypothetical : Activation patterns caused by neural input or active network neurons will cause a specific temporal activation sequence in each neuron . Each neuron may registrate the received activating sequence , for instance in form of ribonucleic-acid -(RNA) molecules , each ribonucleic nucleotid representing a certain degree of activation (Corresponding to an old Quechuan knot - script used in Inca - empire) . The neurons would thereby get filled with lots of such memory - RNA - molecules (The role of RNA and subsequently produced proteins in learning processes is not yet completely understood (Birbaumer and Schmidt , 1996)). If there would be no special input into the net dominating the neural activity, one neuron will begin to send a sequence , it has learned earlier , using the RNA - molecules as " songbook ". (This process mediated by Ion pumps or - channels , being regulated by the sequence of the RNA - nucleotides in the RNA - molecules) . The neurons connected with that neuron may registrate this sequence. They may synthesize in real time an according RNA - molecule and then look by means of detection of homologuous molecules (which is a common task in genetic and immunological processes) for RNA molecules with a homologuous (beginning -) sequence (and most probably best fitting further sequence), they have synthesized earlier .Using these molecules , they could tune in and send activity fitting to the developping pattern into the net . By and by all neurons could be caused " to sing one song " , to reproduce one pattern (which may win upon other competiting starting patterns), using memory molecules in a self - organizing process . Each neuron could make its own contribution to the developping patterns , comparing the produced memory - molecules with earlier produced ones , using those to give the produced pattern its own specific modulation and completion . Thus the memory of the net would be associative . Also the reversed input of recorded temporal activity may reconstruct activity - patterns , as has been shown for wave - transducing mediums (Fink 1996).

Of course , these considerations are speculative . But all means neurons need to perform such procedures are not far from being conceivable .

5 . Resemblances to biological neural networks

Planar and spatial neural networks based on fractal algorithms show morphologic and functional correspondences (in the reflection of fundamental aspects of development and function) to biological neural networks :

In both we may find :

- biomorph symmetrical structures like lobes and hemispheres ,gyri and sulci and cavities similar to ventricles ; the axons of neurons projecting massive parallel , " somatotopic " , some crossing the median line in form of a decussatio (fig. 5b).

- each neuron being connected to almost all other neurons by only few functional interneurons (fig. 4.10).
- periodic and aperiodic orbits as basis of deterministic chaotic activities in biological nervous systems (Freeman 1987) as well as in fractal functions, the visualization of functional parameters in fractal nets resembling to pictures we get by neuroimaging procedures .

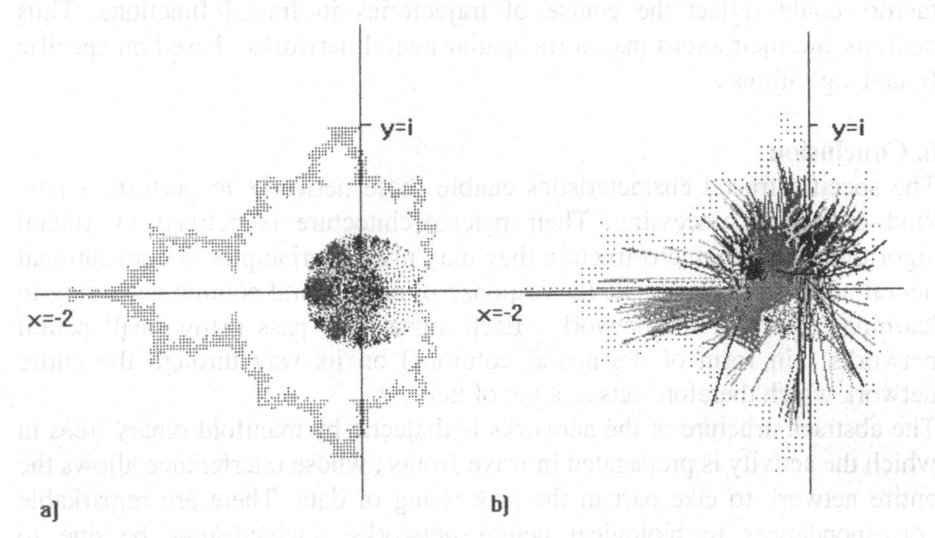

Central structure of the Mandelbrot set as functional " thalamus "

Efferent connections of the " thalamus " to the periphery

Figure 5 : Central structure of the Mandelbrot set (values near zero , which occur in the periodic orbits) in structure and function analogue to a " Thalamus " with distinct "nuclei " (fig. 5a) , specific efferent connections to the periphery shown for "neurons" with y*i > 0 , demonstrating the " Decussatio " of some " Fasciculi " , crossing the median line , resembling to biological thalamic connections (fig.5b).

- activation patterns developing in the course of recurrent projections between peripheral „cortical“ and central „thalamic“ structures . An according central structure we can find in the Mandelbrot set in which each region has its own periodic sequences coming with a certain frequency relatively close to zero. From these " neurons " of the central structure near zero , the activity will be projected to the periphery again (Fig. 5) . Like the Thalamus , this central structure of the Mandelbrot set consists of distinct " nuclei " .
- information processing being a task of the entire network .
- the possibility to activate the entire net from a circumscribed region (near an attractor of f(1) by the recurrent projections of the reversed function in

fractal nets , from some hippocampal regions in human brains in epileptic seizures)
- wave functions playing a role in network function .
- existence (or at least possibility) of associative memory functions .

In ontogenetic development of organisms the orientation of growing axonal fibres in fields of concentration gradients of different neurotrophic growth factors could reflect the course of trajectories in fractal functions. Thus neurons and their axons may form spatial neural networks , based on specific fractal algorithms .

6. Conclusion

The specific fractal characteristics enable these networks to perform a new kind of data - processing. Their macroarchitecture is defined by fractal algorithms. On the microstructure they may use the principles of conventional neural networks to modulate the response of each neural column according to learning history of the network . Each signal will pass many small neural networks (in form of the neural columns) on its way through the entire network which therefore acts as a net of nets .

The abstract structure of the networks is dialectic by manifold binary trees in which the activity is propagated in wave fronts , whose interference allows the entire network to take part in the processing of data .There are remarkable correspondences to biological neural networks , which may be due to reflecting the principles of fractal algorithms in morphogenetic events . Further investigations on these structures promise to offer a chance to produce a new kind of neural networks and to improve our knowledge about function and structure of biological nervous systems .

References :

Babloyantz A , Louren÷o C (1994) Computation with chaos : A paradigm for cortical activity , Proc Natl Acad Sci USA , Vol 91 pp 9027 - 9031 , Biophysics

Birbaumer N , Schmidt RF (1996) Biologische Psychologie, p 598 . Springer Berlin

Destexhe A (1994) Oscillations , complex spatiotemporal behavior , and information transport in networks of excitatory and inhibitory neurons. Physical Review E , Vol 50 Nr 2 : 1594 - 1606

Fink M (1996)Time reversal in acoustics , Contemporary Physics 37 , 2 : 95 - 109

Freeman WJ (1987) Simulation of Chaotic EEG Patterns with a Dynamic Model of the Olfactory System .Biol Cybern 56 : 139-150

Mandelbrot BB (1982) The Fractal Geometry of Nature . Freeman , San Francisco

Pieper H (1985) Die komplexen Zahlen . Deutsch , Frankfurt a. Main

Peitgen HO, Richter PH (1986) The Beauty of Fractals - Images of Complex Dynamical Systems . Springer , Berlin Heidelberg

Wu J , Cohen L, Falk C (1994) , Science 263 , 820 - 823

Adaptive Control Systems Based on Neural Networks

Liliana Dafinca

"Transilvania" University of BRASOV, Electrical Engineering Faculty, 29 Eroilor Bd.
RO-2200 Brasov, Romania
dafinca@unitbv.ro

Abstract. The paper advances the using of the neural networks (NN's) in adaptive control systems. Adaptive control solves the problem of the sensitivity to variation of the plant parameters. In the case of neural adaptive control, the controller parameters are changed by a NN trained off-line. The training patterns are obtained using any design method of the controller for many different values of the plant parameters. A useful tool to train any neural adaptive controller has been developed - a program for Windows '95 that implement the backpropagation algorithm in a general manner.

1. Introduction

The recent rapid and revolutionary progress in power electronics and microelectronics has made it possible to implement and apply modern control theory, well developed over the last decades.

NN's have been proven to be universal approximators of dynamic systems. After they were used for many applications in pattern recognition and in signal or image processing, NN's are now involved in a larger class of scientific disciplines, including system control.

In this context, NN's can be used in adaptive control applications. The term adaptive system implies that the system is capable of accommodating unpredictable environmental changes, whether these changes arise within the system or external to it. This concept has a great deal of appeal to the systems designer since a highly adaptive system, besides accommodating environmental changes, would also accommodate uncertainties and would compensate for the failure of minor system components thereby increasing system reliability.

Among adaptive control techniques, the self-tuning control [1] presented in Fig. 1 is considered in this paper. The system can be thought of as composed of two loops. The inner loop consists of the process and an ordinary feedback controller. The parameters of the controller are adjusted by the outer loop, which is composed of a recursive parameter estimator and a calculator for controller parameters.

The block *controller design* represents an on-line solution to a design problem for a system with known parameters. The self-tuning controller is very flexible with respect to the design method. Virtually any design technique can be accommodated, but on-line design is not always possible because it could be complex and time consuming.

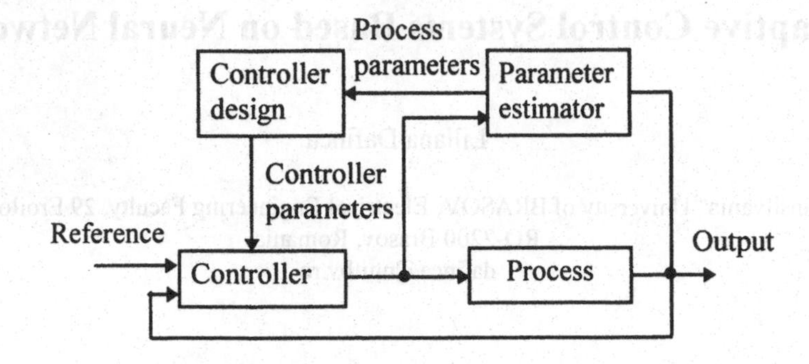

Fig. 1. Block diagram of self-tuning control

If the block *controller design* is replaced by a NN trained off-line, the real time operating becomes possible.

2. Parameter Adaptation of Controller by NN's

2.1 NN Configuration

A three-layer topology allows an efficient training of the NN. Adding one or more intermediate layers does not lead to better results because the hidden nodes have very small variations of weights in the training process, so they learn very slowly [2].

Fig. 2 shows the topology of the NN used in this paper. Every neuron represents a computing element. The output of neuron i at layer l, with $l = 1, 2$ is:

$$y_i^l = f(x_i^l) = \frac{1}{1+e^{-x_i^l \beta}} \quad i = 1,2,\ldots,n_l . \tag{1}$$

$$x_i^l = \sum_{j=0}^{n_l} W_{ij}^l * y_j^{l-1} . \tag{2}$$

Where the notations are:

- f the activation function of the neuron (here it is a sigmoid);
- x_i^l the sum of the inputs to the neuron i at layer $l = 1,2$;
- β the gain of the sigmoid function;
- W_{ij}^l the connection weight from the jth node at layer l-1 to the ith node at layer l.

The neurons from the input layer receive the identified parameters of the process model and yield their outputs through the activation functions. At this layer, the number of neurons (n_0) is imposed by the number of parameters for the process model (ni).

The neurons from the second layer receive the sum of weighted outputs from the input layer and through the activation functions feed their outputs to the output layer.

At this layer, the number of neurons can be chosen by trials. Fewer intermediate neurons will make the training process to become shorter. Following the same procedure as for the previous layer, the parameters of the controller are estimated and outputted from the third layer. The output vector from the third layer contains the estimated parameters of the controller, so the number of neurons at this layer (n_2) is the same with the number of controller parameters (*no*).

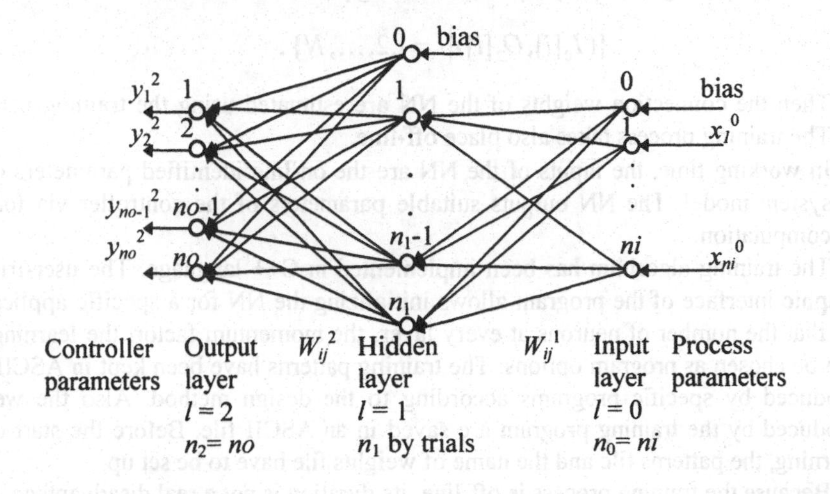

Fig. 2. The neural network configuration

2.2 Neural Adaptive Controller

Replacing the block *controller design* by a NN can be done according to the following procedure.

- Based on the process model and the experiences about the system operating characteristics, the possible ranges of its parameters are defined as:

$$
\begin{aligned}
x_1^0 &\in [x_{1\min}^0, x_{1\max}^0] \\
x_2^0 &\in [x_{2\min}^0, x_{2\max}^0] \\
&\dots \\
x_{ni}^0 &\in [x_{ni\min}^0, x_{ni\max}^0] \, .
\end{aligned}
\tag{3}
$$

Where x_j^0 is the estimated parameter j of the process model.

- The parameter ranges are divided into N sets and the controller is designed off-line for each process parameter set given in (4) according to any standard method considered for the application.

$$I_0[i] = [x_1^0[i], x_2^0[i], \dots, x_{ni}^0[i]], \quad i = 1, 2, \dots, N \, . \tag{4}$$

The parameters of the controller obtained by off-line design for each process parameter set are:

$$O_2 = [o_1^2[i], o_2^2[i], \ldots, o_{no}^2[i]], \quad i = 1, 2, \ldots, N. \tag{5}$$

Where: - $o_k^2\ [i]$ is the parameter k of the controller for process parameters set i.

The data pairs given in (6) represent training patterns for the NN.

$$\{(I_0[i], O_2[i]) \,|\, i = 1,2,\ldots,N\}. \tag{6}$$

- Then the connection weights of the NN are estimated using the training patterns. The training process takes also place off-line.
- In working time, the inputs of the NN are the on-line identified parameters of the system model. The NN outputs suitable parameters of the controller via forward computation.

The training algorithm has been implemented in C++ language. The user-friendly graphic interface of the program allows initializing the NN for a specific application. So that the number of neurons at every layer, the momentum factor, the learning rate can be chosen as program options. The training patterns have been kept in ASCII files produced by specific programs according to the design method. Also the weights produced by the training program are saved in an ASCII file. Before the start of the learning, the patterns file and the name of weights file have to be set up.

Because the training process is off-line, its duration is not a real disadvantage.

The effectiveness of the backpropagation algorithm with momentum for NN training has been tested in several adaptive control systems. One of them is the neural speed control system from Sect. 3.

3. Neural Speed Control for Field-Oriented Induction Motor Drive

In the last decades, many control techniques have been developed and applied to the speed control of electrical drives to obtain high operating performance. However, most of existed controllers cannot lead to good tracking and regulating responses, especially under wide operating ranges. The performances of the system are satisfactory if the controller parameters are changed with the variations of drive parameters. For systematically finding of the controller parameters, a standard design procedure should be used.

3.1. The Speed Control of an Induction Motor Drive

The diagram of an indirect field-oriented induction motor drive is drawn in Fig. 3 [3]. It mainly consists of an induction motor, a hysteresis current-controlled pulse width modulated (PWM) inverter, a slip angular speed estimator, a coordinate translator, and an outer speed feedback control loop.

Parasitic effects such as magnetic saturation, hysteresis, eddy currents and others are generally neglected for control design, so the drive system on Fig.3 can be reasonably represented by the block diagram shown in Fig. 4.

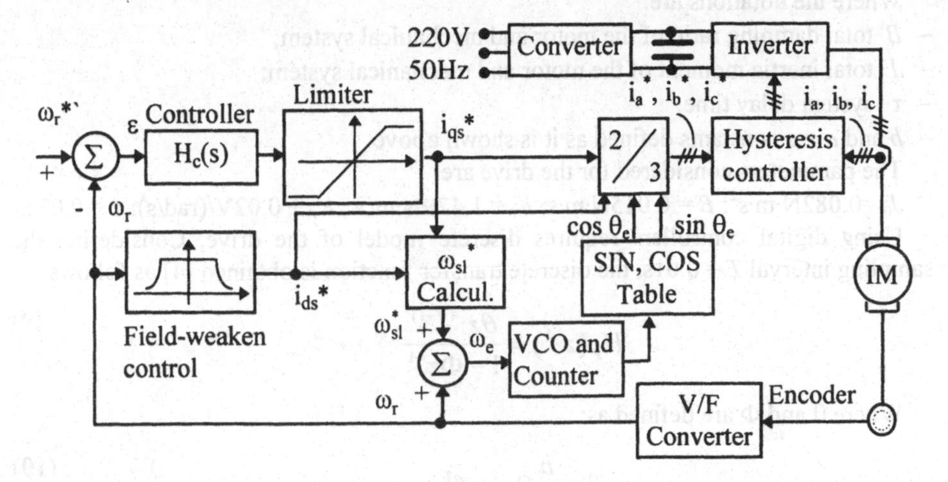

Fig. 3. The diagram of an indirect field oriented induction motor drive

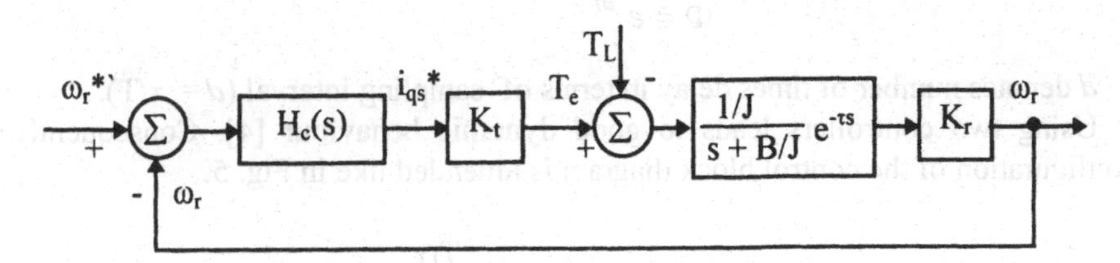

Fig. 4. The dynamic drive model

The notations from Fig. 4 denote:

- K_t the torque constant:

$$K_t = \frac{3p}{4} \cdot \frac{L_m^2}{L_r} \cdot i_{ds}^* . \tag{7}$$

- p number of poles;
- L_m magnetizing inductance per phase;
- L_r rotor inductance per phase referred to stator;
- K_r speed sensor gain factor;
- i_{ds}^*, i_{qs}^* components of the current command;
- T_e, T_L electromagnetic torque and load torque;
- ω_r, $\omega_r^{*\prime}$ machine speed and speed command.

The induction motor is a three-phase squirrel-cage (220V, 50Hz, 4-pole, 750W, Δ-connected, 1000r/min). Its continuous transfer function deduced from the torque equation [3] is given in relation (8).

$$H_p(s) = \frac{1/J}{s + B/J} e^{-\tau s} = \frac{b}{s+a} e^{-\tau s} \,. \tag{8}$$

Where the notations are:

- B total damping ratio of the motor and mechanical system;
- J total inertia moment of the motor and mechanical system;
- τ system delay time;
- b and a are constants defined as it is shown above.

The parameters considered for the drive are:

$J_0 = 0.082\text{N·m·s}^2$; $B = 0.015\text{N·m·s}$; $K_t = 1.479\text{N·m/A}$; $K_r = 0.02\text{V/(rad/s)}$; $\tau = 0.03\text{s}$.

Using digital controllers requires discrete model of the drive. Considering the sampling interval $T = 0.01\text{s}$, the discrete transfer function is obtained [4] as follows:

$$H_P(z^{-1}) = \frac{\theta z^{-(d+1)}}{1 - \Phi z^{-1}} \tag{9}$$

Where θ and Φ are defined as:

$$\theta = \frac{b}{a}(1 - e^{-aT}) \tag{10}$$

$$\Phi = e^{-aT} \,.$$

d denotes number of times delay in terms of sampling interval ($d = \tau/T$).

Using two controllers leads to good dynamic behaviour [4]. Consequently the configuration of the control block diagram is amended like in Fig. 5.

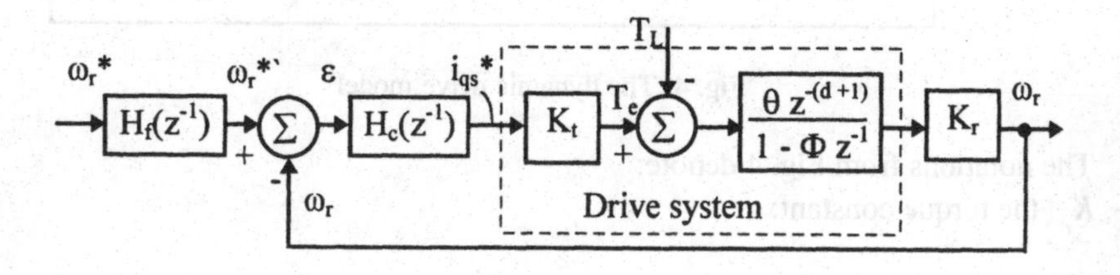

Fig. 5. Drive with two controllers

3.2. A Design Procedure of the Speed Controller

One can chose any design procedure of the speed controller (as in Sect. 2.2). For instance, the direct design method used in digital control systems [4] has been used.

Considering the torque constant K_t and the speed sensor gain factor K_r, the plant model is:

$$H_{pt}(z^{-1}) = \frac{\theta K_t K_r z^{-(d+1)}}{1 - \Phi z^{-1}} = \frac{B(z^{-1})}{A(z^{-1})} \,. \tag{11}$$

Where: $B(z\text{-}1)$, $A(z\text{-}1)$ are polynomial functions defined as it is shown above.
The structures of the controllers have been settled according to [4] as:

$$H_c(z^{-1}) = \frac{z^{-1}E(z^{-1})}{(1-z^{-1})G(z^{-1})} = \frac{z^{-1}(e_0 + e_1 z^{-1})}{(1-z^{-1})(g_0 + g_1 z^{-1} + \cdots + g_{d+1} z^{-(d+1)})} \tag{12}$$

$$H_f(z^{-1}) = \frac{R(z^{-1})}{E(z^{-1})B(z^{-1})} = \frac{r_0 + r_1 z^{-1} + \cdots + r_k z^{-k}}{\theta K_r K_t (e_0 + e_1 z^{-1})} \tag{13}$$

Where $E(z^{-1})$, $G(z^{-1})$ and $R(z^{-1})$ are polynomial functions defined as it has been shown. The order of $R(z^{-1})$ is k that denotes the number of sampling interval in which error in step command tracking response becomes zero.

The design of the feedback controller $H_c(z^{-1})$ is emphasized in obtaining good regulating characteristics and the feedforward controller $H_f(z^{-1})$ is suited to obtain good tracking characteristics.

For the parameters given in the previous section, the discrete drive model of the plant is:

$$H_{pt}(z^{-1}) = \frac{0.0036167z^{-4}}{1 - 0.998203z^{-1}} \tag{14}$$

Following the design procedure introduced in [4], the parameters of the controllers are found as: e_0= 1651.681; e_1= -1374.9984; g_0= 1; g_1= 1.998202; g_2= 2.994609; g_3= 3.989246; g_4= 4.982052; r_0= 0.3; r_1= 0.25; r_2= 0.2; r_3= 0.15; r_4= 0.08; r_5= 0.02.

To test the effectiveness of the designed controller, the simulated rotor speed responses due to a step command change are compared in Fig. 6-8 for three cases: nominal case ($J = J_0$), case 1 ($J = J_0/2$) and case 2 ($J = J_0 \times 2$).

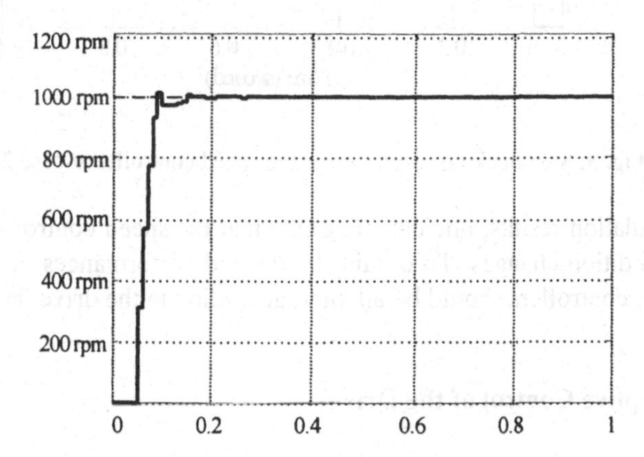

Fig. 6. Step tracking response for the speed controller at nominal case

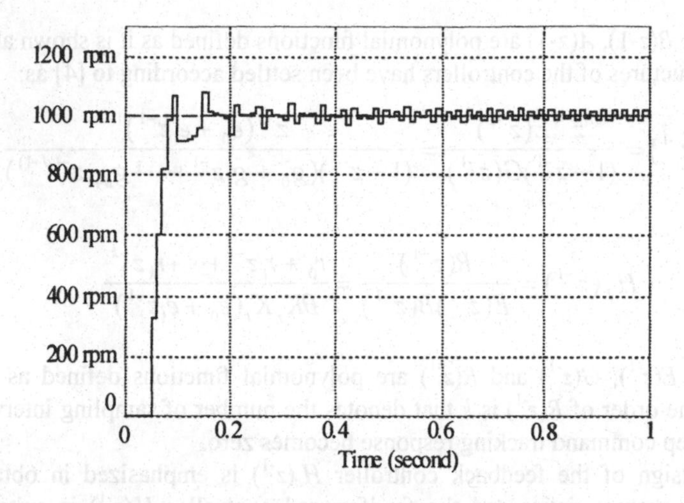

Fig. 7. Step tracking response for the speed controller at case 1

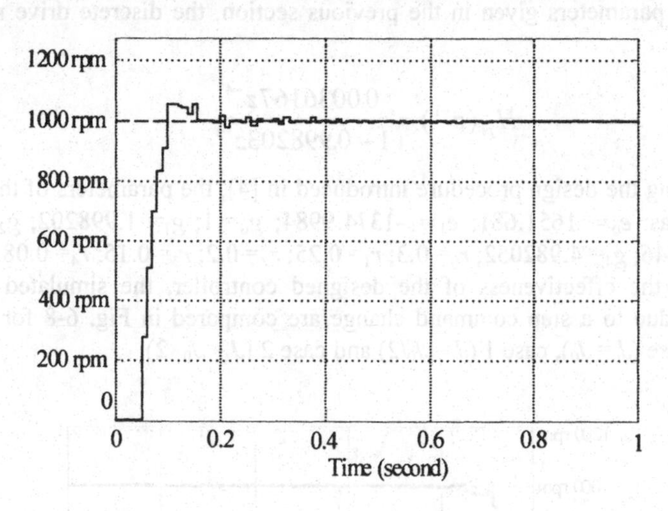

Fig. 8. Step tracking response for the speed controller at case 2

From the simulation results, one can conclude that the speed control is sensitive to the operating condition changes. To obtain the desired performances of the drive, the parameters of the controllers should be adapted according to the drive model changes.

3.3. Neural Adaptive Control of the Drive

Assuming that the mechanical load inertia is significantly changed from the nominal value J_0 to $J_0/2$ (case 1) and to $2xJ_0$ (case 2), according to the estimated drive model, the parameter ranges of the drive model are: $\theta \in [0.061164, 0.244228]$ and $\Phi \in [0.996408, 0.999101]$.

Within these ranges, the parameters are divided into $N = 50$ sets. For each set of drive parameters, a simple program finds the parameters of the controllers.

To enhance the robustness of the speed control, the controller parameters are changed adaptively according to the drive model changes. The parameters of the controllers are modified by a NN as it is shown in Fig. 9.

The neurons from the input layer receive the identified parameters of the drive model and yield their outputs through the activation functions. The inputs to node 1 and 2 at layer 1 are: $x_1^0 = \Phi`$ and $x_2^0 = \theta`$ where $\Phi`$ and $\theta`$ are the estimates of the parameters of the drive model.

The output vector from the third layer contains the estimated parameters of the controller, i.e. $Y^2 = [g_1, g_2, ..., g_{d+1}, e_0, e_1]$, so the number of nodes is $n_2 = d + 2$.

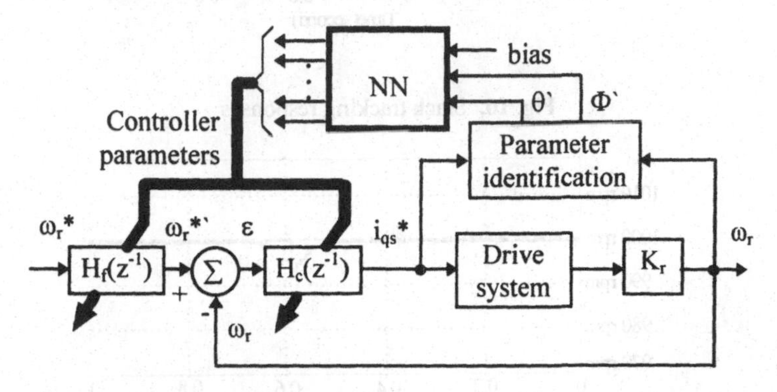

Fig. 9. Neural adaptive control for an induction motor drive

3.4. Simulation Results

For the proposed controller, the tracking responses obtained are rather insensitive to parameter variations as shown in Fig. 10.

Fig. 11 shows the rotor speed response due to step load torque change ($T_L = 1$ N·m applied at 0.5s) for the proposed controller.

Tuning the parameters of the speed controller based on a NN leads to good control performances both in command tracking and load regulation. Also, the control is insensitive to the operating condition changes.

4. Conclusions

This paper describes a new approach of adaptive control based on training ability from examples of NN's and an implementation for Windows '95 of the training algorithm.

The simulation results obtained using SIMULINK (Fig. 10, Fig. 11), for speed control of induction motor drives confirm the effectiveness of neural adaptive control.

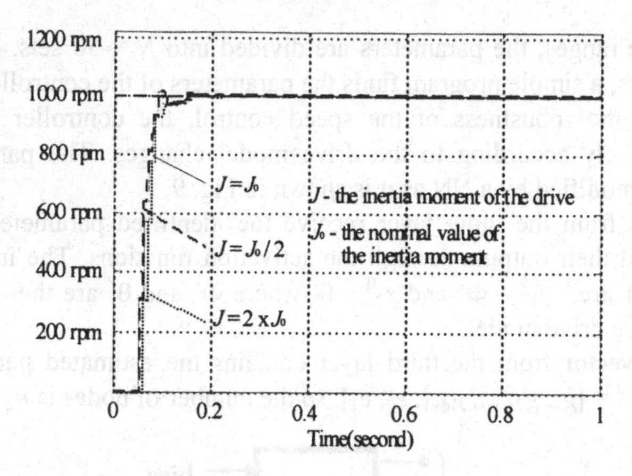

Fig. 10. Stack tracking responses

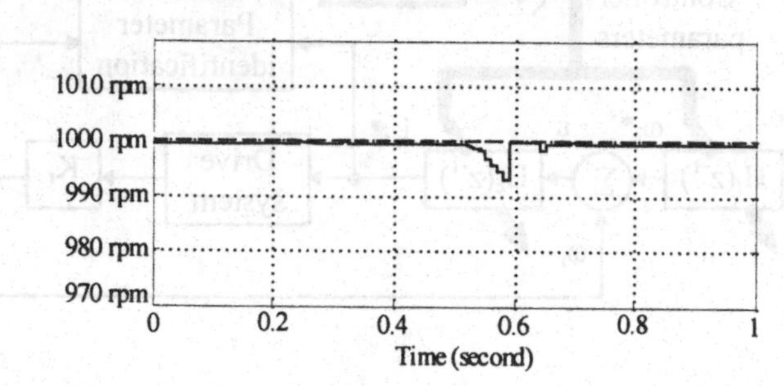

Fig. 11. The rotor speed response due to step load torque change using the proposed controller

The implementation of NN applications can be made by using PC software simulation, by dedicated analog or digital NN chips and by DSP boards. The parallel architecture of the two latter categories are based on multiple processing units that are interconnected to achieve high-speed computation.

Low-cost implementation possibilities and simulation results are arguments for the potential application of NN's to control systems. This motivates to pursue further research in the application of NN's to new types of controllers.

References

1. Dote, Y.: Application of Modern Control Techniques to Motor Control. Proceedings of the IEEE, Vol. 76, No.4 (1988) 438-454
2. Dumitrescu, D., Costin, H.: Retele neuronale. Teorie si aplicatii (Neural Networks. Theory and Applications) Ed. Teora, Bucuresti (1996)
3. Vas, P.: Vector Control of AC Machines. Clarendon Press, Oxford (1990)
4. Calin, S., Petrescu, G., Tabus, I.: Sisteme automate discrete (Digital Control Systems) Ed. Stiintifica si Enciclopedica, Bucuresti (1984)

Function Approximation Using Tensor Product Bernstein Polynomials- Neuro & Evolutionary Approaches

Manuela Buzoianu, Florin Oltean, Alexandru Agapie

National Institute of Microtechnology – IMT Bucharest,
PO BOX 38-160, Bucharest, Romania
E-mail: {manuela, olf, agapie}@oblio.imt.pub.ro

Abstract This paper introduces an approximation technique based on Tensor Product Bernstein Polynomials (TPBPs) and Genetic Algorithms (GAs), res. Neural Networks (NN). First we present the basic model of TPBP, for which suitable control points need to be found, and some of the GA & NN theoretical features. Then we illustrate the efficiency of GAs on multi-parameter optimization in problem of finding optimal control points for TPBPs and the efficiency of NN in our approximation problem. We find these approaches very robust and having good generalization abilities.

1 Introduction

Evolutionary algorithms - [2] receive nowadays a significant attention, especially due to their efficiency in solving multi-modal and multi-parameter optimization problems. One of the most important evolutionary algorithms are Genetic Algorithms.

As for Neural Networks- [5], they provide a robust approach to problems with noisy/incomplete input data. Due to their abstracting ability, NN can also be used in interpolation tasks – this is the subject of our particular case.

2 Tensor Product Bernstein Polynomials

A tensor product Bernstein polynomial is a mapping from $\mathbf{E}^m$ to $\mathbf{R}^n$ defined as a sum of $(w_1+1)(w_2+1)..(w_m+1)$ control points $C_i = \left(C_{i,1}, C_{i,2},...C_{i,n}\right)^T, i = (i_1, i_2,..., i_m)^T$ so that:

$$TPBP^{w_1, w_2, ..., w_m}(t) = \sum_{i_1=0}^{w_1} \sum_{i_2=0}^{w_2} ... \sum_{i_m=0}^{w_m} C_i b_{i_1}^{w_1}(t_1) b_{i_2}^{w_2}(t_2)...b_{i_m}^{w_m}(t_m), \qquad (1)$$

and $b_i^w(t)$ are the Bernstein polynomials:

$$b_i^w(t)=\binom{w}{i}t^i(1-t)^{w-i} \tag{2}$$

For convenience we consider that $w=w_1= w_2=...= w_m$ and call w – the TPBP degree.

Bernstein Polynomials and their tensor products are known in computer graphics. The effects of displacing control points are very intuitive, which enables an easily understanding manual design process. Due to their properties, the general TPBP have proved to be an interesting model for general function approximation. According to [7], TPBPs can also be interpreted as a special class of feed-forward neural networks with two hidden layers.

In order to verify some approximation proprieties of TPBPs we introduce the mean square error *MSE(A)*. Therefore, for a given sample set *A* of pairs of independent values s_k in the source space and dependent values d_k in the destination space, $s_k = (s_{k,1},...,s_{k,m})'$, $d_k = (d_{k,1},...,d_{k,n})'$, $k = 1,...,|A|$, we define

$$MSE(A) = \frac{1}{|A|}\sum_{k=1}^{|A|}\left\|TPBP^{w}(s_k) - d_k\right\|^2 \tag{3}$$

where by $|A|$ we understand the number of set A elements.

We have to minimize this function and to obtain the optimal control point coordinates for TPBPs.

3. Genetic Algorithms and Neural Networks

Genetic Algorithms are probabilistic search algorithms that differ in a fundamental way from traditional optimization methods and search techniques. Basically, a GA requires the definition of five components: the genetic representation of potential problem solutions and its codification, the method for creating an initial population of solutions, the function verifying the fitness of the solution (also called "objective function"), the genetic operators and some constant values for parameters (such as population size, probability of applying an operator).

Thus, the potential solution is represented by a chromosome consisted in real encoded genes which, in our case, describe the control points. So, the length of the chromosome is equal with the number of control points, which varies with the TPBP order. The initial population was randomly generated and its size remains fixed during the algorithm. In the next step, the solutions are evaluated using MSE*(A)* as fitness function and the population is improved in a loop over many generations via the processes of reproduction (selection), mutation and crossover. Besides that, we have taken into consideration other features such as: the crossover probability, $p_c = 0.8$, and the mutation probability, $p_m = 1 / number\ of\ genes$. Moreover, in order to improve the efficiency of GA we employed a *mutation-adaptive* genetic algorithm (this is, a procedure of mutation probability calculus has been added to the canonical GA to assure a better diversity of population from a generation to another).

The characteristics of the operators involved are depicted in Fig.1.

Crossover	*Selection*	*Mutation*	*Next generation selection*
One point	Roulette	Complement	Roulette
		Uniform	Disruptive
Two points	Exponential	Step by step	Elitist
Shuffle	Random	Average	Rank-space

Fig.1. GA operators

All these operators are presented in detail in [6]. For instance the "disruptive next generation" operator works as follows: it preserves the worst and the best chromosome from a generation to another, in order to maintain diversity inside the current population.

When used as an approximation tool, a Neural Network is commonly trained to learn the mapping between the input points from the specific function domain and the output corresponding to the respective function values. In order to achieve a good generalization of function behavior the interpolation methods should have the abstracting ability. As it is well known, NN have this quality and many studies have been dedicated to the problem of function approximation with neural networks.

We designed a special NN, which uses the "information" offered by TPBP polynomial terms. Our concern was not in finding a network that can behave remarkably well with all the test functions. The networks are of feed forward kind with back-propagation of the error. Depending on the function tested we used different back-propagation algorithms depending on the gradient method used in adapting the weights of the network: back-propagation with momentum-[4] and Levenberg-Marquardt back-propagation algorithm.

The network inputs are products of Bernstein polynomials, according to (1):

$$B_{i,j} = b_{i_1}^{w_1}(t_1)b_{i_2}^{w_2}(t_2)...b_{i_m}^{w_m}(t_m)$$

The network is trained to learn these data instead of inputs t_1 and t_2 (the way a simple back-propagation network does). The output layer has one neuron giving the function values to be approximated. Also, the network has one hidden layer, its number of neurons is a parameter to be tuned. The network does not calculate the Bernstein control points directly but a kind of inference.

Let us say that the activating functions are linear and the network is designed for the particular case of a function f:$E^2 \rightarrow$R. Then, the output of the network has the form: $\sum_{j=1}^{m} v_j \sum_{i=1}^{n} w_{ij} x_i$, where n is the number of input neurons, m is the number of hidden neurons, x_i – s are the input points (that is Bernstein products B_{ij}), w is the weight matrix between the input and the hidden layer, and v is the weight vector between the hidden layer and the output. We might say that the coefficient of x_k (after rearranging the sum) is the corresponding Bernstein control point. Thus we could say that in this case the network does calculate directly Bernstein control points.

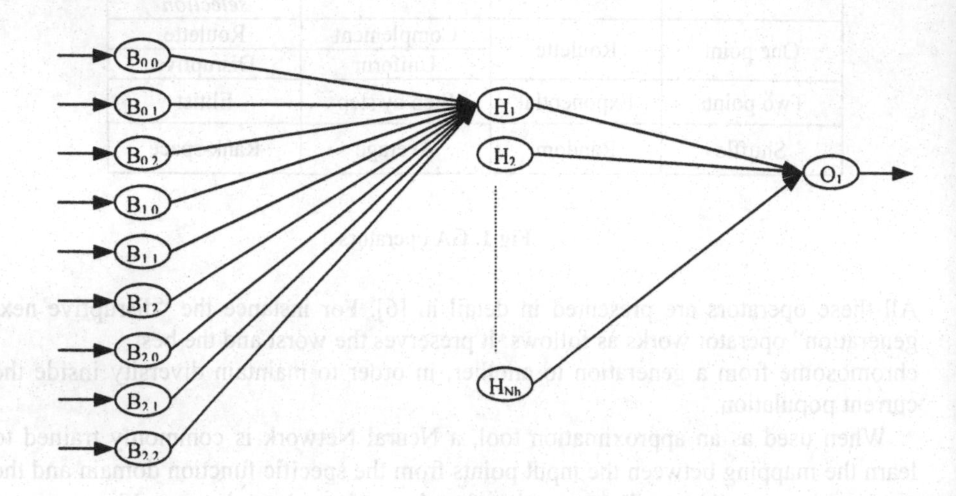

Fig.2. A network model for TPBP of order 2 for a function f:R^2→R to be approximated. (Nh = no. of hidden neurons)

If we introduce sigmoid activating function instead of linear function we could not calculate directly Bernstein control points, but we also might say that the network produce an inference of these control points. This seem reasonably to assume and regarding the fact that the sigmoid function is bijection between R (real numbers) and a segment ([0,1] or [-1,1] depending on the type of sigmoid function chosen).

4 Numerical Examples

We present below the TPBP/GA implementation results, using several sets of test data derived from two test functions. For the GA experiments we used a special GA Toolbox for Excel introduced in [3] with some new improvements, namely a procedure of adaptive genetic algorithm described above. The NN Toolbox in MATLAB obtained the NN results.

Let us introduce the test functions F_1 and F_2 ($[0,1]^2 \rightarrow R$) defined as follows:

$$F_1(x, y) = \sin(10x^2 + 10y^2), \qquad (4)$$

$$F_2(x, y) = -\sin(x)\sin^2(i\pi x^2) - \sin(y)\sin^2(i\pi y^2), \qquad (5)$$

(The second function is a particular case of Michalewicz's function.)

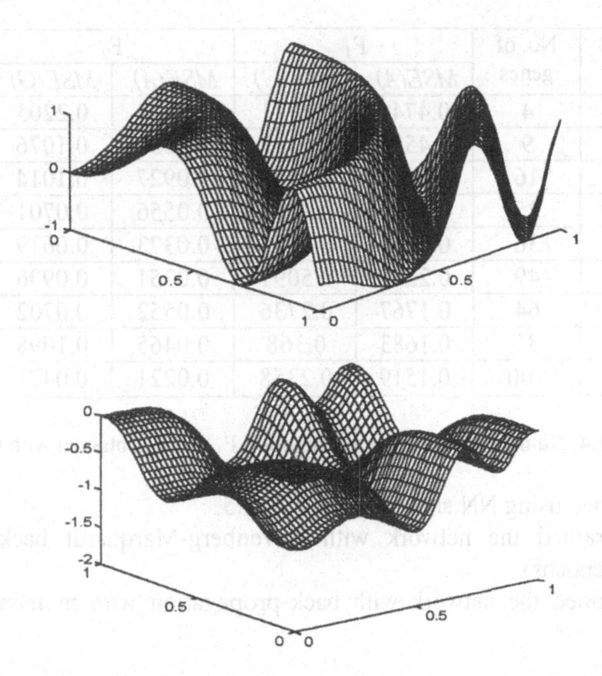

Fig.3. The graphic of function F_1, res. F_2

Let A be a set of 100 randomly chosen pairs from source space. The set A represents the training set.

Generalization abilities of final solutions were measured by using a different set G also made up of 100 samples. Therefore the mean square error $MSE(G)$ is called *generalization error*. (Actually, this corresponds to the usual NN procedure of splitting data into two different sets: *training*, res. *testing* set.) .

We tested both simple and adaptive GA on our optimization problem for every test function introduced above and for TPBP of order $w \in \{1, \ldots, 9\}$ and with two input points. The solution is represented by all control point coordinates C_i, $i=(i_1,i_2)$, where $i_1, i_2 \in \{0, 1, \ldots, w\}$ which can vary between two limits that represent the minimum and, respectively, maximum of test function

We must notice that the best operator setting in the case of simple GA proved to be the following: *shuffle* crossover, *uniform* mutation and *disruptive* next generation. The Adaptive Genetic Algorithms behave very well on our problem, mainly with two mutation types: *complement* and *average*. Almost in all situations, the best results were obtained when we alternate simple GA with adaptive GA in multirun case.

Fig. 4 presents the TPBP/GA numerical results on function F_1 and F_2. The fitting is analyzed for different TPBP degrees. For each degree we run the GA at least two times on the set A and then we averaged the performance of the yielded TPBP on ten randomly chosen testing sets (G). The number of GA's iterations (i.e. generations) is small (increasing from 100 to 20,000, with degree), involving a relatively small running time.

Dg.	No. of genes	F_1		F_2	
		MSE(A)	*MSE(G)*	*MSE(A)*	*MSE(G)*
1	4	0.4745	0.4911	0.2034	0.2205
2	9	0.4524	0.4377	0.1030	0.1076
3	16	0.3944	0.4642	0.0927	0.1014
4	25	0.3230	0.4252	0.0556	0.0701
5	36	0.3067	0.5091	0.0373	0.0619
6	49	0.2232	0.5091	0.0351	0.0996
7	64	0.1767	0.2736	0.0552	0.0702
8	81	0.1683	0.368	0.0465	0.1098
9	100	0.1519	0.2358	0.0221	0.0471

Fig.4. Numerical results on test function F_1 and F_2 obtained with GA

The results obtained using NN are presented in Fig.5.

For F_1 we trained the network with Levenberg-Marquardt back-propagation algorithm (1000 epochs).

For F_2 we trained the network with back-propagation with momentum (10,000 epochs).

Dg.	F_1		F_2	
	MSE(A)	*MSE(G)*	*MSE(A)*	*MSE(G)*
1	0.0051	0.0614	0.0121	0.029
2	0.0042	0.0612	0.0032	0.024
3	0.0023	0.0184	0.0038	0.018
4	0.0061	0.0834	0.0049	0.098

Fig.5. Numerical results on test function F_2 and F_2 obtained with NN

5 Conclusion

As one can see in both tables (fig. 4-5), the GA and NN results - both on training and testing data - confirm the efficiency of TPBP as an interpolation tool.

According to data present in [7], the classical Method of Least Squares (MLS) behave much better on training set than our GA or evolution strategies, but it proved to have not at all generalization abilities. From this point of view, not only GA but also NN perform very well concerning generalization, even though GA do not converge to globally optimal solution as the MLS does. Regarding the special NN designed in our paper, the good approximation achieved might be the result of the extra-information brought by the Bernstein products, B_{ij} (defined above).

Another conclusion is that our NN provide the best results in comparison to GA or to the traditional back-propagation networks used in [7]. This is possible due to NN qualities to be comfortably adapted or to improve solutions intuitively.

References

1. A. Agapie, F. Fagarasan, B. Stanciulescu, *A Genetic Algorithm for a fitting problem*, in Nuclear Instruments & Methods in Physics Research A 389 pp.288-292, 1997.
2. T. Bäck, H.-P. Schwefel, *An overview of evolutionary algorithms for parameter optimization*, Evolutionary Computation, vol.1:1, pp.1-23, 1993.
3. A.H. Dediu, D. Mihaila, *Soft Computing Genetic Tool*, Proc. of the 4th European Congress on Intelligent Techniques and Soft Computing (EUFIT 96), Aachen, Germany, September 1996, pp. 415-418.
4. J.A. Freeman, D.M. Skapura, *Neural Networks, Algorithms, Applications & Programming Techniques*, Springer-Verlag, Berlin.
5. J. Hertz, A. Krogh, R. G. Palmer, *An Introduction to the Theory of Neural Computing*, Addison-Wesley, 1991.
6. Z. Michalewicz, *Genetic Algorithms + Data Structures = Evolution Programs*, 3rd ed., Springer-Verlag, Berlin, 1996.
7. G. Raidl, G. Kodydek, *Evolutionary Optimized Tensor Product Bernstein Polynomials versus Backpropagation Networks*, Proc. of the Int. ICSC/IFAC Symposium on Neural Computation, Vienna, Austria, pp. 885-890, September 1998.

Tuning Considerations above a Fuzzy Controller Used for the Synchronous Generator

Ioan Filip[1], Octavian Prostean[1], Daniel Curiac[1]

[1] "Politehnica" University of Timisoara
Blvd. V.Parvan no.2, Timisoara 1900, Romania
{ifilip, prostean, curiac}@aut.utt.ro

Abstract. This paper presents an on-line tuning algorithm for a PI fuzzy controller and several concrete aspects that can be used for the terminal voltage control of a synchronous generator connected to a power system. The synchronous generator is described by a 6th order nonlinear complex model (proceeding from the Park's equations). Some specific operating conditions (loading and unloading of active power, connecting and disconnecting of power consumers) where considered and simulated using Matlab with Simulink. All results are demonstrated by computer simulation studies.

1 Introduction

The application of fuzzy control described in this paper is referring to the terminal voltage control of a synchronous generator coupled to a power. The control input is the excitation voltage. The transmission network between synchronous generator and power system can be described by equivalent impedance of a long transmission line and by the equivalent impedance of a terminal connected consumer (figure 1.). Tuning the parameters of a fuzzy logic controller is a very important issue that is a prerequisite for any fuzzy control application.

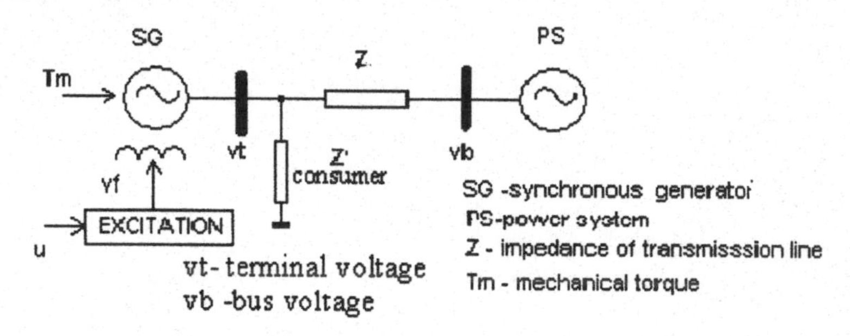

Fig. 1. Synchronous generator connected to a power system.

2 The Designed Fuzzy Control Structure

In figure 2 is depicted the structure of the fuzzy control system taken into consideration, which is build by a proportional-integral (PI) fuzzy controller and a block that models a synchronous generator connected to a power system. The quantities that can disturb this complex process are highlighted and their rejections are one of fuzzy logic controller duties.

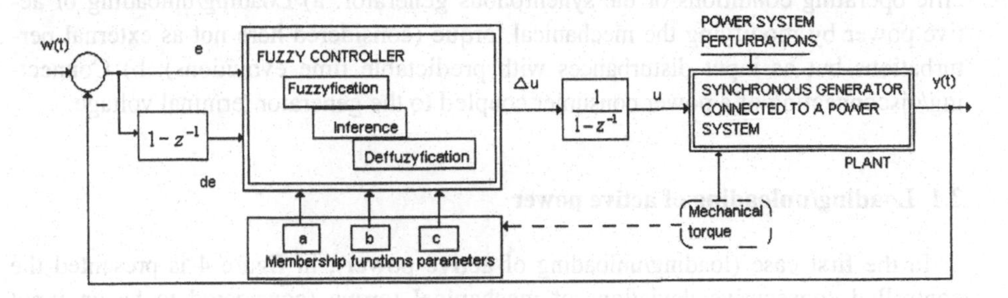

Fig. 2. The Fuzzy Control Structure

Figure 3 presents the membership functions chosen for this application, while the output functions are singletons. As an inference strategy was used Mamdani fuzzy inference method and as a defuzzyfication strategy was used the centroid (center) of area method based on singletons. The relationship between the tuning parameters a,b,*c* used in figure 2 and figure 3 are: $a=a_2=2a_1$, $b=b_2=2b_1$, $c=c_2=2c_1$.

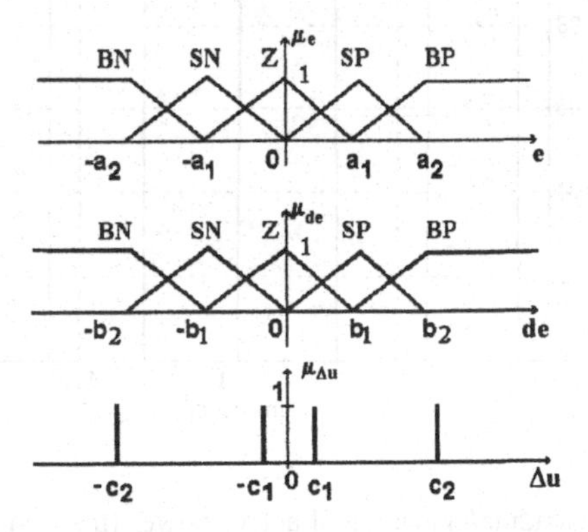

Fig. 3. The membership functions

This shape selection for membership functions (connected with the relation between parameters *a, b, c* mentioned before) takes to a decrease, down to 3, of the fuzzy controller's tuning parameters. The smaller the number of tuning parameters is the easier the tuning process is.

3 Parameter Tuning Algorithm. Simulation and Results

Designing and implementing of a fuzzy controller requires an initial parameters off-line tuning in order to reach the desired performances. Taken into consideration the multitude of GS operating conditions, an on-line parameters tuning becomes almost obligatory.

The tuning process was developed by considering two kinds of tests for two specific operating conditions of the synchronous generator: a) Loading/unloading of active power by modifying the mechanical torque (considered here not as external perturbations but as input disturbances with predictable time evolutions); b) Connecting/disconnecting of a power consumer coupled to the generator terminal voltage.

3.1 Loading/unloading of active power

In the first case (loading/unloading of active power), in figure 4 is presented the controlled consecutive deviations of mechanical torque (considered to be an input signal with a known variation). These consecutive variations are used as test signals for off-line tuning of controller's parameters (*a,b,c*), in order to obtain high performances for controlled output (terminal voltage).

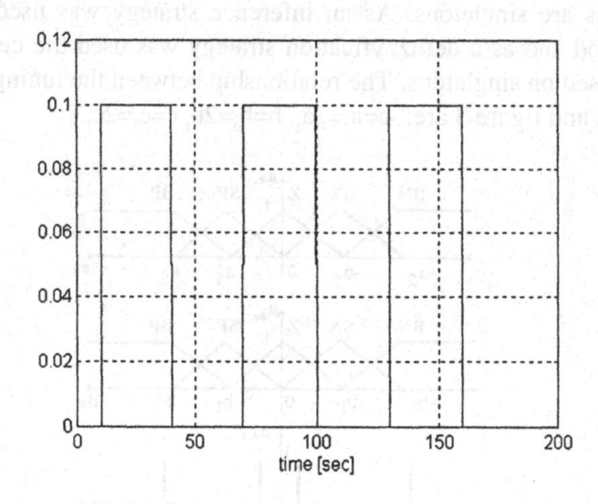

Fig. 4. The consecutive loading/unloading of active power (mechanical torque deviations)

All values of inputs (including disturbances) or outputs (terminal voltage) are in per-unit. Consecutive values of controller's parameters for the three off-line tuning steps (corresponding to the three consecutive variations of mechanical torque) are presented in Table 1.

Table 1. Parameter's variations

Par.	Time moment (t) [sec]		
	t=0	t=60	t=120
a	$a_{(t=0)}=a_0=0.05$	$a_{(t=60)}= a_0+0.01=0.06$	$a_{(t=120)}= a_0+0.015=0.065$
b	$b_{(t=0)}=b_0=0.001$	$b_{(t=60)}= b_0-0.0005=0.0005$	$b_{(t=120)}= b_0-0.00092=0.00008$
c	$c_{(t=0)}=c_0=0.005$	$c_{(t=60)}= c_0+0.005=0.01$	$c_{(t=120)}= c_0+0.0051=0.011$

In figure 5 the output deviation (taken from a set point), and its overshoot/downshoot are decreased with any new controller's parameters tuning step. Consecutive tuning steps, presented in table 1, point to the tendency of variation for controller's parameters. Better performances could be obtained by increasing of the a and c parameters and by a decreasing of b parameter. Final tuning value for (a,b,c) triplet is (0.065,0.00008,0.011) which provides good performances in case of loading/unloading of active power (by modifying the mechanical torque). Other three case studies for tuning parameters (for the same loading/unloading of active power process) had been considered (figure 4). The related results are presented in figures 6, 7 and 8. Each study considers an identical variation with the one described in table 1 for two parameters. For the third parameter only the initial value is modified, but the tuning increment is considered to be the same (Table 1).

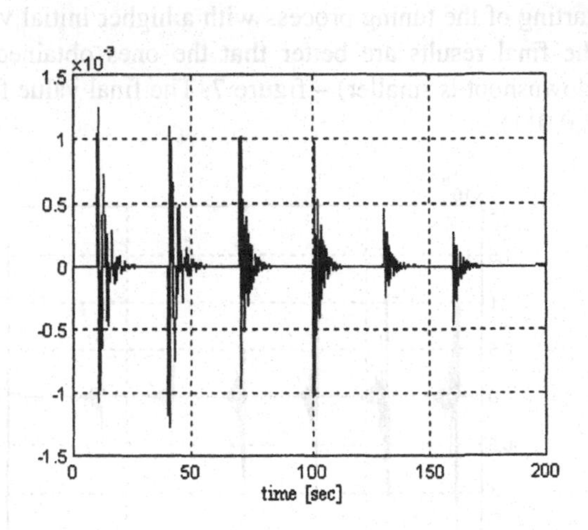

Fig. 5. The consecutive deviations of the output (for $a_{(t=0)}=a_0=0.005$)

Figure 6 presents the output deviation in case that the initial value for a parameter is decreased ($a_{(t=0)}=a_0=0.005$) – the other two parameters are considered to vary like in Table 1. We could see a rise of the oscillations in transient operating conditions.

For the (a,b,c) final triplet taken the value (0.002,0.00008,0.011), the performances obtained are similar with the ones obtained with the triplet (0.065,0.00008,0.011). Thus, the influence of initial modifying of a parameter is not very important.

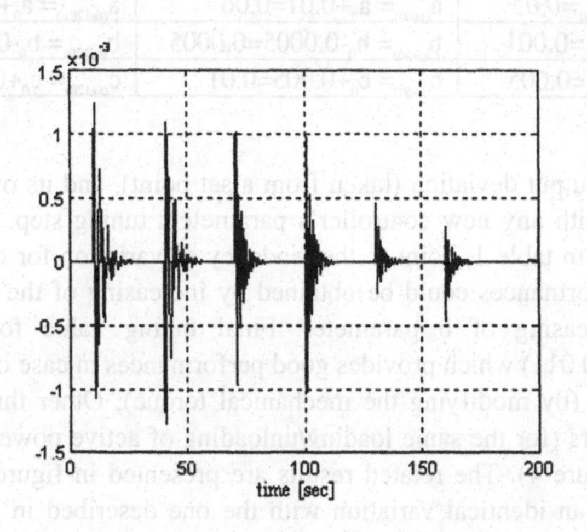

Fig. 6. The consecutive deviations of the output (for $a_{(t=0)}=a_0=0.005$)

In case of restarting of the tuning process with a higher initial value of parameter c ($c_{(t=0)}=c_0=0.01$), the final results are better that the ones obtained in previous cases (final overshoot/downshoot is smaller) – figure 7. The final value for the (a,b,c) triplet is (0.115,0.00008,0.011).

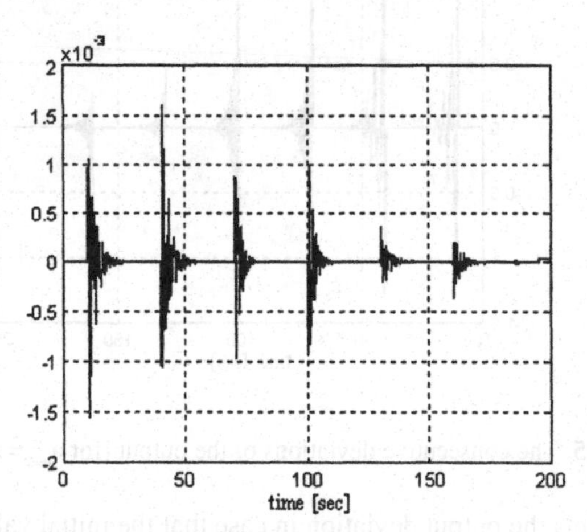

Fig. 7. The consecutive deviations of the output (for $c_{(t=0)}=c_0=0.01$)

In case of restarting of the tuning process with a higher initial value of parameter b ($b_{(t=0)}=b_0=0.002$), the final results are worse that the ones obtained in previous cases and the effect of modification for a and c parameters is useless (figure 8).

A tuning process can be finished after all the imposed performance indicators are reached. The best results had been obtained for the parameter set (a,b,c)= (0.115,0.00008,0.01). It could be possible to find another (a,b,c) set that provide similar performances. Also, we could say that by modifying the parameters a and c around tuned values similar results may be obtained.

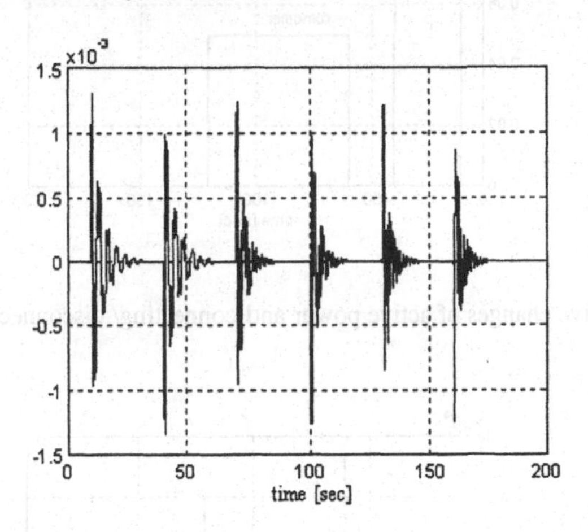

Fig. 8. The consecutive deviations of the output (for $b_{(t=0)}=b_0=0.002$)

3.2 Connecting/disconnecting of a power consumer

By considering the optimum value for the parameters triplet (a,b,c)= (0.115,0.00008,0.011), we made an analysis about the variation of controlled output (terminal voltage) in case of connecting/disconnecting of a power consumer.

The figure 9 presents a sequence of two functioning regimes (operating conditions) for synchronous generator: first a loading/unloading of active power (in this operating conditions the tuning had been made), followed by connecting/disconnecting of a power consumer. Figure 10 shows that in new operating conditions (connecting/disconnecting of a power consumer), the controlled output performances are low (long transient operating conditions, big settling time). It results the necessity of controller's parameters re-tuning for the new operating conditions. This assumption is proved by the figure 11 which presents the output deviation in case that the controller's parameters were re-tuned at the time moment t=60 seconds.

Computer simulation show that by tuning only the b parameter (a and c parameters were considered to have the same values obtained in the off-line tuning process) we could obtain good performances even in the case of power consumer connecting/disconnecting. For example, in figure 11, the b parameter has been raised at t=60 seconds with a small quantity ($b=b_{initial}+0.0004$), so the final (a,b,c) triplet is (0.115,0.00048,0.011). Around this parameters set (a,b,c) an on-line parameter tuning is made for any active power changes resulted from a torque variation.

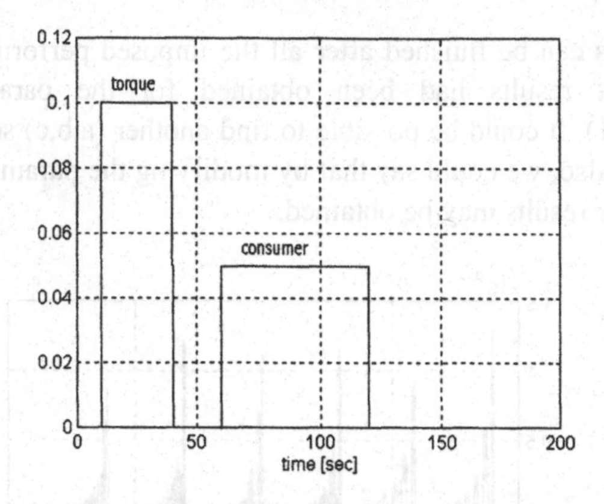

Fig. 9. Consecutive changes of active power and connecting/ disconnecting of a consumer

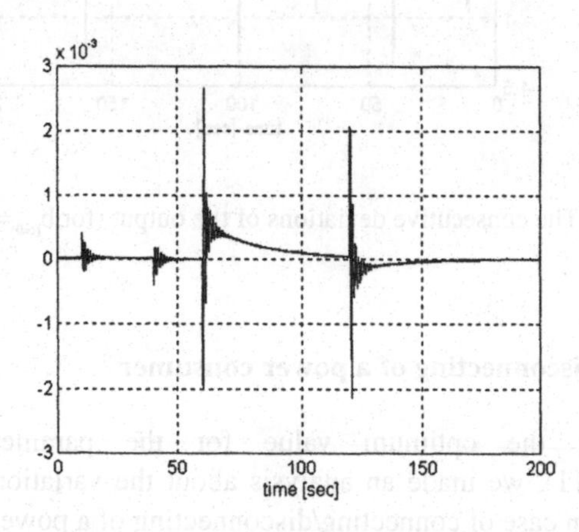

Fig. 10. The consecutive deviations of the output (without parameters re-tuning)

A power consumer unpredictable connecting/disconnecting represents harder operating conditions than an active power change (obtained by a predictable, desired modifying of mechanical torque). For this reason, a controller's parameters set (off-line tuned) is used for consumer unpredictable connecting/disconnecting operating conditions and an on-line parameters tuning method is used for a predictable change of active power.

The objective of o tuning algorithm is to re-shape the membership functions to get the desired system response. The algorithm is based on measuring two indices, which are: the mechanical torque deviation and the system output error. In this case, the algorithm tries to optimize the controlled output performance (in both operating conditions: change of active power -as a desired input disturbance; and connecting/disconnecting of a power consumers - as an unpredictable perturbation) by varying (tuning) the **b** parameter.

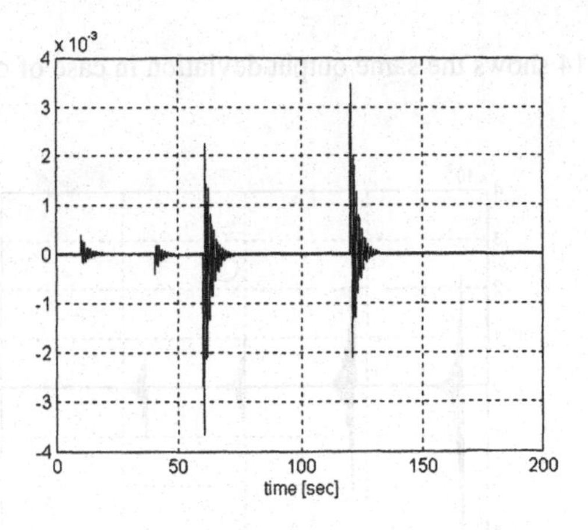

Fig. 11. The consecutive deviations of the output (with re-tuning at time t=60)

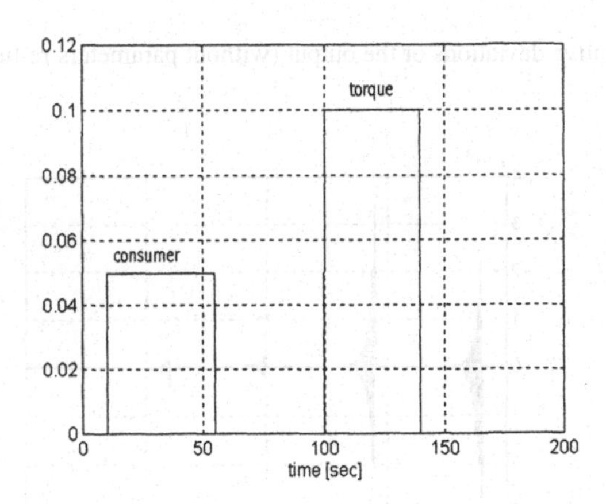

Fig. 12. Consecutive connecting/disconnecting of a consumer and changing of active power.

The basic ideas of on-line tuning method are: a) maintaining the parameters set obtained in off-line tuning process in case of consumers unpredictable connecting or disconnecting ($(a,b,c)_{stationary}$ =(0.115,0.00048,0.011)); b) if a non zero mechanical torque deviation appears, an on-line tuning process will be performed only for b parameter ($b=b_{stationary} -\alpha(\Delta T_m)$, where α is a constant value and (ΔT_m) is mechanical torque variation; c) maintain the new tuned values of b parameter until output error leaves the desired domain; d) if the output error is again in the desired domain the old b parameter is reloaded ($b=b_{stationary}$);

The consecutive modifications of operating condition by connecting/disconnecting of a power consumer and by loading/unloading of active power (modifying mechanical torque) is presented in figure 12. For such a sequence of operating conditions, figure 13 shows the output deviation (terminal voltage error) without parameter re-tuning and figure 14 shows the same output deviation in case of on-line parameters re-tuning.

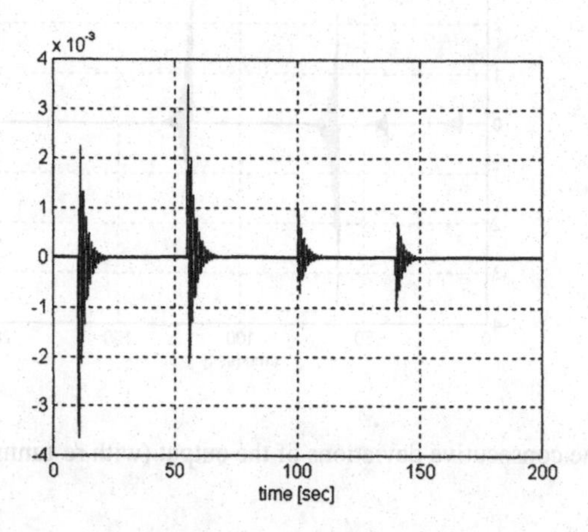

Fig. 13. The consecutive deviations of the output (without parameters re-tuning)

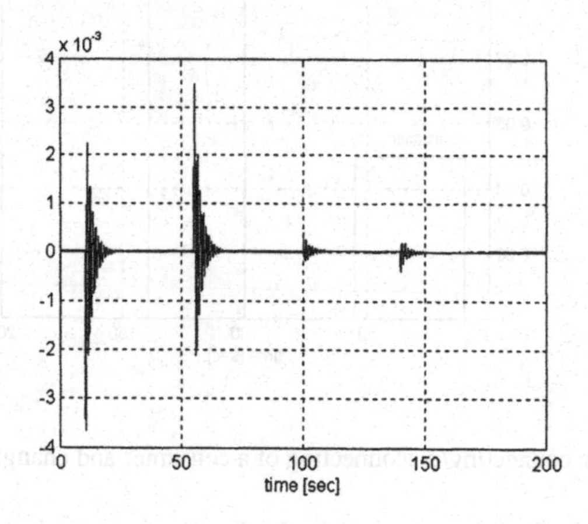

Fig. 14. The consecutive deviations of the output (with on-line re-tuning at time t=100)

4. Conclusion

One can notice the absence of on-line parameters tuning, corresponding to operating conditions caused by connecting/disconnecting of a power consumer, which are practically unpredictable.

Also, the controller's on-line tuning must be done especially for obtaining high performances in operating conditions caused by active power loading/unloading (related to a modification of mechanical torque). This kind of conditions could be anticipated and, simultaneously, a new set of parameters could be obtain by on-line tuning.

The proposed on-line tuning algorithm was validated in different case studies. The computer simulations results, definitively indicates that there is not only one (a,b,c) parameter set for obtaining high performances.

References

1. Zadeh L.A. : Fuzzy Sets, Information and Control, Vol.8, pp.338-353,1965
2. El-Metwally K.A., Hassam M.A., Malik O.P.: Robustness of fuzzy logic controllers , Proceedings, 2nd IASTED International Conference, Computer Applications in Industry, Alexandria, Egypt, pp.386-389, 1992
3. Ollero A., Garcia-Cerezo A.J. : Direct Digital Control, Auto-Tuning and Supervision Using Fuzzy Logic, Fuzzy Sets and System, Vol.30, 1989, pp.135-153
4. Zhao Z. Y., Tomizuka M., Sagara S. : A Fuzzy Tuner for Fuzzy Logic Controllers, Proceedings of the American Control Conference, 1992, pp.2268-2272.
5. Romero D.., Gerald Thomas Heydt : An Adaptive Excitation System Controller in a Stochastic Environment, IEEE Transaction on Power Systems, Vol.PWRS-1, No.1, February 1986
6. Filip I., Prostean O.: Self-Tuning Controller for Synchronous Generator Excitation System, Buletinul Stiintific si Tehnic al UPT, Seria Automatica si Calculatoare, Tom 42(56), 1997

A Sectoring Genetic Algorithm for the Urban Waste Collection Problem

Lamata M.T. [a] Peláez J.I.[b] .Sierra J.C. [b] Bravo J.M.[b]

[a] Dpto. Ciencias de la Computación e Inteligencia Artificial. E.T.S. Ingenieria Informática.
Universidad de Granada. 18071. Granada. España.
E-mail: Mlamata@goliat.ugr.es.
[b] Dpto. Lenguajes y Ciencias de la Computación. Complejo Tecnologico. E.T.S.I. Informática.
Campus de Teatinos. Universidad de Málaga. 29071Málaga. España.
E-mail: jignacio@lcc.uma.es. E-mail: jcs@eva.psi.uma.es.

Abstract. Urban waste collection is a problem that has been widely studied in the last decades. In this work, this problem is analysed and it is shown how traffic direction restrictions may be treated using work levels, so the obtained solutions are similar to those provided by experts. Also a genetic algorithm is introduced, which uses operators based on work levels and a fitness function based on an expansive competitive learning that characterises the optimal solution.

1 Introduction

The Urban Waste Collection Problem (UWCP) is a problem of optimisation, where the problem of sectors is present. The problem consists on a set of geographically distributed nodes connected among themselves. The direction of these connections is that of the traffic. The aim is to find sectors of: (1) shortest distances; (2) adyacent nodes ; (3) greatest load.

Previous works in Operational Research tell us about a great variety of versions and algorithms [1], [2], [3], [4], [5] for different purposes such as: mail service, electrical inspections, waste removal, etc. The generation of efficient sectors that could be found from the demands imposed by experts would imply a saving worth millions of Euros.

The paper is set out as follows: section 2 introduces the l work levels; in section 3 we propose the genetic algorithm; in section 4 provides a practical example; and finally, we present the conclusions.

2 The Representation Problems

The main objection set by urban waste collection companies to the proposed solutions so far is relative to the direction of traffic in the cities. The problem arises when nodes that are apparently neighbouring are in fact distant, due to the traffic direction. The solution to this problem is utterly complicated, since urban distance matrices cannot be applied in an effective way. A possible solution is to use work levels in genetic operators, and so cross and mutation operations take place according to such levels.

We propose the work levels as:

Work levels

We say that nodes C_i and C_{i+1} belong to level *n*, if the shortest route between C_i and C_{i+1} has *n* intermediate nodes.

A feasible solution for sectoring in the UWCP rejects those solutions in which the level between nodes is very high. This causes a simplification of the search space. Similarly, considering only solutions in which nodes belong to level 0 is very restrictive. Therefore, we propose a *work level* that determines the search space for the optimal solution depending on the level between nodes.

3 Genetic Algorithm

We now introduce: (1) the structure of an individual; (2) the evolutionary process; (3) fitness; (4) and the genetic operators.

Individual

In the UWCP each individual represents a feasible solution. An individual is made up of a set of chromosomes, each one of them representing a sector.

Evolutionary process

The evolutionary process used in the genetic algorithm is a "steady-state". Next we illustrates this evolutionary process.

```
Agorithm Evolutionay_process
Begin
        GENERATE Initial Population.
        For i=1 to Num_Generations
            SELECT the two best Individual and
             PUSH in the next generation.
            SELECT fathers and mothers.
            For j = 2  to  MAX_POPULATION
               CROSSING father  and  mother  => Son.
               MUTATING  Son.
                IF Fitness (son) < Worst_Fitness
                ACTUAL the  Population with Son
            End
        End
  End
```

Fitness

The proposed fitness is based on the expansive competitive learning rule, and is composed of two elements: (1) the first one seeks solutions where the sectors' distance to their gravity centre is minimum; (2) the second one attempts to find a global minimum by looking for solutions in which the gravity centres are sufficiently distant from each other.

The first fitness is calculated as follows:

$$Fitness1 = \frac{\sum_{i=1}^{n}\left(\sum_{j=1}^{mi} DE\left(CM_i, C_{ij}\right)\right)}{n} \qquad (1)$$

where: DE is the Euclidean distance, CM_i is the "gravity centre" for chromosome i, C_{ij} represents the coordinates of nodes j for chromosomes i, m_i is the number of nodes of chromosome i, and n the number or cromosomes.

The second fitness is calculated as follows:

$$Fitness2 = \frac{\sum_{i=1}^{n}\left(\overline{CM} - CM_i\right)^2}{n} \quad (2)$$

and finally, the global fitness is calculated as:

$$Fitness = \alpha \bullet Fitness1 - \beta \bullet Fitness2 \quad (3)$$

where $\alpha, \beta \in [0,1]$.

Selection operator

This operator selects the individuals to be crossed, using the roulette method.

Crossing operator

The algorithm shows the crossing operator. The individual *son* is generated by father and mother individuals, which are selected by the selection operator. The functions "Remove_Repeats" and "Put_Pending" carry out their operations considering level restrictions.

```
ALGORITHM Crossing  (Varson: T_Individual;
                              father,m other : T_individual);
VAR chromo: T_chromosome;
Begin
  FOR i = 1  TO Total_Cromos;
    chromo=RAND(CHROMOSOME(father,i),
                CHROMOSOME(mother, i));
    CHROMOSOME(son, i) = chromo;
    REMOVE_ REPEATS(son);
    PUT_PENDING (son),
  End
End
```

Mutation operator

This operator removes links that do not satisfy the level restriction. The algorithm for the mutation operator is:

```
ALGORITHM Mutation (son: T_individual);

VAR  chromo1, chromo2 : T_gen;

    node1, node2 =T_Node;

Begin

        n = RAND (5);

        FOR i = 1 To n do

          chromo1=RAND(Total_Chromosomes);

          chromo2=RAND(Total_Chromosomes);

          node1=NOT_LEVEL (son,chromo1);

          node2= NOT_LEVEL (son,chromo2);

          If CHANGE (node1,node2,son)

              CHANGE (node1,chromo2,chromo1);

              CHANGE(node2,chromo1,chromo2);

          End

      End

End
```

4. Computational Results

Figure 4.1. shows the initial problem, with 246 waste points. Figure 4.2. shows the sectors that are obtained after applying the genetic algorithm. Values α and β used in fitness 1 and fitness 2 are equal to 0.6 and 0.4 respectively, and the work level is set to five..

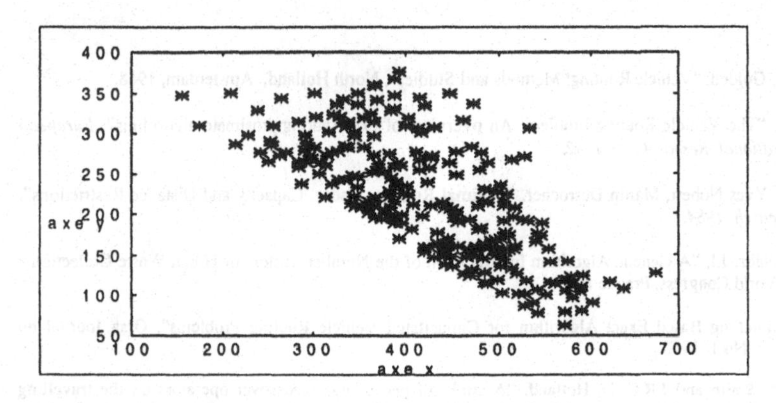

Fig. 4.1. Initial problem

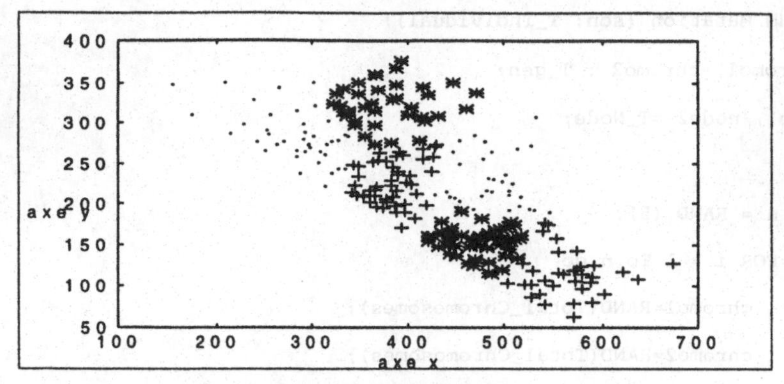

Fig. 4.2. Problem solution

5 Conclusion

The final conclusions are:

- The use of work levels is suitable to solve sectoring problems in the urban waste collection.
- The main factor that influences the urban waste collection solution is representation.
- This algorithm supplies solutions that satisfy the restrictions imposed by the experts.

Acknowledgements:

This work is partially supported by the DGICYT under projects PB92-959 and PB95-95-1181, and the project 8.06/47.1281 between Empresa Malagueña Mixta de Limpieza S.A and Málaga University.

References

1. Assad, A. and B. Golden, "Vehicle Routing: Methods and Studies", North Holland, Amsterdam, 1988.

2. Gilbert Laporte. "The Vehicle Routing Problem: An overview of exact and approsimate algorithms". *European Journal of Operational Research, 59, 1992.*

3. Gilbert Laporte, Yves Nobert, Martin Desrocher. "Optimal Routing under Capacity and Distance Restrictions". *Operations Research. 1984.*

4. Lamata, M.T. Peláez, J.I. "A Genetic Algorithm for Decision of the Number Sectors of Urban Waste Collection". Seventh IFSA World Congress, Prague 1997.

5. Miller D. "A Matching Based Exact Algorithm for Capacitated Vehicle Routing Problems". Orsa Journal on Computing. Vol 7. No 1. 1995.

6. Oliver, I. M,D,J. Smith and J.R.C. D. Holland, "A study of permutation crossover operators on the travelling salesman problem". *Proc. Of the second Int. Conf. On Genetic Algorithms,* Lawrence Erlbaum Assoc. Pp. 224-230, 1987.

7. Pérez J. " Aprendizaje Competitivo para la Cuantificación Vectorial". CAEPIA'97. Málaga. Spain.

8. Sam R. Thangiak, Kendall E. Nygard, Paul L. Juell, "Gideon: Agenetic Algorithm System for Vehicle routing with time windows". IEEE. 1991.

Parameter Optimization of Group Contribution Methods in High Dimensional Solution Spaces

C. Kracht, H. Geyer, P.Ulbig, and S. Schulz

University of Dortmund, Department of Chemical Engineering, Institute for Thermodynamics, 44221 Dortmund, Germany*
kracht@th.chemietechnik.uni-dortmund.de

Abstract. The prediction of certain thermodynamic properties of pure substances and mixtures with calculation methods is a frequent task during the process design in chemical engineering. Group contribution methods divide the molecules into functional groups and if the model parameters for theses groups are known, predictions of thermodynamic properties of compounds that comprise these groups are possible. Their model parameters have to be fitted to experimental data, which usually leads to a multi-parameter multi-modal optimization problem. In this paper, different approaches for the parameter optimization are tested for a certain class of substances. One way to carry out the optimization is to fit only one group interaction at a time, which results in six parameters, that have to be fitted. The downside of this procedure is, that incompatibilities between different parameter sets might occur. The other way is to fit more than one group interaction at a time. This further increases the variable dimension but prevents incompatibilities and leads to thermodynamic more consistent parameters because of a greater data base for their optimization. Therefore, investigations on those different optimization procedures with the help of encapsulated Evolution Strategies are made.

1 Introduction

Group contribution models are used for the prediction of certain thermodynamic properties, such as activity coefficients, excess enthalpies or heats of vaporization to assist in design and simulation of chemical processes. These thermodynamic properties originate in physical interactions between molecules. Group contribution models split the molecules into functional groups and the physical interactions between the functional groups can be determined, if the model parameters for these functional group interactions exist. Figure 1 shows the transition from molecules to groups as an example. The arrows represent the physical interactions between the molecules or between the groups.

* Member of the Collaborative Research Center SFB 531: "Design and Management of Technical Processes and Systems by Using Methods of Computational Intelligence", supported by the Deutsche Forschungsgemeinschaft DFG

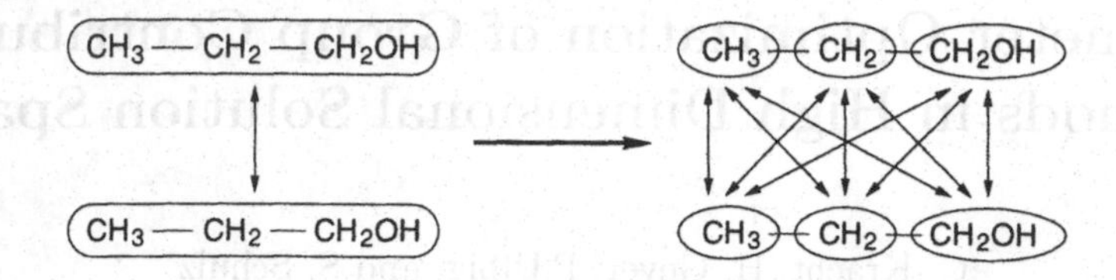

Fig. 1. Group contribution concept

In many modern methods, such as modified UNIFAC [2, 12] and EBGCM [5], the model parameters appear in complex terms such as sums of exponential terms, especially if temperature dependences are described. This leads to a nonlinear regression problem that often results in a multi-modal optimization problem. In this case the search of the global or at least a fairly good optimum becomes rather difficult, although a good prediction of thermodynamic quantities depends on well fitted interaction parameters. Figure 2 shows the topology of the objective function by varying two of the model parameters while the remaining parameters are kept constant.

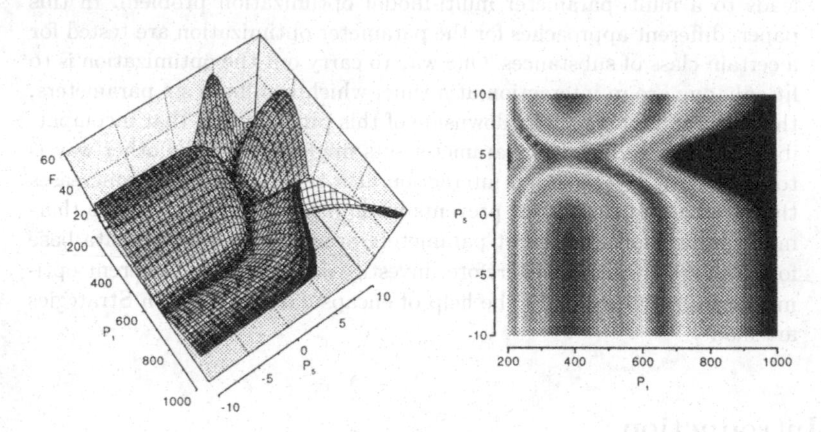

Fig. 2. Adaptive surface of a modified UNIFAC group contribution model by varying two parameters

The parameters of the group contribution models EBGCM and modified UNIFAC are fitted to binary excess enthalpy data sets h^{E} (i. e. the mole fraction of the considered binary mixture and its corresponding excess enthalpy value) at different temperatures. For these models six parameters per one functional group combination have to be optimized [6].

The objective function for the optimization is not the model equation itself in which the model parameter vector $\boldsymbol{p}$ occur but an error criterion which establishes the relation between the calculated value and the respective experimental data point. The fitting is carried out with the mean relative range related error $\mathrm{MRE_R}$ (eq. 1). The range $\Delta h^{\mathrm{E}}_{exp,R_j}$ is defined as the difference between the largest and the smallest h^{E} value of the considered isotherm. By this the deviation between experimental and calculated data points near the zero line (at the edges of the concentration range, for instance) is not weighted too strong.

$$\mathrm{MRE}_{\mathrm{R}}(p) = \frac{1}{Nn} \sum_{j}^{N} \sum_{i}^{n} \frac{\left| h^{\mathrm{E}}_{calc}(x_{ji}, p) - h^{\mathrm{E}}_{exp}(x_{ji}) \right|}{\Delta h^{\mathrm{E}}_{exp,R_j}} \rightarrow \min \tag{1}$$

2 Optimization procedure

Common for all optimization runs is that those group interactions are marked, for which parameters are to be fitted. The optimization procedure itself can be carried out in several ways: It is possible to fit the parameters of one group interaction after the other. By doing so, six interaction parameters have to be fitted at a time. Figure 3 shows extracts of parameter matrices. Every square symbolizes one group interaction with the appropriate six model parameters. For the so called sequential optimization the procedure begins with the interaction in the first row and proceeds row by row to the lower right corner of the matrix.

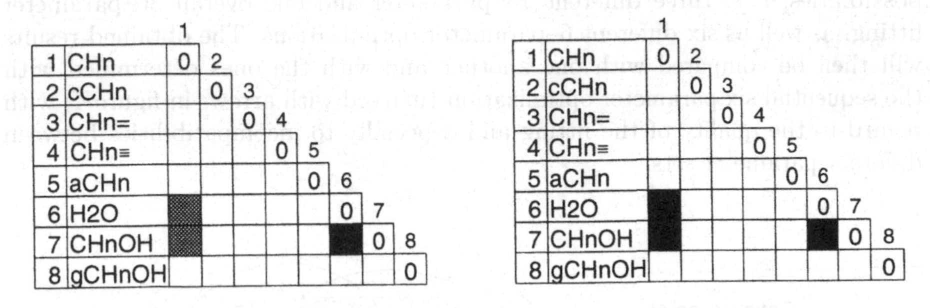

Fig. 3. Parameter matrix for interactions between functional groups

Most of the binary mixtures contain more than two groups (i. e. one group interaction) though, therefore the fitting procedure of the parameters of the group interaction of interest refers to already fitted parameters. If any already fitted parameters of other group interactions are to be taken into account, then these interactions are also marked (in a different way). For example: the optimization of the interaction between H_2O and CH_nOH without referring to other interaction parameters is carried out with experimental data of water/methanol mixtures only. If the interactions $CH_n - CH_nOH$ and $CH_n - H_2O$ are marked in a way that their (previously fitted) parameters apply, the optimization of $H_2O - CH_nOH$ is carried out with experimental data of water/n-alkanol mixtures (including water/methanol systems) (see left matrix of figure 3). Furthermore, there exist some interactions, that can only be optimized, if certain other interactions are taken into account. The downside of this procedure is that incompatibilities between the model parameters might occur. It is possible that no parameter set can be determined that leads to satisfying results, due to the values of the previously fitted parameters of the incoming interactions, which remain constant. Although the parameter sets of the incoming interactions are the sets that yield the best results for the considered group interactions itself, they are evidently not appropriate sets to be used in the combination of different interactions. To avoid this problem, it is possible to optimize the model parameters of more than one interaction at a time. Using the right matrix in figure 3 as an example, three interactions are marked for optimization, i. e. 18 parameters have to be fitted simultaneously. With this optimization procedure, incompatibilities between the different parameter sets cannot occur. Furthermore, the experimental data used for the fitting contains more than one group interaction per mixture, which leads to thermodynamically more consistent parameters. The downside, though, is that the variable dimension increases, which may result in more complex search spaces and thus, higher requirements for the used optimization algorithms.

In this paper, investigations on different multi-parameter optimizations of the EBGC-Model will be made. The interactions between the following groups will be observed: the methyl group CH_n, water H_2O, the alcoholic group CH_nOH and the glycolic group gCH_nOH. Figure 4 shows the different optimization possibilities, i. e. three different 18–parameter and one overall 36–parameter fitting, as well as six different 6–parameter optimizations. The obtained results will then be compared with one another and with the ones determined with the sequential six parameter optimization (marked with arrows in figure 4) with regard to the quality of the fitting and especially to incompatibilities between different parameter sets.

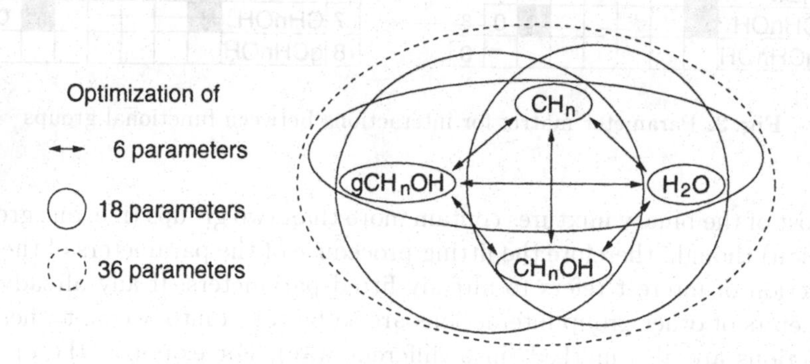

Fig. 4. Considered multi-parameter optimization possibilities

3 Encapsulated Evolution Strategies

Investigations on the considered optimization problem showed that deterministically acting algorithms are not suitable due to the multi-modal character of the fitness landscape. Multi-start algorithms like repeatedly started deterministically acting algorithms or repeatedly started conventional Evolution Strategies lead to better but not satisfying results. In the following, encapsulated Evolution Strategies using only a one dimensional step length and the 1/5 success rule [8], as well as non-encapsulated Evolution Strategies with multi-dimensional and correlated control of step-length [1,9,10] were tested on the problem with different tuning parameters but they also did not deliver satisfying results. Only the combination of both theories, i. e. multimembered $(\mu+\lambda)$– and (μ,λ)–Evolution Strategies with multi-dimensional and correlated step-length control in an encapsulated version lead to good results [3,4,11]. In this case a sequential optimum seeking process on several planes increases a more isolated search in the complex solution space. Notation of the strategy is as follows:

$$\left[r_1^{\boldsymbol{x}} r_1^{\boldsymbol{\sigma}} r_1^{\boldsymbol{\alpha}}\ \mu_1[s_1]\lambda_1\ (r_2^{\boldsymbol{x}} r_2^{\boldsymbol{\sigma}} r_2^{\boldsymbol{\alpha}}\ \mu_2[s_2]\lambda_2)^{\gamma_2}\right]^{\gamma_1} - \mathrm{ES} \quad (2)$$

The 3-number letter codes $r_i^{\boldsymbol{x}} r_i^{\boldsymbol{\sigma}} r_i^{\boldsymbol{\alpha}}$ in the notation give the recombination mechanism for the objective variables $\boldsymbol{x}$, standard deviations $\boldsymbol{\sigma}$, and, if necessary, rotation angles $\boldsymbol{\alpha}$ for each plane. They can be chosen as $r_i^{\boldsymbol{x}} r_i^{\boldsymbol{\sigma}} r_i^{\boldsymbol{\alpha}} \in \{-, d, D, i, I, g, G\}$ referring to Bäck [1,10]. The selection s_i can be chosen as plus [+] or comma [,] for each plane independently.

4 Results with test data

For the determination of optimum tuning parameters for the encapsulated Evolution Strategy, test data were generated with the global optimum exactly known with zero deviation. These test data were predicted by EBGCM with already fitted parameters to ensure that they can be theoretically reproduced during the optimization. At first, a test system for a 6–parameter optimization was generated and different strategy types as well as deterministically acting algorithms were tested and analyzed. Detailed information about those test runs and their results can be obtained elsewhere [3,4]. Using these experiences as a basis, test runs on an 18–parameter test system were carried out. It is important to note, that every parameter has a certain definition area. In the case that parameters leave this range during optimization an exponential penalty function will return extremely high $\mathrm{MRE_R}$ values.

Table 1 shows some results on the 18–parametric test system. The average $\mathrm{MRE_R}$ values, as well as the best and worst results out of 50 test runs are given. Additionally, the arithmetic means of all 50 $\mathrm{MRE_R}$ results in per cent are shown to characterize the reliability and reproducibility of the used algorithms. Not surprisingly, random walks with 900 000 function calls do not deliver any satisfying result with an average $\mathrm{MRE_R}$ of about 400 %. Some test runs with the Simplex method of Nelder and Mead [7] were carried out. The notation is as follows: a $160 \cdot \mathtt{SNM}^{4000}$ is a Simplex run with 4 000 iteration steps, that is repeated 159 times. The best parameter set is kept during the repetitions. It can be seen, that the stochastic part resulting from the repetitions beginning from

Table 1. Optimization results of different deterministically acting algorithms tested on the 18–dimensional test system (MRE_R–values are an average of 50 runs per algorithm)

Algorithm	MRE_R / % average	MRE_R / % best	worst	$\sigma_{(50)}$ / %	function calls
Random walk	**406.42**	215.258	668.072	97.43	900000
160 · SNM^{4000}	**87.36**	48.850	216.438	28.69	890561
320 · SNM^{2000}	**84.98**	35.398	154.461	24.82	883470
650 · SNM^{1000}	**94.34**	49.960	170.785	26.11	900004

different starting points improves the optimization results. Further increase of the repetitions (with decreasing number of iteration steps per Simplex run to maintain a constant overall number of iteration steps of about 900 000) leads again to raising deviations.

Table 2 shows a small extract of all tested Evolution Strategies. The start step length was defined in per cent of each total definition area divided by $\sqrt{n}$ (with n as the number of objective variables), in order to make the standard deviations independent from n and from different definition areas of the parameters. The parent individuals (objective variables) are always chosen by random during their initialization. In case of encapsulated Evolution Strategies, the strategic variables (standard deviations and rotating angles, if used) in the encapsulated plane are also randomly initialized.

Table 2. Optimization results of different Evolution Strategies tested on the 18–dimensional test system (MRE_R–values are an average of 50 runs per strategy type)

ES–notation	MRE_R / % average	start step-width / %	MRE_R / % best	worst	$\sigma_{(50)}$ / %	function calls
(GG 120,800)^{380}	**20.055**	8	15.187	20.245	0.70	304000
(GG 120+800)^{380}	**19.941**	8	15.079	20.236	0.99	304120
[GG 4+8(GG 7+19)^{100}]^{20}	**10.218**	5 / 2	4.480	15.596	2.84	304164
[GG 4+8(GG 7+19)^{400}]^{15}	**3.840**	5 / 2	1.074	8.988	1.93	912124
[GG 4+8(GG 7+19)^{500}]^{12}	**3.392**	5 / 2	0.781	9.190	1.90	912100
[GG 4+8(GG 7+19)^{600}]^{10}	**3.699**	5 / 2	0.440	10.021	2.11	912084
[GG 10,30(SNM^{1000})]^{20}	**17.720**	5	12.909	20.692	2.05	847729
[GG 10+30(SNM^{1000})]^{20}	**15.524**	5	10.391	20.475	2.36	880145

Non-encapsulated (μ+λ)– and (μ,λ)–strategies were found to be unsuitable for this regression problem, as can be seen in the two upper lines of table 2. They did not cope with the 6–parameter problem, either. The encapsulated version shows better but not satisfying results. So the total number of iteration steps had to be increased. The next three lines show those results with the selective pressure on the first plane of about $\lambda/\mu = 2$ and on the encapsulated plane of about $\lambda/\mu \approx 2.7$.

But even with the optimized tuning parameters and as much as 900 000 iteration steps, the mean deviation only reaches about 3.4 %. The 6–parameter test system attained a mean error of about 0.09 % with 300 000 iteration steps [3], which may be a hint for the increasing complexity of the solution space with raising variable dimension. The use of encapsulated Evolution Strategies characterized by an increased locality during the optimum seeking process is necessary to obtain thermodynamically interpretable results for the 6–parameter as well as for the 18–parameter optimization, though.

In the last two lines of table 2 results of an encapsulated Evolution Strategy with a Simplex-Nelder-Mead algorithm running in the encapsulated plane are shown. Both the [,] and the [+] variant did not deliver satisfying results. Although the increased locality through encapsulation leads to good results, as stated above, the locality resulting from the deterministically acting Simplex seems to be too strong.

Test runs for the 36–parameter problem were not carried out. A lot of computing time would be necessary due to the increased number of data points for a test system. So the 36–parameter optimization to real experimental data was carried out with the tuning options obtained for the 18–parameter problem.

5 Results with real–world data

The difference in principle between the 18–parameter test system and real experimental data is, that the test system consists of three group interactions with their data completely independent from each other, i. e. actually a combination of three 6–parameter test systems. Real experimental data consist of systems that contain more than one interaction. This fact causes the problem of incoming parameter sets during optimization, which may result in incompatibilities between different parameter sets, as explained above.

With the determined tuning parameters, optimizations on the above mentioned real-world fitting problems were made. Following the optimization runs with Evolution Strategies, the parameters are usually transferred to a deterministically acting algorithm, like the Simplex method of Nelder and Mead [7] to determine the nearest optimum.

At first, the parameters of the six interactions were optimized in the sequential way, i. e. just one interaction at a time. Table 3 gives a comparison between the results of the 36–parameter and the sequential optimization. The first column gives the group interaction with the number of data points in the second column. In the next column the results of the sequential optimization are displayed. Their parameters were fitted interaction after interaction, six parameters at a time. The last column shows the results of the 36–parameter optimization, that was carried out with the entire data base. To obtain the results for the single interactions, the error values were calculated (but not fitted) with the data bases for the single interactions. The last line gives the number of data points and the results with the entire data base.

The order of the interactions in the table represents the order of the sequential optimization. During the optimization incompatibilities between different parameter sets occurred. The fitting of the parameters of the interaction $CH_n - gCH_nOH$ lead to a deviation greater than 1 000 %. This was due to the incoming parameters of $CH_n - H_2O$ and $H_2O - gCH_nOH$, that were previously fitted. Since the optimization of $CH_nOH - gCH_nOH$ depends on the parameters of $CH_n - gCH_nOH$, no satisfying result was found for this interaction, either. To avoid this problem, the definition area for all six parameters was further restricted, because the values of two parameters of $H_2O - gCH_nOH$ seemed to be too high. That solved the problem for $CH_n - gCH_nOH$ (with a slightly worse result than before), but not for $CH_nOH - gCH_nOH$ with a deviation of about 626 %.

Table 3. Comparison between the 6– and 36–parameter optimizations

Group interaction		Data	MRE$_R$ / % seq.	MRE$_R$ / % 36–par.
CH_n	CH_nOH	4458	14.25	20.47
H_2O	CH_nOH	174	11.85	26.06
CH_n	H_2O	1216	15.84	18.64
H_2O	gCH_nOH	179	9.48	28.48
CH_n	gCH_nOH	218	11.40	24.19
CH_nOH	gCH_nOH	1174	626.04	14.10
All six interactions		7419	102.26	19.56

The results for the 36–parameter optimization show, that interactions with a small database are fitted worse than those with a great database although the deviations of the single interactions are weighted over the number of data points to ensure that all interactions contribute equally to the objective function. (This weighting procedure certainly applies only in case of the simultaneous optimization of more than one interaction.) Moreover, there are some systems, that comprise more than one group interaction. These data are overrated in the weighting process, but this is useful, since they are more valuable from the thermodynamic point of view. This fact may be the cause for the good representation of $CH_nOH - gCH_nOH$, because every data set for this interaction comprises $CH_n - CH_nOH$ and $CH_n - gCH_nOH$. That, on the other hand, caused the incompatibilities during the sequential optimization. Apart from this, all other interactions show better results than those obtained with the 36–parameter optimization, though.

Table 4. Comparison between the sequential and an 18–parameter optimization comprising the groups CH_n, H_2O, and CH_nOH

Group interaction	Data	MRE_R / % seq.	18–par.
CH_n H_2O	1216	15.84	15.20
CH_n CH_nOH	4458	14.25	14.14
H_2O CH_nOH	1447	15.51	16.55
All three interactions	5821	14.29	15.58

Table 5. Comparison between the sequential and two 18–parameter optimizations comprising the groups CH_n, H_2O, and gCH_nOH or CH_n, CH_nOH, and gCH_nOH

Interactions between	Data	MRE_R / % seq.	18–par.
CH_n, H_2O, gCH_nOH	397	10.52	8.05
CH_n, CH_nOH, gCH_nOH	5548	148.07	15.40

Table 4 shows a comparison between the 18–parameter optimization of the groups CH_n, H_2O, and CH_nOH and the appropriate sequential optimization (their results are the same as in the column of the sequential optimization in table 3, of course). The overall deviation is significantly lower for the 6–parameter optimization procedure. Considering the single interactions, it shows that the first two interactions are fitted slightly better simultaneously than sequentially, in contrast to $H_2O - CH_nOH$. One reason may be that the data base for the interaction $H_2O - CH_nOH$ is much smaller than for the two other interactions, although the deviations of the single interactions are weighted over the number of data points, as explained above.

Comparisons of the single deviations between the sequential and the two other 18–parameter optimizations are not possible due to the structure of the experimental data. The sequential parameter optimizations had to be carried out with incoming parameter sets, that were not included in the corresponding 18–parameter fittings. For example, $CH_n - H_2O$ was sequentially fitted with the incoming interactions $CH_n - CH_nOH$ and $H_2O - CH_nOH$, which are not included in the 18–parameter problem with the groups CH_n, H_2O, and gCH_nOH. Table 5 gives a comparison between the results of the remaining two 18–parameter optimizations and the deviation calculated with the 18–parameter data base and the parameters of the sequential optimization. Both 18–parameter optimizations lead to better results than the sequential optimization.

6 Conclusion

The use of Evolutionary Algorithms seems to be indispensable for the successful optimization of parameters of group contribution models by nonlinear regression. Those optimization problems are characterized by multi-modal and complex solution spaces. Conventional multi-membered Evolution Strategies do not cope with this problem. Encapsulated Evolution Strategies are capable of finding quite satisfying results, although the results are significantly worse than for the 6–parameter problem [3, 4]. This fact proves that the complexity of the solution space increases with rising variable dimension.

Different optimizations with real-world data showed that the sequential optimization lead to incompatibilities between different parameter sets, as occurred for the interaction $CH_nOH - gCH_nOH$. Three different 18–parameter optimizations were carried out. It is interesting to note, that the 6–parameter optimizations were worse than the 18–parameter fittings in those cases, when already determined parameter sets applied. The overall fitting of 36 parameters in one delivered quite satisfying results despite the high variable dimension.

Although the runs with an 18–parameter test system did not deliver as good results as with the 6–parameter test system [3, 4], optimizations with real data delivered satisfying results, even with the 36–parameter problem.

To sum up it can be said that sequential parameter optimization of group contribution methods should not be carried out if the fitting refers to already determined parameter sets. Instead, those interactions should be fitted in one.

Acknowledgements

This work is a result of the Collaborative Research Center SFB 531, sponsored by the Deutsche Forschungsgemeinschaft (DFG).

References

1. T. Bäck, *Evolutionary Algorithms in Theory and Practice*, Oxford University Press, New York. **1996**.
2. A. Fredenslund, B. L. Larsen and P. Rasmussen, A Modified UNIFAC Group Contribution Model for Prediction of Phase Equilibria and Heats of Mixing, *Ind. Eng. Chem. Res.* **1987**, 26, 274-286.
3. H. Geyer, P. Ulbig and S. Schulz, Use of Evolutionary Algorithms for the Calculation of Group Contribution Parameters in order to Predict Thermodynamic Properties. Part 2: Encapsulated Evolution Strategies, *Computers Chem. Engng.*, accepted for publication.
4. H. Geyer, P. Ulbig and S. Schulz, Encapsulated Evolution Strategies for the Determination of Group Contribution Model Parameters in order to predict Thermodynamic Properties, in: A. E. Eiben, T. Bäck, M. Schoenauer, H.-P. Schwefel, *Parallel Problem Solving from Nature*, **1998**, 5, 978-987, Springer, Amsterdam.
5. C. Kracht, T. Friese, P. Ulbig and S. Schulz, Development of an Enthalpy Based Group Contribution G_m^E Model, *J. Chem. Thermodynamics*, in press.

6. C. Kracht, H. Geyer, P. Ulbig and S. Schulz, Optimum Tuning Parameters for Encapsulated Evolution Strategies: Results for a Nonlinear Regression Problem, *Technical Reports of the Collaborative Research Center SFB 531: "Design and Management of Technical Processes and Systems by Using Methods of Computational Intelligence"*, **1998**, 42.
7. J. A. Nelder and R. Mead, A Simplex Method for Function Minimization, *Computer Journal* **1965**, 7, 308-313.
8. I. Rechenberg, *Evolutionsstrategie '94, Werkstatt Bionik und Evolutionstechnik*, Band 1, Friedrich Frommann, Stuttgart. **1994**.
9. H.-P. Schwefel, *Numerical Optimization of Computer Models*, Wiley, Chichester. **1981**.
10. H.-P. Schwefel, *Evolution and Optimum Seeking*, Wiley, New York. **1995**.
11. P. Ulbig, T. Friese, H. Geyer, C. Kracht and S. Schulz, Prediction of Thermodynamic Properties for Chemical Engineering with the Aid of Computational Intelligence, *Progress in Connectionist-Based Information Systems - Proceedings of the 1997 International Conference on Neural Information Processing and Intelligent Information Systems*, **1997**, 2, 1259-1262, Springer, New York.
12. U. Weidlich and J. Gmehling, A modified UNIFAC Model, *Ind. Eng. Chem. Res.* **1987**, 26, 1372-1381.

Applying Heuristic Algorithms on Structuring Europeanwide Distribution Networks

Ramin Djamschidi, Andreas Bruckner

Research Institute for Operations Management (FIR) at Aachen University of Technology (RWTH), Pontdriesch 14/16,
D-52062, Germany
{dj, bk}@fir.rwth-aachen.de

Abstract. Due to the recent changes of the European domestic market including the opening up of Eastern Europe and its associated logistical (distribution) consequences, the economically significant question of reorganization of goods distribution networks represents an increasing problem for many companies operating in Europe. The existing concepts indicate several weak points because of insufficient consideration of real, nonlinear transportation and storage costs as well as of underlying strategies behind the distribution structure. In addition, the existing concepts are limited to the optimization of only one warehousing level. The consideration of real conditions incl. simultaneous optimization of multiple distribution echelons increases the complexity enormously, hence a further development of known concepts will practically fail. Contents of the following contribution – promoted by the DFG (Lu 373/18-1) - are based on application of heuristic concepts, i.e. Simulated Annealing, Evolutionary Strategies, Genetic Algorithms, and TABU Search for solving this complex optimization problem.

1 Introduction

The distribution of goods [7], [9], [32], i.e. industrial mass products, forms the termination of the supply chain, which reaches from the raw material procurement over the industrial production of semi- and finished products up to the actual end, i.e. customers. It connects locally settled production with the sales markets which are mostly widespread and can be executed by both the producing enterprises themselves, and -outsourced- by distribution service providers. The physical instrument for the execution of the distribution tasks is the *distribution system/network.* This is characterized by its elements stores, transport and order processing, as well as by the relations between them, i.e. s*tructure* of the distribution network and s*trategy* behind the distribution of goods (warehousing and delivery strategies). Both will be considered in the following contribution.

2 Problem Categories

The following problem categories are regarded in this endeavor [1], [3], [6], [10], [14], [20], [31]:

- "Location Problem",
- "Multifacility Location Problem",
- "Location Allocation Problem", and
- "Warehouse Location Problem".

Each category requires an individual model, target function, distinct concept, and implementation. In addition, each of the heuristic algorithms (Simulated Annealing, Evolutionary Strategies, Genetic Algorithms, and TABU Search) needs to be adapted to the aforementioned concepts. In this paper, the authors will focus -as an example- on applying Simulated Annealing for the Location Allocation Problem, and compare the results with the other heuristic algorithms, especially Genetic Algorithms.

3 Formulation of the Target Function

The target function represents the necessary transportation costs for a given distribution network (and also warehousing costs in the case of Warehouse Location Problem) which is to be minimized. The formulation of the individual functions was developed as close-to-reality as possible through data collection from real companies, both logistic service providers and companies with own distribution divisions.

Based on these investigations concerning type and extent of transportation and warehousing costs, the authors developed a sequence of target functions for each location problem with different settings. The accurate mathematical formulation of the target function proved to be difficult but very important, because the controlling of the developed heuristic procedures and thus the attainable quality of the solutions will primarily depend on the target function value. Without going into details, the target function in the case of the aforementioned Location Allocation Problem is as following:

$$\sum_{s \in M^s} \sum_{j^s=1}^{\overline{j}^s} \sum_{i=1}^{\overline{i}} \sum_{m_i^s=0}^{\overline{m}_i^s} f_{ij^s m_i^s}^{sK} \cdot a_{ij^s m_i^s}^{sK} \cdot \delta_{ij^s} +$$

(representing the costs between distribution stores and customers)

$$\sum_{j^R=1}^{\overline{j}^R} \sum_{j^A=1}^{\overline{j}^A} f_{j^A j^R}^{RA} \cdot a_{j^A j^R}^{RA} \cdot \delta_{j^A j^R} +$$

(... between regional and distribution stores)

$$\sum_{j^Z=1}^{\bar{j}^Z} \sum_{j^R=1}^{\bar{j}^R} f^{ZR}_{j^R j^Z} \cdot a^{ZR}_{j^R j^Z} \cdot \delta_{j^R j^Z} +$$

(... between main and regional stores)

$$\sum_{j^P=1}^{\bar{j}^P} \sum_{j^Z=1}^{\bar{j}^Z} f^{PZ}_{j^z j^P} \cdot a^{PZ}_{j^Z j^P} +$$

(... between factories and main stores)

$$\sum_{\substack{s \in M^s \\ s \neq P}} \sum_{j^s=1}^{\bar{j}^s} f^{L}_{j^s} \cdot BZR + \ldots$$

(... warehousing costs)

"f^{xx}" represent the nonlinear transportation costs, "f^{L}" warehousing costs (both as a driven formula from corresponding cost tables used in different companies) according to each echelon, "*a*" the number of individual deliveries, "*BZR*" the related period, and "δ" a binary value for describing the assignment.

4 Implementation

The basics of the concepts of Simulated Annealing, Evolutionary Strategies, Genetic Algorithms as well as TABU Search used for solving the aforementioned location problems are oriented upon e.g. [4], [5], [11], [13], [15], [16], [17], [23], [24], [25], [30]. Within the field of facility location, there are several applications of heuristic optimization concepts e.g. [29], assuming the *discrete* category. An application of these concepts within the field of the *homogeneous* Location Problem, and also in a combination of both categories as it is in the nature of this optimization problem in the reality - which the authors emphasize on - has not been documented so far.

In order to shorten this paper, the authors will focus on the application of Simulated Annealing (SA), and Genetic Algorithm (GA):

4.1 Simulated Annealing

The ideas that form the core of Simulated Annealing were first published by METROPOLIS [22] realized in an algorithm to simulate the cooling of material in a heat bath - a process known as annealing. KIRKPATRICK [17] and CERNY [4] showed independently that the Metropolis algorithm could be applied to optimization problems. The Simulated Annealing algorithm [19], [21] is similar to the random descent method as far as the neighborhood is sampled at random. It differs in the way that the neighbors giving rise to an increase in the cost function may be accepted and this acceptance will depend on the control parameter called the temperature, and the magnitude of the increase. Thus, there will be temporary worse values in order to come out of possible local minima progressing toward the absolute minimum like the real annealing process.

1. The algorithm loads the following control parameters from a database:

t_start = start temperature
t_end = end temperature
β = parameter for the cooling function
n_iter = max. number of iterations
n_const = max. number of iterations without any decrease
n_rep = max. number of iterations per temperature level
n_acc = max. number of iterations with acceptance per temperature level
ε = wanted upper bound for the cost
p_LG = probability for the change of a connection
r_max = max. distance for one move in terms of *RasterEinheit*

With *RasterEinheit* as a distance unit for moves in radian measure, where 2π is the earth circumference.

2. The temperature *t* is initialized with the value of *t_start*. It will be decreased using the cooling function α defined by $\alpha(t) := t/(1+\beta \cdot t)$. The quality of any location and allocation move during the optimization process depends particularly on the appropriate parameters chosen before. These parameters are e.g. mutation rate or population size for genetic algorithms. In the case of the SA, the most important parameters to be tuned are the "temperature" (*t*) and the "cooling parameter" (β) within the "cooling function".

3. The algorithm runs through the iterations until one of the following conditions is satisfied:

- The number of iterations done is *n_iter*.
- The cost could not be decreased over the last *n_const* iterations.
- The cost is less than ε.
- The temperature *t* is less or equal to *t_end*.

4. A single iteration works in the following way:

- A move is being tried and the decrease δ of the cost is computed.
- If $\delta < 0$ then the move will be accepted. Otherwise it will be accepted with a probability of $\exp(-\delta / t)$.
- Every *n_rep* iterations and *n_acc* acceptances the temperature *t* is cooled down using the cooling function α, i. e. the new *t* becomes $\alpha(t)$.

5. A *move* works in the following way:

– With a probability of *p_LG* a connection will be changed, i.e. an arbitrary customer or store gets a new deliverer.
– Otherwise a store will be moved, i.e. its position will be moved in an arbitrary direction for a distance of *r RasterEinheit* units *r_max(t - t_end)/(t_start - t_end).*

4.2 Genetic Algorithms

As mentioned before the authors would like to focus on describing the application of Simulated Annealing and compare the results with other heuristic algorithms, especially with the Genetic Algorithm. In the following portrait, the authors will only point out the peculiarities of the applied GA.

First of all, it should be reminded that a real distribution problem contains continuous and discrete components. The *continuous* ones are e.g. the locations which can be any mathematical point in Europe (and not at given potential locations). They are calculated by a pair [LG, BG] while LG corresponds to the degree of longitude, and BG to the latitude of a temporary store location. The *discrete* ones are e.g. delivery assignments or the number of stores.

While Genetic Algorithms are coded binary and works usually with only one of the mentioned (continuous or discrete) categories [18], it would lead to too many inconsistent distribution networks which is in the nature of the problem with all its real restrictions and would not be practical. To get consistent results for a distribution network, the manipulation of genes and the crossover procedures are made as a combination of continuous and discrete components. As usual the main processes between two generations are selection, recombination and mutation followed by a valuation of the individuals' fitness which correspond to the reciprocal of the target function values.

For *selection* procedures, there exists different methods, e.g. (Kinnebrock 1994, p.70 f.). In the present application, the Roulette-Wheel-Method is used whereas the individuals (for a recombination procedure) are selected by a probability that is proportional to their fitness. Thus, all individuals can take part at the recombination, whereas the probability of being selected increases with their fitness.

The *recombination* procedure represents the most complex process (in a computation sense), because a new individual represents a whole distribution network of all stores and customers data including delivery assignments, products, and order information and is made arbitrary on the basis of other complex individuals. During the recombination, all allocations between different stores of the distribution network are to be controlled. The implemented procedure starts from the bottom of the distribution network (customers). There exists in analogy to the $(\mu+\lambda)$ strategy, the possibility of keeping some best individuals from each generation, thus, they do not take part at the recombination and mutation.

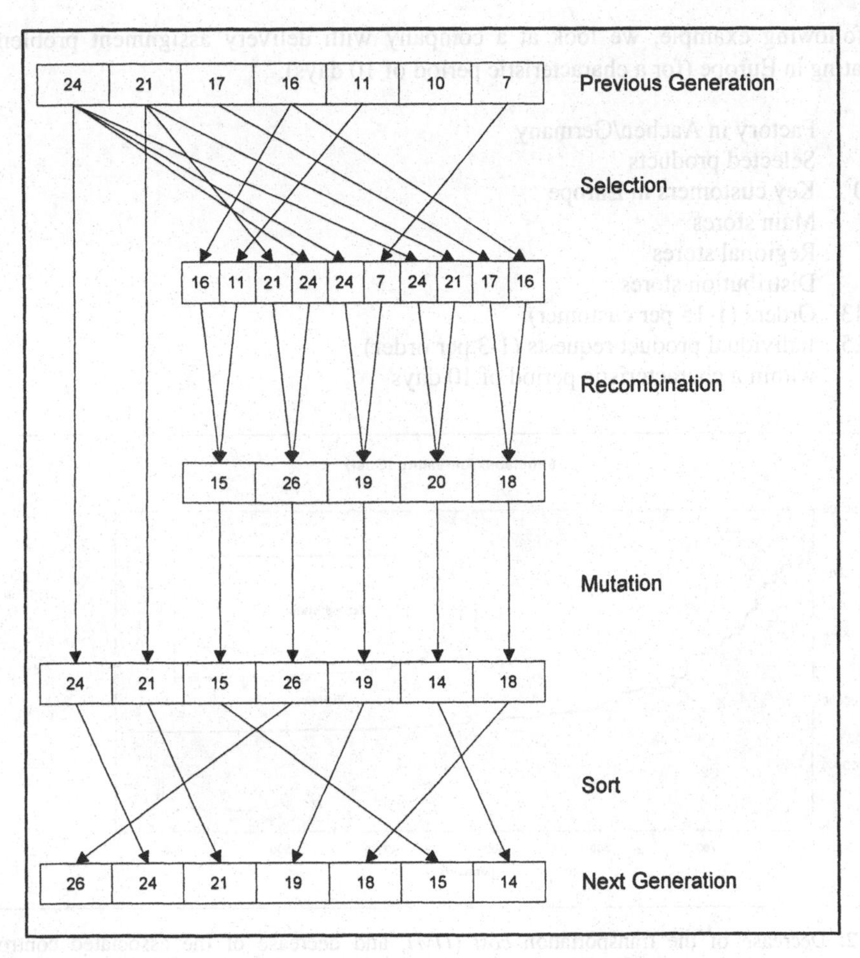

Fig. 1. shows a GA cycle as an example - numbers represent the fitness

Some of the recombined individuals go additionally through a *mutation* procedure (changes of store locations and delivery assignments) by a predetermined probability. The major parameters of the mutation procedure are mutation rate and mutation distance.

The quality of any location and allocation change during the optimization process depends particularly on the appropriate parameters chosen before. All new generated individuals as well as the fixed individuals of the previous generation are sorted corresponding to their fitness. The sorted individuals represent the starting population for the new generation.

5 Results

As mentioned before, the authors focus on applying SA for solving a Location Allocation Problem, and compare the results with the other mentioned algorithms. In the following example, we look at a company with delivery assignment problem operating in Europe (for a characteristic period of 10 days):

1	Factory in Aachen/Germany
7	Selected products
20	Key customers in Europe
3	Main stores
4	Regional stores
5	Distribution stores
143	Orders (1-15 per customer)
325	Individual product requests (1-3 per order) within a characteristic period of 10 days

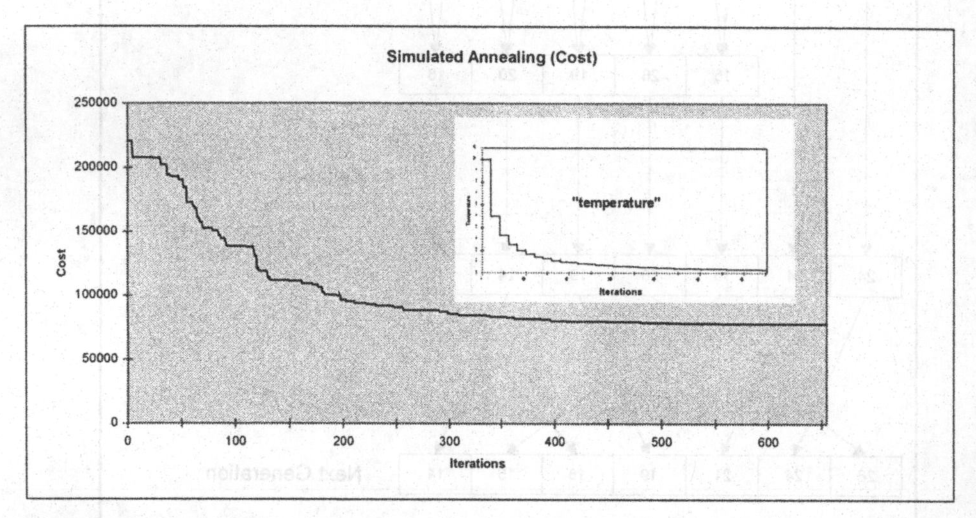

Fig. 2. Decrease of the transportation *cost (DM)*, and decrease of the associated control parameter *temperature* for the aforementioned example using Simulated Annealing

The Simulated Annealing algorithm moved the positions of stores and changed the assignments between stores and customers. In this example, stores were moved - in average - for a distance of 300 km, and 5 out of total 32 connections were changed. The above diagram (Fig. 2) shows the decrease of the cost function.

Table 1. Comparison between the chosen heuristic algorithms for the same company; TA & SA are terminated by the same improvement/iterations; ES and GA by reaching similar values

	Iterations	Improvement
Tabu Search	6.000	31,8 %
Simulated Annealing	643	63,0 %
Evolutionary Strategies	30.000	69,5 %
Genetic Algorithms	15.000	70,6 %

The improvement ratio is calculated by dividing the *costs after applying Heuristic Algorithms* as described before by the *current costs*. For more details concerning the results mentioned in the above table you may read [8] or contact the authors directly.

6 Conclusions

Heuristic concepts which accept temporary deterioration of target function values, proved to be very promising for solving complex distribution problems. Hence, the developed heuristic concepts are the only chance to solve real restructuring problems of goods distribution networks in an acceptable time. The importance of such a solution is immense, since the economically significant question of (re-)configuration of such networks becomes an increasing problem for many companies, due to the changes of the European market and its associated logistical consequences.

7 References

1. Bach, L.: Methoden zur Bestimmung von Standorten und Einzugsbereichen zentraler Einrichtungen. Birkhäuser Verlag, Basel Stuttgart (1978)
2. Battiti, R., Tecchiolli, G.: The Continuous Reactive TABU Search: Blending Combinatorial Optimization and Stochastic Search for Global Optimization – Metaheuristics in Combinatorial Optimization –. In: Annals of Operations Research, 63 (1996) 113-118
3. Blank, U.: Entwicklung eines Verfahrens zur Segmentierung von Warenverteilungssystemen. Dissertation RWTH Aachen (Forschungsinstitut für Rationalisierung - FIR), Aachen (1980)
4. Cerny, V.: A thermodynamical Approach to the Traveling Salesman Problem. In: Journal of Optimization Theory and Applications, 45/1 (1985) 41-51
5. Davis, L.: Handbook of Genetic Algorithms. New York (1991)
6. Djamschidi, R., Friemuth, U.: Distributionslogistik im hochdynamischen Europa. In: Arbeitgeber 15/16 (1997) 483-486
7. Djamschidi, R.: Efficient Algorithms for Structuring Distribution Networks. In: Eufit '98, 6th European Congress on Intelligent Techniques and Soft Computing, Proceedings Volume III. Verlag Mainz Wissenschaftsverlag, Aachen (1998) 1665 - 1670
8. Djamschidi, R., Bruckner, A.: Distributionslogistik in einem sich ändernden Europa. In: Logistikplanung - Methoden, Werkzeuge, Potentiale. Tagungsband der 4. Magdeburger Logistik-Tagung, Magdeburg (1998)
9. Djamschidi, R.: Lagerstandortplanung in Europa. In: Praxishandbuch für Lagerplanung, organisation und -optimierung. WEKA Fachverlag für technische Führungskräfte, Augsburg (1998)
10. Domschke, W., Drexl, A.: Logistik: Standorte. München Wien (1985)
11. Glover, F.: Annals of Operations Research 41. Basel (1993)
12. Glover, F., Laguna, M.: Tabu Search. Kluwer Academic Publishers, Boston (1997)
13. Goldberg, D. E.: Genetic Algorithms in Search, Optimization and Machine Learning. Reading, Massachusetts (1989)

14. Hummelten, W.: Optimierungsmethoden zur betrieblichen Standortwahl. Würzburg Wien (1981)
15. Karaboga, D., Pham, D.: Intelligent Optimization Techniques: Genetic Algorithms, Tabu Search, Simulated Annealing and Neural Networks. Springer-Verlag, London, (1998)
16. Kinnebrock, W.: Optimierung mit genetischen und selektiven Algorithmen. Oldenbourg Verlag, München Wien (1994)
17. Kirkpatrick, S., Gelatt, C. D., Vecchini, M. P.: Optimization by Simulated Annealing. In: Science 220/4598 (1983) 671-680
18. Kitano, H.: A hybrid search for Genetic Algorithms: Combining Genetic Algorithms, Tabu Search and Simulated Annealing. In: Proceedings of the Fifth International Conference on Genetic Algorithms. Morgan Kaufman Publisher (1993)
19. Lee, J.: Faster Simulated Annealing techniques for stochastic optimization problems, with application to queering network simulation. Ph.D. Dissertation, Statistics and Operations Research, North Carolina State University (1995)
20. Love, R.F., Morris, J.G., Wesolowsky, G.O.: Facilities Location: models and methods. North Holland, New York (1988)
21. Malek, M., Guruswamy, M., Pandya, M., Owens, H: Serial and parallel Simulated Annealing and TABU Search algorithms for the travelling salesman problem. In: Annals of Operations Research, 21 (1989) 59-84
22. Metropolis, N., Rosenbluth, A., Teller, A. H.: Equation of state calculation by fast computing machines. In: Journal of Chem. Phys. 21(1953) 1087-1091
23. Murtagh, B. A., Niwattisy, S. R.: An Efficient Method for the Multi-Depot Location-Allocation Problem. In: Journal of the Operational Research Society 33(1982) 629-634
24. Nissen, V.: Evolutionäre Algorithmen. Wiesbaden (1994)
25. Rechenberg, I.: Evolutionsstrategie. Stuttgart (1973)
26. Reeves, C. R.: Modern Heuristic Techniques for Combinatorial Problems. New York (1993)
27. Reeves, C. R., Dowsland, K. A.: Simulated Annealing. In: Modern Heuristic Techniques for Combinatorial Problems. McGraw-Hill (1995)
28. Reeves, C. R., Glover, F., Laguna, M.: TABU Search. In: Modern Heuristic Techniques for Combinatorial Problems. McGraw-Hill (1995)
29. Schildt, B.: Strategische Produktions- und Distributionsplanung. Wiesbaden (1994)
30. Schwefel, H.-P.: Numerische Optimierung von Computer-Modellen mittels der Evolutionsstrategie. Basel Stuttgart (1977)
31. Sivazlian, B. D., Stanfel, L. E.: Analysis of Systems in Operations Research. Englewood Cliffs (1975)
32. Tempelmeier, H.: Quantitative Marketing-Logistik. Berlin (1983)

The Control of a Nonlinear System Using Neural Networks

Ovidiu Grigore[1], Octavian Grigore[2]

[1] Polytechnic University of Bucharest, Electronic and Telecommunications Department
Splaiul Independentei 313, sect.6, Bucharest Romania
ovgrig@alpha.imag.pub.ro

[2] Polytechnic University of Bucharest, Aerospace Engineering Department
Splaiul Independentei 313, sect.6, Bucharest Romania
ogrigore@aeronet.propulsion.pub.ro

Abstract: This paper presents a neural network based controller used in commanding time varying systems with uncertainties task First, a reduction procedure of the initial set of parameters using an unsupervised pattern recognition technique was applied. After this a feed-forward neural network was trained using the minimized set of data. The advantage of this method is over-passing of the difficulties implied by the direct solving of the differential models, which are necessary in a classical approach. An application of a missile-target tracking was implemented using the mentioned method, and the results are compared with those obtain in a classical approach.

1 Introduction

When all apriori information about the controlled process is known, there can be designed an optimal controller using deterministic optimization techniques [1],[2]. When the apriori information required is unknown or partially known an optimal design, like: dynamical programming, maximum principle (Pontriaghin principle) or variational calculus, is used. In this case we can discuss two different approaches in solving the problem. One approach is to design a controller based only on the amount of information available, the unknown information being ignored or approximated with a known value given by a performance criterion. In this case the systems designed are, in general, suboptimal [3],[4]. The second approach is designing a controller able to estimate the unknown information, so if it is possible to approach gradually the true information, then the designed controller approaches the optimal one. In this case, it could be said that the controller learns during the process what decisions will make in the following. This quality of *learning* may be viewed as a problem of estimating or successive approximating the unknown values or function. Therefore, as the controller accumulates more information about the unknown function or parameters, the control law will be altered according to the

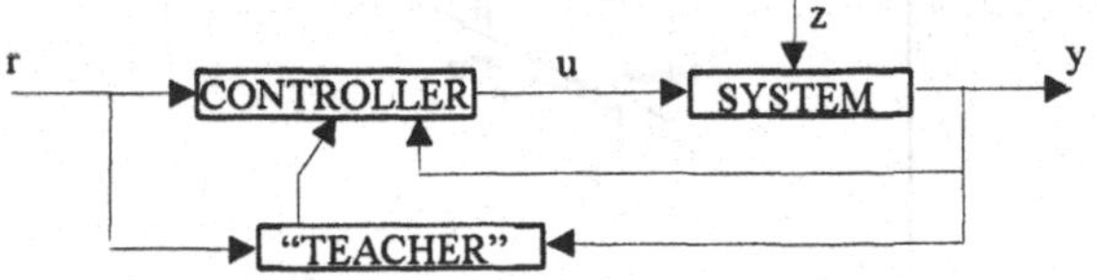

Fig.1 Block diagram of a learning control system

updated information in order to improve the system's performance. A basic block diagram for a learning control system is shown in Fig. 1, where the dynamical system under control **u** is disturbed by the perturbation z, assumed to be unknown or partially known. In classical theory the unknown values are estimated to decrease the error between the true value and the designed one, therefore in the *learning control theory* exists a *teacher* which, using a certain rule, teaches controller the commands necessary in reaching the expected goal.

2 NN Applied to Control Uncertain Nonlinear Dynamical Systems

For implement a controller we used a feed-forward neural network (NN), with a *back propagation* algorithm in the training step. Designing a controller means to find the values of the actual control using the input values, state values and the previous control. So, in this case the network's output must be the control at present time and the inputs are the state, the inputs and the previous control (Fig. 2).

Suppose the inputs $x_1 \ldots x_K$, represented by a k-dimensional vector **X** in the feature space Ω_X [17], and $\omega_1 \ldots \omega_M$ be the M classes of the control situations, the control operation can be interpreted as a partition of the k-dimensional space Ω_X into M decision regions corresponding to the M control classes. To determine the M control classes a clustering method based on the minimum Euclidean distances between vectors and classes' prototype [8] was applied.

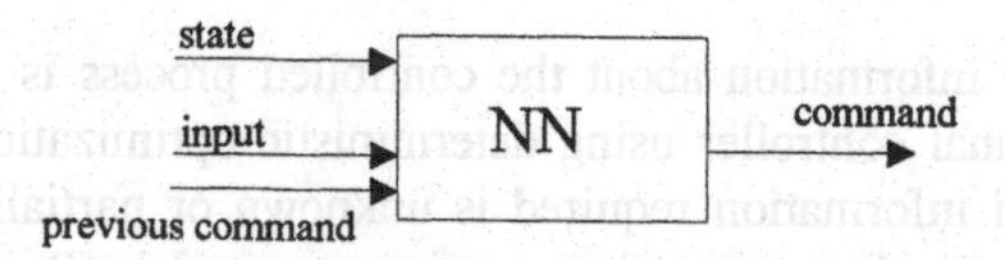

Fig.2 Neural network controller scheme

We can say that all vectors of the same class have approximately the same features and we can approximate their controls with the control corresponding to the center of the class.

3 Application

In this section we design the neural control of a missile which track a target with an unknown dynamics and evolution. The system's parameters and axes are attached as shown in Fig.3. The general case of missile's movement is described by a differential

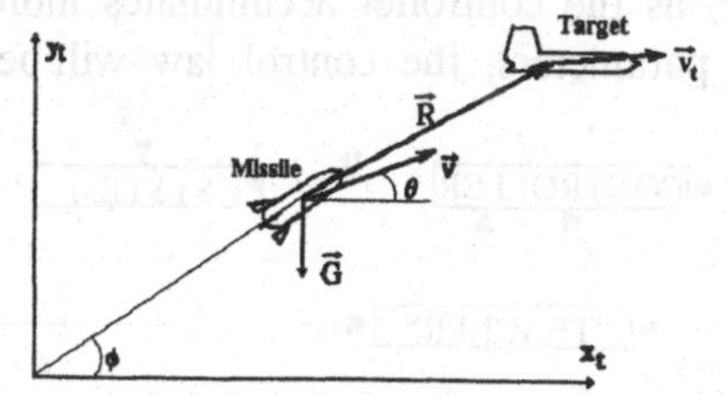

Fig.3 System's parameters and attached axes

system with 23-equation [9]. Therefore, designing a controller using classical methods consists of a numerical real-time solve of this differential system. A NN based controller remove this disadvantage.

The distance missile-target R, line of sight angle Φ and tilt angle of the trajectory θ are the inputs of the NN, so the network has 3 input nodes. The NN's output is the next command with values between -0.56 and 0.56 (due to the maximum admissible overload condition n = ± 10g). Instead of training the NN with one output varying in a continuous domain, the output range was divided into 11 intervals, for each of them having a corresponding output neuron of the NN. Good results were obtained using 7 neurons in hidden layer of the NN.

The learning set used in NN training contains 20000 experimental values [10]. Due to the great amount of data and the longer time used to learn the NN we reduced the number of input - output vectors by "Isodata" clustering method, finally obtaining a set with only 470 representative vectors.

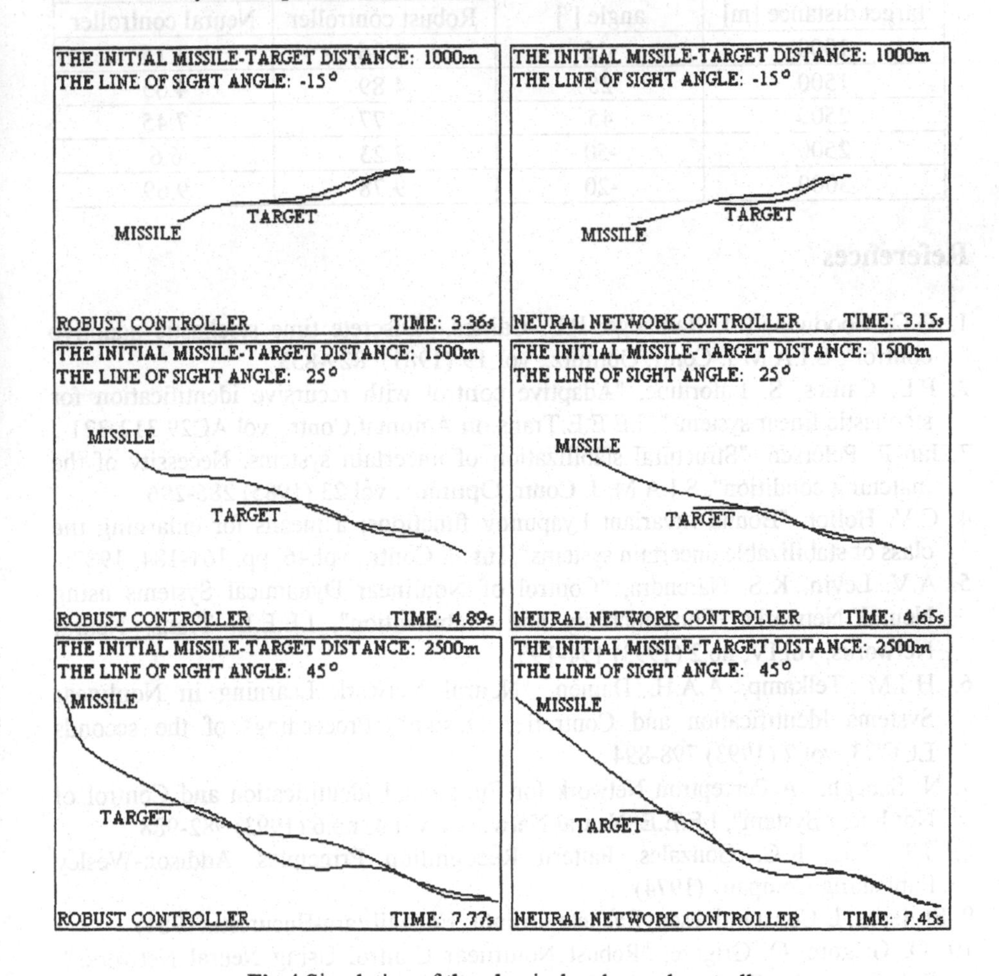

Fig.4 Simulation of the classical and neural controller

4 Conclusions

Controller's design by classical methods has to solve in real time a differential multiequation system with many unknowns. This disadvantage is eliminated using the neuronal network because there is a directly correspondence between the input and the output values after it was trained.

In Fig.4 and Table 1 are presented the results of the simulation using a controller with the neural network described above. The results are compared with those obtained using a robust controller [23]. Speaking from the quality point of view (catch time of target) we can see that our method (with neural network) is better than classical one (the robust controller)

Table 1. Comparison of the classical and neural controller

The initial missile-target distance [m]	The line of sight angle [°]	Time [s]	
		Robust controller	Neural controller
1000	-15	3.36	3.15
1500	25	4.89	4.65
2500	45	7.77	7.45
2500	-50	7.23	6.6
3000	-20	9.78	9.69

References

1. G.C. Goodwin, P.J. Ramadge, P.E. Caines, "Discrete time stochastic adaptive control", S.I.A.M J. Contr. Optimiz., vol.19 (1981) 829-853
2. P.E. Caines, S. Lafortune, "Adaptive control with recursive identification for stochastic linear systems", I.E.E.E.Trans.on Automat.Contr., vol.AC29 312-321
3. Ian R. Petersen, "Structural stabilization of uncertain systems. Necessity of the matching condition", S.I.A.M. J. Contr. Optimiz., vol.23 (1985) 286-296
4. C.V. Hollot, "Bound invariant Lyapunov functions: a means for enlarging the class of stabilizable uncertain systems", Int. J. Contr., vol.46, pp. 161-184, 1987;
5. A.V. Levin, K.S. Narendra, "Control of Nonlinear Dynamical Systems using Neural Networks: Controllability and Stabilization", I.E.E.E. Trans. Neural Networks, vol.IV, no.2 (1993) 194-195
6. H.J.M. Telkamp, A.A.H. Damen, "Neural Network Learning in Nonlinear Systems Identification and Controller Design", Proceedings of the seconds ECC'93, vol.2 (1993) 798-894
7. N. Sadegh, "A Perceptron Network for Functional Identification and Control of Nonlinear System", I.E.E.E. Neural Networks, vol.4, no.6 (1993) 982-988
8. J.T. Tou, R.C. Gonzales, Pattern Recognition Principles, Addison-Wesley Publishing Company (1974)
9. St. Ispas, L Constantinescu, Racheta dirijata, Ed. Militara-Bucuresti (1984)
10. O. Grigore, O. Grigore, "Robust Nonlinear Control Using Neural Networks", Rev. Roum Sci. Techn., 3, vol.40 (1995) 367-390.

Demonstration of the Use of Hyperinference: Position Control Using an Electric Disc-Motor

Peter Krause

Faculty of Electrical Engineering, University of Dortmund, Otto-Hahn-Str. 4,
D-44221 Dortmund, Germany
Krause@esr.e-technik.uni-dortmund.de

Abstract. This paper demonstrates the use of two-way fuzzy controllers with hyperinference [1] as it is realized in *fuzzy*TECH. The plant is an electrical disc-motor, which is highly nonlinear because of static friction. This paper illustrates the design procedure and the advantages gained by the possibility of processing negative rules by hyperinference.

1 Introduction

Fuzzy control aims at transferring qualitative knowledge in the form of IF-THEN rules directly into a fuzzy controller and making use of it there. Conventional fuzzy controllers have only one path for processing rules and, therefore, can process only *positive* rules. These are rules that describe experience or knowledge in the form of *recommendations*. However, in many applications, it is also advisable or necessary to consider experience in the form of warnings or vetoes. This kind of knowledge can be described by *negative* rules. However, a conventional fuzzy controller cannot process negative rules. In contrast, the patented *two-way* fuzzy controller with a *hyperinference* unit can process both positive and negative rules [1, 2, 3, 4, 5].

For fast real-time applications, a simplified version of hyperinference has been developed [6]. In this paper we demonstrate that, although the current version has restricted functionality to ensure adequate performance in real-time applications, it is possible to solve complex design problems. As an example application, we consider the problem of controlling an electric disc-motor. In particular, we focus on the aspect that the disc-motor exhibits considerable static friction. The fuzzy controllers used in this example are designed and realized using *fuzzy*TECH [7]. For simulation, our tool *DORA for Windows* is used.

2 The plant and the control problem

As an example of a plant, we consider a DC servo-motor (disc-motor), which is used to assist the steering of a motor car. The input to the servo-motor is a voltage u. The controlled variable is the turning angle φ of the motor, which determines the angle φ' of the front axle. The angular velocity of the servo-motor is designated as ω.

The voltage u produces a torque τ_{Motor}, which is proportional to u. when a voltage u is applied to the resting motor, the motor begins to rotate only if $|\tau_{Motor}|$ is larger than a critical value τ_C, which is determined by static friction. In addition, a braking torque τ_S due to sliding friction is present as long as the servo-motor rotates and the absolute value of τ_S is independent of the angular velocity.

The problem is to design a fuzzy positioning control system. The input variables of the fuzzy controller are the control error $\Delta\varphi$ (difference between reference value φ_{ref} and actual value φ of the turning angle) and the angular velocity ω. The output value of the fuzzy controller is the voltage u [2, 4].

3 Design of a conventional one-way fuzzy controller

For each of the chosen input variables $\Delta\varphi$ and ω, we provide five linguistic values covering the working domain of the input values and for the output variable u we provide seven linguistic values.

The following heuristics are considered for specifying the rules.

1. If $\Delta\varphi$ and ω are both positive or negative, the servo-motor should be accelerated in the opposite direction.

2. If $\Delta\varphi$ and ω have opposite signs, the brake of the servo-motor should be applied slightly.

These heuristics are refined by making the absolute values of the acceleration and braking dependent on the absolute values of the input values. Each of the resulting five by five rules supplies a recommendation for acting if the premise of the rule is met. Consequently, all these rules are *positive* rules.

To analyse the above controller, a reference function $\varphi_{ref}(t)$ is applied to the resulting control system. Considering the actual changes $\varphi(t)$ and $u(t)$ of the angle and the voltage, we find that the performance of the control system shows the following two drawbacks.

Drawback 1

After changing the reference value, we find a *residual control error* $\Delta\varphi \neq 0$. The reason for this undesired remaining control error lies in the observation that the output voltage u of the controller is not big enough to overcome the static friction, if the absolute values of the control error and the angular velocity are small.

Drawback 2

If the servo-motor does not rotate, the controller output does not take exactly the value $u = 0$. This means that an undesired voltage (rest voltage) is permanently applied to the servo-motor. This effect is also caused by the static friction.

4 Design of a two-way fuzzy controller

In the following, the modified hyperinference unit for real-time applications is described [6].

The positive rules are first processed conventionally. The resulting output membership function $\mu^{+}(u)$, consisting of individual singletons located at positions u_i with corresponding activations μ_i, is fed into the hyperinference module and *preprocessed* there.

The first step of the *preprocessing* consists of determining the position u_{COG} of the centre of gravity corresponding to $\mu^{+}(u)$. The second step consists of constructing a parabola $\hat{\mu}^{+}(u)$: its top coincides with the position u_{COG} and the height of the top is determined by the sum of the activations μ_i of the individual singletons, bounded to the maximum value 1 (Fig. 1). The width Δu of the parabola is determined according to the following principles: Δu is proportional to the extension u_{max}-u_{min} of the chosen working domain and Δu is bigger if more singletons located at different positions are activated simultaneously.

The functional values of this parabola $\hat{\mu}^{+}(u)$ indicate the degree to which it is recommended by the positive rules, for each potential output value u. The value u_{COG} that is recommended most strongly is that which corresponds to the conventional COG-defuzzification. Values $u \neq u_{COG}$ in the neighbourhood of u_{COG} are also recommended. However, they are recommended less as the distance between u and u_{COG} increases. Values u with $|u - u_{COG}| \geq \Delta u/2$ are not recommended at all. The magnitude of Δu determines the amount of free play that can be used for considering warnings and vetoes.

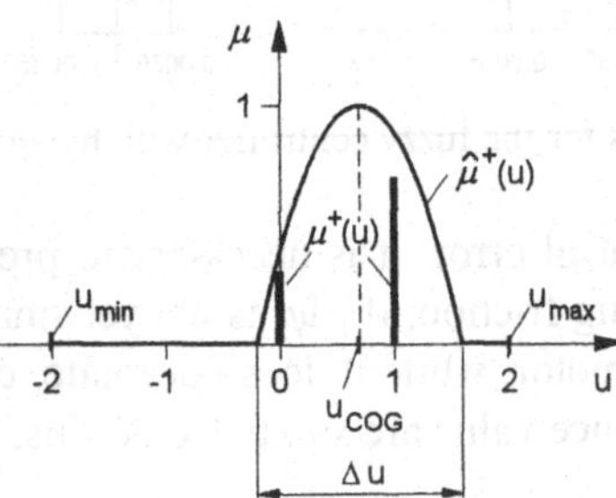

Fig. 1. Construction of the parabola $\hat{\mu}^{+}(u)$ from the function $\mu^{+}(u)$, which is produced by the positive rules

The path that processes the negative rules produces an output membership function $\mu^{-}(u)$. It describes, for each potential output value u, the degree to which all the negative rules together warn against taking this value u as the output value.

The heart of the hyperinference module is the hyperinference strategy. It combines the functions $\hat{\mu}^{+}(u)$ and $\mu^{-}(u)$ according to the prescription

$$\mu(u) = \max\{\hat{\mu}^{+}(u) - \mu^{-}(u), 0\}.$$

The resulting function $\mu(u)$ indicates, for each potential output value u, the degree to which it is supported altogether, i. e., considering all positive and negative rules. Consequently, that value for which $\mu(u)$ takes its maximum is chosen as output value u_D.

In the following, we show that the drawbacks of the conventional one-way fuzzy controller described above can be overcome by the use of a two-way fuzzy controller. In order to prevent a rest voltage $u \neq 0$ while $\omega = 0$, we must forbid any output value $u_D \neq u$ if $\omega \approx 0$ holds and $\Delta\varphi$ is small enough. To do this, the negative rule

IF ω = *infinitesimal* AND $\Delta\varphi$ = *enough* THEN u = *unequal zero* FORBIDDEN

with membership functions according to Fig. 2 is introduced.

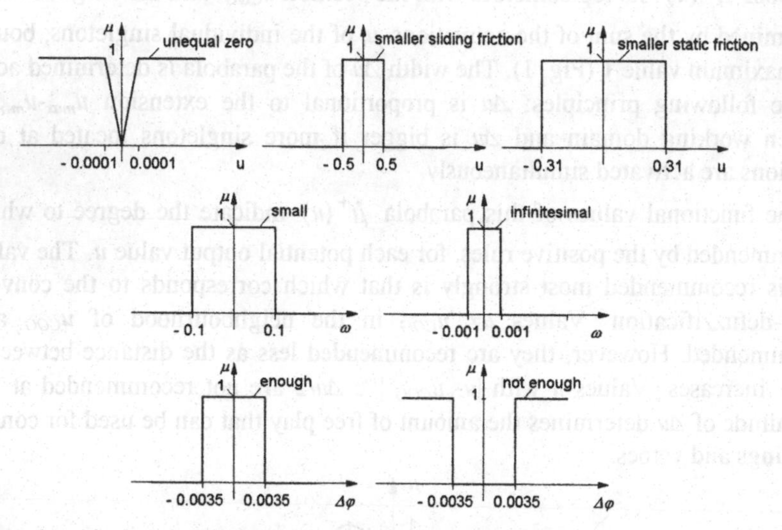

Fig. 2. Membership functions for the fuzzy controller with hyperinference

To reduce the residual control error, it is necessary to prevent the servo-motor from coming to rest due to sliding friction, if $|\Delta\varphi|$ is not yet small enough. Also it must be guaranteed that the servo-motor, while it does not rotate, can be set into motion even if the changes of the reference value are small. To do this, the negative rules

IF ω = *small* AND $\Delta\varphi$ = *not enough* FORBID u = *less than sliding friction*

IF ω = *infinitesimal* AND $\Delta\varphi$ = not *enough* FORBID u = *less than static friction*

are added. The shapes of the additional membership functions are shown in Fig 2.

When the above 25 positive rules are applied together with the negative rules the performance improves markedly, as can be seen in Fig. 3.

Remark: More details of the hyperinference version used here and the controller design are available at:

http://esr.e-technik.uni-dortmund.de/for_exa_hyp.htm

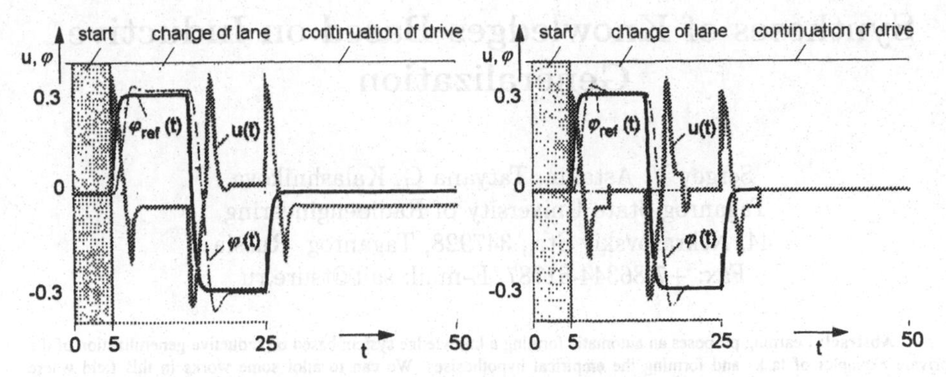

Fig. 3. Changes in $\varphi_{ref}(t)$, $\varphi(t)$ and $u(t)$ for the control system with a conventional fuzzy controller (left) and with the two-way fuzzy controller (right).

5. Conclusions

This example illustrates that a simplified version of hyperinference that allows fast real-time applications has enough flexibility to solve complex control problems. The results obtained demonstrate that the advantages of hyperinference can now be utilized by a commercial tool.

Literature

[1] Kiendl, H.: Verfahren zur Erzeugung von Stellgrößen am Ausgang eines Fuzzy-Reglers und Fuzzy-Regler hierfür. Patent DE 4808083, 1993

[2] Kiendl, H.: Sytem of controlling or monitoring processes or industrial plants employing a dual-line fuzzy unit, U. S. Patent 08/513,987, 1998

[3] Kiendl, H.: Fuzzy Control methodenorientier. R. Oldenbourg Verlag München Wien, 1997

[4] Kiendl, H., Knicker, R., Niewels, F.: Two-way fuzzy controllers based on hyperinference and inference filter. in: Jamshidi, M., Yuh, J., Danchez, P. (eds.), Proceedings Second World Automation Congress, Vol. 4, Intelligent Automation and Control, TSI Press, Montpellier 1996, pp. 387 – 394.

[5] Krone, A., Frenck, Ch., Russak, O.: Design of a Fuzzy Controller for an Alkoxylation Process using the ROSA Method for Automatic Rule Generation, EUFIT'95, Aachen, S. 760-764, August 1995

[6] Kiendl, H.: Fast Realisation of Two-way Fuzzy Controllers with Hyperinference for Real-Time Applications. Forschungsbericht 0299 der Fakultät für Elektrotechnik der Universität Dortmund, 1999, ISSN 0941-4169

[7] Inform GmbH Aachen, www.fuzzytech.com

Syntheses of Knowledges Based on Inductive Generalization

Sergey V. Astanin, Tatyana G. Kalashnikova
Taganrog State University of Radioengineering,
44, Nekrasovskiy Str., 347928, Taganrog, Russia
Fax: +7 86344-61787, E-mail: sait@tsure.ru

Abstract: Learning proposes an automatic forming a knowledge system based on inductive generalization of the private examples of tasks and forming the empirical hypothesises. We can to allot some works in this field where formalized schemes of inductive reasoning are presented. Haek and Havranek and Finn proposed its methods of plausible hypothesises construction The difference of this work consist of operator behaviour anticipation (outrunning reflection) registration at the forming hypothesises. In this paper the problem is formulated, and the mechanism of inductive reasoning is described. This approach was applied practically in medical-biological experiment at the man-operator parameters interdependence (correlation) reveality.

Index terms: inductive conclusion, anticipation of behaviour of the object, fuzzy variable, plausible hypothesises, basis set of fuzzy variable (BSFV).

INTRODUCTION

Synthesis of knowledges means an automatic forming of a knowledge system based on inductive generalization of the private examples of tasks, and forming the empirical hypothesises. As an induction we used to understand a process of hypothesises forming in the form of general regularities on the base of observing the private events. To find the regularities it takes to do the following

- revealing resemblance of private events;
- forming a suggestion for general features which reflect resemblance;
- forming a confirmation of suggestion on sufficient amount of examples;
- and finally, generalizing a suggestion in the form of regularities.

The methods of deductive reasoning are based on powerful formal systems. However, the theory of non-standard logical calculations is passing a stage of its formation. At present, the elaboration in this field is associated with the study, analysis and formalization of different schemes of inductive conclusions. So P.Hajek and T.Havranek [1] developed GUHA-method for construction of plausible hypothesises based on the theories of mathematical logic and mathematical statistics. V.Finn [2] developed JSM-method based on induction schemes offered by J.Mille and on using of monotonous and nonmonotonous strategies to raise and prove the hypothesises. The given methods allow to solve the following problems. The objects and their properties are considered. It is assumed that laws, which control presence or absence of these properties are studied not enough or unknown. The selection of objects is considered and it is known for each object whether – it has some property or not. The objects from other selection are presented. It is required to get plausible hypothesises, which suppose that these objects have the earlier fixed properties.

II. THE FORMULATION OF THE PROBLEM

We are interested in the following problem. The properties of object are pointed out. In the order to simplify the formulation of the problem, we will consider that object is described by two characteristics. The set of certainly properties is known for characteristic A. Other characteristic B is presented only by measurements. Further we will call the set of characteristic's properties as a state of object. The dependency of characteristic's properties is expected and object has the following peculiarity: changing a property of one characteristic is possible before changing a property of the other characteristic. The object seems to adjust for changing its state. Such a peculiarity of the object's behaviour is known in physiology and is named a principle of the behaviour anticipation (outrunning reflection)[3]. It is necessary to select unknown properties of characteristic B and influence of properties of characteristic B on properties of characteristic A in the process of simultaneous observing the values of both characteristics paying attention to the anticipation of object's behaviour. Such a formulation of the problem is distinguished from above considered by the necessity to pay attention to the anticipation of object's behaviour while forming the hypothesises. Such peculiarity of behaviour corresponds to the identity law of dialectical logic: any Q is Q and it is not Q at the same time. The similar statement of the problem is distinctive in events of correspondence determination between reports of intellect agents in the contact cession, and in events of revealing a dependency between characteristics of situations at the synthesis of situational network.

III. THE RELATION DEPENDENCE CONSTRUCTION OF THE OBJECT CHARACTERISTICS PROPERTIES ON THE BASE INDUCTIVE GENERALIZATION

The relation of dependences of the object characteristics properties, based on the inductive generalization, is constructed. Let A and B be the characteristics of some object x, each of which have properties A and B. In the process of observing these properties can be measured by means of values a_i and b_i, where $a_i \in A$, $b_i \in B$, $i \in T=\{1, 2, \ldots, N\}$, and T defines consequent moments of the observation's time. The sets of subsets $\{A_j\}$, $\{B_j\}$, $j=\overline{1,n}$ will be assigned on sets of values A and B by means of methods of experimental optimization. Assume that a_i, a_{i+1}, b_i, b_{i+1} are the values of properties A and B fixed at consequent moments of time, and $b_i=b_{i+1}$. We shall consider values a_i, a_{i+1} as points of experimental plan on the interval $[c_j, d_j]$ the limits of which are unknown. In accordance with the methods of experimental optimization, efficient planning is built by using Fibonacci numbers, which are defined by the sequence $F_0=F_1=1$, $F_k=F_{k-1}+F_{k-2}$, $k>1$. Coordinates of two points on the interval [c, d] are assumed equal $a_i=c+(d-c)\xi$, $a_{i+1}=d-(d-c)\xi$ with $\xi = F_{k-2}/F_k$, where k is chosen as sufficiently big. The problem is to calculate membership function of the fuzzy variables on the scale of characteristic A. Each fuzzy variable is linguistic value of characteristic A. Fuzzy variable is presented by a triplet $<\alpha, A, C(\alpha)>$, where α is a name of fuzzy variable, $A=\{a\}$ is an area of its determination (base set), $C(\alpha)=\{<\mu_{C(\alpha)}(a)/a>\}(a \in A)$ is a fuzzy subset of set A which describe the restrictions on possible values α. If a_i, a_{i+1}, b_i, b_{i+1} are values of properties A and B, fixed at consequent moments of time when $b_i=b_{i+1}$ then the interval $A_1=[c_1, d_1] \subset A$ is named as a base set of fuzzy variable (BSFV), where c_j and d_j are defined by solving an equation system:

$$\begin{cases} a_i = c_j + (d_j - c_j)\xi \\ a_{i+1} = d_j - (d_j - c_j)\xi \end{cases},$$

where on each step limit value ξ is used at $k \to \infty$. Values a_i, a_{i+1} are considered as experimental plan points the forming of which consists in the calculation of limits of interval, the values of it are similar relatively to b_i (b_{i+1})[4].

Definition 1. Expression of type $A_k \to_1 b_s$, where A_k is BSFV and $b_s \in B$ is observed simultaneously with $a_1 \in A_k$, is named an empirical hypothesis of 1-st kind.

Example 1. An example of forming the empirical hypothesises of 1-st kind is considered. Let A - "pulse", B - "risk activity breakdown" and measurements $a_1=60$, b_1="low", $a_2=70$, b_2="low" are fixed in the observing base. It is necessary to define BSFV as $A_1=[c_1, d_1]$ solving the equation system at $\xi=0.382$.

There is:

$$\begin{cases} 60 = c_1 + (d_1 - c_1)0.382 \\ 70 = d_1 - (d_1 - c_1)0.382 \end{cases}.$$

As a solution we get: $A_1=[44, 86]$ and $A_1 \to_1$"low".

Definition 2. Let $\{a_t\}$ and $\{b_t\}$ are observing the values of properties fixed at consequent moments of time and $A_k \to_1 b_s$ is an empirical hypothesis of 1-st kind for some a_t at $1 \le t \le r$. Changes of the values a_t for $t>(r+1)$ such as $a_t \notin A_k$ and $b_t=b_s$ will be called anticipation relatively to b_p, if a value b_p, $(p > r+2)$ different from b_s for all $b_t=b_s$ at $t < p$ will be found. At the same time, if a_t is anticipation then we will conditionally consider that a_t and b_p are fixed at the same moment of time.

Definition 3. If a_t is anticipation relatively to some b_p then the expression of type $a_t \to_2 b_p$ will be named an empirical hypothesis of 2-nd kind. It is easy to see that if empirical hypothesis 1-st kind $A_k \to_1 b_s$ is formed then it is possible to convert it into empirical hypothesis 2-nd kind when among $a_t \in A_k$ it will be found such value which is anticipation relatively to b_s.

Theorem 1. If a_t is the anticipation relatively to b_p then the empirical hypothesis of 2-nd kind has an interval character of type $A_k \to_2 b_p$ and $a_t \in A_k$.

Proof. According to the definition 2 we have the following pair of observing a_t, b_p, a_p, b_p. Then it is possible to form BSFV as $A_k=[c, d] \subset A$ where a_t, $a_p \in A_k$. Consequently, it is $A_k \to_2 b_p$.

Example 2. The values of characteristics of preceding example are considered. The observing base has the form of: $a_1=60$, b_1="low", $a_2=70$, b_2="low", $a_3=90$, b_3="low", $a_4=100$, b_4="high". In accordance with the definition 2 pulse value equal to 90 is anticipation relatively to the value "risk activity breakdown" - "high". Let's form an empirical hypothesis of 2-nd kind, taking to consideration the following observing sequence: $a_3=90$, b_3="high", $a_4=100$, b_4="high". Let's build BSFV as $A_1=[74, 116]$. Then empirical hypothesis 2-nd kind will have the form of: $[74, 116] \to_2$"high".

Theorem 2. If in consequent observing $1 \le t \le r$ values are absent which are anticipation, and on the interval [l, r] BSFV as $A_k=[c, d]$ is formed with the empirical hypothesis 1-st kind $A_k \to_1 b_s$, and the value $a_{r+1} \notin A_k$ at $b_{r+1}=b_s$ then BSFV can be extended to $A_k'=[c', d']$, with $A_k \subset A_k'$.

Proof. Since one of the experiment plan points is a value a_{r+1} then the second point is limited by the interval [c, d]. From the given interval as the second of the experiment plan points we will take an average value of interval equal to $(c+d)/2$. Calculate new BSFV as $A_k'=[c', d']$. Consider two possible variants of value a_{r+1}.

1. $a_{r+1} > d$.

In this case we have:

$$c^{l}=\frac{\frac{(c+d)}{2}-\left(\frac{(c+d)}{2}+a_{r+1}\right)\xi}{(1-2\xi)}=2c+2d-2\xi d-1.6a_{r+1}=c+d-1.6a_{r+1}.$$

Let a_{r+1}=d+ε. Then c^{l}=c + d - 1.6d - 1.6ε. Obviously, that $\lim\limits_{\varepsilon\to 0} c^{l}=c-0.6d$, i.e. $c^{l}<c$.

Define the value of right the limit A_k^{l}:

$$d^{l}=\frac{\frac{(c+d)}{2}-c^{l}(1-\xi)}{\xi}=2.3d-0.3c+2.5\varepsilon.$$

Obviously that $\lim\limits_{\substack{c\to d\\ \varepsilon\to 0}} d^{l}>d$ and $\lim\limits_{\varepsilon\to\infty} d^{l}\to\infty$.

Thus both the former interval [c, d] and the value a_{r+1} belong to the newly formed BSFV.

2. $a_{r+1}<c$.

There is:

$$c^{l}=\frac{a_{r+1}-\left(a_{r+1}+\frac{(c+d)}{2}\right)\xi}{(1-2\xi)}=4.2a_{r+1}-4.2a_{r+1}\xi-0.8c-0.8d=2.6a_{r+1}-0.8c-0.8d.$$

Let a_{r+1}=c - ε. Then c^{l}=2.6c - 2.6ε - 0.8c – 0.8d = 1.8c – 0.8d - 2.6ε and $\lim\limits_{\substack{\varepsilon\to 0\\ d\to c}} c^{l}\to c$, $\lim\limits_{\varepsilon\to\infty} c^{l}\to-\infty$.

Define the value of right the limit:

$$d^{l}=\frac{a_{r+1}-\frac{c}{(1-\xi)}}{\xi}=2.6a_{r+1}-1.6c^{l}=1.3c+1.3d-1.4a_{r+1}.$$

Let a_{r+1}=c - ε. Then d^{l}=1.3c+1.3d–1.4c+1.4ε =1.3d–0.1c+1.4ε and $\lim\limits_{\substack{\varepsilon\to 0\\ c\to d}} d^{l}\to 1.2d$, $\lim\limits_{\varepsilon\to\infty} d^{l}\to\infty$. In both cases d^{l}> d.

Thus at a_{r+1}<c $A_k\subset A_k^{l}$ and $a_{r+1}\in A_k^{l}$.

Obviously, that fuzzy set $A_j=\langle\mu_v/\{a_j(t)\}\rangle$ is a fuzzy subset of set A, and $a_j(t)$ are values of BSFV as $A_j=\{a_j(t)\}$ with the function of membership $\mu_v(a_j(t))=n/n^*$; n - an amount of $a_j(t)$, which satisfy to predicate $a_j(t)\to_1 b_s(t)$; n^* - the total amount of observing of $b_s(t)$. Function μ_v is the private determination of predicate of type "$\to_1$". It will similarly define the private determinations of predicates of type "$\to_2$" calculating them as $\vartheta_w=\frac{\sum_v n_v}{n^*}$, where $\sum n_v$ is amount $a_j(t)\in A_j$, which satisfy to predicate $a_j(t)\to_1 b_s(t)$. The fuzzy variable formed according to the method described above allows to go over to more general predicates descriptions "$\to_1$"and "$\to_2$", which will have the form of: $x\in A_j$ $x\to_1 b_s$, $x\in A_j$ $x\to_2 b_s$, $x\in A_j$ $b_s\to_2 x$. When turning from the private determination ϑ_w to quantifiers then predicates of type "$\to_2$" will have the form of: $\Delta x\in A_j$ $x\to_2 b_s$, $\Delta x\in A_j$ $b_s\to_2 x$, where Δ is fuzzy quantifier. Value A_j is the interval generalization of the observing interpreted as an fuzzy variable.

Fuzzy variables A_k and A_{k+1} will be named coinciding, if $A_k\cong A_{k+1}$ for all $A_k\to_1 b_s$ and $A_{k+1}\to_1 b_s$. Here "≅" is an operation of fuzzy equality. Operation of combining fuzzy variables A_k and A_{k+1} will be named an operation of crossing the fuzzy subsets $A_k\cap A_{k+1}$, if fuzzy variables A_k and A_{k+1} are coinciding. These determinations allow to reveal surplus in observing or in other words, if $A_k\approx A_{k+1}$ then $A_k\cap A_{k+1}\to_1 b_s$. Using operation of crossing the fuzzy subsets A_k and A_{k+1} as operation of combining to fuzzy variables is stipulated by the circumstance that in this case nonsignificant observations are not taken into account, i.e. those observations for which function ϑ_w has values less than some thresholdous value. Using an operation of crossing the fuzzy subsets allows to dispose of surplus information in observations and to reduce a number of analysed hypothesises. Let's enter a threshold of determination ϑ_w to form the plausible hypothesises. If $\vartheta_w\in[0.6, 1]$ then the observation is considered as hypothesises with the plausibility equal to ϑ_w. If $\vartheta_w\in[0, 0.6]$ then we will consider that data is insufficient for forming a hypothesis about the influence of properties of one characteristic on properties of the another.

IV. CALCULUS OF PLAUSIBLE HYPOTHESIS FORMING

Consider a general case, when observing is in the form of: $b_l^{i}(k),\ldots,b_l^{u}(k)$. We are interested in twin mutual influences of characteristics properties by means of determination of partly determined predicates of type "$\to_2$". Let's build the language for automatic hypothesises forming for general case.

Sorts of terms:

1. Characteristics of object x: $A_1,\ldots,A_n$.
2. Properties of characteristics: $b_l^{i},\ldots,b_l^{u}$.
3. Base sets of fuzzy variables.
4. Natural numbers: k, l,...

5. Individual constants: $b_i^i(k),...,b_i^{ll}(k)$.

Predicates:

1. $b_i^i \rightarrow_1 b_r^s$ is "the property b_i^i of i characteristic is observed jointly with the property b_r^s of s characteristic";
2. $b_i^i(k) = b_p^i(k+p)$ is "the property $b_i^i(k)$ fixed on k step of observing coincides with the property $b_p^i(k+p)$, fixed on (k+p) step";
3. $x \in B_i^i$ is "the variable belongs to the class (fuzzy set) B_i^i";
4. $b_i^i(k) \in B_i^i(p)$ is "the value of observation $b_i^i(k)$ fixed on k-step belongs to BSFV $B_i^i(p)$, formed on p-step of observation".

Partly determined predicates:

$b_i^i \rightarrow_2 b_r^s$ is the property b_i^i influences the property b_r^s. The area of values of truth (validity) predicates of type "$\rightarrow_2$" are values of special fuzzy logic from [0, 1]. Values of truth are appropriated to formulas of type: $x \in B_i^i y \in B_r^s \alpha(x \rightarrow_2 y)$, where α is validity of hypothesis denominated by this formula.

Production rules:

$$\frac{b_i^i(k), b_r^s(k), b_i^i(k+1), b_r^s(k+1), b_r^s(k) = b_r^s(k+1)}{B_i^i(k+1) \rightarrow_1 b_r^s}; \tag{4.7}$$

$$\frac{b_i^i(k), b_i^i(k) \in B_i^i(t) / t \le k, b_r^s(k)}{B_i^i(k) \rightarrow_1 b_r^s(v) / v \ge k}; \tag{4.8}$$

$$\frac{B_i^i(k), b_r^s(k) = b_r^s(k+1), b_i^i(k+1) \rightarrow_1 b_r^s(k+1), b_i^i(k+1) \notin B_i^i(k)}{B_i^i(k+1) \rightarrow_1 b_r^s(k) / B_i^i(k) \subseteq B_i^i(k+1) \,\&\, b_i^i(k+1) \in B_i^i(k+1)}; \tag{4.9}$$

$$\frac{b_i^i(k+1) \in B_i^i(k), b_s^r(k+1) = b_s^r(k)}{B_i^i(k) \rightarrow_2 b_s^r} \tag{4.10}$$

$$\frac{\begin{matrix} B_i^i(k) \rightarrow_1 b_r^s(k), b_i^i(k+1) \notin B_i^i(k), b_r^s(k+1) = b_r^s(k), \\ b_i^i(k+p) = b_i^i(k+1), b_r^s(k+p) \ne b_r^s(k) \end{matrix}}{b_i^i(k+1) \rightarrow_2 b_r^s(k+p)}; \tag{4.11}$$

$$\frac{B_i^i(k) \rightarrow_2 b_r^s(k)}{x \in B_i^i x \rightarrow_2 b_r^s}; \tag{4.12}$$

$$\frac{x \in B_i y \in B_p x \rightarrow_2 y}{\alpha(x \in B_i y \in B_p x \rightarrow_2 y)}. \tag{4.13}$$

Rules (4.7-4.13) allow turn from quantitative estimations of properties to qualitative ones, formally expressed in the manner of ensembles B_i^i of rule (4.12). Rule (4.11) expresses the properties of anticipation, and rule (4.13) is a plausible conclusion rule.

V. THE ALGORITHM OF HYPOTHESIS AND KNOWLEDGE BASE FORMING ANALYSIS

The computer-oriented algorithm of hypothesis and knowledge base forming analysis consist of following steps:

1. The selection (k+1) record from database.
2. If the situation S_{k+1} coincides with the situation S_k then go to (turn to) step 3 else go to step 8.
3. If the BSFV were formed then go to step 4 else go to step 12.
4. If the (k+1) value belongs the BSFV then go to step 5 else go to step 12.
5. If the self-estimation value C_{k+1} coincides with the value C_k then go to step 6 else go to step 13.
6. It is forming of the hypothesis of form: $S_k \rightarrow_1$ BSFV(f_k); $S_k \rightarrow_1 C_k$.
7. If the record (i+1) is last then the end of algorithm else go to step 1.
8. If the value f_{k+1} coincides with the value f_k or $f_{k+1} \in$ BSFV(f_k) then go to step 9 else go to step 10.
9. It is forming of the plausible hypothesis of type: $C_{k+1} \rightarrow_2 S_{k+1}$; $C_{k+1} \rightarrow_2 f_{k+1}$. Go to step 7.
10. If the value C_{k+1} is anticipation then go to step 9 else go to step 11.
11. It is forming of the hypothesis of type: $S_{k+1} \rightarrow_2 C_{k+1} \& f_{k+1}$. Go to step 1.

12. It is forming of the BSFV for the quantitative factors. Go to step 3.
13. It is forming of the hypothesis of type: "C_{k+1} is anticipation".
14. If the hypothesis "C_{k+1} is anticipation" is confirmed then go to step 15 else go to step 16.
15. It is forming of the hypothesis $C_{k+1} \rightarrow_2 S_{k+1}$. Go to step 1.
16. It is forming of the hypothesis $S_{k+1} \rightarrow_2 C_{k+1}$.

VI. RESULTS

The experimental research of the external factors influence on man-operator state was conducted. The experimental observing data processing is realized on above considered algorithm. The experiment goal was to reveal causal relations between parameters of psycho-physiological components of the man behavior. These components were selected as following:

- the state of physiological organism subsystems;
- the psychological portrait of man-operator;
- the man-operator self-estimation of his state regarding to the solving task results;
- the sate (results) of the solving task.

The state of physiological organism subsystems was estimated by using following factors:

- the heart abridgements frequency (HAF);
- the tension index (TI);
- the breathing amplitude;
- the breathing frequency;
- the relation between the time of breath and the time of exhalation, and other.

The psychological portrait of man-operator was constructed on the base of the tests collection:

- Spilberg-Hanin;
- General state – activity – mood;
- Lusher;
- Supos-8;
- the distributing, selecting and attention switching, and other.

The self-estimate variations are as following:

- the certainty in successful task decision is weak (1);
- the certainty in successful task decision is not very weak (2);
- the certainty in successful task decision is middle (3);
- the certainty in successful task decision is high (4);
- the certainty in successful task decision is very high (0).

The state of solving task was estimated as following:

- the activity flowing according to optimal strategy (1);
- the activity flowing with the low task decision risk break (2);
- the activity flowing with the sufficiently high task decision risk break(3);
- the activity flowing with the very high task decision risk break (3);
- the activity was disrupted on i-step of the task decision (5).

The three from four selected elements are characterized by the qualitative estimations, and the one from four is characterized by the quantitative estimations. The experimental data and processing results by using inductive reasoning are presented in the table.

Let F is a set of the physiological factors, C_i ($i=\overline{0,4}$) is a set of the self-estimate values, S_j ($j=\overline{1,5}$) is a set of the situation estimations. Consider that one of the physiological factors values is F(TI) named as tension index. Define causal relations between three kinds of the factors by means of using experimental data analysis by inductive reasoning. Let S_k, C_k, f_k are current value of situations, self-estimations, and tension index. The formed hypothesises collection is presented in Exhibit. The analysis shows that four types of the causal relations between elements of selected structure exist: self-estimation -- situation, self-estimation – physiology, situation – self-estimation, physiology – (situation & self-estimation), physiology – situation. Concrete type of these causal relations for defined element values is presented below.

The fragment of experimental data and processing results by using inductive reasoning are presented in the table.

THE EXPERIMENTAL DATA

Rec.№ in DataBase	The activity disruption risk	Physiological parameters (F)				Code of self-estimation (C)	Code of situation (S)
		The heart abridgements frequency (HAF)	The dispersion of the HAF	The tension index (TI)	The breathing amplitude		
1	0,00	90,13	4,04	324,32	3,51	3	1
2	0,00	93,65	4,45	216,52	3,46	3	1
3	0,00	92,00	7,34	216,52	3,91	3	1
4	0,25	89,00	5,64	216,52	4,19	0	2
5	0,25	90,92	5,53	122,91	4,07	0	2
6	0,67	91,29	3,50	122,91	4,19	4	3
7	0,67	79,92	9,45	122,91	4,11	4	3
8	0,67	91,32	3,71	450,45	4,06	4	3
9	0,50	82,68	5,33	450,45	4,19	0	3
10	0,50	80,45	4,35	450,45	4,29	0	3
11	0,00	81,97	4,10	298,25	4,35	0	1
12	0,00	82,33	2,74	299,25	4,51	0	1

The Hypothesis reliability estimation:

$C_0 \rightarrow S_1 - 0.46$;

$S_2 \rightarrow C_0 - 0.3$;

$S_3 \rightarrow C_4 - 0.45$;

$F(TI)=[280\text{-}350] \rightarrow C_0 - 0.33$;

$F(TI)=[175\text{-}216] \rightarrow C_0 - 0.6$;

$F(TI)=[370\text{-}450] \rightarrow C_4 - 0.3$;

$F(TI)=[110\text{-}175] \rightarrow S_3 - 0.4$;

$F(TI)=[130\text{-}210] \rightarrow S_1 - 0.57$;

$C_4 \rightarrow F(TI)=[50\text{-}200] - 0.67$.

VI. CONCLUSION

The considered approach to forming the plausible hypothesises and inductive generalization of sequence of measurements in the form of fuzzy sets (fuzzy variables) allows to realize an analysis of observations databases of the objects characteristics values, which possess the property of an behaviour anticipation. The approach was practically approved in the course of conducting a medical-biological experiment when solving a problem of revealing an interdependence of person-operator parameters.

The automation intelligent system of a medical-biological experiment implementation was realized in EDB "RITM" of Taganrog State University of RadioEngineering. The publications analysis shows absence similar tool systems. At the research more 100 experiments was conducted, that it is a sufficient large set. These researches shown the receipt possibility of the more qualitative conclusions regarding the man-machine-operator activity estimation when man is considered as an organizational system modeled by computer environment. If early the number of erroneous conclusions by using only the man-machine-operator activity analysis was equal 40-50% then the number of errors by using considered technology reduced to 15%.

REFERENCES

[1] Hajek P., Havranek T. Mechanizing hypothesis formation. Mathematical Foundation for a General Theory. Springer-Verlag Berlin Heidelberg, 1978.

[2] Finn V.K. 1995. JSM – Reasoning for control problem in open (±)-world/ Proceedings of the 1995 ISIC Workshop. 10th IEEE International Symposium on Intelligent Control, Monterey, CA, USA. PP 75-79.

[3] Lomov B.F., Surkov E.N. Anticipation in the structure activity.-M.: Nauka, 1980.-278p.

[4] Hartmann K., Lezki E., Schafer W. Statistische Versuchsplanung und-auswertung in der Stoffwirtschaft. Veb Deut Scher Verlag fur Grundstoffindustrie. Leipzig. 1974.

ANN-Based Fault Type and Location Identification System for Autonomous Preventive-Restoration Control of Complex Electrical Power System Plants

A. Halinka, M. Szewczyk, B. Witek
Silesian Technical University of Gliwice

Introduction

In modern electrical power systems (EPS) an increasing importance of generating nodes, such as: pump storage plants or combined power plants including gas-turbine as well as steam-turbine generators, is observed. Such objects posess composed but highly redundant structure offering a significant opertion flexibility e.g. the adjacent start-up system use in a fault case in the dedicated start-up system, or two generating sets co-operation with one step-up transformer. The essential features of these objects are: very good power regulation abilities, short start-up time and internal structure modification possibility. The above features determine the growing importance of such objects in the EPS operation control process, aiming at the optimal operation mode as well as fast and effective restoration after a system fault occurrence, which may be ensured by the adaptive protection and control system (APCS) - sufficiently reliable both in normal operating conditions and in a fault case. One of the basic functional modules included in APCS is the autonomous preventive restoration control module (APRCM), developing the object control rules and preventing it from a fault or limiting possible fault effects via fast restoration of generation and regulation abilities.

Alarms analysis and fault location system dedicated to reversible hydro-turbine generating sets (FID)

The main part of APRCM makes alarms analysis system and fault type and location estimation FID (fault identification). Information developed in the FID system is a basic knowledge source for APRCM and is used for signals supply for power generation and voltage regulator control.

In case of disturbance in the protected object or in the adjacent systems - such as: out-of-step, frequency deviation, overload or transformer cooling system failure - FID system analyses the alarms, and the results are used by APRCM to initiate the preventive actions, leading to the further object operation in emergency conditions.

If the object (or its element) operation is no longer possible and it has to be switched-off, the algorithm of fast autonomous post fault restoration is started, replacing the faulted element by the redundant system. As an example of the composed generating plant the system comprising two reversible hydro-tubine sets has been taken, for which the FID and APRCM - based adaptive protection/control system has been proposed.

Due to the object complexity the adaptive protection/control system has been decentralized and divided into modules, dedicated to the specified functional blocks as well as to the protected object elements (machine, step-up transformer etc.). Decentralized APCS structure allows to break up the by particular blocks realized functions among the representative modules, which increases the number of processors but also considerably decreases the processor load degree and the reaction time of the entire system. The parallel data processing - insensitive to any information error - is essential for the correct alarms interpretation as well as fault location and identification in FID. Since in practice an infinitive number of fault types is possible, the FID should be able to generalize events and classify the properly. An implementation of ANN in FID structure seem to be the best way to comply with the above specified requirements. The FID of the particular hydro-turbine generating set consists of four modules (FID-T, FID-SUT, FID-M1, FID-M2) assigned to the object elements and co-operating with them APRCMs. The particular modules are built on the multi-layer perceptron (MLP) structures, and error back-propagation algorithm is applied for the net supervised training.

Such an approach ensures any non-linear problem solution. It has been assumed that the number of neurons in the first layer is 15 and in the second 10, which was optimal to represent the complex identification relationships. The output layer contains as many neurons as the number of FID outputs. The hyperbolic tangent activation function has been used for the first and second layer, and for the third layer the linear function has been applied.

The following signals for the particular modules neural nets have been used:

- binary signals representing the elements topology (switches positions),
- measuring signals for binary information verification,
- warning signals from regulation systems (e.g. AVRS),
- alarms indicating protective functions activation or protection operation, supplied by the protection-measuring block modules,
- output information from the object operating mode identification block,
- adjacent FID modules output signals.

In Fig.1 the FID scheme dedicated to the synchronous machine is presented.

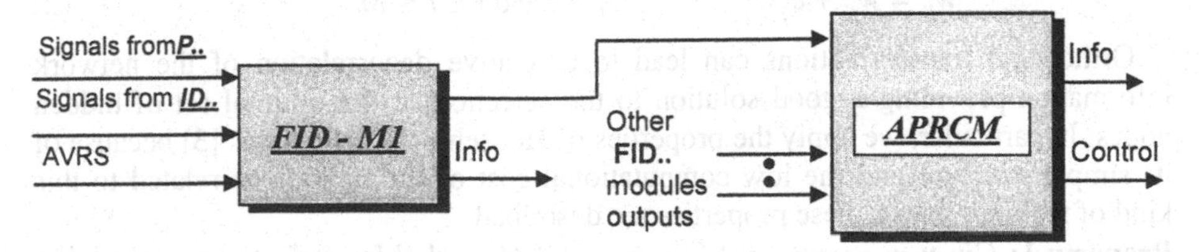

Fig.1 Pictorial diagram of the **FID - M1** module dedicated to the synchronous machine

Optimal Hidden Structure for Feedforward Neural Networks

P. Bachiller, R.M. Pérez, P. Martínez, P.L. Aguilar, P. Díaz

Department of Computer Science. University of Extremadura. Escuela Politécnica. 10071 Cáceres. Spain.
{pilarb, rosapere, pablomar, paguilar, pdiaz}@unex.es

Abstract. The selection of an adequate hidden structure of a feedforward neural network is a very important issue of its design. When the hidden structure of the network is too large and complex for the model being developed, the network may tend to memorize input and output sets rather than learning relationships between them. In addition, training time will significantly increase when the network is unnecessarily large. We propose two methods to optimize the size of feedforward neural networks using orthogonal transformations. These two approaches avoid the retraining process of the reduced-size network, which is necessary in any pruning technique.

1 Selection of the Optimal Hidden Structure

The basic principle of network size reduction is to detect and to eliminate collinearity between the input data sets at the hidden layers. Consider a feedforward neural network with N nodes in its single hidden layer and M output nodes. Let h_i be a vector formed by the outputs of the i-th hidden node for all the training patterns. If h_i can be computed as a linear combination of h_j's (expression 1), the i-th hidden unit can be eliminated and equation (2) provides a direct adjustment for the weights of the hidden links connected to the nodes to be retained (w_{lj}) that preserves the initial behavior.

$$h_i = \sum_{j=1\, j\neq i}^{N} c_j h_j \qquad (1)$$

$$w_{lj} = w_{lj} + c_j w_{li} \qquad \forall j \neq i \text{ and } 1 \leq l \leq M \qquad (2)$$

Orthogonal transformations can lead to a relative decorrelation of the network information providing a good solution to the selection of the optimal set of hidden nodes. In particular, we apply the properties of Householder reflections [3] because of its simple structure and the low computational cost of the operations related to this kind of matrices. Next, these properties are described:

Property 1: *Given m vectors* $\in \Re^n$ $[x_1, x_2, \ldots, x_m]$*, Householder reflections are used to determine which of them are linearly independent.*

Property 2: *If a vector* $x \in \Re^n$ *is a linear combination of m linearly independent vectors* $(x_1, x_2, \ldots, x_m)$*, Householder reflections provide the coefficients of such combination by solving a square triangular equation system.*

The first method we propose uses the above properties to compute the optimal hidden structure of a given network. It starts with a large trained network and selects the optimum hidden units applying the property 1 on the vectors formed by the hidden outputs for the whole set of training patterns. Then, the coefficients c_i of expression 1 are determined by the second property and the weights of the optimal links are adjusted using equation 2.

2 The Optimizing and Training Algorithm

When a pruning technique is used to compute the optimal set of hidden nodes of a given network, it is always necessary to train an overparameterized network [2,4]. The training time increases significantly when the hidden structure of the network is too large and complex for the model being developed. In order to both, accelerate the training process and improve the network generalization, we proposed an optimizing and training algorithm (OTA). This method is based on the idea of determining, in each iteration of the training process, the set of linearly independent outputs of the hidden layer(s) and then updating only the weights of the links connected to those hidden units.

The method starts with a *NxMxO* neural network and *P* training patterns. In each iteration of the algorithm, the set of optimum hidden nodes is computed applying the first property of Householder matrices to the *M* vectors obtained by the outputs of the hidden nodes for the *P* training patterns. For all the hidden nodes belonging to the obtained set, the weights of their links are adjusted using the *Backpropagation* algorithm. The algorithm iterates until the network reaches the convergence.

To test the effectiveness of our algorithms, experimental results on different test problems have been obtained. These results show that the optimizing and training algorithm reduces significantly the network training time with regards to the original Backpropagation algorithm [1]. Moreover, the generalization capability of the network is improved when both optimizing algorithms are used.

References

1. P. Bachiller, R.M. Pérez, P. Martínez and P.L. Aguilar: *A method based on orthogonal transformation for the design of optimal feedforward network architecture*. Proceedings of the 3rd International Meeting on Vector and Parallel Processing (VECPAR'98), 1998, pp. 541-552.

2. G. Castellano, A. M. Fanelli and M. Pelillo: *An iterative pruning algorithm for feedforward neural networks*. IEEE Transactions on Neural Networks, vol. 8, no. 3, 1997, pp. 519-531.

3. G. H. Golub and C. F. Van Loan: *Matrix computations*. Baltimore, MD: John Hopkins Univ. Press, 1989.

4. P. Kanjilal and N. Banerjee: *On the application of orthogonal transformation for the design and analysis of feedforward networks*. IEEE Transactions on Neural Networks, vol. 5, no. 5, 1995, pp. 1061-1070.

Relation between the IADL and Physical Fitness Tests Focusing on Uncertainty of Answering Questionnaire in Elderly Women

Hayato Uchida [1], Yutaka Hata [1] Kensaku Suei [1], Hiroshi Nakagawa [1] and Hideyasu Aoyama [2]

[1]Himeji Institute of Technology, 2167, Shosha, Himeji, 671-2201, Japan.
uchida@hept.himeji-tech.ac.jp
[2]Okayama University Medical School, 2-5-1, Shikata-cho, Okayama, 700-8558

Abstract. This paper examines the uncertainty of answering questionnaire in elderly women. We derive a membership function demonstrating a robustness of the knee-raising test and height related to decreased instrumental activities of daily living (IADL) for the uncertainty of answering questionnaire. This result could be maintained if the number of movement were within range from −10 to +5 persons.

1 Introduction

This paper concerns the uncertainty of answering questionnaire in elderly women, and then analyzes the relation between the result and physical fitness tests. Consequently, we obtain a membership function, which demonstrates a robustness of our analysis result for the uncertainty on answering questionnaire.

2 Subjects and Methods on Uncertainty for Answering Questionnaire

The survey was carried out in June and July of 1993. The subjects included 128 women aged 65 years or over, who lived at home in Y Town, near Himeji City in Hyogo Prefecture, Japan, where 12.8 % of the population in this town is 65 years old and over, similar to the national average for the entire Japanese in 1992 [1]. According to the results of a medical check, the subjects did not have serious paralysis or a decrease of muscular strength due to cerebrovascular or musculoskeletal disease.

IADL status was assessed by questionnaire, using an index of competence created by the Tokyo Metropolitan Institute of Gerontology [2], consisting of five items [Question (Q)1, going out to use a train or bus; Q2, going shopping; Q3, preparing meals; Q4, making payments of bills; and Q5, obtaining receipts of postal savings and/or payments from bank accounts]. The elderly women answer every items of the questionnaire. In determining 'Yes' or 'No' in the questionnaire, when the elderly women can not choice either, "They throw the dice/coin". This is, the answer of the questionnaire includes uncertainty and fuzziness [3]. If we add third value 'unknown' into 'Yes' or 'No' in the questionnaire, it is impossible to analyze on odds ratio (OR)-based statistical analysis[4].

We carried out various physical fitness tests including measurement of height, body weight, grip strength, sitting stepping, standing on one leg with open eyes, standing on one leg with closed eyes, sit-and-reach test, jumping reaction time and knee-raising test.

Body mass index was calculated as body weight (kg) / height (m)^2.

Our analysis method is as follows:

Step (1) employ questionnaire.

Step (2) classify into two groups according to the disability level competence: the dependent in IADL (Dep) group consisted of 49 persons (38.3%) who were dependent in regard to at least one of the five functions, and the independent in IADL (Ind) group consisted of 79 persons (61.7%) who were independent with regard to all five functions.

Step (3) do χ^2 test of the data obtained in Step 2.

Step (4) select from 5 to 40 persons in Dep group by random sampling on a computer, and move them to Ind group.

Step (5) select from 5 to 40 persons in Ind group by random sampling on a computer, and move them to Dep group.

Step (6) do χ^2 test of the data obtained in Step 4 and 5.

Step (7) repeat Step 4 to Step 6 on different three random sampling (cases A to C).

Steps 4 and 5 mean movements of "No" to "Yes" and "Yes" to "No" in answering questionnaire, respectively. A 0.05 probability level was taken as the level of significance. All physical fitness tests were divided into two categories according to their median value. To aid in the calculation of measures of association, epidemiologic data are often presented in the form of a two-by-two table, also called a four-fold or contingency table [4]. The two-by- two table derives its name from the fact that it contains two rows and two columns, each representing the presence or absence of the exposure or disease [4]. In this study, this creates four cells, labeled *a*, *b*, *c*, and *d*, each of which represents the number of individuals having that particular combination of fitness and IADL status, then: *a* = the number of individuals who are low fitness level and dependent in IADL; *b* = the number who are low fitness level and independent in IADL; *c* = the number who are high fitness level and dependent in IADL; *d* = the number who are high fitness level and independent in IADL. The odds ratio (OR) was calculated to assess the magnitude of the association of decrease in IADL with the physical fitness tests. The OR is calculated by "*ad/bc*". Disability of IADL was found to be correlated most with advanced age in the elderly at home. Therefore, we adjusted the results in the two-by-two table analysis for age (grouped by 5-year intervals), using the Mantel-Haenszel method. The data were analyzed using the HALBAU statistical software.

3 Results and Discussions

We examine the relationship between IADL questionnaire result and physical fitness tests in elderly persons. Reference [5,6] concludes the results obtained from Step 1 to Step 3 as follows. Result 1: The following five fitness tests showed a significant relation to decreased IADL; knee-raising test, height, grip strength, sit-and-reach test, and standing on one leg with closed eyes. Result 2: Low values for the knee-raising test and height were strongly associated with decreased IADL. Figures 1 and 2 illustrated the results obtained in Step 7. Figure 1 shows the relation between the number of movement and the number of fitness tests which were found to be significantly associated with a decrease in IADL.

Figure 2 shows the relation between the number of movement and two tests, knee-raising test and height. On considering these results, we can derive a membership function shown in Figure 3. The function shows a robustness of Results 1 and 2 for uncertainty on answering questionnaire. We conclude that Results 1 and 2 could be maintained if the number of movement were within range from –10 to +5 persons.

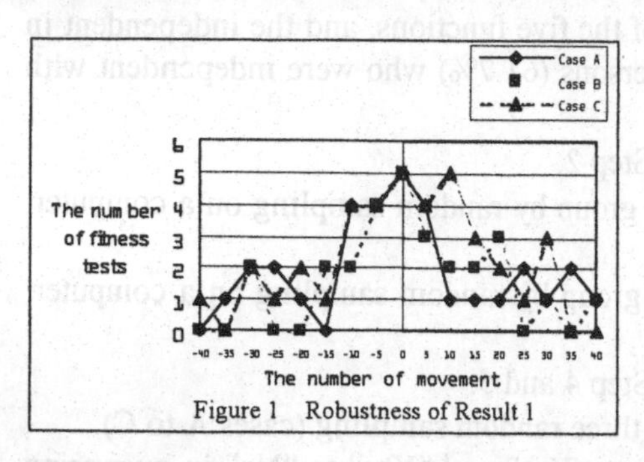

Figure 1 Robustness of Result 1

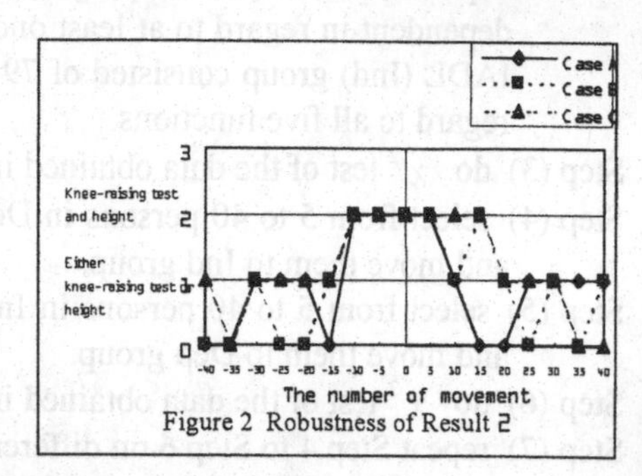

Figure 2 Robustness of Result 2

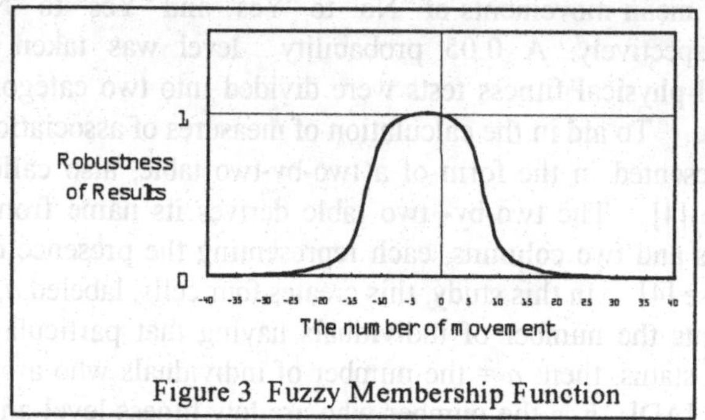

Figure 3 Fuzzy Membership Function

References

[1] Health and Welfare Statistics Association ed : Vital Statistics of Japan 1996, Health and Welfare Statistics Association, Tokyo (1996) pp37-42. (in Japanese)

[2] Koyano W, Shibata H, Haga H and Suyama Y: Measurement of competence in the elderly living at home: development of an index of competence. Jpn J Public Health (1987) 34, 109-114. (in Japanese)

[3] Zadeh LA: Fuzzy sets. Inform Control (1965) 8, 338-353.

[4] Charles HH and Julie EB: Epidemiology in Medicine, Little, Brown and Company, Boston (1987) pp.54-98.

[5] Uchida H, Mino Y, Tsuda T, Babazono A, Kawada Y, Araki H, Ogawa T and Aoyama H: Relation between the instrumental activities of daily living and physical fitness tests in elderly women. Acta Med Okayama (1996) 50, 325-333.

[6] Uchida H, Hata Y and Aoyama H: Fuzzy Analysis between Instrumental Activities of Daily Living and Physical Fitness Tests in Elderly Women, Proc. of 5th Intern. Conf. on Soft Computing and Information/Intelligent Systems, Iizuka, Japan (1998) pp367-370.

A Toolbox of Artificial Brain Cells to Simulate Classical and Operant Learning Behavior

Gerd Doeben-Henisch[1], Joachim P. Hasebrook[2]

The recent discussion of Artificial Intelligence (AI) is strongly influenced by controversial concepts of 'meaning' and 'intelligence'. Alan M. Turing gave the problem a specific format when asking: 'What are necessary and sufficient conditions that a human being responds to a machine as if it would be (partly) intelligent?' This point of view introduced what later on has been called Turing test: A machine or program can be considered to be intelligent, if human subjects cannot decide whether she or he is communicating with a machine or another human being.

The restriction to the surface of meaning, i.e. to the overt observable behavior like in a Turing test, says not too much about the underlying mechanisms which are responsible for the generation of the observable behavior. Taking into account these underlying structures the fields of neurobiology and phenomenology are entered, the two main and methodologically different sources of information about inner states of an organism. If the additional assumption is made that the subjective experience within a phenomenological approach is caused by certain physiological body states, neurobiology becomes the primary source of information. Although neurobiological facts cannot be interpreted without a link to the actual behavior or subjective experience, they represent the core of our knowledge about the mechanisms underlying the observable behavior.

Most programs which mimic intelligent behavior are based on logical oriented knowledge-based techniques which proofed to be prescriptive and too inflexible to represent even primitive forms of learning. Moreover, they elicit a number of paradox behavior when applied to support human learning. In classical AI different forms of logical based representational schemes are used and in connectionism researchers adhere to different types of artificial neural networks (ANN).

ANNs have achieved some success in non-linear forecasting, pattern matching and in artificial life paradigms. But ANNs still lack many of the vital features of biological neural networks (BNN), such as the ability of real neurons to allow some kind of self-modification with regard to short term learning and long term learning. The general linear model of an ANN transforming a numerical input vector into an output vector applying one general transformation rule to all 'neurons' can not account for the flexibility of a BNN.

The simulation of BNNs developed by neurobiologists does not seem to be promising either because recent attempts have shown that exact simulations of neuron brain cells consumes a vast amount of computer resources (e. g. 18.2h computing time on 5 connected sun sparc-2 workstations to simulate one second of activity of a single neuron, Goodard & Hood, 1997).

Within the Knowbotic Interface Project (KIP) we tried to combine the behavioral perspective with the physiological perspective, both embedded in concepts of learning and sign based communications (Semiotics). We call self learning and sign-using systems 'Knowbots'(cf. figure 1). We assume that the physiological structure is the main cause for observable behavior and therefore we focus on neurobiology. Thus, we have to find a model of the human brain neuron. We claimed that the neuron models should be empirically more sound than the classical ANNs, but they should also be still practical feasible on 'ordinary' PCs.

Analyzing the recent neurobiological literature we finally were able to extract a model of the human brain cell which encompasses all main features of biological neuronal cells especially those which are considered to be the key features of learning (Sheperd, 1998). The INM neuron model allows an arbitrary number of membranes and synapses, and for the membranes one can use three chemical substances simultaneously together with the necessary ionic channels, pumps, receptors, and transmitters. This chemical mechanism is used as basis for the computation of the corresponding electric potentials. Assuming a mean signaling frequency of about 100 Hz and a mean membrane density of 11 membranes per neuron, it is possible to simulate about 350 neurons in real-time with an

---

[1] Dr. Gerd Döben-Henisch, Institut für Neue Medien, Daimlerstr. 32, 60314 Frankfurt (eMail: doeb@inm.de)

[2] Dr. Joachim Hasebrook, Bankakademie e.V., Oeder Weg 16-18, 60318 Frankfurt (eMail: hasebrook@bankakademie.de)

Alpha chip working at 600 MHZ. The model simulates neurons almost perfectly with respect to the height of the potentials, the timing of the processes and the concentrations of chemical substances involved. Moreover, our neural network can model the local and global influence of hormones and neuropharmaca on brain cells.

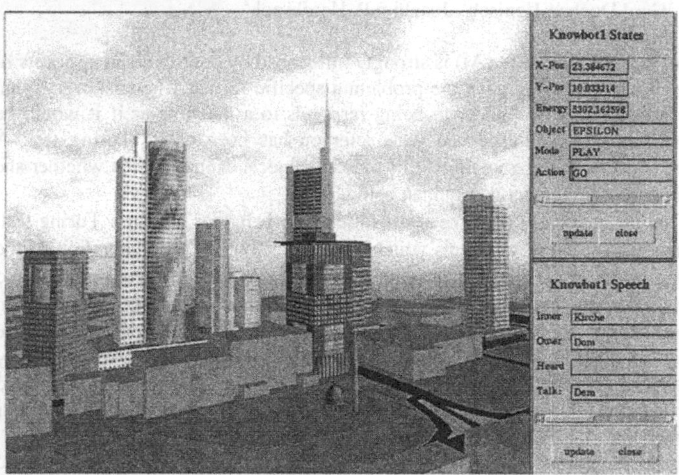

Figure 1: A Knowbot demonstrator build to guide avatars representing human users through the city of Frankfurt. The Knowbot adopts terms entered by the user and links it to its knowledge map about the buildings in the city.

Maass (1997) classifies the existing neural network models according to the following three generations:

- *First Generation:* McCulloch-Pitts neurons, Perceptrons, and threshold gates are used to implement multi-layered perceptrons, Hopfield nets, and Boltzmann machines. They can compute boolean functions as well as digital functions.
- *Second Generation:* Applying activation functions with a continuous set of possible output values to a weighted sum of the inputs (e.g. sigmoid functions, linear saturated functions, piecewise exponential functions) enabled the implementation of feed forward and recurrent sigmoid neural networks and networks of radial basics function units. These networks can compute boolean function, digital functions, and continuous functions with a compact domain. Moreover, they support learning algorithms based on gradient descent.
- *Third Generation:* Spiking neurons, such as integrate and fire neurons, form networks of spiking neurons in which timing of individual computational steps plays a key-role for the computation.

Maass (1997) proofs that the third generation model has at least the same computational power as neural nets from the first two generations of a similar size. Furthermore, many functions require significantly *fewer* spiking neurons than neurons of the first or second generation. We´d like to emphasize that our neuron model fits the definition of a spiking neuron mentioned above. Therefore, all of the results reported by Maass (1997) are also valid for our neural networks.

An outstanding feature of our neuron model is the fact that within this neuron model the response function and the threshold functions are given as concrete functions simulating the given chemical mechanism of a brain cell, that is: our model simulates the mechanism of an axon with compartments followed by a complex synapse with possibly several membranes and finally a post-synaptic membrane; all these parts are forming the response function epsilon. Similarly, the action potential and threshold function both are realized as a complex membrane model which can be triggered by potentials and react according to it's actual state of receptors, ionic channels, ionic pumps and substance concentrations.

We are aiming to represent only those properties of biological cells which are functionally important for the processing of information and especially for the learning of new behavior patterns. Along this

line of research we are looking for neuronal circuits which represent possible neuronal correlates of learning and shed some light on behavioral learning concepts.
Only a few BNNs underlying learning, however, have been identified, yet. As a first test case we have chosen a classical conditioning circuit and several candidates that might be responsible for operant conditioning. In a first experiment we implemented the network which represents the eye blink reflex of a rabbit. The network matches the neuropsychological data almost perfectly (cf. figure 2): The connection of the unconditioned stimulus (US = air flow) and the conditioned stimulus (CS = sound) is learned in a few trials, if the CS is given slightly before the US. Several runs presenting the CS without the US extinguish the connection. It is reestablished very quickly, if the CS and the US are displayed together again (cf. Menzel, 1996).

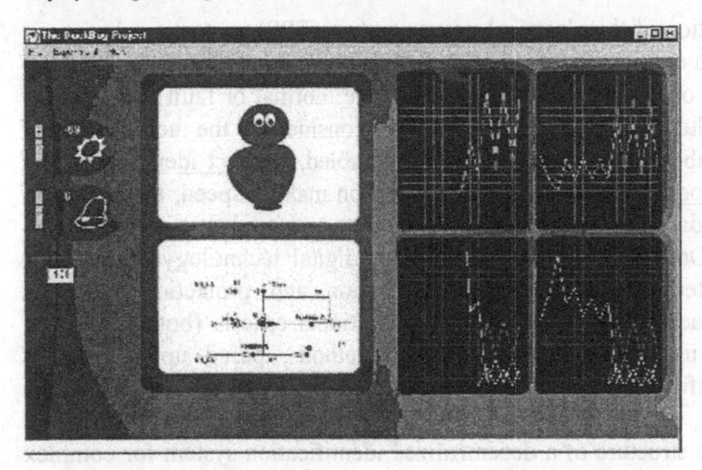

Figure 2: Test environment for (classical and operant) conditioning experiments with Knowbots based on artificial biological neural networks implemented with Java.

Up to now, no clear candidate acting as an neurological correlate of operant conditioning has been identified. In order to overcome this shortcoming, we set up an (artificial) BNN with our neurons testing in a defined experimental environment instances of operant conditioning. While writing this article, we have identified first candidates but have not completed the experiments, yet.
The Knowbots with embedded BNNs based on our neuron model can deepen our understanding of basic learning behavior. And they will be useful to implement intelligent agent software. One aim of the KIP is to develop a set of 'software brain tools' combined with 'environment test suits' which enable the user to implement the basic operations of brains that have been identified or suggested so far, such as signal detection, habituation, classical and operant conditioning, orientation and movement in three-dimensional environments.
In the near future we are planning to simulate the neural system of primitive organism that are well described in the research literature, as well as to construct formalized neuronal circuits which prove to have the capability of learning within clearly defined test environments.

References

Goddard, N. & Hood, N. (1997) Parallel GENESIS for large scale modeling. In J. M. Bower (Ed.), Computational Neuroscience '96. New York: Plenum.
Sheperd, G. M. (Ed)(1998). The Synaptic Organization of the Brain (4th ed.). Oxford: Oxford University Press.
Maass, W. (1997). Networks of Spiking Neurons: The Third Generation of Neural Network Models. Graz: Institute for Theoretical Computer Science, TU Graz.
Menzel, R. 1996. Neuronale Plastizität, Lernen und Gedächtnis. In J. Dudel, R. Menzel & R.F. Schmidt (Eds.), Neurowissenschaft. Vom Molekül zur Kognition (pp.485 – 518). New York: Springer.

Intelligent, ANN-Based Identification System of Complex Generating Sets Operating Mode

A. Halinka, M. Szewczyk, B. Witek
Silesian Technical University of Gliwice, Poland

Introduction

The effective operation of the electrical power system (EPS) protection devices is mainly determined by a correctness of generated decisions, classifying actual state of the protected object to one of the two event classes (i.e. normal or fault), which may be realized only if the following conditions are considered: the acquisition of extended information about the protected object is enabled, correct identification of the object operating mode, data processing and decision making speed, the ability of the protective system adaptation to the actual object state and its „insensitivity" to any kind of interference. On the other hand the use of digital technology allows new approach to the protected object state identification and protection functions adaptation. It may be achieved by implementing additional criteria (both measuring and logical) and by use of new computational methods, based upon *artificial intelligence* such as artificial neural networks (ANNs).

The paper presents a structure of a decentralized identification system for complex EPS objects, applied to their protection and control arrangements. Such system appears as a basic source of information - essential for an adequate realization of a *structure criterion*, which enables the adjustment of the protective system functions to the protected object varying conditions - understood as a change of interconnections between the object elements (topology change) as well as the input parameters (e.g. frequency) variation. As an example of complex generating system the so called „combined arrangement" (equipped with cooperating gas-turbine and steam turbine generators) has been taken.

EPS generating node operating mode identification

The protected object has been divided into basic functional sections: gas turbine generators with adjacent cooperating systems (unit transformer, frequency start-up system, excitation system); steam turbine generator with accompanying equipment as well as general auxiliaries system (GAS). For each elementary functional block (consisting of gas turbine and generator) a three-level autonomous identification module has been defined.

By dividing the identification modules into levels the reduction of decision development time is achieved (via multiprocessor conditioning) whereby a certain level „specialization" is created. Chosen functional blocks of the protected object have been divided into „elementary systems" consisting a single device or a group of devices (e.g. generator, frequency start-up system etc.). Each level initially identifies -

by using logical or measuring-logical (i.e. considering measuring algorithms outputs) information about connectors positions - the actual configuration of the nearest „elementary system" functional block. In the following „main" level (B1) the necessary exchange of information generated in the preceding levels (BP1-A, BP1-B). The final decision level (BF1-A, BF1-B) uses information pre-processed in the preceding levels as well as in the adjacent blocks BPO and GAS, which is meant as the identification process verification and increases the certainty of output decision.

The proposed identification system is based upon a multi-layer perceptron (MLP) ANN structure. It is worth noticing that the application of such structures does not require a direct definition of relations between specified information (signals) essential for the identification system. During the learning process the nets are optimizing their structures by adequate weights and biases values adjustment for correct identification and state classification pre defined in a learning base as well as states being their generalization. Particular advantage of such a structure is a possibility of certain errors elimination e.g. missing or incomplete sets of binary signals (a situation frequently occurring in practice), which in the boolean logic - based systems causes the appearance of an output error signal and the identification process termination. As mentioned, the most important problem for the ANN-based identification system is a learning base selection considering the network learning with possibly maximal number of cases assigned to all possible events i.e. to all operating states of the protected plant.

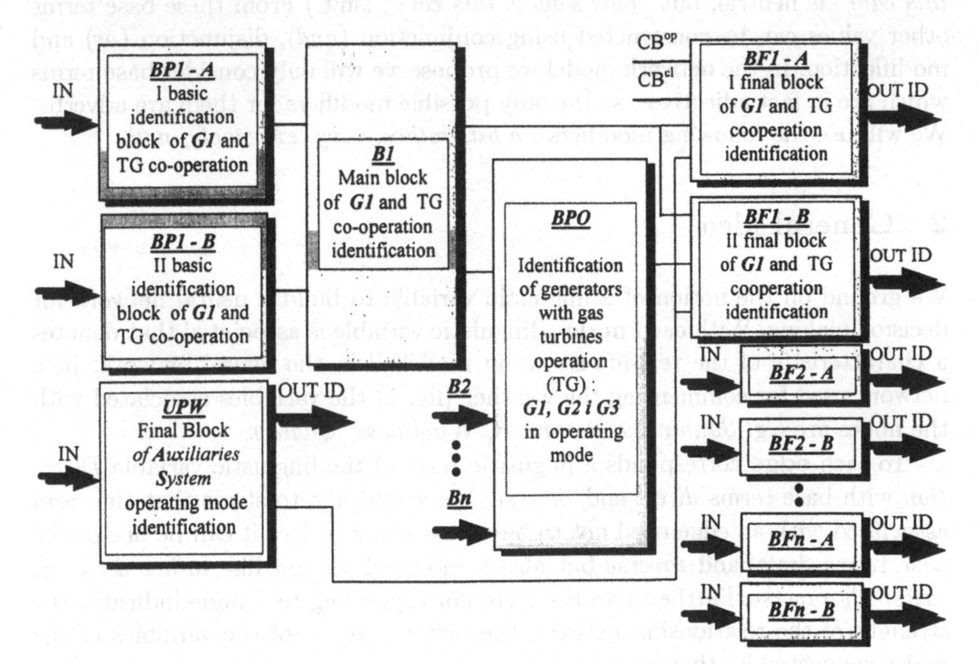

Fig.1 Functional block „gas-turbine generator" operating state identification system scheme for protection and control purposes

A Neural Network Based on Linguistic Modifiers

Martine De Cock, Frederik Vynckier, and Etienne E. Kerre

Fuzziness and Uncertainty Modelling**
Department of Applied Mathematics and Computer Science
University of Gent, Krijgslaan 281 (S9), B–9000 Gent, Belgium

key-words : **neural networks, fuzzy neural networks, linguistic variable, linguistic modifiers, approximate reasoning**

1 Basic concept : a linguistic variable

While the values of a numerical variable are numbers, the values of a **linguistic variable** are linguistic terms [2]. The term set consisting of all values of a linguistic variable (e.g. *Speed*) contains one or more base terms : a primary term (e.g. *fast*), most often also its antonym (e.g. *slow*), and sometimes a third primary term such as *medium*. (We want to emphasize that the concepts "primary term" and "antonym" are not interchangeable : the question *"how fast is this car?"* is neutral, but *"how slow is this car?"* isn't.) From these base terms other values can be constructed using conjunction (*and*), disjunction (*or*) and modification. In the network model we propose we will only consider base terms which are in fact adjectives, so the only possible modifiers for them are adverbs. We will use the following modifiers : *a bit, rather, very, extremely, not*[1].

2 General idea

We ground on the notion of a linguistic variable to build a neural network for decision making. With each **node** a linguistic variable is associated that denotes a characteristic of the real-life situation modelled in the neural network. In a network used for commenting the weather (fig. 1) the variables associated with the nodes are e.g. *Sunniness, Raininess, Windiness, Quality*.

To each **edge** corresponds a linguistic term of the linguistic variable *Direction* with base terms *direct* and *inverse*. We would like to stress that the term associated with an edge need not to be a base term; in fact it can be one of the base terms *direct* and *inverse* but also a modified version like *rather direct* or *extremely inverse*. Furthermore the term corresponding to a node indicates the strength of the relationship between the *primary terms* of the variables of the nodes connected by that edge.

** Martine De Cock and Frederik Vynckier would like to thank the Fund for Scientific Research – Flanders (Belgium) for funding the research reported on in this abstract.

[1] For convenience we consider *not* as a modifier too, instead of as a logical operator for negation.

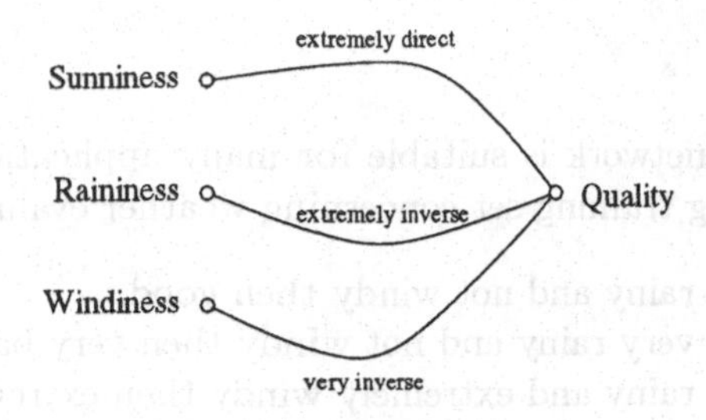

Fig. 1. A neural network for commenting the weather

The user is asked to comment on the variables of the first layer using linguistic descriptions. So the actual input signals of the network are linguistic terms. E.g. when you're glad you didn't forget your umbrella because it's pouring, you could give *very rainy* as an input to the node *Raininess*. Intuitively one might expect that when *"it's very sunny, not rainy and a bit windy"*, then *"it's very good weather"*. This is because there is an *extremely direct* connection between *sunny* and *good weather*. The connection between *windy* and *good weather* is *very inverse*, but the fact that *"it's very sunny"* dominates and makes us forget that little bit of wind. Furthermore the fact that it's *not rainy* and that there is an *inverse* connection between *rainy* and *good weather* helps to reinforce the idea that *"the weather is good"*. How does the network make such a conclusion? Starting from the linguistic input terms, a linguistic term of the variable of the output layer is calculated (see below for details). In the example mentioned, this could be *very good*.

3 Realisation

The primary term of a linguistic variable is mapped onto the number 1. If there's also an antonym in the term set, this is mapped onto -1. If it occurs, the term *medium* is mapped onto 0. Furthermore each modifier is mapped onto a real number, e.g. *extremely* (3), *very* (2), *rather* (0.5), *a bit* (0.33), *not* (0), so all input signals, output signals and weights are real numbers. The output value y of a neuron with incoming inputs $x_1, x_2, ..., x_n$ $(n \in \mathbb{N})$ and weights $w_1, w_2, ..., w_n$ on the corresponding edges is calculated by

$$y = approx(\frac{1}{n}\sum_{i=1}^{n}(w_i x_i))$$

with $approx$ a $\mathbb{R} - \{-3, -2, -1, -0.5, -0.33, 0, 0.33, 0.5, 1, 2, 3\}$ mapping satisfying $(\forall x \in \mathbb{R})(\forall z \in \{-3, -2, -1, -0.5, -0.33, 0, 0.33, 0.5, 1, 2, 3\})$
$(|x - approx(x)| \leq |x - z|)$. The initial weights are arbitrary chosen. The neural network learns from a training set using the delta learning rule [1].

4 Test results

This kind of neural network is suitable for many applications. As an example consider the following training set concerning weather evaluation :

- If sunny and not rainy and not windy then good.
- If not sunny and very rainy and not windy then very bad.
- If not sunny and rainy and extremely windy then extremely bad.
- If very sunny and a bit rainy and not windy then very good.

Using the delta learning rule, the following weights were obtained : Sunniness – Quality : extremely direct; Raininess – Quality : extremely inverse; Windiness – Quality : very inverse. Now the network is able to comment on descriptions of other weather situations; it has actually "learned" the concepts "good" and "bad weather".

- For very sunny, not rainy, rather windy, it says : very good.
- For not sunny, rather rainy, rather windy, it says : bad.
- ...

References

1. L. FAUSETT, **Fundamentals of Neural Networks : Architectures, Algorithms, and Applications**, Prentice Hall, International Editions, 1994
2. L. A. ZADEH, **The Concept of a Linguistic Variable and its Application to Approximate Reasoning I, II, III**, Information Sciences, **8**(1975), 199-249, 301-357, **9(1975)**, 43-80

Fuzzy Computing in a MultiPurpose Neural Network Implementation

Ciprian-Daniel Neagu, Vasile Palade

University "Dunarea de Jos" of Galati, Department of Applied Informatics,
6200 Galati, Romania
{dneagu, pvasile}@cs.ugal.ro

Abstract. A family of connectionist tools is developed as a collection of fuzzy processing operators to model logic oriented computation of fuzzy sets. A generalized neuron with fuzzy capabilities (developed using the MAPI structure proposed in [4]) could be a useful tool to add another level of programmability. These combinations of generalized fuzzy computation, the expanded MAPI model and specific distributed architecture, are used as a powerful processing tool in financial forecasting, particularly in a portfolio problem. The neural reasoning engine is accorded to fuzzy rules, which model a real world portfolio evaluation process, taken from the current Romanian financial context.

1 Introduction

A special focus in fuzzy reasoning is to develop some universal computing models, easy customizing to meet wide subjects of particular specifications. For this purpose, it is indispensable to identify generic-processing modules, performing general computations on fuzzy sets. In this paper, a connectionist model based on MAPI formal neuron [4] is proposed to solve a portfolio problem. In section 2, MAPI formalism, and fuzzy operators implemented using MAPI structure are presented. In section 3, it is proposed a structure of MPNN equivalent with a set of explicit fuzzy rules used in portfolio application. The paper is ending with conclusions and ideas on future work.

2 Fuzzy Logic and MAPI Neuron

The extended version of Modus Ponens is proposed in [6]:

$$\begin{array}{c} \text{IF } X_1 \text{ is } A_1 \wedge \ldots \wedge X_j \text{ is } A_j \text{ THEN } Y \text{ is } B \\ \underline{(X_1 \text{ is } A'_1) \wedge \ldots \wedge (X_j \text{ is } A'_j)} \\ Y \text{ is } B' \end{array} \tag{1}$$

This process is performed in four steps: 1) Matching (the compatibility σ between A' and A), 2) Aggregation (based on triangular norm), 3) Projection: the compatibility σ_c

of the consequent is obtained as function of the aggregated value σ_a, measuring the compatibility of (Y is B') with (Y is B), and 4) Inverse-Matching and Defuzzification, in MAPI context performed at the axonic terminals. ANN performing fuzzy computation [2], [3], [5] are parallel structures aimed at logic-based processing of fuzzy data. Aggregation and projection are performed by generalized aggregative OR and AND neurons, involving triangular norms or co-norms:

$$y=\text{T-conorm}_{i=1,..,n}\ [x_i\ \text{T-norm}\ w_i]. \quad (2)$$

$$y=\text{T-norm}_{i=1,..,n}\ [x_i\ \text{T-conorm}\ w_i], \quad (3)$$

where x_i and w_i are the inputs and the weights of the MAPI neuron. The neural structure is an equivalent system [1], generating so-called MPNN [4].

3 Portfolio Evaluation Using MPNN

The capabilities of MPNN to perform fuzzy computing [4] are tested to solve a portfolio evaluation process in current Romanian financial context described by fuzzy rules: (R1) *If USD is weak against DEM and USD is strong against JPY and USD is weak against ROL then our portfolio value is positive*, (R2) *If USD is medium against DEM and USD is medium against JPY and USD is medium against ROL then our portfolio value is about zero*, (R3) *If USD is strong against DEM and USD is strong against JPY and USD is strong against ROL then our portfolio value is negative.*

3.1 Reasoning with a Fuzzy Rule Base

Let be considered a single rule with three antecedents described as:

$$\text{IF X is A AND Y is B AND U is C THEN Z is D.} \quad (4)$$

A, B, C, D are fuzzy sets having associated matching functions μ_A, μ_B, μ_C, μ_D. Let the matching function $\mu_A(\xi)$ be described by a vector X of size Nx, so that:

$$x_i=\mu_A(\xi),\ \text{if}\ \alpha_i<\xi\leq\alpha_{i+1},\ i=1,2,.., Nx-1. \quad (5)$$

Thus, the fuzzy set A is: $A=[x_1\ ..\ x_{Nx}]$.
Similarly, fuzzy sets B, C and D are described in discrete forms as follows:

$$B=[y_1\ ..\ y_{Ny}],\ y_i=\mu_B(\psi),\ \text{if}\ \beta_i<\psi\leq\beta_{i+1},\ i=1,2,.., Ny-1\ . \quad (6)$$

$$C=[u_1\ ..\ u_{Nu}],\ u_i=\mu_C(\upsilon),\ \text{if}\ \gamma_i<\upsilon\leq\gamma_{i+1},\ i=1,2,.., Nu-1\ . \quad (7)$$

$$D=[z_1\ ..\ z_{Nz}],\ z_i=\mu_D(\zeta),\ \text{if}\ \delta_i<\zeta\leq\delta_{i+1},\ i=1,2,.., Nz-1\ . \quad (8)$$

$$R{:}A\times B\times C\times D\rightarrow[0,1]\ ,\ \mu_R(x,y,u,z)=(\mu_A(\xi)\wedge\mu_B(\psi)\wedge\mu_C(\upsilon))\Gamma\mu_D(\zeta). \quad (9)$$

The fuzzy relation (9) defines the implication according to (4), where ^ is a conjunctive T-norm and Γ is an associative T-norm; given A', B', C' (° is max-Γ operator):

$$D'=(A'\wedge B'\wedge C')\circ R . \tag{10}$$

3.2 Mapping Fuzzy Rules into MPNN

Application uses triangular membership functions for primary fuzzy sets $\{L_i,M_i,H_i\}$, i=1,2,3, (fig. 1) each one described by vectors as in (6),(7) of size N_{Li}, N_{Mi}, N_{Hi}.

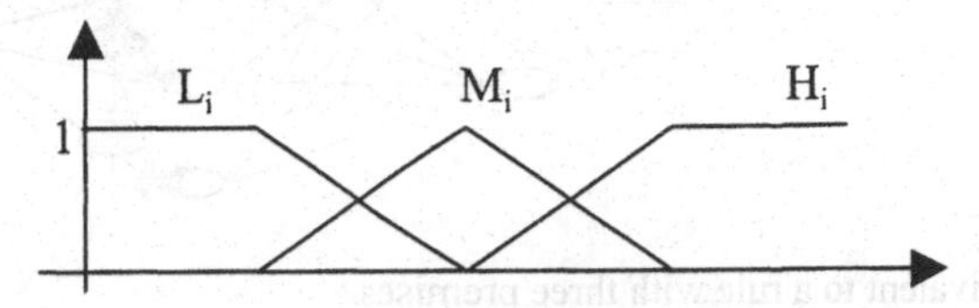

Fig. 1. Membership functions for "x_i is low", "x_i is medium", "x_i is high", i=1,2,3.

The MPNN shown in fig. 2 is equivalent with a discrete fuzzy rule base described by (10) if the neurons N_i are used to convert the current values of entries X, Y, and U to correspondent values $\mu_{A'}(\xi)$, $\mu_{B'}(\psi)$, $\mu_{C'}(\upsilon)$, the weights between input and associative neurons are set to 1, the associative neurons H_{ijk} process $(x'_i \wedge y'_j \wedge u'_k)$ if the encoding function is ^ T-norm, the weights between associative and output neurons are $w_{ijkl}=(\mu_A(x)\wedge\mu_B(y)\wedge\mu_C(u))\Gamma\mu_D(z)$. The setting neurons S_i, i=0,..,N_x*N_y*N_u, provide the synchronism $H_{111}<H_{Nx11}<H_{1Ny1}< H_{NxNy1}<H_{11Nu}<H_{Nx1Nu}<H_{1NyNu}<H_{NxNyNu}$ (where "<" means *fires before*). The vectors are considered of size N_x, N_y, N_u, N_z, respectively. In these conditions, D'=(A'^B'^C')°R. The defuzzification of the output signals follows the methods proposed in [4], using a final controlling neuron.

3.3 Portfolio explicit knowledge implementation

The resulted MPNN structure for the module of explicit knowledge is a combination of NR modules presented in fig. 2 (NR indicates the number of rules), being a part of a neural expert system used to evaluate portfolio values in the Romanian financial market. The knowledge base mapped in the MPNN structure is:

IF (X is L_1) AND (Y is H_2) AND (U is L_3) THEN (Z is PO) .
IF (X is M_1) AND (Y is M_2) AND (U is M_3) THEN (Z is ZE) . **(11)**
IF (X is H_1) AND (Y is H_2) AND (U is H_3) THEN (Z is NE) .

where the antecedents using X ("USD/DM"), Y ("USD/100JPY"), and U ("USD/1000ROL") are described by the parity values of the US dollar, Japanese yen and German mark for the financial year 1997. The portfolio values derived from the fuzzy rule base are obtained using the MPNN proposed.

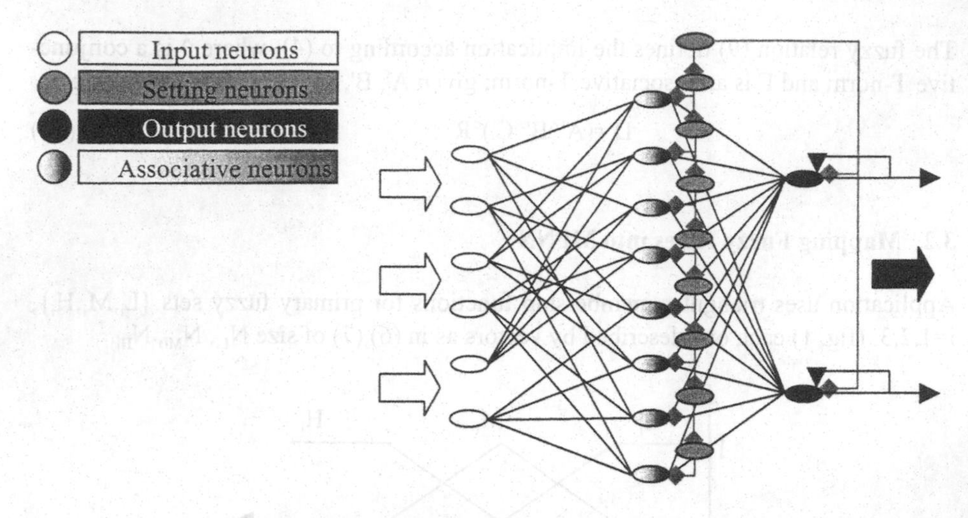

Fig. 2. MPNN equivalent to a rule with three premises.

4. Conclusions and Future Work

This paper described a neural approach of fuzzy reasoning using MPNN structure based on generalized MAPI model, equivalent with the discrete fuzzy rule base. The raison of implementing fuzzy rules using MPNN consists in both, the necessity of combining explicit and implicit knowledge, and the property of MAPI neurons to be fuzzy processors [4]. Future research will focus on using neural MAPI operators and MPNN structures for knowledge acquisition, in order to develop empty expert systems. This aim supposes to clarify the problems of combining fuzzy neural modules, to identify methods to extract fuzzy rules from trained neural networks and to describe strategies of using neural expert systems in knowledge acquisition.

References

1. Buckley, J.J., Hayashi, Y., Neural nets for fuzzy systems, Fuzzy Sets and Systems 71, Elsevier (1995) 265-276
2. Pedrycz, W., Fuzzy Neural Networks and Neurocomputations, Fuzzy Sets and Systems, 56 (1993) 1-28
3. Pedrycz, W., Rocha, A.F., Fuzzy Set Based Models of Neurons and Knowledge Based Networks, IEEE Transactions on Fuzzy Systems, Vol.1, no.5 (1993) 254-266
4. Rocha, A.F., Neural Nets– a Theory for Brains and Machines, Springer-Verlag (1992)
5. Rocha, A.F., Yager, R.R., Neural nets and fuzzy logic, in Intelligent Hybrid Systems, CRC Press Inc (1992)
6. Zadeh, L.A., The role of fuzzy logic in the management of uncertainty in expert systems, Fuzzy Sets and Systems,11/3 (1983) 199-227

Fuzzy Logic Control of Child Blood Pressure During Anesthesia

Ahmet Yardimci[1], A.Sukru Onural[2]

[1]Akdeniz University Technician Training Centre, Kampüs, 07059, Antalya-Turkey
yardimci@ufuk.lab.akdeniz.edu.tr
[2]Akdeniz University Engineering Faculty, Topcular, Antalya-Turkey

Abstract. This study developed a multivariable, computer-based fuzzy logic method to control of child blood pressure during anaesthesia. Fuzzy logic reasoning, a powerful new form of information processing, will be employed to infer depth of anaesthesia from on-line measurements of cerebral activity (EEG) and cardiovascular function (hearth rate and arterial blood pressure). reflexes of a patient break down under general anaesthesia. Many electronic monitoring systems continuously give information about vital human parameters (blood pressure, hearth frequency, etc.). Other, unmeasured information can only be monitored by an anesthesist. Develop an intelligent integrated control system that can take over some of the routine tasks of anaesthesia after the induction.

1 Introduction

Anesthesia is used to make surgery possible and to keep your child comfortable and pain free. There are various forms of anaesthesia which your child may receive, depending on the operation your child is to have. In this study a fuzzy logic controller was used to control mean arterial pressure (MAP), which was taken as a measure of the depth of anaesthesia. The main reason for automating the control of depth anaesthesia is to release the anesthesist so that he or she can devote attention to other tasks as well (controlling fluid balance, ventilation, and drug application) that cannot yet be adequately automated and thus to increase the patient's safety.

1.1 General Anaesthesia and Blood Pressure During Anaesthesia

General Anaesthesia causes loss of consciousness, meaning your child will be asleep during the operation. This may be started either by way of injection of medicines into an intraveneous needle/drip, or by breathing anaesthetic gases. This will then continue until end of the operation. There are two different systems available to deliver anaesthetic gases and vapours to the patient[1]. In the draw-over system, air used as the carrier gas to which volatile agents or compressed medical gases are added. In the continuous - flow system air is not used, but compressed medical gases, usually nitrous oxide and oxygen, pass through flow meters (rotameters) and vaporizers to supply anaesthetic to the patient[2]. The aim is to provide a pleasant induction and lack of awareness for the patient, using a technique that is safe for both patient and anaesthetist and that provides good operating conditions for the surgeon. Unfortunately, the ideal anaesthetic drug with all the desired qualities does not exist[3]. In anaesthetic practice, the systolic pressure has greater significance than the diastolic pressure, which is frequently not recorded, particularly if access to the arm is difficult No "normal" blood pressure can be specified for the anaesthetized patient; in general the systolic pressure should be stable in the range 90-140 mmHg (12.0-18-7 kPa).[4] Systolic arterial blood pressure is normally lower in children (45-75 mmHg or 6.0-10.0 kPa in neonates). Blood pressure value always contrast to depth of anaetshesia. If the depth of anaesthesia in deep blood pressure goes to down. In other case, it goes to up.

2 Application

Because a biological process like anaesthesia has a nonlinear, time-varying structure and time-varying parameters, modeling it suggests the use of rule-based controllers like fuzzy controllers. The design process here was iterative, and the reference points of the membership functions as well as the linguistic rules were determined by trial and error[5]. The control rules made use of the error between the desired and the actual values of MAP as well as the integral of the error. Depth of anaestesia is controlled by using a mixture of drugs that are injected intravenously or inhaled gases, most often in a mixture of 0 to 2 percent by volume of isofluorane in oxygen and/or nitrous oxide[6]. The linguistic rules that describe the anaesthetist's actions and the fam table for these rules are given in Table.1 The abbreviations NB, NS, ZE, PS, PB, and PV refer to the linguistic variables "negative big," "negative small," "zero," "positive small," "positive big," and "positive very big," respectively. In The membership functions used in this case are bell-shaped functions that can be described by the following exponential equation, where ξ is the input value and λ the shifting of the function in relation to zero:

$$\eta = \exp[-K(\xi-\lambda)^2] \qquad (1)$$

Table.1. Linguistic Rules and FAM Table

Rule Number	Input e	Input ∫e	Output u
1	NS	-	PB
2	PS	-	PS
3	NB	-	PV
4	PB	-	ZE
5	ZE	ZE	PM
6	ZE	PS	PS
7	ZE	NS	PB
8	-	NB	PV
9	-	PB	ZE

e\∫e	NB	NS	ZE	PS	PB
NB	PV	PV	PV	PV	PV\ZE
NS	PB\PV	PB\PV	PB	PB	PB\ZE
ZE	PV	PB	PM	PS	ZE
PS	PS\PV	PS	PS	PS	PS\ZE
PB	ZE\PV	ZE	ZE	ZE	ZE

The factor K determines the width of the bell. The reference points for the membership functions are in Table 2. The resulting membership functions are shown in Fig.1., where a value of K=5 used for all the membership functions for *e* and *u*. The results for a simulation run using the linguistic rules and refrence points provided in Tables 1-2 are shown Figure 2.

Table 2. The reference points of membership functions

	Input e, (mm Hg)	Input ė, (mm Hg.s)	Output	Output u, (%)
NB	-4	-100	ZE	0
NS	-2	-50	PS	0,5
ZE	0	0	PM	1
PS	2	50	PB	1,5
PB	4	100	PV	2

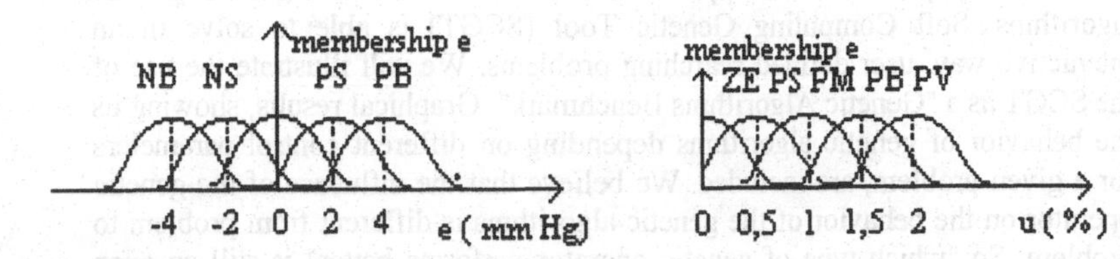

Fig 1. The resulting membership functions

A max-min composition and a center of gravity defuzzification method were used here to evaluate the linguistic rules.

3 Conclusion

Child's anaesthetist is either a fully qualified specialist doctor in anaesthesia, resuscitation and pain control, or a specialist in training. He or she is responsible for child's overall care during the time of surgery. An intelligent integrated control system can take over some of the routine tasks of anaesthesia. This application controls the blood pressure. If the other changable measurement values control by the other intelligent control sytems it will be full automated.

References

1. Dobson M.B., "Anaesthesia At The District Hospital", World Health Organization,Geneva 1988
2. World Health Organization, "Drugs Used in Anaesthesia", Geneva, 1989
3. Erengül A., "Lokal Anestezi",İstanbul 1980
4. Carfi M.A, Öktem Y., Çetin O., "Anestiyoloji", İzmir, 1978
5. Ross J.T, "Fuzzy Logic With Engineering Applications"; McGraw-Hill, USA,1995.
6. Yardimci A., Comlekci S., Aydogan T., Cakir A.," Fuzzy Logic Control of Blood Pressure with Isofluorane During Anaesthesia" Intelligent Manufacturing Systems Symphosium Page:154-161 6-7 August, Adapazarı, Turkey, 1998

Soft Computing Genetic Tool V3.0 - Applications

Adrian Horia Dediu, Alexandru Agapie, Nicolae Varachiu

Institute of Microtechnology, Erou Iancu Nicolae 32B, 72225, Bucharest, Romania
Hd@imt.ro, agapie@imt.pub.ro, nichi@imt.pub.ro

Abstract. We present some applications of a software tool based on genetic algorithms. Soft Computing Genetic Tool (SCGT) is able to solve in an interactive way, user defined searching problems. We will illustrate the use of the SCGT as a "Genetic Algorithms Benchmark". Graphical results, showing us the behavior of genetic algorithms depending on different control parameters for a given problem, are included. We believe that the influence of the genetic operator on the behavior of the genetic algorithms is different from problem to problem. So "which type of genetic operator performs better" is still an open problem.

Comparing the performances of the genetic algorithms

Because of the lack of theoretical demonstrations in the Genetic Algorithm area, we have problems when we have to choose the genetic operators and parameters, for a real world problem.

De Jong proposed a method to compare the performances of two different strategies. He defined the on-line and offline performances of a strategy as the average of all fitness functions, respectively the average of maximum fitness function until one moment.

In 1996, Hugues Bersini, Marco Doringo, Luca Gambardella, Stefan Langerman, Gregory Seront, organized the first contest on evolutionary optimization during the IEEE International Conference on Evolutionary Computation (ICEC'96) and they adopted a more sophisticated method for comparing the performances of the genetic algorithms.

Since the genetic algorithms are probabilistic techniques, in order to be able to compare the performances of two different algorithms we need to test the average results after a given number of runs.

We used the "Multiple Run" option for the examples that we studied with 20 repeated executions. The average values of 20 maximum fitness functions are presented as results. The examples illustrate the influence of different genetic algorithm parameters on the genetic evolution, on F1 problem: "find $\overline{x_0} \in ([0,256) \cap N)^{10}$ so that, the function $f(\overline{x}) = \sum_{i=1}^{10} (x_i - 200)^2$ may attain its minimum". We studied the influence of the coding scheme on the evolution of the fitness function, thus we used three coding schemata: integers, binary and Gray code.

Results for the F1 problem

Using the same genetic operators and control parameters, we find out that for the F1 problem, *Gray coding* performs better than *binary coding* that performs better than *integer coding.*

If we compare the influence of the crossover operator we can say that for the F1 problem the *shuffle crossover* is better than the *two points crossover* or *the one point crossover.*

Finally, when we used an integer type coding, with an average mutation operator the best selection operator was the rank space selection (see **Fig. 1**). The results were even better than the results obtained using the Gray coding

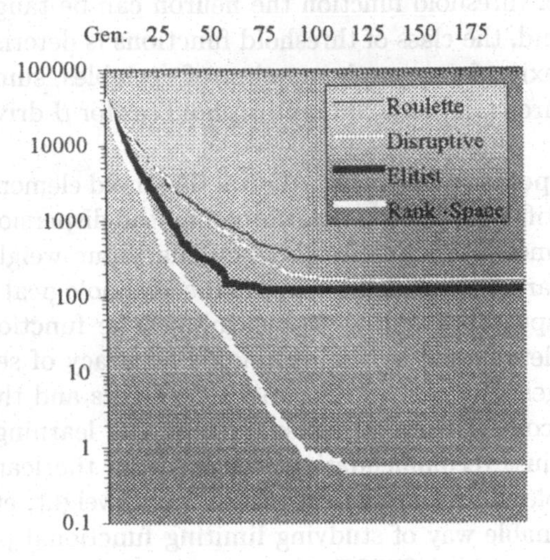

Fig. 1. the selection type influence on the genetic algorithms performances

References

1 A.-H. Dediu, D. Mihaila, Soft Computing Genetic Tool, Proceedings of First European Congress on Intelligent Techniques and oft Computing EUFIT'96, Aachen 2-5 September 1996, Vol. 2,pp. 415-418

2 A. Agapie, A.-H. Dediu, GA for Deceptive problems: Inverting Schemata by statistical approach Proceedings of IEEE International Conference on Evolutionary Computation, ICEC'96 Nagoya, Japan 22 May 1996, pp. 336-340

3 Z. Michalewicz. Genetic Algorithms + Data Structures = Evolution Programs. Springer Verlag, 1992

4 M.Srinivas, and L. M. Patnoik, Adaptive Probabilities of Crossover and Mutation in Genetic Algorithms, IEEE Transactions on Systems, Man and Cybernetics, Vol. 24, No. 4, April 1994, pp. 656-668

5 Patrick Henry Winston, "Artificial Intelligence" third ed., Addison-Wesley, June 1992.

Learning Experiments with CMOS Artificial Neuron

Victor Varshavsky and Vyacheslav Marakhovsky

The University of Aizu, Aizu-Wakamatsu City, 965-8580 Japan
victor@u-aizu.ac.jp and marak@u-aizu.ac.jp

One of the key problems in designing analog/digital implementations of artificial neuron is the problem of its limiting functional power, i.e. the question about what class of threshold function the neuron can be taught to produce. In problems of this kind, the class of threshold functions is determined by a certain criterion of complexity, for example, number of variables, sum of input weights (as for νCMOS neuron [1,2]), maximum threshold (as for β-driven CMOS neuron [3,4]), etc.

The functional power of the analog/digital threshold element what is the base for building the artificial neuron is determined by the dispersion of technological and physical parameters of the circuit when the input weights and threshold are fixed. For a learnable artificial neuron, the technological dispersion of the parameters is compensated during the learning. The functional power of the neuron is mainly determined by the attainable accuracy of setting (during the leaning) the voltages that control the synaptic weights and threshold. Analytic study of how the control voltages behave during the learning is very difficult, providing at best just recommendations on choosing the learning parameters, such as, for example, a unit increment of the input weight, etc. It looks today like the only reasonable way of studying limiting functional power of learnable artificial neuron is computer SPICE simulation.

This approach may look simple and attractive, but there is one principal difficulty: simulation time. It depends on two factors associated with the teaching (test) function. First, on the number of variables determining the number of synapses and, hence, the number of components in the simulated circuit. Second, on the length of a single teaching sequence (number of input variable combinations specifying the test threshold function). Taking into account that the learning process consists in multiple (hundreds of times and more) showing of the single teaching sequence, the experiment for threshold functions of limiting complexity can be as long as tens of hours, if the test function was not chosen well.

A single teaching sequence containing a number of variable combinations smallest for a given threshold function will be called bearing teaching sequence.

Let T and F be the sets of value combinations of variables X, such that for threshold function $Y(X)$ $Y(X{\in}T) = 1$, $Y(X{\in}F) = 0$. Then the bearing sets T_0 and F_0 of the threshold function are such maximum subsets of T and F respectively that if $X_j \in T_0$, then for any comparable with X_j combination $X_i < X_j$ it is true that $X_i \in F$, and conversely for $X_i \in F_0$ and any comparable with it combination $X_j > X_i$ it is true that $X_j \in T$. Combinations in T_0 and

F_0 are determined by terms of minimum forms of Boolean functions representing the threshold function and its inversion respectively. Combinations from T_0 and F_0 form the bearing teaching sequence. The length of the bearing teaching sequence for a threshold function of n variables varies in wide limits: from $n+1$ to $2C_n^{(n-1)/2} = \frac{2 \cdot n!}{((n-1)/2)!((n+1)/2)!}$ (for odd n).

As test functions, we consider threshold functions the Boolean expression for which can be represented as Gorner's scheme:

$G_1(X) = \text{Sign}\left(\sum_{j=0}^{n-1} w_j x_j - \nu_1\right) = x_{n-1}(x_{n-2} \vee x_{n-3}(x_{n-4} \vee ...)),$

$G_2(X) = \text{Sign}\left(\sum_{j=0}^{n-1} w_j x_j - \nu_2\right) = x_{n-1} \vee x_{n-2}(x_{n-3} \vee x_{n-4}(x_{n-5} \vee ...)).$

We will refer to them as Gorner's functions of the first and second kinds respectively. In respect to these functions, the following statements are proved.

1. The input weights and threshold for Gorner's functions form Fibonacci sequence:

$w_{n-1} = \frac{1}{\sqrt{5}}\left[\left(\frac{1+\sqrt{5}}{2}\right)^n - \left(\frac{1-\sqrt{5}}{2}\right)^n\right];$

$\nu_1(n) = w_n;\ \nu_2(n) = w_{n-1};\ \sum_{j=0}^{n-1} w_j = w_{n+1} - 1.$

2. Among all threshold functions of n variables Gorner's function has the shortest teaching sequence. The length of the sequence is determined as $L_{min}(n) = |T_0(n)| + |F_0(n)| = n+1$ were $|T_0(n)|$ and $|F_0(n)|$ are equal to the numbers of terms in the minimum representations of the threshold function $Y_n(X)$ as Gorner's function $Y_n(X) = G_2(n)$ and $\overline{Y}_n(X) = G_1(n)$ respectively.

We apply the results of the experiments on teaching a beta-driven artificial neuron to producing Gorner's threshold functions of 10 and 11 variables with sums of the input weights 143 and 232 respectively:

$Y_{10} = \text{Sign}(x_0 + x_1 + 2x_2 + 3x_3 + 5x_4 + 8x_5 + 13x_6 + 21x_7 + 34x_8 + 55x_9 - 55),$
$T_0(10) = \{341, 342, 344, 352, 384, 512\},\ F_0(10) = \{255, 319, 335, 339, 340\};$
$Y_{11} = \text{Sign}(x_0 + x_1 + 2x_2 + 3x_3 + 5x_4 + 8x_5 + 13x_6 + 21x_7 + 34x_8 + 55x_9 + 89x_{10} - 89),$
$T_0(11) = \{683, 684, 688, 704, 768, 1024\},\ F_0(11) = \{511, 639, 671, 679, 681, 682\}.$
Here the elements of the sets F_0 and T_0 are the decimal equivalents of binary numbers corresponding to the variable value combinations.

It is easy to see that Gorner's threshold functions cover wide range of input weights. This increases the efficiency of their usage as test functions.

References

1. Shibata, T., Ohmi, T.: Neuron MOS binary-logic integrated circuits: Part 1, Design fundamentals and soft-hardware logic circuit implementation. IEEE Trans. Electron Devices Vol.**40**, No.**5** (1993) 974–979
2. Ohmi, T., Shibata, T., Kotani, K.: Four-Terminal Device Concept for Intelligence Soft Computing on Silicon Integrated Circuits. Proc. Of IIZUKA'96 (1996) 49–59
3. Varshavsky, V.: Simple CMOS Learnable Threshold Element. International ICSC/IFAC Symposium on Neural Computation, Vienna, Austria, (1998)
4. Varshavsky, V.: CMOS Artificial Neuron on the Base of Beta-Driven Threshold Elements. IEEE International Conference on Systems, Man and Cybernetics, San Diego, CA (1998) 1897–1861

Author Index

Lecture Notes in Computer Science

For information about Vols. 1–1520
please contact your bookseller or Springer-Verlag

Vol. 1521: B. Rovan (Ed.), SOFSEM'98: Theory and Practice of Informatics. Proceedings, 1998. XI, 453 pages. 1998.

Vol. 1522: G. Gopalakrishnan, P. Windley (Eds.), Formal Methods in Computer-Aided Design. Proceedings, 1998. IX, 529 pages. 1998.

Vol. 1524: G.B. Orr, K.-R. Müller (Eds.), Neural Networks: Tricks of the Trade. VI, 432 pages. 1998.

Vol. 1525: D. Aucsmith (Ed.), Information Hiding. Proceedings, 1998. IX, 369 pages. 1998.

Vol. 1526: M. Broy, B. Rumpe (Eds.), Requirements Targeting Software and Systems Engineering. Proceedings, 1997. VIII, 357 pages. 1998.

Vol. 1527: P. Baumgartner, Theory Reasoning in Connection Calculi. IX, 283. 1999. (Subseries LNAI).

Vol. 1528: B. Preneel, V. Rijmen (Eds.), State of the Art in Applied Cryptography. Revised Lectures, 1997. VIII, 395 pages. 1998.

Vol. 1529: D. Farwell, L. Gerber, E. Hovy (Eds.), Machine Translation and the Information Soup. Proceedings, 1998. XIX, 532 pages. 1998. (Subseries LNAI).

Vol. 1530: V. Arvind, R. Ramanujam (Eds.), Foundations of Software Technology and Theoretical Computer Science. XII, 369 pages. 1998.

Vol. 1531: H.-Y. Lee, H. Motoda (Eds.), PRICAI'98: Topics in Artificial Intelligence. XIX, 646 pages. 1998. (Subseries LNAI).

Vol. 1096: T. Schael, Workflow Management Systems for Process Organisations. Second Edition. XII, 229 pages. 1998.

Vol. 1532: S. Arikawa, H. Motoda (Eds.), Discovery Science. Proceedings, 1998. XI, 456 pages. 1998. (Subseries LNAI).

Vol. 1533: K.-Y. Chwa, O.H. Ibarra (Eds.), Algorithms and Computation. Proceedings, 1998. XIII, 478 pages. 1998.

Vol. 1534: J.S. Sichman, R. Conte, N. Gilbert (Eds.), Multi-Agent Systems and Agent-Based Simulation. Proceedings, 1998. VIII, 237 pages. 1998. (Subseries LNAI).

Vol. 1535: S. Ossowski, Co-ordination in Artificial Agent Societies. XV; 221 pages. 1999. (Subseries LNAI).

Vol. 1536: W.-P. de Roever, H. Langmaack, A. Pnueli (Eds.), Compositionality: The Significant Difference. Proceedings, 1997. VIII, 647 pages. 1998.

Vol. 1537: N. Magnenat-Thalmann, D. Thalmann (Eds.), Modelling and Motion Capture Techniques for Virtual Environments. Proceedings, 1998. IX, 273 pages. 1998. (Subseries LNAI).

Vol. 1538: J. Hsiang, A. Ohori (Eds.), Advances in Computing Science – ASIAN'98. Proceedings, 1998. X, 305 pages. 1998.

Vol. 1539: O. Rüthing, Interacting Code Motion Transformations: Their Impact and Their Complexity. XXI,225 pages. 1998.

Vol. 1540: C. Beeri, P. Buneman (Eds.), Database Theory – ICDT'99. Proceedings, 1999. XI, 489 pages. 1999.

Vol. 1541: B. Kågström, J. Dongarra, E. Elmroth, J. Waśniewski (Eds.), Applied Parallel Computing. Proceedings, 1998. XIV, 586 pages. 1998.

Vol. 1542: H.I. Christensen (Ed.), Computer Vision Systems. Proceedings, 1999. XI, 554 pages. 1999.

Vol. 1543: S. Demeyer, J. Bosch (Eds.), Object-Oriented Technology ECOOP'98 Workshop Reader. 1998. XXII, 573 pages. 1998.

Vol. 1544: C. Zhang, D. Lukose (Eds.), Multi-Agent Systems. Proceedings, 1998. VII, 195 pages. 1998. (Subseries LNAI).

Vol. 1545: A. Birk, J. Demiris (Eds.), Learning Robots. Proceedings, 1996. IX, 188 pages. 1998. (Subseries LNAI).

Vol. 1546: B. Möller, J.V. Tucker (Eds.), Prospects for Hardware Foundations. Survey Chapters, 1998. X, 468 pages. 1998.

Vol. 1547: S.H. Whitesides (Ed.), Graph Drawing. Proceedings 1998. XII, 468 pages. 1998.

Vol. 1548: A.M. Haeberer (Ed.), Algebraic Methodology and Software Technology. Proceedings, 1999. XI, 531 pages. 1999.

Vol. 1549: M. Pettersson, Compiling Natural Semantics. XVII, 240 pages. 1999.

Vol. 1550: B. Christianson, B. Crispo, W.S. Harbison, M. Roe (Eds.), Security Protocols. Proceedings, 1998. VIII, 241 pages. 1999.

Vol. 1551: G. Gupta (Ed.), Practical Aspects of Declarative Languages. Proceedings, 1999. VIII, 367 pgages. 1999.

Vol. 1552: Y. Kambayashi, D.L. Lee, E.-P. Lim, M.K. Mohania, Y. Masunaga (Eds.), Advances in Database Technologies. Proceedings, 1998. XIX, 592 pages. 1999.

Vol. 1553: S.F. Andler, J. Hansson (Eds.), Active, Real-Time, and Temporal Database Systems. Proceedings, 1997. VIII, 245 pages. 1998.

Vol. 1554: S. Nishio, F. Kishino (Eds.), Advanced Multimedia Content Processing. Proceedings, 1998. XIV, 454 pages. 1999.

Vol. 1555: J.P. Müller, M.P. Singh, A.S. Rao (Eds.), Intelligent Agents V. Proceedings, 1998. XXIV, 455 pages. 1999. (Subseries LNAI).

Vol. 1556: S. Tavares, H. Meijer (Eds.), Selected Areas in Cryptography. Proceedings, 1998. IX, 377 pages. 1999.

Vol. 1557: P. Zinterhof, M. Vajteršic, A. Uhl (Eds.), Parallel Computation. Proceedings, 1999. XV, 604 pages. 1999.

Vol. 1558: H. J.v.d. Herik, H. Iida (Eds.), Computers and Games. Proceedings, 1998. XVIII, 337 pages. 1999.

Vol. 1559: P. Flener (Ed.), Logic-Based Program Synthesis and Transformation. Proceedings, 1998. X, 331 pages. 1999.

Vol. 1560: K. Imai, Y. Zheng (Eds.), Public Key Cryptography. Proceedings, 1999. IX, 327 pages. 1999.

Vol. 1561: I. Damgård (Ed.), Lectures on Data Security.VII, 250 pages. 1999.

Vol. 1562: C.L. Nehaniv (Ed.), Computation for Metaphors, Analogy, and Agents. X, 389 pages. 1999. (Subseries LNAI).

Vol. 1563: Ch. Meinel, S. Tison (Eds.), STACS 99. Proceedings, 1999. XIV, 582 pages. 1999.

Vol. 1565: P. P. Chen, J. Akoka, H. Kangassalo, B. Thalheim (Eds.), Conceptual Modeling. XXIV, 303 pages. 1999.

Vol. 1567: P. Antsaklis, W. Kohn, M. Lemmon, A. Nerode, S. Sastry (Eds.), Hybrid Systems V. X, 445 pages. 1999.

Vol. 1568: G. Bertrand, M. Couprie, L. Perroton (Eds.), Discrete Geometry for Computer Imagery. Proceedings, 1999. XI, 459 pages. 1999.

Vol. 1569: F.W. Vaandrager, J.H. van Schuppen (Eds.), Hybrid Systems: Computation and Control. Proceedings, 1999. X, 271 pages. 1999.

Vol. 1570: F. Puppe (Ed.), XPS-99: Knowledge-Based Systems. VIII, 227 pages. 1999. (Subseries LNAI).

Vol. 1571: P. Noriega, C. Sierra (Eds.), Agent Mediated Electronic Commerce. Proceedings, 1998. IX, 207 pages. 1999. (Subseries LNAI).

Vol. 1572: P. Fischer, H.U. Simon (Eds.), Computational Learning Theory. Proceedings, 1999. X, 301 pages. 1999. (Subseries LNAI).

Vol. 1574: N. Zhong, L. Zhou (Eds.), Methodologies for Knowledge Discovery and Data Mining. Proceedings, 1999. XV, 533 pages. 1999. (Subseries LNAI).

Vol. 1575: S. Jähnichen (Ed.), Compiler Construction. Proceedings, 1999. X, 301 pages. 1999.

Vol. 1576: S.D. Swierstra (Ed.), Programming Languages and Systems. Proceedings, 1999. X, 307 pages. 1999.

Vol. 1577: J.-P. Finance (Ed.), Fundamental Approaches to Software Engineering. Proceedings, 1999. X, 245 pages. 1999.

Vol. 1578: W. Thomas (Ed.), Foundations of Software Science and Computation Structures. Proceedings, 1999. X, 323 pages. 1999.

Vol. 1579: W.R. Cleaveland (Ed.), Tools and Algorithms for the Construction and Analysis of Systems. Proceedings, 1999. XI, 445 pages. 1999.

Vol. 1580: A. Včkovski, K.E. Brassel, H.-J. Schek (Eds.), Interoperating Geographic Information Systems. Proceedings, 1999. XI, 329 pages. 1999.

Vol. 1581: J.-Y. Girard (Ed.), Typed Lambda Calculi and Applications. Proceedings, 1999. VIII, 397 pages. 1999.

Vol. 1582: A. Lecomte, F. Lamarche, G. Perrier (Eds.), Logical Aspects of Computational Linguistics. Proceedings, 1997. XI, 251 pages. 1999. (Subseries LNAI).

Vol. 1584: G. Gottlob, E. Grandjean, K. Seyr (Eds.), Computer Science Logic. Proceedings, 1998. X, 431 pages. 1999.

Vol. 1586: J. Rolim et al. (Eds.), Parallel and Distributed Processing. Proceedings, 1999. XVII, 1443 pages. 1999.

Vol. 1587: J. Pieprzyk, R. Safavi-Naini, J. Seberry (Eds.), Information Security and Privacy. Proceedings, 1999. XI, 327 pages. 1999.

Vol. 1590: P. Atzeni, A. Mendelzon, G. Mecca (Eds.), The World Wide Web and Databases. Proceedings, 1998. VIII, 213 pages. 1999.

Vol. 1592: J. Stern (Ed.), Advances in Cryptology – EUROCRYPT '99. Proceedings, 1999. XII, 475 pages. 1999.

Vol. 1593: P. Sloot, M. Bubak, A. Hoekstra, B. Hertzberger (Eds.), High-Performance Computing and Networking. Proceedings, 1999. XXIII, 1318 pages. 1999.

Vol. 1594: P. Ciancarini, A.L. Wolf (Eds.), Coordination Languages and Models. Proceedings, 1999. IX, 420 pages. 1999.

Vol. 1596: R. Poli, H.-M. Voigt, S. Cagnoni, D. Corne, G.D. Smith, T.C. Fogarty (Eds.), Evolutionary Image Analysis, Signal Processing and Telecommunications. Proceedings, 1999. X, 225 pages. 1999.

Vol. 1597: H. Zuidweg, M. Campolargo, J. Delgado, A. Mullery (Eds.), Intelligence in Services and Networks. Proceedings, 1999. XII, 552 pages. 1999.

Vol. 1598: R. Poli, P. Nordin, W.B. Langdon, T.C. Fogarty (Eds.), Genetic Programming. Proceedings, 1999. X, 283 pages. 1999.

Vol. 1599: T. Ishida (Ed.), Multiagent Platforms. Proceedings, 1998. VIII, 187 pages. 1999. (Subseries LNAI).

Vol. 1601: J.-P. Katoen (Ed.), Formal Methods for Real-Time and Probabilistic Systems. Proceedings, 1999. X, 355 pages. 1999.

Vol. 1602: A. Sivasubramaniam, M. Lauria (Eds.), Network-Based Parallel Computing. Proceedings, 1999. VIII, 225 pages. 1999.

Vol. 1605: J. Billington, M. Diaz, G. Rozenberg (Eds.), Application of Petri Nets to Communication Networks. IX, 303 pages. 1999.

Vol. 1609: Z. W. Raś, A. Skowron (Eds.), Foundations of Intelligent Systems. Proceedings, 1999. XII, 676 pages. 1999. (Subseries LNAI).

Vol. 1610: G. Cornuéjols, R.E. Burkard, G.J. Woeginger (Eds.), Integer Programming and Combinatorial Optimization. Proceedings, 1999. IX, 453 pages. 1999.

Vol. 1615: C. Polychronopoulos, K. Joe, A. Fukuda, S. Tomita (Eds.), High Performance Computing. Proceedings, 1999. XIV, 408 pages. 1999.

Vol. 1621: D. Fensel, R. Studer (Eds.), Knowledge Acquisition Modeling and Management. Proceedings, 1999. XI, 404 pages. 1999. (Subseries LNAI).

Vol. 1625: B. Reusch (Ed.), Computational Intelligence. Proceedings, 1999. XIV, 710 pages. 1999.